UNITED STATES DEPARTMENT OF COMMERCE ● C. R. Smith, *Secretary*

NATIONAL BUREAU OF STANDARDS ● A. V. Astin, *Director*

MICROWAVE SPECTRAL TABLES

Volume IV. Polyatomic Molecules Without Internal Rotation

Marian S. Cord, Jean D. Petersen, Matthew S. Lojko,
and Rudolph H. Haas

Institute for Basic Standards, National Bureau of Standards, Boulder, Colorado

National Bureau of Standards Monograph 70 — Volume IV

Issued October 1968

For sale by the Superintendent of Documents, U.S. Government Printing Office
Washington, D.C. 20402 - Price $5.50

245976

Microwave Spectral Tables

This five-volume NBS Monograph will present a comprehensive compilation of microwave spectra including measured frequencies, assigned molecular species, assigned quantum numbers, and molecular constants determined from these data. Four volumes are now available and the other one will be available in about two months.

Titles of the volumes

   I. Diatomic Molecules, $2.00   *O.P. (2/70)*

  II. Line Strengths of Asymmetric Rotors, $3.00 *O.P. (2/70)*

 III. Polyatomic Molecules With Internal Rotation (~~In press~~) *June, 1969, $4.75*

 IV. Polyatomic Molecules Without Internal Rotation, $5.50

  V. Spectral Line Listing, $4.75

The volumes are available by purchase from the Superintendent of Documents, Government Printing Office, Washington, D.C. 20402. Orders from foreign countries should include an additional one-fourth of the price of the publication to cover the postage.

Library of Congress Catalog Card Number: 63–62235

Foreword

Volume IV concludes the present compilation of data under the title of Microwave Spectral Tables—NBS Monograph 70, begun in 1953 by Dr. Paul F. Wacker. Volume V, published prior to this volume, contains a numerical listing of spectral lines given in Volumes I, III, and IV. Volume I, Diatomic Molecules, and Volume II, Line Strengths of Asymmetric Rotors, were completed under the supervision of Dr. Wacker and published in December, 1964.

In May, 1964, general supervision of the project was assumed by Dr. Yardley Beers. Subsequently, Mrs. Marian Cord was appointed project leader for the completion of Volume III, Polyatomic Molecules with Internal Rotation, Volume IV, Polyatomic Molecules Without Internal Rotation, and Volume V, Spectral Line Listing.

Extensive efforts have been made to reduce the number of errors in recording the data to a minimum. Any corrections, criticisms, or suggestions will be appreciated by the authors.

Acknowledgments

The authors wish to express appreciation for much assistance in the preparation of this volume; for preliminary reviewing: Masataka Mizushima, Allen Garrison and William Dambeck; for preliminary programming: Raymond Kukol and Jean Troyer; for checking: Emmett MacKenzie, Judy Stephenson and Carol Nielsen; for the INTRODUCTION, general consultation and valuable advice: Paul Wacker and Yardley Beers. To this list should be added the names of many scientists in the field of microwave spectroscopy who readily and graciously complied to requests for information.

Contents

Microwave Spectral Tables

Volume IV. Polyatomic Molecules Without Internal Rotation

Marian S. Cord, Jean D. Petersen, Matthew S. Lojko, and Rudolph H. Haas

Measured frequencies, assigned molecular species, and assigned quantum numbers are given for abour 14,000 spectral lines of polyatomic molecules without internal rotation observed by coherent radiation techniques. Molecular data, such as rotational constants, dipole moments, and various coupling constants, determined by such techniques, are also tabulated. References are given for all included data.

Key words: Centrifugal distortion constant; coherent radiation technique; dipole moment; l-doubling; microwave spectra; molecular spectra; polyatomic molecules; quadrupole coupling constants; rotational constant; spectral lines.

1. Introduction

This volume contains data on the microwave spectra of 166 polyatomic molecules incapable of exhibiting internal rotation. These data are based upon a systematic search of the literature up to January 1961 and include some information of later dates. Molecules capable of exhibiting internal rotation are those whose atoms can be arranged in groups which may rotate separately about a common axis. This rotation is known to take place, or supposed to be observable, when the bond between adjacent atoms in two groups is a single one. However, the splittings due to torsional oscillation about a multiple bond are so very small that it is usually more suitable to treat such phenomena with techniques developed for vibrational spectra. The microwave spectra of molecules capable of exhibiting internal rotation about single bonds are tabulated in Volume III of this series. The present volume contains data on all the other polyatomic molecules covered by this survey, including linear polyatomic molecules.

One special topic which deserves some brief mention, as it is not encountered extensively in other volumes of this series, is l-type doubling. This phenomenon occurs with molecules having degenerate modes of vibration. The linear motions of the nuclei belonging to the two modes are in orthogonal directions. Therefore, if both modes are excited, the combined effect generally is to cause the nuclei to appear to travel in elliptical orbits. Associated with this motion is the concept of "vibrational angular momentum," which is quantized with the quantum number "l". If the molecule as a whole has rotation as well, however, the two normal mode vibrations are not completely equivalent, since the directions of linear vibrations generally have different angles with the vector representing the other rotation. Therefore, the Coriolis forces corresponding to the two modes are different, and generally there results a splitting of the otherwise degenerate levels. The class of molecules in which the effect has been most extensively studied is that of linear triatomic molecules. For a linear molecule, the angular momentum is $lh/2\pi$, where l is an integer no greater in absolute value than the corresponding vibrational quantum number v, and is even or odd as is the latter. Neglecting the bending which occurs if the amplitudes of the two component modes are unequal, the linear molecule is similar to an asymmetric top with l analogous to the prolate quantum number K_{-1}. Commonly only the absolute value of l is given in both the microwave and infrared literature. If the sign is definite, a + sign or a − sign is used; if the sign is known relative to that of the axial quantum number K in a symmetric top molecule, then $\pm$ or $\mp$ is used with K and l to indicate this correspondence. If the quantum number l is non-zero, the energy for a given J value is dependent upon the sign of l, i.e., upon whether the vibrational rotation is clockwise or counterclockwise [see H. H. Nielsen and W. H. Shaffer, J. Chem. Phys., **11**, 142 (1943)].

The l-splitting in the case of $|l| \geqslant 1$, is of the order of magnitude of $B_e(B_e/\omega_2)^{2|l|-1}$ for singlet states, where B_e is the rotational constant and ω_2 is the bending frequency in comparable units [H. H. Nielsen in Encyclopedia of Physics XXXVII/1, page 280 (1959)]. Hence, the splitting is so small for $|l| > 1$ that attention is confined here to the usual $|l| = 1$ case. Then, to a first approximation, which is adequate for most purposes, the splitting (frequency units) is given by

$$\Delta W_l = \frac{q_l}{2}(v_2+1)J(J+1),$$

where q_l is a constant, J is a total angular momentum ($J \geq |l|$), and v_2 is quantum number pertaining to the bending vibration. Transitions can take place according to either of the following selection rules: (1) $\Delta J = \pm 1$, $\Delta l = 0$, or (2) $\Delta J = 0$ and $\Delta |l| = 0$. As a consequence of the first set of selection rules, each $\Delta J = \pm 1$ rotational line originating from the ground vibrational state is accompanied by a vibrational satellite or pair, displaced slightly from it for each finite value of v_2 (and $|l|$). The second set of selection rules give rise to the "l-type" spectra mentioned earlier, and the frequencies of the lines, to the first approximation, are given by the above formula. This formula is to be considered as an approximation adequate for nearly all cases which have been investigated. However, for a few molecules, especially HCN and ClCN, there have been obtained data of sufficient quantity and accuracy to show a slight variation of q_l with J. In such cases the tables will list an approximate value and in some cases a footnote will give the best available formula for q_l as a function of J along with a suitable reference.

An extensive discussion and bibliography pertaining to "l-type doubling" may be found in "Microwave Spectroscopy" by C. H. Townes and A. L. Schawlow [McGraw-Hill Book Company, Inc., New York, 1955], especially Section 2–2, for which the following suggestions are given:

    (A) The factor h should be added to (2–13) and to the second and third terms of (2–11);

    (B) The denominator of (2–16) should be $J(J+1)$ as in their (3–41) rather than $(J+1)(2J+1)$;

    (C) The order of magnitude of the l-splitting is that given by the first sentence of our previous paragraph rather than $B_e(B_e/\omega_2)^l$.

2. Explanation of Tables

Data for these Microwave Spectral Tables have been compiled from a review of articles found in worldwide journals as well as in private communications regarding unpublished work. Generally speaking, date and accuracy of data measurement were the criteria by which any required selection was made. An explanation of the format and contents of the Tables follows.

The first line of data contains an identification number and the Chemical Abstracts name. The lines immediately following give alternate names, if any. The last line of the heading contains the empirical formula, the symmetry point group of the molecule, and, where possible, the "quasi"-structural formula. The point group given assumes only one isotopic form of each element and the molecular configuration believed to have the lowest potential energy. Molecules are ordered first according to empirical formula, as in Chemical Abstracts, then by name. A numbering system has been established for identification of the molecules and their isotopic species for convenience in programming and referencing. Each molecule was given a number ending in zero in order of its occurrence on the list. Its species have been numbered consecutively from that point, skipping the next zero when the number of isotopic species exceeds nine. Every species is given for which any molecular constants or spectral lines from microwave sources have been reported up to January 1961. However, a glance at the bibliography will show that a considerable number of later references have been included. In the formulas for the species, mass numbers have been placed as superscripts to the right of the atomic symbols. Hydrogen is the one exception—for the normal mass of "1" the superscript is omitted, and for deuterium (H^2) the symbol D is used.

In the second column of the molecular constant table, the overall symmetry point group (Pt. Gp.) has been given for each isotopic species of the molecule, as the analysis permits. Some consideration of structural information was necessarily a part of these assignments.

Most of the rotational constants tabulated have been derived from microwave sources and this is indicated by the M occurring after the values. In a few instances, other experimental sources have been designated by the code letters N or F (for near or far infrared measurements). Use of symmetry has been made in reporting rotational constants A, B, and C (with $A \geq B \geq C$). If, for example, the axis of symmetry of a symmetric top is known to be the axis of least moment of inertia, then the constants B and C are equal and are so recorded. All moments of inertia have been converted to the corresponding constant, A, B or C (in megahertz) by multiplication with the appropriate conversion factor. These factors are given later following the list of symbols and abbreviations.

Centrifugal distortion effects are indicated by the constants D_J, D_{JK}, and D_K which are dependent upon the rotational constants and the vibrational force constants. The first two of these are given in the molecular constant tables, but the third, D_K, when reported, is to be found in the data given following the molecular constant table for each molecule.

With planar molecules, the inertial defect, $\Delta = I_c - I_a - I_b$, is frequently of interest. Classically, for a rigid planar molecule, this is zero, but in actuality it is non-zero because of the existence of zero point vibrations.

Ray's asymmetry parameter is a measure of the variation from the symmetric top. The value of κ $\left(\text{defined as } \dfrac{2B-A-C}{A-C}\right)$ ranges between -1 for a prolate top and $+1$ for an oblate top. In cases where the original papers have used other asymmetry parameters, these have been converted to κ values. Formulas which were used for these conversions are as follows:

$$\kappa = 2\delta - 1 = \frac{3b_p + 1}{b_p - 1} = \frac{1 + 3b_0}{1 + b_0} = \frac{6\epsilon - 1}{2\epsilon + 1},$$

where

$$b_p = \frac{C-B}{2A-B-C}, \; b_0 = \frac{A-B}{2C-B-A},$$

$$\delta = \frac{B-C}{A-C}, \text{ and } \epsilon = \frac{B-C}{2(2A-B-C)}$$

In many instances, the total dipole moment μ was found in the literature, but it had not been resolved into components parallel to the axes of least, intermediate, and greatest moments of inertia, respectively (denoted by the subscripts a, b, and c). Symmetric top molecules have two equal principal moments of inertia, designated by I_b. The dipole moment of such molecules lies along the third principal axis. Some of the original papers did not specify whether this moment of inertia was smaller or larger than the two others and, therefore, did not indicate how this third axis should be labelled. An effort was made to resolve the question by making use of available structural information to calculate the moment of inertia with respect to the symmetry axis as described in Volume III. Using the convention $I_a \leqslant I_b \leqslant I_c$, if the computed value was smaller than the given I_b, then the figure axis was considered to be "a" and, if greater than I_b, then the figure axis was considered to be "c". A symbol M follows the value if the dipole moment was measured by microwave techniques, usually the Stark effect. In the case of asymmetric molecules, if the components of the dipole moment were not specified, the reported dipole moment was entered under the component which was adjudged the major one (according to transitions assigned to spectral lines, or, if these were lacking, according to structural information available). In addition, the letter "u" designates the unknown, but possible, second component. The fact that certain dipole moment components are known to be zero by virtue of established symmetries is indicated by an X, after

the data (for example, a component perpendicular to the plane of symmetry, or one in a plane perpendicular to the axis of symmetry, would always be zero). Effects of isotopic substitution on symmetry of the molecule have been taken into account. Where dipole moments are reported which were obtained by other than microwave techniques, different code letter designations are used. If the moments are reportedly obtained from the Debye equation with temperature-variation for the molar polarization of the gas or liquid, the code letters G or L, respectively, are used. A code letter P, denotes that the value was reported in the literature as computed from the molar polarization α. In all other cases of non-microwave sources for these data the code letter T is used.

Quadrupole coupling constants, eQq, obtained from hyperfine structure measurements are reported as found in the literature. In some cases authors have reported these as referred to particular bonds or axes, and if so, they are recorded thus. To facilitate later description of transitions involving excited vibrational states, each vibrational mode excited in any observed microwave transition is assigned a letter subscript and is identified by the frequency, ω, in cm^{-1}, and the degeneracy d.

For a limited number of molecules some additional molecular constant data, not described in the foregoing paragraphs, were available. This has been condensed, where practicable, and inserted after the regular molecular constant tables for each molecule. No effort is made here to define these miscellaneous constants. The interested reader is therefore referred to the particular paper from which the data were abstracted.

The table of reported spectral lines for each molecule starts on a new page following the molecular constant tables. Immediately following the heading for the spectral lines, the formula and identification number of each isotopic molecular species are printed. Lines are arranged in the following pattern: ascending order of rotational quantum numbers according to (1) J', (2) J, (3) K'_{-1}, (4) K_{-1}, (5) K'_{+1}, (6) K_{+1}, then in ascending magnitude of the frequency. For asymmetric tops the rotational quantum numbers are arranged in the pattern J', K'_{-1}, $K'_{+1} \leftarrow J$, K_{-1}, K_{+1}, where primes denote upper state. For prolate or oblate symmetric tops the arrangement is J', $K'_{-1} \leftarrow J$, K_{-1} or J', $K'_{+1} \leftarrow J$, K_{+1}, respectively. Blanks occur where the identification is not complete in the literature. Unassigned lines are indicated by the term "not reported." It should be recognized that lines for which rotational quantum numbers have not been assigned may be due to impurities.

It should be mentioned that in a few $\Delta J = \pm 1$ transitions of very asymmetric top molecules it is difficult to ascertain which is the upper state, even if quantum numbers have been assigned to both states. There is indication that some authors have not clearly stated the existence of this uncertainty or that they have been careless with their notation in this regard. In some instances clarification has been obtained from the author, but in others this was not possible. In any event the quantum numbers on the left denote the state which is believed to be the upper state.

There is considerable non-uniformity in the listing of the hyperfine components, mainly because many times only a few of the rotational transitions of a given molecule may have their hyperfine splittings analyzed. If the hyperfine splitting is resolved, the individual lines are ordinarily reported separately with the total angular momenta F' and F of the upper and lower states, respectively. If the splitting is partially resolved, the assigned F values are given for the resolved lines. For large J values, the $\Delta F = \Delta J$ components are relatively strong and displaced but little from the frequency the line would have if there were no hyperfine splitting. If the hyperfine splitting is quite small (less than 5 MHz), the frequency of the hypothetical line for zero quadrupole moment is commonly listed and shifts from these frequencies are given in the text or in footnotes.

Efforts were made to determine from the author's statements, the accuracies to which spectral lines were measured, though, in many cases, these were not well defined in the literature. References are indicated for all data, but wherever possible, footnotes have been given to obviate the necessity of referring to these. References are arranged in order of year of publication, then alphabetically according to first author and numbered consecutively. A few exceptions occur in the case of books and of articles which were not found until after the numbering had been completed and a change in reference number would have necessitated extensive changes throughout the entire volume.

3. List of Symbols, Abbreviations, and Conversion Factors

3.1. Symbols and Abbreviations

The following symbols and abbreviations are listed in the order in which they may first occur in the tables.

Molecular Constant Table

$\vdots$	Double bond.
$\vdots$	Triple bond.
...	Extra (cyclic) bond between two atoms preceding these symbols.
$a-$	Asymmetric.
b	As a superscript to an atomic symbol, indicates isotope not stated.
$c-$	Cis structure.
$g-$	Gauche structure.
$t-$	Trans structure.
$s-$	Symmetric.
Pt. Gp.	Symmetry point group of the molecular species.
Id. No.	Identification number of the molecule or molecular isotopic species.
A, B, C	Rotation constants (MHz). $A \geqslant B \geqslant C$.
M	Microwave determination.
N	Near infrared determination.
F	Far infrared determination.
L	Liquid temperature variation procedure used for determination.
G	Gas temperature variation procedure used for determination.
P	Value reported as computed from molar polarization α.
$D_{J, JK, K}$	Centrifugal distortion constants (MHz).
κ	Ray's asymmetry parameter $(2B-A-C)/(A-C)$.
Δ	Inertial defect (amu Å²).
a, b, c	Principal axes corresponding to A, B, and C, respectively.
$\mu_{a, b, c}$	Components of the dipole moment along the principal axes (Debye).
D	Abbreviation for Debye units.
u	Value not known.
X	Value precisely zero due to symmetry.
Acc	Accuracy of the previous quantity.
eQq	For a symmetric top, the electric quadrupole coupling constant along the symmetry axis (MHz).
$\omega_{a, b, \ldots}$	Vibrational frequencies of modes, $a, b, \ldots$ excited in observed microwave spectral lines (cm⁻¹).
d	Degeneracy of the preceding vibrational mode.
Ref.	Numbers are reference numbers as in the bibliography.
$eQq_{aa}, \ldots$ $\chi_{aa}, \ldots$	Electric quadrupole coupling constant along indicated principal axis (MHz).
I_α	Coupling parameter, i.e., moment of inertia of rotatable group (amu Å²).
$I_{a, b, c}$	Moments of inertia of the whole molecule with respect to the indicated principal axis.

Spectral Line Table

$v_a; v^l; \ldots$	Quantum number of the vibrational mode identified as $a, b, \ldots$ in the molecular constant table. The l quantum number is given after a comma rather than as a superscript.
$F_1', \ldots, F$	Total angular momentum quantum number. Primed values are for the upper state, unprimed values for the lower state. Those with subscript "1" include nuclear spin only for nucleus with largest eQq.

3.2. Conversion Factors

The following formulas were used to convert the literature values x, to the values y, of the tables. The respective units are given in parentheses. It should be pointed out that the conversion factor given here for amu-Å² to MHz, is based on the atomic weight scale in which $C^{12} = 12$. However, most data covered in these tables were based on the older scale in which $O^{16} = 16$. The formulas used are:

amu Å² to MHz: $y(\text{MHz}) = \dfrac{(8.391420 \times 10^5)}{x(\text{amu Å}^2)\ (1.6604345)} = \dfrac{(5.0537497 \times 10^5)}{x(\text{amu Å}^2)}$

cm⁻¹ to MHz: $y(\text{MHz}) = (2.997925 \times 10^4)\ x\ (\text{cm}^{-1})$

gm-cm² to MHz: $y(\text{MHz}) = \dfrac{(8.3914204 \times 10^5)}{x(\text{gm-cm}^2)}.$

4. Molecular Constants and Spectral Line Tabulation

AsCl₃ .... C$_{3v}$ .... AsCl₃

Isotopic Species	Pt. Gp.	Id. No.	A MHz		B MHz		C MHz	D$_J$ MHz	D$_{JK}$ MHz	Δ Amu A²	κ
As^{75}Cl$_3^{35}$	C$_{3v}$	11	2 147.2	M	2 147.2	M					
As^{75}Cl$_3^{37}$	C$_{3v}$	14	2 044.7	M	2 044.7	M					

Id. No.	μ$_a$ Debye	μ$_b$ Debye	μ$_c$ Debye	eQq Value(MHz) Rel.		eQq Value(MHz) Rel.	eQq Value(MHz) Rel.	ω$_a$ d 1/cm	ω$_b$ d 1/cm	ω$_c$ d 1/cm	ω$_d$ d 1/cm
11	0. X	0. X	2.1 G	−173	As			193 1			

References:

ABC: 206,446   μ: 446   eQq: 285   ω: 1028

Add. Ref. 285,534

Arsenic Trichloride .... Spectral Line Table

Isotopic Species	Id. No.	Rotational Quantum Nos.	Vib. State v$_a$	Hyperfine F$_1'$	F'	F$_1$	F	Frequency MHz	Acc. ±MHz	Ref.
As^{75}Cl$_3^{35}$	11	5, ← 4,	1					21 426.	10.	446
	11	5, ← 4,	Ground					21 472.	10.	446
	11	6, ← 5,	2					25 675.	10.	446
	11	6, ← 5,	2					25 724.	10.	446
	11	6, ← 5,	Ground					25 767.	2.	446
As^{75}Cl$_2^{35}$Cl^{37}	12	6, , ← 5, ,	Ground					25 308.	10.	446
	12	Not Reported	Ground					25 354.	10.	446
	12	Not Reported	Ground					25 381.	10.	446
	12	Not Reported	Ground					25 393.	10.	446
	12	Not Reported	Ground					25 411.	10.	446
As^{75}Cl^{35}Cl$_2^{37}$	13	6, , ← 5, ,	Ground					24 932.	10.	446
	13	Not Reported	Ground					24 973.	10.	446
As^{75}Cl$_3^{37}$	14	6, ← 5,	Ground					24 536.	2.	446

Isotopic Species	Pt. Gp.	Id. No.	A MHz	B MHz	C MHz	D_J MHz	D_{JK} MHz	Δ Amu A²	κ
As^{75}F$_3^{19}$	C$_{3v}$	21	5 878.971 M	5 878.971 M			−.009		

Id. No.	μ_a Debye	μ_b Debye	μ_c Debye	eQq Value(MHz)	Rel.	eQq Value(MHz)	Rel.	eQq Value(MHz)	Rel.	ω_a d 1/cm	ω_b d 1/cm	ω_c d 1/cm	ω_d d 1/cm
21			2.815 M	−236.23	As					341 1	274 2		

References:
ABC: 447 D_{JK}: 447 μ: 236,534   eQq: 447 ω: 1029

Add. Ref. 77,81,173,236,446

Arsenic Trifluoride

Spectral Line Table

Isotopic Species	Id. No.	Rotational Quantum Nos.	Vib. State v_a ; v_b	F_1'	F'	F_1	F	Frequency MHz	Acc. ±MHz	Ref.
As^{75}F$_3^{19}$	21	2, 0← 1, 0	Ground		3/2		1/2	23 457.		447
	21	2, 0← 1, 0	Ground		5/2		5/2	23 461.905	.01	447
	21	2, 0← 1, 0	1; 0		5/2		5/2	23 482.		447
	21	2, 0← 1, 0	Ground		1/2		1/2	23 515.865	.01	447
	21	2, 0← 1, 0	Ground		7/2		5/2	23 520.965	.01	447
	21	2, 0← 1, 0	Ground		5/2		3/2	23 520.965	.01	447
	21	2, 0← 1, 0	0; 1		5/2		3/2	23 521.		447
	21	2, 0← 1, 0	0; 1		7/2		5/2	23 521.		447
	21	2, 0← 1, 0	1; 0		5/2		3/2	23 543.		447
	21	2, 0← 1, 0	1; 0		7/2		5/2	23 543.		447
	21	2, 0← 1, 0	Ground		3/2		3/2	23 563.222	.01	447
	21	2, 0← 1, 0	1; 0		3/2		3/2	23 584.		447
	21	2, 0← 1, 0	Ground		1/2		3/2	23 622.2		447
	21	2, 1← 1, 1	Ground		5/2		3/2	23 471.334	.01	447
	21	2, 1← 1, 1	Ground		3/2		3/2	23 492.423	.01	447
	21	2, 1← 1, 1	Ground		5/2		5/2	23 500.834	٭.01	447
	21	2, 1← 1, 1	1; 0		3/2		3/2	23 513.		447
	21	2, 1← 1, 1	Ground		7/2		5/2	23 530.318	.01	447
	21	2, 1← 1, 1	0; 1		7/2		5/2	23 531.		447
	21	2, 1← 1, 1	Ground		3/2		1/2	23 545.596	.015	447
	21	2, 1← 1, 1	1; 0		7/2		5/2	23 553.		447
	21	2, 1← 1, 1	Ground		1/2		1/2	23 575.		447
	21	2, 1← 1, 1	1; 0		1/2		1/2	23 595.		447
	21	2, 1← 2, 1						23 509.		447

Arsenic Trihydride, Hydrogen Arsenide

AsH$_3$ C$_{3v}$ AsH$_3$

Isotopic Species	Pt. Gp.	Id. No.	A MHz		B MHz		C MHz	D$_J$ MHz	D$_{JK}$ MHz	Δ Amu A²	κ
As^{75}H$_3$	C$_{3v}$	31	112 468.5	M	112 468.5	M		2.237	−2.220		
As^{75}H$_2$D	C$_s$	32									−.8625
As^{75}D$_3$	C$_{3v}$	33	57 477.15	M	57 477.15	M		.569	−.201		

Id. No.	μ$_a$ Debye		μ$_b$ Debye		μ$_c$ Debye		eQq Value(MHz)	Rel.	eQq Value(MHz)	Rel.	eQq Value(MHz)	Rel.	ω$_a$ d 1/cm	ω$_b$ d 1/cm	ω$_c$ d 1/cm	ω$_d$ d 1/cm
31	0.	X	0.	X	.22	M	−160.1	As^{75}								
32							−164	As^{75}								
33							−165.9	As^{75}								

References:

ABC: 586   D$_J$: 879   D$_{JK}$: 879   κ: 289   μ: 289   eQq: 289,586

Add. Ref. 6,213,338

For species 31, D$_K$ = 2.147 MHz; species 33, D$_K$ = 0.174 MHz. Ref. 879.

Arsine

Isotopic Species	Id. No.	Rotational Quantum Nos.	Vib. State	F$_1'$	F'	F$_1$	F	Frequency MHz	Acc. ±MHz	Ref.
As^{75}H$_3$	31	1, 0← 0, 0	Ground		3/2		3/2	224 895.91	.5	586
	31	1, 0← 0, 0	Ground		5/2		3/2	224 936.78	.5	586
	31	1, 0← 0, 0	Ground		1/2		3/2	224 967.56	.5	586
As^{75}H$_2$D	32	1, 1, 1← 1, 0, 1	Ground		1/2		3/2	35 413.79		289
	32	1, 1, 1← 1, 0, 1	Ground		5/2		3/2	35 430.10		289
	32	1, 1, 1← 1, 0, 1	Ground		5/2		5/2	35 435.19		289
	32	1, 1, 1← 1, 0, 1	Ground		3/2		3/2	35 450.51		289
	32	1, 1, 1← 1, 0, 1	Ground		3/2		5/2	35 455.64		289
	32	1, 1, 1← 1, 0, 1	Ground		3/2		1/2	35 460.08		289
	32	3, 1, 3← 3, 0, 3	Ground		3/2		5/2	29 498.52		289
	32	3, 1, 3← 3, 0, 3	Ground		9/2		7/2	29 507.91		289
	32	3, 1, 3← 3, 0, 3	Ground		3/2		3/2	29 507.91		289
	32	3, 1, 3← 3, 0, 3	Ground		9/2		9/2	29 516.96		289
	32	3, 1, 3← 3, 0, 3	Ground		5/2		7/2	29 522.64		289
	32	3, 1, 3← 3, 0, 3	Ground		5/2		5/2	29 526.69		289
	32	3, 1, 3← 3, 0, 3	Ground		7/2		7/2	29 535.30		289
	32	3, 1, 3← 3, 0, 3	Ground		5/2		3/2	29 535.30		289
	32	3, 1, 3← 3, 0, 3	Ground		7/2		5/2	29 539.38		289
	32	3, 1, 3← 3, 0, 3	Ground		7/2		9/2	29 545.05		289
As^{75}D$_3$	33	1, 0← 0, 0	Ground		3/2		3/2	114 918.94	.3	586
	33	1, 0← 0, 0	Ground		5/2		3/2	114 960.57	.3	586
	33	1, 0← 0, 0	Ground		1/2		3/2	114 993.50	.3	586

B$_2$BrH$_5$ C$_s$ H$_2$BH$_2$BHBr

Isotopic Species	Pt. Gp.	Id. No.	A MHz	B MHz	C MHz	D$_J$ MHz	D$_{JK}$ MHz	Δ Amu A^2	κ
H$_2$B^{11}H$_2$B^{11}HBr^{79}	C$_s$	41		3 369.65 M	3 141.48 M				
H$_2$B^{11}H$_2$B^{11}HBr^{81}	C$_s$	42		3 350.75 M	3 124.95 M				
H$_2$B^{10}H$_2$B^{11}HBr^{79}	C$_s$	43		3 523.72 M	3 278.42 M				
H$_2$B^{11}H$_2$B^{10}HBr^{79}	C$_s$	45		3 398.62 M	3 176.05 M				
H$_2$B^{11}H$_2$B^{10}HBr^{81}	C$_s$	46		3 379.95 M	3 159.85 M				

Id. No.	μ$_a$ Debye	μ$_b$ Debye	μ$_c$ Debye	eQq Value(MHz)	Rel.	eQq Value(MHz)	Rel.	eQq Value(MHz)	Rel.	ω$_a$ 1/cm	d	ω$_b$ 1/cm	d	ω$_c$ 1/cm	d	ω$_d$ 1/cm	d
41				293	aa												
42				244	aa												

References:

ABC: 191   eQq: 191

For H$_2$B^{10}H$_2$B^{11}HBr^{81}, B$_0$ + C$_0$ = 6766.4 MHz. Ref. 191.

Bromodiborane   Spectral Line Table

Isotopic Species	Id. No.	Rotational Quantum Nos.	Vib. State	F$_1'$	F'	F$_1$	F	Frequency MHz	Acc. ±MHz	Ref.
H$_2$B^{11}H$_2$B^{11}HBr^{79}	41	3, 0, 3 ← 2, 0, 2	Ground		7/2		5/2	19 524.7		190
	41	3, 0, 3 ← 2, 0, 2	Ground		9/2		7/2	19 524.7		190
	41	3, 0, 3 ← 2, 0, 2	Ground		3/2		1/2	19 543.2		190
	41	3, 0, 3 ← 2, 0, 2	Ground		5/2		3/2	19 543.2		190
	41	3, 1, 2 ← 2, 1, 1	Ground		9/2		7/2	19 867.6		190
	41	3, 1, 2 ← 2, 1, 1	Ground		7/2		5/2	19 885.8		190
	41	3, 1, 3 ← 2, 1, 2	Ground		9/2		7/2	19 182.0		190
	41	3, 1, 3 ← 2, 1, 2	Ground		7/2		5/2	19 200.6		190
	41	3, 2, 1 ← 2, 2, 0	Ground		9/2		7/2	19 517.5		190
	41	3, 2, 1 ← 2, 2, 0	Ground		7/2		5/2	19 589.8		190
	41	3, 2, 2 ← 2, 2, 1	Ground		9/2		7/2	19 511.0		190
	41	3, 2, 2 ← 2, 2, 1	Ground		7/2		5/2	19 586.1		190
	41	4, 0, 4 ← 3, 0, 3	Ground		9/2		7/2	26 029.3		190
	41	4, 0, 4 ← 3, 0, 3	Ground		11/2		9/2	26 029.3		190
	41	4, 0, 4 ← 3, 0, 3	Ground		5/2		3/2	26 037.3		190
	41	4, 0, 4 ← 3, 0, 3	Ground		7/2		5/2	26 037.3		190
	41	4, 1, 4 ← 3, 1, 3	Ground		11/2		9/2	25 581.2		190
	41	4, 1, 4 ← 3, 1, 3	Ground		9/2		7/2	25 587.5		190
	41	4, 2, 2 ← 3, 2, 1	Ground		11/2		9/2	26 046.5		190
	41	4, 2, 2 ← 3, 2, 1	Ground		7/2		5/2	26 065.		190
H$_2$B^{11}H$_2$B^{11}HBr^{81}	42	3, 0, 3 ← 2, 0, 2	Ground		7/2		5/2	19 419.1		190
	42	3, 0, 3 ← 2, 0, 2	Ground		9/2		7/2	19 419.1		190
	42	3, 0, 3 ← 2, 0, 2	Ground		3/2		1/2	19 434.4		190
	42	3, 0, 3 ← 2, 0, 2	Ground		5/2		3/2	19 434.4		190
	42	3, 1, 2 ← 2, 1, 1	Ground		9/2		7/2	19 759.0		190

Isotopic Species	Id. No.	Rotational Quantum Nos.	Vib. State	Hyperfine				Frequency MHz	Acc. ±MHz	Ref.
				F_1'	F'	F_1	F			
H$_2$B^{11}H$_2$B^{11}HBr^{81}	42	3, 1, 2← 2, 1, 1	Ground		7/2		5/2	19 774.5		190
	42	3, 1, 3← 2, 1, 2	Ground		9/2		7/2	19 080.7		190
	42	3, 1, 3← 2, 1, 2	Ground		7/2		5/2	19 095.8		190
	42	3, 2, 1← 2, 2, 0	Ground		9/2		7/2	19 412.5		190
	42	3, 2, 1← 2, 2, 0	Ground		7/2		5/2	19 475.9		190
	42	3, 2, 2← 2, 2, 1	Ground		9/2		7/2	19 412.5		190
	42	3, 2, 2← 2, 2, 1	Ground		7/2		5/2	19 470.1		190
	42	4, 0, 4← 3, 0, 3	Ground		9/2		7/2	25 889.0		190
	42	4, 0, 4← 3, 0, 3	Ground		11/2		9/2	25 889.0		190
	42	4, 0, 4← 3, 0, 3	Ground		7/2		5/2	25 895.5		190
	42	4, 0, 4← 3, 0, 3	Ground		5/2		3/2	25 895.5		190
	42	4, 1, 3← 3, 1, 2	Ground		9/2		7/2	26 353.2		190
	42	4, 1, 4← 3, 1, 3	Ground		11/2		9/2	25 444.2		190
	42	4, 1, 4← 3, 1, 3	Ground		9/2		7/2	25 450.2		190
	42	4, 2, 2← 3, 2, 1	Ground		11/2		9/2	25 906.7		190
	42	4, 2, 2← 3, 2, 1	Ground		7/2		5/2	25 921.8		190
	42	4, 2, 3← 3, 2, 2	Ground		7/2		5/2	25 906.7		190
	42	4, 2, 3← 3, 2, 2	Ground		9/2		7/2	25 919.0		190
H$_2$B^{10}H$_2$B^{11}HBr^{79}	43	3, 0, 3← 2, 0, 2	Ground		9/2		7/2	20 396.9		190
	43	3, 0, 3← 2, 0, 2	Ground		7/2		5/2	20 396.9		190
	43	3, 1, 2← 2, 1, 1	Ground		9/2		7/2	20 766.4		190
	43	3, 1, 3← 2, 1, 2	Ground		9/2		7/2	20 029.4		190
	43	3, 1, 3← 2, 1, 2	Ground		7/2		5/2	20 047.4		190
	43	4, 0, 4← 3, 0, 3	Ground		9/2		7/2	27 191.6		190
	43	4, 0, 4← 3, 0, 3	Ground		11/2		9/2	27 191.6		190
H$_2$B^{10}H$_2$B^{11}HBr^{81}	44	3, 0, 3← 2, 0, 2	Ground		7/2		5/2	20 290.3		190
	44	3, 0, 3← 2, 0, 2	Ground		9/2		7/2	20 290.3		190
	44	4, 0, 4← 3, 0, 3	Ground		9/2		7/2	27 049.0		190
	44	4, 0, 4← 3, 0, 3	Ground		11/2		9/2	27 049.0		190
H$_2$B^{11}H$_2$B^{10}HBr^{79}	45	3, 0, 3← 2, 0, 2	Ground		7/2		5/2	19 715.7		190
	45	3, 0, 3← 2, 0, 2	Ground		9/2		7/2	19 715.7		190
	45	3, 1, 2← 2, 1, 1	Ground		9/2		7/2	20 049.5		190
	45	3, 1, 2← 2, 1, 1	Ground		7/2		5/2	20 067.6		190
	45	3, 1, 3← 2, 1, 2	Ground		9/2		7/2	19 380.6		190
	45	3, 1, 3← 2, 1, 2	Ground		7/2		5/2	19 398.8		190
	45	4, 0, 4← 3, 0, 3	Ground		9/2		7/2	26 284.7		190
	45	4, 0, 4← 3, 0, 3	Ground		11/2		9/2	26 284.7		190
	45	4, 0, 4← 3, 0, 3	Ground		5/2		3/2	26 292.4		190
	45	4, 0, 4← 3, 0, 3	Ground		7/2		5/2	26 292.4		190
H$_2$B^{11}H$_2$B^{10}HBr^{81}	46	3, 0, 3← 2, 0, 2	Ground		7/2		5/2	19 611.7		190
	46	3, 0, 3← 2, 0, 2	Ground		9/2		7/2	19 611.7		190
	46	3, 1, 2← 2, 1, 1	Ground		9/2		7/2	19 942.8		190
	46	3, 1, 2← 2, 1, 1	Ground		7/2		5/2	19 958.6		190
	46	3, 1, 3← 2, 1, 2	Ground		9/2		7/2	19 281.		190
	46	3, 1, 3← 2, 1, 2	Ground		7/2		5/2	19 297.		190

B₅H₉ $\qquad$ C$_{4v}$ $\qquad$ B₅H₉

Isotopic Species	Pt. Gp.	Id. No.	A MHz		B MHz		C MHz		D$_J$ MHz	D$_{JK}$ MHz	Δ Amu A²	κ
B$_5^{11}$H₉	C$_{4v}$	51	7 002.9	M	7 002.9	M	4 890.	M				
B^{10}B$_4^{11}$H₉	C$_{4v}$	52	7 089.8	M	7 089.8	M						
B$_5^{11}$D₉	C$_{4v}$	61	5 211.35	M	5 211.35	M	3 700.	M				

Id. No.	μ$_a$ Debye		μ$_b$ Debye		μ$_c$ Debye		eQq Value(MHz)	Rel.	eQq Value(MHz)	Rel.	eQq Value(MHz)	Rel.	ω$_a$ 1/cm	d	ω$_b$ 1/cm	d	ω$_c$ 1/cm	d	ω$_d$ 1/cm	d
51	0.	X	0.	X	2.13	M														
61	0.	X	0.	X	2.16	M														

References:

ABC: 1029 $\qquad$ μ: 529

Add. Ref. 358

Isotopic Species	Id. No.	Rotational Quantum Nos.	Vib. State	Hyperfine				Frequency MHz	Acc. ±MHz	Ref.
				F_1'	F'	F_1	F			
$B_5^{11}H_9$	51	2, ← 1,	Ground					28 011.4	.1	529
	51	3, ← 2,	Ground					42 017.3	.1	529
$B^{10}B_4^{11}H_9$	52	2, ← 1,	Ground					28 359.7	.2	529
	52	3, ← 2,	Ground					42 539.8	.2	529
$B^{11}B^{10}B_3^{11}H_9$	53	2, 1, 1← 1, 0, 1	Ground					28 187.1	.2	529
	53	2, 2, 0← 1, 1, 0	Ground					28 359.7	.2	529
	53	2, 2, 1← 1, 1, 1	Ground					28 513.8	.2	529
	53	3, 1, 2← 2, 0, 2	Ground					42 488.1	.2	529
	53	3, 2, 1← 2, 1, 1	Ground					42 286.5	.2	529
	53	3, 2, 2← 2, 1, 2	Ground					42 525.3	.4	529
	53	3, 3, 0← 2, 2, 0	Ground					42 563.1	.2	529
	53	3, 3, 1← 2, 2, 1	Ground					42 776.1	.2	529
$B_2^{10}B_3^{11}H_9$	54	2, 1, 1← 1, 0, 1	Ground					28 537.7	.2	529
	54	2, 2, 0← 1, 1, 0	Ground					28 715.	1.0	529
	54	2, 2, 1← 1, 1, 1	Ground					28 872.	1.0	529
$B^{11}B^{10}B^{11}B^{10}B^{11}H_9$	55	2, 1, 1← 1, 0, 1	Ground					28 364.5	.5	529
	55	2, 2, 0← 1, 1, 0	Ground					28 732.	1.0	529
	55	2, 2, 1← 1, 1, 1	Ground					29 024.	1.0	529
$B_3^{11}B_2^{10}H_9$	56	2, 1, 1← 1, 0, 1	Ground					28 686.5	1.0	529
	56	2, 2, 0← 1, 1, 0	Ground					28 689.5	.2	529
	56	2, 2, 1← 1, 1, 1	Ground					28 694.0	.5	529
$B_2^{11}B_3^{10}H_9$	57	2, 1, 1← 1, 0, 1	Ground					29 045.	1.0	529
	57	2, 2, 0← 1, 1, 0	Ground					29 052.	1.0	529
	57	2, 2, 1← 1, 1, 1	Ground					29 057.	1.0	529
$B_3^{10}B_2^{11}H_9$	58	2, 1, 1← 1, 0, 1	Ground					28 870.	1.0	529
	58	2, 2, 0← 1, 1, 0	Ground					29 045.	1.0	529
$B^{10}B^{11}B^{10}B^{11}B^{10}H_9$	59	2, 1, 1← 1, 0, 1	Ground					28 715.	1.0	529
	59	2, 2, 0← 1, 1, 0	Ground					29 095.	1.0	529
$B_5^{11}D_9$	61	2, ← 1,	Ground					20 845.4	.2	529
$B_4^{11}B^{10}D_9$	62	2, ← 1,	Ground					21 051.8	.2	529
$B_2^{10}B_3^{11}D_9$	63	2, 1, 1← 1, 0, 1	Ground					21 146.	1.0	529
	63	2, 2, 0← 1, 1, 0	Ground					21 241.	1.0	529
	63	2, 2, 1← 1, 1, 1	Ground					21 329.	1.0	529
$B_3^{11}B_2^{10}D_9$	64	2, 2, 0← 1, 1, 0	Ground					21 211.	1.0	529
$B_5^{b}H_9$	65	Not Reported	Ground					28 392.	.1	529
	65	Not Reported	Ground					28 989.	.1	529
	65	Not Reported	Ground					29 063.	.1	529
$B^{11}B^{10}B_3^{11}D_9$	66	2, 1, 1← 1, 0, 1	Ground					20 938.8	.2	529
	66	2, 2, 0← 1, 1, 0	Ground					21 032.1	.2	529
	66	2, 2, 1← 1, 1, 1	Ground					21 117.2	.2	529

BiCl$_3$ C$_{3v}$ BiCl$_3$

Isotopic Species	Pt. Gp.	Id. No.	A MHz	B MHz	C MHz	D$_J$ MHz	D$_{JK}$ MHz	Δ Amu A^2	κ
Bi^{209}Cl$_3^{35}$	C$_{3v}$	71		1 497. M		<0.03			

References:

ABC: 759   D$_J$: 759

Bismuth Trichloride   Spectral Line Table

Isotopic Species	Id. No.	Rotational Quantum Nos.	Vib. State	F$_1'$	F'	F$_1$	F	Frequency MHz	Acc. ±MHz	Ref.
Bi^{209}Cl$_3^{35}$	71	3. ← 2.	Ground					8 982.	1.	759
	71	6. ← 5.	Ground					17 944.	1.	759

BrF₃   C₂ᵥ   BrF₃

Isotopic Species	Pt. Gp.	Id. No.	A MHz		B MHz		C MHz		D_J MHz	D_JK MHz	Δ Amu A²	κ
Br⁷⁹F₃¹⁹	C₂ᵥ	81	10 841.25	M	4 077.57	M	2 958.59	M			.260	−.71609
Br⁸¹F₃¹⁹	C₂ᵥ	82	10 806.99	M	4 077.21	M	2 956.01	M			.250	−.71438

Id. No.	μ_a Debye		μ_b Debye		μ_c Debye		eQq Value(MHz)	Rel.	eQq Value(MHz)	Rel.	eQq Value(MHz)	Rel.	ω_a d 1/cm	ω_b d 1/cm	ω_c d 1/cm	ω_d d 1/cm
81	0.	X	1.0	M	0.	X	607.57	aa	501.78	bb	−1109.35	cc				
82							506.13	aa	419.21	bb	−925.34	cc				

References:

ABC: 765   Δ: 765   κ: 765   μ: 765   eQq: 765

Bromine Trifluoride

Spectral Line Table

Isotopic Species	Id. No.	Rotational Quantum Nos.	Vib. State	F_1'	F'	F_1	F	Frequency MHz	Acc. ±MHz	Ref.
Br⁷⁹F₃¹⁹	81	1, 1, 0← 1, 0, 1	Ground		3/2		5/2	7 691.42		765
	81	1, 1, 0← 1, 0, 1	Ground		3/2		1/2	7 812.84		765
	81	1, 1, 0← 1, 0, 1	Ground		5/2		3/2	7 816.22		765
	81	1, 1, 0← 1, 0, 1	Ground		5/2		5/2	7 967.96		765
	81	1, 1, 0← 1, 0, 1	Ground		1/2		3/2	8 039.00		765
	81	2, 1, 2← 1, 0, 1	Ground		7/2		5/2	19 667.92		765
	81	2, 1, 2← 1, 0, 1	Ground		5/2		3/2	19 793.83		765
	81	2, 1, 1← 2, 0, 2	Ground		5/2		5/2	8 911.82		765
	81	2, 1, 1← 2, 0, 2	Ground		7/2		7/2	9 215.06		765
	81	3, 1, 2← 3, 0, 3	Ground		7/2		7/2	11 022.16		765
	81	3, 1, 2← 3, 0, 3	Ground		5/2		5/2	11 162.14		765
	81	3, 1, 2← 3, 0, 3	Ground		9/2		9/2	11 319.44		765
	81	3, 1, 2← 3, 0, 3	Ground		3/2		3/2	11 462.68		765
Br⁸¹F₃¹⁹	82	1, 1, 0← 1, 0, 1	Ground		3/2		5/2	7 691.42		765
	82	1, 1, 0← 1, 0, 1	Ground		3/2		1/2	7 792.44		765
	82	1, 1, 0← 1, 0, 1	Ground		5/2		3/2	7 795.73		765
	82	1, 1, 0← 1, 0, 1	Ground		5/2		5/2	7 922.30		765
	82	1, 1, 0← 1, 0, 1	Ground		1/2		3/2	7 981.36		765
	82	2, 1, 2← 1, 0, 1	Ground		7/2		5/2	19 633.95		765
	82	2, 1, 2← 1, 0, 1	Ground		5/2		3/2	19 739.27		765
	82	2, 1, 1← 2, 0, 2	Ground		5/2		5/2	8 819.04		765
	82	2, 1, 1← 2, 0, 2	Ground		7/2		7/2	9 173.13		765
	82	3, 1, 2← 3, 0, 3	Ground		7/2		7/2	11 033.69		765
	82	3, 1, 2← 3, 0, 3	Ground		5/2		5/2	11 150.02		765
	82	3, 1, 2← 3, 0, 3	Ground		9/2		9/2	11 282.56		765
	82	3, 1, 2← 3, 0, 3	Ground		3/2		3/2	11 401.82		765

BrF₃Si — C_{3v} — SiF₃Br

Isotopic Species	Pt. Gp.	Id. No.	A MHz	B MHz	C MHz	D_J MHz	D_JK MHz	Δ Amu A²	κ
$Si^{28}F_3^{19}Br^{79}$	C_{3v}	91		1 549.98 M	1 549.98 M		.0008		
$Si^{28}F_3^{19}Br^{81}$	C_{3v}	92		1 534.14 M	1 534.14 M		.0008		

Id. No.	μ_a Debye	μ_b Debye	μ_c Debye	eQq Value(MHz)	Rel.	eQq Value(MHz)	Rel.	eQq Value(MHz)	Rel.	ω_a d 1/cm	ω_b d 1/cm	ω_c d 1/cm	ω_d d 1/cm
91				440	Br^{79}								
92				370	Br^{81}								

References:

ABC: 311 D_{JK}: 311   eQq: 311

Add. Ref. 233,493

Bromotrifluorosilane

Spectral Line Table

Isotopic Species	Id. No.	Rotational Quantum Nos.	Vib. State	F_1'	F'	F_1	F	Frequency MHz	Acc. ±MHz	Ref.
$Si^{28}F_3^{19}Br^{79}$	91	16, 0←15, 0	Ground	29/2		27/2		49 597.80		311
	91	16, 0←15, 0	Ground	35/2		33/2		49 599.22		311
	91	16, 0←15, 0	Ground	31/2		31/2		49 599.22		311
	91	16, 0←15, 0	Ground	31/2		29/2		49 599.86		311
	91	16, 0←15, 0	Ground	29/2		27/2		49 599.86		311
	91	16, 1←15, 1	Ground	35/2		33/2		49 599.22		311
	91	16, 1←15, 1	Ground	33/2		31/2		49 599.22		311
	91	16, 1←15, 1	Ground	31/2		29/2		49 599.86		311
	91	16, 1←15, 1	Ground	29/2		27/2		49 599.86		311
	91	16, 2←15, 2	Ground	31/2		29/2		49 599.22		311
	91	16, 2←15, 2	Ground	29/2		27/2		49 599.22		311
	91	16, 2←15, 2	Ground	35/2		33/2		49 599.22		311
	91	16, 2←15, 2	Ground	31/2		29/2		49 599.86		311
	91	16, 3←15, 3	Ground	35/2		33/2		49 598.42		311
	91	16, 3←15, 3	Ground	29/2		27/2		49 599.22		311
	91	16, 3←15, 3	Ground	33/2		31/2		49 599.86		311
	91	16, 3←15, 3	Ground	31/2		29/2		49 599.86		311
	91	16, 4←15, 4	Ground	35/2		33/2		49 597.80		311
	91	16, 4←15, 4	Ground	33/2		31/2		49 599.86		311
	91	16, 4←15, 4	Ground	31/2		29/2		49 600.90		311
	91	16, 5←15, 5	Ground	29/2		27/2		49 596.60		311
	91	16, 5←15, 5	Ground	35/2		33/2		49 596.60		311
	91	16, 6←15, 6	Ground	35/2		33/2		49 595.72		311
	91	16, 6←15, 6	Ground	29/2		27/2		49 595.72		311
	91	16, 6←15, 6	Ground	31/2		29/2		49 601.56		311
	91	16, 6←15, 6	Ground	33/2		31/2		49 601.56		311
	91	16, 7←15, 7	Ground	35/2		33/2		49 594.38		311
	91	16, 7←15, 7	Ground	29/2		27/2		49 594.38		311
	91	16, 8←15, 8	Ground	29/2		27/2		49 592.25		311
	91	16, 8←15, 8	Ground	35/2		33/2		49 593.00		311

Isotopic Species	Id. No.	Rotational Quantum Nos.	Vib. State	F_I'	F'	F_I	F	Frequency MHz	Acc. ±MHz	Ref.
Si^{28}F$_3^{19}$Br^{79}	91	16, 8←15, 8	Ground		33/2		31/2	49 603.30		311
	91	16, 9←15, 9	Ground		35/2		33/2	49 591.66		311
	91	16, 9←15, 9	Ground		31/2		29/2	49 603.30		311
	91	16, 5←16, 5	Ground		31/2		29/2	49 600.90		311
	91	16, 5←16, 5	Ground		33/2		31/2	49 600.90		311
Si^{28}F$_3^{19}$Br^{81}	92	16, 0←15, 0	Ground		33/2		31/2	49 092.38		311
	92	16, 0←15, 0	Ground		35/2		33/2	49 092.38		311
	92	16, 0←15, 0	Ground		31/2		29/2	49 092.88		311
	92	16, 0←15, 0	Ground		29/2		27/2	49 092.88		311
	92	16, 1←15, 1	Ground		35/2		33/2	49 092.38		311
	92	16, 1←15, 1	Ground		33/2		31/2	49 092.38		311
	92	16, 1←15, 1	Ground		29/2		27/2	49 092.88		311
	92	16, 1←15, 1	Ground		31/2		29/2	49 092.88		311
	92	16, 2←15, 2	Ground		33/2		31/2	49 092.38		311
	92	16, 2←15, 2	Ground		29/2		27/2	49 092.38		311
	92	16, 2←15, 2	Ground		31/2		29/2	49 092.88		311
	92	16, 3←15, 3	Ground		35/2		33/2	49 091.70		311
	92	16, 3←15, 3	Ground		31/2		29/2	49 092.88		311
	92	16, 3←15, 3	Ground		33/2		31/2	49 092.88		311
	92	16, 4←15, 4	Ground		29/2		27/2	49 091.00		311
	92	16, 4←15, 4	Ground		35/2		33/2	49 091.00		311
	92	16, 4←15, 4	Ground		33/2		31/2	49 092.88		311
	92	16, 4←15, 4	Ground		31/2		29/2	49 093.60		311
	92	16, 5←15, 5	Ground		29/2		27/2	49 090.22		311
	92	16, 5←15, 5	Ground		35/2		33/2	49 090.22		311
	92	16, 5←15, 5	Ground		33/2		31/2	49 093.60		311
	92	16, 5←15, 5	Ground		31/2		29/2	49 093.60		311
	92	16, 6←15, 6	Ground		35/2		33/2	49 089.18		311
	92	16, 6←15, 6	Ground		29/2		27/2	49 089.18		311
	92	16, 6←15, 6	Ground		31/2		29/2	49 094.20		311
	92	16, 6←15, 6	Ground		33/2		31/2	49 094.20		311
	92	16, 7←15, 7	Ground		35/2		33/2	49 088.01		311
	92	16, 7←15, 7	Ground		29/2		27/2	49 088.01		311

BrGeH₃ — C₃ᵥ — GeH₃Br

Isotopic Species	Pt. Gp.	Id. No.	A MHz	B MHz	C MHz	D_J MHz	D_{JK} MHz	Δ Amu A²	κ
Ge⁷⁰H₃Br⁷⁹	C₃ᵥ	101		2 438.57 M					
Ge⁷⁰H₃Br⁸¹	C₃ᵥ	102		2 410.17 M					
Ge⁷²H₃Br⁷⁹	C₃ᵥ	103		2 406.42 M					
Ge⁷²H₃Br⁸¹	C₃ᵥ	104		2 378.01 M					
Ge⁷⁴H₃Br⁷⁹	C₃ᵥ	105		2 375.88 M					
Ge⁷⁴H₃Br⁸¹	C₃ᵥ	106		2 347.46 M					
Ge⁷⁶H₃Br⁷⁹	C₃ᵥ	107		2 346.84 M					
Ge⁷⁶H₃Br⁸¹	C₃ᵥ	108		2 318.37 M					

Id. No.	μ_a Debye	μ_b Debye	μ_c Debye	eQq Value(MHz)	Rel.	eQq Value(MHz)	Rel.	eQq Value(MHz)	Rel.	ω_a d 1/cm	ω_b d 1/cm	ω_c d 1/cm	ω_d d 1/cm
101				380	Br⁷⁹								
102				321	Br⁸¹								

References:

ABC: 230   eQq: 230

Add. Ref. 375

Bromogermane — Spectral Line Table

Isotopic Species	Id. No.	Rotational Quantum Nos.	Vib. State	F₁'	F'	F₁	F	Frequency MHz	Acc. ±MHz	Ref.
Ge⁷⁰H₃Br⁷⁹	101	5, ← 4,	Ground					24 385.54	1.	230
Ge⁷⁰H₃Br⁸¹	102	5, ← 4,	Ground					24 101.61	1.	230
Ge⁷²H₃Br⁷⁹	103	5, ← 4,	Ground					24 064.35	1.	230
Ge⁷²H₃Br⁸¹	104	5, ← 4,	Ground					23 780.11	1.	230
Ge⁷⁴H₃Br⁷⁹	105	5, ← 4,	Ground					23 758.99	1.	230
Ge⁷⁴H₃Br⁸¹	106	5, ← 4,	Ground					23 474.75	1.	230
Ge⁷⁶H₃Br⁷⁹	107	5, ← 4,	Ground					23 468.0	1.	230
Ge⁷⁶H₃Br⁸¹	108	5, ← 4,	Ground					23 183.8	1.	230

Bromosilicane, Silyl Bromide

BrH$_3$Si   C$_{3v}$ SiH$_3$Br

Isotopic Species	Pt. Gp.	Id. No.	A MHz	B MHz	C MHz	D$_J$ MHz	D$_{JK}$ MHz	Δ Amu A^2	κ
Si^{28}H$_3$Br^{79}	C$_{3v}$	111		4 321.72 M	4 321.72 M				
Si^{28}H$_3$Br^{81}	C$_{3v}$	112		4 292.64 M	4 292.64 M				
Si^{29}H$_3$Br^{79}	C$_{3v}$	113		4 232.96 M	4 232.96 M				
Si^{29}H$_3$Br^{81}	C$_{3v}$	114		4 203.70 M	4 203.70 M				
Si^{30}H$_3$Br^{79}	C$_{3v}$	115		4 149.39 M	4 149.39 M				
Si^{30}H$_3$Br^{81}	C$_{3v}$	116		4 120.09 M	4 120.09 M				

Id. No.	μ$_a$ Debye	μ$_b$ Debye	μ$_c$ Debye	eQq Value(MHz)	Rel.	eQq Value(MHz)	Rel.	eQq Value(MHz)	Rel.	ω$_a$ d 1/cm	ω$_b$ d 1/cm	ω$_c$ d 1/cm	ω$_d$ d 1/cm
111	1.32 M	0. X	0. X	336	Br^{79}								
112				278	Br^{81}								

References:

ABC: 159   μ: 375   eQq: 159

Add. Ref. 230,493

Bromosilane

Isotopic Species	Id. No.	Rotational Quantum Nos.	Vib. State	F$_1'$	F'	F$_1$	F	Frequency MHz	Acc. ±MHz	Ref.
Si^{28}H$_3$Br^{79}	111	2, ← 1,	Ground					17 287.30	.08	159
	111	3, ← 2,	Ground					25 930.32	.08	159
Si^{28}H$_3$Br^{81}	112	2; ← 1,	Ground					17 170.45	.08	159
	112	3, ← 2,	Ground					25 755.89	.08	159 [1]
Si^{29}H$_3$Br^{79}	113	3, ← 2,	Ground					25 397.80	.08	159
Si^{29}H$_3$Br^{81}	114	3, ← 2,	Ground					25 222.21	.08	159
Si^{30}H$_3$Br^{79}	115	3, ← 2,	Ground					24 896.33	.08	159
Si^{30}H$_3$Br^{81}	116	3, ← 2,	Ground					24 720.57	.08	159

1. This line is possibly the same one for which the Stark effect was measured by Mays and Dailey, as reported in JCP *20*, 1965 (1952).

BrNO C_s NOBr

Isotopic Species	Pt. Gp.	Id. No.	A MHz		B MHz		C MHz		D_J MHz	D_{JK} MHz	Δ Amu A^2	κ
$N^{14}O^{16}Br^{79}$	C_s	121	83 340.	M	3 747.24	M	3 586.00	M				
$N^{14}O^{16}Br^{81}$	C_s	122	83 340.	M	3 722.49	M	3 563.34	M				

Id. No.	μ_a Debye		μ_b Debye		μ_c Debye		eQq Value(MHz)	Rel.	eQq Value(MHz)	Rel.	eQq Value(MHz)	Rel.	ω_a d 1/cm	ω_b d 1/cm	ω_c d 1/cm	ω_d d 1/cm
121	1.80	M	.50	¹M	0.	X	388.3	aa	−239.5	bb	−148.8	cc				
122							325.5	aa	−200.2	bb	−125.3	cc				

1. Calculated, since $\mu_a = 1.80$ Debye and $\mu = 1.87$ Debye.

References:
ABC: 727 μ: 856   eQq: 727

Add. Ref. 647,669,798
For species 121, eQq(xx) = −290.2 MHz, eQq(yy) = −148.8 MHz.
For species 122, eQq(xx) = −242.6 MHz, eQq(yy) = −125.3 MHz. Ref. 727.

Nitrosyl Bromide   Spectral Line Table

Isotopic Species	Id. No.	Rotational Quantum Nos.	Vib. State	F_1'	F'	F_1	F	Frequency MHz	Acc. ±MHz	Ref.
$N^{14}O^{16}Br^{79}$	121	3, 0, 3 ← 2, 0, 2	Ground		9/2		7/2	21 993.86	.25	727
	121	3, 0, 3 ← 2, 0, 2	Ground		7/2		5/2	21 993.86	.25	727
	121	3, 0, 3 ← 2, 0, 2	Ground		3/2		1/2	22 016.97	.25	727
	121	3, 0, 3 ← 2, 0, 2	Ground		5/2		3/2	22 016.97	.25	727
	121	3, 1, 2 ← 2, 1, 1	Ground		9/2		7/2	22 230.38	.25	727
	121	3, 1, 2 ← 2, 1, 1	Ground		3/2		1/2	22 233.60	.25	727
	121	3, 1, 2 ← 2, 1, 1	Ground		7/2		5/2	22 254.57	.25	727
	121	3, 1, 2 ← 2, 1, 1	Ground		5/2		3/2	22 258.34	.25	727
	121	3, 1, 3 ← 2, 1, 2	Ground		5/2		5/2	21 742.93	.25	727
	121	3, 1, 3 ← 2, 1, 2	Ground		3/2		1/2	21 745.74	.25	727
	121	3, 1, 3 ← 2, 1, 2	Ground		9/2		7/2	21 747.91	.25	727
	121	3, 1, 3 ← 2, 1, 2	Ground		5/2		3/2	21 769.84	.25	727
	121	3, 1, 3 ← 2, 1, 2	Ground		7/2		5/2	21 771.95	.25	727
	121	3, 2, 1 ← 2, 2, 0	Ground		5/2		7/2	21 972.29	.25	727
	121	3, 2, 1 ← 2, 2, 0	Ground		7/2		7/2	21 972.29	.25	727
	121	3, 2, 1 ← 2, 2, 0	Ground		9/2		7/2	21 972.29	.25	727
	121	3, 2, 1 ← 2, 2, 0	Ground		3/2		3/2	22 001.07	.25	727
	121	3, 2, 1 ← 2, 2, 0	Ground		5/2		3/2	22 001.07	.25	727
	121	3, 2, 1 ← 2, 2, 0	Ground		3/2		5/2	22 069.77	.25	727
	121	3, 2, 1 ← 2, 2, 0	Ground		7/2		5/2	22 069.77	.25	727
	121	3, 2, 1 ← 2, 2, 0	Ground		5/2		5/2	22 069.77	.25	727
	121	3, 2, 2 ← 2, 2, 1	Ground		5/2		7/2	21 972.29	.25	727
	121	3, 2, 2 ← 2, 2, 1	Ground		9/2		7/2	21 972.29	.25	727
	121	3, 2, 2 ← 2, 2, 1	Ground		7/2		7/2	21 972.29	.25	727
	121	3, 2, 2 ← 2, 2, 1	Ground		3/2		3/2	22 001.07	.25	727
	121	3, 2, 2 ← 2, 2, 1	Ground		5/2		3/2	22 001.07	.25	727
	121	3, 2, 2 ← 2, 2, 1	Ground		3/2		5/2	22 069.77	.25	727
	121	3, 2, 2 ← 2, 2, 1	Ground		7/2		5/2	22 069.77	.25	727
	121	3, 2, 2 ← 2, 2, 1	Ground		5/2		5/2	22 069.77	.25	727
$N^{14}O^{16}Br^{81}$	122	3, 0, 3 ← 2, 0, 2	Ground		9/2		7/2	21 852.23	.25	727
	122	3, 0, 3 ← 2, 0, 2	Ground		7/2		5/2	21 852.23	.25	727
	122	3, 0, 3 ← 2, 0, 2	Ground		5/2		3/2	21 871.10	.25	727
	122	3, 0, 3 ← 2, 0, 2	Ground		3/2		1/2	21 871.10	.25	727
	122	3, 1, 2 ← 2, 1, 1	Ground		9/2		7/2	22 086.65	.25	727

Isotopic Species	Id. No.	Rotational Quantum Nos.	Vib. State	Hyperfine				Frequency MHz	Acc. ±MHz	Ref.
				F_1'	F'	F_1	F			
$N^{14}O^{16}Br^{81}$	122	3, 1, 2← 2, 1, 1	Ground		3/2		1/2	22 089.42	.25	727
	122	3, 1, 2← 2, 1, 1	Ground		7/2		5/2	22 106.69	.25	727
	122	3, 1, 2← 2, 1, 1	Ground		5/2		3/2	22 110.25	.25	727
	122	3, 1, 3← 2, 1, 2	Ground		5/2		5/2	21 606.22	.25	727
	122	3, 1, 3← 2, 1, 2	Ground		3/2		1/2	21 608.49	.25	727
	122	3, 1, 3← 2, 1, 2	Ground		9/2		7/2	21 609.91	.25	727
	122	3, 1, 3← 2, 1, 2	Ground		5/2		3/2	21 628.84	.25	727
	122	3, 1, 3← 2, 1, 2	Ground		7/2		5/2	21 630.32	.25	727
	122	3, 2, 1← 2, 2, 0	Ground		5/2		7/2	21 835.19	.25	727
	122	3, 2, 1← 2, 2, 0	Ground		9/2		7/2	21 835.19	.25	727
	122	3, 2, 1← 2, 2, 0	Ground		7/2		7/2	21 835.19	.25	727
	122	3, 2, 1← 2, 2, 0	Ground		5/2		3/2	21 858.69	.25	727
	122	3, 2, 1← 2, 2, 0	Ground		3/2		3/2	21 858.69	.25	727
	122	3, 2, 1← 2, 2, 0	Ground		3/2		5/2	21 916.77	.25	727
	122	3, 2, 1← 2, 2, 0	Ground		5/2		5/2	21 916.77	.25	727
	122	3, 2, 1← 2, 2, 0	Ground		7/2		5/2	21 916.77	.25	727
	122	3, 2, 2← 2, 2, 1	Ground		7/2		7/2	21 835.19	.25	727
	122	3, 2, 2← 2, 2, 1	Ground		5/2		7/2	21 835.19	.25	727
	122	3, 2, 2← 2, 2, 1	Ground		9/2		7/2	21 835.19	.25	727
	122	3, 2, 2← 2, 2, 1	Ground		5/2		3/2	21 858.69	.25	727
	122	3, 2, 2← 2, 2, 1	Ground		3/2		3/2	21 858.69	.25	727
	122	3, 2, 2← 2, 2, 1	Ground		5/2		5/2	21 916.77	.25	727
	122	3, 2, 2← 2, 2, 1	Ground		3/2		5/2	21 916.77	.25	727
	122	3, 2, 2← 2, 2, 1	Ground		7/2		5/2	21 916.77	.25	727
$N^{14}O^{16}Br^b$	123	4, , ← 3, ,						28 820.1		798
	123	4, , ← 3, ,						28 823.9		798
	123	4, , ← 3, ,						28 827.4		798
	123	4, , ← 3, ,						28 829.4		798
	123	4, , ← 3, ,						28 831.7		798
	123	4, , ← 3, ,						29 007.1		798
	123	4, , ← 3, ,						29 014.9		798
	123	4, , ← 3, ,						29 122.8		798
	123	4, , ← 3, ,						29 132.6		798
	123	4, , ← 3, ,						29 154.0		798
	123	4, , ← 3, ,						29 196.0		798
	123	4, , ← 3, ,						29 266.4		798
	123	4, , ← 3, ,						29 307.9		798
	123	4, , ← 3, ,						29 320.0		798
	123	4, , ← 3, ,						29 326.3		798
	123	4, , ← 3, ,						29 330.4		798
	123	4, , ← 3, ,						29 359.1		798
	123	4, , ← 3, ,						29 463.3		798
	123	4, , ← 3, ,						29 468.4		798
	123	4, , ← 3, ,						29 504.3		798
	123	4, , ← 3, ,						29 517.8		798
	123	4, , ← 3, ,						29 618.2		798
	123	4, , ← 3, ,						29 657.8		798
	123	4, , ← 3, ,						29 664.0		798
	123	5, , ← 4, ,						35 870.		798
	123	5, , ← 4, ,						36 030.		798
	123	5, , ← 4, ,						36 280.		798
	123	5, , ← 4, ,						36 420.		798
	123	5, , ← 4, ,						36 480.		798
	123	5, , ← 4, ,						36 630.		798
	123	5, , ← 4, ,						36 710.		798
	123	5, , ← 4, ,						36 810.		798
	123	5, , ← 4, ,						36 830.		798
	123	5, , ← 4, ,						37 100.		798

Br_3P C_{3v} PBr_3

Isotopic Species	Pt. Gp.	Id. No.	A MHz	B MHz		C MHz		D_J MHz	D_{JK} MHz	Δ Amu A²	κ
$P^bBr_3^{79}$	C_{3v}	131		996.4	M	996.4	M				
$P^bBr_3^{81}$	C_{3v}	132		974.4	M	974.4	M				

Id. No.	μ_a Debye		μ_b Debye		μ_c Debye		eQq Value(MHz)	Rel.	eQq Value(MHz)	Rel.	eQq Value(MHz)	Rel.	ω_a d 1/cm	ω_b d 1/cm	ω_c d 1/cm	ω_d d 1/cm
131	.61	M	0.	X	0.	X										

References:

ABC: 252 μ: 534

No Spectral Lines

CBrN $C_{\infty v}$ BrCN

Isotopic Species	Pt. Gp.	Id. No.	A MHz	B MHz		C MHz		D_J MHz	D_{JK} MHz	Δ Amu A²	κ
$Br^{79}C^{12}N^{14}$	$C_{\infty v}$	141		4120.230	M	4120.230	M	.00088			
$Br^{81}C^{12}N^{14}$	$C_{\infty v}$	142		4096.804	M	4096.804	M	.00087			
$Br^{79}C^{13}N^{14}$	$C_{\infty v}$	143		4073.373	M	4073.373	M				
$Br^{81}C^{13}N^{14}$	$C_{\infty v}$	144		4049.608	M	4049.608	M				
$Br^{79}C^{12}N^{15}$	$C_{\infty v}$	145		3944.846	M	3944.846	M				
$Br^{81}C^{12}N^{15}$	$C_{\infty v}$	146		3921.787	M	3921.787	M				

Id. No.	μ_a Debye		μ_b Debye		μ_c Debye		eQq Value(MHz)	Rel.	eQq Value(MHz)	Rel.	eQq Value(MHz)	Rel.	ω_a d 1/cm	ω_b d 1/cm	ω_c d 1/cm	ω_d d 1/cm
141							−3.83	N^{14}	685.9	Br^{79}			580 1	368 2	2187 1	
142	2.94	M	0.	X	0.	X			572.5	Br^{81}						

References:

ABC: 240,401 D_J: 663 μ: 112   eQq: 112,401 ω: 1029

Add. Ref. 42,59,61,62,63,64,70,81,97,105,200,569,601,636

Excited Mode:	(0, 1, 0)	(1, 0, 0)	(0, 2, 0)	
Species	eQq(Br) in MHz			q_l(MHz)
141	682.84	688.5	680.9	3.912
142	570.44	575.2	568.1	3.845
Ref.	401	773	773	112

For a discussion of Fermi resonance as exhibited by this molecule, and calculated values of related parameters, see references 663 and 401.

Isotopic Species	Id. No.	Rotational Quantum Nos.	Vib. State v_a ; v_b^l ; v_c	Hyperfine				Frequency MHz	Acc. ±MHz	Ref.
				F_1'	F'	F_1	F			
$Br^{79}C^{12}N^{14}$	141	$1 \leftarrow 0$	1; 0, 0; 0	1/2		3/2		8 047.08	.1	773
	141	$1 \leftarrow 0$	Ground	1/2		3/2		8 070.45	.1	773
	141	$1 \leftarrow 0$	0; 2, 0; 0	1/2		3/2		8 109.78	.1	773
	141	$1 \leftarrow 0$	1; 0, 0; 0	5/2		3/2		8 183.87	.1	773
	141	$1 \leftarrow 0$	Ground	5/2	5/2	3/2	5/2	8 206.18	.1	773
	141	$1 \leftarrow 0$	Ground	5/2	7/2	3/2	5/2	8 206.92	.1	773
	141	$1 \leftarrow 0$	Ground	5/2	3/2	3/2	5/2	8 207.39	.1	773
	141	$1 \leftarrow 0$	0; 2, 0; 0	5/2		3/2		8 245.29	.1	773
	141	$1 \leftarrow 0$	1; 0, 0; 0	3/2		3/2		8 356.48	.1	773
	141	$1 \leftarrow 0$	Ground	3/2	5/2	3/2	3/2	8 377.95	.1	773
	141	$1 \leftarrow 0$	Ground	3/2	3/2	3/2	3/2	8 378.58	.1	773
	141	$1 \leftarrow 0$	Ground	3/2	1/2	3/2	3/2	8 379.38	.1	773
	141	$1 \leftarrow 0$	0; 2, 0; 0	3/2		3/2		8 416.30	.1	773
	141	$2 \leftarrow 1$	Ground	1/2	3/2	3/2	3/2	16 172.30	.1	773
	141	$2 \leftarrow 1$	Ground	1/2	3/2	3/2	5/2	16 172.90	.1	773
	141	$2 \leftarrow 1$	Ground	1/2	3/2	3/2	1/2	16 173.40	.1	773
	141	$2 \leftarrow 1$	1; 0, 0; 0	3/2		3/2		16 299.07	.1	773
	141	$2 \leftarrow 1$	Ground	3/2	5/2	3/2	5/2	16 345.38	.1	773
	141	$2 \leftarrow 1$	Ground	3/2	3/2	3/2	3/2	16 346.17	.1	773
	141	$2 \leftarrow 1$	1; 0, 0; 0	7/2		5/2		16 420.97	.1	773
	141	$2 \leftarrow 1$	1; 0, 0; 0	5/2		3/2		16 420.97	.1	773
	141	$2 \leftarrow 1$	1; 0, 0; 0	1/2		1/2		16 435.3	.1	773
	141	$2 \leftarrow 1$	Ground	7/2		5/2		16 466.35	.1	773
	141	$2 \leftarrow 1$	Ground	5/2		3/2		16 466.35	.1	773
	141	$2 \leftarrow 1$	Ground	1/2		1/2		16 480.92	.1	773
	141	$2 \leftarrow 1$	Ground	3/2		5/2		16 517.8	.1	773
	141	$2 \leftarrow 1$	0; 2, 0; 0	7/2		5/2		16 542.45	.1	773
	141	$2 \leftarrow 1$	0; 2, 0; 0	5/2		3/2		16 542.45	.1	773
	141	$2 \leftarrow 1$	0; 2, 0; 0	1/2		1/2		16 556.67	.1	773
	141	$2 \leftarrow 1$	1; 0, 0; 0	5/2		5/2		16 593.38	.1	773
	141	$2 \leftarrow 1$	1; 0, 0; 0	3/2		1/2		16 608.6	.1	773
	141	$2 \leftarrow 1$	Ground	5/2		5/2		16 638.40	.1	773
	141	$2 \leftarrow 1$	Ground	3/2		1/2		16 653.58	.1	773
	141	$2 \leftarrow 1$	0; 2, 0; 0	5/2		5/2		16 712.9	.1	773
	141	$2 \leftarrow 1$	0; 2, 0; 0	3/2		1/2		16 727.97	.1	773
	141	$3 \leftarrow 2$	Ground	3/2		3/2		24 583.00		112
	141	$3 \leftarrow 2$	Ground	5/2		5/2		24 633.71		112
	141	$3 \leftarrow 2$	1; 0, 0; 0	3/2		1/2		24 687.11		112
	141	$3 \leftarrow 2$	1; 0, 0; 0	5/2		3/2		24 687.11		112
	141	$3 \leftarrow 2$	Ground	7/2		5/2		24 713.05		112
	141	$3 \leftarrow 2$	Ground	9/2		7/2		24 713.05		112
	141	$3 \leftarrow 2$	Ground	3/2		1/2		24 755.22		112
	141	$3 \leftarrow 2$	Ground	5/2		3/2		24 755.22		112
	141	$3 \leftarrow 2$	0; 1, −1; 0	3/2		1/2		24 760.76		112
	141	$3 \leftarrow 2$	0; 1, −1; 0	9/2		7/2		24 760.76		112
	141	$3 \leftarrow 2$	0; 1, +1; 0	3/2		1/2		24 784.02		112
	141	$3 \leftarrow 2$	0; 1, +1; 0	9/2		7/2		24 784.02		112
	141	$3 \leftarrow 2$	1; 1, −1; 0	5/2		3/2		24 803.00		112
	141	$3 \leftarrow 2$	0; 1, −1; 0	7/2		5/2		24 803.00		112
	141	$3 \leftarrow 2$	0; 1, +1; 0	5/2		3/2		24 826.70		112

Isotopic Species	Id. No.	Rotational Quantum Nos.	Vib. State v_a ; v_b^l ; v_c	Hyperfine				Frequency MHz	Acc. ±MHz	Ref.
				F_1'	F'	F_1	F			
Br^{79}C^{12}N^{14}	141	3 ← 2	0; 1, +1; 0	7/2		5/2		24 826.70		112
	141	3 ← 2	0; 2, ±2; 0	3/2		3/2		24 860.6		1020
	141	3 ← 2	0; 2, ±2; 0	5/2		3/2		24 860.6		1020
	141	3 ← 2	Ground	7/2		7/2		24 884.57		112
	141	3 ← 2	0; 1, −1; 0	7/2		7/2		24 890.0		1020
	141	3 ← 2	0; 2, ±2; 0	5/2		5/2		24 981.5		1020
	141	3 ← 2	0; 2, ±2; 0	3/2		5/2		24 981.5		1020
	141	3 ← 2	0; 2, ±2; 0	7/2		5/2		24 981.5		1020
	141	3 ← 2	0; 1, +1; 0	7/2		7/2		25 006.0		1020
	141	4 ← 3	Ground	5/2		5/2		32 804.56		106
	141	4 ← 3	Ground	11/2		9/2		32 956.68		106
	141	4 ← 3	Ground	9/2		7/2		32 956.68		106
	141	4 ← 3	Ground	7/2		5/2		32 976.40		106
	141	4 ← 3	Ground	5/2		3/2		32 976.40		106
	141	6 ← 5	Ground	9/2		9/2		49 274.99	.10	401
	141	6 ← 5	1; 0, 0; 0	15/2		13/2		49 302.27	.10	401
	141	6 ← 5	1; 0, 0; 0	13/2		11/2		49 302.27	.10	401
	141	6 ← 5	1; 0, 0; 0	9/2		7/2		49 310.06	.10	401
	141	6 ← 5	1; 0, 0; 0	11/2		9/2		49 310.06	.10	401
	141	6 ← 5	Ground	11/2		11/2		49 398.90	.10	401
	141	6 ← 5	0; 1, −1; 0	9/2		9/2		49 403.74	.15	401
	141	6 ← 5	Ground	13/2	13/2	11/2	13/2	49 438.01	.15	401
	141	6 ← 5	Ground	15/2	15/2	13/2	15/2	49 438.01	.15	401
	141	6 ← 5	Ground	13/2		11/2		49 439.12	.10	401
	141	6 ← 5	Ground	15/2		13/2		49 439.12	.10	401
	141	6 ← 5	Ground	13/2	11/2	11/2	11/2	49 440.40	.15	401
	141	6 ← 5	Ground	15/2	13/2	13/2	13/2	49 440.40	.15	401
	141	6 ← 5	1; 1, −1; 0	15/2		13/2		49 443.83	.20	401
	141	6 ← 5	Ground	9/2	9/2	7/2	9/2	49 445.65	.15	401
	141	6 ← 5	Ground	11/2	11/2	9/2	11/2	49 445.65	.15	401
	141	6 ← 5	Ground	11/2		9/2		49 446.90	.10	401
	141	6 ← 5	Ground	9/2		7/2		49 446.90	.10	401
	141	6 ← 5	Ground	9/2	7/2	7/2	7/2	49 448.31	.15	401
	141	6 ← 5	1; 1, −1; 0	13/2		11/2		49 448.31	.20	401
	141	6 ← 5	Ground	11/2	9/2	9/2	9/2	49 448.31	.15	401
	141	6 ← 5	0; 1, +1; 0	9/2		9/2		49 452.59	.15	401
	141	6 ← 5	1; 1, +1; 0	9/2		7/2		49 496.11	.20	401
	141	6 ← 5	1; 1, +1; 0	11/2		9/2		49 501.09	.20	401
	141	6 ← 5	0; 1, −1; 0	11/2		11/2		49 520.03	.15	401
	141	6 ← 5	0; 1, −1; 0	15/2		13/2		49 552.62	.10	401
	141	6 ← 5	0; 1, −1; 0	13/2		11/2		49 557.49	.10	401
	141	6 ← 5	0; 1, −1; 0	9/2		7/2		49 558.32	.10	401
	141	6 ← 5	0; 1, −1; 0	11/2		9/2		49 563.28	.10	401
	141	6 ← 5	0; 1, +1; 0	11/2		11/2		49 567.45	.15	401
	141	6 ← 5	0; 1, +1; 0	15/2		13/2		49 599.57	.10	401
	141	6 ← 5	0; 1, +1; 0	13/2		11/2		49 604.35	.10	401
	141	6 ← 5	0; 1, +1; 0	9/2		7/2		49 605.29	.10	401
	141	6 ← 5	Ground	13/2		13/2		49 610.43	.15	401
	141	6 ← 5	0; 1, +1; 0	11/2		9/2		49 610.43	.15	401
	141	6 ← 5	0; 2, 0; 0	13/2		11/2		49 666.19	.10	401

Isotopic Species	Id. No.	Rotational Quantum Nos.	Vib. State v_a ; v_b^l ; v_c	F_1'	F'	F_1	F	Frequency MHz	Acc. ±MHz	Ref.
$Br^{79}C^{12}N^{14}$	141	6← 5	0; 2, 0; 0	15/2		13/2		49 666.19	.10	401
	141	6← 5	0; 2, 0; 0	9/2		7/2		49 673.93	.10	401
	141	6← 5	0; 2, 0; 0	11/2		9/2		49 673.93	.10	401
	141	6← 5	0; 2,±2; 0	9/2		7/2		49 709.00	.10	401
	141	6← 5	0; 2,±2; 0	15/2		13/2		49 709.00	.10	401
	-141-	6← 5	0; 1,−1; 0	13/2		13/2		49 712.00	.15	401
	141	6← 5	0; 2,±2; 0	13/2		11/2		49 728.46	.10	401
	141	6← 5	0; 2,±2; 0	11/2		9/2		49 728.46	.10	401
	141	6← 5	0; 1,+1; 0	13/2		13/2		49 757.39	.15	401
	141	9← 8	Ground	15/2		13/2		74 159.48	.18	240
	141	9← 8	Ground	17/2		15/2		74 159.48	.18	240
	141	9← 8	Ground	19/2		17/2		74 162.76	.18	240
	141	9← 8	Ground	21/2		19/2		74 162.76	.18	240
	141	9← 8	Ground					74 165.		165
	141	10← 9	Ground					82 405.		165
	141	12←11	Ground					98 879.19	.20	663
	141	12←11	0; 1,−1; 0					99 108.45		663 [1]
	141	12←11	0; 1,−1; 0					99 109.06		663 [1]
	141	12←11	0; 1,−1; 0					99 110.12		663 [1]
	141	12←11	0; 1,−1; 0					99 110.75		663 [1]
	141	12←11	0; 1,+1; 0					99 203.06		663 [1]
	141	12←11	0; 1,+1; 0					99 204.09		663 [1]
	141	12←11	0; 1,+1; 0					99 204.70		663 [1]
	141	15←14	Ground					123 594.73	.25	663
	141	15←14	0; 1,−1; 0					123 881.78		663 [1]
	141	15←14	0; 1,−1; 0					123 882.90		663 [1]
	141	15←14	0; 1,+1; 0					123 999.22		663 [1]
	141	15←14	0; 1,+1; 0					124 000.30		663 [1]
	141	18←17	Ground					148 307.36	.30	663
	141	18←17	0; 1,−1; 0					148 652.01		663 [1]
	141	18←17	0; 1,−1; 0					148 652.83		663 [1]
	141	18←17	0; 1,+1; 0					148 792.99		663 [1]
	141	18←17	0; 1,+1; 0					148 793.79		663 [1]
	141	21←20	Ground					173 016.45	.35	663
	141	21←20	0; 1,−1; 0					173 418.77		663
	141	21←20	0; 1,+1; 0					173 583.26		663
	141	24←23	Ground					197 721.69	.40	663
	141	24←23	0; 1,−1; 0					198 181.25		663
	141	24←23	0; 1,+1; 0					198 368.79		663
	141	27←26	0; 1,+1; 0					222 150.44		663
	141	27←26	Ground					222 422.34	.45	663
	141	27←26	0; 1,−1; 0					222 939.11		663
	141	30←29	Ground					247 117.75	.50	663
	141	30←29	0; 1,−1; 0					247 691.48		663
	141	30←29	0; 1,+1; 0					247 925.95		663
	141	33←32	Ground					271 807.59	.55	663
	141	36←35	Ground					296 490.86	.60	663
	141	39←38	Ground					321 167.1	.65	663
	141	42←41	Ground					345 837.0	1.0	663

1. Some quadrupole splitting was observed in the high J transitions in the first excited bending mode, however they were not assigned F values. These are listed as given in the 1956 article by Burrus and Gordy.

Isotopic Species	Id. No.	Rotational Quantum Nos.	Vib. State v_a ; v_b^l ; v_c	Hyperfine F_1'	F'	F_1	F	Frequency MHz	Acc. ±MHz	Ref.
$Br^{81}C^{12}N^{14}$	142	1← 0	1; 0, 0; 0	1/2		3/2		8 028.04	.1	773
	142	1← 0	Ground	1/2		3/2		8 051.48	.1	773
	142	1← 0	0; 2, 0; 0	1/2		3/2		8 090.25	.1	773
	142	1← 0	1; 0, 0; 0	5/2		3/2		8 142.89	.1	773
	142	1← 0	Ground	5/2	5/2	3/2	5/2	8 164.96	.1	773
	142	1← 0	Ground	5/2	7/2	3/2	5/2	8 165.68	.1	773
	142	1← 0	1; 0, 0; 0	3/2		3/2		8 286.87	.1	773
	142	1← 0	Ground	3/2	5/2	3/2	3/2	8 308.27	.1	773
	142	1← 0	Ground	3/2	3/2	3/2	3/2	8 308.89	.1	773
	142	1← 0	Ground	3/2	1/2	3/2	3/2	8 309.67	.1	773
	142	1← 0	0; 2, 0; 0	3/2		3/2		8 346.07	.1	773
	142	2← 1	Ground	1/2		3/2		16 129.96	.1	773
	142	2← 1	1; 0, 0; 0	3/2		3/2		16 227.73	.1	773
	142	2← 1	Ground	3/2	5/2	3/2	5/2	16 273.00	.1	773
	142	2← 1	Ground	3/2	5/2	3/2	3/2	16 273.62	.1	773
	142	2← 1	Ground	3/2	3/2	3/2	1/2	16 274.38	.1	773
	142	2← 1	1; 0, 0; 0	7/2		5/2		16 329.72	.1	773
	142	2← 1	1; 0, 0; 0	5/2		3/2		16 329.72	.1	773
	142	2← 1	Ground	5/2		3/2		16 375.43	.1	773
	142	2← 1	Ground	7/2		5/2		16 375.43	.1	773
	142	2← 1	Ground	1/2		1/2		16 387.20	.1	773
	142	2← 1	Ground	3/2		5/2		16 417.0	.1	773
	142	2← 1	0; 2, 0; 0	7/2		5/2		16 451.27	.1	773
	142	2← 1	0; 2, 0; 0	5/2		3/2		16 451.27	.1	773
	142	2← 1	1; 0, 0; 0	5/2		5/2		16 473.20	.1	773
	142	2← 1	1; 0, 0; 0	3/2		1/2		16 486.04	.1	773
	142	2← 1	Ground	5/2		5/2		16 518.62	.1	773
	142	2← 1	Ground	3/2		1/2		16 531.31	.1	773
	142	2← 1	0; 2, 0; 0	3/2		1/2		16 605.92	.1	773
	142	3← 2	Ground	3/2		3/2		24 465.33		112
	142	3← 2	1; 0, 0; 0	9/2		7/2		24 506.75		112
	142	3← 2	1; 0, 0; 0	7/2		5/2		24 506.75		112
	142	3← 2	Ground	5/2		5/2		24 507.38		112
	142	3← 2	1; 0, 0; 0	3/2		1/2		24 541.18		112
	142	3← 2	1; 0, 0; 0	5/2		3/2		24 541.18		112
	142	3← 2	Ground	7/2		5/2		24 573.86		112
	142	3← 2	Ground	9/2		7/2		24 573.86		112
	142	3← 2	Ground	3/2		1/2		24 608.92		112
	142	3← 2	Ground	5/2		3/2		24 608.92		112
	142	3← 2	0; 1, −1; 0	9/2		7/2		24 622.93		112
	142	3← 2	0; 1, −1; 0	3/2		1/2		24 622.93		112
	142	3← 2	0; 1, +1; 0	9/2		7/2		24 645.82		112
	142	3← 2	0; 1, +1; 0	3/2		1/2		24 645.82		112
	142	3← 2	0; 1, −1; 0	5/2		3/2		24 658.89		112
	142	3← 2	0; 1, −1; 0	7/2		5/2		24 658.89		112
	142	3← 2	0; 1, +1; 0	5/2		3/2		24 682.13		112
	142	3← 2	0; 1, +1; 0	7/2		5/2		24 682.13		112
	142	3← 2	Ground	7/2		7/2		24 717.19		112
	142	4← 3	Ground	5/2		5/2		32 643.13		106
	142	4← 3	Ground	7/2		7/2		32 720.28		106

Isotopic Species	Id. No.	Rotational Quantum Nos.	Vib. State v_a ; v_b^l ; v_c	F_1'	F'	F_1	F	Frequency MHz	Acc. ±MHz	Ref.
$Br^{81}C^{12}N^{14}$	142	4 ← 3	Ground	11/2		9/2		32 770.13		106
	142	4 ← 3	Ground	9/2		7/2		32 770.13		106
	142	4 ← 3	Ground	7/2		5/2		32 786.65		106
	142	4 ← 3	Ground	5/2		3/2		32 786.65		106
	142	4 ← 3	Ground	9/2		9/2		32 913.24		106
	142	6 ← 5	1; 0, 0; 0	15/2		13/2		49 021.91	.10	401
	142	6 ← 5	Ground	9/2		9/2		49 021.91	.10	401
	142	6 ← 5	1; 0, 0; 0	13/2		11/2		49 021.91	.10	401
	142	6 ← 5	1; 0, 0; 0	9/2		7/2		49 028.32	.10	401
	142	6 ← 5	1; 0, 0; 0	11/2		9/2		49 028.32	.10	401
	142	6 ← 5	Ground	11/2		11/2		49 125.04	.10	401
	142	6 ← 5	0; 1, −1; 0	9/2		9/2		49 147.47	.15	401
	142	6 ← 5	Ground	15/2	15/2	13/2	15/2	49 157.27	.15	401
	142	6 ← 5	Ground	13/2	13/2	11/2	13/2	49 157.27	.15	401
	142	6 ← 5	Ground	15/2		13/2		49 158.64	.10	401
	142	6 ← 5	Ground	13/2		11/2		49 158.64	.10	401
	142	6 ← 5	Ground	15/2	13/2	13/2	13/2	49 159.85	.15	401
	142	6 ← 5	Ground	13/2	11/2	11/2	11/2	49 159.85	.15	401
	142	6 ← 5	Ground	11/2	11/2	9/2	11/2	49 163.82	.15	401
	142	6 ← 5	Ground	9/2	9/2	7/2	9/2	49 163.82	.15	401
	142	6 ← 5	Ground	11/2		9/2		49 165.10	.10	401
	142	6 ← 5	Ground	9/2		7/2		49 165.10	.10	401
	142	6 ← 5	Ground	11/2	9/2	9/2	9/2	49 166.56	.15	401
	142	6 ← 5	Ground	9/2	7/2	7/2	7/2	49 166.56	.15	401
	142	6 ← 5	0; 1, +1; 0	9/2		9/2		49 195.39	.15	401
	142	6 ← 5	0; 1, −1; 0	11/2		11/2		49 244.47	.15	401
	142	6 ← 5	0; 1, −1; 0	15/2		13/2		49 271.73	.10	401
	142	6 ← 5	0; 1, −1; 0	13/2		11/2		49 275.67	.10	401
	142	6 ← 5	0; 1, −1; 0	9/2		7/2		49 276.54	.10	401
	142	6 ← 5	0; 1, −1; 0	11/2		9/2		49 280.69	.10	401
	142	6 ← 5	0; 1, +1; 0	11/2		11/2		49 291.25	.15	401
	142	6 ← 5	Ground	13/2		13/2		49 301.67	.10	401
	142	6 ← 5	0; 1, +1; 0	15/2		13/2		49 318.15	.10	401
	142	6 ← 5	0; 1, +1; 0	13/2		11/2		49 322.08	.10	401
	142	6 ← 5	0; 1, +1; 0	9/2		7/2		49 323.03	.10	401
	142	6 ← 5	0; 1, +1; 0	11/2		9/2		49 327.08	.10	401
	142	6 ← 5	0; 2, 0; 0	13/2		11/2		49 385.48	.10	401
	142	6 ← 5	0; 2, 0; 0	15/2		13/2		49 385.48	.10	401
	142	6 ← 5	0; 2, 0; 0	9/2		7/2		49 391.95	.10	401
	142	6 ← 5	0; 2, 0; 0	11/2		9/2		49 391.95	.10	401
	142	6 ← 5	0; 1, −1; 0	13/2		13/2		49 404.69	.15	401
	142	6 ← 5	0; 2, ±2; 0	9/2		7/2		49 427.87	.10	401
	142	6 ← 5	0; 2, ±2; 0	15/2		13/2		49 427.87	.10	401
	142	6 ← 5	0; 2, ±2; 0	11/2		9/2		49 444.30	.10	401
	142	6 ← 5	0; 2, ±2; 0	13/2		11/2		49 444.30	.10	401
	142	6 ← 5	0; 1, +1; 0	13/2		13/2		49 449.98	.15	401
	142	9 ← 8	Ground	15/2		13/2		73 738.42	.18	240
	142	9 ← 8	Ground	17/2		15/2		73 738.42	.18	240
	142	9 ← 8	Ground	19/2		17/2		73 741.20	.18	240
	142	9 ← 8	Ground	21/2		19/2		73 741.20	.18	240

Isotopic Species	Id. No.	Rotational Quantum Nos.	Vib. State $v_a : v_b^l : v_c$	F_1'	F'	F_1	F	Frequency MHz	Acc. ±MHz	Ref.
$Br^{81}C^{12}N^{14}$	142	$9 \leftarrow 8$	Ground					73 742.		165
	142	$10 \leftarrow 9$	Ground					81 936.		165
	142	$12 \leftarrow 11$	Ground					98 317.37	.20	663
	142	$12 \leftarrow 11$	$0; 1, -1; 0$					98 545.65		663 [1]
	142	$12 \leftarrow 11$	$0; 1, -1; 0$					98 546.18		663 [1]
	142	$12 \leftarrow 11$	$0; 1, -1; 0$					98 547.08		663 [1]
	142	$12 \leftarrow 11$	$0; 1, -1; 0$					98 547.63		663 [1]
	142	$12 \leftarrow 11$	$0; 1, +1; 0$					98 638.49		663 [1]
	142	$12 \leftarrow 11$	$0; 1, +1; 0$					98 639.01		663 [1]
	142	$12 \leftarrow 11$	$0; 1, +1; 0$					98 639.95		663 [1]
	142	$12 \leftarrow 11$	$0; 1, +1; 0$					98 640.42		663 [1]
	142	$15 \leftarrow 14$	Ground					122 892.50	.25	663
	142	$15 \leftarrow 14$	$0; 1, -1; 0$					123 178.33		663 [1]
	142	$15 \leftarrow 14$	$0; 1, -1; 0$					123 179.40		663 [1]
	142	$15 \leftarrow 14$	$0; 1, +1; 0$					123 294.38		663 [1]
	142	$15 \leftarrow 14$	$0; 1, +1; 0$					123 295.31		663 [1]
	142	$18 \leftarrow 17$	Ground					147 464.71	.30	663
	142	$18 \leftarrow 17$	$0; 1, -1; 0$					147 807.77		663 [1]
	142	$18 \leftarrow 17$	$0; 1, -1; 0$					147 808.49		663 [1]
	142	$18 \leftarrow 17$	$0; 1, +1; 0$					147 947.11		663 [1]
	142	$18 \leftarrow 17$	$0; 1, +1; 0$					147 947.85		663 [1]
	142	$21 \leftarrow 20$	Ground					172 033.53	.35	663
	142	$21 \leftarrow 20$	$0; 1, -1; 0$					172 434.07		663
	142	$21 \leftarrow 20$	$0; 1, +1; 0$					172 596.70		663
	142	$24 \leftarrow 23$	Ground					196 598.50	.40	663
	142	$24 \leftarrow 23$	$0; 1, -1; 0$					197 055.84		663
	142	$24 \leftarrow 23$	$0; 1, +1; 0$					197 241.70		663
	142	$27 \leftarrow 26$	Ground					221 158.86	.45	663
	142	$27 \leftarrow 26$	$0; 1, -1; 0$					221 673.17		663
	142	$27 \leftarrow 26$	$0; 1, +1; 0$					221 881.84		663
	142	$30 \leftarrow 29$	Ground					245 714.32	.50	663
	142	$33 \leftarrow 32$	Ground					270 263.95	.55	663
	142	$36 \leftarrow 35$	Ground					294 807.52	.60	663
	142	$39 \leftarrow 38$	Ground					319 345.52	1.0	663
	142	$42 \leftarrow 41$	Ground					343 873.0	1.5	663
	142	$48 \leftarrow 47$	Ground					392 907.0	.9	897
	142	$52 \leftarrow 51$	Ground					425 575.9	.9	897
	142	$56 \leftarrow 55$	Ground					458 226.2	.9	897
$Br^{79}C^{13}N^{14}$	143	$4 \leftarrow 3$	Ground	11/2		9/2		32 581.73		106
	143	$4 \leftarrow 3$	Ground	9/2		7/2		32 581.73		106
	143	$4 \leftarrow 3$	Ground	5/2		3/2		32 601.46		106
	143	$4 \leftarrow 3$	Ground	7/2		5/2		32 601.46		106
	143	$6 \leftarrow 5$	Ground	15/2		13/2		48 877.11	.10	401
	143	$6 \leftarrow 5$	Ground	13/2		11/2		48 877.11	.10	401
	143	$6 \leftarrow 5$	Ground	9/2		7/2		48 884.82	.10	401
	143	$6 \leftarrow 5$	Ground	11/2		9/2		48 884.82	.10	401
$Br^{81}C^{13}N^{14}$	144	$4 \leftarrow 3$	Ground	11/9		9/2		32 392.59		106
	144	$4 \leftarrow 3$	Ground	9/2		7/2		32 392.59		106
	144	$4 \leftarrow 3$	Ground	5/2		3/2		32 409.06		106

1. Some quadrupole splitting was observed in the high J transitions in the first excited bending mode, however they were not assigned F values. These are listed as given in the 1956 article by Burrus and Gordy.

Isotopic Species	Id. No.	Rotational Quantum Nos.	Vib. State $v_a ; v_b^1 ; v_c$	F_1'	F'	F_1	F	Frequency MHz	Acc. ±MHz	Ref.
$Br^{81}C^{13}N^{14}$	144	$4 \leftarrow 3$	Ground	7/2		5/2		32 409.06		106
	144	$6 \leftarrow 5$	Ground	15/2		13/2		48 592.37	.10	401
	144	$6 \leftarrow 5$	Ground	13/2		11/2		48 592.37	.10	401
	144	$6 \leftarrow 5$	Ground	11/2		9/2		48 598.93	.10	401
	144	$6 \leftarrow 5$	Ground	9/2		7/2		48 598.93	.10	401
$Br^{79}C^{12}N^{15}$	145	$6 \leftarrow 5$	Ground	15/2		13/2		47 334.84	.10	401
	145	$6 \leftarrow 5$	Ground	13/2		11/2		47 334.84	.10	401
	145	$6 \leftarrow 5$	Ground	9/2		7/2		47 342.44	.10	401
	145	$6 \leftarrow 5$	Ground	11/2		9/2		47 342.44	.10	401
$Br^{81}C^{12}N^{15}$	146	$6 \leftarrow 5$	Ground	13/2		11/2		47 058.19	.1	401
	146	$6 \leftarrow 5$	Ground	15/2		13/2		47 058.19	.1	401
	146	$6 \leftarrow 5$	Ground	9/2		7/2		47 064.86	.15	401
	146	$6 \leftarrow 5$	Ground	11/2		9/2		47 064.86	.15	401

150 − Chloro-Fluorocarbonyl   Molecular Constant Table

CClFO C_s ClFCO

Isotopic Species	Pt. Gp.	Id. No.	A MHz		B MHz		C MHz		D_J MHz	D_{JK} MHz	Δ Amu A²	κ
$Cl^{35}F^{19}C^{12}O^{16}$	C_s	151	11 830.31	M	5 286.95	M	3 648.59	M	.0014	.0103	.17	−.59951
$Cl^{37}F^{19}C^{12}O^{16}$	C_s	152	11 829.42	M	5 127.73	M	3 572.54	M			.18	−.62330

Id. No.	μ_a Debye	μ_b Debye	μ_c Debye	eQq Value(MHz)	Rel.	eQq Value(MHz)	Rel.	eQq Value(MHz)	Rel.	ω_a d 1/cm	ω_b d 1/cm	ω_c d 1/cm	ω_d d 1/cm
151				−73.7	aa	44.5	bb						
152				−58.04	aa	29.68	bb						

References:

ABC: 976 D_J: 976 D_{JK}: 976 Δ: 976 κ: 976   eQq: 976,977

For species 151, $D_K = 0.699$ MHz, $\delta_J = -0.00028$ MHz, $R_5 = 0.0234$ MHz, $R_6 = -0.00079$ MHz, $R_{10} = 0.0158$ MHz. Ref. 976.

Isotopic Species	Id. No.	Rotational Quantum Nos.	Vib. State	F_I'	F'	F_1	F	Frequency MHz	Acc. ±MHz	Ref.
$Cl^{35}F^{19}C^{12}O^{16}$	151	1, 1, 1← 0, 0, 0	Ground		1/2		3/2	15 467.68	.05	976
	151	1, 1, 1← 0, 0, 0	Ground		5/2		3/2	15 476.73	.05	976
	151	1, 1, 1← 0, 0, 0	Ground		3/2		3/2	15 487.73	.05	976
	151	2, 0, 2← 1, 0, 1	Ground		3/2		1/2	17 582.10	.1	976
	151	2, 0, 2← 1, 0, 1	Ground		5/2		5/2	17 584.14	.07	976
	151	2, 0, 2← 1, 0, 1	Ground		7/2		5/2	17 601.44	.05	976
	151	2, 0, 2← 1, 0, 1	Ground		3/2		3/2	17 615.12	.07	976
	151	2, 1, 2← 1, 0, 1	Ground		5/2		3/2	22 785.82	.05	976
	151	2, 1, 2← 1, 1, 1	Ground		5/2		3/2	16 218.39	.05	976
	151	2, 1, 2← 1, 1, 1	Ground		7/2		5/2	16 237.11	.05	976
	151	2, 1, 2← 1, 1, 1	Ground		3/2		1/2	16 243.64	.05	976
	151	2, 1, 2← 1, 1, 1	Ground		1/2		1/2	16 251.05	.05	976
	151	2, 2, 1← 1, 1, 0	Ground		7/2		5/2	39 135.12	.05	976
	151	4, 3, 2← 4, 2, 3	Ground					37 647.41	.05	976
	151	5, 3, 3← 5, 2, 4	Ground					38 919.47	.05	976
	151	6, 3, 3← 5, 3, 2	Ground					55 833.34	.05	976
	151	6, 2, 4← 6, 1, 5	Ground					21 533.58	.05	976
	151	6, 3, 3← 6, 2, 4	Ground					29 663.01	.05	976
	151	6, 4, 2← 6, 3, 3	Ground					49 072.09	.05	976
	151	6, 4, 3← 6, 3, 4	Ground					51 109.94	.05	976
	151	7, 3, 4← 7, 2, 5	Ground					27 960.91	.05	976
	151	7, 3, 5← 7, 2, 6	Ground					43 850.54	.05	976
	151	8, 2, 7← 7, 2, 6	Ground					68 433.00	.05	976
	151	8, 3, 5← 7, 3, 4	Ground					76 741.27	.05	976
	151	8, 3, 6← 7, 3, 5	Ground					72 250.03	.05	976
	151	8, 4, 4← 7, 4, 3	Ground					73 816.70	.05	976
	151	8, 5, 4← 7, 5, 3	Ground					72 805.86	.05	976
	151	8, 4, 5← 8, 3, 6	Ground					52 320.30	.05	976
	151	9, 3, 6← 9, 3, 7	Ground					15 753.04	.05	976
	151	10, 3, 7←10, 3, 8	Ground					23 880.16	.05	976
$Cl^{37}F^{19}C^{12}O^{16}$	152	1, 1, 1← 0, 0, 0	Ground		3/2		3/2	15 407.90	.05	976
	152	2, 0, 2← 1, 0, 1	Ground		7/2		5/2	17 161.28	.05	976
	152	2, 1, 2← 1, 0, 1	Ground		7/2		5/2	22 546.22	.05	976
	152	2, 1, 1← 1, 1, 0	Ground		7/2		5/2	18 959.00	.07	976
	152	2, 1, 2← 1, 1, 1	Ground		7/2		5/2	15 848.84	.05	976
	152	2, 1, 2← 1, 1, 1	Ground		1/2		1/2	15 859.84	.05	976
	152	6, 3, 3← 6, 2, 4	Ground					30 639.42	.05	976
	152	6, 3, 3← 6, 4, 2	Ground					50 212.06	.05	976
	152	7, 3, 4← 7, 2, 5	Ground					28 761.07	.05	976
	152	7, 3, 4← 7, 4, 3	Ground					48 346.65	.05	976
	152	7, 3, 5← 7, 4, 4	Ground					52 207.15	.05	976
	152	8, 3, 5← 8, 4, 4	Ground					45 659.90	.05	976

CClF$_3$ C$_{3v}$ CF$_3$Cl

Isotopic Species	Pt. Gp.	Id. No.	A MHz	B MHz	C MHz	D$_J$ MHz	D$_{JK}$ MHz	Δ Amu A^2	κ
C^{12}F$^{19}_3$Cl^{35}	C$_{3v}$	161		3 335.56 M	3 335.56 M				
C^{12}F$^{19}_3$Cl^{37}	C$_{3v}$	162		3 251.51 M	3 251.51 M				

Id. No.	μ$_a$ Debye	μ$_b$ Debye	μ$_c$ Debye	eQq Value(MHz)	Rel.	eQq Value(MHz)	Rel.	eQq Value(MHz)	Rel.	ω$_a$ 1/cm	d	ω$_b$ 1/cm	d	ω$_c$ 1/cm	d	ω$_d$ 1/cm	d
161	.50 M	0. X	0. X	78.05	Cl^{35}												
162				61.44	Cl^{37}												

References:

ABC: 125   μ: 741   eQq: 125

Add. Ref. 502,810

Isotopic Species	Id. No.	Rotational Quantum Nos.	Vib. State	Hyperfine				Frequency MHz	Acc. ±MHz	Ref.
				F_1'	F'	F_1	F			
$C^{12}F_3^{19}Cl^{35}$	161	3, 1← 2, 1	Ground		5/2		3/2	20 010.84		1020
	161	3, 1← 2, 1	Ground		7/2		5/2	20 010.84		1020
	161	3, 1← 2, 1	Ground		9/2		7/2	20 015.77		1020
	161	3, 1← 2, 1	Ground		3/2		1/2	20 015.77		1020
	161	3, 2← 2, 2	Ground		7/2		5/2	19 999.66		1020
	161	3, 2← 2, 2	Ground		5/2		5/2	19 999.66		1020
	161	3, 2← 2, 2	Ground		5/2		3/2	20 013.68		1020
	161	3, 2← 2, 2	Ground		3/2		3/2	20 013.68		1020
	161	3, 2← 2, 2	Ground		9/2		7/2	20 019.17		1020
	161	3, 2← 2, 2	Ground		7/2		7/2	20 019.17		1020
	161	4, ← 3,	Ground					26 669.78		1020
	161	4, 1← 3, 1	Ground		7/2		5/2	26 682.81		1020
	161	4, 1← 3, 1	Ground		9/2		7/2	26 683.78		1020
	161	4, 1← 3, 1	Ground		5/2		3/2	26 684.69		1020
	161	4, 1← 3, 1	Ground		11/2		9/2	26 685.73		1020
	161	4, 2← 3, 2	Excited		9/2		7/2	26 631.58		1020
	161	4, 2← 3, 2	Excited		7/2		5/2	26 634.31		1020
	161	4, 2← 3, 2	Excited		11/2		9/2	26 639.40		1020
	161	4, 2← 3, 2	Excited		5/2		3/2	26 642.22		1020
	161	4, 2← 3, 2	Ground		9/2		7/2	26 679.62		1020
	161	4, 2← 3, 2	Ground		7/2		5/2	26 682.30		1020
	161	4, 2← 3, 2	Ground		11/2		9/2	26 687.38		1020
	161	4, 2← 3, 2	Ground		5/2		3/2	26 690.14		1020
	161	4, 3← 3, 3	Ground		7/2		7/2	26 670.19		1020
	161	4, 3← 3, 3	Ground		9/2		7/2	26 672.59		1020
	161	4, 3← 3, 3	Ground		5/2		5/2	26 674.77		1020
	161	4, 3← 3, 3	Ground		7/2		5/2	26 681.52		1020
	161	4, 3← 3, 3	Ground		11/2		9/2	26 690.14		1020
	161	4, 3← 3, 3	Ground		9/2		9/2	26 697.04		1020
	161	4, 3← 3, 3	Ground		5/2		3/2	26 699.14		1020
$C^{12}F_3^{19}Cl^{37}$	162	4, ← 3,	Ground					26 000.77		1020
	162	4, 2← 3, 2	Ground		9/2		7/2	26 008.55		1020
	162	4, 2← 3, 2	Ground		7/2		5/2	26 010.73		1020
	162	4, 2← 3, 2	Ground		11/2		9/2	26 014.69		1020
	162	4, 2← 3, 2	Ground		5/2		3/2	26 016.84		1020
	162	4, 3← 3, 3	Ground		7/2		7/2	26 001.20		1020
	162	4, 3← 3, 3	Ground		9/2		7/2	26 003.04		1020
	162	4, 3← 3, 3	Ground		7/2		5/2	26 010.09		1020
	162	4, 3← 3, 3	Ground		11/2		9/2	26 016.84		1020
	162	4, 3← 3, 3	Ground		9/2		9/2	26 022.20		1020
	162	4, 3← 3, 3	Ground		5/2		3/2	26 023.94		1020

CClN $C_{\infty v}$ ClCN

Isotopic Species	Pt. Gp.	Id. No.	A MHz	B MHz	C MHz	D_J MHz	D_{JK} MHz	Δ Amu A^2	κ
Cl^{35}C^{12}N^{14}	$C_{\infty v}$	171		5 970.831 M	5 970.831 M	.00166			
Cl^{35}C^{13}N^{14}	$C_{\infty v}$	172		5 941.03 M	5 941.03 M				
Cl^{37}C^{12}N^{14}	$C_{\infty v}$	173		5 847.243 M	5 847.243 M	.00161			
Cl^{37}C^{13}N^{14}	$C_{\infty v}$	174		5 815.92 M	5 815.92 M				
Cl^{36}C^{12}N^{14}	$C_{\infty v}$	175		5 907.31 M	5 907.31 M				

Id. No.	μ_a Debye	μ_b Debye	μ_c Debye	eQq Value(MHz)	Rel.	eQq Value(MHz)	Rel.	eQq Value(MHz)	Rel.	ω_a 1/cm	d	ω_b 1/cm	d	ω_c 1/cm	d	ω_d 1/cm	d
171 173 175	2.802 M	0. X	0. X	−83.33 −65.7 −42.2	Cl^{35} Cl^{37} Cl^{36}	−3.63	N^{14}			714	1	396	2	2213	1		

References:

ABC: 106,663,1029 D_J: 663 μ: 1030   eQq: 112,348,800,1020 ω: 1028

Add. Ref. 30,59,61,62,63,64,81,97,105,116,236,732,934

For species 171, $\alpha_2 = -16.39$ MHz, ref. 112; $q_1 = 7.467467 - 1.327 \times 10^{-5} J(J+1)$, with values for various J's given in ref. 732.

Cyanogen Chloride   Spectral Line Table

Isotopic Species	Id. No.	Rotational Quantum Nos.	Vib. State v_a^l	Hyperfine F_1'	F'	F_1	F	Frequency MHz	Acc. ±MHz	Ref.
Cl^{35}C^{12}N^{14}	171	1 ← 0	Ground					11 879.		298
	171	2 ← 1	Ground					23 759.		298
	171	2 ← 1	Ground	3/2		1/2		23 862.57		112
	171	2 ← 1	Ground	1/2		1/2		23 883.30		112
	171	2 ← 1	Ground	7/2	5/2	5/2	3/2	23 885.16		112
	171	2 ← 1	Ground	7/2	9/2	5/2	7/2	23 885.16		112
	171	2 ← 1	Ground	3/2	1/2	3/2	3/2	23 899.59		112
	171	2 ← 1	Ground	3/2	5/2	3/2	3/2	23 899.59		112
	171	2 ← 1	Ground	3/2	3/2	3/2	3/2	23 899.59		112
	171	2 ← 1	Ground	3/2	5/2	3/2	5/2	23 900.20		112
	171	2 ← 1	Ground	3/2	3/2	3/2	5/2	23 900.20		112
	171	2 ← 1	1,±1	5/2		3/2		23 917.9		112
	171	2 ← 1	Ground	1/2		3/2		23 920.91		112
	171	2 ← 1	1,±1	3/2		3/2		23 925.5		112
	171	2 ← 1	1,±1	5/2		5/2		23 928.7		112
	171	2 ← 1	1,±1	7/2		5/2		23 938.6		112
	171	2 ← 1	1,±1	3/2		1/2		23 944.4		112
	171	2 ← 1	1,±1	5/2		3/2		23 948.2		112
	171	2 ← 1	1,±1	3/2		3/2		23 954.5		112
	171	2 ← 1	1,±1	1/2		1/2		23 954.5		112
	171	2 ← 1	1,±1	5/2		5/2		23 958.4		112
	171	2 ← 1	1,±1	7/2		5/2		23 968.6		112
	171	2 ← 1	1,±1	3/2		1/2		23 974.4		112
	171	2 ← 1	1,±1	1/2		1/2		23 984.6		112
	171	3 ← 2	Ground					35 638.		298

Isotopic Species	Id. No.	Rotational Quantum Nos.	Vib. State v_a^1	Hyperfine				Frequency MHz	Acc. ±MHz	Ref.
				F_1'	F'	F_1	F			
$Cl^{35}C^{12}N^{14}$	171	$3 \leftarrow 2$	Ground	7/2		7/2		35 805.09		106
	171	$3 \leftarrow 2$	Ground	3/2		1/2		35 820.65		106
	171	$3 \leftarrow 2$	Ground	5/2		3/2		35 820.65		106
	171	$3 \leftarrow 2$	Ground	7/2		5/2		35 825.95		106
	171	$3 \leftarrow 2$	Ground	9/2		7/2		35 825.95		106
	171	$3 \leftarrow 2$	Ground	5/2		5/2		35 835.74		106
	171	$8 \leftarrow 7$	Ground					95 529.86	.20	663
	171	$8 \leftarrow 7$	$1,\pm1$					95 731.52		663
	171	$8 \leftarrow 7$	$1,\pm1$					95 850.88		663
	171	$10 \leftarrow 9$	Ground					119 409.82	.25	663
	171	$10 \leftarrow 9$	$1,\pm1$					119 662.15		663
	171	$10 \leftarrow 9$	$1,\pm1$					119 811.35		663
	171	$12 \leftarrow 11$	Ground					143 288.45	.30	663
	171	$12 \leftarrow 11$	$1,\pm1$					143 591.01		663
	171	$12 \leftarrow 11$	$1,\pm1$					143 770.23		663
	171	ι $14 \leftarrow 13$	Ground					167 165.15	.35	663
	171	$14 \leftarrow 13$	$1,\pm1$					167 517.88		663
	171	$14 \leftarrow 13$	$1,\pm1$					167 726.78		663
	171	$16 \leftarrow 15$	Ground					191 039.44	.40	663
	171	$16 \leftarrow 15$	$1,\pm1$					191 442.66		663
	171	$16 \leftarrow 15$	$1,\pm1$					191 681.13		663
	171	$17 \leftarrow 17$	$1,\pm1$					2 283.698	.010	800
	171	$17 \leftarrow 17$	$1,\pm1$					2 283.906	.010	800
	171	$18 \leftarrow 17$	Ground					214 911.20	.45	663
	171	$18 \leftarrow 17$	$1,\pm1$					215 364.77		663
	171	$18 \leftarrow 17$	$1,\pm1$					215 633.04		663
	171	$18 \leftarrow 18$	$1,\pm1$					2 552.224	.010	800
	171	$18 \leftarrow 18$	$1,\pm1$					2 552.448	.010	800
	171	$19 \leftarrow 19$	$1,\pm1$					2 835.622	.010	800
	171	$19 \leftarrow 19$	$1,\pm1$					2 835.829	.010	800
	171	$20 \leftarrow 19$	Ground					238 780.22	.50	663
	171	$20 \leftarrow 19$	$1,\pm1$					239 283.85		663
	171	$20 \leftarrow 19$	$1,\pm1$					239 582.25		663
	171	$22 \leftarrow 21$	Ground					262 645.82	.55	663
	171	$23 \leftarrow 23$	$1,\pm1$					4 117.869	.020	800
	171	$23 \leftarrow 23$	$1,\pm1$					4 118.067	.020	800
	171	$24 \leftarrow 23$	Ground					286 507.95	.60	663
	171	$26 \leftarrow 25$	Ground					310 365.90	.65	663
	171	$28 \leftarrow 27$	Ground					334 219.5	1.5	663
$Cl^{35}C^{13}N^{14}$	172	$1 \leftarrow 0$	Ground					11 941.		298
	172	$2 \leftarrow 1$	Ground	7/2	9/2	5/2	7/2	23 760.98		112
	172	$2 \leftarrow 1$	Ground					23 883.		298
	172	$3 \leftarrow 2$	Ground	7/2		7/2		35 618.81		106
	172	$3 \leftarrow 2$	Ground	3/2		1/2		35 634.85		106
	172	$3 \leftarrow 2$	Ground	5/2		3/2		35 634.85		106
	172	$3 \leftarrow 2$	Ground	9/2		7/2		35 639.78		106
	172	$3 \leftarrow 2$	Ground	7/2		5/2		35 639.78		106
	172	$3 \leftarrow 2$	Ground	5/2		5/2		35 649.56		106
	172	$3 \leftarrow 2$	Ground					35 824.		298

Isotopic Species	Id. No.	Rotational Quantum Nos.	Vib. State v_a^l	Hyperfine				Frequency MHz	Acc. ±MHz	Ref.
				F_1'	F'	F_1	F			
$Cl^{37}C^{12}N^{14}$	173	2← 1	Ground	3/2	5/2	1/2	3/2	23 372.72		112
	173	2← 1	Ground	1/2		1/2		23 389.00		112
	173	2← 1	Ground	5/2	5/2	3/2	3/2	23 389.61		112
	173	2← 1	Ground	7/2	9/2	5/2	7/2	23 390.53		112
	173	2← 1	Ground	7/2	5/2	5/2	3/2	23 390.53		112
	173	2← 1	Ground	3/2	5/2	3/2	5/2	23 402.47		112
	173	3← 2	Ground	7/2		7/2		35 067.99		106
	173	3← 2	Ground	5/2		3/2		35 080.39		106
	173	3← 2	Ground	3/2		1/2		35 080.39		106
	173	3← 2	Ground	7/2		5/2		35 084.15		106
	173	3← 2	Ground	9/2		7/2		35 084.15		106
	173	3← 2	Ground	5/2		5/2		35 091.97		106
	173	8← 7	Ground					93 552.59	.20	663
	173	8← 7	1,±1					93 751.28		663
	173	8← 7	1,±1					93 865.98		663
	173	10← 9	Ground					116 938.45	.25	663
	173	10← 9	1,±1					117 186.54		663
	173	10← 9	1,±1					117 329.90		663
	173	12←11	Ground					140 322.75	.30	663
	173	12←11	1,±1					140 620.55		663
	173	12←11	1,±1					140 792.59		663
	173	14←13	Ground					163 705.31	.35	663
	173	14←13	1,±1					164 052.55		663
	173	14←13	1,±1					164 253.26		663
	173	16←15	Ground					187 085.58	.40	663
	173	16←15	1,±1					187 482.50		663
	173	16←15	1,±1					187 711.52		663
	173	18←17	Ground					210 463.24	.45	663 [1]
	173	18←17	1,−1					215 769.51		663
	173	18←17	1,+1					216 027.44		663
	173	20←19	Ground					233 838.30	.50	663
	173	22←21	Ground					257 210.28	.55	663
	173	24←23	Ground					280 578.92	.60	663
	173	26←25	Ground					303 943.87	.65	663
$Cl^{37}C^{13}N^{14}$	174	2← 1	Ground	7/2	9/2	5/2	7/2	23 260.31		112
	174	3← 2	Ground	9/2		7/2		34 889.05		106
$Cl^{36}C^{12}N^{14}$	175	2← 1	Ground					23 635.		171

1. In the 1956 article by Burrus and Gordy, the ground state line 210463.24 is not consistent with the l-type doublets 215769.51 and 216027.44.

CCl$_2$O $\qquad\qquad\qquad\qquad\qquad$ C$_{2v}$ $\qquad\qquad\qquad\qquad\qquad$ COCl$_2$

Isotopic Species	Pt. Gp.	Id. No.	A MHz		B MHz		C MHz		D$_J$ MHz	D$_{JK}$ MHz	Δ Amu A²	κ
C^{12}O^{16}Cl$_2^{35}$	C$_{2v}$	181	7 918.14	M	3 474.72	M	2 412.07	M			.251	−.61401
C^{12}O^{16}Cl^{35}Cl^{37}	C$_s$	182	7 867.16	M	3 379.68	M	2 361.30	M			.253	−.63007

Id. No.	μ_a Debye		μ_b Debye		μ_c Debye		eQq Value(MHz)	Rel.	eQq Value(MHz)	Rel.	eQq Value(MHz)	Rel.	ω_a d 1/cm	ω_b d 1/cm	ω_c d 1/cm	ω_d d 1/cm
181	0.	X	1.17	M	0.	X	−37.20	aa	10.13	bb	27.07	cc				
182											24.20	Cl^{37}				

References:

ABC: 471 $\qquad$ Δ: 471 $\qquad$ κ: 471 $\qquad$ μ: 965 $\qquad$ eQq: 471

Add. Ref. 472

Isotopic Species	Id. No.	Rotational Quantum Nos.	Vib. State	Hyperfine				Frequency MHz	Acc. ±MHz	Ref.
				F_1'	F'	F_1	F			
$C^{12}O^{16}Cl_2^{35}$	181	2, 2, 1← 1, 1, 0	Ground	7/2	3	3/2	3	26 158.02	.1	471
	181	2, 2, 1← 1, 1, 0	Ground	3/2	2	3/2	3	26 160.72	.1	471
	181	2, 2, 1← 1, 1, 0	Ground	7/2	3	3/2	2	26 161.73	.1	471
	181	2, 2, 1← 1, 1, 0	Ground		2		1	26 163.24	.1	471
	181	2, 2, 1← 1, 1, 0	Ground	3/2	2	3/2	2	26 164.10	.1	471
	181	2, 2, 1← 1, 1, 0	Ground	5/2	1	3/2	0	26 165.52	.1	471
	181	2, 2, 1← 1, 1, 0	Ground	1/2	1	5/2	2	26 165.52	.1	471
	181	2, 2, 1← 1, 1, 0	Ground	7/2	5	5/2	4	26 165.95	.1	471
	181	2, 2, 1← 1, 1, 0	Ground	1/2	2	5/2	2	26 167.72	.1	471
	181	2, 2, 1← 1, 1, 0	Ground	5/2	4	3/2	3	26 168.50	.1	471
	181	2, 2, 1← 1, 1, 0	Ground	5/2	3	3/2	3	26 171.10	.1	471
	181	2, 2, 1← 1, 1, 0	Ground		2		1	26 174.12	.1	471
	181	2, 2, 1← 1, 1, 0	Ground	5/2	3	3/2	2	26 174.81	.1	471
	181	2, 2, 1← 1, 1, 0	Ground	7/2	3	5/2	2	26 176.10	.1	471
	181	2, 2, 1← 1, 1, 0	Ground	3/2	2	5/2	2	26 178.62	.1	471
	181	2, 2, 1← 1, 1, 0	Ground	5/2	4	5/2	4	26 179.44	.1	471
	181	3, 3, 0← 3, 2, 1	Ground	3/2	2	3/2	1	24 159.62	.1	471
	181	3, 3, 0← 3, 2, 1	Ground	3/2	3	3/2	3	24 161.42	.1	471
	181	3, 3, 0← 3, 2, 1	Ground	9/2	6	9/2	6	24 165.82	.1	471
	181	3, 3, 0← 3, 2, 1	Ground	5/2	3	3/2	2	24 165.82	.1	471
	181	3, 3, 0← 3, 2, 1	Ground	3/2	1	3/2	2	24 166.92	.1	471
	181	3, 3, 0← 3, 2, 1	Ground	9/2	4	9/2	4	24 171.70	.1	471
	181	3, 3, 0← 3, 2, 1	Ground	5/2	3	5/2	3	24 178.75	.1	471
	181	3, 3, 0← 3, 2, 1	Ground	7/2	4	5/2	3	24 178.75	.1	471
	181	3, 3, 0← 3, 2, 1	Ground	3/2	2	5/2	3	24 179.92	.1	471
	181	3, 3, 0← 3, 2, 1	Ground	7/2	5	7/2	5	24 180.63	.1	471
	181	3, 3, 0← 3, 2, 1	Ground	7/2	2	7/2	2	24 181.28	.1	471
	181	3, 3, 0← 3, 2, 1	Ground	7/2	4	7/2	4	24 186.65	.1	471
	181	3, 3, 1← 3, 2, 2	Ground	9/2	4	9/2	3	24 961.48	.1	471
	181	3, 3, 1← 3, 2, 2	Ground	9/2	5	9/2	5	24 973.78	.1	471
	181	3, 3, 1← 3, 2, 2	Ground	5/2	2	5/2	2	24 973.78	.1	471
	181	3, 3, 1← 3, 2, 2	Ground	5/2	4	5/2	4	24 973.78	.1	471
	181	3, 3, 1← 3, 2, 2	Ground	5/2	1	5/2	1	24 973.78	.1	471
	181	3, 3, 1← 3, 2, 2	Ground	7/2	3	7/2	3	24 985.83	.1	471
	181	4, 1, 4← 3, 0, 3	Ground					23 637.1	.3	471
	181	4, 3, 1← 4, 2, 2	Ground					23 218.36	.1	471
	181	4, 3, 2← 4, 2, 3	Ground	5/2	4	5/2	4	25 391.24	.1	471
	181	4, 3, 2← 4, 2, 3	Ground	11/2	7	11/2	7	25 393.35	.1	471
	181	4, 3, 2← 4, 2, 3	Ground	5/2	3	5/2	3	25 396.48	.1	471
	181	4, 3, 2← 4, 2, 3	Ground	11/2	5	11/2	5	25 397.98	.1	471
	181	4, 3, 2← 4, 2, 3	Ground	9/2	6	9/2	6	25 402.67	.1	471
	181	4, 3, 2← 4, 2, 3	Ground	7/2	4	7/2	4	25 403.90	.1	471
	181	4, 3, 2← 4, 2, 3	Ground	9/2	3	9/2	3	25 403.90	.1	471
	181	4, 3, 2← 4, 2, 3	Ground	9/2	5	9/2	5	25 407.53	.1	471
	181	5, 0, 5← 4, 1, 4	Ground					25 304.87	.1	471
	181	5, 1, 5← 4, 0, 4	Ground					27 764.0	.3	471
	181	5, 2, 4← 5, 1, 5	Ground	13/2	5	13/2	5	23 359.62	.1	471
	181	5, 2, 4← 5, 1, 5	Ground	13/2	7	13/2	7	23 364.80	.1	471
	181	5, 2, 4← 5, 1, 5	Ground	9/2	3	9/2	3	23 364.80	.1	471
	181	5, 2, 4← 5, 1, 5	Ground	9/2	4	9/2	4	23 364.80	.1	471

Isotopic Species	Id. No.	Rotational Quantum Nos.	Vib. State	Hyperfine				Frequency MHz	Acc. ±MHz	Ref.
				F_1'	F'	F_1	F			
$C^{12}O^{16}Cl_2^{35}$	181	5, 2, 4← 5, 1, 5	Ground	9/2	6	9/2	6	23 364.80	.1	471
	181	5, 2, 4← 5, 1, 5	Ground	11/2	5	11/2	5	23 369.93	.1	471
	181	5, 3, 3← 5, 2, 4	Ground					26 196.98	.1	471
	181	6, 2, 5← 6, 1, 6	Ground					26 765.1	.3	471
	181	6, 3, 4← 6, 2, 5	Ground					27 476.1	.3	471
	181	7, 2, 5← 6, 3, 4	Ground					25 719.8	.3	471
	181	9, 2, 7← 9, 1, 8	Ground					26 375.1	.3	471
	181	9, 4, 5← 9, 3, 6	Ground					27 977.6	.3	471
	181	10, 4, 6←10, 3, 7	Ground					25 998.02	.1	471
$C^{12}O^{16}Cl^{35}Cl^{37}$	182	2, 2, 1← 1, 1, 0	Ground	1/2	1	5/2	2	25 962.1	.3	471
	182	2, 2, 1← 1, 1, 0	Ground	7/2	5	5/2	4	25 962.5	.3	471
	182	2, 2, 1← 1, 1, 0	Ground	7/2	4	5/2	3	25 964.6	.3	471
	182	2, 2, 1← 1, 1, 0	Ground	5/2	4	3/2	3	25 964.9	.3	471
	182	2, 2, 1← 1, 1, 0	Ground	3/2	3	5/2	3	25 964.9	.3	471
	182	2, 2, 1← 1, 1, 0	Ground	3/2	3	1/2	2	25 964.9	.3	471
	182	2, 2, 1← 1, 1, 0	Ground	5/2	2	3/2	1	25 966.0	.3	471
	182	2, 2, 1← 1, 1, 0	Ground	5/2	3	3/2	3	25 967.0	.3	471
	182	2, 2, 1← 1, 1, 0	Ground	5/2	3	3/2	2	25 970.3	.3	471
	182	3, 3, 0← 3, 2, 1	Ground					24 340.3	.3	471
	182	3, 3, 1← 3, 2, 2	Ground	3/2	3	3/2	3	25 063.76	.1	471
	182	3, 3, 1← 3, 2, 2	Ground	9/2	3	9/2	3	25 064.68	.1	471
	182	3, 3, 1← 3, 2, 2	Ground	9/2	6	9/2	6	25 068.66	.1	471
	182	3, 3, 1← 3, 2, 2	Ground	9/2	4	9/2	4	25 073.68	.1	471
	182	3, 3, 1← 3, 2, 2	Ground	9/2	5	9/2	5	25 075.21	.1	471
	182	3, 3, 1← 3, 2, 2	Ground	5/2	4	5/2	4	25 075.21	.1	471
	182	3, 3, 1← 3, 2, 2	Ground	5/2	3	5/2	3	25 080.68	.1	471
	182	3, 3, 1← 3, 2, 2	Ground	7/2	5	7/2	5	25 082.38	.1	471
	182	3, 3, 1← 3, 2, 2	Ground	7/2	3	7/2	3	25 085.93	.1	471
	182	3, 3, 1← 3, 2, 2	Ground	7/2	4	7/2	4	25 088.13	.1	471
	182	4, 1, 4← 3, 0, 3	Ground					23 261.0	.3	471
	182	4, 3, 1← 4, 2, 2	Ground					23 449.9	.3	471
	182	4, 3, 2← 4, 2, 3	Ground					25 466.8	.3	471
	182	5, 0, 5← 4, 1, 4	Ground					24 653.6	.3	471
	182	5, 1, 5← 4, 0, 4	Ground					27 285.5	.3	471
	182	5, 2, 4← 5, 1, 5	Ground					23 066.28	.1	471
	182	5, 3, 3← 5, 2, 4	Ground					26 201.9	.3	471
	182	6, 2, 5← 6, 1, 6	Ground					26 328.5	.3	471
	182	6, 3, 4← 6, 2, 5	Ground					27 387.0	.3	471
	182	7, 2, 5← 6, 3, 4	Ground					24 132.7	.3	471
	182	7, 1, 6← 7, 0, 7	Ground					26 146.5	.3	471
	182	9, 2, 7← 9, 1, 8	Ground					25 297.0	.3	471
	182	10, 4, 6←10, 3, 7	Ground					26 557.5	.3	471

CCl₃F

C₃ᵥ

CCl₃F

Isotopic Species	Pt. Gp.	Id. No.	A MHz		B MHz		C MHz	D_J MHz	D_JK MHz	Δ Amu A²	κ
C¹²Cl₃³⁵F¹⁹	C₃ᵥ	191	2 465.39	M	2 465.39	M			−.196		
C¹²Cl₂³⁵Cl³⁷F¹⁹	Cₛ	192	2 463.22	M	2 398.50	M					

Id. No.	μ_a Debye		μ_b Debye		μ_c Debye		eQq Value(MHz)	Rel.	eQq Value(MHz)	Rel.	eQq Value(MHz)	Rel.	ω_a d 1/cm	ω_b d 1/cm	ω_c d 1/cm	ω_d d 1/cm
191	0.	X	0.	X	.49	P	37.3	zz								

References:

ABC: 914   D_JK: 914   μ: 707   eQq: 914

Trichlorofluoromethane

Spectral Line Table

Isotopic Species	Id. No.	Rotational Quantum Nos.	Vib. State	F₁′	F′	F₁	F	Frequency MHz	Acc. ±MHz	Ref.
C¹²Cl₃³⁵F¹⁹	191	2, 1← 1, 1	Ground					9 853.68	.08	914
	191	2, 1← 1, 1	Ground					9 855.60	.20	914
	191	2, 1← 1, 1	Ground					9 859.30	.04	914
	191	2, 1← 1, 1	Ground					9 860.60	.10	914
	191	2, 1← 1, 1	Ground					9 861.80	.30	914
	191	2, 1← 1, 1	Ground					9 863.96	.05	914
	191	2, 1← 1, 1	Ground					9 865.00	.20	914
	191	2, 1← 1, 1	Ground					9 866.69	.11	914
	191	2, 1← 1, 1	Ground					9 870.50	.20	914
	191	2, 1← 1, 1	Ground					9 875.00	.25	914
	191	2, 1← 1, 1	Ground					9 877.00	.25	914
	191	3, 1← 2, 1	Ground					14 790.46	.08	914
	191	3, 1← 2, 1	Ground					14 792.80	.06	914
	191	3, 1← 2, 1	Ground					14 793.92	.08	914
	191	3, 1← 2, 1	Ground					14 795.26	.08	914
	191	3, 1← 2, 1	Ground					14 799.50	.30	914
	191	3, 2← 2, 2	Ground					14 784.75	.10	914
	191	3, 2← 2, 2	Ground					14 794.66	.08	914
	191	3, 2← 2, 2	Ground					14 796.25	.10	914
	191	3, 2← 2, 2	Ground					14 797.12	.10	914
	191	3, 2← 2, 2	Ground					14 803.40	.30	914
	191	4, ← 3,	Ground					19 725.17	.01	914
	191	5, ← 4,	Ground					24 657.26	.03	914
	191	6, ← 5,	Ground					29 588.95	.01	914
	191	7, ← 6,	Ground					34 520.42	.01	914

CFN $\qquad$ C$_{\infty v}$ $\qquad$ FCN

Isotopic Species	Pt. Gp.	Id. No.	A MHz	B MHz	C MHz	D_J MHz	D_{JK} MHz	Δ Amu A^2	κ
F^{19}C^{12}N^{14}	C$_{\infty v}$	201		10 554.20 M	10 554.20 M	.0053			

Id. No.	μ_a Debye	μ_b Debye	μ_c Debye	eQq Value(MHz)	Rel.	eQq Value(MHz)	Rel.	eQq Value(MHz)	Rel.	ω_a d 1/cm	ω_b d 1/cm	ω_c d 1/cm	ω_d d 1/cm
201	1.68 M	0. X	0. X	−2.67	N^{14}								

References:

ABC: 926 $\qquad$ D$_J$: 926 $\qquad$ μ: 926 $\qquad$ eQq: 926

No Spectral Lines

CF$_2$O

C$_{2v}$

COF$_2$

Isotopic Species	Pt. Gp.	Id. No.	A MHz		B MHz		C MHz		D$_J$ MHz	D$_{JK}$ MHz	Δ Amu A^2	κ
C^{12}O^{16}F$_2^{19}$	C$_{2v}$	211	11 813.45	M	11 752.99	M	5 880.91	M				
C^{13}O^{16}F$_2^{19}$	C$_{2v}$	212	11 814.66	M	11 747.27	M	5 879.81	M				
C^{12}O^{18}F$_2^{19}$	C$_{2v}$	213	11 813.48	M	10 878.54	M	5 653.32	M				

Id. No.	μ$_a$ Debye		μ$_b$ Debye		μ$_c$ Debye		eQq Value(MHz) Rel.	eQq Value(MHz) Rel.	eQq Value(MHz) Rel.	ω$_a$ d 1/cm	ω$_b$ d 1/cm	ω$_c$ d 1/cm	ω$_d$ d 1/cm
211	.951	M	0.	X	0.	X							

References:

ABC: 972   μ: 972

Carbonyl Fluoride

Isotopic Species	Id. No.	Rotational Quantum Nos.	Vib. State	F$_1'$	F'	F$_1$	F	Frequency MHz	Acc. ±MHz	Ref.
C^{12}O^{16}F$_2^{19}$	211	1, 0, 1← 0, 0, 0	Ground					17 633.95	.10	972
	211	2, 0, 2← 1, 0, 1	Ground					29 455.70	.10	972
	211	2, 1, 2← 1, 1, 1	Ground					29 395.71	.10	972
	211	2, 1, 1← 2, 1, 2	Ground					17 616.26	.10	972
	211	2, 2, 1← 2, 0, 2	Ground					17 798.14	.10	972
	211	3, 1, 2← 3, 1, 3	Ground					29 509.27	.10	972
	211	3, 2, 2← 3, 0, 3	Ground					29 512.10	.10	972
	211	3, 2, 1← 3, 3, 3	Ground					17 525.47	.10	972
	211	3, 3, 1← 3, 1, 2	Ground					17 890.51	.10	972
	211	4, 2, 2← 4, 2, 3	Ground					29 504.93	.10	972
	211	4, 3, 2← 4, 1, 3	Ground					29 513.80	.10	972
	211	4, 3, 1← 4, 3, 2	Ground					17 404.77	.10	972
	211	4, 4, 1← 4, 2, 2	Ground					18 015.94	.10	972
	211	5, 3, 2← 5, 3, 3	Ground					29 497.00	.10	972
	211	5, 4, 2← 5, 2, 3	Ground					29 517.37	.10	972
	211	5, 4, 1← 5, 4, 2	Ground					17 254.35	.10	972
	211	5, 5, 1← 5, 3, 2	Ground					18 176.22	.10	972
	211	6, 4, 2← 6, 4, 3	Ground					29 483.85	.10	972
	211	6, 5, 2← 6, 3, 3	Ground					29 523.74	.10	972
	211	6, 5, 1← 6, 5, 2	Ground					17 074.52	.10	972
	211	6, 6, 1← 6, 4, 2	Ground					18 373.76	.10	972
	211	7, 5, 2← 7, 5, 3	Ground					29 463.51	.10	972
	211	7, 6, 2← 7, 4, 3	Ground					29 533.95	.10	972
	211	7, 6, 1← 7, 6, 2	Ground					16 865.76	.10	972
	211	7, 7, 1← 7, 5, 2	Ground					18 611.18	.10	972
	211	8, 6, 2← 8, 6, 3	Ground					29 433.64	.10	972
	211	8, 7, 2← 8, 5, 3	Ground					29 549.17	.10	972
	211	8, 7, 1← 8, 7, 2	Ground					16 628.49	.10	972
	211	8, 8, 1← 8, 6, 2	Ground					18 891.66	.10	972
	211	9, 7, 2← 9, 7, 3	Ground					29 391.88	.10	972

Isotopic Species	Id. No.	Rotational Quantum Nos.	Vib. State	Hyperfine				Frequency MHz	Acc. ±MHz	Ref.
				F_1'	F'	F_1	F			
$C^{12}O^{16}F_2^{19}$	211	9, 8, 2← 9, 6, 3	Ground					29 570.75	.10	972
	211	9, 8, 1← 9, 8, 2	Ground					16 363.41	.10	972
	211	9, 9, 1← 9, 7, 2	Ground					19 218.49	.10	972
	211	10, 8, 2←10, 8, 3	Ground					29 335.23	.10	972
	211	10, 9, 2←10, 7, 3	Ground					29 600.24	.10	972
	211	10, 9, 1←10, 9, 2	Ground					16 071.05	.10	972
	211	10,10, 1←10, 8, 2	Ground					19 595.29	.10	972
	211	11, 9, 2←11, 9, 3	Ground					29 261.23	.10	972
	211	11,10, 2←11, 8, 3	Ground					29 639.29	.10	972
	211	11,10, 1←11,10, 2	Ground					15 752.33	.10	972
	211	11,11, 1←11, 9, 2	Ground					20 025.87	.10	972
	211	12,10, 2←12,10, 3	Ground					29 166.78	.10	972
	211	12,11, 2←12, 9, 3	Ground					29 689.77	.10	972
	211	12,11, 1←12,11, 2	Ground					15 408.00	.10	972
	211	12,12, 1←12,10, 2	Ground					20 514.14	.10	972
$C^{13}O^{16}F_2^{19}$	212	1, 0, 1← 0, 0, 0	Ground					17 627.05	.10	972
	212	2, 0, 2← 1, 0, 1	Ground					29 453.50	.10	972
	212	2, 1, 2← 1, 1, 1	Ground					29 386.69	.10	972
	212	2, 1, 1← 2, 1, 2	Ground					17 602.40	.10	972
	212	2, 2, 1← 2, 0, 2	Ground					17 805.26	.10	972
	212	3, 2, 1← 3, 2, 2	Ground					17 501.02	.10	972
	212	3, 3, 1← 3, 1, 2	Ground					17 908.80	.10	972
	212	4, 3, 1← 4, 3, 2	Ground					17 366.77	.10	972
	212	4, 4, 1← 4, 2, 2	Ground					18 049.35	.10	972
	212	5, 4, 1← 5, 4, 2	Ground					17 199.10	.10	972
	212	5, 5, 1← 5, 3, 2	Ground					18 229.66	.10	972
	212	6, 5, 1← 6, 5, 2	Ground					16 998.78	.10	972
	212	6, 6, 1← 6, 4, 2	Ground					18 452.36	.10	972
	212	7, 6, 1← 7, 6, 2	Ground					16 766.23	.10	972
	212	7, 7, 1← 7, 5, 2	Ground					18 720.94	.10	972
	212	8, 7, 1← 8, 7, 2	Ground					16 502.08	.10	972
	212	8, 8, 1← 8, 6, 2	Ground					19 039.51	.10	972
	212	9, 8, 1← 9, 8, 2	Ground					16 206.91	.10	972
	212	9, 9, 1← 9, 7, 2	Ground					19 411.99	.10	972
	212	10, 9, 1←10, 9, 2	Ground					15 881.70	.10	972
	212	10,10, 1←10, 8, 2	Ground					19 843.06	.10	972
	212	11,10, 1←11,10, 2	Ground					15 527.27	.10	972
	212	11,11, 1←11, 9, 2	Ground					20 337.14	.10	972
	212	12,11, 1←12,11, 2	Ground					15 143.78	.10	972
$C^{12}O^{18}F_2^{19}$	213	1, 0, 1← 0, 0, 0	Ground					16 531.91	.10	972
	213	2, 0, 2← 1, 0, 1	Ground					28 658.85	.10	972
	213	2, 1, 2← 1, 1, 1	Ground					27 838.50	.10	972
	213	2, 1, 2← 2, 1, 1	Ground					15 675.67	.10	972
	213	3, 2, 1← 3, 2, 2	Ground					14 350.10	.10	972
	213	4, 2, 2← 4, 2, 3	Ground					27 207.95	.10	972
	213	5, 3, 2← 5, 3, 3	Ground					25 869.20	.10	972

CBrF$_3$ C$_{3v}$ CF$_3$Br

Isotopic Species	Pt. Gp.	Id. No.	A MHz	B MHz	C MHz	D$_J$ MHz	D$_{JK}$ MHz	Δ Amu A²	κ
C^{12}F$_3^{19}$Br^{79}	C$_{3v}$	221		2 098.06 M	2 098.06 M		.00126		
C^{12}F$_3^{19}$Br^{81}	C$_{3v}$	222		2 078.50 M	2 078.50 M		.00122		

Id. No.	μ$_a$ Debye	μ$_b$ Debye	μ$_c$ Debye	eQq Value(MHz)	Rel.	eQq Value(MHz)	Rel.	eQq Value(MHz)	Rel.	ω$_a$ d 1/cm	ω$_b$ d 1/cm	ω$_c$ d 1/cm	ω$_d$ d 1/cm
221	.65 G	0. X	0. X	619	Br^{79}								
222				517	Br^{81}								

References:

ABC: 229,391   D$_{JK}$: 391   μ: 593   eQq: 391

Add. Ref. 232

Bromotrifluoromethane   Spectral Line Table

Isotopic Species	Id. No.	Rotational Quantum Nos.	Vib. State	F$_1'$	F'	F$_1$	F	Frequency MHz	Acc. ±MHz	Ref.
C^{12}F$_3^{19}$Br^{79}	221	5, 0← 4, 0	Ground		13/2		11/2	20 978.2		780
	221	5, 0← 4, 0	Ground		11/2		9/2	20 978.2		780
	221	5, 0← 4, 0	Ground		9/2		7/2	20 988.5		780
	221	5, 0← 4, 0	Ground		7/2		5/2	20 988.5		780
	221	5, 1← 4, 1	Ground		9/2		7/2	20 990.1		780
	221	5, 2← 4, 2	Ground		7/2		5/2	20 963.2		780
	221	5, 2← 4, 2	Ground		13/2		11/2	20 968.2		780
	221	5, 2← 4, 2	Ground		9/2		7/2	20 994.7		780
	221	5, 2← 4, 2	Ground		11/2		9/2	20 998.8		780
	221	5, 3← 4, 3	Ground		7/2		5/2	20 931.7		780
	221	5, 3← 4, 3	Ground		13/2		11/2	20 955.9		780
	221	5, 3← 4, 3	Ground		9/2		7/2	21 001.9		780
	221	5, 3← 4, 3	Ground		11/2		9/2	21 025.0		780
	221	6, 0← 5, 0	Ground		13/2		11/2	25 175.2		780
	221	6, 1← 5, 1	Ground		15/2		13/2	25 173.3		780
	221	6, 2← 5, 2	Ground		13/2		11/2	25 186.7		780
	221	6, 3← 5, 3	Ground		15/2		13/2	25 161.3		780
	221	6, 3← 5, 3	Ground		11/2		9/2	25 192.5		780
	221	11, 0←10, 0	Ground		23/2		21/2	46 156.60		391
	221	11, 0←10, 0	Ground		25/2		23/2	46 156.60		391
	221	11, 0←10, 0	Ground		19/2		17/2	46 158.56		391
	221	11, 0←10, 0	Ground		21/2		19/2	46 158.56		391
	221	11, 1←10, 1	Ground		25/2		23/2	46 156.28		391
	221	11, 1←10, 1	Ground		23/2		21/2	46 156.98		391
	221	11, 1←10, 1	Ground		19/2		17/2	46 158.06		391
	221	11, 1←10, 1	Ground		21/2		19/2	46 158.80		391
	221	11, 2←10, 2	Ground		25/2		23/2	46 155.32		391
	221	11, 2←10, 2	Ground		19/2		17/2	46 156.60		391
	221	11, 2←10, 2	Ground		23/2		21/2	46 158.06		391
	221	11, 2←10, 2	Ground		21/2		19/2	46 159.50		391

Isotopic Species	Id. No.	Rotational Quantum Nos.	Vib. State	F_1'	F'	F_1	F	Frequency MHz	Acc. ±MHz	Ref.
$C^{12}F_3^{19}Br^{79}$	221	11, 3←10, 3	Ground		25/2		23/2	46 153.72		391
	221	11, 3←10, 3	Ground		19/2		17/2	46 154.28		391
	221	11, 3←10, 3	Ground		23/2		21/2	46 160.02		391
	221	11, 3←10, 3	Ground		21/2		19/2	46 160.70		391
	221	11, 4←10, 4	Ground		19/2		17/2	46 151.04		391
	221	11, 4←10, 4	Ground		25/2		23/2	46 151.52		391
	221	11, 4←10, 4	Ground		21/2		19/2	46 162.40		391
	221	11, 4←10, 4	Ground		23/2		21/2	46 162.68		391
	221	11, 5←10, 5	Ground		19/2		17/2	46 146.92		391
	221	11, 5←10, 5	Ground		25/2		23/2	46 148.76		391
	221	11, 5←10, 5	Ground		21/2		19/2	46 164.58		391
	221	11, 5←10, 5	Ground		23/2		21/2	46 166.14		391
	221	11, 6←10, 6	Ground		19/2		17/2	46 141.80		391
	221	11, 6←10, 6	Ground		25/2		23/2	46 145.34		391
	221	11, 6←10, 6	Ground		21/2		19/2	46 167.24		391
	221	11, 6←10, 6	Ground		23/2		21/2	46 170.44		391
	221	11, 7←10, 7	Ground		19/2		17/2	46 135.84		391
	221	11, 7←10, 7	Ground		25/2		23/2	46 141.18		391
	221	11, 7←10, 7	Ground		21/2		19/2	46 170.44		391
	221	11, 7←10, 7	Ground		23/2		21/2	46 175.44		391
	221	11, 8←10, 8	Ground		19/2		17/2	46 128.96		391
	221	11, 8←10, 8	Ground		25/2		23/2	46 136.42		391
	221	11, 8←10, 8	Ground		21/2		19/2	46 173.94		391
	221	11, 8←10, 8	Ground		23/2		21/2	46 181.24		391
	221	11, 9←10, 9	Ground		19/2		17/2	46 121.20		391
	221	11, 9←10, 9	Ground		25/2		23/2	46 130.54		391
	221	11, 9←10, 9	Ground		21/2		19/2	46 177.98		391
	221	11, 9←10, 9	Ground		21/2		19/2	46 182.44		391
	221	11, 9←10, 9	Ground		23/2		21/2	46 187.96		391
$C^{12}F_3^{19}Br^{81}$	222	5, 0← 4, 0	Ground		13/2		11/2	20 782.9		780
	222	5, 0← 4, 0	Ground		11/2		9/2	20 782.9		780
	222	5, 0← 4, 0	Ground		9/2		7/2	20 791.8		780
	222	5, 0← 4, 0	Ground		7/2		5/2	20 791.8		780
	222	5, 1← 4, 1	Ground		9/2		7/2	20 793.0		780
	222	5, 2← 4, 2	Ground		7/2		5/2	20 770.7		780
	222	5, 2← 4, 2	Ground		13/2		11/2	20 774.3		780
	222	5, 2← 4, 2	Ground		9/2		7/2	20 796.8		780
	222	5, 2← 4, 2	Ground		11/2		9/2	20 800.0		780
	222	5, 3← 4, 3	Ground		7/2		5/2	20 744.2		780
	222	5, 3← 4, 3	Ground		13/2		11/2	20 764.2		780
	222	5, 3← 4, 3	Ground		9/2		7/2	20 803.0		780
	222	5, 3← 4, 3	Ground		11/2		9/2	20 822.2		780
	222	5, 4← 4, 4	Ground		13/2		11/2	20 749.6		780
	222	5, 4← 4, 4	Ground		9/2		7/2	20 811.2		780
	222	5, 4← 4, 4	Ground		11/2		9/2	20 852.8		780
	222	6, 0← 5, 0	Ground		15/2		13/2	24 941.2		780
	222	6, 0← 5, 0	Ground		13/2		11/2	24 941.2		780
	222	6, 1← 5, 1	Ground		15/2		13/2	24 939.6		780
	222	6, 1← 5, 1	Ground		13/2		11/2	24 943.3		780

Isotopic Species	Id. No.	Rotational Quantum Nos.	Vib. State	Hyperfine				Frequency MHz	Acc. ±MHz	Ref.
				F_1'	F'	F_1	F			
$C^{12}F_3^{19}Br^{81}$	222	6, 1 ← 5, 1	Ground		11/2		9/2	24 947.6		780
	222	6, 2 ← 5, 2	Ground		15/2		13/2	24 935.7		780
	222	6, 2 ← 5, 2	Ground		9/2		7/2	24 935.7		780
	222	6, 2 ← 5, 2	Ground		13/2		11/2	24 950.4		780
	222	6, 3 ← 5, 3	Ground		11/2		9/2	24 955.4		780
	222	11, 0 ← 10, 0	Ground		25/2		23/2	45 726.26		391
	222	11, 0 ← 10, 0	Ground		23/2		21/2	45 726.26		391
	222	11, 0 ← 10, 0	Ground		21/2		19/2	45 727.88		391
	222	11, 0 ← 10, 0	Ground		19/2		17/2	45 727.88		391
	222	11, 1 ← 10, 1	Ground		25/2		23/2	45 726.02		391
	222	11, 1 ← 10, 1	Ground		23/2		21/2	45 726.60		391
	222	11, 1 ← 10, 1	Ground		19/2		17/2	45 727.50		391
	222	11, 1 ← 10, 1	Ground		21/2		19/2	45 728.12		391
	222	11, 2 ← 10, 2	Ground		25/2		23/2	45 725.18		391
	222	11, 2 ← 10, 2	Ground		19/2		17/2	45 726.26		391
	222	11, 2 ← 10, 2	Ground		23/2		21/2	45 727.50		391
	222	11, 2 ← 10, 2	Ground		21/2		19/2	45 728.70		391
	222	11, 3 ← 10, 3	Ground		25/2		23/2	45 723.84		391
	222	11, 3 ← 10, 3	Ground		19/2		17/2	45 724.34		391
	222	11, 3 ← 10, 3	Ground		23/2		21/2	45 729.10		391
	222	11, 3 ← 10, 3	Ground		21/2		19/2	45 729.66		391
	222	11, 4 ← 10, 4	Ground		19/2		17/2	45 721.66		391
	222	11, 4 ← 10, 4	Ground		25/2		23/2	45 722.12		391
	222	11, 4 ← 10, 4	Ground		21/2		19/2	45 731.06		391
	222	11, 4 ← 10, 4	Ground		23/2		21/2	45 731.36		391
	222	11, 5 ← 10, 5	Ground		19/2		17/2	45 718.16		391
	222	11, 5 ← 10, 5	Ground		25/2		23/2	45 719.66		391
	222	11, 5 ← 10, 5	Ground		21/2		19/2	45 732.88		391
	222	11, 5 ← 10, 5	Ground		23/2		21/2	45 734.22		391
	222	11, 6 ← 10, 6	Ground		19/2		17/2	45 713.80		391
	222	11, 6 ← 10, 6	Ground		25/2		23/2	45 716.70		391
	222	11, 6 ← 10, 6	Ground		21/2		19/2	45 735.00		391
	222	11, 6 ← 10, 6	Ground		23/2		21/2	45 737.66		391
	222	11, 7 ← 10, 7	Ground		19/2		17/2	45 708.80		391
	222	11, 7 ← 10, 7	Ground		25/2		23/2	45 713.24		391
	222	11, 7 ← 10, 7	Ground		21/2		19/2	45 737.56		391
	222	11, 7 ← 10, 7	Ground		23/2		21/2	45 741.90		391
	222	11, 8 ← 10, 8	Ground		19/2		17/2	45 702.98		391
	222	11, 8 ← 10, 8	Ground		25/2		23/2	45 709.20		391
	222	11, 8 ← 10, 8	Ground		21/2		19/2	45 740.64		391
	222	11, 8 ← 10, 8	Ground		23/2		21/2	45 746.74		391
	222	11, 9 ← 10, 9	Ground		19/2		17/2	45 696.34		391
	222	11, 9 ← 10, 9	Ground		25/2		23/2	45 704.58		391
	222	11, 9 ← 10, 9	Ground		21/2		19/2	45 743.94		391
	222	11, 9 ← 10, 9	Ground		23/2		21/2	45 752.20		391
	222	11,10 ← 10,10	Ground		25/2		23/2	45 699.46		391
	222	11,10 ← 10,10	Ground		21/2		19/2	45 747.52		391

CF_3I C_{3v} CF_3I

Isotopic Species	Pt. Gp.	Id. No.	A MHz	B MHz	C MHz	D_J MHz	D_{JK} MHz	Δ Amu A²	κ
$C^{12}F_3^{19}I^{127}$	C_{3v}	231		1 523.23 M	1 523.23 M	.0006			

Id. No.	μ_a Debye	μ_b Debye	μ_c Debye	eQq Value(MHz) Rel.	eQq Value(MHz) Rel.	eQq Value(MHz) Rel.	ω_a d 1/cm	ω_b d 1/cm	ω_c d 1/cm	ω_d d 1/cm
231	1.0 M	0. X	0. X	−2142.5						

References:

ABC: 391 D_J: 391 μ: 637   eQq: 638

Add. Ref. 232,502,528,593

Trifluoroiodomethane   Spectral Line Table

Isotopic Species	Id. No.	Rotational Quantum Nos.	Vib. State	F_1'	F'	F_1	F	Frequency MHz	Acc. ±MHz	Ref.
$C^{12}F_3^{19}I^{127}$	231	1, 0← 0, 0	Ground		5/2		5/2	2 715.1	.3	565
	231	1, 0← 0, 0	Ground		7/2		5/2	3 162.9	.25	565
	231	1, 0← 0, 0	Ground		3/2		5/2	3 362.9	.4	565
	231	3, 3← 3, 3	Ground		9/2		11/2	500.4	.4	638
	231	10, ← 9,	Ground					30 386.08		391
	231	10, ← 9,	Ground					30 386.82		391
	231	10, ← 9,	Ground					30 387.43		391
	231	10, ← 9,	Ground					30 390.29		391
	231	10, ← 9,	Ground					30 391.03		391
	231	10, ← 9,	Ground					30 392.19		391
	231	10, ← 9,	Ground					30 395.37		391
	231	10, ← 9,	Ground					30 397.85		391
	231	10, ← 9,	Ground					30 414.44		391
	231	10, ← 9,	Ground					30 421.04		391
	231	10, ← 9,	Ground					30 425.23		391
	231	10, ← 9,	Ground					30 426.25		391
	231	10, ← 9,	Ground					30 434.89		391
	231	10, ← 9,	Ground					30 436.74		391
	231	10, ← 9,	Ground					30 446.96		391
	231	10, ← 9,	Ground					30 467.09		391
	231	10, 0← 9, 0	Ground		17/2		15/2	30 455.63		391
	231	10, 0← 9, 0	Ground		19/2		17/2	30 459.00		391
	231	10, 0← 9, 0	Ground		15/2		13/2	30 459.51		391
	231	10, 0← 9, 0	Ground		21/2		19/2	30 465.76		391
	231	10, 0← 9, 0	Ground		25/2		23/2	30 468.32		391
	231	10, 0← 9, 0	Ground		23/2		21/2	30 480.14		391
	231	10, 1← 9, 1	Ground		17/2		15/2	30 456.13		391
	231	10, 1← 9, 1	Ground		19/2		17/2	30 458.15		391
	231	10, 1← 9, 1	Ground		15/2		13/2	30 462.16		391
	231	10, 1← 9, 1	Ground		21/2		19/2	30 464.26		391
	231	10, 1← 9, 1	Ground		23/2		21/2	30 480.14		391
	231	10, 2← 9, 2	Ground		21/2		21/2	30 369.52		391
	231	10, 2← 9, 2	Ground		19/2		17/2	30 455.46		391
	231	10, 2← 9, 2	Ground		17/2		15/2	30 457.42		391
	231	10, 2← 9, 2	Ground		21/2		19/2	30 459.83		391

Isotopic Species	Id. No.	Rotational Quantum Nos.	Vib. State	Hyperfine				Frequency MHz	Acc. ±MHz	Ref.
				F_1'	F'	F_1	F			
$C^{12}F_3^{19}I^{127}$	231	10, 2← 9, 2	Ground		15/2		13/2	30 467.72		391
	231	10, 2← 9, 2	Ground		25/2		23/2	30 473.55		391
	231	10, 2← 9, 2	Ground		23/2		21/2	30 476.54		391
	231	10, 2← 9, 2	Ground		19/2		19/2	30 514.83		391
	231	10, 3← 9, 3	Ground		19/2		17/2	30 451.31		391
	231	10, 3← 9, 3	Ground		21/2		19/2	30 452.49		391
	231	10, 3← 9, 3	Ground		17/2		15/2	30 459.51		391
	231	10, 3← 9, 3	Ground		23/2		21/2	30 469.63		391
	231	10, 3← 9, 3	Ground		15/2		13/2	30 477.65		391
	231	10, 3← 9, 3	Ground		25/2		23/2	30 480.14		391
	231	10, 3← 9, 3	Ground		19/2		19/2	30 500.73		391
	231	10, 4← 9, 4	Ground		21/2		19/2	30 442.17		391
	231	10, 4← 9, 4	Ground		19/2		17/2	30 445.32		391
	231	10, 4← 9, 4	Ground		23/2		21/2	30 461.76		391
	231	10, 4← 9, 4	Ground		17/2		15/2	30 462.46		391
	231	10, 4← 9, 4	Ground		19/2		19/2	30 476.54		391
	231	10, 4← 9, 4	Ground		25/2		23/2	30 489.17		391
	231	10, 4← 9, 4	Ground		15/2		13/2	30 491.56		391
	231	10, 5← 9, 5	Ground		21/2		21/2	30 412.09		391
	231	10, 5← 9, 5	Ground		21/2		19/2	30 428.97		391
	231	10, 5← 9, 5	Ground		23/2		21/2	30 436.24		391
	231	10, 5← 9, 5	Ground		19/2		17/2	30 437.54		391
	231	10, 5← 9, 5	Ground		23/2		21/2	30 449.49		391
	231	10, 5← 9, 5	Ground		17/2		15/2	30 466.29		391
	231	10, 5← 9, 5	Ground		17/2		17/2	30 500.73		391
	231	10, 5← 9, 5	Ground		25/2		23/2	30 500.73		391
	231	10, 5← 9, 5	Ground		15/2		13/2	30 509.58		391
	231	10, 6← 9, 6	Ground		21/2		19/2	30 412.64		391
	231	10, 6← 9, 6	Ground		19/2		19/2	30 413.84		391
	231	10, 6← 9, 6	Ground		19/2		17/2	30 427.99		391
	231	10, 6← 9, 6	Ground		21/2		21/2	30 432.06		391
	231	10, 6← 9, 6	Ground		17/2		17/2	30 432.06		391
	231	10, 6← 9, 6	Ground		17/2		15/2	30 470.95		391
	231	10, 6← 9, 6	Ground		15/2		15/2	30 476.54		391
	231	10, 6← 9, 6	Ground		23/2		23/2	30 509.58		391
	231	10, 6← 9, 6	Ground		25/2		23/2	30 514.83		391
	231	10, 6← 9, 6	Ground		15/2		13/2	30 532.05		391
	231	10, 7← 9, 7	Ground		21/2		19/2	30 393.47		391
	231	10, 7← 9, 7	Ground		19/2		17/2	30 416.42		391
	231	10, 7← 9, 7	Ground		23/2		21/2	30 420.49		391
	231	10, 7← 9, 7	Ground		21/2		21/2	30 459.00		391
	231	10, 7← 9, 7	Ground		25/2		23/2	30 470.02		391
	231	10, 7← 9, 7	Ground		17/2		15/2	30 477.65		391
	231	10, 7← 9, 7	Ground		25/2		23/2	30 531.33		391
	231	10, 8← 9, 8	Ground		21/2		19/2	30 371.50		391
	231	10, 8← 9, 8	Ground		23/2		21/2	30 400.97		391
	231	10, 8← 9, 8	Ground		19/2		17/2	30 402.76		391
	231	10, 8← 9, 8	Ground		21/2		21/2	30 489.17		391
	231	10, 9← 9, 9	Ground		21/2		19/2	30 346.49		391
	231	10, 9← 9, 9	Ground		19/2		17/2	30 387.87		391
	231	10, 9← 9, 9	Ground		17/2		15/2	30 489.17		391

CHBr$_3$ C$_{3v}$ CHBr$_3$

Isotopic Species	Pt. Gp.	Id. No.	A MHz		B MHz		C MHz	D$_J$ MHz	D$_{JK}$ MHz	Δ Amu A^2	κ
C^{12}HBr$_3^{79}$	C$_{3v}$	241	1 247.61	M	1 247.61	M					
C^{12}HBr$_3^{81}$	C$_{3v}$	242	1 217.30	M	1 217.30	M					
C^{12}DBr$_3^{79}$	C$_{3v}$	243	1 239.45	M	1 239.45	M					
C^{12}DBr$_3^{81}$	C$_{3v}$	244	1 209.51	M	1 209.51	M					

Id. No.	μ$_a$ Debye		μ$_b$ Debye		μ$_c$ Debye		eQq Value(MHz)	Rel.	eQq Value(MHz)	Rel.	eQq Value(MHz)	Rel.	ω$_a$ d 1/cm	ω$_b$ d 1/cm	ω$_c$ d 1/cm	ω$_d$ d 1/cm
241	0.	X	0.	X	.99	L	577	Br^{79}								
242					--		482	Br^{81}								

References:

ABC: 410   μ: 1032   eQq: 368

Add. Ref. 252,528,536

Tribromomethane[1]   Spectral Line Table

Isotopic Species	Id. No.	Rotational Quantum Nos.	Vib. State	Hyperfine F$_1'$	F'	F$_1$	F	Frequency MHz	Acc. ±MHz	Ref.
C^{12}HBr$_3^{79}$	241	11, ←10,	Ground					27 447.9	.5	410
	241	12, ←11,	Ground					29 942.7	.5	410
	241	13, ←12,	Ground					32 438.2	.5	410
	241	15, ←14,	Ground					37 427.8	.5	410
C^{12}HBr$_3^{81}$	242	11, ←10,	Ground					26 781.1	.5	410
	242	12, ←11,	Ground					29 215.4	.5	410
	242	13, ←12,	Ground					31 649.9	.5	410
	242	14, ←13,	Ground					34 084.1	.5	410
C^{12}DBr$_3^{79}$	243	12, ←11,	Ground					29 747.2	.5	410
	243	13, ←12,	Ground					32 225.6	.5	410
	243	14, ←13,	Ground					34 704.6	.5	410
C^{12}DBr$_3^{81}$	244	12, ←11,	Ground					29 028.2	.5	410
	244	13, ←12,	Ground					31 447.3	.5	410
	244	14, ←13,	Ground					33 866.2	.5	410

1. Lines in the reference by Kojima, et al., were not included since discrepancies in their assignments were pointed out in a later reference by Hermann.

CHClF$_2$ C$_s$ CHClF$_2$

Isotopic Species	Pt. Gp.	Id. No.	A MHz	B MHz	C MHz	D$_J$ MHz	D$_{JK}$ MHz	Δ Amu A^2	κ
C^{12}HCl^{35}F$_2^{19}$	C$_s$	251	10 234.68 M	4 861.22 M	3 507.415 M				−.5975
C^{12}HCl^{37}F$_2^{19}$	C$_s$	252	10 233.82 M	4 717.12 M	3 431.812 M				−.6221
C^{13}HCl^{35}F$_2^{19}$	C$_s$	253	10 204.16 M	4 845.998 M	3 503.335 M				

Id. No.	μ$_a$ Debye	μ$_b$ Debye	μ$_c$ Debye	eQq Value(MHz)	Rel.	eQq Value(MHz)	Rel.	eQq Value(MHz)	Rel.	ω$_a$ d 1/cm	ω$_b$ d 1/cm	ω$_c$ d 1/cm	ω$_d$ d 1/cm
251	.12 M	0. X	1.43 M	−64.96	aa	35.61	bb	29.35	cc				
252				−51.07	aa	27.92	bb	23.15	cc				

References:

ABC: 974   κ: 1006   μ: 1006   eQq: 1006

Add. Ref. 151,619,626

Chlorodifluoromethane   Spectral Line Table

Isotopic Species	Id. No.	Rotational Quantum Nos.	Vib. State	F$_1'$	F'	F$_1$	F	Frequency MHz	Acc. ±MHz	Ref.
C^{12}HCl^{35}F$_2^{19}$	251	1, 1, 0← 0, 0, 0	Ground		1/2		3/2	15 088.57		1006
	251	1, 1, 0← 0, 0, 0	Ground		5/2		3/2	15 094.42		1006
	251	1, 1, 0← 0, 0, 0	Ground					15 095.94	.02	974
	251	1, 1, 0← 0, 0, 0	Ground		3/2		3/2	15 101.91		1006
	251	2, 1, 1← 1, 0, 1	Ground		3/2		1/2	24 801.9	.05	706
	251	2, 1, 1← 1, 0, 1	Ground		3/2		1/2	24 802.17		1006
	251	2, 1, 1← 1, 0, 1	Ground		5/2		5/2	24 808.7	.05	706
	251	2, 1, 1← 1, 0, 1	Ground		1/2		1/2	24 810.8	.05	706
	251	2, 1, 1← 1, 0, 1	Ground		1/2		1/2	24 810.96		1006
	251	2, 1, 1← 1, 0, 1	Ground		7/2		5/2	24 817.4	.05	706
	251	2, 1, 1← 1, 0, 1	Ground		7/2		5/2	24 817.64		1006
	251	2, 1, 1← 1, 0, 1	Ground					24 818.34	.03	974
	251	2, 1, 1← 1, 0, 1	Ground		5/2		3/2	24 824.9	.05	706
	251	2, 1, 1← 1, 0, 1	Ground		5/2		3/2	24 825.06		1006
	251	2, 1, 1← 1, 0, 1	Ground		3/2		3/2	24 831.2	.05	706
	251	2, 1, 1← 1, 0, 1	Ground		3/2		3/2	24 831.37		1006
	251	2, 1, 1← 1, 0, 1	Ground		5/2		5/2	24 808.76		1006
	251	2, 2, 0← 1, 1, 0	Ground		1/2		1/2	34 428.02		1006
	251	2, 2, 0← 1, 1, 0	Ground		3/2		3/2	34 430.59		1006
	251	2, 2, 0← 1, 1, 0	Ground		7/2		5/2	34 433.37		1006
	251	2, 2, 0← 1, 1, 0	Ground					34 436.37		1006
	251	2, 2, 0← 1, 1, 0	Ground		5/2		3/2	34 441.75		1006
	251	2, 2, 0← 1, 1, 0	Ground		3/2		1/2	34 443.60		1006
	251	2, 2, 0← 1, 1, 0	Ground		5/2		5/2	34 449.00		1006
	251	2, 2, 1← 1, 1, 1	Ground		1/2		3/2	35 541.81		1006
	251	2, 2, 1← 1, 1, 1	Ground		3/2		3/2	35 557.98		1006
	251	2, 2, 1← 1, 1, 1	Ground		7/2		5/2	35 562.30		1006
	251	2, 2, 1← 1, 1, 1	Ground					35 565.3	.2	974
	251	2, 2, 1← 1, 1, 1	Ground		5/2		5/2	35 578.59		1006
	251	3, 1, 2← 2, 0, 2	Ground		7/2		7/2	35 275.12		1006

43

Isotopic Species	Id. No.	Rotational Quantum Nos.	Vib. State	F_1'	F'	F_1	F	Frequency MHz	Acc. $\pm$MHz	Ref.
$C^{12}HCl^{35}F_2^{19}$	251	3, 1, 2← 2, 0, 2	Ground		3/2		1/2	35 282.61		1006
	251	3, 1, 2← 2, 0, 2	Ground		5/2		3/2	35 285.66		1006
	251	3, 1, 2← 2, 0, 2	Ground		9/2		7/2	35 288.02		1006
	251	3, 1, 2← 2, 0, 2	Ground					35 288.25	.05	974
	251	3, 1, 2← 2, 0, 2	Ground		7/2		5/2	35 290.94		1006
	251	3, 1, 2← 2, 0, 2	Ground		5/2		5/2	35 296.90		1006
	251	3, 1, 2← 2, 0, 2	Ground		3/2		3/2	35 298.48		1006
	251	3, 2, 1← 3, 1, 3	Ground					23 429.85	.03	974
	251	3, 3, 0← 3, 2, 2	Ground		3/2		3/2	30 389.4	.05	706
	251	3, 3, 0← 3, 2, 2	Ground		9/2		9/2	30 398.8	.05	706
	251	3, 3, 0← 3, 2, 2	Ground					30 405.68	.10	974
	251	3, 3, 0← 3, 2, 2	Ground		5/2		5/2	30 409.4	.05	706
	251	3, 3, 0← 3, 2, 2	Ground		7/2		7/2	30 418.9	.05	706
	251	3, 3, 1← 3, 2, 1	Ground		3/2		3/2	29 277.84		1006
	251	3, 3, 1← 3, 2, 1	Ground		9/2		9/2	29 287.5	.05	706
	251	3, 3, 1← 3, 2, 1	Ground					29 294.80	.03	974
	251	3, 3, 1← 3, 2, 1	Ground		5/2		5/2	29 298.9	.05	706
	251	3, 3, 1← 3, 2, 1	Ground		7/2		7/2	29 308.7	.05	706
	251	4, 3, 2← 4, 2, 2	Ground					27 906.05	.05	974
	251	5, 3, 2← 5, 2, 4	Ground		7/2		7/2	32 630.4	.05	706
	251	5, 3, 2← 5, 2, 4	Ground		13/2		13/2	32 632.4	.05	706
	251	5, 3, 2← 5, 2, 4	Ground		9/2		9/2	32 637.6	.05	706
	251	5, 3, 2← 5, 2, 4	Ground		11/2		11/2	32 639.6	.05	706
	251	5, 3, 3← 5, 2, 3	Ground		13/2		13/2	25 625.6	.05	706
	251	5, 3, 3← 5, 2, 3	Ground					25 629.03	.46	974
	251	5, 3, 3← 5, 2, 3	Ground		9/2		9/2	25 632.7	.05	706
	251	5, 3, 3← 5, 2, 3	Ground		11/2		11/2	25 635.3	.05	706
	251	6, 3, 4← 6, 2, 4	Ground		9/2		9/2	22 545.9	.05	706
	251	6, 3, 4← 6, 2, 4	Ground					22 550.41	.03	974
	251	6, 3, 4← 6, 2, 4	Ground		11/2		11/2	22 552.9	.05	706
	251	6, 3, 4← 6, 2, 4	Ground		13/2		13/2	22 554.5	.05	706
	251	6, 3, 4← 6, 2, 4	Ground		15/2		15/2	25 547.5	.05	706
	251	8, 4, 5← 8, 3, 5	Ground					35 209.96	.04	974
	251	9, 4, 6← 9, 3, 6	Ground		15/2		15/2	31 194.9	.05	706
	251	9, 4, 6← 9, 3, 6	Ground		21/2		21/2	31 195.8	.05	706
	251	9, 4, 6← 9, 3, 6	Ground		17/2		17/2	31 200.1	.05	706
	251	9, 4, 6← 9, 3, 6	Ground		19/2		19/2	31 201.0	.05	706
	251	10, 4, 7←10, 3, 7	Ground		17/2		17/2	26 439.4	.05	706
	251	10, 4, 7←10, 3, 7	Ground		23/2		23/2	26 440.0	.05	706
	251	10, 4, 7←10, 3, 7	Ground		19/2		19/2	26 443.8	.05	706
	251	10, 4, 7←10, 3, 7	Ground		21/2		21/2	26 444.4	.05	706
$C^{12}HCl^{37}F_2^{19}$	252	1, 1, 0← 0, 0, 0	Ground		1/2		3/2	14 945.25		1006
	252	1, 1, 0← 0, 0, 0	Ground		5/2		3/2	14 949.85		1006
	252	1, 1, 0← 0, 0, 0	Ground					14 950.94	.03	974
	252	1, 1, 0← 0, 0, 0	Ground		3/2		3/2	14 955.67		1006
	252	2, 1, 1← 1, 0, 1	Ground		3/2		1/2	24 372.39		1006
	252	2, 1, 1← 1, 0, 1	Ground		5/2		5/2	24 377.65		1006
	252	2, 1, 1← 1, 0, 1	Ground		1/2		1/2	24 379.38		1006
	252	2, 1, 1← 1, 0, 1	Ground		7/2		5/2	24 384.64		1006
	252	2, 1, 1← 1, 0, 1	Ground		7/2		5/2	24 384.7	.05	706
	252	2, 1, 1← 1, 0, 1	Ground					24 385.21	.03	974

Isotopic Species	Id. No.	Rotational Quantum Nos.	Vib. State	Hyperfine				Frequency MHz	Acc. ±MHz	Ref.
				F_1'	F'	F_1	F			
$C^{12}HCl^{37}F_2^{19}$	252	2, 1, 1← 1, 0, 1	Ground		5/2		3/2	24 390.43		1006
	252	2, 1, 1← 1, 0, 1	Ground		3/2		3/2	24 395.56		1006
	252	2, 2, 0← 1, 1, 0	Ground		1/2		1/2	34 325.80		1006
	252	2, 2, 0← 1, 1, 0	Ground		7/2		5/2	34 330.43		1006
	252	2, 2, 0← 1, 1, 0	Ground		5/2		3/2	34 337.10		1006
	252	2, 2, 0← 1, 1, 0	Ground		5/2		5/2	34 342.73		1006
	252	2, 2, 1← 1, 1, 1	Ground		3/2		3/2	35 412.78		1006
	252	2, 2, 1← 1, 1, 1	Ground		7/2		5/2	35 416.22		1006
	252	2, 2, 1← 1, 1, 1	Ground					35 418.51	.10	974
	252	2, 2, 1← 1, 1, 1	Ground		5/2		3/2	35 422.15		1006
	252	2, 2, 1← 1, 1, 1	Ground		3/2		1/2	35 425.55		1006
	252	2, 2, 1← 1, 1, 1	Ground		5/2		5/2	35 429.16		1006
	252	3, 1, 2← 2, 0, 2	Ground		7/2		7/2	34 515.29		1006
	252	3, 1, 2← 2, 0, 2	Ground		3/2		1/2	34 521.01		1006
	252	3, 1, 2← 2, 0, 2	Ground		9/2		7/2	34 525.30		1006
	252	3, 1, 2← 2, 0, 2	Ground		7/2		5/2	34 527.65		1006
	252	3, 2, 1← 3, 1, 3	Ground					23 417.65	.10	974
	252	3, 3, 0← 3, 2, 2	Ground					30 932.07	.10	974
	252	3, 3, 1← 3, 2, 1	Ground		3/2		3/2	29 931.86		1006
	252	3, 3, 1← 3, 2, 1	Ground		9/2		9/2	29 939.30		1006
	252	3, 3, 1← 3, 2, 1	Ground					29 944.7	.3	974
	252	3, 3, 1← 3, 2, 1	Ground		5/2		5/2	29 948.23		1006
	252	3, 3, 1← 3, 2, 1	Ground		7/2		7/2	29 955.81		1006
	252	4, 3, 2← 4, 2, 2	Ground		5/2		5/2	28 681.8	.05	706
	252	4, 3, 2← 4, 2, 2	Ground		11/2		11/2	28 685.5	.05	706
	252	4, 3, 2← 4, 2, 2	Ground					28 689.19	.05	974
	252	4, 3, 2← 4, 2, 2	Ground		7/2		7/2	28 692.6	.05	706
	252	4, 3, 2← 4, 2, 2	Ground		9/2		9/2	28 695.6	.05	706
	252	6, 3, 4← 6, 2, 4	Ground		9/2		9/2	23 704.6	.05	706
	252	6, 3, 4← 6, 2, 4	Ground		15/2		15/2	23 705.8	.05	706
	252	6, 3, 4← 6, 2, 4	Ground					23 708.20	.02	974
	252	6, 3, 4← 6, 2, 4	Ground		11/2		11/2	23 710.0	.05	706
	252	6, 3, 4← 6, 2, 4	Ground		13/2		13/2	23 711.4	.05	706
	252	9, 4, 6← 9, 3, 6	Ground		15/2		15/2	33 167.4	.05	706
	252	9, 4, 6← 9, 3, 6	Ground		21/2		21/2	33 168.1	.05	706
	252	9, 4, 6← 9, 3, 6	Ground		17/2		17/2	33 171.4	.05	706
	252	9, 4, 6← 9, 3, 6	Ground		19/2		19/2	33 172.1	.05	706
	252	10, 4, 7←10, 3, 7	Ground		17/2		17/2	28 697.7	.05	706
	252	10, 4, 7←10, 3, 7	Ground		23/2		23/2	28 698.2	.05	706
	252	10, 4, 7←10, 3, 7	Ground		19/2		19/2	28 701.1	.05	706
	252	10, 4, 7←10, 3, 7	Ground		21/2		21/2	28 701.6	.05	706
	252	11, 4, 8←11, 3, 8	Ground		19/2		19/2	23 728.2	.05	706
	252	11, 4, 8←11, 3, 8	Ground		25/2		25/2	23 728.6	.05	706
	252	11, 4, 8←11, 3, 8	Ground		21/2		21/2	23 731.0	.05	706
	252	11, 4, 8←11, 3, 8	Ground		23/2		23/2	23 731.4	.05	706
$C^{13}HCl^{35}F_2^{19}$	253	2, 1, 1← 1, 0, 1	Ground					24 742.15	.05	974
	253	3, 1, 2← 2, 0, 2	Ground					35 175.15	.05	974
	253	3, 2, 1← 3, 1, 3	Ground					23 318.61	.15	974
	253	3, 3, 0← 3, 2, 2	Ground					30 299.2	.2	974
	253	3, 3, 1← 3, 2, 1	Ground					29 202.44	.10	974
	253	4, 3, 2← 4, 2, 2	Ground					27 830.2	.3	974
	253	6, 3, 4← 6, 2, 4	Ground					22 525.15	.10	974

CHCl₃ — $CHCl_3$ C₃ᵥ $CHCl_3$

Isotopic Species	Pt. Gp.	Id. No.	A MHz		B MHz		C MHz		D_J MHz	D_{JK} MHz	Δ Amu A²	κ
$C^{12}HCl_3^{35}$	C_{3v}	261	3 301.94	M	3 301.94	M			.00412	.055		
$C^{12}HCl_2^{35}Cl^{37}$	C_s	262	3 302.20	M	3 187.19	M	1 682.67	M				
$C^{12}HCl_3^{37}$	C_{3v}	263	3 129.51	M	3 129.51	M						
$C^{12}DCl_3^{35}$	C_{3v}	264	3 250.17	M	3 250.17	M						
$C^{13}DCl_3^{35}$	C_{3v}	265	3 296.38	M	3 296.38	M						

Id. No.	μ_a Debye		μ_b Debye		μ_c Debye		eQq Value(MHz)	Rel.	eQq Value(MHz)	Rel.	eQq Value(MHz)	Rel.	ω_a 1/cm	d	ω_b 1/cm	d	ω_c 1/cm	d	ω_d 1/cm	d
261	0.	X	0.	X	1.2	M	28.70	zz													

References:
ABC: 350,729,967 D_J: 729 D_{JK}: 729 μ: 1029   eQq: 729

Add. Ref. 55,406,475

Trichloromethane   Spectral Line Table

Isotopic Species	Id. No.	Rotational Quantum Nos.	Vib. State	F_1'	F'	F_1	F	Frequency MHz	Acc. ±MHz	Ref.
$C^{12}HCl_3^{35}$	261	3, ← 2,	Ground					19 812.92	.10	729
	261	3, ← 2,	Ground					19 814.95	.10	729
	261	3, ← 2,	Ground					19 816.76	.14	729
	261	3, ← 2,	Ground					19 817.66	.13	729
	261	3, ← 2,	Ground					19 818.73	.17	729
	261	3, ← 2,	Ground					19 819.65	.17	729
	261	3, 1← 2, 1	Ground					19 808.28	.12	729
	261	3, 2← 2, 2	Ground					19 801.23		729
	261	3, 2← 2, 2	Ground					19 802.67	.14	729
	261	3, 2← 2, 2	Ground					19 803.70	.16	729
	261	3, 2← 2, 2	Ground					19 804.43	.16	729
	261	3, 2← 2, 2	Ground					19 810.57	.05	729
	261	3, 2← 2, 2	Ground					19 821.00	.10	729
	261	3, 2← 2, 2	Ground					19 822.19	.12	729
	261	3, 2← 2, 2	Ground					19 823.08	.17	729
	261	4, ← 3,	Ground					26 417.		166
	261	5, ← 4,	Ground					33 020.0	.4	967
	261	7, ← 6,	Ground					46 227.2	.15	249
	261	Not Reported	Ground					19 805.57	.14	729
	261	Not Reported	Ground					19 806.21		729
$C^{12}HCl_2^{35}Cl^{37}$	262	3, 2, 1← 2, 1, 1	Ground					19 298.31	.21	729
	262	3, 3, 0← 2, 2, 0	Ground					19 492.81	.17	729
	262	3, 3, 1← 2, 2, 1	Ground					19 643.13	.01	729
	262	5, 4, 1← 4, 3, 1	Ground					32 187.3	.4	967
	262	5, 5, 0← 4, 4, 0	Ground					32 565.3	.4	967
	262	5, 5, 1← 4, 4, 1	Ground					32 756.4	.4	967
$C^{12}HCl_3^{37}$	263	5, ← 4,	Ground					31 295.4	.4	967
	263	6, ← 5,	Ground					37 554.11	.40	350
$C^{12}DCl_3^{35}$	264	5, ← 4,	Ground					32 502.1	.4	967
	264	7, ← 6,	Ground					45 502.4	.15	249
$C^{13}DCl_3^{35}$	265	5, ← 4,	Ground					32 973.9	.4	967

CHFO C_s **FCHO**

Isotopic Species	Pt. Gp.	Id. No.	A MHz	B MHz	C MHz	D_J MHz	D_{JK} MHz	Δ Amu A²	κ
$F^{19}HC^{12}O^{16}$	C_s	271	91 153.57 M	11 760.37 M	10 396.79 M			.0918	−.96623
$F^{19}DC^{12}O^{16}$	C_s	272	65 096.59 M	11 761.74 M	9 941.71 M			.1020	
$F^{19}HC^{13}O^{16}$	C_s	273	88 505.1 M	11 755.2 M	10 357.3 M			.0904	
$F^{19}HC^{12}O^{18}$	C_s	274	89 769.5 M	11 102.9 M	9 863.4 M				

Id. No.	μ_a Debye	μ_b Debye	μ_c Debye	eQq Value(MHz) Rel.	eQq Value(MHz) Rel.	eQq Value(MHz) Rel.	ω_a d 1/cm	ω_b d 1/cm	ω_c d 1/cm	ω_d d 1/cm
271	.595 M	1.934 M	0. X							

References:

ABC: 858,900,947   Δ: 858,913   κ: 858   μ: 913

Add. Ref. 899,924

Formyl Fluoride   Spectral Line Table

Isotopic Species	Id. No.	Rotational Quantum Nos.	Vib. State	F_1'	F'	F_1	F	Frequency MHz	Acc. ±MHz	Ref.
$F^{19}HC^{12}O^{16}$	271	1, 0, 1← 0, 0, 0						22 156.8	.1	913
	271	1, 1, 1← 0, 0, 0						101 550.36	.05	858
	271	1, 1, 1← 2, 0, 2						35 097.5	.1	913
	271	2, 0, 2← 1, 0, 1						44 296.0	.1	913
	271	2, 0, 2← 1, 1, 1						35 097.26	.04	962
	271	2, 1, 2← 1, 0, 1						122 343.95	.05	858
	271	2, 1, 1← 1, 1, 0						45 676.8	.1	913
	271	2, 1, 2← 1, 1, 1						42 950.7	.1	913
	271	2, 1, 2← 3, 0, 3						11 647.9	.1	913
	271	3, 0, 3← 2, 1, 2						11 648.01	.04	962
	271	3, 1, 3← 2, 0, 2						142 462.44	.05	858
	271	4, 0, 4← 3, 1, 3						12 388.8	.1	913
	271	4, 1, 4← 3, 0, 3						161 927.56	.05	858
	271	4, 2, 3← 5, 1, 4						119 298.55	.05	858
	271	5, 0, 5← 4, 1, 4						36 956.0	.1	913
	271	5, 1, 4← 5, 0, 5						90 786.64	.05	858
	271	5, 2, 4← 6, 1, 5						93 164.04	.05	858
	271	6, 1, 5← 6, 0, 6						95 339.16	.05	858
	271	6, 1, 5← 6, 1, 6						28 613.9	.1	913
	271	7, 0, 7← 6, 0, 6						154 120.02	.05	858
	271	7, 1, 6← 6, 1, 5						159 618.91	.04	900
	271	7, 1, 7← 6, 1, 6						150 098.73	.05	858
	271	7, 2, 5← 6, 2, 4						155 923.97	.04	900
	271	7, 2, 6← 6, 2, 5						154 955.55	.04	900
	271	7, 3, 4← 6, 3, 3						155 251.93	.04	900
	271	7, 3, 5← 6, 3, 4						155 234.17	.04	900
	271	7, 4, ← 6, 4,						155 195.15	.04	900
	271	7, 5, ← 6, 5,						155 183.56	.04	900
	271	7, 6, ← 6, 6,						155 186.37	.04	900
	271	7, 1, 6← 7, 0, 7						100 838.05	.05	858

Isotopic Species	Id. No.	Rotational Quantum Nos.	Vib. State	Hyperfine				Frequency MHz	Acc. ±MHz	Ref.
				F_1'	F'	F_1	F			
$F^{19}HC^{12}O^{16}$	271	7, 2, 6← 8, 1, 7						39 044.3	.1	913
	271	8, 1, 7← 7, 2, 6						39 039.23	.04	962
	271	8, 1, 7← 8, 0, 8						107 362.11	.05	858
	271	9, 0, 9← 8, 1, 8						138 988.38	.05	858
	271	9, 1, 8← 9, 0, 9						114 993.80	.05	858
	271	10, 1, 9←10, 0,10						123 812.35	.05	858
$F^{19}DC^{12}O^{16}$	272	2, 0, 2← 1, 0, 1						43 361.9	.1	913
	272	2, 1, 2← 1, 0, 1						94 920.60	.04	900
	272	2, 1, 1← 1, 1, 0						45 226.9	.1	913
	272	2, 1, 2← 1, 1, 1						41 586.8	.1	913
	272	2, 2, 1← 2, 1, 2						165 448.32	.04	900
	272	2, 2, 1← 3, 1, 2						92 176.92	.04	900
	272	3, 1, 3← 2, 0, 2						113 911.30	.04	900
	272	3, 2, 1← 3, 1, 2						157 515.96	.04	900
	272	3, 2, 2← 3, 1, 3						168 210.00	.04	900
	272	4, 0, 4← 3, 1, 3						37 370.8	.1	913
	272	4, 1, 4← 3, 0, 3						132 067.57	.04	900
	272	4, 1, 3← 4, 1, 4						18 193.7	.1	913
	272	4, 2, 2← 4, 1, 3						154 391.16	.04	900
	272	4, 2, 3← 5, 1, 4						40 852.1	.1	913
	272	5, 1, 5← 4, 0, 4						149 483.49	.04	900
	272	5, 2, 3← 5, 1, 4						150 856.20	.04	900
	272	5, 3, 2← 6, 2, 5						141 567.45	.04	900
	272	6, 2, 4← 6, 1, 5						147 150.05	.04	900
	272	7, 0, 7← 6, 0, 6						149 439.56	.04	900
	272	7, 1, 6← 6, 1, 5						157 633.14	.04	900
	272	7, 1, 7← 6, 1, 6						144 988.68	.04	900
	272	7, 2, 5← 6, 2, 4						154 029.84	.04	900
	272	7, 2, 6← 6, 2, 5						151 558.74	.04	900
	272	7, 3, 4← 6, 3, 3						152 343.06	.04	900
	272	7, 3, 5← 6, 3, 4						152 253.06	.04	900
	272	7, 4, 3← 6, 4, 2						152 147.81	.04	900
	272	7, 4, 4← 6, 4, 3						152 147.81	.04	900
	272	7, 5, ← 6, 5,						152 088.00	.04	900
	272	7, 6, ← 6, 6,						152 059.08	.04	900
	272	7, 2, 5← 7, 1, 6						143 547.24	.04	900
	272	8, 1, 7← 8, 0, 8						93 881.72	.04	900
	272	8, 2, 6← 8, 1, 7						140 340.46	.04	900
	272	9, 1, 8← 9, 0, 9						105 584.79	.04	900
	272	10, 1, 9←10, 0,10						119 163.65	.04	900
	272	11, 1,10←11, 0,11						134 614.08	.04	900
	272	12, 1,11←12, 0,12						151 855.87	.04	900
$F^{19}HC^{13}O^{16}$	273	1, 0, 1← 0, 0, 0						22 112.5	.1	913
	273	1, 1, 1← 2, 0, 2						32 514.1	.1	913
	273	1, 1, 1← 2, 0, 2						32 544.1	.2	947
	273	2, 1, 1← 1, 1, 0						45 623.0	.1	913
	273	2, 1, 2← 1, 1, 1						44 827.2	.1	913
	273	4, 0, 4← 3, 1, 3						14 919.9	.2	947
	273	7, 2, 6← 8, 1, 7						30 947.4	.2	947
$F^{19}HC^{12}O^{18}$	274	1, 0, 1← 0, 0, 0	Ground					20 966.32	.2	947
	274	1, 1, 1← 2, 0, 2	Ground					36 748.3	.2	947
	274	2, 1, 1← 1, 0, 1	Ground					43 172.2	.4	947
	274	2, 1, 2← 1, 1, 1	Ground					40 692.6	.4	947
	274	2, 1, 2← 3, 0, 3	Ground					14 601.9	.2	947
	274	4, 0, 4← 3, 1, 3	Ground					8 085.9	.2	947
	274	5, 0, 5← 4, 1, 4	Ground					31 267.2	.2	947

CHF$_3$ C$_{3v}$ CHF$_3$

Isotopic Species	Pt. Gp.	Id. No.	A MHz		B MHz		C MHz	D$_J$ MHz	D$_{JK}$ MHz	Δ Amu A^2	κ
C^{12}HF$_3^{19}$	C$_{3v}$	281	10 348.74	M	10 348.74	M		.0113	−.0181		
C^{13}HF$_3^{19}$	C$_{3v}$	282	10 422.00	M	10 422.00	M					
C^{12}DF$_3^{19}$	C$_{3v}$	283	9 921.35	M	9 921.35	M					

Id. No.	μ$_a$ Debye		μ$_b$ Debye		μ$_c$ Debye		eQq Value(MHz)	Rel.	eQq Value(MHz)	Rel.	eQq Value(MHz)	Rel.	ω$_a$ d 1/cm	ω$_b$ d 1/cm	ω$_c$ d 1/cm	ω$_d$ d 1/cm
281	0.	X	0.	X	1.645	M										

References:

ABC: 350   D$_J$: 745   D$_{JK}$: 745   μ: 312

Add. Ref. 130,593,611,666,741,810

The following parameters have been reported in ref. 959: for species 281, $\alpha_1 = 7.49$ MHz, $\alpha_4 = 62.96$ MHz, α_4 (perturbed) = 10.7 MHz, $\alpha_3 = 19.53$ MHz; $\alpha_6 = -4.275$ MHz, $\gamma_6 = 0.080$ MHz; for species 283, $\alpha_1 = 39.6$ MHz, $\alpha_2 = 55.46$ MHz, $\alpha_3 = 20.63$ MHz, $\alpha_4 = 12.78$ MHz, $\alpha_5 = 9.42$ MHz, $\alpha_6 = -1.98$ MHz.

Trifluoromethane   Spectral Line Table

Isotopic Species	Id. No.	Rotational Quantum Nos.	Vib. State v$_a$; v$_b$; v$_c$; v$_d$;v$_e^l$	F$_1'$	F'	F$_1$	F	Frequency MHz	Acc. ±MHz	Ref.
C^{12}HF$_3^{19}$	281	1, ← 0,	0; 0; 0; 1; 0, 0					20 650.59	.10	959
	281	1, ← 0,	0; 1; 0; 0; 0, 0					20 658.67	.05	959
	281	1, ← 0,	0; 1; 0; 0; 1, 0					20 667.26	.10	959
	281	1, ← 0,	0; 0; 1; 0; 0, 0					20 677.36	.10	959
	281	1, ← 0,	Ground					20 697.73	.05	959
	281	1, ← 0,	0; 0; 0; 0; 1, 0					20 705.82	.05	959
	281	1, ← 0,	0; 0; 0; 0; 2, 0					20 713.54	.10	959
	281	1, ← 0,	0; 0; 0; 0; 3, 0					20 720.99	.10	959
	281	1, ← 0,	1; 0; 0; 0; 0, 0					20 872.78	.10	959
	281	2, ← 1,	0; 1; 0; 0; 0, 0					41 316.72	.05	959
	281	2, ← 1,	Ground					41 394.95	.18	129
	281	2, ← 1,	0; 0; 0; 0; 3, 0					41 441.00	.10	959
	281	2, 0← 1, 0	Excited					41 426.22	.10	959
	281	2, 1← 1, 1	0; 1; 0; 0; 1, 0					41 334.0	.10	959
	281	2, 1← 1, 1	0; 1; 0; 0; 1,−1					41 338.99	.10	959
	281	2, 1← 1, 1	0; 0; 0; 0; 1, 0					41 411.52	.05	959
	281	2, 1← 1, 1	0; 0; 0; 0; 1,+1					41 483.60	.10	959
	281	4, 0← 3, 0	Ground					82 788.00	.20	745
	281	4, 1← 3, 1	Ground					82 788.00	.20	745
	281	4, 2← 3, 2	Ground					82 788.59	.20	745

Isotopic Species	Id. No.	Rotational Quantum Nos.	Vib. State v_a ; v_b ; v_c ; v_d ; v_e	F_1'	F'	F_1	F	Frequency MHz	Acc. ±MHz	Ref.
$C^{12}HF_3^{19}$	281	4, 3← 3, 3	Ground					82 789.28	.20	745
	281	7, 0← 6, 0	Ground					144 868.57	.30	745
	281	7, 1← 6, 1	Ground					144 868.82	.30	745
	281	7, 2← 6, 2	Ground					144 869.58	.30	745
	281	7, 3← 6, 3	Ground					144 870.83	.30	745
	281	7, 4← 6, 4	Ground					144 872.60	.30	745
	281	7, 5← 6, 5	Ground					144 874.89	.30	745
	281	7, 6← 6, 6	Ground					144 877.70	.30	745
	281	9, 0← 8, 0	Ground					186 246.50	.40	745
	281	9, 1← 8, 1	Ground					186 246.87	.40	745
	281	9, 2← 8, 2	Ground					186 247.81	.40	745
	281	9, 3← 8, 3	Ground					186 249.42	.40	745
	281	9, 4← 8, 4	Ground					186 251.70	.40	745
	281	9, 5← 8, 5	Ground					186 254.65	.40	745
	281	9, 6← 8, 6	Ground					186 258.21	.40	745
	281	9, 7← 8, 7	Ground					186 262.43	.40	745
	281	9, 8← 8, 8	Ground					186 267.30	.40	745
$C^{13}HF_3^{19}$	282	1, ← 0,	Ground					20 643.93	.10	959
	282	2, ← 1,	Ground					41 287.45	.10	959
$C^{12}DF_3^{19}$	283	1, ← 0,	1; 0; 0; 0; 0, 0					19 731.29	.10	959
	283	1, ← 0,	0; 1; 0; 0; 0, 0					19 800.96	.05	959
	283	1, ← 0,	0; 1; 0; 0; 1, 0					19 810.30	.10	959
	283	1, ← 0,	0; 0; 1; 0; 0, 0					19 811.26	.10	959
	283	1, ← 0,	0; 0; 0; 1; 0, 0					19 823.40	.05	959
	283	1, ← 0,	Ground					19 842.21	.05	959
	283	1, ← 0,	0; 0; 0; 0; 1, 0					19 846.16	.05	959
	283	1, ← 0,	0; 0; 0; 0; 2, 0					19 850.13	.10	959
	283	1, ← 0,	0; 0; 0; 0; 2, 0					19 851.33	.10	959
	283	1, ← 0,	0; 0; 0; 0; 3, 0					19 854.13	.10	959
	283	2, ← 1,	1; 0; 0; 0; 0, 0					39 462.03	.10	959
	283	2, ← 1,	0; 1; 0; 0; 0, 0					39 601.68	.05	959
	283	2, ← 1,	0; 1; 0; 0; 1, 0					39 620.32	.10	959
	283	2, ← 1,	0; 0; 1; 0; 0, 0					39 622.05	.10	959
	283	2, ← 1,	Excited					39 641.50	.10	959
	283	2, ← 1,	0; 0; 0; 1; 0, 0					39 646.51	.05	959
	283	2, ← 1,	Excited					39 651.45	.10	959
	283	2, ← 1,	Ground					39 684.21	.05	959
	283	2, ← 1,	0; 0; 0; 0; 2, 0					39 700.07	.10	959
	283	2, ← 1,	0; 0; 0; 0; 2, 0					39 701.76	.10	959
	283	2, ← 1,	0; 0; 0; 0; 3, 0					39 708.18	.10	959
	283	2, 0← 1, 0	0; 0; 0; 0; 1, −1					39 606.04	.10	959
	283	2, 0← 1, 0	0; 0; 0; 0; 1, +1					39 762.87	.10	959
	283	2, 1← 1, 1	0; 0; 0; 0; 1, −1					39 628.39	.10	959
	283	2, 1← 1, 1	0; 0; 0; 0; 1, 0					39 691.98	.10	959
	283	2, 1← 1, 1	0; 0; 0; 0; 1, +1					39 755.60	.10	959
$C^{13}DF_3^{19}$	284	1, ← 0,	Ground					19 798.67	.05	959
	284	2, ← 1,	Ground					39 597.13	.05	959
$C^bHF_3^b$	285	1, ← 0,						20 621.38	.10	959
	285	2, ← 1,						41 241.52	.10	959
	285	2, ← 1,						41 345.49	.10	959
	285	2, ← 1,						41 350.02	.10	959
	285	2, ← 1,						41 759.73	.10	959
	285	2, ← 1,						42 073.64	.10	959

CHN $C_{\infty v}$ HCN

Isotopic Species	Pt. Gp.	Id. No.	A MHz	B MHz		C MHz		D_J MHz	D_{JK} MHz	Δ Amu A²	κ
$HC^{12}N^{14}$	$C_{\infty v}$	291		44 315.99	M	44 315.99	M	.09040			
$DC^{12}N^{14}$	$C_{\infty v}$	292		36 207.42	M	36 207.42	M	.05738			
$HC^{13}N^{14}$	$C_{\infty v}$	293		43 169.83	M	43 169.83	M				
$DC^{13}N^{14}$	$C_{\infty v}$	294		35 587.56	M	35 587.56	M				

Id. No.	μ_a Debye	μ_b Debye	μ_c Debye	eQq Value(MHz)	Rel.	eQq Value(MHz)	Rel.	eQq Value(MHz)	Rel.	ω_a d 1/cm	ω_b d 1/cm	ω_c d 1/cm	ω_d d 1/cm
291	2.986 M	0. X	0. X	−4.714	N^{14}					727 2			
292										569 2			

References:

ABC: 241,394,663 D_J: 663 μ: 895   eQq: 895 ω: 590,922

Add. Ref. 221,223,239,323,357,383,384,385,404,434,488,508,521,550,571,628,644,667,690,730,731,740,776,803,822

For species 291, eQq(N^{14}) for $v_2 = 1$ is −4.80 MHz, the I · J interaction constant is .012 MHz, ref. 733. For species 292, eQq(D)=±.290 MHz, ref. 645.

Species	α_1	α_2	α_3	Ref.
291	+.0103	−.0037	+.0099	667
292	+.0066	−.00415	+.0104	590

Isotopic Species	Id. No.	Rotational Quantum Nos.	Vib. State v_a^1	Hyperfine				Frequency MHz	Acc. ±MHz	Ref.
				F_1'	F'	F_1	F			
HC¹²N¹⁴	291	1← 0	Ground		1		1	88 630.431	.005	895
	.291	1← 0	Ground		2		1	88 631.871	.005	895
	291	1← 0	Ground		0		1	88 633.954	.005	895
	291	1← 0	Ground					88 671.		165
	291	1← 1	1,±1		1		1	448.967	.010	733
	291	1← 1	1,±1		2		2	448.967	.010	733
	291	2← 1	Ground					177 260.99	.40	663
	291	2← 2	1,±1		2		2	1 346.677	.005	733
	291	2← 2	1,±1		1		1	1 346.796	.005	733
	291	2← 2	1,±1		3		3	1 346.796	.005	733
	291	3← 2	Ground					265 886.18	.55	663
	291	3← 3	1,±1		3		2	2 691.757	.008	733
	291	3← 3	1,±1		3		4	2 692.071	.006	733
	291	3← 3	1,±1		3		3	2 693.250	.009	733
	291	3← 3	1,±1		4		4	2 693.395	.006	733
	291	3← 3	1,±1		2		2	2 693.395	.006	733
	291	3← 3	1,±1		4		3	2 694.582	.009	733
	291	3← 3	1,±1		2		3	2 694.954	.009	733
	291	3← 3	5,±1					8 557.50	.1	953
	291	4← 4	1,±1		4		3	4 486.762	.013	733
	291	4← 4	1,±1		4		5	4 487.000	.006	733
	291	4← 4	1,±1		4		4	4 488.381	.020	733
	291	4← 4	1,±1		3		3	4 488.522	.020	733
	291	4← 4	1,±1		5		5	4 488.522	.020	733
	291	4← 4	3,±1					9 242.20	.1	953
	291	4← 4	5,±1					14 224.60	.1	953
	291	5← 5	1,±1		5		5	6 731.793	.011	733
	291	5← 5	1,±1		4		4	6 731.925	.009	733
	291	5← 5	1,±1		6		6	6 731.925	.009	733
	291	5← 5	3,±1					13 861.45	.1	953
	291	6← 6	1,±1					9 423.32	.02	953
	291	6← 6	3,±1					19 402.20	.1	953
	291	7← 7	1,±1					12 562.32	.03	953
	291	7← 7	3,±1					25 863.35	.1	953
	291	8← 8	1,±1					16 148.55	.05	953
	291	9← 9	1,±1					20 181.40	.05	953
	291	10←10	1,±1					24 660.37	.05	953
	291	10←10	1,±1					24 689.96	.1	237
	291	11←11	1,±1					29 585.12	.20	405
	291	11←11	1,±1					29 650.	30.	237
	291	12←12	1,±1					34 953.5		570
	291	12←12	1,±1					35 043.24	.1	237
DC¹²N¹⁴	292	1← 0	Ground		1		1	72 413.25	.20	241
	292	1← 0	Ground		2		1	72 414.62	.20	241
	292	1← 0	Ground		0		1	72 416.68	.20	241
	292	2← 1	Ground					144 827.86	.30	663
	292	3← 2	Ground					217 238.40	.45	663
	292	3← 3	5,±1					7 050.92	.1	953
	292	4← 3	Ground					286 644.67	.60	663

Isotopic Species	Id. No.	Rotational Quantum Nos.	Vib. State v_a^1	Hyperfine F_1'	F'	F_1	F	Frequency MHz	Acc. ±MHz	Ref.
DC^{12}N^{14}	292	4← 4	1,±1					3 722.98	.02	708
	292	4← 4	3,±1					7 634.45	.1	953
	292	4← 4	5,±1					11 680.55	.1	953
	292	5← 5	1,±1					5 583.85	.02	708
	292	5← 5	3,±1					11 449.55	.1	953
	292	6← 6	1,±1					7 816.20	.02	708
	292	6← 6	3,±1					16 025.35	.1	953
	292	7← 7	1,±1					10 419.88	.03	685
	292	7← 7	3,±1					21 360.15	.1	953
	292	8← 8	1,±1					13 394.50	.03	685
	292	9← 9	1,±1					16 739.42	.03	685
	292	10←10	1,±1					20 454.40	.05	953
	292	11←11	1,±1					24 538.92	.05	953
	292	12←12	1,±1					28 992.55	.20	405
HC^{13}N^{14}	293	1← 0	Ground	1			1	86 338.12	.30	241
	293	1← 0	Ground	2			1	86 339.49	.30	241
	293	1← 0	Ground	0			1	86 341.54	.30	241
	293	6← 6	1,±1					9 018.87	.05	953
	293	7← 7	1,±1					12 023.25	.05	953
	293	8← 8	1,±1					15 455.64	.1	953
	293	9← 9	1,±1					19 315.70	.1	953
	293	10←10	1,±1					23 602.60	.1	953
DC^{13}N^{14}	294	1← 0	Ground	1			1	71 173.58	.20	241
	294	1← 0	Ground	2			1	71 174.96	.20	241
	294	1← 0	Ground	0			1	71 177.02	.20	241
	294	6← 6	1,±1					7 652.70	.05	953
	294	7← 7	1,±1					10 201.95	.05	953
	294	8← 8	1,±1					13 114.35	.1	953
	294	9← 9	1,±1					16 389.63	.1	953
	294	10←10	1,±1					20 027.10	.1	953
	294	11←11	1,±1					24 026.60	.1	953
HC^{12}N^{15}	295	6← 6	1,±1					8 897.20	.1	953
	295	7← 7	1,±1					11 861.0	.1	953
	295	8← 8	1,±1					15 247.1	.2	953
	295	9← 9	1,±1					19 055.4	.3	953
	295	10←10	1,±1					23 284.1	.3	953
DC^{12}N^{15}	296	6← 6	1,±1					7 391.80	.1	953
	296	7← 7	1,±1					9 854.15	.1	953
	296	8← 8	1,±1					12 667.25	.1	953
	296	9← 9	1,±1					15 830.90	.1	953
	296	10←10	1,±1					19 344.30	.1	953
	296	11←11	1,±1					23 207.45	.2	953

238-605 O-68-5

CHNO C_s HNCO

Isotopic Species	Pt. Gp.	Id. No.	A MHz		B MHz		C MHz		D_J MHz	D_{JK} MHz	Δ Amu A²	κ
HN¹⁴C¹²O¹⁶	C_s	301	956 400.	M	11 071.02	M	10 910.58	M	3.55	.8370	.1429	−.99966
DN¹⁴C¹²O¹⁶	C_s	303	534 500.	M	10 313.61	M	10 079.67	M	2.95	−.2271	.1918	−.99911

Id. No.	μ_a Debye		μ_b Debye		μ_c Debye		eQq Value(MHz)	Rel.	eQq Value(MHz)	Rel.	eQq Value(MHz)	Rel.	ω_a 1/cm	d	ω_b 1/cm	d	ω_c 1/cm	d	ω_d 1/cm	d
301	1.592	M	0.	X	0.	X	2.00						572	1	670	1				
303	1.619	M	0.	X	0.	X														

References:

ABC: 993 D_J: 993 D_{JK}: 993   Δ: 993 κ: 993 μ: 204   eQq: 993 ω: 1028

Add. Ref. 312

Isotopic Species	Id. No.	Rotational Quantum Nos.	Vib. State v_a ; v_b	F_1'	F'	F_1	F	Frequency MHz	Acc. ±MHz	Ref.
HN^{14}C^{12}O^{16}	301	1, 0, 1← 0, 0, 0	Ground					21 981.7	.02	314
	301	1, 0, 1← 0, 0, 0	Excited					21 993.0	.02	314
	301	1, 0, 1← 0, 0, 0	Excited					22 017.3	.02	314
	301	4, , ← 3, ,	Excited					87 913.08	1.0	993
	301	4, , ← 3, ,	0; 1					87 969.72	1.0	993
	301	4, , ← 3, ,	Excited					88 623.90	1.0	993
	301	4, 0, ← 3, 0,	1; 0					88 067.49	1.0	993
	301	4, 0, ← 3, 0,	1; 0					88 083.84	1.0	993
	301	4, 0, 4← 3, 0, 3	Ground					87 925.45	.5	993
	301	4, 1, ← 3, 1,	0; 1					88 131.78	1.0	993
	301	4, 1, ← 3, 1,	1; 0					88 334.64	1.0	993
	301	4, 1, 3← 3, 1, 2	Ground					88 239.03	.5	993
	301	4, 1, 4← 3, 1, 3	Ground					87 597.03	.5	993
	301	4, 2, ← 3, 2,	1; 0					88 069.80	1.0	993
	301	4, 2, 2← 3, 2, 1	Ground					87 898.53	.5	993
	301	4, 2, 3← 3, 2, 2	Ground					87 898.53	.5	993
	301	4, 3, 1← 3, 3, 0	Ground		5		4	87 866.82	1.0	993
	301	4, 3, 1← 3, 3, 0	Ground		3		2	87 866.82	1.0	993
	301	4, 3, 1← 3, 3, 0	Ground		4		3	87 867.60	1.0	993
	301	4, 3, 2← 3, 3, 1	Ground		3		2	87 866.82	1.0	993
	301	4, 3, 2← 3, 3, 1	Ground		5		4	87 866.82	1.0	993
	301	4, 3, 2← 3, 3, 1	Ground		4		3	87 867.60	1.0	993
	301	5, , ← 4, ,	Excited					109 890.69	1.0	993
	301	5, , ← 4, ,	0; 1					109 958.67	1.0	993
	301	5, 0, ← 4, 0,	1; 0					110 083.80	1.0	993
	301	5, 0, ← 4, 0,	1; 0					110 104.11	1.0	993
	301	5, 0, 5← 4, 0, 4	Ground					109 905.90	.5	993
	301	5, 1, ← 4, 1,	1; 0					109 776.42	1.0	993
	301	5, 1, ← 4, 1,	0; 1					110 164.08	1.0	993
	301	5, 1, ← 4, 1,	1; 0					110 419.08	1.0	993
	301	5, 1, 4← 4, 1, 3	Ground					110 297.82	.5	993
	301	5, 1, 5← 4, 1, 4	Ground					109 495.71	.5	993
	301	5, 2, ← 4, 2,	1; 0					110 086.08	1.0	993
	301	5, 2, ← 4, 2,	1; 0					110 105.52	1.0	993
	301	5, 2, 3← 4, 2, 2	Ground					109 872.99	.5	993
	301	5, 2, 4← 4, 2, 3	Ground					109 872.60	.5	993
	301	5, 3, ← 4, 3,	1; 0					110 089.41	1.0	993
	301	5, 3, 2← 4, 3, 1	Ground		4		3	109 833.45	1.0	993
	301	5, 3, 2← 4, 3, 1	Ground		6		5	109 833.45	1.0	993
	301	5, 3, 2← 4, 3, 1	Ground		5		4	109 833.84	1.0	993
	301	5, 3, 3← 4, 3, 2	Ground		4		3	109 833.45	1.0	993
	301	5, 3, 3← 4, 3, 2	Ground		6		5	109 833.45	1.0	993
	301	5, 3, 3← 4, 3, 2	Ground		5		4	109 833.84	1.0	993
	301	5, 4, 1← 4, 4, 0	Ground		4		3	109 778.37	1.0	993
	301	5, 4, 1← 4, 4, 0	Ground		6		5	109 778.37	1.0	993
	301	5, 4, 1← 4, 4, 0	Ground		5		4	109 778.94	1.0	993
	301	5, 4, 2← 4, 4, 1	Ground		4		3	109 778.37	1.0	993
	301	5, 4, 2← 4, 4, 1	Ground		6		5	109 778.37	1.0	993
	301	5, 4, 2← 4, 4, 1	Ground		5		4	109 778.94	1.0	993
	301	6, , ← 5, ,	Excited					131 863.33	1.0	993

Isotopic Species	Id. No.	Rotational Quantum Nos.	Vib. State $v_a ; v_b$	F'_1	F'	F_1	F	Frequency MHz	Acc. ±MHz	Ref.
HN¹⁴C¹²O¹⁶	301	6, , ← 5, ,	0; 1					131 952.11	1.0	993
	301	6, 0, ← 5, 0,	1; 0					132 099.82	1.0	993
	301	6, 0, ← 5, 0,	1; 0					132 115.04	1.0	993
	301	6, 0, 6← 5, 0, 5	Ground					131 885.52	.5	993
	301	6, 1, ← 5, 1,	1; 0					131 730.84	1.0	993
	301	6, 1, ← 5, 1,	0; 1					132 195.68	1.0	993
	301	6, 1, ← 5, 1,	1; 0					132 500.92	1.0	993
	301	6, 1, 5← 5, 1, 4	Ground					132 356.76	.5	993
	301	6, 1, 6← 5, 1, 5	Ground					131 394.40	.5	993
	301	6, 2, ← 5, 2,	1; 0					132 102.88	1.0	993
	301	6, 2, ← 5, 2,	1; 0					132 125.44	1.0	993
	301	6, 2, 4← 5, 2, 3	Ground					131 846.28	.5	993
	301	6, 2, 5← 5, 2, 4	Ground					131 845.56	.5	993
	301	6, 3, ← 5, 3,	1; 0					132 106.60	1.0	993
	301	6, 3, 3← 5, 3, 2	Ground					131 799.12	1.0	993
	301	6, 3, 4← 5, 3, 3	Ground					131 799.12	1.0	993
	301	6, 4, 2← 5, 4, 1	Ground		7		6	131 733.64	1.0	993
	301	6, 4, 2← 5, 4, 1	Ground		5		4	131 733.64	1.0	993
	301	6, 4, 2← 5, 4, 1	Ground		6		5	131 734.00	1.0	993
	301	6, 4, 3← 5, 4, 2	Ground		7		6	131 733.64	1.0	993
	301	6, 4, 3← 5, 4, 2	Ground		5		4	131 733.64	1.0	993
	301	6, 4, 3← 5, 4, 2	Ground		6		5	131 734.00	1.0	993
	301	6, 5, 1← 5, 5, 0	Ground					131 640.60	1.0	993
	301	6, 5, 2← 5, 5, 1	Ground					131 640.60	1.0	993
HN¹⁵C¹²O¹⁶	302	1, , ← 0, ,	Ground					21 323.5		204
DN¹⁴C¹²O¹⁶	303	1, , ← 0, ,	Ground					20 394.7		314
	303	4, 0, 4← 3, 0, 3	Ground					81 571.53	.5	993
	303	4, 1, 3← 3, 1, 2	Ground					82 042.23	.5	993
	303	4, 1, 4← 3, 1, 3	Ground					81 106.32	.5	993
	303	4, 2, 2← 3, 2, 1	Ground					81 579.61	.5	993
	303	4, 2, 3← 3, 2, 2	Ground					81 578.97	.5	993
	303	4, 3, 1← 3, 3, 0	Ground					81 586.95	1.0	993
	303	4, 3, 2← 3, 3, 1	Ground					81 586.95	1.0	993
	303	5, 0, 5← 4, 0, 4	Ground					101 963.49	.5	993
	303	5, 1, 4← 4, 1, 3	Ground					102 551.73	.5	993
	303	5, 1, 5← 4, 1, 4	Ground					101 382.24	.5	993
	303	5, 2, 3← 4, 2, 2	Ground					101 975.19	.5	993
	303	5, 2, 4← 4, 2, 3	Ground					101 973.63	.5	993
	303	5, 3, 2← 4, 3, 1	Ground					101 981.97	1.0	993
	303	5, 3, 3← 4, 3, 2	Ground					101 981.97	1.0	993
	303	5, 4, 1← 4, 4, 0	Ground					101 988.27	1.0	993
	303	5, 4, 2← 4, 4, 1	Ground					101 988.27	1.0	993
	303	7, 0, 7← 6, 0, 6	Ground					142 744.44	.5	993
	303	7, 1, 6← 6, 1, 5	Ground					143 569.77	.5	993
	303	7, 1, 7← 6, 1, 6	Ground					141 932.08	.5	993
	303	7, 2, 5← 6, 2, 4	Ground					142 764.97	.5	993
	303	7, 2, 6← 6, 2, 5	Ground					142 760.57	.5	993
	303	7, 3, 4← 6, 3, 3	Ground					142 773.02	1.0	993
	303	7, 3, 5← 6, 3, 4	Ground					142 773.02	1.0	993
	303	7, 4, 3← 6, 4, 2	Ground					142 782.23	1.0	993
	303	7, 4, 4← 6, 4, 3	Ground					142 782.23	1.0	993
	303	7, 5, 2← 6, 5, 1	Ground					142 785.65	1.0	993
	303	7, 5, 3← 6, 5, 2	Ground					142 785.65	1.0	993

CHNS C_s HNCS

Isotopic Species	Pt. Gp.	Id. No.	A MHz		B MHz		C MHz		D_J MHz	D_{JK} MHz	Δ Amu A²	κ
$HN^{14}C^{12}S^{32}$	C_s	311	483 000.	M	5 883.42	M	5 845.62	M	1.17		.2148	−.99995
$DN^{14}C^{12}S^{32}$	C_s	312	723 400.	M	5 500.51	M	5 445.26	M	1.22		.2337	−.99985
$HN^{14}C^{12}S^{33,}$	C_s	315										
$HN^{14}C^{12}S^{34}$	C_s	316			5 744.81	M	5 708.73	M				

Id. No.	μ_a Debye	μ_b Debye	μ_c Debye	eQq Value(MHz)	Rel.	eQq Value(MHz)	Rel.	eQq Value(MHz)	Rel.	ω_a d 1/cm	ω_b d 1/cm	ω_c d 1/cm	ω_d d 1/cm
311 315	1.72 M		0. X	−27.5	S^{33}					469 1	600 1		

References:

ABC: 993 D_J: 993   Δ: 993 κ: 993 μ: 118   eQq: 429 ω: 1028

Add. Ref. 81,314,970

For species 311 the rotational constants for the excited state are: A = 900000 MHz; B = 5884.3 MHz; C = 5846.8 MHz. Ref. 257.

Isothiocyanic Acid   Spectral Line Table

Isotopic Species	Id. No.	Rotational Quantum Nos.	Vib. State v_a ; v_b	F_1'	F'	F_1	F	Frequency MHz	Acc. ±MHz	Ref.
$HN^{14}C^{12}S^{32}$	311	2, , ← 1, ,	Ground					23 464.		26
	311	2, 0, ← 1, 0,	0; 1					23 475.	.5	993
	311	2, 0, 2← 1, 0, 1	Ground					23 458.		182
	311	2, 1, 1← 1, 1, 0	Ground					23 499.5		257
	311	2, 1, 1← 1, 1, 0	0; 1					23 520.	.5	993
	311	2, 1, 1← 1, 1, 1	Ground					23 537.		182 [1]
	311	2, 1, 2← 1, 1, 0	Ground					23 387.		182 [1]
	311	2, 1, 2← 1, 1, 1	Ground					23 424.5		257
	311	8, , ← 7, ,	Excited					93 695.0	1.0	993
	311	8, , ← 7, ,	Excited					93 851.2	1.0	993
	311	8, , ← 7, ,	Excited					93 895.6	1.0	993
	311	8, , ← 7, ,	Excited					93 899.3	1.0	993
	311	8, , ← 7, ,	Excited					93 902.9	1.0	993
	311	8, , ← 7, ,	Excited					93 953.3	1.0	993
	311	8, , ← 7, ,	Excited					93 983.7	1.0	993
	311	8, , ← 7, ,	Excited					94 053.4	1.0	993
	311	8, , ← 7, ,	Excited					94 073.3	1.0	993
	311	8, , ← 7, ,	Excited					94 086.9	1.0	993
	311	8, 0, ← 7, 0,	1; 0					94 003.9	1.0	993
	311	8, 0, ← 7, 0,	1; 0					94 012.2	1.0	993
	311	8, 0, 8← 7, 0, 7	Ground					93 829.91	.5	993
	311	8, 1, 7← 7, 1, 6	Ground					93 994.96	.5	993
	311	8, 1, 8← 7, 1, 7	Ground					93 692.76	.5	993
	311	8, 2, 6← 7, 2, 5	Ground					93 863.28	.5	993
	311	8, 2, 7← 7, 2, 6	Ground					93 863.28	.5	993

1. Two lines in the ground state could not actually be seen at zero field, but their frequencies were fixed by extrapolation. The transitions given are normally forbidden, but they can be identified in the presence of the Stark field.

Isotopic Species	Id. No.	Rotational Quantum Nos.	Vib. State v_a ; v_b	F_1'	F'	F_1	F	Frequency MHz	Acc. ±MHz	Ref.
HN^{14}C^{12}S^{32}	311	8, 3, 5← 7, 3, 4	Ground					93 875.28	1.0	993
	311	8, 3, 6← 7, 3, 5	Ground					93 875.28	1.0	993
	311	9, , ← 8, ,	Excited					105 406.7	1.0	993
	311	9, , ← 8, ,	Excited					105 585.4	1.0	993
	311	9, , ← 8, ,	Ground					105 631.7	1.0	993 [2]
	311	9, , ← 8, ,	Excited					105 640.1	1.0	993
	311	9, , ← 8, ,	Excited					105 672.8	1.0	993
	311	9, , ← 8, ,	Excited					105 698.6	1.0	993
	311	9, , ← 8, ,	Excited					105 730.5	1.0	993
	311	9, , ← 8, ,	Excited					105 802.2	1.0	993
	311	9, , ← 8, ,	Excited					105 810.0	1.0	993
	311	9, , ← 8, ,	Excited					105 831.6	1.0	993
	311	9, , ← 8, ,	Excited					105 847.5	1.0	993
	311	9, 0, ← 8, 0,	1; 0					105 756.8	1.0	993
	311	9, 0, ← 8, 0,	1; 0					105 765.8	1.0	993
	311	9, 0, 9← 8, 0, 8	Ground					105 558.08	.5	993
	311	9, 1, 8← 8, 1, 7	Ground					105 743.77	.5	993
	311	9, 1, 9← 8, 1, 8	Ground					105 403.63	.5	993
	311	9, 2, ← 8, 2,	1; 0					105 749.3	1.0	993
	311	9, 2, ← 8, 2,	1; 0					105 764.9	1.0	993
	311	9, 2, 7← 8, 2, 6	Ground					105 595.61	.5	993
	311	9, 2, 8← 8, 2, 7	Ground					105 595.61	.5	993
	311	9, 3, ← 8, 3,	1; 0					105 748.3	1.0	993
	311	9, 3, 6← 8, 3, 5	Ground					105 609.18	1.0	993
	311	9, 3, 7← 8, 3, 6	Ground					105 609.18	1.0	993
	311	10, , ← 9, ,	Excited					117 315.4	1.0	993
	311	10, , ← 9, ,	Excited					117 371.3	1.0	993
	311	10, , ← 9, ,	Excited					117 413.8	1.0	993
	311	10, , ← 9, ,	Excited					117 444.9	1.0	993
	311	10, , ← 9, ,	Excited					117 479.9	1.0	993
	311	10, , ← 9, ,	Excited					117 562.5	1.0	993
	311	10, , ← 9, ,	0; 1					117 589.8	1.0	993
	311	10, , ← 9, ,	Excited					117 606.8	1.0	993
	311	10, , ← 9, ,	Excited					117 615.4	1.0	993
	311	10, 0, ← 9, 0,	0; 1					117 377.2	1.0	993
	311	10, 0, ← 9, 0,	1; 0					117 506.6	1.0	993
	311	10, 0, ← 9, 0,	1; 0					117 516.8	1.0	993
	311	10, 0, ← 9, 0,	1; 0					117 523.4	1.0	993
	311	10, 0,10← 9, 0, 9	Ground					117 285.45	.5	993
	311	10, 1, 9← 9, 1, 8	Ground					117 491.95	.5	993
	311	10, 1,10← 9, 1, 9	Ground					117 114.04	.5	993
	311	10, 1,10← 9, 1, 9	0; 1					117 117.5	1.0	993
	311	10, 2, 8← 9, 2, 7	Ground					117 327.47	.5	993
	311	10, 2, 9← 9, 2, 8	Ground					117 327.47	.5	993
	311	10, 3, 7← 9, 3, 6	Ground					117 342.63	1.0	993
	311	10, 3, 8← 9, 3, 7	Ground					117 342.63	1.0	993
	311	10, 4, ← 9, 4,	Ground					117 367.94	1.0	993 [2]
	311	16, , ←15, ,	Excited					188 131.7	1.0	993
	311	16, , ←15, ,	Excited					188 181.5	1.0	993
	311	16, 0, ←15, 0,	1; 0					188 001.0	1.0	993
	311	16, 0, ←15, 0,	1; 0					188 015.4	1.0	993
	311	16, 0,16←15, 0,15	Ground					187 645.15	.5	993
	311	16, 1,15←15, 1,14	Ground					187 976.13	.5	993
	311	16, 1,16←15, 1,15	Ground					187 370.57	.5	993
	311	16, 2, ←15, 2,	1; 0					188 010.9	1.0	993

2. These may be K=4, ground vibrational state.

Isotopic Species	Id. No.	Rotational Quantum Nos.	Vib. State $v_a ; v_b$	F_1'	F'	F_1	F	Frequency MHz	Acc. ±MHz	Ref.
HN^{14}C^{12}S^{32}	311	16, 2,14←15, 2,13	Ground					187 712.225	.5	993
	311	16, 2,15←15, 2,14	Ground					187 711.730	.5	993
	311	16, 3,13←15, 3,12	Ground					187 736.68	1.0	993
	311	16, 3,14←15, 3,13	Ground					187 736.68	1.0	993
DN^{14}C^{12}S^{32}	312	2, 0, 2← 1, 0, 1	Ground					21 891.66	.5	429
	312	2, 1, 1← 1, 1, 0	Ground					21 951.82	.5	429
	312	2, 1, 2← 1, 1, 1	Ground					21 841.04	.5	429
	312	8, , ← 7, ,	Excited					87 531.0	1.0	993
	312	8, , ← 7, ,	Excited					87 555.0	1.0	993
	312	8, , ← 7, ,	Excited					87 631.1	1.0	993
	312	8, , ← 7, ,	Excited					87 671.8	1.0	993
	312	8, , ← 7, ,	1; 0					87 730.7	1.0	993
	312	8, , ← 7, ,	1; 0					87 734.2	1.0	993
	312	8, , ← 7, ,	1; 0					87 742.8	1.0	993
	312	8, , ← 7, ,	Excited					87 777.1	1.0	993
	312	8, , ← 7, ,	Excited					87 799.3	1.0	993
	312	8, , ← 7, ,	Excited					87 870.4	1.0	993
	312	8, 0, 8← 7, 0, 7	Ground					87 563.05	.5	993
	312	8, 1, 7← 7, 1, 6	Ground					87 803.28	.5	993
	312	8, 1, 8← 7, 1, 7	Ground					87 361.71	.5	993
	312	8, 2, 6← 7, 2, 5	Ground					87 626.37	.5	993
	312	8, 2, 7← 7, 2, 6	Ground					87 626.37	.5	993
	312	8, 3, 5← 7, 3, 4	Ground					87 679.66	1.0	993
	312	8, 3, 6← 7, 3, 5	Ground					87 679.66	1.0	993
	312	9, , ← 8, ,	Excited					98 440.8	1.0	993
	312	9, , ← 8, ,	Excited					98 472.6	1.0	993
	312	9, , ← 8, ,	Excited					98 504.1	1.0	993
	312	9, , ← 8, ,	Excited					98 584.5	1.0	993
	312	9, , ← 8, ,	Excited					98 628.9	1.0	993
	312	9, , ← 8, ,	1; 0					98 691.0	1.0	993
	312	9, , ← 8, ,	1; 0					98 697.6	1.0	993
	312	9, , ← 8, ,	1; 0					98 701.6	1.0	993
	312	9, , ← 8, ,	1; 0					98 710.2	1.0	993
	312	9, , ← 8, ,	Excited					98 773.9	1.0	993
	312	9, , ← 8, ,	Excited					98 854.3	1.0	993
	312	9, 0, 9← 8, 0, 8	Ground					98 507.94	.5	993
	312	9, 1, 8← 8, 1, 7	Ground					98 778.22	.5	993
	312	9, 1, 9← 8, 1, 8	Ground					98 281.15	.5	993
	312	9, 2, 7← 8, 2, 6	Ground					98 579.41	.5	993
	312	9, 2, 8← 8, 2, 7	Ground					98 578.99	.5	993
	312	9, 3, 6← 8, 3, 5	Ground					98 638.64	1.0	993
	312	9, 3, 7← 8, 3, 6	Ground					98 638.64	1.0	993
	312	10, , ← 9, ,	Excited					109 239.9	1.0	993
	312	10, , ← 9, ,	Excited					109 413.3	1.0	993 [3]
	312	10, , ← 9, ,	Excited					109 445.9	1.0	993
	312	10, , ← 9, ,	Excited					109 588.8	1.0	993
	312	10, , ← 9, ,	1; 0					109 656.9	1.0	993
	312	10, , ← 9, ,	1; 0					109 663.2	1.0	993
	312	10, , ← 9, ,	1; 0					109 664.5	1.0	993
	312	10, , ← 9, ,	1; 0					109 667.4	1.0	993
	312	10, , ← 9, ,	Excited					109 679.3	1.0	993
	312	10, , ← 9, ,	Excited					109 710.6	1.0	993
	312	10, , ← 9, ,	Excited					109 721.4	1.0	993
	312	10, , ← 9, ,	Excited					109 748.5	1.0	993
	312	10, , ← 9, ,	Excited					109 837.2	1.0	993
	312	10, 0, ← 9, 0,	0; 1					109 537.4	.5	993

3. This line may be either a K=1 line of J=10 ←9 for $v_5 = 1$ or $v_4 = 1$.

Isotopic Species	Id. No.	Rotational Quantum Nos.	Vib. State $v_a ; v_b$	Hyperfine F_1'	F'	F_1	F	Frequency MHz	Acc. $\pm$MHz	Ref.
$DN^{14}C^{12}S^{32}$	312	10, 1,10← 9, 0, 9	Ground					109 452.06	.5	993
	312	10, 1, 9← 9, 1, 8	Ground					109 752.49	.5	993
	312	10, 1,10← 9, 1, 9	Ground					109 200.75	.5	993
	312	10, 1,10← 9, 1, 9	0; 1					117 117.5	.5	993
	312	10, 2, 8← 9, 2, 7	Ground					109 532.36	.5	993
	312	10, 2, 9← 9, 2, 8	Ground					109 531.83	.5	993
	312	10, 3, 7← 9, 3, 6	Ground					109 597.60	1.0	993
	312	10, 3, 8← 9, 3, 7	Ground					109 597.60	1.0	993
	312	Not Reported	Ground					21 897.		182
$HN^{14}C^{13}S^{32}$	313	Not Reported	Ground					23 389.		182 [4]
$DN^{14}C^{13}S^{32}$	314	Not Reported	Ground					21 839.		182 [4]
$HN^{14}C^{12}S^{33}$	315	2, 0, 2← 1, 0, 1	Ground		3/2		1/2	23 167.89	.5	429
	315	2, 0, 2← 1, 0, 1	Ground		5/2		5/2	23 167.89	.5	429
	315	2, 0, 2← 1, 0, 1	Ground		5/2		3/2	23 174.56	.5	429
	315	2, 0, 2← 1, 0, 1	Ground		7/2		5/2	23 174.56	.5	429
	315	2, 0, 2← 1, 0, 1	Ground		1/2		1/2	23 174.56	.5	429
	315	2, 0, 2← 1, 0, 1	Ground		3/2		3/2	23 179.70	.5	429
$HN^{14}C^{12}S^{34}$	316	2, 0, 2← 1, 0, 1	Ground					22 906.79	.5	429
	316	2, 1, 1← 1, 1, 0	Ground					22 946.47	.5	429
	316	2, 1, 2← 1, 1, 1	Ground					22 874.59	.5	429
	316	8, 0, 8← 7, 0, 7	Ground					91 635.9	.5	993
	316	8, 1, 7← 7, 1, 6	Ground					91 784.3	.5	993
	316	8, 1, 8← 7, 1, 7	Ground					91 495.6	.5	993
	316	8, 2, 6← 7, 2, 5	Ground					91 658.7	.5	993
	316	8, 2, 7← 7, 2, 6	Ground					91 658.7	.5	993
	316	8, 3, 5← 7, 3, 4	Ground					91 671.7	1.0	993
	316	8, 3, 6← 7, 3, 5	Ground					91 671.7	1.0	993
	316	Not Reported	Ground					22 915.		182

4. These lines are averages between two different transitions, therefore they have not been entered under either.

CH₂Br₂ C_{2v} **CH₂Br₂**

Isotopic Species	Pt. Gp.	Id. No.	A MHz	B MHz	C MHz	D_J MHz	D_{JK} MHz	Δ Amu A²	κ
$C^{12}H_2Br_2^b$	C_{2v}	321							

Id. No.	μ_a Debye	μ_b Debye	μ_c Debye	
321	1.5 M	0. X	0. X	

References:

μ: 1029

For species 321, $A - (B + C)/2 = 0.821$ MHz. Ref. 1028.

Dibromomethane   Spectral Line Table

Isotopic Species	Id. No.	Rotational Quantum Nos.	Vib. State	F_1'	F'	F_1	F	Frequency MHz	Acc. ±MHz	Ref.
$C^{12}H_2Br_2^b$	321	Not Reported	Ground					24 908.		155
	321	Not Reported	Ground					24 943.		155
	321	Not Reported	Ground					24 972.		155
	321	Not Reported	Ground					24 982.		155
	321	Not Reported	Ground					25 002.		155
	321	Not Reported	Ground					25 013.		155
	321	Not Reported	Ground					25 042.		155
	321	Not Reported	Ground					25 056.	5.	113
	321	Not Reported	Ground					25 072.		155
	321	Not Reported	Ground					25 090.		155
	321	Not Reported	Ground					25 128.		155
	321	Not Reported	Ground					25 147.		155
	321	Not Reported	Ground					25 152.		155
	321	Not Reported	Ground					25 160.		155
	321	Not Reported	Ground					25 170.		155
	321	Not Reported	Ground					25 203.		155
	321	Not Reported	Ground					25 223.		155

CH$_2$ClF   C$_s$ CH$_2$ClF

Isotopic Species	Pt. Gp.	Id. No.	A MHz		B MHz		C MHz		D$_J$ MHz	D$_{JK}$ MHz	Δ Amu A^2	κ
C^{12}H$_2$Cl^{35}F^{19}	C$_s$	331	41 810.1	M	5 715.7	M	5 194.6	M				
C^{12}H$_2$Cl^{37}F^{19}	C$_s$	332	41 738.2	M	5 580.5	M	5 081.6	M				

Id. No.	μ$_a$ Debye		μ$_b$ Debye	μ$_c$ Debye		eQq Value(MHz)	Rel.	eQq Value(MHz)	Rel.	eQq Value(MHz)	Rel.	ω$_a$ 1/cm	d	ω$_b$ 1/cm	d	ω$_c$ 1/cm	d	ω$_d$ 1/cm	d
331	1.82	G		0.	X	−52.18	aa	38.83	bb										

References:

ABC: 463   μ: 995   eQq: 463

Isotopic Species	Id. No.	Rotational Quantum Nos.	Vib. State	Hyperfine				Frequency MHz	Acc. ±MHz	Ref.
				F_1'	F'	F_1	F			
$C^{12}H_2Cl^{35}F^{19}$	331	1, 1, 0← 1, 0, 1	Ground					36 592.71	1.0	463
	331	1, 1, 0← 1, 0, 1	Ground					36 611.22	.1	463
	331	1, 1, 0← 1, 0, 1	Ground					36 616.50	.1	463
	331	1, 1, 0← 1, 0, 1	Ground					36 620.92	.1	463
	331	1, 1, 0← 1, 0, 1	Ground					36 624.40	.1	463
	331	2, 1, 1← 2, 0, 2	Ground					37 129.14	1.0	463
	331	2, 1, 1← 2, 0, 2	Ground					37 132.67	1.0	463
	331	2, 1, 1← 2, 0, 2	Ground					37 136.25	1.0	463
	331	2, 1, 1← 2, 0, 2	Ground					37 140.00	1.0	463
	331	2, 1, 1← 2, 0, 2	Ground					37 146.66	1.0	463
	331	2, 1, 1← 2, 0, 2	Ground					37 149.50	.1	463
	331	2, 1, 1← 2, 0, 2	Ground					37 153.50	1.0	463
	331	3, 1, 2← 3, 0, 3	Ground					37 940.52	1.0	463
	331	5, 0, 5← 4, 1, 4	Ground					20 619.17	.1	463
	331	5, 0, 5← 4, 1, 4	Ground					20 620.26	.1	463
	331	5, 0, 5← 4, 1, 4	Ground					20 621.30	.1	463
	331	6, 0, 6← 5, 1, 5	Ground					32 659.51	1.0	463
	331	6, 0, 6← 5, 1, 5	Ground					32 660.83	.1	463
	331	6, 0,·6← 5, 1, 5	Ground					32 662.11	.1	463
	331	7, 1, 6← 6, 2, 5	Ground					25 518.46	.1	463
	331	8, 1, 8← 7, 2, 5	Ground					32 035.40	.1	463
	331	8, 1, 8← 7, 2, 5	Ground					32 041.58	.1	463
	331	9, 1, 9← 8, 2, 6	Ground					24 020.97	.1	463
	331	9, 1, 9← 8, 2, 6	Ground					24 021.75	.1	463
	331	9, 1, 9← 8, 2, 6	Ground					24 026.92	.1	463
	331	9, 1, 9← 8, 2, 6	Ground					24 027.57	.1	463
	331	10, 1,10← 9, 2, 7	Ground					16 487.12	.1	463
	331	10, 1,10← 9, 2, 7	Ground					16 492.71	.1	463
	331	11, 1,10←10, 2, 9	Ground					27 498.50	.1	463
	331	11, 1,10←10, 2, 9	Ground					27 500.30	.1	463
	331	15, 2,14←14, 3,11	Ground					21 834.20	.1	463
	331	15, 2,14←14, 3,11	Ground					21 836.00	.1	463
	331	17, 1,17←16, 2,14	Ground					18 961.29	.1	463
	331	17, 1,17←16, 2,14	Ground					18 967.52	.1	463
	331	17, 2,15←16, 3,14	Ground					18 916.70	.1	463
	331	18, 1,18←17, 2,15	Ground					21 084.05	.1	463
	331	18, 1,18←17, 2,15	Ground					21 090.31	.1	463
	331	18, 2,16←17, 3,15	Ground					31 816.40	.1	463
	331	18, 2,16←17, 3,15	Ground					31 818.03	.1	463
	331	19, 1,19←18, 2,16	Ground					22 395.57	.1	463
	331	19, 1,19←18, 2,16	Ground					22 402.23	.1	463
	331	20, 1,20←19, 2,17	Ground					22 889.15	.1	463
	331	20, 1,20←19, 2,17	Ground					22 895.76	.1	463
	331	20, 2,19←19, 3,16	Ground					24 048.17	.1	463
	331	20, 2,19←19, 3,16	Ground					24 050.16	.1	463
	331	21, 1,21←20, 2,18	Ground					22 413.	1.0	463
	331	21, 1,21←20, 2,18	Ground					22 419.25	.1	463
	331	22, 1,22←21, 2,19	Ground					21 429.80	.1	463
	331	22, 1,22←21, 2,19	Ground					21 437.05	.1	463

Isotopic Species	Id. No.	Rotational Quantum Nos.	Vib. State	Hyperfine F_1'	F'	F_1	F	Frequency MHz	Acc. ±MHz	Ref.
$C^{12}H_2Cl^{37}F^{19}$	332	1, 1, 0← 1, 0, 1	Ground					36 652.77	1.0	463
	332	1, 1, 0← 1, 0, 1	Ground					36 657.47	1.0	463
	332	1, 1, 0← 1, 0, 1	Ground					36 660.75	1.0	463
	332	1, 1, 0← 1, 0, 1	Ground					36 663.42	1.0	463
	332	2, 1, 1← 2, 0, 2	Ground					37 158.82	1.0	463
	332	2, 1, 1← 2, 0, 2	Ground					37 166.20	1.0	463
	332	3, 1, 2← 3, 0, 3	Ground					37 924.77	1.0	463
	332	5, 0, 5← 4, 1, 4	Ground					19 232.16	.1	463
	332	6, 0, 6← 5, 1, 5	Ground					30 984.37	.1	463
	332	7, 1, 6← 6, 2, 5	Ground					27 711.27	.1	463
	332	8, 1, 8← 7, 2, 5	Ground					33 703.27	.1	463
	332	8, 1, 8← 7, 2, 5	Ground					33 708.06	.1	463
	332	9, 1, 9← 8, 2, 6	Ground					25 791.65	.1	463
	332	9, 1, 9← 8, 2, 6	Ground					25 796.15	.1	463
	332	11, 1,10←10, 2, 9	Ground					23 937.44	.1	463
	332	14, 2,12←13, 3,11	Ground					26 215.43	.1	463
	332	14, 2,12←13, 3,11	Ground					26 216.84	.1	463
	332	15, 2,14←14, 3,11	Ground					25 494.60	.1	463
	332	17, 1,17←16, 2,14	Ground					17 648.07	.1	463
	332	17, 1,17←16, 2,14	Ground					17 652.92	.1	463
	332	18, 1,18←17, 2,15	Ground					20 011.22	.1	463
	332	18, 1,18←17, 2,15	Ground					20 016.47	.1	463
	332	18, 2,16←17, 3,15	Ground					25 878.70	.1	463
	332	19, 1,19←18, 2,16	Ground					21 596.35	.1	463
	332	19, 1,19←18, 2,16	Ground					21 600.0	1.0	463
	332	20, 2,19←19, 3,16	Ground					19 917.48	.1	463
	332	20, 2,19←19, 3,16	Ground					19 919.07	.1	463
	332	22, 1,22←21, 2,19	Ground					21 603.23	.1	463
	332	22, 1,22←21, 2,19	Ground					21 608.55	.1	463
$C^bH_2Cl^bF^b$	333	Not Reported						16 117.91	.1	463
	333	Not Reported						16 166.58	.1	463
	333	Not Reported						16 296.9	1.	463
	333	Not Reported						16 419.80	.1	463
	333	Not Reported						17 228.62	.1	463
	333	Not Reported						17 331.76	.1	463
	333	Not Reported						17 416.50	.1	463
	333	Not Reported						17 782.62	.1	463
	333	Not Reported						17 783.68	.1	463
	333	Not Reported						17 978.22	.1	463
	333	Not Reported						18 229.29	.1	463
	333	Not Reported						18 381.82	.1	463
	333	Not Reported						18 382.50	.1	463
	333	Not Reported						18 584.59	.1	463
	333	Not Reported						19 076.34	.1	463
	333	Not Reported						19 273.78	.1	463
	333	Not Reported						19 316.96	.1	463
	333	Not Reported						19 400.74	.1	463
	333	Not Reported						19 447.26	.1	463
	333	Not Reported						19 624.74	.1	463

Isotopic Species	Id. No.	Rotational Quantum Nos.	Vib. State	Hyperfine				Frequency MHz	Acc. ±MHz	Ref.
				F_1'	F'	F_1	F			
$C^bH_2^bCl^bF^b$	333	Not Reported						19 640.67	.1	463
	333	Not Reported						19 687.12	.1	463
	333	Not Reported						19 868.76	.1	463
	333	Not Reported						20 042.38	.1	463
	333	Not Reported						20 143.55	1.	463
	333	Not Reported						20 285.53	1.	463
	333	Not Reported						20 469.5	1.	463
	333	Not Reported						21 053.4	1.	463
	333	Not Reported						21 298.60	.1	463
	333	Not Reported						21 626.6	1.	463
	333	Not Reported						21 915.40	.1	463
	333	Not Reported						21 965.24	.1	463
	333	Not Reported						21 966.23	.1	463
	333	Not Reported						22 234.17	.1	463
	333	Not Reported						22 251.93	.1	463
	333	Not Reported						22 253.07	.1	463
	333	Not Reported						22 443.0	1.	463
	333	Not Reported						22 446.6	1.	463
	333	Not Reported						22 564.64	.1	463
	333	Not Reported						22 892.11	.1	463
	333	Not Reported						22 919.28	.1	463
	333	Not Reported						23 088.58	.1	463
	333	Not Reported						23 154.47	.1	463
	333	Not Reported						23 158.02	.1	463
	333	Not Reported						23 408.54	.1	463
	333	Not Reported						23 450.31	.1	463
	333	Not Reported						23 452.00	.1	463
	333	Not Reported						23 602.58	.1	463
	333	Not Reported						23 703.	1.	463
	333	Not Reported						23 711.04	.1	463
	333	Not Reported						23 727.50	1.	463
	333	Not Reported						23 883.90	.1	463
	333	Not Reported						24 105.38	.1	463
	333	Not Reported						24 250.	1.	463
	333	Not Reported						24 345.15	.1	463
	333	Not Reported						24 349.08	.1	463
	333	Not Reported						24 349.84	.1	463
	333	Not Reported						24 928.22	.1	463
	333	Not Reported						25 082.26	.1	463
	333	Not Reported						25 088.20	.1	463
	333	Not Reported						25 197.90	.1	463
	333	Not Reported						25 214.92	.1	463
	333	Not Reported						25 619.82	.1	463
	333	Not Reported						25 628.67	.1	463
	333	Not Reported						25 707.52	.1	463
	333	Not Reported						25 708.51	.1	463
	333	Not Reported						26 257.63	.1	463
	333	Not Reported						26 368.30	.1	463
	333	Not Reported						26 429.08	.1	463
	333	Not Reported						26 551.16	.1	463

Isotopic Species	Id. No.	Rotational Quantum Nos.	Vib. State	Hyperfine				Frequency MHz	Acc. ±MHz	Ref.
				F_1'	F'	F_1	F			
$C^bH_2^bCl^bF^b$	333	Not Reported						26 616.38	.1	463
	333	Not Reported						26 617.97	.1	463
	333	Not Reported						26 920.64	.1	463
	333	Not Reported						27 138.5	1.	463
	333	Not Reported						27 781.00	.1	463
	333	Not Reported						27 869.96	.1	463
	333	Not Reported						28 006.22	.1	463
	333	Not Reported						28 007.90	.1	463
	333	Not Reported						28 462.5	1.	463
	333	Not Reported						28 526.33	.1	463
	333	Not Reported						28 746.70	.1	463
	333	Not Reported						28 887.99	.1	463
	333	Not Reported						29 252.29	.1	463
	333	Not Reported						29 358.15	.1	463
	333	Not Reported						29 463.66	.1	463
	333	Not Reported						29 658.40	.1	463
	333	Not Reported						29 925.4	1.	463
	333	Not Reported						29 926.	1.	463
	333	Not Reported						29 997.2	1.	463
	333	Not Reported						30 100.	1.	463
	333	Not Reported						30 219.4	1.	463
	333	Not Reported						30 220.7	1.	463
	333	Not Reported						30 290.67	.1	463
	333	Not Reported						30 292.	1.	463
	333	Not Reported						30 580.73	.1	463
	333	Not Reported						30 645.	1.	463
	333	Not Reported						30 659.9	1.	463
	333	Not Reported						30 837.8	1.	463
	333	Not Reported						30 843.59	.1	463
	333	Not Reported						30 992.18	.1	463
	333	Not Reported						31 023.24	.1	463
	333	Not Reported						31 509.67	.1	463
	333	Not Reported						31 535.33	.1	463
	333	Not Reported						31 887.8	1.	463
	333	Not Reported						31 922.92	.1	463
	333	Not Reported						31 924.32	.1	463
	333	Not Reported						32 479.50	.1	463
	333	Not Reported						33 051.68	.1	463
	333	Not Reported						33 057.8	1.	463
	333	Not Reported						33 162.40	.1	463
	333	Not Reported						33 224.05	.1	463
	333	Not Reported						33 404.40	.1	463
	333	Not Reported						33 406.00	.1	463
	333	Not Reported						33 633.04	.1	463
	333	Not Reported						33 677.98	.1	463
	333	Not Reported						33 733.21	.1	463
	333	Not Reported						33 769.71	.1	463
	333	Not Reported						33 787.00	.1	463
	333	Not Reported						36 583.05	1.	463
	333	Not Reported						36 585.2	1.	463

Isotopic Species	Id. No.	Rotational Quantum Nos.	Vib. State	Hyperfine				Frequency MHz	Acc. ±MHz	Ref.
				F_1'	F'	F_1	F			
$C^bH_2^bCl^bF^b$	333	Not Reported						36 587.57	1.	463
	333	Not Reported						36 631.	1.	463
	333	Not Reported						37 914.	.1	463
	333	Not Reported						37 930.0	1.	463
	333	Not Reported						37 938.	1.	463
	333	Not Reported						37 950.	1.	463
	333	Not Reported						37 952.	1.	463
	333	Not Reported						37 957.	1.	463
	333	Not Reported						37 962.76	1.	463
	333	Not Reported						38 198.	1.	463
	333	Not Reported						38 199.27	1.	463
	333	Not Reported						38 423.18	1.	463
	333	Not Reported						38 928.22	1.	463
	333	Not Reported						39 021.44	1.	463
	333	Not Reported						39 028.46	1.	463

340 — Dichloromethane
Methylene Chloride

CH_2Cl_2 C_{2v} CH_2Cl_2

Isotopic Species	Pt. Gp.	Id. No:	A MHz		B MHz		C MHz		D_J MHz	D_{JK} MHz	Δ Amu A²	κ
$C^{12}H_2Cl_2^{35}$	C_{2v}	341	32 001.8	M	3 320.4	M	3 065.2	M				
$C^{12}H_2Cl^{35}Cl^{37}$	C_s	342	31 878.25	M	3 231.5	M	2 988.25	M				
$C^{12}H_2Cl_2^{37}$	C_{2v}	343	31 754.	M	3 143.	M	2 912.	M				
$C^{12}DHCl_2^{35}$	C_s	344	27 198.	M	3 305.	M	3 027.	M				
$C^{12}DHCl^{35}Cl^{37}$	C_1	345	27 090.5	M	3 217.5	M	2 951.5	M				
$C^{12}D_2Cl_2^{35}$	C_{2v}	346	23 676.5	M	3 284.	M	2 993.5	M				
$C^{12}D_2Cl^{35}Cl^{37}$	C_s	347	23 582.	M	3 197.5	M	2 920.	M				

Id. No.	μ_a Debye		μ_b Debye		μ_c Debye		eQq Value(MHz)	Rel.	eQq Value(MHz)	Rel.	eQq Value(MHz)	Rel.	ω_a 1/cm	d	ω_b 1/cm	d	ω_c 1/cm	d	ω_d 1/cm	d
341	0.	X	1.618	M	0.	X	−41.8	aa	2.6	bb	39.2	cc	282	1						
342	0.	u	1.623	M	0.	X														
344	0.	X	1.616	M	0.	u														
345	0.	u	1.625	M	0.	u														
346	0.	X	1.644	M	0.	X														
347	0.	u	1.640	M	0.	X														

References:

ABC: 381 μ: 381   eQq: 381 ω: 406

Add. Ref. 622

Isotopic Species	Id. No.	Rotational Quantum Nos.	Vib. State v_a	Hyperfine				Frequency MHz	Acc. ±MHz	Ref.
				F_1'	F'	F_1	F			
$C^{12}H_2Cl_2^{35}$	341	1, 1, 1← 0, 0, 0	Ground					35 067.0	.2	381
	341	1, 1, 1← 0, 0, 0	1					35 258.	2.	381 [1]
	341	1, 1, 1← 0, 0, 0	2					35 451.	5.	381 [1]
	341	1, 1, 0← 1, 0, 1	Ground					28 936.6	.1	381
	341	2, 1, 2← 1, 0, 1	Ground					41 197.8	.2	381
	341	2, 1, 1← 2, 0, 2	Ground					29 193.4	.3	381
	341	3, 1, 2← 3, 0, 3	Ground					29 582.0	.3	381
	341	4, 1, 3← 4, 0, 4	Ground					30 105.	5.	381
	341	5, 1, 4← 5, 0, 5	Ground					30 769.	5.	381
	341	6, 1, 5← 6, 0, 6	Ground					31 577.	5.	381
	341	7, 0, 7← 6, 1, 6	Ground					18 375.	5.	381
	341	7, 1, 6← 7, 0, 7	Ground					32 546.	2.	381
	341	7, 2, 6← 8, 1, 7	Ground					30 812.	5.	381
	341	8, 0, 8← 7, 1, 7	Ground					25 533.	5.	381
	341	8, 1, 7← 8, 0, 8	Ground					33 672.	2.	381
	341	8, 2, 7← 9, 1, 8	Ground					23 312.	5.	381
	341	9, 0, 9← 8, 1, 8	Ground					32 769.	5.	381
	341	9, 1, 8← 9, 0, 9	Ground					34 970.	5.	381
	341	10, 1, 9←10, 0,10	Ground					36 451.3	.3	381
$C^{12}H_2Cl^{35}Cl^{37}$	342	1, 1, 1← 0, 0, 0	Ground					34 866.5	.2	381
	342	1, 1, 0← 1, 0, 1	Ground					28 890.0	.1	381
	342	2, 1, 2← 1, 0, 1	Ground					40 842.	1.	381
	342	2, 1, 1← 2, 0, 2	Ground					29 135.	2.	381
	342	3, 1, 2← 3, 0, 3	Ground					29 509.	5.	381
	342	4, 1, 3← 4, 0, 4	Ground					30 005.	5.	381
	342	5, 1, 4← 5, 0, 5	Ground					30 635.	5.	381
	342	6, 1, 5← 6, 0, 6	Ground					31 405.	5.	381
	342	7, 1, 6← 7, 0, 7	Ground					32 325.	5.	381
	342	7, 2, 6← 8, 1, 7	Ground					32 229.	5.	381
	342	8, 0, 8← 7, 1, 7	Ground					24 100.	5.	381
	342	8, 1, 7← 8, 0, 8	Ground					33 393.	2.	381
	342	8, 2, 7← 9, 1, 8	Ground					24 947.	5.	381
	342	9, 0, 9← 8, 1, 8	Ground					31 140.	5.	381
	342	9, 1, 8← 9, 0, 9	Ground					34 623.	5.	381
	342	10, 1, 9←10, 0,10	Ground					36 026.	5.	381
$C^{12}H_2Cl_2^{37}$	343	1, .1, 1← 0, 0, 0	Ground					34 665.3	.2	381
	343	4, 1, 3← 4, 0, 4	Ground					29 905.	5.	381
	343	5, 1, 4← 5, 0, 5	Ground					30 502.	5.	381
	343	6, 1, 5← 6, 0, 6	Ground					31 225.	5.	381
	343	7, 1, 6← 7, 0, 7	Ground					32 100.	5.	381
	343	8, 1, 7← 8, 0, 8	Ground					33 121.	5.	381
	343	9, 1, 8← 9, 0, 9	Ground					34 286.	5.	381
	343	10, 1, 9←10, 0,10	Ground					35 614.	5.	381
$C^{12}DHCl_2^{35}$	344	1, 1, 1← 0, 0, 0	Ground					30 224.7	.2	381
	344	1, 1, 0← 1, 0, 1	Ground					24 171.	2.	381
	344	2, 1, 2← 1, 0, 1	Ground					36 283.	2.	381
	344	2, 1, 1← 2, 0, 2	Ground					24 450.	5.	381
	344	3, 1, 2← 3, 0, 3	Ground					24 878.	5.	381

1. The assignment of these lines to $C^{12}H_2Cl_2^{35}$ is somewhat doubtful.

Isotopic Species	Id. No.	Rotational Quantum Nos.	Vib. State v_a	Hyperfine				Frequency MHz	Acc. ±MHz	Ref.
				F_1'	F'	F_1	F			
$C^{12}DHCl_2^{35}$	344	4, 1, 3← 4, 0, 4	Ground					25 453.	5.	381
	344	5, 1, 4← 5, 0, 5	Ground					26 183.	5.	381
	344	6, 1, 5← 6, 0, 6	Ground					27 090.	5.	381
	344	7, 1, 6← 7, 0, 7	Ground					28 167.	5.	381
	344	8, 1, 7← 8, 0, 8	Ground					29 434.	5.	381
	344	9, 1, 8← 9, 0, 9	Ground					30 900.	5.	381
	344	10, 1, 9←10, 0,10	Ground					32 583.	2.	381
$C^{12}DHCl^{35}Cl^{37}$	345	1, 1, 1← 0, 0, 0	Ground					30 042.2	.2	381
	345	1, 1, 0← 1, 0, 1	Ground					24 139.	5.	381
	345	2, 1, 2← 1, 0, 1	Ground					35 945.	2.	381
	345	2, 1, 1← 2, 0, 2	Ground					24 405.	5.	381
	345	3, 1, 2← 3, 0, 3	Ground					24 820.	5.	381
	345	4, 1, 3← 4, 0, 4	Ground					25 365.	5.	381
	345	5, 1, 4← 5, 0, 5	Ground					26 063.	5.	381
	345	6, 1, 5← 6, 0, 6	Ground					26 920.	5.	381
	345	7, 1, 6← 7, 0, 7	Ground					27 943.	5.	381
	345	8, 1, 7← 8, 0, 8	Ground					29 143.	5.	381
	345	9, 1, 8← 9, 0, 9	Ground					30 535.	5.	381
	345	10, 1, 9←10, 0,10	Ground					32 127.	5.	381
$C^{12}D_2Cl_2^{35}$	346	1, 1, 1← 0, 0, 0	Ground					26 670.2	.2	381
	346	2, 1, 2← 1, 0, 1	Ground					32 656.	2.	381
	346	2, 1, 1← 2, 0, 2	Ground					20 976.7	.2	381
	346	3, 1, 3← 2, 0, 2	Ground					38 502.	2.	381
	346	3, 1, 2← 3, 0, 3	Ground					21 425.	5.	381
	346	4, 1, 3← 4, 0, 4	Ground					22 030.	5.	381
	346	5, 1, 4← 5, 0, 5	Ground					22 805.	5.	381
	346	6, 1, 5← 6, 0, 6	Ground					23 761.	5.	381
	346	7, 1, 6← 7, 0, 7	Ground					24 908.	5.	381
	346	8, 1, 7← 8, 0, 8	Ground					26 268.	5.	381
	346	9, 1, 8← 9, 0, 9	Ground					27 850.	5.	381
	346	10, 1, 9←10, 0,10	Ground					29 667.7	.2	381
$C^{12}D_2Cl^{35}Cl^{37}$	347	1, 1, 1← 0, 0, 0	Ground					26 501.5	.2	381
	347	2, 1, 1← 2, 0, 2	Ground					20 942.	5.	381
	347	3, 1, 3← 2, 0, 2	Ground					38 043.	2.	381
	347	3, 1, 2← 3, 0, 3	Ground					21 370.	5.	381
	347	4, 1, 3← 4, 0, 4	Ground					21 948.	5.	381
	347	5, 1, 4← 5, 0, 5	Ground					22 687.	5.	381
	347	6, 1, 5← 6, 0, 6	Ground					23 590.	5.	381
	347	7, 1, 6← 7, 0, 7	Ground					24 687.	5.	381
	347	8, 1, 7← 8, 0, 8	Ground					25 974.	5.	381
	347	9, 1, 8← 9, 0, 9	Ground					27 473.	5.	381
	347	10, 1, 9←10, 0,10	Ground					29 196.	5.	381

Methylene Fluoride

CH_2F_2 C_{2v} CH_2F_2

Isotopic Species	Pt. Gp.	Id. No.	A MHz		B MHz		C MHz		D_J MHz	D_{JK} MHz	Δ Amu A²	κ
$C^{12}H_2F_2^{19}$	C_{2v}	351	49 138.4	M	10 603.89	M	9 249.20	M				−.932077
$C^{13}H_2F_2^{19}$	C_{2v}	352	47 720.	M	10 604.	M	9 198.	M				−.9270

Id. No.	μ_a Debye		μ_b Debye		μ_c Debye		eQq Value(MHz)	Rel.	eQq Value(MHz)	Rel.	eQq Value(MHz)	Rel.	ω_a d 1/cm	ω_b d 1/cm	ω_c d 1/cm	ω_d d 1/cm
351	0.	X	1.96	M	0.	X										

References:
ABC: 370 κ: 370 μ: 370
Add. Ref. 371,620

Difluoromethane   Spectral Line Table

Isotopic Species	Id. No.	Rotational Quantum Nos.	Vib. State	F_1'	F'	F_1	F	Frequency MHz	Acc. ±MHz	Ref.
$C^{12}H_2F_2^{19}$	351	3, 0, 3← 2, 1, 2	Ground					22 204.18	.10	370
	351	3, 2, 2← 4, 1, 3	Ground					31 543.75	.10	370
	351	4, 2, 2← 5, 1, 5	Ground					29 268.90	.10	370
	351	7, 3, 5← 8, 2, 6	Ground					31 777.75	.10	370
	351	8, 3, 5← 9, 2, 8	Ground					20 237.69	.10	370
	351	9, 1, 9← 8, 2, 6	Ground					21 980.68	.10	370
	351	10, 1,10← 9, 2, 7	Ground					30 679.01	.10	370
	351	15, 3,13←14, 4,10	Ground					24 760.40	.10	370
	351	15, 5,11←16, 4,12	Ground					30 962.08	.10	370
	351	19, 4,16←18, 5,13	Ground					29 624.66	.10	370
	351	19, 6,13←20, 5,16	Ground					29 630.87	.10	370
	351	19, 6,14←20, 5,15	Ground					29 339.38	.10	370
	351	23, 7,16←24, 6,19	Ground					27 603.66	.10	370
	351	23, 7,17←24, 6,18	Ground					27 516.98	.10	370
	351	27, 8,19←28, 7,22	Ground					25 694.13	.10	370
	351	27, 8,20←28, 7,21	Ground					25 669.29	.10	370
	351	31, 9,22←32, 8,25	Ground					23 871.82	.10	370
	351	31, 9,23←32, 8,24	Ground					23 864.92	.10	370
	351	34, 8,26←33, 9,25	Ground					17 429.88	.10	370
	351	34, 8,27←33, 9,24	Ground					17 411.86	.10	370
	351	35,10,25←36, 9,28	Ground					22 137.35	.10	370
	351	35,10,26←36, 9,27	Ground					22 135.44	.10	370
	351	38, 9,29←37,10,28	Ground					19 039.76	.10	370
	351	38, 9,30←37,10,27	Ground					19 034.89	.10	370
	351	39,11,28←40,10,31	Ground					20 500.38	.10	370
	351	39,11,29←40,10,30	Ground					20 499.85	.10	370
	351	42,10,32←41,11,31	Ground					20 571.99	.10	370
	351	42,10,33←41,11,30	Ground					20 570.63	.10	370
	351	43,12,31←44,11,34	Ground					18 972.43	.10	370
	351	43,12,32←44,11,33	Ground					18 972.43	.10	370
	351	46,11,35←45,12,34	Ground					21 934.33	.10	370
	351	46,11,36←45,12,33	Ground					21 934.33	.10	370
	351	50,12,38←49,13,37	Ground					23 188.45	.10	370
	351	50,12,39←49,13,36	Ground					23 188.45	.10	370
	351	54,13,41←53,14,40	Ground					24 297.52	.10	370
	351	54,13,42←53,14,39	Ground					24 297.52	.10	370
	351	Not Reported	Ground					21 423.20	.10	370
	351	Not Reported	Ground					22 579.40	.10	370
$C^{13}H_2F_2^{19}$	352	3, 0, 3← 2, 1, 2	Ground					23 501.2	.4	370
	352	4, 2, 2← 5, 1, 5	Ground					25 829.9	.4	370

360 — Cyanamide
  Carbamonitrile

CH$_2$N$_2$ C$_s$ H$_2$NCN

Isotopic Species	Pt. Gp.	Id. No.	A MHz	B MHz	C MHz	D$_J$ MHz	D$_{JK}$ MHz	Δ Amu A^2	κ
H$_2$N^{14}C^{12}N^{14}	C$_s$	361		10 129.2 M	9 865.8 M				
HDN^{14}C^{12}N^{14}	C$_1$	362		9 604.0 M	9 257.1 M				
D$_2$N^{14}C^{12}N^{14}	C$_s$	363		9 155.3 M	8 743.5 M				

Id. No.	μ_a Debye	μ_b Debye	μ_c Debye	eQq Value(MHz) Rel.	eQq Value(MHz) Rel.	eQq Value(MHz) Rel.	ω_a d 1/cm	ω_b d 1/cm	ω_c d 1/cm	ω_d d 1/cm
361	4.24 M		0. X				638 1	429 1		
362	4.28 M						545 1			
363	4.24 M		0. X							

References:

ABC: 975 μ: 982 ω: 975

Add. Ref. 885,986,987

For species 361, the inversion barrier is given as 370 cm^{-1}. Ref. 975.

Cyanamide   Spectral Line Table

Isotopic Species	Id. No.	Rotational Quantum Nos.	Vib. State v$_a$; v$_b$	F$_1'$	F'	F$_1$	F	Frequency MHz	Acc. ±MHz	Ref.
H$_2$N^{14}C^{12}N^{14}	361	1, 0, ← 0, 0,	1; 0					19 980.0	.2	951
	361	1, 0, ← 0, 0,	Ground					19 995.8	.2	951
	361	2, 0, 2← 1, 0, 1	1; 0					39 958.2	.5	975
	361	2, 0, 2← 1, 0, 1	Ground					39 991.0	.5	975
	361	2, 0, 2← 1, 0, 1	1; 1					40 068.0	.5	975
	361	2, 0, 2← 1, 0, 1	0; 1					40 120.0	.5	975
	361	2, 1, 1← 1, 1, 0	1; 0					40 203.0	.5	975
	361	2, 1, 1← 1, 1, 0	Ground					40 252.2	.5	975
	361	2, 1, 1← 1, 1, 0	1; 1					40 315.9	.5	975
	361	2, 1, 1← 1, 1, 0	0; 1					40 394.2	.5	975
	361	2, 1, 2← 1, 1, 1	1; 0					39 712.3	.5	975
	361	2, 1, 2← 1, 1, 1	Ground					39 725.4	.5	975
	361	2, 1, 2← 1, 1, 1	1; 1					39 820.		975
	361	2, 1, 2← 1, 1, 1	0; 1					39 845.8	.5	975
	361	3, 0, ← 2, 0,	1; 0					59 937.6	.4	951
	361	3, 0, ← 2, 0,	Ground					59 986.0	.4	951
	361	3, 1, ← 2, 1,	1; 0					59 569.6	.4	951
	361	3, 1, ← 2, 1,	Ground					59 586.8	.4	951
	361	3, 1, ← 2, 1,	1; 0					60 308.0	.4	951
	361	3, 1, ← 2, 1,	Ground					60 379.2	.4	951
	361	3, 2, ← 2, 2,	1; 0					59 848.8	.4	951
	361	3, 2, ← 2, 2,	Ground					59 973.0	.4	951
	361	7, 1, 6← 7, 1, 7	Ground					7 384.82	.5	975
	361	8, 1, 7← 8, 1, 8	Ground					9 490.48	.5	975
	361	8, 1, 7← 8, 1, 8	Ground					9 491.29	.5	975

Isotopic Species	Id. No.	Rotational Quantum Nos.	Vib. State $v_a : v_b$	Hyperfine F_1'	F'	F_1	F	Frequency MHz	Acc. ±MHz	Ref.
$H_2N^{14}C^{12}N^{14}$	361	8, 1, 7← 8, 1, 8	Ground					9 493.00	.5	975
	361	8, 1, 7← 8, 1, 8	Ground					9 493.77	.5	975
	361	14, 1,13←14, 1,14	Ground					27 676.0	.5	975
	361	14, 1,13←14, 1,14	Ground					27 678.6	.5	975
	361	15, 1,14←15, 1,15	Ground					31 627.4	.5	975
	361	15, 1,14←15, 1,15	Ground					31 630.6	.5	975
	361	16, 1,15←16, 1,16	Ground					35 841.7	.5	975
	361	16, 1,15←16, 1,16	Ground					35 844.4	.5	975
$HDN^{14}C^{12}N^{14}$	362	1, 0, 1← 0, 0, 0	1; 0					18 861.7	.1	982
	362	1, 0, 1← 0, 0, 0	Ground					18 861.7	.1	982
	362	2, 0, 2← 1, 0, 1	1; 0					37 723.8	.5	975
	362	2, 0, 2← 1, 0, 1	Ground					37 723.8	.5	975
	362	2, 1, 1← 1, 1, 0	1; 0					38 056.8	.5	975
	362	2, 1, 1← 1, 1, 0	Ground					38 069.1	.5	975
	362	2, 1, 2← 1, 1, 1	Ground					37 375.4	.5	975
	362	2, 1, 2← 1, 1, 1	1; 0					37 384.7	.5	975
	362	3, 0, 3← 2, 0, 2	Ground					56 583.1	.3	982
	362	3, 0, 3← 2, 0, 2	1; 0					56 583.1	.3	982
	362	3, 1, 2← 2, 1, 1	1; 0					57 086.0	.3	982
	362	3, 1, 2← 2, 1, 1	Ground					57 105.2	.3	982
	362	3, 1, 3← 2, 1, 2	Ground					56 061.0	.3	982
	362	3, 1, 3← 2, 1, 2	1; 0					56 083.8	.3	982
	362	3, 2, 1← 2, 2, 0	1; 0					56 542.0	.3	982
	362	3, 2, 1← 2, 2, 0	Ground					56 576.0	.3	982
	362	3, 2, 2← 2, 2, 1	1; 0					56 542.0	.3	982
	362	3, 2, 2← 2, 2, 1	Ground					56 576.0	.3	982
$D_2N^{14}C^{12}N^{14}$	363	1, 0, 1← 0, 0, 0	Ground					17 899.7	.1	982
	363	1, 0, 1← 0, 0, 0	1; 0					17 905.2	.2	982
	363	2, 0, 2← 1, 0, 1	Ground					35 797.8	.5	975
	363	2, 0, 2← 1, 0, 1	1; 0					35 809.7	.5	975
	363	2, 1, 1← 1, 1, 0	Ground					36 209.0	.5	975
	363	2, 1, 1← 1, 1, 0	1; 0					36 223.6	.5	975
	363	2, 1, 1← 1, 1, 0	1; 0					36 229.0	.5	982
	363	2, 1, 2← 1, 1, 1	Ground					35 385.5	.5	975
	363	2, 1, 2← 1, 1, 1	1; 0					35 422.0	.5	975
	363	2, 1, 2← 1, 1, 1	1; 0					35 425.8	.2	982
	363	3, 0, 3← 2, 0, 2	Ground					53 694.8	.3	982
	363	3, 0, 3← 2, 0, 2	1; 0					53 712.2	.3	982
	363	3, 1, 2← 2, 1, 1	Ground					54 313.6	.3	982
	363	3, 1, 2← 2, 1, 1	1; 0					54 334.4	.3	982
	363	3, 1, 3← 2, 1, 2	Ground					53 077.6	.3	982
	363	3, 1, 3← 2, 1, 2	1; 0					53 132.4	.3	982
	363	3, 2, 1← 2, 2, 0	1; 0					53 652.0	.3	982
	363	3, 2, 1← 2, 2, 0	Ground					53 672.8	.3	982
	363	3, 2, 2← 2, 2, 1	Ground					53 672.8	.3	982
	363	3, 2, 2← 2, 2, 1	1; 0					53 702.1	.1	982

CH₂N₂ C_{2v} H_2CNN

Isotopic Species	Pt. Gp.	Id. No.	A MHz	B MHz		C MHz		D_J MHz	D_{JK} MHz	Δ Amu A²	κ
$H_2C^{12}N^{14}N^{14}$	C_{2v}	371		11 305.5	M	10 845.3	M	.006	.39		−.9966
$HDC^{12}N^{14}N^{14}$	C_s	372		10 609.5	M	10 031.2	M	.005	.27		−.9936
$D_2C^{12}N^{14}N^{14}$	C_{2v}	373		10 042.6	M	9 346.4	M		.22		−.9891
$H_2C^{12}N^{15}N^{14}$	C_{2v}	374		10 952.47	M	10 519.95	M				
$H_2C^{13}N^{14}N^{14}$	C_{2v}	375		10 946.89	M	10 514.85	M				
$D_2C^{13}N^{14}N^{14}$	C_{2v}	376		9 792.90	M	9 129.77	M				
$H_2C^{13}N^{15}N^{14}$	C_{2v}	377		10 600.03	M	10 194.38	M				

Id. No.	μ_a Debye		μ_b Debye		μ_c Debye		eQq Value(MHz)	Rel.	eQq Value(MHz)	Rel.	eQq Value(MHz)	Rel.	ω_a d 1/cm	ω_b d 1/cm	ω_c d 1/cm	ω_d d 1/cm
371	1.45	M	0.	X	0.	X	−1.19	aa	1.04	bb	.15	cc				

References:

ABC: 816,981 D_J: 981 D_{JK}: 981 κ: 981 μ: 816   eQq: 981

Add. Ref. 545

No Spectral Lines

CH$_2$O   C$_{2v}$ HCHO

Isotopic Species	Pt. Gp.	Id. No.	A MHz	B MHz	C MHz	D$_J$ MHz	D$_{JK}$ MHz	Δ Amu A^2	κ
HC^{12}HO^{16}	C$_{2v}$	381	282 106. M	38 834. M	34 004. M	.0826	1.311	.0574	−.961067
DC^{12}DO^{16}	C$_{2v}$	384						.0777	
HC^{12}DO^{16}	C$_s$	385						.0679	

Id. No.	μ$_a$ Debye	μ$_b$ Debye	μ$_c$ Debye	eQq Value(MHz) Rel.	eQq Value(MHz) Rel.	eQq Value(MHz) Rel.	ω$_a$ d 1/cm	ω$_b$ d 1/cm	ω$_c$ d 1/cm	ω$_d$ d 1/cm
381	2.340 M	0. X	0. X							

References:
ABC: 288,671   D$_J$: 671   D$_{JK}$: 671   Δ: 979   κ: 288   μ: 312

Add. Ref. 86,121,142,143,210,211,479,624,712,713,746,852,881,883

Formaldehyde   Spectral Line Table

Isotopic Species	Id. No.	Rotational Quantum Nos.	Vib. State	F$_1'$	F'	F$_1$	F	Frequency MHz	Acc. ±MHz	Ref.
HC^{12}HO^{16}	381	1, 0, 1← 0, 0, 0	Ground					72 838.14		288
	381	1, 1, 0← 1, 1, 1	Ground					4 829.73	.01	684
	381	2, 0, 2← 1, 0, 1	Ground					145 603.1	.73	671
	381	2, 1, 1← 1, 1, 0	Ground					150 498.2	.75	671
	381	2, 1, 2← 1, 1, 1	Ground					140 839.3	.70	671
	381	2, 1, 1← 2, 1, 2	Ground					14 488.65		288
	381	3, 0, 3← 2, 0, 2	Ground					218 221.6	1.1	671
	381	3, 1, 2← 2, 1, 1	Ground					225 698.2	1.1	671
	381	3, 1, 3← 2, 1, 2	Ground					211 210.6	1.1	671
	381	3, 2, 0← 2, 2, 0	Ground					218 759.4	1.1	671
	381	3, 2, 2← 2, 2, 1	Ground					218 475.1	1.1	671
	381	3, 1, 2← 3, 1, 3	Ground					28 974.85		288
	381	3, 2, 1← 3, 2, 2	Ground					355.586	.005	989
	381	4, 1, 3← 4, 1, 4	Ground					48 284.60		288
	381	4, 2, 2← 4, 2, 3	Ground					1 065.85	.02	989
	381	5, 1, 4← 5, 1, 5	Ground					72 409.35		288
	381	6, 2, 4← 6, 2, 5	Ground					4 954.76	.01	684
	381	7, 2, 5← 7, 2, 6	Ground					8 884.87		288
	381	8, 2, 6← 8, 2, 7	Ground					14 726.74		288
	381	8, 3, 5← 8, 3, 6	Ground					301.10	.01	989
	381	9, 2, 7← 9, 2, 8	Ground					22 965.71		288
	381	9, 3, 6← 9, 3, 7	Ground					601.07	.005	989
	381	11, 2, 9←11, 2,10	Ground					48 612.70	.1	551
	381	12, 3, 9←12, 3,10	Ground					3 225.58	.01	684
	381	13, 3,10←13, 3,11	Ground					5 136.58	.01	684
	381	14, 3,11←14, 3,12	Ground					7 892.03		288
	381	15, 3,12←15, 3,13	Ground					11 753.13		288
	381	16, 3,13←16, 3,14	Ground					17 027.60		288
	381	17, 3,14←17, 3,15	Ground					24 068.31		288
	381	19, 3,16←19, 3,17	Ground					45 063.10	.1	551
	381	20, 4,16←20, 4,17	Ground					3 518.85	.5	684
	381	21, 4,17←21, 4,18	Ground					5 138.57	.5	684
	381	22, 4,18←22, 4,19	Ground					7 362.60		288
	381	23, 4,19←23, 4,20	Ground					10 366.51		288

Isotopic Species	Id. No.	Rotational Quantum Nos.	Vib. State	Hyperfine				Frequency MHz	Acc. ±MHz	Ref.
				F_1'	F'	F_1	F			
HC^{12}HO^{16}	381	24, 4,20←24, 4,21	Ground					14 361.54		288
	381	25, 4,21←25, 4,22	Ground					19 595.23		288
	381	26, 4,22←26, 4,23	Ground					26 358.82		288
	381	28, 4,24←28, 4,25	Ground					45 835.58		631
	381	31, 5,26←31, 5,27	Ground					7 833.20		288
	382	1, 1, 0← 1, 1, 1	Ground					4 593.26	.5	684
	382	2, 1, 1← 2, 1, 2	Ground					13 778.86		209
	382	3, 1, 2← 3, 1, 3	Ground					27 555.73		209
	382	4, 1, 3← 4, 1, 4	Ground					45 920.08		631
	382	7, 2, 5← 7, 2, 6	Ground					8 012.56	.5	684
	382	9, 2, 7← 9, 2, 8	Ground					20 736.30		209
	382	14, 3,11←14, 3,12	Ground					6 752.31	.5	684
	382	16, 3,13←16, 3,14	Ground					14 592.44		209
	382	17, 3,14←17, 3,15	Ground					20 649.30		209
	382	18, 3,15←18, 3,16	Ground					28 582.40		209
HC^{12}HO^{18}	383	1, 1, 0← 1, 1, 1	Ground					4 388.85	.5	684
DC^{12}DO^{16}	384	1, 1, 0← 1, 1, 1	Ground					6 096.10	.02	686
	384	2, 1, 1← 2, 1, 2	Ground					18 287.90		686
	384	4, 2, 2← 4, 2, 3	Ground					3 687.28	.04	686
	384	5, 2, 3← 5, 2, 4	Ground					8 519.10		686
	384	6, 2, 4← 6, 2, 5	Ground					16 759.64		686
	384	8, 3, 5← 8, 3, 6	Ground					2 850.62	.03	686
	384	9, 3, 6← 9, 3, 7	Ground					5 636.98		686
	384	10, 3, 7←10, 3, 8	Ground					10 304.64		686
	384	13, 4, 9←13, 4,10	Ground					3 079.48	.03	686
	384	14, 4,10←14, 4,11	Ground					5 461.54		686
	384	15, 4,11←15, 4,12	Ground					9 259.88		686
	384	16, 4,12←16, 4,13	Ground					15 080.34		686
	384	19, 5,14←19, 5,15	Ground					4 508.39	.04	686
HC^{12}DO^{16}	385	1, 1, 0← 1, 1, 1	Ground					5 346.64	.03	686
	385	2, 1, 1← 2, 1, 2	Ground					16 038.06		686
	385	3, 2, 1← 3, 2, 2	Ground					644.893	.005	989
	385	5, 2, 3← 5, 2, 4	Ground					4 489.08	.03	686
	385	6, 2, 4← 6, 2, 5	Ground					8 922.59		686
	385	7, 2, 5← 7, 2, 6	Ground					15 907.38		686
	385	10, 3, 7←10, 3, 8	Ground					3 283.09	.03	686
	385	11, 3, 8←11, 3, 9	Ground					5 702.6		686
	385	12, 3, 9←12, 3,10	Ground					9 412.51		686
	385	13, 3,10←13, 3,11	Ground					14 873.02		686
	385	16, 4,12←16, 4,13	Ground					2 946.67	.03	686
	385	17, 4,13←17, 4,14	Ground					4 713.90		686
	385	18, 4,14←18, 4,15	Ground					7 322.35		686
	385	19, 4,15←19, 4,16	Ground					11 074.30		686
	385	23, 5,18←23, 5,19	Ground					3 330.66	.04	686
	385	24, 5,19←24, 5,20	Ground					5 018.25		686
HC^{13}DO^{16}	386	1, 1, 0← 1, 1, 1	Ground					5 156.19	.10	686

CH$_3$BO

C$_{3v}$

H$_3$BCO

Isotopic Species	Pt. Gp.	Id. No.	A MHz	B MHz		C MHz		D$_J$ MHz	D$_{JK}$ MHz	Δ Amu A^2	κ
H$_3$B^{10}C^{12}O^{16}	C$_{3v}$	391		8 979.94	M	8 979.94	M	.177	.39		
H$_3$B^{11}C^{12}O^{16}	C$_{3v}$	392		8 657.22	M	8 657.22	M		.36		
D$_3$B^{10}C^{12}O^{16}	C$_{3v}$	393		7 530.34	M	7 530.34	M		.29		
D$_3$B^{11}C^{12}O^{16}	C$_{3v}$	394		7 336.56	M	7 336.56	M		.24		

Id. No.	μ_a Debye	μ_b Debye	μ_c Debye	eQq Value(MHz)	Rel.	eQq Value(MHz)	Rel.	eQq Value(MHz)	Rel.	ω_a d 1/cm	ω_b d 1/cm	ω_c d 1/cm	ω_d d 1/cm
391 392	1.795 M	0. X	0. X	3.4 1.55	B^{10} B^{11}								

References:

ABC: 198   D$_J$: 168   D$_{JK}$: 198 μ: 168   eQq: 198

Add. Ref. 81,82,134

For species 391, B$_{e(v1)}$ = 9002.66 MHz, B$_{e(v2)}$ = 8985.80 MHz.

Carbonyl Borane

Spectral Line Table

Isotopic Species	Id. No.	Rotational Quantum Nos.	Vib. State	F$_I'$	F'	F$_I$	F	Frequency MHz	Acc. ±MHz	Ref.
H$_3$B^{10}C^{12}O^{16}	391	1, 0 ← 0, 0	Ground		2		3	17 959.67	.05	250
	391	1, 0 ← 0, 0	Ground		4		3	17 959.91	.05	250
	391	1, 0 ← 0, 0	Ground		3		3	17 960.60	.05	250
	391	2, 0 ← 1, 0	Ground		2		3	35 919.08		198
	391	2, 0 ← 1, 0	Ground		5		4	35 919.60		198
	391	2, 0 ← 1, 0	Ground		4		3	35 919.60		198
	391	2, 0 ← 1, 0	Ground		1		2	35 919.60		198
	391	2, 0 ← 1, 0	Ground		3		3	35 919.60		198
	391	2, 0 ← 1, 0	Ground		2		2	35 919.95		198
	391	2, 0 ← 1, 0	Ground		4		4	35 920.22		198
	391	2, 0 ← 1, 0	Ground		3		4	35 920.22		198
	391	2, 0 ← 1, 0	Ground		3		2	35 920.40		198
	391	2, 1 ← 1, 1	Ground		1		2	35 917.66		198
	391	2, 1 ← 1, 1	Ground		5		4	35 917.96		198
	391	2, 1 ← 1, 1	Ground		2		2	35 917.96		198
	391	2, 1 ← 1, 1	Ground		4		4	35 918.29		198
	391	2, 1 ← 1, 1	Ground		3		4	35 918.29		198
	391	2, 1 ← 1, 1	Ground		2		3	35 918.29		198
	391	2, 1 ← 1, 1	Ground		3		3	35 918.55		198
	391	2, 1 ← 1, 1	Ground		4		3	35 918.55		198
H$_3$B^{11}C^{12}O^{16}	392	2, 0 ← 1, 0	Ground		1/2		3/2	34 628.16		198
	392	2, 0 ← 1, 0	Ground		3/2		3/2	34 628.58		198
	392	2, 0 ← 1, 0	Ground		5/2		3/2	34 628.85		198
	392	2, 0 ← 1, 0	Ground		7/2		5/2	34 628.85		198
	392	2, 0 ← 1, 0	Ground		3/2		5/2	34 628.85		198
	392	2, 0 ← 1, 0	Ground		1/2		1/2	34 628.85		198
	392	2, 0 ← 1, 0	Ground		3/2		1/2	34 629.27		198
	392	2, 0 ← 1, 0	Ground		5/2		5/2	34 629.27		198
	392	2, 1 ← 1, 1	Ground		1/2		1/2	34 627.16		198

Isotopic Species	Id. No.	Rotational Quantum Nos.	Vib. State	Hyperfine				Frequency MHz	Acc. ±MHz	Ref.
				F_1'	F'	F_1	F			
$H_3B^{11}C^{12}O^{16}$	392	2, 1← 1, 1	Ground		3/2		1/2	34 627.42		198
	392	2, 1← 1, 1	Ground		3/2		5/2	34 627.42		198
	392	2, 1← 1, 1	Ground		7/2		5/2	34 627.42		198
	392	2, 1← 1, 1	Ground		1/2		3/2	34 627.42		198
	392	2, 1← 1, 1	Ground		5/2		5/2	34 627.64		198
	392	2, 1← 1, 1	Ground		3/2		3/2	34 627.64		198
	392	2, 1← 1, 1	Ground		5/2		3/2	34 627.81		198
$D_3B^{10}C^{12}O^{16}$	393	2, 0← 1, 0	Ground		2		3	30 120.86		198
	393	2, 0← 1, 0	Ground		3		3	30 121.21		198
	393	2, 0← 1, 0	Ground		4		3	30 121.21		198
	393	2, 0← 1, 0	Ground		1		2	30 121.21		198
	393	2, 0← 1, 0	Ground		5		4	30 121.21		198
	393	2, 0← 1, 0	Ground		2		2	30 121.56		198
	393	2, 0← 1, 0	Ground		4		4	30 121.86		198
	393	2, 0← 1, 0	Ground		3		4	30 121.86		198
	393	2, 0← 1, 0	Ground		3		2	30 121.86		198
	393	2, 1← 1, 1	Ground		1		2	30 119.91		198
	393	2, 1← 1, 1	Ground		2		2	30 120.21		198
	393	2, 1← 1, 1	Ground		5		4	30 120.21		198
	393	2, 1← 1, 1	Ground		3		4	30 120.56		198
	393	2, 1← 1, 1	Ground		2		3	30 120.56		198
	393	2, 1← 1, 1	Ground		4		4	30 120.56		198
	393	2, 1← 1, 1	Ground		4		3	30 120.86		198
	393	2, 1← 1, 1	Ground		3		3	30 120.86		198
$D_3B^{11}C^{12}O^{16}$	394	2, 0← 1, 0	Ground		3/2		3/2	29 345.93		198
	394	2, 0← 1, 0	Ground		5/2		3/2	29 346.24		198
	394	2, 0← 1, 0	Ground		7/2		5/2	29 346.24		198
	394	2, 0← 1, 0	Ground		3/2		5/2	29 346.24		198
	394	2, 0← 1, 0	Ground		1/2		1/2	29 346.24		198
	394	2, 0← 1, 0	Ground		3/2		1/2	29 346.65		198
	394	2, 0← 1, 0	Ground		5/2		5/2	29 346.65		198
	394	2, 1← 1, 1	Ground		1/2		1/2	29 345.03		198
	394	2, 1← 1, 1	Ground		7/2		5/2	29 345.28		198
	394	2, 1← 1, 1	Ground		3/2		1/2	29 345.28		198
	394	2, 1← 1, 1	Ground		1/2		3/2	29 345.28		198
	394	2, 1← 1, 1	Ground		3/2		5/2	29 345.28		198
	394	2, 1← 1, 1	Ground		3/2		3/2	29 345.52		198
	394	2, 1← 1, 1	Ground		5/2		5/2	29 345.52		198
	394	2, 1← 1, 1	Ground		5/2		3/2	29 345.68		198

CH$_3$Br   C$_{3v}$ CH$_3$Br

Isotopic Species	Pt. Gp.	Id. No.	A MHz		B MHz		C MHz		D$_J$ MHz	D$_{JK}$ MHz	Δ Amu A^2	κ
C^{12}H$_3$Br^{79}	C$_{3v}$	401	152 354.5	M	9 568.20	M	9 568.20	M	.0099	.1283		
C^{12}H$_3$Br^{81}	C$_{3v}$	402	152 354.5	M	9 531.82	M	9 531.82	M	.0097	.1274		
C^{12}D$_3$Br^{79}	C$_{3v}$	403			7 714.57	M	7 714.57	M		.039		
C^{12}D$_3$Br^{81}	C$_{3v}$	404			7 681.23	M	7 681.23	M		.039		
C^{13}H$_3$Br^{79}	C$_{3v}$	405			9 119.507	M	9 119.507	M				
C^{13}H$_3$Br^{81}	C$_{3v}$	406			9 082.860	M	9 082.866	M				

Id. No.	μ$_a$ Debye	μ$_b$ Debye	μ$_c$ Debye	eQq Value(MHz)	Rel.	eQq Value(MHz)	Rel.	eQq Value(MHz)	Rel.	ω$_a$ d 1/cm	ω$_b$ d 1/cm	ω$_c$ d 1/cm	ω$_d$ d 1/cm
401	1.797 M	0. X	0. X	577.3	Br^{79}					610 1			
402	.			482.4	Br^{81}								
403				574.6	Br^{79}					577 1			
404				479.8	Br^{81}								

References:

ABC: 242,395,537,568,1028   D$_J$: 568   D$_{JK}$: 395,568   μ: 375   eQq: 395,568   ω: 1028,1029

Add. Ref. 39,102,103,185,208,245,282,538,723,741,810,827,1015

Species	2(B − C) MHz	B in MHz for v$_3$ = 1
407	317.70	9495.43
408	314.44	9454.51
Ref.	376	537

Isotopic Species	Id. No.	Rotational Quantum Nos.	Vib. State v_a	Hyperfine				Frequency MHz	Acc. ±MHz	Ref.
				F_1'	F'	F_1	F			
$C^{12}H_3Br^{79}$	401	1, ← 0,	Ground					19 136.73	.1	161
	401	1, 0← 0, 0	l		1/2		3/2	18 846.88	.05	537
	401	1, 0← 0, 0	l		5/2		3/2	18 962.19	.05	537
	401	1, 0← 0, 0	Ground		1/2		3/2	18 992.47	.05	537
	401	1, 0← 0, 0	l		3/2		3/2	19 106.60	.10	537
	401	1, 0← 0, 0	Ground		5/2		3/2	19 107.72	.05	537
	401	1, 0← 0, 0	Ground		3/2		3/2	19 252.11	.05	537
	401	2, 0← 1, 0	Ground		3/2		3/2	38 157.30	.08	83
	401	2, 0← 1, 0	Ground		7/2		5/2	38 260.10	.08	83
	401	2, 0← 1, 0	Ground		5/2		3/2	38 260.10	.08	83
	401	2, 0← 1, 0	Ground		1/2		1/2	38 272.40	.08	83
	401	2, 0← 1, 0	Ground		5/2		5/2	38 404.49	.08	83
	401	2, 0← 1, 0	Ground		3/2		1/2	38 417.09	.08	83
	401	2, 1← 1, 1	Ground		1/2		1/2	38 128.40	.08	83
	401	2, 1← 1, 1	Ground		3/2		1/2	38 200.52	.08	83
	401	2, 1← 1, 1	Ground		7/2		5/2	38 237.14	.08	83
	401	2, 1← 1, 1	Ground		5/2		5/2	38 309.45	.08	83
	401	2, 1← 1, 1	Ground		3/2		3/2	38 330.25	.08	83
	401	2, 1← 1, 1	Ground		5/2		3/2	38 381.70	.08	83
	401	4, 0← 3, 0	Ground		9/2		7/2	76 538.02	.18	240
	401	4, 0← 3, 0	Ground		11/2		9/2	76 538.02	.18	240
	401	4, 0← 3, 0	Ground		5/2		3/2	76 554.82	.18	240
	401	4, 0← 3, 0	Ground		7/2		5/2	76 554.82	.18	240
	401	4, 1← 3, 1	Ground		11/2		9/2	76 532.88	.18	240
	401	4, 1← 3, 1	Ground		5/2		3/2	76 540.20	.18	240
	401	4, 1← 3, 1	Ground		9/2		7/2	76 547.24	.18	240
	401	4, 1← 3, 1	Ground		7/2		5/2	76 554.82	.18	240
	401	4, 2← 3, 2	Ground		5/2		3/2	76 496.60	.18	240
	401	4, 2← 3, 2	Ground		11/2		9/2	76 517.36	.18	240
	401	4, 2← 3, 2	Ground		7/2		5/2	76 554.82	.18	240
	401	4, 2← 3, 2	Ground		9/2		7/2	76 575.22	.18	240
	401	4, 3← 3, 3	Ground		5/2		3/2	76 425.18	.18	240
	401	4, 3← 3, 3	Ground		11/2		9/2	76 491.36	.18	240
	401	4, 3← 3, 3	Ground		7/2		5/2	76 554.82	.18	240
	401	4, 3← 3, 3	Ground		9/2		7/2	76 621.78	.18	240
	401	5, 0← 4, 0	Ground		11/2		9/2	95 673.51	.27	240
	401	5, 0← 4, 0	Ground		13/2		11/2	95 673.51	.27	240
	401	5, 0← 4, 0	Ground		7/2		5/2	95 683.62	.27	240
	401	5, 0← 4, 0	Ground		9/2		7/2	95 683.62	.27	240
	401	5, 1← 4, 1	Ground		13/2		11/2	95 669.97	.27	240
	401	5, 1← 4, 1	Ground		7/2		5/2	95 676.39	.27	240
	401	5, 1← 4, 1	Ground		11/2		9/2	95 677.20	.27	240
	401	5, 1← 4, 1	Ground		9/2		7/2	95 683.62	.27	240
	401	5, 2← 4, 2	Ground		7/2		5/2	95 654.73	.27	240
	401	5, 2← 4, 2	Ground		13/2		11/2	95 659.20	.27	240
	401	5, 2← 4, 2	Ground		9/2		7/2	95 683.62	.27	240
	401	5, 2← 4, 2	Ground		11/2		9/2	95 688.27	.27	240
	401	5, 3← 4, 3	Ground		7/2		5/2	95 619.24	.27	240
	401	5, 3← 4, 3	Ground		13/2		11/2	95 640.87	.27	240
	401	5, 3← 4, 3	Ground		9/2		7/2	95 683.62	.27	240

Isotopic Species	Id. No.	Rotational Quantum Nos.	Vib. State v_a	F_1'	F'	F_1	F	Frequency MHz	Acc. ±MHz	Ref.
$C^{12}H_3Br^{79}$	401	5, 3← 4, 3	Ground		11/2		9/2	95 706.12	.27	240
	401	5, 4← 4, 4	Ground		13/2		11/2	95 615.73	.27	240
	401	8, 0← 7, 0	Ground		17/2		15/2	153 069.61		568
	401	8, 0← 7, 0	Ground		19/2		17/2	153 069.61		568
	401	8, 0← 7, 0	Ground		13/2		11/2	153 073.09		568
	401	8, 0← 7, 0	Ground		15/2		13/2	153 073.09		568
	401	8, 1← 7, 1	Ground		19/2		17/2	153 066.92		568
	401	8, 1← 7, 1	Ground		17/2		15/2	153 068.60		568
	401	8, 1← 7, 1	Ground		15/2		13/2	153 071.60		568
	401	8, 2← 7, 2	Ground		19/2		17/2	153 058.75		568
	401	8, 2← 7, 2	Ground		17/2		15/2	153 065.61		568
	401	8, 3← 7, 3	Ground		13/2		11/2	153 044.21		568
	401	8, 3← 7, 3	Ground		19/2		17/2	153 045.22		568
	401	8, 3← 7, 3	Ground		15/2		13/2	153 059.73		568
	401	8, 3← 7, 3	Ground		17/2		15/2	153 060.67		568
	401	8, 4← 7, 4	Ground		13/2		11/2	153 021.82		568
	401	8, 4← 7, 4	Ground		19/2		17/2	153 026.32		568
	401	8, 4← 7, 4	Ground		15/2		13/2	153 049.30		568
	401	8, 4← 7, 4	Ground		17/2		15/2	153 053.70		568
	401	8, 5← 7, 5	Ground		13/2		11/2	152 992.98		568
	401	8, 5← 7, 5	Ground		19/2		17/2	153 001.93		568
	401	8, 5← 7, 5	Ground		15/2		13/2	153 035.98		568
	401	8, 6← 7, 6	Ground		15/2		13/2	153 019.63		568
	401	8, 6← 7, 6	Ground		17/2		15/2	153 034.03		568
$C^{12}H_3Br^{81}$	402	1, ← 0,	Ground					19 064.40	.1	161
	402	1, 0← 0, 0	1		1/2		3/2	18 798.72	.05	537
	402	1, 0← 0, 0	1		5/2		3/2	18 895.04	.05	537
	402	1, 0← 0, 0	Ground		1/2		3/2	18 943.38	.05	537
	402	1, 0← 0, 0	1		3/2		3/2	19 015.66	.05	537
	402	1, 0← 0, 0	Ground		5/2		3/2	19 039.69	.05	537
	402	1, 0← 0, 0	Ground		3/2		3/2	19 160.30	.05	537
	402	2, 0← 1, 0	Ground		3/2		3/2	38 030.77	.08	83
	402	2, 0← 1, 0	Ground		7/2		5/2	38 116.65	.08	83
	402	2, 0← 1, 0	Ground		5/2		3/2	38 116.65	.08	83
	402	2, 0← 1, 0	Ground		1/2		1/2	38 126.97	.08	83
	402	2, 0← 1, 0	Ground		5/2		5/2	38 237.14	.08	83
	402	2, 0← 1, 0	Ground		3/2		1/2	38 247.77	.08	83
	402	2, 1← 1, 1	Ground		1/2		1/2	38 006.47	.08	83
	402	2, 1← 1, 1	Ground		3/2		1/2	38 066.72	.08	83
	402	2, 1← 1, 1	Ground		7/2		5/2	38 097.45	.08	83
	402	2, 1← 1, 1	Ground		5/2		5/2	38 157.70	.08	83
	402	4, 0← 3, 0	Ground		9/2		7/2	76 248.32	.18	240
	402	4, 0← 3, 0	Ground		11/2		9/2	76 248.32	.18	240
	402	4, 0← 3, 0	Ground		5/2		3/2	76 261.96	.18	240
	402	4, 0← 3, 0	Ground		7/2		5/2	76 261.96	.18	240
	402	4, 1← 3, 1	Ground		11/2		9/2	76 243.66	.18	240
	402	4, 1← 3, 1	Ground		5/2		3/2	76 249.94	.18	240
	402	4, 1← 3, 1	Ground		9/2		7/2	76 255.68	.18	240
	402	4, 1← 3, 1	Ground		7/2		5/2	76 261.96	.18	240

Isotopic Species	Id. No.	Rotational Quantum Nos.	Vib. State v_a	Hyperfine				Frequency MHz	Acc. ±MHz	Ref.
				F_1'	F'	F_1	F			
$C^{12}H_3Br^{81}$	402	4, 2← 3, 2	Ground		5/2		3/2	76 213.16	.18	240
	402	4, 2← 3, 2	Ground		11/2		9/2	76 230.18	.18	240
	402	4, 2← 3, 2	Ground		7/2		5/2	76 261.96	.18	240
	402	4, 2← 3, 2	Ground		9/2		7/2	76 278.16	.18	240
	402	4, 3← 3, 3	Ground		5/2		3/2	76 152.28	.18	240
	402	4, 3← 3, 3	Ground		11/2		9/2	76 207.66	.18	240
	402	5, 0← 4, 0	Ground		13/2		11/2	95 310.78	.27	240
	402	5, 0← 4, 0	Ground		11/2		9/2	95 310.78	.27	240
	402	5, 0← 4, 0	Ground		9/2		7/2	95 319.12	.27	240
	402	5, 0← 4, 0	Ground		7/2		5/2	95 319.12	.27	240
	402	5, 1← 4, 1	Ground		13/2		11/2	95 307.48	.27	240
	402	5, 1← 4, 1	Ground		9/2		7/2	95 319.12	.27	240
	402	5, 2← 4, 2	Ground		7/2		5/2	95 253.89	.27	240
	402	5, 2← 4, 2	Ground		13/2		11/2	95 297.55	.27	240
	402	5, 2← 4, 2	Ground		11/2		9/2	95 322.15	.27	240
	402	5, 3← 4, 3	Ground		7/2		5/2	95 263.47	.27	240
	402	5, 3← 4, 3	Ground		13/2		11/2	95 281.53	.27	240
	402	5, 3← 4, 3	Ground		11/2		9/2	95 336.01	.27	240
	402	5, 4← 4, 4	Ground		13/2		11/2	95 259.24	.27	240
	402	8, 0← 7, 0	Ground		19/2		17/2	152 488.14		568
	402	8, 0← 7, 0	Ground		17/2		15/2	152 488.14		568
	402	8, 0← 7, 0	Ground		15/2		13/2	152 491.10		568
	402	8, 0← 7, 0	Ground		13/2		11/2	152 491.10		568
	402	8, 1← 7, 1	Ground		15/2		15/2	152 466.19		568
	402	8, 1← 7, 1	Ground		19/2		17/2	152 485.61		568
	402	8, 1← 7, 1	Ground		17/2		15/2	152 487.03		568
	402	8, 1← 7, 1	Ground		13/2		11/2	152 488.14		568
	402	8, 1← 7, 1	Ground		15/2		13/2	152 489.56		568
	402	8, 2← 7, 2	Ground		19/2		17/2	152 477.87		568
	402	8, 2← 7, 2	Ground		13/2		11/2	152 479.20		568
	402	8, 2← 7, 2	Ground		17/2		15/2	152 483.60		568
	402	8, 2← 7, 2	Ground		15/2		13/2	152 484.89		568
	402	8, 3← 7, 3	Ground		19/2		17/2	152 464.97		568
	402	8, 3← 7, 3	Ground		15/2		13/2	152 477.06		568
	402	8, 3← 7, 3	Ground		17/2		15/2	152 477.87		568
	402	8, 4← 7, 4	Ground		13/2		11/2	152 443.13		568
	402	8, 4← 7, 4	Ground		19/2		17/2	152 446.88		568
	402	8, 4← 7, 4	Ground		15/2		13/2	152 466.19		568
	402	8, 4← 7, 4	Ground		17/2		15/2	152 469.84		568
	402	8, 4← 7, 4	Ground		17/2		17/2	152 487.03		568
	402	8, 5← 7, 5	Ground		15/2		13/2	152 452.02		568
	402	8, 5← 7, 5	Ground		17/2		15/2	152 459.44		568
	402	8, 6← 7, 6	Ground		17/2		15/2	152 446.88		568
	402	Not Reported	Ground		3/2		3/2	38 175.08	.08	83
	402	Not Reported	Ground		5/2		3/2	38 218.21	.08	83
$C^{12}D_3Br^{79}$	403	2, 0← 1, 0	Ground		3/2		3/2	30 743.99	.10	395
	403	2, 0← 1, 0	Ground		7/2		5/2	30 846.00	.10	395
	403	2, 0← 1, 0	Ground		5/2		3/2	30 846.00	.10	395
	403	2, 0← 1, 0	Ground		1/2		1/2	30 858.24	.10	395
	403	2, 0← 1, 0	Ground		5/2		5/2	30 898.82	.10	395

Isotopic Species	Id. No.	Rotational Quantum Nos.	Vib. State v_a	F_1'	F'	F_1	F	Frequency MHz	Acc. $\pm$MHz	Ref.
$C^{12}D_3Br^{79}$	403	2, 0← 1, 0	Ground		3/2		1/2	31 002.43	.10	395
	403	2, 1← 1, 1	Ground		7/2		5/2	30 823.44	.10	395
	403	2, 1← 1, 1	Ground		5/2		5/2	30 895.58	.10	395
	403	2, 1← 1, 1	Ground		3/2		3/2	30 916.21	.10	395
	403	2, 1← 1, 1	Ground		5/2		3/2	30 967.53	.10	395
$C^{12}D_3Br^{81}$	404	1, 0← 1, 0	Ground		5/2		5/2	30 834.73	.10	395
	404	2, 0← 1, 0	Ground		3/2		3/2	30 629.28	.10	395
	404	2, 0← 1, 0	Ground		7/2		5/2	30 714.74	.10	395
	404	2, 0← 1, 0	Ground		5/2		3/2	30 714.74	.10	395
	404	2, 0← 1, 0	Ground		1/2		1/2	30 724.89	.10	395
	404	2, 1← 1, 1	Ground		3/2		1/2	30 665.35	.10	395
	404	2, 1← 1, 1	Ground		7/2		5/2	30 695.83	.10	395
	404	2, 1← 1, 1	Ground		5/2		5/2	30 756.12	.10	395
	404	2, 1← 1, 1	Ground		3/2		3/2	30 773.17	.10	395
	404	2, 1← 1, 1	Ground		5/2		3/2	30 815.96	.10	395
$C^{13}H_3Br^{79}$	405	2, ← 1,	Ground					36 477.67	.09	242
$C^{13}H_3Br^{81}$	406	2, ← 1,	Ground					36 331.10	.09	242
$C^{12}HD_2Br^{79}$	407	2, 1, 1← 1, 1, 0	Ground		7/2		5/2	33 057.1		376
	407	2, 1, 2← 1, 1, 1	Ground		7/2		5/2	32 739.4		376
$C^{12}HD_2Br^{81}$	408	2, 1, 1← 1, 1, 0	Ground		7/2		5/2	32 925.1		376
	408	2, 1, 2← 1, 1, 1	Ground		7/2		5/2	32 610.6		376

CH₃BrHg

C_{3v}

CH₃HgBr

Isotopic Species	Pt. Gp.	Id. No.	A MHz	B MHz		C MHz		D_J MHz	D_{JK} MHz	Δ Amu A²	κ
C¹²H₃Hg¹⁹⁸Br⁷⁹	C_{3v}	411		1 142.86	M	1 142.86	M		.0082		
C¹²H₃Hg¹⁹⁸Br⁸¹	C_{3v}	412		1 125.28	M	1 125.28	M		.0080		
C¹²H₃Hg¹⁹⁹Br⁷⁹	C_{3v}	413		1 142.10	M	1 142.10	M				
C¹²H₃Hg¹⁹⁹Br⁸¹	C_{3v}	414		1 124.51	M	1 124.51	M				
C¹²H₃Hg²⁰⁰Br⁷⁹	C_{3v}	415		1 141.36	M	1 141.36	M				
C¹²H₃Hg²⁰⁰Br⁸¹	C_{3v}	416		1 123.76	M	1 123.76	M				
C¹²H₃Hg²⁰²Br⁷⁹	C_{3v}	417		1 139.88	M	1 139.88	M				
C¹²H₃Hg²⁰²Br⁸¹	C_{3v}	418		1 122.27	M	1 122.27	M				

Id. No.	μ_a Debye	μ_b Debye	μ_c Debye	eQq Value(MHz)	Rel.	eQq Value(MHz)	Rel.	eQq Value(MHz)	Rel.	ω_a d 1/cm	ω_b d 1/cm	ω_c d 1/cm	ω_d d 1/cm
411				350	Br⁷⁹								
412				290	Br⁸¹								

References:
ABC: 522 D_{JK}: 522   eQq: 522

Methyl Mercuric Bromide

Isotopic Species	Id. No.	Rotational Quantum Nos.	Vib. State	Hyperfine F₁′	F′	F₁	F	Frequency MHz	Acc. ±MHz	Ref.
C¹²H₃Hg¹⁹⁸Br⁷⁹	411	16, ←15,	Ground					36 571.55		522
C¹²H₃Hg¹⁹⁸Br⁸¹	412	16, ←15,	Ground					36 008.79		522
C¹²H₃Hg¹⁹⁹Br⁷⁹	413	16, ←15,	Ground					36 547.33		522
C¹²H₃Hg¹⁹⁹Br⁸¹	414	16, ←15,	Ground					35 984.46		522
C¹²H₃Hg²⁰⁰Br⁷⁹	415	16, ←15,	Ground					36 523.48		522
C¹²H₃Hg²⁰⁰Br⁸¹	416	16, ←15,	Ground					35 960.40		522
C¹²H₃Hg²⁰²Br⁷⁹	417	16, ←15,	Ground					36 476.28		522
C¹²H₃Hg²⁰²Br⁸¹	418	16, ←15,	Ground					35 912.50		522

CH₃CL   C$_{3v}$ CH₃Cl

Isotopic Species	Pt. Gp.	Id. No.	A MHz	B MHz	C MHz	D$_J$ MHz	D$_{JK}$ MHz	Δ Amu A²	κ
C^{12}H$_3$Cl^{35}	C$_{3v}$	421		13292.84 M	13292.84 M	.0181	.198		
C^{12}H$_3$Cl^{36}	C$_{3v}$	422		13187.60 M	13187.60 M				
C^{12}H$_3$Cl^{37}	C$_{3v}$	423		13088.13 M	13088.13 M	.0270	.185		
C^{12}D$_3$Cl^{35}	C$_{3v}$	424		10841.88 M	10841.88 M				
C^{12}D$_3$Cl^{37}	C$_{3v}$	425		10658.43 M	10658.43 M				
C^{13}H$_3$Cl^{35}	C$_{3v}$	426		12799.18 M	12799.18 M				
C^{13}H$_3$Cl^{37}	C$_{3v}$	427		12592.13 M	12592.13 M				
C^{12}HD$_2$Cl^{35}	C$_s$	431		11679. ¹M	11370. ¹M				

1. Calculations for B and C for C^{12}HD$_2$Cl^{35} were made from ref. 296 by adding the quantities for 3B + C and 3C + B to obtain B + C, then combining this with the given value for 2(B − C).

Id. No.	μ$_a$ Debye	μ$_b$ Debye	μ$_c$ Debye	eQq Value (MHz) Rel.		ω$_a$ d 1/cm	ω$_b$ d 1/cm
421	1.869 M	0. X	0. X	−74.77	Cl^{35}	725 1	1012 2
422				−15.83	Cl^{36}		
423				−58.93	Cl^{37}		
424				−74.41	Cl^{35}		
425				−58.58	Cl^{37}		

References
ABC: 218, 296, 395, 537, 576   D$_J$: 240, 568   D$_{JK}$: 240, 568   μ: 236   eQq: 537, 576   ω:406
Add. Ref.: 39, 83, 89, 90, 102, 103, 116, 119, 127, 141, 164, 185, 299, 306, 316, 341, 498, 499, 510, 599, 741, 810, 827, 1022

Species	eQq (MHz) v$_3$ = 1	eQq (MHz) v$_6$ = 1	α$_3^B$(MHz)	α$_6^B$(MHz)	B + C (MHz)	Ref.
421 (Cl^{35})	−74.87	−74.89	115.21	49.01		537
423 (Cl^{37})	−58.89	−58.76	112.30	48.19		537
428					24658	218
429					24266	218
430					22674	218

238-605 O-68—7

Isotopic Species	Id. No.	Rotational Quantum Nos.	Vib. State v_a ; v_b	Hyperfine				Frequency MHz	Acc. ±MHz	Ref.
				F_1'	F'	F_1	F			
$C^{12}H_3Cl^{35}$	421	1, 0← 0, 0	1; 0		3/2		3/2	26 340.27	.05	537
	421	1, 0← 0, 0	1; 0		5/2		3/2	26 359.01	.05	537
	421	1, 0← 0, 0	1; 0		1/2		3/2	26 373.96	.05	537
	421	1, 0← 0, 0	0; 1		3/2		3/2	26 472.68	.10	537
	421	1, 0← 0, 0	0; 1		5/2		3/2	26 491.39	.10	537
	421	1, 0← 0, 0	0; 1		1/2		3/2	26 506.38	.10	537
	421	1, 0← 0, 0	Ground		3/2		3/2	26 570.73	.05	537
	421	1, 0← 0, 0	Ground		5/2		3/2	26 589.40	.05	537
	421	1, 0← 0, 0	Ground		1/2		3/2	26 604.38	.05	537
	421	3, 0← 2, 0	Ground		7/2		7/2	79 736.96	.16	240
	421	3, 0← 2, 0	Ground		5/2		3/2	79 751.44		568
	421	3, 0← 2, 0	Ground		3/2		1/2	79 751.44		568
	421	3, 0← 2, 0	Ground		9/2		7/2	79 756.00		568
	421	3, 0← 2, 0	Ground		7/2		5/2	79 756.00		568
	421	3, 0← 2, 0	Ground		5/2		5/2	79 764.56	.16	240
	421	3, 0← 2, 0	Ground		3/2		3/2	79 769.94	.16	240
	421	3, 1← 2, 1	Ground		7/2		5/2	79 751.44		568
	421	3, 1← 2, 1	Ground		9/2		7/2	79 756.00		568
	421	3, 1← 2, 1	Ground		3/2		1/2	79 756.00		568
	421	3, 2← 2, 2	Ground		5/2		5/2	79 736.96	.16	240
	421	3, 2← 2, 2	Ground		7/2		5/2	79 736.96	.16	240
	421	3, 2← 2, 2	Ground		7/2		7/2	79 756.00		568
	421	3, 2← 2, 2	Ground		9/2		7/2	79 756.00		568
	421	3, 2← 2, 2	Ground		5/2		7/2	79 756.00		568
	421	3, 2← 2, 2	Ground		3/2		1/2	79 768.98	.16	240
	421	4, 0← 3, 0	Ground		9/2		9/2	106 320.08		568
	421	4, 0← 3, 0	Ground		5/2		3/2	106 336.59		568
	421	4, 0← 3, 0	Ground		7/2		5/2	106 336.59		568
	421	4, 0← 3, 0	Ground		9/2		7/2	106 338.75		568
	421	4, 0← 3, 0	Ground		11/2		9/2	106 338.75		568
	421	4, 0← 3, 0	Ground		7/2		7/2	106 345.33		568
	421	4, 0← 3, 0	Ground		5/2		5/2	106 355.35		568
	421	4, 1← 3, 1	Ground		9/2		7/2	106 335.79		568
	421	4, 1← 3, 1	Ground		5/2		3/2	106 336.59		568
	421	4, 1← 3, 1	Ground		11/2		9/2	106 337.70		568
	421	4, 1← 3, 1	Ground		7/2		7/2	106 341.52		568
	421	4, 1← 3, 1	Ground		5/2		5/2	106 350.77		568
	421	4, 2← 3, 2	Ground		9/2		7/2	106 327.02		568
	421	4, 2← 3, 2	Ground		9/2		9/2	106 327.02		568
	421	4, 2← 3, 2	Ground		11/2		9/2	106 334.53		568
	421	4, 3← 3, 3	Ground		7/2		7/2	106 310.31		568
	421	4, 3← 3, 3	Ground		9/2		7/2	106 312.37		568
	421	4, 3← 3, 3	Ground		5/2		5/2	106 314.52		568
	421	4, 3← 3, 3	Ground		7/2		5/2	106 321.24		568
	421	4, 3← 3, 3	Ground		9/2		9/2	106 335.79		568
	421	4, 3← 3, 3	Ground		11/2		9/2	106 392.24		568
	421	6, 0← 5, 0	Ground		13/2		13/2	159 480.29		568
	421	6, 0← 5, 0	Ground		9/2		7/2	159 498.25		568
	421	6, 0← 5, 0	Ground		11/2		9/2	159 498.25		568
	421	6, 0← 5, 0	Ground		15/2		13/2	159 499.02		568

Isotopic Species	Id. No.	Rotational Quantum Nos.	Vib. State v_a ; v_b	Hyperfine F_1'	F'	F_1	F	Frequency MHz	Acc. ±MHz	Ref.
$C^{12}H_3Cl^{35}$	421	6, 0← 5, 0	Ground		13/2		11/2	159 499.02		568
	421	6, 1← 5, 1	Ground		7/2		7/2	106 310.31		568
	421	6, 1← 5, 1	Ground					159 495.99		568
	421	6, 2← 5, 2	Ground		11/2		9/2	159 488.13		568
	421	6, 2← 5, 2	Ground		13/2		11/2	159 488.13		568
	421	6, 2← 5, 2	Ground		9/2		7/2	159 490.27		568
	421	6, 2← 5, 2	Ground		15/2		13/2	159 490.27		568
	421	6, 3← 5, 3	Ground		13/2		11/2	159 474.48		568
	421	6, 3← 5, 3	Ground		11/2		9/2	159 475.47		568
	421	6, 3← 5, 3	Ground		15/2		13/2	159 479.25		568
	421	6, 3← 5, 3	Ground		9/2		7/2	159 480.29		568
	421	6, 4← 5, 4	Ground		13/2		11/2	159 455.37		568
	421	6, 4← 5, 4	Ground		15/2		13/2	159 463.92		568
	421	6, 4← 5, 4	Ground		9/2		7/2	159 466.53		568
$C^{12}H_3Cl^{36}$	422	1, 0← 0, 0	Ground		2		2	26 372.42	.025	576
	422	1, 0← 0, 0	Ground		2		3	26 376.03	.035	576
	422	1, 0← 0, 0	Ground		2		1	26 377.96	.025	576
$C^{12}H_3Cl^{37}$	423	1, 0← 0, 0	1; 0		3/2		3/2	25 939.87	.05	537
	423	1, 0← 0, 0	1; 0		5/2		3/2	25 954.60	.05	537
	423	1, 0← 0, 0	1; 0		1/2		3/2	25 966.37	.05	537
	423	1, 0← 0, 0	0; 1		3/2		3/2	26 068.83	.10	537
	423	1, 0← 0, 0	0; 1		5/2		3/2	26 082.83	.10	537
	423	1, 0← 0, 0	0; 1		1/2		3/2	26 094.56	.10	537
	423	1, 0← 0, 0	Ground		3/2		3/2	26 164.48	.05	537
	423	1, 0← 0, 0	Ground		5/2		3/2	26 179.18	.05	537
	423	1, 0← 0, 0	Ground		1/2		3/2	26 191.00	.05	537
	423	3, 0← 2, 0	Ground		7/2		7/2	78 512.80	.16	240
	423	3, 0← 2, 0	Ground		3/2		1/2	78 523.32	.16	240
	423	3, 0← 2, 0	Ground		5/2		3/2	78 523.32	.16	240
	423	3, 0← 2, 0	Ground		9/2		7/2	78 527.10	.16	240
	423	3, 0← 2, 0	Ground		7/2		5/2	78 527.10	.16	240
	423	3, 1← 2, 1	Ground		5/2		3/2	78 523.32	.16	240
	423	3, 1← 2, 1	Ground		7/2		5/2	78 523.32	.16	240
	423	3, 1← 2, 1	Ground		9/2		7/2	78 527.10	.16	240
	423	3, 1← 2, 1	Ground		3/2		1/2	78 527.10	.16	240
	423	3, 2← 2, 2	Ground		7/2		5/2	78 511.68	.16	240
	423	3, 2← 2, 2	Ground		5/2		5/2	78 511.68	.16	240
	423	3, 2← 2, 2	Ground		3/2		3/2	78 522.00	.16	240
	423	3, 2← 2, 2	Ground		5/2		3/2	78 522.00	.16	240
	423	3, 2← 2, 2	Ground		9/2		7/2	78 526.14	.16	240
	423	3, 2← 2, 2	Ground		7/2		7/2	78 526.14	.16	240
$C^{12}D_3Cl^{35}$	424	1, 0← 0, 0	Ground		3/2		3/2	21 668.88	.08	395
	424	1, 0← 0, 0	Ground		5/2		3/2	21 687.46	.08	395
	424	1, 0← 0, 0	Ground		1/2		3/2	21 702.36	.08	395
$C^{12}D_3Cl^{37}$	425	1, 0← 0, 0	Ground		3/2		3/2	21 305.15	.08	395
	425	1, 0← 0, 0	Ground		5/2		3/2	21 319.79	.08	395
	425	1, 0← 0, 0	Ground		1/2		3/2	21 331.51	.08	395
$C^{13}H_3Cl^{35}$	426	1, ← 0,	Ground					25 577.40	.1	218

Isotopic Species	Id. No.	Rotational Quantum Nos.	Vib. State v_a ; v_b	F_1'	F'	F_1	F	Frequency MHz	Acc. ±MHz	Ref.
$C^{13}H_3Cl^{35}$	426	1, ← 0,	Ground					25 596.19	.1	218
	426	1, ← 0,	Ground					25 611.09	.1	218
$C^{13}H_3Cl^{37}$	427	1, ← 0,	Ground					25 167.68	.1	218
	427	1, ← 0,	Ground					25 182.50	.1	218
	427	1, ← 0,	Ground					25 194.20	.1	218
$C^{12}H_2DCl^{35}$	428	1, 0, 1← 0, 0, 0	Ground					24 641.70		218
	428	1, 0, 1← 0, 0, 0	Ground					24 660.33		218
	428	1, 0, 1← 0, 0, 0	Ground					24 675.25		218
$C^{12}H_2DCl^{37}$	429	1, 0, 1← 0, 0, 0	Ground					24 252.00		218
	429	1, 0, 1← 0, 0, 0	Ground					24 266.68		218
	429	1, 0, 1← 0, 0, 0	Ground					24 278.33		218
$C^{12}HD_2Cl^{35}$	431	1, 0, 1← 0, 0, 0	Ground					23 035.00		218
	431	1, 0, 1← 0, 0, 0	Ground					23 053.62		218
	431	1, 0, 1← 0, 0, 0	Ground					23 068.51		218
	431	2, 1, 2← 1, 0, 1	Ground					46 099.4		296
	431	2, 1, 1← 1, 1, 0	Ground		7/2		5/2	46 407.		376
	431	2, 1, 2← 1, 1, 1	Ground		7/2		5/2	45 789.		376
$C^{12}HD_2Cl^{37}$	432	1, 0, 1← 0, 0, 0	Ground					22 659.29		218
	432	1, 0, 1← 0, 0, 0	Ground					22 673.80		218
	432	1, 0, 1← 0, 0, 0	Ground					22 685.60		218

CH$_3$ClHg   C$_{3v}$ CH$_3$HgCl

Isotopic Species	Pt. Gp.	Id. No.	A MHz	B MHz		C MHz		D$_J$ MHz	D$_{JK}$ MHz	Δ Amu A^2	κ
C^{12}H$_3$Hg^{198}Cl^{35}	C$_{3v}$	441		2 077.48	M	2 077.48	M	.00024	.0210		
C^{12}H$_3$Hg^{198}Cl^{37}	C$_{3v}$	442		2 006.14	M	2 006.14	M		.0195		
C^{12}H$_3$Hg^{199}Cl^{35}	C$_{3v}$	443		2 077.18	M	2 077.18	M	.00026	.0210		
C^{12}H$_3$Hg^{199}Cl^{37}	C$_{3v}$	444		2 005.79	M	2 005.79	M		.0195		
C^{12}H$_3$Hg^{200}Cl^{35}	C$_{3v}$	445		2 076.86	M	2 076.86	M	.00026	.0211		
C^{12}H$_3$Hg^{200}Cl^{37}	C$_{3v}$	446		2 005.45	M	2 005.45	M		.0195		
C^{12}H$_3$Hg^{202}Cl^{35}	C$_{3v}$	447		2 076.24	M	2 076.24	M	.00025	.0211		
C^{12}H$_3$Hg^{202}Cl^{37}	C$_{3v}$	448		2 004.76	M	2 004.76	M		.0195		
C^{12}H$_3$Hg^{204}Cl^{35}	C$_{3v}$	449		2 075.59	M	2 075.59	M				
C^{12}H$_3$Hg^{204}Cl^{37}	C$_{3v}$	451		2 004.09	M	2 004.09	M				

Id. No.	μ$_a$ Debye		μ$_b$ Debye		μ$_c$ Debye		eQq Value(MHz)	Rel.	eQq Value(MHz)	Rel.	eQq Value(MHz)	Rel.	ω$_a$ d 1/cm	ω$_b$ d 1/cm	ω$_c$ d 1/cm	ω$_d$ d 1/cm
441	3.36	L	0.	X	0.	X	−42	Cl^{35}								
442							−33	Cl^{37}								

References:

ABC: 522,592   D$_J$: 592   D$_{JK}$: 522,592   μ: 995   eQq: 522

Add. Ref. 199

Isotopic Species	Id. No.	Rotational Quantum Nos.	Vib. State	Hyperfine				Frequency MHz	Acc. ±MHz	Ref.
				F_1'	F'	F_1	F			
$C^{12}H_3Hg^{198}Cl^{35}$	441	9, ← 8,	Ground					37 394.00		522
	441	17, 0←16, 0	Ground					70 629.71		592
	441	17, 1←16, 1	Ground					70 629.03		592
	441	17, 2←16, 2	Ground					70 626.86		592
	441	17, 3←16, 3	Ground					70 623.21		592
	441	17, 4←16, 4	Ground					70 618.31		592
$C^{12}H_3Hg^{198}Cl^{37}$	442	9, ← 8,	Ground					36 110.53		522
$C^{12}H_3Hg^{199}Cl^{35}$	443	9, ← 8,	Ground					37 388.40		522
	443	17, 0←16, 0	Ground					70 618.92		592
	443	17, 1←16, 1	Ground					70 618.32		592
	443	17, 2←16, 2	Ground					70 616.02		592
	443	17, 3←16, 3	Ground					70 612.46		592
	443	17, 4←16, 4	Ground					70 607.56		592
$C^{12}H_3Hg^{199}Cl^{37}$	444	9, ← 8,	Ground					36 104.30		522
$C^{12}H_3Hg^{200}Cl^{35}$	445	9, ← 8,	Ground					37 382.80		522
	445	17, 0←16, 0	Ground					70 608.29		592
	445	17, 1←16, 1	Ground					70 607.58		592
	445	17, 2←16, 2	Ground					70 605.45		592
	445	17, 3←16, 3	Ground					70 601.86		592
	445	17, 4←16, 4	Ground					70 596.92		592
	445	17, 5←16, 5	Ground					70 590.40		592
$C^{12}H_3Hg^{200}Cl^{37}$	446	9, ← 8,	Ground					36 098.07		522
$C^{12}H_3Hg^{202}Cl^{35}$	447	9, ← 8,	Ground					37 371.60		522
	447	17, 0←16, 0	Ground					70 587.33		592
	447	17, 1←16, 1	Ground					70 586.55		592
	447	17, 2←16, 2	Ground					70 584.44		592
	447	17, 3←16, 3	Ground					70 580.86		592
	447	17, 4←16, 4	Ground					70 575.74		592
$C^{12}H_3Hg^{202}Cl^{37}$	448	9, ← 8,	Ground					36 085.75		522
$C^{12}H_3Hg^{204}Cl^{35}$	449	9, ← 8,	Ground					37 360.62		522
$C^{12}H_3Hg^{204}Cl^{37}$	451	9, ← 8,	Ground					36 073.62		522

CH₃F   C$_{3v}$ CH₃F

Isotopic Species	Pt. Gp.	Id. No.	A MHz		B MHz		C MHz		D$_J$ MHz	D$_{JK}$ MHz	Δ Amu A²	κ
C¹²H₃F¹⁹	C$_{3v}$	461	152 793.2	N	25 530.59	M	25 530.59	M	.0593	.445		
C¹²D₃F¹⁹	C$_{3v}$	462	76 858.39	N	20 445.57	N	20 445.57	N				
C¹³H₃F¹⁹	C$_{3v}$	463			24 857.20	M	24 857.20	M				
C¹³D₃F¹⁹	C$_{3v}$	464			20 111.88	M	20 111.88	M	.033			
C¹²H₂DF¹⁹	C$_s$	465	.		24 043.	M	22 959.	M	.05	.35		

Id. No.	μ$_a$ Debye	μ$_b$ Debye	μ$_c$ Debye	eQq Value(MHz) Rel.	eQq Value(MHz) Rel.	eQq Value(MHz) Rel.	ω$_a$ d 1/cm	ω$_b$ d 1/cm	ω$_c$ d 1/cm	ω$_d$ d 1/cm
461	1.8555 M	0. X	0. X							

References:

ABC: 568,652,781   D$_J$: 568,781   D$_{JK}$: 568,781   μ: 994

Add. Ref. 83,103,130,196,426,434,437,533,666,670,741,810

Fluoromethane   Spectral Line Table

Isotopic Species	Id. No.	Rotational Quantum Nos.	Vib. State	F$_1'$	F'	F$_1$	F	Frequency MHz	Acc. ±MHz	Ref.
C¹²H₃F¹⁹	461	1, 0← 0, 0	Ground					51 071.98		568
	461	2, 0← 1, 0	Ground					102 142.56	.20	533
	461	2, 1← 1, 1	Ground					102 140.86	.20	533
	461	3, 0← 2, 0	Ground					153 210.44	.30	533
	461	3, 1← 2, 1	Ground					153 207.65	.30	533
	461	3, 2← 2, 2	Ground					153 199.58	.30	533
	461	4, 0← 3, 0	Ground					204 273.69	.40	533
	461	4, 1← 3, 1	Ground					204 270.09	.40	533
	461	4, 2← 3, 2	Ground					204 259.48	.40	533
	461	4, 3← 3, 3	Ground					204 241.71	.40	533
C¹²D₃F¹⁹	462	3, 0← 2, 0	Ground					122 695.50	.30	282
	462	3, 1← 2, 1	Ground					122 694.20	.30	282
	462	3, 2← 2, 2	Ground					122 690.02	.30	282
C¹³H₃F¹⁹	463	1, 0← 0, 0	Ground					49 724.73	.18	129
C¹³D₃F¹⁹	464	1, 0← 0, 0	Ground					40 223.64		781
C¹²H₂DF¹⁹	465	1, 0, 1← 0, 0, 0	Ground					47 002.52	.1	781
	465	2, 0, 2← 1, 0, 1	Ground					93 994.7	1.	781
	465	2, 1, 1← 1, 1, 0	Ground					95 086.0	1.	781
	465	2, 1, 2← 1, 1, 1	Ground					92 918.7	1.	781
C¹²HD₂F¹⁹	466	1, 0, 1← 0, 0, 0	Ground					43 689.82	.1	781

CH₃HgI C_{3v} CH₃HgI

Isotopic Species	Pt. Gp.	Id. No.	A MHz	B MHz	C MHz	D_J MHz	D_{JK} MHz	Δ Amu A²	κ
$C^{12}H_3Hg^bI^b$	C_{3v}	471		788.0 M					

1. The J = 28 ← 27 and 29 ← 28 transitions were observed, but spectral lines were not given.

References:

ABC: 1020

Add. Ref. 199

·No Spectral Lines

CH₃I C_{3v} CH₃I

Isotopic Species	Pt. Gp.	Id. No.	A MHz	B MHz	C MHz	D_J MHz	D_{JK} MHz	Δ Amu A²	κ
$C^{12}H_3I^{127}$	C_{3v}	481	152 570.9 M	7 501.30 M	7 501.30 M	.00628	.0985		
$C^{12}D_3I^{127}$	C_{3v}	482		6 040.285 M	6 040.285 M	.00358	.04832		
$C^{13}H_3I^{127}$	C_{3v}	483	152 570.9 M	7 119.04 M	7 119.04 M				

Id. No.	μ_a Debye	μ_b Debye	μ_c Debye	eQq Value(MHz) Rel.	eQq Value(MHz) Rel.	eQq Value(MHz) Rel.	ω_a d 1/cm	ω_b d 1/cm	ω_c d 1/cm	ω_d d 1/cm
481	1.647 M	0. X	0. X	−1933.99 I^{127}						

References:

ABC: 582,940,1010 D_J: 568,1010 D_{JK}: 568,1010 μ: 236   eQq: 638

Add. Ref. 40,41,79,102,103,144,151,164,416,454,741,753,754,807,810

For species 484: 2(B − C) = 195.44 MHz, Ref. 376.

For species $C^{12}H_3I^{129}$: nuclear g-factor = 0.783, Ref. 133.

For species $C^{12}H_3I^{131}$: eQq (I^{131}) = − 973 MHz, Ref. 455.

Isotopic Species	Id. No.	Rotational Quantum Nos.	Vib. State	Hyperfine				Frequency MHz	Acc. ±MHz	Ref.
				F_1'	F'	F_1	F			
$C^{12}H_3I^{127}$	481	1, 0← 0, 0	Ground		5/2		5/2	14 695.22	.05	806
	481	1, 0← 0, 0	Ground		7/2		5/2	15 100.74	.05	806
	481	1, 0← 0, 0	Ground		3/2		5/2	15 275.87	.05	806
	481	1, 1← 1, 1	Ground		5/2		3/2	292.5	.04	638
	481	2, 0← 1, 0	Ground		5/2		3/2	29 598.95	.08	83
	481	2, 0← 1, 0	Ground		7/2		7/2	29 673.95	.08	83
	481	2, 0← 1, 0	Ground		5/2		7/2	29 773.95	.08	83
	481	2, 0← 1, 0	Ground		3/2		3/2	29 872.52	.08	83
	481	2, 0← 1, 0	Ground		9/2		7/2	30 046.99	.08	83
	481	2, 0← 1, 0	Ground		7/2		5/2	30 079.72	.08	83
	481	2, 0← 1, 0	Ground		1/2		3/2	30 121.32	.08	83
	481	2, 0← 1, 0	Ground		5/2		5/2	30 179.71	.08	83
	481	2, 0← 1, 0	Ground		3/2		5/2	30 453.46	.08	83
	481	2, 1← 1, 1	Ground		7/2		5/2	29 735.71	.08	83
	481	2, 1← 1, 1	Ground		5/2		5/2	29 782.71	.08	83
	481	2, 1← 1, 1	Ground		3/2		5/2	29 923.50	.08	83
	481	2, 1← 1, 1	Ground		7/2		7/2	29 939.87	.08	83
	481	2, 1← 1, 1	Ground		5/2		7/2	29 986.84	.08	83
	481	2, 1← 1, 1	Ground		5/2		3/2	30 075.08	.08	83
	481	2, 1← 1, 1	Ground		9/2		7/2	30 123.64	.08	83
	481	2, 1← 1, 1	Ground		3/2		3/2	30 215.95	.08	83
	481	2, 2← 2, 2	Ground		7/2		9/2	375.0	.3	638
	481	3, 3← 3, 3	Ground		9/2		11/2	444.76	.10	638
	481	4, 1← 3, 1	Ground		7/2		5/2	59 955.05	.2	775
	481	4, 1← 3, 1	Ground		5/2		3/2	59 965.	1.	775
	481	4, 1← 3, 1	Ground		9/2		7/2	59 975.55	.2	775
	481	4, 1← 3, 1	Ground		11/2		9/2	60 010.55	.2	775
	481	4, 1← 3, 1	Ground		13/2		11/2	60 036.35	.2	775
	481	4, 2← 3, 2	Ground		9/2		7/2	59 922.75	.2	775
	481	4, 2← 3, 2	Ground		11/2		9/2	59 937.75	.2	775
	481	4, 2← 3, 2	Ground		7/2		5/2	59 975.55	.2	775
	481	4, 2← 3, 2	Ground		5/2		3/2	60 059.75	.2	775
	481	4, 2← 3, 2	Ground		13/2		11/2	60 076.05	.2	775
	481	4, 3← 3, 3	Ground		11/2		9/2	59 819.35	.2	775
	481	4, 3← 3, 3	Ground		9/2		7/2	59 832.75	.2	775
	481	4, 3← 3, 3	Ground		7/2		5/2	60 011.95	.2	775
	481	4, 3← 3, 3	Ground		13/2		11/2	60 141.55	.2	775
	481	4, 4← 4, 4	Ground		11/2		13/2	481.05	.10	638
	481	5, 0← 4, 0	Ground		7/2		5/2	74 967.66	.16	240
	481	5, 0← 4, 0	Ground		9/2		7/2	74 977.62	.16	240
	481	5, 0← 4, 0	Ground		5/2		3/2	74 986.14	.16	240
	481	5, 0← 4, 0	Ground		11/2		9/2	75 004.28	.16	240
	481	5, 0← 4, 0	Ground		15/2		13/2	75 019.28	.16	240
	481	5, 0← 4, 0	Ground		13/2		11/2	75 027.58	.16	240
	481	5, 1← 4, 1	Ground		9/2		7/2	74 976.22	.16	240
	481	5, 1← 4, 1	Ground		7/2		5/2	74 977.62	.16	240
	481	5, 1← 4, 1	Ground		11/2		9/2	74 993.28	.16	240
	481	5, 1← 4, 1	Ground		13/2		11/2	75 016.20	.16	240
	481	5, 1← 4, 1	Ground		15/2		13/2	75 026.20	.16	240
	481	5, 2← 4, 2	Ground		11/2		9/2	74 960.76	.16	240

Isotopic Species	Id. No.	Rotational Quantum Nos.	Vib. State	Hyperfine F_1'	F'	F_1	F	Frequency MHz	Acc. $\pm$MHz	Ref.
$C^{12}H_3I^{127}$	481	5, 2← 4, 2	Ground		9/2		7/2	74 971.76	.16	240
	481	5, 2← 4, 2	Ground		13/2		11/2	74 982.18	.16	240
	481	5, 2← 4, 2	Ground		7/2		5/2	75 007.62	.16	240
	481	5, 2← 4, 2	Ground		15/2		13/2	75 046.48	.16	240
	481	5, 3← 4, 3	Ground		13/2		11/2	74 926.04	.16	240
	481	5, 3← 4, 3	Ground		9/2		7/2	74 964.36	.16	240
	481	5, 3← 4, 3	Ground		15/2		13/2	75 081.02	.16	240
	481	5, 4← 4, 4	Ground		11/2		9/2	74 829.54	.16	240
	481	5, 4← 4, 4	Ground		13/2		11/2	74 849.92	.16	240
	481	5, 5← 5, 5	Ground		13/2		15/2	503.05	.15	638
	481	6, 6← 6, 6	Ground		15/2		17/2	517.3	.4	638
	481	8, 0← 7, 0	Ground		17/2		17/2	119 919.48		568
	481	8, 0← 7, 0	Ground		13/2		11/2	119 994.36		568
	481	8, 0← 7, 0	Ground		15/2		13/2	119 998.90		568
	481	8, 0← 7, 0	Ground		11/2		9/2	120 000.34		568
	481	8, 0← 7, 0	Ground		17/2		15/2	120 008.43		568
	481	8, 0← 7, 0	Ground		21/2		19/2	120 012.42		568
	481	8, 0← 7, 0	Ground		19/2		17/2	120 015.95		568
	481	8, 1← 7, 1	Ground		17/2		17/2	119 920.02		568
	481	8, 1← 7, 1	Ground		13/2		11/2	119 993.97		568
	481	8, 1← 7, 1	Ground		15/2		13/2	119 996.05		568
	481	8, 1← 7, 1	Ground		11/2		9/2	120 002.44		568
	481	8, 1← 7, 1	Ground		17/2		15/2	120 004.25		568
	481	8, 1← 7, 1	Ground		19/2		17/2	120 012.42		568
	481	8, 1← 7, 1	Ground		21/2		19/2	120 012.99		568
	481	8, 1← 7, 1	Ground		15/2		15/2	120 070.89		568
	481	8, 2← 7, 2	Ground		17/2		17/2	119 921.72		568
	481	8, 2← 7, 2	Ground		15/2		13/2	119 987.78		568
	481	8, 2← 7, 2	Ground		17/2		15/2	119 991.88		568
	481	8, 2← 7, 2	Ground		13/2		11/2	119 992.78		568
	481	8, 2← 7, 2	Ground		19/2		17/2	120 002.44		568
	481	8, 2← 7, 2	Ground		11/2		9/2	120 009.33		568
	481	8, 2← 7, 2	Ground		21/2		19/2	120 014.89		568
	481	8, 3← 7, 3	Ground		17/2		15/2	119 971.10		568
	481	8, 3← 7, 3	Ground		15/2		13/2	119 973.48		568
	481	8, 3← 7, 3	Ground		19/2		17/2	119 985.10		568
	481	8, 3← 7, 3	Ground		13/2		11/2	119 990.79		568
	481	8, 3← 7, 3	Ground		21/2		19/2	120 017.88		568
	481	8, 3← 7, 3	Ground		11/2		9/2	120 020.67		568
	481	8, 4← 7, 4	Ground		17/2		15/2	119 941.94		568
	481	8, 4← 7, 4	Ground		15/2		13/2	119 953.94		568
	481	8, 4← 7, 4	Ground		19/2		17/2	119 961.23		568
	481	8, 4← 7, 4	Ground		13/2		11/2	119 987.78		568
	481	8, 4← 7, 4	Ground		13/2		13/2	120 014.89		568
	481	8, 4← 7, 4	Ground		21/2		19/2	120 022.18		568
	481	8, 4← 7, 4	Ground		11/2		9/2	120 036.54		568
	481	8, 5← 7, 5	Ground		17/2		15/2	119 904.58		568
	481	8, 5← 7, 5	Ground		15/2		13/2	119 928.43		568
	481	8, 5← 7, 5	Ground		19/2		17/2	119 930.63		568
	481	8, 5← 7, 5	Ground		13/2		11/2	119 984.47		568

Isotopic Species	Id. No.	Rotational Quantum Nos.	Vib. State	Hyperfine				Frequency MHz	Acc. ±MHz	Ref.
				F_1'	F'	F_1	F			
$C^{12}H_3I^{127}$	481	8, 5← 7, 5	Ground		21/2		19/2	120 027.60		568
	481	8, 5← 7, 5	Ground		11/2		9/2	120 057.13		568
	481	8, 6← 7, 6	Ground		17/2		15/2	119 858.79		568
	481	8, 6← 7, 6	Ground		19/2		17/2	119 893.31		568
	481	8, 6← 7, 6	Ground		15/2		13/2	119 897.38		568
	481	8, 6← 7, 6	Ground		17/2		17/2	119 941.94		568
	481	8, 6← 7, 6	Ground		13/2		11/2	119 980.18		568
	481	8, 6← 7, 6	Ground		21/2		19/2	120 034.26		568
	481	8, 6← 7, 6	Ground		11/2		9/2	120 082.48		568
$C^{12}D_3I^{127}$	482	2, 0← 1, 0	Ground		7/2		7/2	23 831.16	.08	395
	482	2, 0← 1, 0	Ground		3/2		3/2	24 029.16	.08	395
	482	2, 0← 1, 0	Ground		9/2		7/2	24 203.19	.08	395
	482	2, 0← 1, 0	Ground		7/2		5/2	24 235.67	.08	395
	482	2, 0← 1, 0	Ground		1/2		3/2	24 277.13	.08	395
	482	2, 0← 1, 0	Ground		5/2		5/2	24 336.26	.08	395
	482	2, 0← 1, 0	Ground		3/2		5/2	24 608.61	.08	395
	482	2, 1← 1, 1	Ground		7/2		5/2	23 893.51	.08	395
	482	2, 1← 1, 1	Ground		5/2		5/2	23 940.01	.08	395
	482	2, 1← 1, 1	Ground		3/2		5/2	24 081.25	.08	395
	482	2, 1← 1, 1	Ground		7/2		7/2	24 097.49	.08	395
	482	2, 1← 1, 1	Ground		5/2		3/2	24 232.16	.08	395
	482	2, 1← 1, 1	Ground		9/2		7/2	24 279.89	.08	395
	482	2, 1← 1, 1	Ground		3/2		3/2	24 373.36	.08	395
	482	2, 1← 1, 1	Ground		1/2		3/2	24 493.36	.08	395
$C^{13}H_3I^{127}$	483	2, 0← 1, 0	Ground		5/2		3/2	28 069.99	.08	83
	483	2, 0← 1, 0	Ground		7/2		7/2	28 145.01	.08	83
	483	2, 0← 1, 0	Ground		3/2		3/2	28 343.64	.08	83
	483	2, 0← 1, 0	Ground		9/2		7/2	28 518.14	.08	83
	483	2, 0← 1, 0	Ground		7/2		5/2	28 550.86	.08	83
	483	2, 0← 1, 0	Ground		5/2		5/2	28 650.91	.08	83
	483	2, 1← 1, 1	Ground		7/2		5/2	28 206.90	.08	83
	483	2, 1← 1, 1	Ground		5/2		5/2	28 253.84	.08	83
	483	2, 1← 1, 1	Ground		7/2		7/2	28 411.19	.08	83
	483	2, 1← 1, 1	Ground		9/2		7/2	28 594.74	.08	83
	483	2, 1← 1, 1	Ground		3/2		3/2	28 687.21	.08	83
$C^{12}HD_2I^{127}$	484	2, 0, ← 1, 1,	Ground		9/2		7/2	26 022.2		376
	484	2, 1, ← 1, 0,	Ground		9/2		7/2	25 826.9		376

CH$_3$NSSi C$_{3v}$ SiH$_3$NCS˙

Isotopic Species	Pt. Gp.	Id. No.	A MHz	B MHz	C MHz	D$_J$ MHz	D$_{JK}$ MHz	Δ Amu A²	κ
Si^{28}H$_3$N^{14}C^{12}S^{32}	C$_{3v}$	491		1516.018 M	1516.018 M	<.0003	.0419		
Si^{28}D$_3$N^{14}C^{12}S^{32}	C$_{3v}$	492		1412.403 M	1412.403 M		.0314		
Si^{30}D$_3$N^{14}C^{12}S^{32}	C$_{3v}$	493		1377.047 M	1377.047 M				
Si^{29}H$_3$N^{14}C^{12}S^{32}	C$_{3v}$	494		1493.389 M	1493.389 M				
Si^{30}H$_3$N^{14}C^{12}S^{32}	C$_{3v}$	495		1471.902 M	1471.902 M				
Si^{28}H$_3$N^{14}C^{12}S^{34}	C$_{3v}$	496		1473.39 M	1473.39 M				
Si^{28}H$_2$DN^{14}C^{12}S^{32}	C$_s$	497		1483.326 M	1474.844 M		.0405		
Si^{28}HD$_2$N^{14}C^{12}S^{32}	C$_s$	498		1448.947 M	1440.283 M		.0370		

References:

ABC: 968   D$_J$: 968   D$_{JK}$: 968

For species 491, excited state: D$_{JK}$ = 0.0451 MHz, A$_v$ = 69,200 MHz, B$_v$ = 1526.28 MHz, and ζ = .99. Ref. 968.

Isotopic Species	Id. No.	Rotational Quantum Nos.	Vib. State v_a^l	F_1'	F'	F_1	F	Frequency MHz	Acc. ±MHz	Ref.
$Si^{28}H_3N^{14}C^{12}S^{32}$	491	6, ← 5,	Ground					18 192.17		968 [2]
	491	7, ← 6,	Ground					21 224.56		968 [2]
	491	8, 1← 7, 1	Ground					24 255.75	.1	968 [1]
	491	8, 1← 7, 1	1,−1					24 399.00		968
	491	8, 1← 7, 1	1,∓1					24 418.65		968
	491	8, 1← 7, 1	1,+1					24 444.32		968
	491	8, 2← 7, 2	Ground					24 253.81	.1	968 [1]
	491	8, 2← 7, 2	1,∓1					24 414.46		968
	491	8, 2← 7, 2	1,±1					24 420.40		968
	491	8, 2← 7, 2	1,∓1					24 422.33		968
	491	8, 3← 7, 3	Ground					24 250.15	.1	968 [1]
	491	8, 3← 7, 3	1,∓1					24 409.80		968
	491	8, 3← 7, 3	1,±1					24 419.31		968
	491	8, 4← 7, 4	Ground					24 245.53	.1	968 [1]
	491	8, 4← 7, 4	1,∓1					24 403.55		968
	491	8, 4← 7, 4	1,±1					24 416.85		968
	491	8, 5← 7, 5	Ground					24 239.39	.1	968 [1]
	491	8, 5← 7, 5	1,∓1					24 396.19		968
	491	8, 5← 7, 5	1,±1					24 412.92		968
	491	8, 6← 7, 6	Ground					24 232.05	.1	968 [1]
	491	8, 6← 7, 6	1,±1					24 406.83		968
	491	8, 7← 7, 7	Ground					24 223.56	.1	968 [1]
	491	8, 7← 7, 7	1,±1					24 401.33		968
	491	10, ← 9,	Ground					30 320.35		968 [2]
$Si^{28}D_3N^{14}C^{12}S^{32}$	492	8, ← 7,	Ground					22 598.75		968 [2]
	492	9, ← 8,	Ground					25 423.26		968 [2]
$Si^{30}D_3N^{14}C^{12}S^{32}$	493	9, ← 8,	Ground					24 786.84		968 [2]
$Si^{29}H_3N^{14}C^{12}S^{32}$	494	8, ← 7,	Ground					23 894.22		968 [2]
$Si^{30}H_3N^{14}C^{12}S^{32}$	495	8, ← 7,	Ground					23 550.43		968 [2]
$Si^{28}H_3N^{14}C^{12}S^{34}$	496	8, ← 7,	Ground					23 573.42		968 [2]
$Si^{28}H_2DN^{14}C^{12}S^{32}$	497	8, 1, ← 7, 1,	Ground					23 630.62	.1	968 [1]
	497	8, 1, ← 7, 1,	Ground					23 698.48	.1	968 [1]
	497	8, 2, ← 7, 2,	Ground					23 662.43	.1	968 [1]
	497	8, 3, ← 7, 3,	Ground					23 659.41	.1	968 [1]
	497	8, 4, ← 7, 4,	Ground					23 655.15	.1	968 [1]
	497	8, 5, ← 7, 5,	Ground					23 649.25	.1	968 [1]
	497	8, 6, ← 7, 6,	Ground					23 642.29	.1	968 [1]
	497	8, 7, ← 7, 7,	Ground					23 633.38	.1	968 [1]
$Si^{28}HD_2N^{14}C^{12}S^{32}$	498	8, 1, ← 7, 1,	Ground					23 078.48	.1	968 [1]
	498	8, 1, ← 7, 1,	Ground					23 147.76	.1	968 [1]
	498	8, 2, ← 7, 2,	Ground					23 111.59	.1	968 [1]
	498	8, 3, ← 7, 3,	Ground					23 108.25	.1	968 [1]
	498	8, 4, ← 7, 4,	Ground					23 104.24	.1	968 [1]
	498	8, 5, ← 7, 5,	Ground					23 099.36	.1	968 [1]
	498	8, 6, ← 7, 6,	Ground					23 092.61	.1	968 [1]
	498	8, 7, ← 7, 7,	Ground					23 084.74	.1	968 [1]

1. A typographical error in Table 1 of Jenkins, Kewley, and Sugden's article has been corrected to read $J = 8 \leftarrow 7$.
2. These lines for the transition origin (ν for $K = 0$) were stated to have been found graphically.

CH₃NSi C_{3v} H₃SiCN

Isotopic Species	Pt. Gp.	Id. No.	A MHz	B MHz	C MHz	D_J MHz	D_{JK} MHz	Δ Amu A²	κ
$H_3Si^{28}C^{12}N^{14}$	C_{3v}	501		4 972.7 M	4 972.7 M				
$D_3Si^{28}C^{12}N^{14}$	C_{3v}	505		4 535.0 M	4 535.0 M				

Id. No.	μ_a Debye	μ_b Debye	μ_c Debye	eQq Value(MHz) Rel.		eQq Value(MHz) Rel.		eQq Value(MHz) Rel.		ω_a 1/cm d	ω_b 1/cm d	ω_c 1/cm d	ω_d 1/cm d
·501				−4.7	N¹⁴								

References:

ABC: 916   eQq: 925

No Spectral Lines

Iodine Cyanide

CIN $C_{\infty v}$ ICN

Isotopic Species	Pt. Gp.	Id. No.	A MHz	B MHz	C MHz	D_J MHz	D_{JK} MHz	Δ Amu A²	κ
$I^{127}C^{12}N^{14}$	$C_{\infty v}$	511		3 225.578 M	3 225.578 M	.00088			
$I^{127}C^{13}N^{14}$	$C_{\infty v}$	512		3 177.69 M	3 177.69 M				

Id. No.	μ_a Debye	μ_b Debye	μ_c Debye	eQq Value(MHz) Rel.		eQq Value(MHz) Rel.		eQq Value(MHz) Rel.		ω_a 1/cm d	ω_b 1/cm d	ω_c 1/cm d	ω_d 1/cm d
511	3.71 M	0. X	0. X	−3.80	N¹⁴	−2420.	I¹²⁷			470 1	321 2	2158 1	

References:

ABC: 106,112,240,774 D_J: 240   μ: 112   eQq: 104,774   ω: 1028

Add. Ref. 42,79,104,105,361

For species 511 in excited states:

	$v_1 = 1$	$v_2 = 1$	$v_2 = 2$
eQq (I¹²⁷) in MHz	−2425.9	−2410.85	−2303.2
Ref.	774	615	774

Also given for species 511 are the quantities: $q_1 = 2.643$ MHz, Ref. 615; $\alpha_1 = 9.33$ MHz, Ref. 112; $\alpha_2 = -9.50$ MHz, Ref. 112.

Isotopic Species	Id. No.	Rotational Quantum Nos.	Vib. State v_a; v_b; v_c	Hyperfine				Frequency MHz	Acc. ±MHz	Ref.
				F′	F′	F₁	F			
$I^{127}C^{12}N^{14}$	511	1← 0	1; 0, 0; 0		5/2		5/2	6 051.36	.5	774
	511	1← 0	Ground		5/2		5/2	6 070.66	.5	774
	511	1← 0	Ground		5/2		5/2	6 071.61	.5	774
	511	1← 0	0; 2, 0; 0		5/2		5/2	6 106.71	.5	774
	511	1← 0	1; 0, 0; 0		7/2		5/2	6 559.63	.5	774
	511	1← 0	Ground		7/2		5/2	6 577.06	.5	774
	511	1← 0	Ground		7/2		5/2	6 577.95	.5	774
	511	1← 0	0; 2, 0; 0		7/2		5/2	6 610.24	.5	774
	511	.1← 0	1; 0, 0; 0		3/2		5/2	6 782.16	.5	774
	511	1← 0	Ground		3/2		5/2	6 799.79	.5	774
	511	1← 0	0; 2, 0; 0		3/2		5/2	6 830.47	.5	774
	511	2← 1	Ground		5/2		3/2	12 400.35	.5	774
	511	2← 1	Ground		5/2		3/2	12 401.33	.5	774
	511	2← 1	1; 0, 0; 0		7/2		7/2	12 451.36	.5	774
	511	2← 1	Ground		7/2		7/2	12 489.27	.5	774
	511	2← 1	Ground		7/2		7/2	12 489.88	.5	774
	511	2← 1	0; 2, 0; 0		7/2		7/2	12 558.38	.5	774
	511	2← 1	Ground		5/2		7/2	12 622.19	.5	774
	511	2← 1	1; 0, 0; 0		3/2		3/2	12 699.69	.5	774
	511	2← 1	Ground		3/2		3/2	12 737.22	.5	774
	511	2← 1	Ground		9/2		7/2	12 956.38	.5	774
	511	2← 1	1; 0, 0; 0		7/2		5/2	12 959.17	.5	774
	511	2← 1	Ground		7/2		5/2	12 994.98	.5	774
	511	2← 1	Ground		7/2		5/2	12 996.50	.5	774
	511	2← 1	Ground		7/2		5/2	12 996.93	.5	774
	511	2← 1	1; 0, 0; 0		1/2		3/2	13 011.34	.5	774
	511	2← 1	0; 2, 0; 0		9/2		7/2	13 021.92	.5	774
	511	2← 1	Ground		1/2		3/2	13 048.33	.5	774
	511	2← 1	Ground		5/2		5/2	13 129.36	.5	774
	511	2← 1	Ground		5/2		5/2	13 129.68	.5	774
	511	2← 1	0; 2, 0; 0		5/2		5/2	13 193.32	.5	774
	511	2← 1	1; 0, 0; 0		3/2		5/2	13 429.80	.5	774
	511	2← 1	Ground		3/2		5/2	13 465.77	.5	774
	511	2← 1	0; 2, 0; 0		3/2		5/2	13 527.88	.5	774
	511	4← 3	Ground	11/2	9/2	11/2	9/2	25 393.517		615
	511	4← 3	Ground	11/2	13/2	11/2	13/2	25 393.517		615
	511	4← 3	Ground	11/2	11/2	11/2	11/2	25 393.776		615
	511	4← 3	0; 1, −1; 0	11/2	13/2	11/2	13/2	25 542.856		615
	511	4← 3	0; 1, −1; 0	11/2	9/2	11/2	9/2	25 542.856		615
	511	4← 3	0; 1, −1; 0	11/2	11/2	11/2	11/2	25 542.856		615
	511	4← 3	0; 1, +1; 0	11/2	11/2	11/2	11/2	25 567.571		615
	511	4← 3	0; 1, +1; 0	11/2	13/2	11/2	13/2	25 567.571		615
	511	4← 3	0; 1, +1; 0	11/2	9/2	11/2	9/2	25 567.571		615
	511	4← 3	Ground	5/2		3/2		25 711.50	.1	112
	511	4← 3	Ground	7/2		5/2		25 728.77	.1	112
	511	4← 3	1; 0, 0; 0	13/2		11/2		25 748.18	.1	112
	511	4← 3	Ground	3/2		1/2		25 752.65	.1	112
	511	4← 3	1; 0, 0; 0	11/2		9/2		25 763.23	.1	112
	511	4← 3	Ground	9/2		7/2		25 783.50	.1	112
	511	4← 3	Ground	9/2		9/2		25 789.85	.1	112

Isotopic Species	Id. No.	Rotational Quantum Nos.	Vib. State v_a ; v_b ; v_c	F'_1	F'	F_1	F	Frequency MHz	Acc. ±MHz	Ref.
$I^{127}C^{12}N^{14}$	511	4← 3	0; 1,−1; 0	7/2		5/2		25 802.92	.1	112
	511	4← 3	0; 1,−1; 0	5/2		3/2		25 815.34	.1	112
	511	4← 3	Ground	13/2		11/2		25 823.08	.1	112
	511	4← 3	0; 1,−1; 0	9/2		7/2		25 829.31	.1	112
	511	4← 3	Ground	11/2		9/2		25 837.64	.1	112
	511	4← 3	0; 1,+1; 0	9/2		7/2		25 850.78	.1	112
	511	4← 3	0; 1,−1; 0	11/2		9/2		25 872.24	.1	112
	511	4← 3	0; 1,+1; 0	11/2		9/2		25 893.73	.1	112
	511	4← 3	0; 1,−1; 0	13/2		11/2		25 906.28	.1	112
	511	4← 3	0; 1,+1; 0	13/2		11/2		25 927.66	.1	112
	511	4← 3	Ground	7/2		7/2		25 954.36	.1	112
	511	4← 3	Ground	3/2		3/2		25 969.58	.1	112
	511	4← 3	0; 2, 0; 0	13/2			11/2	25 979.72		618
	511	4← 3	Ground	5/2		5/2		25 991.92	.1	112
	511	4← 3	0; 2, 2; 0	13/2		11/2		26 046.32	.1	112
	511	4← 3	0; 1,−1; 0			5/2	7/2	26 196.540		615
	511	4← 3	0; 1,−1; 0			7/2	9/2	26 196.931		615
	511	4← 3	0; 1,−1; 0			5/2	7/2	26 197.588		615
	511	4← 3	0; 1,+1; 0			5/2	7/2	26 216.380		615
	511	4← 3	0; 1,+1; 0			7/2	9/2	26 216.771		615
	511	4← 3	Ground	5/2	11/2	7/2	11/2	26 217.022		615
	511	4← 3	0; 1,+1; 0			5/2	7/2	26 217.428		615
	511	4← 3	0; 1,−1; 0	3/2	5/2	5/2	7/2	26 247.900		615
	511	4← 3	0; 1,−1; 0	3/2	5/2	5/2	5/2	26 247.900		615
	511	4← 3	0; 1,−1; 0	3/2	3/2	5/2	5/2	26 248.238		615
	511	4← 3	0; 1,−1; 0	3/2	3/2	5/2	3/2	26 248.238		615
	511	4← 3	0; 1,−1; 0	3/2	1/2	5/2	3/2	26 248.300		615
	511	4← 3	Ground	3/2	3/2	5/2	5/2	26 248.971		615
	511	4← 3	0; 1,+1; 0	3/2	5/2	5/2	7/2	26 265.210		615
	511	4← 3	0; 1,+1; 0	3/2	5/2	5/2	5/2	26 265.210		615
	511	4← 3	0; 1,+1; 0	3/2	3/2	5/2	3/2	26 265.548		615
	511	4← 3	0; 1,+1; 0	3/2	3/2	5/2	5/2	26 265.548		615
	511	4← 3	0; 1,+1; 0	3/2	1/2	5/2	3/2	26 265.610		615
	511	5← 4	Ground			13/2	13/2	31 848.77		106
	511	5← 4	Ground			11/2	11/2	32 200.58		106
	511	5← 4	Ground			7/2	5/2	32 203.57		106
	511	5← 4	Ground			9/2	7/2	32 215.56		106
	511	5← 4	Ground			5/2	3/2	32 226.85		106
	511	5← 4	Ground			11/2	9/2	32 248.52		106
	511	5← 4	Ground			15/2	13/2	32 268.33		106
	511	5← 4	Ground			13/2	11/2	32 278.55		106
	511	5← 4	Ground			9/2	9/2	32 386.29		106
	511	11←10	Ground			19/2	17/2	70 949.66	.18	240
	511	11←10	Ground			23/2	21/2	70 959.14	.18	240
	511	11←10	Ground			27/2	25/2	70 961.30	.18	240
	511	11←10	Ground			25/2	23/2	70 963.90	.18	240
	511	12←11	Ground					77 413.		165
	511	13←12	Ground					83 864.		165
	511	20←19	Ground					129 000.		196
$I^{127}C^{13}N^{14}$	512	1← 0	Ground			5/2	5/2	5 974.43	.5	774
	512	1← 0	Ground			7/2	5/2	6 480.72	.5	774
	512	1← 0	Ground			3/2	5/2	6 703.25	.5	774
	512	5← 4	Ground			7/2	5/2	31 718.28		106
	512	5← 4	Ground			9/2	7/2	31 730.50		106
	512	5← 4	Ground			5/2	3/2	31 741.50		106
	512	5← 4	Ground			11/2	9/2	31 763.34		106
	512	5← 4	Ground			15/2	13/2	31 783.31		106
	512	5← 4	Ground			13/2	11/2	31 793.46		106

COS

$C_{\infty v}$

OCS

Isotopic Species	Pt. Gp.	Id. No.	A MHz	B MHz	C MHz	D_J MHz	D_{JK} MHz	Δ Amu A²	κ
$O^{16}C^{12}S^{32}$	$C_{\infty v}$	521		6 081.480 M	6 081.480 M	.00131			
$O^{16}C^{12}S^{33}$	$C_{\infty v}$	522		6 004.918 M	6 004.918 M				
$O^{16}C^{12}S^{34}$	$C_{\infty v}$	523		5 932.843 M	5 932.843 M				
$O^{16}C^{13}S^{32}$	$C_{\infty v}$	526		6 061.923 M	6 061.923 M				
$O^{16}C^{13}S^{34}$	$C_{\infty v}$	527		5 911.730 M	5 911.730 M				
$O^{16}C^{14}S^{32}$	$C_{\infty v}$	528		6 043.25 M	6 043.25 M				
$O^{17}C^{12}S^{32}$	$C_{\infty v}$	529							
$O^{18}C^{12}S^{32}$	$C_{\infty v}$	531		5 704.825 M	5 704.825 M				

Id. No.	μ_a Debye	μ_b Debye	μ_c Debye	eQq Value(MHz)	Rel.	eQq Value(MHz)	Rel.	eQq Value(MHz)	Rel.	ω_a 1/cm	d	ω_b 1/cm	d	ω_c 1/cm	d	ω_d 1/cm	d
521 522 529	.7124 M	0. X	0. X	−29.07 −1.32	S^{33} O^{17}					859	1	522	2	2050	1		

References:

ABC: 112 D_J: 533 μ: 766   eQq: 344,349 ω: 1028

Add. Ref. 15, 43, 55, 61, 86, 97, 147, 163, 188, 193, 202, 223, 238, 269, 270, 272, 281, 312, 347, 362, 366, 367, 432, 492, 521, 579, 636, 686, 672, 715, 721, 854, 855, 887, 1017, 1021, 1026

For species 521, $v_2 = 1$, $\mu = .700$ D, Ref: 238; $q_1 = 6.344$ MHz, $\alpha_1 = 20.56$ MHz, $\alpha_2 = 10.563$ MHz, $\alpha_3 = 52.6$ MHz.

For this molecule Fermi resonance interactions occur between various vibrationally excited states. Values given here for α_i are those which will allow accurate prediction of the rotational frequencies in the lowest excited vibrational state according to the usual formula; see Ref. 402.

Experimental B values for excited vibrational states are given in the same reference.

For species 529, ref. 349 gives eQq $(O^{17}) = -1.32$ MHz.

Isotopic Species	Id. No.	Rotational Quantum Nos.	Vib. State v_a ; v_b^l ; v_c	F_1'	F'	F_1	F	Frequency MHz	Acc. ±MHz	Ref.
$O^{16}C^{12}S^{32}$	521	1← 0	Ground					12 162.97		894
	521	2← 1	2; 0, 0; 0					24 179.62		618
	521	2← 1	0; 0, 0; 1					24 180.47	.2	1008
	521	2← 1	1; 0, 0; 0					24 253.51		618
	521	2← 1	1; 1,−1; 0					24 289.97		618
	521	2← 1	1; 1,+1; 0					24 316.76		618
	521	2← 1	Ground					24 325.921	.002	331
	521	2← 1	0; 1,−1; 0					24 355.50		112
	521	2← 1	0; 1,+1; 0					24 381.07		112
	521	2← 1	0; 2, 0; 0					24 401.0		618
	521	2← 1	0; 3,−1; 0					24 411.	2.	618
	521	2← 1	0; 3,+1; 0					24 459.	2.	618
	521	3← 2	Ground					36 488.82	.03	170
	521	3← 2	0; 1,−1; 0					36 532.47		618
	521	3← 2	0; 1,+1; 0					36 570.83		618
	521	3← 2	0; 2, 0; 0					36 600.81		618
	521	3← 2	0; 2,±2; 0					36 615.3		618
	521	4← 3	1; 0, 0; 0					48 506.24	.10	402
	521	4← 3	Ground					48 651.40	.10	402
	521	4← 3	0; 1,−1; 0					48 710.80	.10	402
	521	4← 3	0; 1,+2; 0					48 761.55	.10	402
	521	4← 3	0; 2, 0; 0					48 801.08	.10	402
	521	4← 3	0; 2,±2; 0					48 819.92	.10	402
	521	5← 4	Ground					60 814.08	.05	170
	521	6← 5	Ground					72 976.80		578
	521	8← 7	Ground					97 301.19	.20	445
	521	10← 9	Ground					121 624.63	.25	445
	521	12←11	Ground					145 946.79	.30	445
	521	14←13	Ground					170 267.49	.35	445
	521	16←15	Ground					194 586.44	.40	533
	521	18←17	Ground					218 903.27	.45	533
	521	20←19	Ground					243 218.09	.50	533
	521	22←21	Ground					267 529.56	.55	533
	521	24←23	Ground					291 839.22	.60	533
	521	26←25	Ground					316 144.7	1.0	504
	521	28←27	Ground					340 449.2	1.0	504
	521	30←29	Ground					364 747.5	1.5	504
	521	32←31	Ground					389 041.	2.0	504
	521	36←35	Ground					486 184.2		1007
	521	42←41	Ground					510 457.3		1007
$O^{16}C^{12}S^{33}$	522	2← 1	1; 0, 0; 0		7/2		5/2	23 947.4		1008
	522	2← 1	1; 0, 0; 0		5/2		3/2	23 947.4		1008
	522	2← 1	1; 0, 0; 0		1/2		1/2	23 947.4		1008
	522	2← 1	Ground		3/2		1/2	24 012.33	.02	344
	522	2← 1	Ground		5/2		5/2	24 012.94	.02	344
	522	2← 1	Ground		3/2		5/2	24 018.13	.02	344
	522	2← 1	Ground		1/2		1/2	24 019.59	.02	344
	522	2← 1	Ground		5/2		3/2	24 020.23	.02	344
	522	2← 1	Ground		7/2		5/2	24 020.23	.02	344
	522	2← 1	Ground		3/2		3/2	24 025.42	.02	344

Isotopic Species	Id. No.	Rotational Quantum Nos.	Vib. State v_a ; v_b^l ; v_c	F_1'	F'	F_1	F	Frequency MHz	Acc. ±MHz	Ref.
$O^{16}C^{12}S^{33}$	522	2← 1	Ground		1/2		3/2	24 032.68	.02	344
	522	2← 1	0; 1, −1; 0		5/2		3/2	24 044.0		1008
	522	2← 1	0; 1, −1; 0		5/2		5/2	24 046.9		1008
	522	2← 1	0; 1, −1; 0		3/2		3/2	24 046.9		1008
	522	2← 1	0; 1, −1; 0		7/2		5/2	24 051.2		1008
	522	2← 1	0; 1, +1; 0		5/2		3/2	24 069.2		1008
	522	2← 1	0; 1, +1; 0		5/2		5/2	24 072.0		1008
	522	2← 1	0; 1, +1; 0		3/2		3/2	24 072.0		1008
	522	2← 1	0; 1, +1; 0		7/2		5/2	24 075.7		1008
	522	2← 1	0; 2, 0; 0		1/2		1/2	24 092.4		1008
	522	2← 1	0; 2, 0; 0		7/2		5/2	24 092.4		1008
	522	2← 1	0; 2, 0; 0		5/2		3/2	24 092.4		1008
	522	2← 1	Ground		5/2		3/2	48 038.19	.10	402
	522	4← 3	Ground		7/2		5/2	48 038.19	.10	402
	522	4← 3	Ground		11/2		9/2	48 039.13	.10	402
	522	4← 3	Ground		9/2		7/2	48 039.13	.10	402
$O^{16}C^{12}S^{34}$	523	2← 1	Ground					23 661.		146
	523	2← 1	Ground					23 731.299	.003	331
	523	4← 3	Ground					47 462.40	.05	170
	523	8← 7	Ground					97 301.31	.20	282
	523	10← 9	Ground					121 624.79	.25	282
$O^{16}C^{12}S^{35}$	524	2← 1	Ground		3/2		3/2	23 453.323	.011	331
	524	2← 1	Ground		7/2		5/2	23 456.963	.011	331
	524	2← 1	Ground		5/2		3/2	23 456.963	.011	331
	524	2← 1	Ground		5/2		5/2	23 462.343	.011	331
	524	2← 1	Ground		3/2		1/2	23 462.343	.011	331
$O^{16}C^{12}S^{36}$	525	2← 1	Ground					23 198.66		146
$O^{16}C^{13}S^{32}$	526	2← 1	1; 0, 0; 0					24 176.07		1020
	526	2← 1	Ground					24 247.82	.03	170
	526	2← 1	0; 1, ±1; 0					24 274.84	.03	170
	526	2← 1	0; 1, ±1; 0					24 300.58	.03	170
	526	4← 3	Ground					48 494.76	.10	402
$O^{16}C^{13}S^{34}$	527	2← 1	Ground					23 646.92		112
$O^{16}C^{14}S^{32}$	528	2← 1	Ground					24 173.0	1.0	96
	528	2← 1	0; 1, ±1; 0					24 197.0	1.0	170
	528	2← 1	0; 1, ±1; 0					24 224.0	1.0	170
$O^{17}C^{12}S^{32}$	529	2← 1	Ground		5/2		3/2	23 534.101	.014	349
	529	2← 1	Ground		7/2		7/2	23 534.164	.012	349
	529	2← 1	Ground		1/2		3/2	23 534.308	.012	349
	529	2← 1	Ground		9/2		7/2	23 534.422		349
	529	2← 1	Ground		7/2		5/2	23 534.422		349
	529	2← 1	Ground		3/2		3/2	23 534.481	.014	349
	529	2← 1	Ground		5/2		5/2	23 534.481	.014	349
$O^{18}C^{12}S^{32}$	531	2← 1	1; 0, 0; 0					22 754.6	.2	119
	531	2← 1	Ground					22 819.30		112
	531	2← 1	0; 1, ±1; 0					22 848.70	.1	119
	531	2← 1	0; 1, ±1; 0					22 871.28	.1	119
$O^{18}C^{12}S^{34}$	532	2← 1	Ground					22 239.6	.2	119
$O^{18}C^{13}S^{32}$	533	2← 1	Ground					22 763.8	.2	119

COSe $\qquad$ $C_{\infty v}$ $\qquad$ OCSe

Isotopic Species	Pt. Gp.	Id. No.	A MHz	B MHz	C MHz	D_J MHz	D_{JK} MHz	Δ Amu A^2	κ
O^{16}C^{12}Se^{74}	$C_{\infty v}$	541		4 095.786 M	4 095.786 M				
O^{16}C^{12}Se^{75}	$C_{\infty v}$	542		4 081.926 M	4 081.926 M				
O^{16}C^{12}Se^{76}	$C_{\infty v}$	543		4 068.438 M	4 068.438 M	.00068			
O^{16}C^{12}Se^{77}	$C_{\infty v}$	544		4 055.241 M	4 055.241 M	.00068			
O^{16}C^{12}Se^{78}	$C_{\infty v}$	545		4 042.413 M	4 042.413 M	.00068			
O^{16}C^{12}Se^{79}	$C_{\infty v}$	546							
O^{16}C^{12}Se^{80}	$C_{\infty v}$	547		4 017.649 M	4 017.649 M	.00067			
O^{16}C^{12}Se^{82}	$C_{\infty v}$	548		3 994.064 M	3 994.064 M	.00066			
O^{16}C^{13}Se^{78}	$C_{\infty v}$	549		4 005.112 M	4 005.112 M				
O^{16}C^{13}Se^{80}	$C_{\infty v}$	551		3 980.045 M	3 980.045 M				

Id. No.	μ_a Debye	μ_b Debye	μ_c Debye	eQq Value(MHz)	Rel.	eQq Value(MHz)	Rel.	eQq Value(MHz)	Rel.	ω_a 1/cm	d	ω_b 1/cm	d	ω_c 1/cm	d	ω_d 1/cm	d
542				946.0													
545	.754 M	0. X	0. X														
546				752.09													
547	.754 M	0. X	0. X							642	1	466	2				

References:

ABC: 169,575,663 $\qquad$ D$_J$: 663 $\qquad$ μ: 169 $\qquad$ eQq: 436,575 $\qquad$ ω: 618

Add. Ref. 86, 115, 195, 217, 355, 491, 1030

For species 547, the dipole moments for excited states are given in ref. 169: for (1, 0, 0), $\mu = .728$D.; for (0, 1, 0), $\mu = .730$ D.

For the molecule in general, ref. 217 gives the "Fermi resonance" interaction energy $W_{ni} = 45.7$ cm^{-1} and (unperturbed) $\alpha_1 = 14.01$ MHz, $\alpha_2 = -6.88$ MHz.

Isotopic Species	Id. No.	Rotational Quantum Nos.	Vib. State v_a ; v_b^l	Hyperfine F_1'	F'	F_1	F	Frequency MHz	Acc. ±MHz	Ref.
$O^{16}C^{12}Se^{74}$	541	3← 2	Ground					24 514.67	.03	169
	541	3← 2	Ground					24 574.86		194
$O^{16}C^{12}Se^{75}$	542	3← 2	Ground		5/2		5/2	24 429.58	.05	575
	542	3← 2	Ground		3/2		3/2	24 455.21	.05	575
	542	3← 2	Ground		9/2		7/2	24 471.31	.05	575
	542	3← 2	Ground		11/2		9/2	24 480.45	.05	575
	542	3← 2	Ground		7/2		5/2	24 517.93	.05	575
	542	3← 2	Ground		5/2		3/2	24 565.87	.03	575
$O^{16}C^{12}Se^{76}$	543	3← 2	Ground					24 410.58		194
	543	3← 2	0; 1,−1					24 442.98	.03	169
	543	3← 2	0; 1,+1					24 462.42	.03	169
	543	12←11	Ground					97 637.78	.20	663
	543	15←14	Ground					122 043.90	.25	663
	543	18←17	Ground					146 447.90	.30	663
	543	21←20	Ground					170 849.06	.35	663
	543	24←23	Ground					195 247.17	.40	663
	543	27←26	Ground					219 641.79	.45	663
	543	30←29	Ground					244 032.33	.50	663
$O^{16}C^{12}Se^{77}$	544	3← 2	1; 0, 0					24 250.84	.03	169
	544	3← 2	Ground					24 331.38		194
	544	3← 2	0; 1,−1'					24 363.97	.03	169
	544	3← 2	0; 1,+1					24 383.21	.03	169
	544	12←11	Ground					97 321.07	.20	663
	544	15←14	Ground					121 647.98	.25	663
	544	18←17	Ground					145 972.74	.30	663
	544	21←20	Ground					170 294.80	.35	663
	544	24←23	Ground					194 613.75	.40	663
	544	27←26	Ground					218 929.21	.45	663
$O^{16}C^{12}Se^{78}$	545	3← 2	1; 0, 0					24 174.30	.03	169
	545	3← 2	Ground					24 254.43		194
	545	3← 2	0; 1,−1					24 286.82	.03	169
	545	3← 2	0; 1,+1					24 305.95	.03	169
	545	6← 5	Ground					48 508.88	.03	169
	545	7← 6	Ground					56 593.16	.03	169
	545	12←11	Ground					97 013.24	.20	663
	545	15←14	Ground					121 263.28	.25	663
	545	18←17	Ground					145 511.08	.30	663
	545	21←20	Ground					169 756.27	.35	663
	545	24←23	Ground					193 998.34	.40	663
	545	27←26	Ground					218 236.97	.45	663
	545	30←29	Ground					242 471.47	.50	663
	545	33←32	Ground					266 701.93	.55	663
$O^{16}C^{12}Se^{79}$	546	3← 2	Ground		7/2		7/2	24 153.204		436
	546	3← 2	Ground		1/2		3/2	24 158.9		354
	546	3← 2	Ground		11/2		9/2	24 158.9		354
	546	3← 2	Ground		13/2		11/2	24 170.194		436
	546	3← 2	Ground		9/2		7/2	24 190.787		436
	546	3← 2	Ground		3/2		3/2	24 204.692		436

Isotopic Species	Id. No.	Rotational Quantum Nos.	Vib. State v_a ; v_b^l	Hyperfine				Frequency MHz	Acc. ±MHz	Ref.
				F_1'	F'	F_1	F			
$O^{16}C^{12}Se^{79}$	546	$3 \leftarrow 2$	Ground		7/2		5/2	24 234.329		436
$O^{16}C^{12}Se^{80}$	547	$3 \leftarrow 2$	1; 0, 0					24 026.39	.03	169
	547	$3 \leftarrow 2$	Ground					24 105.85		194
	547	$3 \leftarrow 2$	0; 1,−1					24 138.05	.03	169
	547	$3 \leftarrow 2$	0; 1,+1					24 156.93	.03	169
	547	$3 \leftarrow 2$	0; 2, 0					24 183.97		618
	547	$3 \leftarrow 2$	0; 2,±2					24 188.18		618
	547	$6 \leftarrow 5$	Ground					48 211.46	.03	169
	547	$7 \leftarrow 6$	Ground					56 246.47	.03	169
	547	$12 \leftarrow 11$	Ground					96 418.95	.20	663
	547	$12 \leftarrow 11$	0; 1,−1					96 546.60		663
	547	$12 \leftarrow 11$	0; 1,+1					96 622.76		663
	547	$15 \leftarrow 14$	Ground					120 520.40	.25	663
	547	$15 \leftarrow 14$	0; 1,−1					120 679.98		663
	547	$15 \leftarrow 14$	0; 1,+1					120 775.11		663
	547	$18 \leftarrow 17$	Ground					144 619.81	.30	663
	547	$18 \leftarrow 17$	0; 1,−1					144 811.06		663
	547	$18 \leftarrow 17$	0; 1,+1					144 925.26		663
	547	$21 \leftarrow 20$	Ground					168 716.41	.35	663
	547	$21 \leftarrow 20$	0; 1,−1					168 939.60		663
	547	$21 \leftarrow 20$	0; 1,+1					169 072.81		663
	547	$24 \leftarrow 23$	Ground					192 810.17	.40	663
	547	$24 \leftarrow 23$	0; 1,−1					193 065.08		663
	547	$24 \leftarrow 23$	0; 1,±1					193 217.26		663
	547	$27 \leftarrow 26$	Ground					216 900.38	.45	663
	547	$27 \leftarrow 26$	0; 1,−1					217 186.90		663
	547	$27 \leftarrow 26$	0; 1,+1					217 358.18		663
	547	$30 \leftarrow 29$	Ground					240 986.62	.50	663
	547	$33 \leftarrow 32$	Ground					265 068.60	.55	663
	547	$36 \leftarrow 35$	Ground					289 145.50	.60	663
	547	$39 \leftarrow 38$	Ground					313 217.57	.65	663
$O^{16}C^{12}Se^{82}$	548	$3 \leftarrow 2$	1; 0, 0					23 885.76	.03	169
	548	$3 \leftarrow 2$	Ground					23 964.33		194
	548	$3 \leftarrow 2$	0; 1,−1					23 996.26	.03	169
	548	$3 \leftarrow 2$	0; 1,+1					24 014.97	.03	169
	548	$7 \leftarrow 6$	Ground					55 916.19	.03	169
	548	$12 \leftarrow 11$	Ground					95 852.94	.20	663
	548	$15 \leftarrow 14$	Ground					119 812.89	.25	663
	548	$18 \leftarrow 17$	Ground					143 770.82	.30	663
	548	$21 \leftarrow 20$	Ground					167 726.08	.35	663
	548	$24 \leftarrow 23$	Ground					191 678.34	.40	663
	548	$27 \leftarrow 26$	Ground					215 627.18	.45	663
	548	$30 \leftarrow 29$	Ground					239 572.08	.50	663
	548	$33 \leftarrow 32$	Ground					263 512.90	.55	663
$O^{16}C^{13}Se^{78}$	549	$3 \leftarrow 2$	Ground					24 030.58	.03	169
$O^{16}C^{13}Se^{80}$	551	$3 \leftarrow 2$	Ground					23 880.18	.03	169

CSSe
$C_{\infty V}$
SCSe

Isotopic Species	Pt. Gp.	Id. No.	A MHz	B MHz	C MHz	D_J MHz	D_{JK} MHz	Δ Amu A²	κ
$S^{32}C^{12}Se^{82}$	$C_{\infty V}$	561		2 001.56 M	2 001.56 M				
$S^{32}C^{12}Se^{80}$	$C_{\infty V}$	562		2 016.74 M	2 016.74 M				
$S^{32}C^{12}Se^{78}$	$C_{\infty V}$	563		2 031.16 M	2 031.16 M				
$S^{32}C^{12}Se^{77}$	$C_{\infty V}$	564		2 042.16 M	2 042.16 M				
$S^{32}C^{12}Se^{76}$	$C_{\infty V}$	565		2 049.95 M	2 049.95 M				

References:

ABC: 181

Thiocarbonyl Selenide

Spectral Line Table

Isotopic Species	Id. No.	Rotational Quantum Nos.	Vib. State v_a^l	F_1'	F'	F_1	F	Frequency MHz	Acc. ±MHz	Ref.
$S^{32}C^{12}Se^{82}$	561	6 ← 5	Ground					24 021.	4.	181
	561	6 ← 5	1,±1					24 048.	4.	181
	561	6 ← 5	1,±1					24 075.	4.	181
$S^{32}C^{12}Se^{80}$	562	6 ← 5	Ground					24 203.	4.	181
	562	6 ← 5	1,±1					24 214.	4.	181
	562	6 ← 5	1,±1					24 230.	4.	181
$S^{32}C^{12}Se^{78}$	563	6 ← 5	Ground					24 376.	4.	181
	563	6 ← 5	1,±1					24 386.	4.	181
	563	6 ← 5	1,±1					24 406.	4.	181
$S^{32}C^{12}Se^{77}$	564	6 ← 5	Ground					24 508.	4.	181
	564	6 ← 5	1,±1					24 521.	4.	181
	564	6 ← 5	1,±1					24 527.	4.	181
$S^{32}C^{12}Se^{76}$	565	6 ← 5	Ground					24 602.	4.	181
	565	6 ← 5	1,±1					24 614.	4.	181
	565	6 ← 5	1,±1					24 627.	4.	181

CSTe $C_{\infty V}$ SCTe

Isotopic Species	Pt. Gp.	Id. No.	A MHz	B MHz	C MHz	D_J MHz	D_{JK} MHz	Δ Amu A²	κ
$S^{32}C^{12}Te^{122}$	$C_{\infty V}$	571		1 584.122 M	1 584.122 M				
$S^{32}C^{12}Te^{123}$	$C_{\infty V}$	572		1 580.926 M	1 580.926 M				
$S^{32}C^{12}Te^{124}$	$C_{\infty V}$	573		1 577.790 M	1 577.790 M				
$S^{32}C^{12}Te^{125}$	$C_{\infty V}$	574		1 574.692 M	1 574.692 M				
$S^{32}C^{12}Te^{126}$	$C_{\infty V}$	575		1 571.652 M	1 571.652 M				
$S^{32}C^{12}Te^{128}$	$C_{\infty V}$	576		1 565.702 M	1 565.702 M				
$S^{32}C^{12}Te^{130}$	$C_{\infty V}$	577		1 559.930 M	1 559.930 M				

Id. No.	μ_a Debye	μ_b Debye	μ_c Debye	eQq Value(MHz) Rel.	eQq Value(MHz) Rel.	eQq Value(MHz) Rel.	ω_a d 1/cm	ω_b d 1/cm	ω_c d 1/cm	ω_d d 1/cm
571	.172 M	0. X	0. X							

References:

ABC: 526 μ: 526

Add. Ref. 393

The following parameters for the various species have been reported in ref. 526:

Species No.	B_0 (MHz)	α_2 (MHz)	q_l (MHz)
571	1584.1224	3.2870	.6786
572	1580.9261	3.2818	.6776
573	1577.7898	3.2764	.6752
574	1574.6925	3.2712	.6728
575	1571.6524	3.2657	.6706
576	1565.7022	3.2551	.6649
577	1559.9303	3.2446	.6599

Thiocarbonyl Telluride   Spectral Line Table

Isotopic Species	Id. No.	Rotational Quantum Nos.	Vib. State v_a^l	F_1'	F'	F_1	F	Frequency MHz	Acc. ±MHz	Ref.
$S^{32}C^{12}Te^{122}$	571	8← 7	1,−1					25 392.929	.010	526
	571	8← 7	1,+1					25 403.788	.010	526
$S^{32}C^{12}Te^{123}$	572	8← 7	1,−1					25 341.714	.010	526
	572	8← 7	1,+1					25 352.555	.010	526
$S^{32}C^{12}Te^{124}$	573	8← 7	1,−1					25 291.465	.010	526
	573	8← 7	1,+1					25 302.268	.010	526
$S^{32}C^{12}Te^{125}$	574	8← 7	1,−1					25 241.844	.010	526
	574	8← 7	1,+1					25 252.608	.010	526
$S^{32}C^{12}Te^{126}$	575	8← 7	1,−1					25 193.132	.010	526
	575	8← 7	1,+1					25 203.861	.010	526
$S^{32}C^{12}Te^{128}$	576	8← 7	1,−1					25 097.805	.010	526
	576	8← 7	1,+1					25 108.444	.010	526
$S^{32}C^{12}Te^{130}$	577	8← 7	1,−1					25 005.326	.010	526
	577	8← 7	1,+1					25 015.884	.010	526

C₂Cl₃N   C₃ᵥ   CCl₃CN

Isotopic Species	Pt. Gp.	Id. No.	A MHz		B MHz		C MHz		D$_J$ MHz	D$_{JK}$ MHz	Δ Amu A²	κ
C¹²Cl₃³⁵C¹²N¹⁴	C₃ᵥ	581			1667.3	M	1667.3	M				
C¹²Cl³⁵Cl₂³⁷C¹²N¹⁴	Cₛ	583	1690.2	M	1659.9	M	1634.0	M				
C¹²Cl₃³⁷C¹²N¹⁴	C₃ᵥ	584			1613.8	M	1613.8	M				

Id. No.	μ$_a$ Debye		μ$_b$ Debye		μ$_c$ Debye		eQq Value(MHz)	Rel.	eQq Value(MHz)	Rel.	eQq Value(MHz)	Rel.	ω$_a$ 1/cm	d	ω$_b$ 1/cm	d	ω$_c$ 1/cm	d	ω$_d$ 1/cm	d
581	1.93	L	0.	X	0.	X							163	1						

References:

ABC: 738,950   μ: 995   ω: 950

Add. Ref. 735,802

Trichloroacetonitrile   Spectral Line Table

Isotopic Species	Id. No.	Rotational Quantum Nos.	Vib. State v$_a$	F$_I'$	F'	F$_I$	F	Frequency MHz	Acc. ±MHz	Ref.
C¹²Cl₃³⁵C¹²N¹⁴	581	5, ← 4,	Ground					16 667.		738
	581	6, ← 5,	1					20 017.1	1.0	950
	581	6, ← 5,	2					20 028.9	1.0	950
	581	7, ← 6,	1					23 353.2	1.0	950
	581	8, ← 7,	1					26 691.9	1.0	950
	581	8, ← 7,	2					26 705.4	1.0	950
	581	9, ← 8,	Ground					30 009.2	1.0	950
	581	9, ← 8,	1					30 027.4	1.0	950
	581	9, ← 8,	2					30 035.0	1.0	950
	581	10, ← 9,	Ground					33 346.5	1.0	950
	581	10, ← 9,	1					33 363.7	1.0	950
	581	11, ←10,	Ground					36 681.9	1.0	950
	581	11, ←10,	1					36 701.7	1.0	950
	581	12, ←11,	Ground					40 050.0	20.	950
C¹²Cl₂³⁵Cl³⁷C¹²N¹⁴	582	7, , ← 6, ,	Ground					23 111.9	1.0	950
	582	7, , ← 6, ,	1					23 127.3	1.0	950
	582	7, 0, 7← 6, 0, 6	Ground					22 914.		738
	582	7, 1, 6← 6, 1, 5	Ground					22 992.		738
	582	7, 1, 7← 6, 1, 6	Ground					22 914.		738
	582	7, 2, 6← 6, 2, 5	Ground					22 992.		738
	582	7, 3, 4← 6, 3, 3	Ground					23 138.		738
	582	7, 3, 5← 6, 3, 4	Ground					23 055.		738
	582	7, 4, 3← 6, 4, 2	Ground					23 138.		738
	582	8, , ← 7, ,	Ground					26 394.8	1.0	950
	582	8, , ← 7, ,	1					26 426.0	1.0	950
	582	9, , ← 8, ,	Ground					29 695.1	1.0	950
	582	10, , ← 9, ,	Ground					32 999.9	1.0	950
	582	11, , ←10, ,	Ground					36 301.8	1.0	950
	582	11, , ←10, ,	1					36 335.2	1.0	950
	582	12, , ←11, ,	Ground					39 600.0	20.	950
C¹²Cl³⁵Cl₂³⁷C¹²N¹⁴	583	8, , ← 7, ,	Ground					26 115.9	1.0	950
	583	8, , ← 7, ,	1					26 142.3	1.0	950
	583	9, , ← 8, ,	Ground					29 400.0	20.	950
	583	10, , ← 9, ,	Ground					32 650.0	20.	950
	583	11, , ←10, ,	Ground					35 900.0	20.	950
	583	12, , ←11, ,	Ground					39 150.0	20.	950
C¹²Cl₃³⁷C¹²N¹⁴	584	8, ← 7,	Ground					25 818.5	1.0	950
	584	9, ← 8,	Ground					29 050.0	20.	950
	584	10, ← 9,	Ground					32 279.5	1.0	950
	584	11, ←10,	Ground					35 499.4	1.0	950

Trifluoroethanenitrile, Trifluoromethyl Cyanide

C_2F_3N C_{3v} CF_3CN

Isotopic Species	Pt. Gp.	Id. No.	A MHz	B MHz	C MHz	D_J MHz	D_{JK} MHz	Δ Amu A²	κ
$C^{12}F_3^{19}C^{12}N^{14}$	C_{3v}	591		2 945.528 M	2 945.528 M	.00031	.00581		
$C^{13}F_3^{19}C^{12}N^{14}$	C_{3v}	592		2 944.23 M	2 944.23 M				
$C^{12}F_3^{19}C^{13}N^{14}$	C_{3v}	593		2 921.86 M	2 921.86 M				
$C^{12}F_3^{19}C^{12}N^{15}$	C_{3v}	594		2 855.859 M	2 855.859 M	.0004	.0056		

References:

ABC: 391,745,1013 D_J: 391,745 D_{JK}: 391,745

Add. Ref. 232

For molecule 590, first vibrational state, $B_v = 2950.52$ MHz, $\alpha = -4.98$ MHz, and $q_l = 3.60$ MHz. Ref. 1013.

Isotopic Species	Id. No.	Rotational Quantum Nos.	Vib. State	Hyperfine F_1'	F'	F_1	F	Frequency MHz	Acc. ±MHz	Ref.
$C^{12}F_3^{19}C^{12}N^{14}$	591	6, 0← 5, 0	Ground		6		5	35 346.03		391
	591	6, 0← 5, 0	Ground		7		6	35 346.03		391
	591	6, 0← 5, 0	Ground		5		4	35 346.03		391
	591	6, 1← 5, 1	Ground		6		5	35 346.03		391
	591	6, 1← 5, 1	Ground		5		4	35 346.03		391
	591	6, 1← 5, 1	Ground		7		6	35 346.03		391
	591	6, 2← 5, 2	Ground		6		5	35 345.60		391
	591	6, 2← 5, 2	Ground		6		5	35 345.90		391
	591	6, 2← 5, 2	Ground		5		4	35 345.90		391
	591	6, 3← 5, 3	Ground		6		5	35 345.15		391
	591	6, 3← 5, 3	Ground		6		5	35 345.60		391
	591	6, 3← 5, 3	Ground		7		6	35 345.60		391
	591	6, 4← 5, 4	Ground		6		5	35 344.44		391
	591	6, 4← 5, 4	Ground		5		4	35 345.15		391
	591	6, 4← 5, 4	Ground		7		6	35 345.15		391
	591	6, 5← 5, 5	Ground		6		5	35 343.50		391
	591	6, 5← 5, 5	Ground		7		6	35 344.65		391
	591	6, 5← 5, 5	Ground		5		4	35 344.91		391
	591	8, 0← 7, 0	Ground		9		8	47 127.74		391
	591	8, 0← 7, 0	Ground		8		7	47 127.74		391
	591	8, 0← 7, 0	Ground		7		6	47 127.74		391
	591	8, 1← 7, 1	Ground		9		8	47 127.74		391
	591	8, 1← 7, 1	Ground		8		7	47 127.74		391
	591	8, 1← 7, 1	Ground		7		6	47 127.74		391
	591	8, 2← 7, 2	Ground		7		6	47 127.44		391
	591	8, 2← 7, 2	Ground		9		8	47 127.44		391
	591	8, 2← 7, 2	Ground		8		7	47 127.44		391
	591	8, 3← 7, 3	Ground		8		7	47 126.83		391
	591	8, 3← 7, 3	Ground		7		6	47 127.02		391
	591	8, 3← 7, 3	Ground		9		8	47 127.02		391
	591	8, 4← 7, 4	Ground		8		7	47 126.07		391
	591	8, 4← 7, 4	Ground		9		8	47 126.46		391
	591	8, 4← 7, 4	Ground		7		6	47 126.46		391
	591	8, 5← 7, 5	Ground		7		6	47 125.62		391
	591	8, 5← 7, 5	Ground		9		8	47 125.62		391
	591	8, 6← 7, 6	Ground		8		7	47 123.88		391
	591	8, 6← 7, 6	Ground		9		8	47 124.58		391
	591	8, 6← 7, 6	Ground		7		6	47 124.68		391
	591	8, 7← 7, 7	Ground		8		7	47 122.48		391
	591	8, 7← 7, 7	Ground		9		8	47 123.48		391
	591	8, 7← 7, 7	Ground		7		6	47 123.60		391
	591	16, 0←15, 0	Ground					94 251.93	.20	745
	591	16, 1←15, 1	Ground					94 251.78	.20	745
	591	16, 2←15, 2	Ground					94 251.20	.20	745
	591	16, 3←15, 3	Ground					94 250.24	.20	745
	591	16, 4←15, 4	Ground					94 248.88	.20	745
	591	16, 5←15, 5	Ground					94 247.19	.20	745
	591	16, 6←15, 6	Ground					94 245.17	.20	745
	591	16, 7←15, 7	Ground					94 242.76	.20	745
	591	16, 8←15, 8	Ground					94 239.97	.20	745

Isotopic Species	Id. No.	Rotational Quantum Nos.	Vib. State	Hyperfine				Frequency MHz	Acc. ±MHz	Ref.
				F_1'	F'	F_1	F			
$C^{12}F_3^{19}C^{12}N^{14}$	591	16, 9←15, 9	Ground					94 236.82	.20	745
	591	16,10←16,10	Ground					94 233.31	.20	745
	591	16,11←16,11	Ground					94 229.43	.20	745
	591	16,12←16,12	Ground					94 225.12	.20	745
	591	25, 0←24, 0	Ground					147 257.28	.30	745
	591	25, 1←24, 1	Ground					147 257.03	.30	745
	591	25, 2←24, 2	Ground					147 256.03	.30	745
	591	25, 3←24, 3	Ground					147 254.55	.30	745
	591	25, 4←24, 4	Ground					147 252.50	.30	745
	591	25, 5←24, 5	Ground					147 249.92	.30	745
	591	25, 6←24, 6	Ground					147 246.77	.30	745
	591	25, 8←24, 8	Ground					147 238.71	.30	745
	591	25, 9←24, 9	Ground					147 233.76	.30	745
	591	25,10←24,10	Ground					147 228.26	.30	745
	591	25,11←24,11	Ground					147 222.19	.30	745
	591	25,12←24,12	Ground					147 215.48	.30	745
	591	25,13←24,13	Ground					147 208.22	.30	745
	591	25,14←24,14	Ground					147 200.36	.30	745
	591	25,15←24,15	Ground					147 191.93	.30	745
	591	25,17←24,17	Ground					147 173.28	.30	745
	591	25,18←24,18	Ground					147 163.10	.30	745
	591	25,19←24,19	Ground					147 152.20	.30	745
	591	25,20←24,20	Ground					147 140.97	.30	745
	591	25,21←24,21	Ground					147 129.08	.30	745
	591	25,22←24,22	Ground					147 116.55	.30	745
	591	25,23←24,23	Ground					147 103.56	.30	745
	591	25,24←24,24	Ground					147 089.91	.30	745
	591	33, 0←32, 0	Ground					194 360.70	.40	745
	591	33, 2←32, 2	Ground					194 359.19	.40	745
	591	33, 3←32, 3	Ground					194 357.22	.40	745
	591	33, 4←32, 4	Ground					194 354.51	.40	745
	591	33, 5←32, 5	Ground					194 351.15	.40	745
	591	33, 6←32, 6	Ground					194 346.88	.40	745
	591	33, 7←32, 7	Ground					194 342.10	.40	745
	591	33, 8←32, 8	Ground					194 336.36	.40	745
	591	33, 9←32, 9	Ground					194 329.69	.40	745
	591	33,10←32,10	Ground					194 322.46	.40	745
	591	33,11←32,11	Ground					194 314.42	.40	745
	591	33,12←32,12	Ground					194 305.68	.40	745
$C^{12}F_3^{19}C^{12}N^{15}$	594	6, 0← 5, 0	Ground					34 269.81		391
	594	6, 1← 5, 1	Ground					34 269.81		391
	594	6, 2← 5, 2	Ground					34 269.81		391
	594	6, 3← 5, 3	Ground					34 269.30		391
	594	6, 4← 5, 4	Ground					34 268.83		391
	594	6, 5← 5, 5	Ground					34 268.24		391
	594	8, 0← 7, 0	Ground					45 692.94		391
	594	8, 1← 7, 1	Ground					45 692.94		391
	594	8, 2← 7, 2	Ground					45 692.74		391
	594	8, 3← 7, 3	Ground					45 692.16		391
	594	8, 4← 7, 4	Ground					45 691.56		391
	594	8, 6← 7, 6	Ground					45 690.72		391
	594	8, 7← 7, 7	Ground					45 689.82		391
	594	8, 8← 7, 8	Ground					45 688.54		391

C₂HCl — C_2HCl $C_{\infty v}$ HC : CCl

Isotopic Species	Pt. Gp.	Id. No.	A MHz	B MHz	C MHz	D_J MHz	D_JK MHz	Δ Amu A²	κ
$HC^{12}:C^{12}Cl^{35}$	$C_{\infty v}$	601		5 684.24 M	5 684.24 M				
$HC^{12}:C^{12}Cl^{37}$	$C_{\infty v}$	602		5 572.38 M	5 572.38 M				
$DC^{12}:C^{12}Cl^{35}$	$C_{\infty v}$	603		5 187.01 M	5 187.01 M				
$DC^{12}:C^{12}Cl^{37}$	$C_{\infty v}$	604		5 084.24 M	5 084.24 M				

Id. No.	μ_a Debye	μ_b Debye	μ_c Debye	eQq Value(MHz) Rel.	eQq Value(MHz) Rel.	eQq Value(MHz) Rel.	ω_a d 1/cm	ω_b d 1/cm	ω_c d 1/cm	ω_d d 1/cm
601	.44 M	0. X	0. X	−79.67 Cl^{35}						
602				−62.75 Cl^{37}						
603				−79.66 Cl^{35}						
604				−63.12 Cl^{37}						

References:

ABC: 175 μ: 175   eQq: 175

Add. Ref. 176,571

For species 603 and 604, ref. 645 gives eQq(D) = .175 MHz.

Chloroacetylene   Spectral Line Table

Isotopic Species	Id. No.	Rotational Quantum Nos.	Vib. State	F_1'	F'	F_1	F	Frequency MHz	Acc. ±MHz	Ref.
$HC^{12}:C^{12}Cl^{35}$	601	2← 1	Ground		3/2		1/2	22 717.07	.1	175
	601	2← 1	Ground		5/2		5/2	22 718.80	.1	175
	601	2← 1	Ground		3/2		5/2	22 732.90	.1	175
	601	2← 1	Ground		1/2		1/2	22 737.00	.1	175
	601	2← 1	Ground		7/2		5/2	22 738.68	.1	175
	601	2← 1	Ground		5/2		3/2	22 738.68	.1	175
	601	2← 1	Ground		3/2		3/2	22 752.95	.1	175
	601	2← 1	Ground		1/2		3/2	22 772.82	.1	175
$HC^{12}:C^{12}Cl^{37}$	602	2← 1	Ground		3/2		1/2	22 273.90	.1	175
	602	2← 1	Ground		5/2		5/2	22 275.10	.1	175
	602	2← 1	Ground		1/2		1/2	22 289.55	.1	175
	602	2← 1	Ground		5/2		3/2	22 290.85	.1	175
	602	2← 1	Ground		7/2		5/2	22 290.85	.1	175
	602	2← 1	Ground		3/2		3/2	22 302.10	.1	175
$DC^{12}:C^{12}Cl^{35}$	603	2← 1	Ground		3/2		1/2	20 728.03	.1	175
	603	2← 1	Ground		5/2		5/2	20 729.79	.1	175
	603	2← 1	Ground		3/2		5/2	20 744.00	.1	175
	603	2← 1	Ground		1/2		1/2	20 748.02	.1	175
	603	2← 1	Ground		7/2		5/2	20 749.76	.1	175
	603	2← 1	Ground		5/2		3/2	20 749.76	.1	175
	603	2← 1	Ground		3/2		3/2	20 763.96	.1	175
	603	2← 1	Ground		1/2		3/2	20 783.80	.1	175
$DC^{12}:C^{12}Cl^{37}$	604	2← 1	Ground		3/2		1/2	20 321.12	.1	175
	604	2← 1	Ground		5/2		5/2	20 322.50	.1	175
	604	2← 1	Ground		1/2		1/2	20 336.84	.1	175
	604	2← 1	Ground		7/2		5/2	20 338.29	.1	175
	604	2← 1	Ground		5/2		3/2	20 338.29	.1	175
	604	2← 1	Ground		3/2		3/2	20 349.48	.1	175

1,1-Difluorovinyl Chloride, Difluorochloroethylene

C_2HClF_2 C_s $CHCl:CF_2$

Isotopic Species	Pt. Gp.	Id. No.	A MHz		B MHz		C MHz		D_J MHz	D_{JK} MHz	Δ Amu A^2	κ
$C^{12}HCl^{35}:C^{12}F_2^{19}$	C_s	611	10 710.4	M	2 296.6	M	1 890.2	M			.16	−.907725
$C^{12}HCl^{37}:C^{12}F_2^{19}$	C_s	612	10 710.8	M	2 232.8	M	1 846.6	M			.16	−.912859

Id. No.	μ_a Debye	μ_b Debye	μ_c Debye	eQq Value(MHz)	Rel.	eQq Value(MHz)	Rel.	eQq Value(MHz)	Rel.	ω_a 1/cm	d	ω_b 1/cm	d	ω_c 1/cm	d	ω_d 1/cm	d
611				−51.7	aa	18.2	bb	33.5	cc								
612				−51.7	aa	18.2	bb	33.5	cc								

References:

ABC: 865 Δ: 865 κ: 865   eQq: 865

For species 611 and 612, ref. 865 gives: eQq(zz) = −84.3 MHz, eQq(xx) = 50.9 MHz, eQq(yy) = 33.5 MHz.

1,1-Difluoro-2-Chloroethene   Spectral Line Table

Isotopic Species	Id. No.	Rotational Quantum Nos.	Vib. State	F_1'	F'	F_1	F	Frequency MHz	Acc. ±MHz	Ref.
$C^{12}HCl^{35}:C^{12}F_2^{19}$	611	2, 2, 0 ← 2, 1, 1	Ground		7/2		7/2	25 249.3	.3	865
	611	2, 2, 0 ← 2, 1, 1	Ground		3/2		3/2	25 254.3	.5	865
	611	2, 2, 0 ← 2, 1, 1	Ground		3/2		5/2	25 257.2	.5	865
	611	2, 2, 0 ← 2, 1, 1	Ground		5/2		3/2	25 263.7	.5	865
	611	2, 2, 0 ← 2, 1, 1	Ground		5/2		5/2	25 266.9	.5	865
	611	3, 1, 3 ← 2, 0, 2	Ground		5/2		3/2	19 961.4	.5	865
	611	3, 1, 3 ← 2, 0, 2	Ground		9/2		7/2	19 964.0	.5	865
	611	3, 1, 3 ← 2, 0, 2	Ground		7/2		5/2	19 965.3	.5	865
	611	3, 1, 3 ← 2, 0, 2	Ground		5/2		5/2	19 970.6	.5	865
	611	3, 2, 1 ← 3, 1, 2	Ground		3/2		3/2	24 704.2	.5	865
	611	3, 2, 1 ← 3, 1, 2	Ground		9/2		9/2	24 708.2	.3	865
	611	3, 2, 1 ← 3, 1, 2	Ground		5/2		5/2	24 712.3	.5	865
	611	3, 2, 1 ← 3, 1, 2	Ground		7/2		7/2	24 716.3	.5	865
	611	4, 1, 4 ← 3, 0, 3	Ground		7/2		5/2	23 365.2	.1	865
	611	4, 1, 4 ← 3, 0, 3	Ground		5/2		3/2	23 365.2	.1	865
	611	4, 1, 4 ← 3, 0, 3	Ground		11/2		9/2	23 366.8	.1	865
	611	4, 1, 4 ← 3, 0, 3	Ground		9/2		7/2	23 366.8	.1	865
	611	4, 2, 2 ← 4, 1, 3	Ground		5/2		5/2	24 055.5	.3	865
	611	4, 2, 2 ← 4, 1, 3	Ground		11/2		11/2	24 057.1	.3	865
	611	4, 2, 2 ← 4, 1, 3	Ground		7/2		7/2	24 059.8	.3	865
	611	4, 2, 2 ← 4, 1, 3	Ground		9/2		9/2	24 061.3	.3	865
	611	5, 2, 3 ← 5, 1, 4	Ground		7/2		7/2	23 357.1	.1	865
	611	5, 2, 3 ← 5, 1, 4	Ground		9/2		9/2	23 360.5	.1	865
	611	6, 2, 4 ← 6, 1, 5	Ground		15/2		15/2	22 680.8	.1	865
	611	6, 2, 4 ← 6, 1, 5	Ground		9/2		9/2	22 680.8	.1	865
	611	6, 2, 4 ← 6, 1, 5	Ground		13/2		13/2	22 682.6	.1	865
	611	6, 2, 4 ← 6, 1, 5	Ground		11/2		11/2	22 682.6	.1	865
	611	7, 0, 7 ← 6, 1, 6	Ground					23 434.2	.1	865
	611	7, 2, 5 ← 7, 1, 6	Ground		17/2		17/2	22 107.3	.1	865
	611	7, 2, 5 ← 7, 1, 6	Ground		11/2		11/2	22 107.3	.1	865

Isotopic Species	Id. No.	Rotational Quantum Nos.	Vib. State	Hyperfine				Frequency MHz	Acc. ±MHz	Ref.
				F_1'	F'	F_1	F			
$C^{12}HCl^{35}{:}C^{12}F_2^{19}$	611	7, 2, 5← 7, 1, 6	Ground		13/2		13/2	22 108.6	.1	865
	611	7, 2, 5← 7, 1, 6	Ground		15/2		15/2	22 108.6	.1	865
	611	8, 2, 6← 8, 1, 7	Ground		17/2		17/2	21 715.0	.1	865
	611	8, 2, 6← 8, 1, 7	Ground		15/2		15/2	21 715.0	.1	865
	611	8, 2, 6← 8, 1, 7	Ground		19/2		19/2	21 715.9	.1	865
	611	8, 2, 6← 8, 1, 7	Ground		13/2		13/2	21 715.9	.1	865
	611	9, 1, 8← 8, 2, 7	Ground					20 128.1	.1	865
	611	9, 2, 7← 9, 1, 8	Ground		15/2		15/2	21 576.1	.1	865
	611	9, 2, 7← 9, 1, 8	Ground		21/2		21/2	21 576.1	.1	865
	611	9, 2, 7← 9, 1, 8	Ground		17/2		17/2	21 577.0	.1	865
	611	9, 2, 7← 9, 1, 8	Ground		19/2		19/2	21 577.0	.1	865
	611	10, 2, 8←10, 1, 9	Ground		21/2		21/2	21 754.4	.1	865
	611	10, 2, 8←10, 1, 9	Ground		19/2		19/2	21 754.4	.1	865
	611	10, 2, 8←10, 1, 9	Ground		23/2		23/2	21 755.5	.1	865
	611	10, 2, 8←10, 1, 9	Ground		17/2		17/2	21 755.5	.1	865
	611	11, 2, 9←11, 1,10	Ground		21/2		21/2	22 304.1	.1	865
	611	11, 2, 9←11, 1,10	Ground		23/2		23/2	22 304.1	.1	865
	611	11, 2, 9←11, 1,10	Ground		25/2		25/2	22 305.3	.1	865
	611	11, 2, 9←11, 1,10	Ground		19/2		19/2	22 305.3	.1	865
	611	12, 2,10←12, 1,11	Ground		25/2		25/2	23 270.9	.1	865
	611	12, 2,10←12, 1,11	Ground		23/2		23/2	23 270.9	.1	865
	611	12, 2,10←12, 1,11	Ground		27/2		27/2	23 272.4	.1	865
	611	12, 2,10←12, 1,11	Ground		21/2		21/2	23 272.4	.1	865
	611	13, 2,11←12, 3,10	Ground					21 450.	.1	865
	611	13, 2,11←13, 1,12	Ground					24 691.6	.1	865
	611	13, 2,11←13, 1,12	Ground					24 693.2	.1	865
$C^{12}HCl^{37}{:}C^{12}F_2^{19}$	612	6, 2, 4← 6, 1, 5	Ground					22 948.4	.3	865
	612	6, 2, 4← 6, 1, 5	Ground					22 949.6	.3	865
	612	7, 0, 7← 6, 1, 6	Ground					22 562.0	.1	865
	612	7, 2, 5← 7, 1, 6	Ground					22 361.6	.3	865
	612	7, 2, 5← 7, 1, 6	Ground					22 362.7	.3	865
	612	8, 2, 6← 8, 1, 7	Ground					21 931.0	.1	865
	612	8, 2, 6← 8, 1, 7	Ground					21 932.0	.1	865
	612	9, 2, 7← 9, 1, 8	Ground					21 725.5	.1	865
	612	9, 2, 7← 9, 1, 8	Ground					21 726.3	.1	865
	612	10, 2, 8←10, 1, 9	Ground					21 805.4	.1	865
	612	11, 2, 9←11, 1,10	Ground					22 222.5	.1	865
	612	11, 2, 9←11, 1,10	Ground					22 223.5	.1	865
	612	12, 2,10←12, 1,11	Ground					23 022.5	.1	865
	612	12, 2,10←12, 1,11	Ground					23 023.5	.1	865
	612	13, 2,11←13, 1,12	Ground					24 241.7	.1	865
	612	13, 2,11←13, 1,12	Ground					24 242.8	.1	865

C₂HF $\quad\quad$ C_∞ᵥ $\quad\quad$ HC:CF

Isotopic Species	Pt. Gp.	Id. No.	A MHz	B MHz	C MHz	D_J MHz	D_JK MHz	Δ Amu A²	κ
HC¹²:C¹²F¹⁹	C_∞ᵥ	621		9 706.22 M					
HC¹²:C¹³F¹⁹	C_∞ᵥ	622		9 700.71 M					
HC¹³:C¹²F¹⁹	C_∞ᵥ	623		9 373.95 M					
DC¹²:C¹²F¹⁹	C_∞ᵥ	624		8 736.09 M					
DC¹²:C¹³F¹⁹	C_∞ᵥ	625		8 733.94 M					
DC¹³:C¹²F¹⁹	C_∞ᵥ	626		8 486.33 M					

Id. No.	μ_a Debye	μ_b Debye	μ_c Debye	eQq Value(MHz) Rel.	eQq Value(MHz) Rel.	eQq Value(MHz) Rel.	ω_a d 1/cm	ω_b d 1/cm	ω_c d 1/cm	ω_d d 1/cm
621	.75 M	0. X	0. X							

References:

ABC: 931 $\quad$ μ: 931

No Spectral Lines

C₂H₂ClF $\quad\quad$ C_s $\quad\quad$ CH₂:CFCl

Isotopic Species	Pt. Gp.	Id. No.	A MHz	B MHz	C MHz	D_J MHz	D_JK MHz	Δ Amu A²	κ
C¹²H₂:C¹²F¹⁹Cl³⁵	C_s	631	10 681.62 M		3 448.38 M				−.542724
C¹²H₂:C¹²F¹⁹Cl	C_s	632	10 681.33 M		3 380.49 M				−.568678

Id. No.	μ_a Debye	μ_b Debye	μ_c Debye	eQq Value(MHz) Rel.	eQq Value(MHz) Rel.	eQq Value(MHz) Rel.	ω_a d 1/cm	ω_b d 1/cm	ω_c d 1/cm	ω_d d 1/cm
631				−73.3 aa	39.8 bb					

References:

ABC: 186 $\quad$ κ: 186 $\quad$ eQq: 186

Add. Ref. 74

1,1-Chlorofluoroethene $\quad\quad\quad\quad\quad\quad\quad\quad\quad\quad\quad\quad\quad\quad\quad\quad\quad\quad\quad\quad\quad\quad\quad\quad$ Spectral Line Table

Isotopic Species	Id. No.	Rotational Quantum Nos.	Vib. State	F_I'	F'	F_I	F	Frequency MHz	Acc. ±MHz	Ref.
C¹²H₂:C¹²F¹⁹Cl³⁵	631	2, 1, 2← 1, 0, 1	Ground					21 026.70		186
	631	2, 2, 1← 2, 1, 2	Ground					21 699.70		186
	631	3, 0, 3← 2, 1, 2	Ground					20 214.29		186
	631	3, 2, 2← 3, 1, 3	Ground					24 362.48		186
	631	5, 1, 4← 5, 0, 5	Ground					24 601.24		186
	631	6, 2, 4← 6, 1, 5	Ground					20 391.51		186
	631	6, 3, 3← 6, 2, 4	Ground					24 895.46		186
	631	7, 3, 4← 6, 4, 3	Ground					22 419.13		186
	631	7, 2, 5← 7, 1, 6	Ground					25 656.30		186
	631	7, 3, 4← 7, 2, 5	Ground					23 896.29		186
C¹²H₂:C¹²F¹⁹Cl	632	2, 1, 2← 1, 0, 1	Ground					20 822.8		186
	632	2, 2, 1← 2, 1, 2	Ground					21 902.50		186
	632	3, 2, 2← 3, 1, 3	Ground					24 427.38		186
	632	9, 4, 5← 8, 5, 4	Ground					22 852.40		186

C₂H₂ClF $\qquad$ C_s $\qquad$ CHF:CHCl

Isotopic Species	Pt. Gp.	Id. No.	A MHz		B MHz		C MHz		D_J MHz	D_{JK} MHz	Δ Amu A²	κ
c-C^{12}HF^{19}:C^{12}HCl^{35}	C$_s$	641	16 405.9	M	3 756.05	M	3 052.67	M			−.63	−.89465
c-C^{12}HF^{19}:C^{12}HCl^{37}	C$_s$	642	16 346.6	M	3 662.49	M	2 988.61	M				−.89911

Id. No.	μ_a Debye	μ_b Debye	μ_c Debye	eQq Value(MHz)	Rel.	eQq Value(MHz)	Rel.	eQq Value(MHz)	Rel.	ω_a d 1/cm	ω_b d 1/cm	ω_c d 1/cm	ω_d d 1/cm
641		1		−22.46	aa	−10.88	bb	56.7	ab				
642				−17.31	aa	−7.96	bb						

1. Presence of quadrupole coupling renders interpretation of Stark effect data difficult, hence no dipole moment is given. However, a value of $\mu_b = 1.6$ Debye is indicated in a preliminary treatment.

References:

ABC: 941 $\qquad$ Δ: 941 $\qquad$ κ: 941 $\qquad$ μ: 941 $\qquad$ eQq: 941

cis-1-Chloro-2-Fluoroethene $\qquad$ Spectral Line Table

Isotopic Species	Id. No.	Rotational Quantum Nos.	Vib. State	F_1'	F'	F_1	F	Frequency MHz	Acc. ±MHz	Ref.
c-C^{12}HF^{19}:C^{12}HCl^{35}	641	1, 1, 1← 0, 0, 0	Ground		3/2		3/2	19 456.39		941
	641	1, 1, 1← 0, 0, 0	Ground		5/2		3/2	19 459.04		941
	641	1, 1, 1← 0, 0, 0	Ground		1/2		3/2	19 461.25		941
	641	2, 1, 2← 1, 0, 1	Ground		5/2		3/2	25 563.1		941
	641	2, 1, 2← 1, 0, 1	Ground		7/2		5/2	25 565.7		941
	641	3, 1, 2← 3, 0, 3	Ground		3/2		3/2	15 229.76		941
	641	3, 1, 2← 3, 0, 3	Ground		9/2		9/2	15 233.21		941
	641	3, 1, 2← 3, 0, 3	Ground		5/2		5/2	15 237.39		941
	641	3, 1, 2← 3, 0, 3	Ground		9/2		7/2	15 239.60		941
	641	3, 1, 2← 3, 0, 3	Ground		5/2		7/2	15 240.14		941
	641	3, 1, 2← 3, 0, 3	Ground		7/2		7/2	15 240.67		941
	641	4, 1, 3← 4, 0, 4	Ground		5/2		5/2	16 860.48		941
	641	4, 1, 3← 4, 0, 4	Ground		11/2		11/2	16 863.18		941
	641	4, 1, 3← 4, 0, 4	Ground		7/2		7/2	16 868.13		941
	641	4, 1, 3← 4, 0, 4	Ground		9/2		9/2	16 870.66		941
	641	5, 1, 4← 5, 0, 5	Ground		7/2		7/2	19 043.40		941
	641	5, 1, 4← 5, 0, 5	Ground		13/2		13/2	19 045.89		941
	641	5, 1, 4← 5, 0, 5	Ground		9/2		9/2	19 051.40		941
	641	5, 1, 4← 5, 0, 5	Ground		11/2		11/2	19 053.22		941
c-C^{12}HF^{19}:C^{12}HCl^{37}	642	1, 1, 1← 0, 0, 0	Ground		3/2		3/2	19 333.62		941
	642	1, 1, 1← 0, 0, 0	Ground		5/2		3/2	19 335.66		941
	642	1, 1, 1← 0, 0, 0	Ground		1/2		3/2	19 337.18		941
	642	3, 1, 2← 3, 0, 3	Ground		9/2		9/2	15 154.71		941
	642	3, 1, 2← 3, 0, 3	Ground		5/2		5/2	15 157.89		941
	642	3, 1, 2← 3, 0, 3	Ground		7/2		7/2	15 160.37		941
	642	4, 1, 3← 4, 0, 4	Ground		5/2		5/2	16 705.23		941
	642	4, 1, 3← 4, 0, 4	Ground		11/2		11/2	16 707.20		941
	642	4, 1, 3← 4, 0, 4	Ground		7/2		7/2	16 710.90		941
	642	4, 1, 3← 4, 0, 4	Ground		9/2		9/2	16 712.87		941
	642	5, 1, 4← 5, 0, 5	Ground		7/2		7/2	18 780.34		941
	642	5, 1, 4← 5, 0, 5	Ground		13/2		13/2	18 781.91		941

Chloroethanenitrile, Chloromethyl Cyanide

C_2H_2ClN C_s CH_2ClCN

Isotopic Species	Pt. Gp.	Id. No.	A MHz	B MHz	C MHz	D_J MHz	D_{JK} MHz	Δ Amu A²	κ
$C^{12}H_2Cl^{35}C^{12}N^{14}$	C_s	651	25 284.77 M	3 151.61 M	2 849.90 M				−.9731
$C^{12}H_2Cl^{37}C^{12}N^{14}$	C_s	652	25 135.17 M	3 081.77 M	2 790.84 M				

Id. No.	μ_a Debye	μ_b Debye	μ_c Debye	eQq Value(MHz)	Rel.	eQq Value(MHz)	Rel.	eQq Value(MHz)	Rel.	ω_a d 1/cm	ω_b d 1/cm	ω_c d 1/cm	ω_d d 1/cm
651		3.00 L	0. X	−76.36									

References:

ABC: 905   κ: 905   μ: 995   eQq: 905

For species 651, $\mu_a/\mu_b = .65$. Ref. 905.

Monochloroacetonitrile

Isotopic Species	Id. No.	Rotational Quantum Nos.	Vib. State	F_1'	F'	F_1	F	Frequency MHz	Acc. ±MHz	Ref.
$C^{12}H_2Cl^{35}C^{12}N^{14}$	651	3, 1, 3← 2, 1, 2	Ground					17 550.	1.	905
	651	4, 0, 4← 3, 0, 3	Ground					23 974.	1.	905
	651	4, 2, 2← 3, 2, 1	Ground					24 036.	1.	905
	651	4, 2, 3← 3, 2, 2	Ground					24 002.	1.	905
	651	4, 3, 1← 3, 3, 0	Ground					24 011.	1.	905
	651	4, 3, 2← 3, 3, 1	Ground					24 011.	1.	905
	651	4, 1, 3← 4, 0, 4	Ground					23 829.80	.2	905
	651	6, 0, 6← 5, 1, 5	Ground					15 794.54	.2	905
	651	7, 0, 7← 6, 1, 6	Ground					22 554.00	.2	905
	651	9, 2, 7←10, 1,10	Ground					16 347.33	.2	905
	651	12, 1,11←11, 2,10	Ground					16 469.00	.2	905
$C^{12}H_2Cl^{37}C^{12}N^{14}$	652	3, 1, 3← 2, 1, 2	Ground					17 177.	1.	905
	652	6, 0, 6← 5, 1, 5	Ground					15 038.34	.2	905
	652	9, 2, 7←10, 1,10	Ground					17 004.60	.2	905
	652	12, 1,11←11, 2,10	Ground					14 789.70	.2	905

1,1-Dichloroethylene, Vinylidene Chloride

$C_2H_2Cl_2$ C_{2v} $H_2C:CCl_2$

Isotopic Species	Pt. Gp.	Id. No.	A MHz	B MHz	C MHz	D_J MHz	D_{JK} MHz	Δ Amu A²	κ
$H_2C^{12}:C^{12}Cl_2^{35}$	C_{2v}	661	7 466.8 M	3 411.3 M	2 339.0 M			.23	−.58178
$H_2C^{12}:C^{12}Cl^{35}Cl^{37}$	C_{2v}	662	7 423.8 M	3 319.2 M	2 291.4 M			.22	−.59448

Id. No.	μ_a Debye	μ_b Debye	μ_c Debye	eQq Value(MHz)	Rel.	eQq Value(MHz)	Rel.	eQq Value(MHz)	Rel.	ω_a d 1/cm	ω_b d 1/cm	ω_c d 1/cm	ω_d d 1/cm
661	0. X	1.34 M	0. X	−78.7	Cl^{35}								
662				−78.7	Cl^{35}								

References:

ABC: 778   Δ: 778   κ: 778   μ: 965   eQq: 778

Add. Ref. 341

Isotopic Species	Id. No.	Rotational Quantum Nos.	Vib. State	Hyperfine				Frequency MHz	Acc. ±MHz	Ref.
				F_1'	F'	F_1	F			
$H_2C^{12}:C^{12}Cl_2^{35}$	661	3, 1, 3← 2, 0, 2	Ground					18 703.9	1.	778
	661	4, 1, 4← 3, 0, 3	Ground					22 701.2	.5	778
	661	4, 3, 2← 4, 2, 3	Ground					23 530.7	1.	778
	661	5, 0, 5← 4, 1, 4	Ground					24 739.8	1.	778
	661	5, 2, 4← 5, 1, 5	Ground					22 314.	1.	778
	661	5, 2, 4← 5, 1, 5	Ground					22 318.	1.	778
	661	5, 3, 2← 5, 2, 3	Ground					19 747.0	1.	778
	661	6, 3, 3← 6, 2, 4	Ground					18 281.0	1.	778
	661	6, 3, 4← 6, 2, 5	Ground					25 771.6	.5	778
	661	9, 3, 6← 9, 2, 7	Ground					18 657.3	.5	778
	661	9, 4, 5← 9, 3, 6	Ground					24 960.08	.5	778
	661	9, 4, 5← 9, 3, 6	Ground					24 961.40	.5	778
	661	9, 4, 5← 9, 3, 6	Ground					24 962.72	.5	778
	661	10, 3, 7←10, 2, 8	Ground					21 490.1	.5	778
	661	10, 3, 7←10, 2, 8	Ground					21 492.3	.5	778
	661	10, 3, 7←10, 2, 8	Ground					21 494.8	.5	778
	661	10, 4, 6←10, 3, 7	Ground					23 303.7	.5	778
	661	11, 3, 8←11, 2, 9	Ground					25 785.7	.5	778
	661	11, 4, 7←11, 3, 8	Ground					22 650.6	.5	778
	661	12, 4, 8←12, 3, 9	Ground					23 428.5	.5	778
	661	13, 4, 9←13, 3,10	Ground					25 859.5	.5	778
	661	13, 4, 9←13, 3,10	Ground					25 861.5	.5	778
	661	13, 4, 9←13, 3,10	Ground					25 863.3	.5	778
$H_2C^{12}:C^{12}Cl^{35}Cl^{37}$	662	4, 1, 4← 3, 0, 3	Ground					22 351.3	.5	778
	662	5, 0, 5← 4, 1, 4	Ground					24 131.4	1.	778
	662	5, 1, 5← 4, 0, 4	Ground					26 287.3	.5	778
	662	6, 3, 4← 6, 2, 5	Ground					25 692.5	1.	778
	662	9, 2, 7← 9, 1, 8	Ground					25 497.8	1.	778
	662	9, 3, 6← 9, 2, 7	Ground					18 367.5	.5	778
	662	9, 4, 5← 9, 3, 6	Ground					25 573.4	1.	778
	662	10, 3, 7←10, 2, 8	Ground					20 799.9	.5	778
	662	10, 3, 7←10, 2, 8	Ground					20 801.9	.5	778
	662	10, 3, 7←10, 2, 8	Ground					20 804.2	.5	778
	662	10, 4, 6←10, 3, 7	Ground					23 796.8	.5	778
	662	11, 3, 8←11, 2, 9	Ground					24 672.6	1.	778
	662	11, 4, 7←11, 3, 8	Ground					22 858.7	.5	778
	662	12, 4, 8←12, 3, 9	Ground					23 203.8	.5	778
	662	13, 4, 9←13, 3,10	Ground					25 093.7	.5	778
$H_2C^{12}:C^{12}Cl_2^b$	663	Not Reported	Ground					21 428.5	1.	778
	663	Not Reported	Ground					21 461.5	1.	778
	663	Not Reported	Ground					21 483.0	1.5	778
	663	Not Reported	Ground					21 560.3	1.	778
	663	Not Reported	Ground					21 562.5	1.	778
	663	Not Reported	Ground					21 564.	1.	778
	663	Not Reported	Ground					21 601.5	1.5	778
	663	Not Reported	Ground					22 234.1	1.	778
	663	Not Reported	Ground					22 311.8	1.5	778
	663	Not Reported	Ground					22 526.9	1.5	778
	663	Not Reported	Ground					22 578.6	.5	778

Isotopic Species	Id. No.	Rotational Quantum Nos.	Vib. State	Hyperfine				Frequency MHz	Acc. ±MHz	Ref.
				F_1'	F'	F_1	F			
$H_2C^{12}:C^{12}Cl_2^b$	663	Not Reported	Ground					22 597.	1.5	778
	663	Not Reported	Ground					22 631.	1.5	778
	663	Not Reported	Ground					22 727.2	1.5	778
	663	Not Reported	Ground					22 752.7	1.	778
	663	Not Reported	Ground					22 782.8	1.	778
	663	Not Reported	Ground					22 874.0	1.	778
	663	Not Reported	Ground					22 958.7	1.	778
	663	Not Reported	Ground					23 103.8	1.	778
	663	Not Reported	Ground					23 153.9	1.	778
	663	Not Reported	Ground					23 240.0	1.5	778
	663	Not Reported	Ground					23 262.7	1.5	778
	663	Not Reported	Ground					23 315.1	.5	778
	663	Not Reported	Ground					23 326.2	1.5	778
	663	Not Reported	Ground					23 350.2	1.5	778
	663	Not Reported	Ground					23 380.3	1.	778
	663	Not Reported	Ground					23 386.5	1.5	778
	663	Not Reported	Ground					23 400.0	1.5	778
	663	Not Reported	Ground					23 418.5	1.5	778
	663	Not Reported	Ground					23 539.4	.5	778
	663	Not Reported	Ground					23 558.5	1.5	778
	663	Not Reported	Ground					23 618.6	1.5	778
	663	Not Reported	Ground					23 643.3	1.	778
	663	Not Reported	Ground					23 660.0	1.5	778
	663	Not Reported	Ground					23 754.6	1.5	778
	663	Not Reported	Ground					23 896.5	1.5	778
	663	Not Reported	Ground					24 366.5	1.	778
	663	Not Reported	Ground					24 403.2	1.	778
	663	Not Reported	Ground					24 469.0	1.5	778
	663	Not Reported	Ground					24 491.2	.5	778
	663	Not Reported	Ground					24 563.5	.5	778
	663	Not Reported	Ground					24 607.2	1.5	778
	663	Not Reported	Ground					24 661.2	1.	778
	663	Not Reported	Ground					25 063.5	.5	778
	663	Not Reported	Ground					25 075.0	1.	778
	663	Not Reported	Ground					25 215.5	.5	778
	663	Not Reported	Ground					25 328.2	1.5	778
	663	Not Reported	Ground					25 734.8	1.	778
	663	Not Reported	Ground					25 736.1	1.5	778
	663	Not Reported	Ground					25 740.9	1.5	778
	663	Not Reported	Ground					25 766.9	1.	778
	663	Not Reported	Ground					25 783.1	1.5	778
	663	Not Reported	Ground					25 918.4	1.	778
	663	Not Reported	Ground					25 991.1	.5	778
	663	Not Reported	Ground					26 011.8	1.	778
	663	Not Reported	Ground					26 236.0	1.	778

$C_2H_2Cl_2$ C_{2v} HClC:CHCl

Isotopic Species	Pt. Gp.	Id. No.	A MHz		B MHz		C MHz		D_J MHz	D_{JK} MHz	Δ Amu A^2	κ
c-HCl^{35}C^{12}:C^{12}HCl^{35}	C_{2v}	671	11 518.33	M	2 545.15	M	2 082.57	M			.229	−.90195
c-HCl^{35}C^{12}:C^{12}HCl^{37}	C_s	672	10 774.7	M	2 663.3	M	2 135.5	M			.01	−.87782

Id. No.	μ_a Debye		μ_b Debye		μ_c Debye		eQq Value(MHz)	Rel.	eQq Value(MHz)	Rel.	eQq Value(MHz)	Rel.	ω_a d 1/cm	ω_b d 1/cm	ω_c d 1/cm	ω_d d 1/cm
671	0.	X	2.95	G	0.	X	3.7	aa	−35.6	bb	31.9	cc				

References:

ABC: 928,963 Δ: 928,963 κ: 928,963 μ: 995   eQq: 963

cis-1,2-Dichloroethene   Spectral Line Table

Isotopic Species	Id. No.	Rotational Quantum Nos.	Vib. State	F_1'	F'	F_1	F	Frequency MHz	Acc. ±MHz	Ref.
c-HCl^{35}C^{12}:C^{12}HCl^{35}	671	1, 1, 1← 0, 0, 0	Ground					13 600.90		963
	671	3, 1, 2← 3, 0, 3	Ground					10 668.00		963
	671	4, 0, 4← 3, 1, 3	Ground					13 445.60		963
	671	4, 1, 3← 4, 0, 4	Ground					11 730.06		963
	671	4, 2, 2← 4, 1, 3	Ground					22 833.4	.5	928
	671	5, 1, 4← 5, 0, 5	Ground					13 147.17		963
	671	5, 2, 3← 5, 1, 4	Ground					22 049.0	1.	928
	671	6, 0, 6← 5, 1, 5	Ground					20 962.10		963
	671	6, 1, 5← 6, 0, 6	Ground					14 968.41		963
	671	6, 2, 4← 6, 1, 5	Ground					21 399.2	1.	928
	671	6, 2, 4← 6, 1, 5	Ground					24 090.1		963
	671	8, 1, 7← 7, 2, 6	Ground					23 236.9	.5	928
	671	8, 1, 7← 8, 0, 8	Ground					19 977.83		963
	671	8, 1, 7← 8, 0, 8	Ground					21 916.2	.5	928
	671	9, 1, 8← 9, 0, 9	Ground					24 943.5	1.	928
	671	9, 2, 7← 9, 1, 8	Ground					21 485.3	.5	928
	671	10, 2, 8←10, 1, 9	Ground					22 507.3	.5	928
	671	11, 2, 9←11, 1,10	Ground					24 137.0	1.	928
	671	11, 2, 9←11, 1,10	Ground					24 139.42		963
	671	11, 2, 9←11, 1,10	Ground					24 140.1	.5	928
	671	11, 2, 9←11, 1,10	Ground					24 143.2	1.	928
	671	11, 2, 9←11, 3, 8	Ground					22 501.8	.5	928
c-HCl^{35}C^{12}:C^{12}HCl^{37}	672	6, 0, 6← 5, 1, 5	Ground					22 869.4	.5	928
	672	6, 0, 6← 5, 1, 5	Ground					22 873.9	.5	928
	672	8, 1, 7← 7, 2, 6	Ground					21 772.3	1.	928
	672	8, 1, 7← 8, 0, 8	Ground					21 161.6	1.	928
	672	10, 2, 8←10, 1, 9	Ground					22 181.0	1.	928
	672	10, 2, 8←10, 1, 9	Ground					22 183.1	1.	928
	672	11, 2, 9←11, 1,10	Ground					23 620.5	1.	928

680 — 1,1-Difluoroethene
     1,1-Difluoroethylene

C₂H₂F₂   C₂ᵥ   H₂C:CF₂

Isotopic Species	Pt. Gp.	Id. No.	A MHz		B MHz		C MHz		D_J MHz	D_{JK} MHz	Δ Amu A²	κ
H₂C¹²:C¹²F₂¹⁹	C₂ᵥ	681	11 000.8	M	10 428.8	M	5 345.6	M			.234	−.79771
HDC¹²:C¹²F₂¹⁹	Cₛ	682	10 926.9	M	9 545.5	M	5 086.7	M			.262	−.52692
D₂C¹²:C¹²F₂¹⁹	C₂ᵥ	683	10 590.5	M	8 994.1	M	4 855.9	M			.235	−.44326

Id. No.	μ_a Debye		μ_b Debye		μ_c Debye		eQq Value(MHz)	Rel.	eQq Value(MHz)	Rel.	eQq Value(MHz)	Rel.	ω_a 1/cm	d	ω_b 1/cm	d	ω_c 1/cm	d	ω_d 1/cm	d
681	1.366	M	0.	X	0.	X														

References:

ABC: 747   Δ: 747   κ: 747   μ: 156

Add. Ref. 157

1,1-Difluoroethene   Spectral Line Table

Isotopic Species	Id. No.	Rotational Quantum Nos.	Vib. State	F₁′	F′	F₁	F	Frequency MHz	Acc. ±MHz	Ref.
H₂C¹²:C¹²F₂¹⁹	681	2, 0, 2← 1, 0, 1	Ground					26 991.7	.3	747
	681	2, 1, 1← 1, 1, 0	Ground					36 633.	1.	747
	681	2, 1, 2← 1, 1, 1	Ground					26 465.6	.3	747
	681	3, 0, 3← 2, 0, 2	Ground					37 461.	1.	747
	681	3, 1, 3← 2, 1, 2	Ground					37 417.	1.	747
	681	3, 1, 2← 3, 1, 3	Ground					26 646.2	.3	747
	681	3, 2, 2← 3, 0, 3	Ground					26 881.2	.3	747
	681	3, 3, 1← 3, 1, 2	Ground					18 075.0	.3	747
	681	4, 2, 2← 4, 2, 3	Ground					26 627.6	.3	747
	681	4, 3, 2← 4, 1, 3	Ground					27 014.7	.3	747
	681	4, 4, 1← 4, 2, 2	Ground					19 710.9	.3	747
	681	5, 3, 2← 5, 3, 3	Ground					25 725.2	.3	747
	681	5, 4, 2← 5, 2, 3	Ground					27 297.0	.3	747
	681	6, 4, 1← 6, 4, 2	Ground					25 245.7	.3	747
	681	6, 4, 2← 6, 4, 3	Ground					24 771.0	.3	747 [1]
	681	6, 5, 2← 6, 3, 3	Ground					27 821.2	.3	747
	681	7, 5, 2← 7, 5, 3	Ground					23 435.7	.3	747
	681	7, 6, 2← 7, 4, 3	Ground					28 696.	1.	747
	681	8, 6, 2← 8, 6, 3	Ground					21 745.2	.3	747
	681	Not Reported						17 008.	1.	747
	681	Not Reported						17 229.3	.3	747
	681	Not Reported						17 843.8	.3	747
	681	Not Reported						17 892.1	.3	747
	681	Not Reported						18 907.5	.3	747
	681	Not Reported						19 098.9	.3	747
	681	Not Reported						19 254.	1.	747
	681	Not Reported						19 660.9	.3	747
	681	Not Reported						19 723.3	.3	747
	681	Not Reported						19 726.5	.3	747
	681	Not Reported						20 038.2	.3	747

1. The transition given in the reference appears to be a misprint. It has been entered as given in an earlier article co-authored by Edgell.

Isotopic Species	Id. No.	Rotational Quantum Nos.	Vib. State	F_1'	F'	F_1	F	Frequency MHz	Acc. ±MHz	Ref.
$H_2C^{12}{:}C^{12}F_2^{19}$	681	Not Reported						20 345.0	.3	747
	681	Not Reported						20 957.1	.3	747
	681	Not Reported						21 484.0	.3	747
	681	Not Reported						21 572.6	.3	747
	681	Not Reported						21 689.0	.3	747
	681	Not Reported						22 236.3	.3	747
	681	Not Reported						22 273.4	.3	747
	681	Not Reported						22 386.2	.3	747
	681	Not Reported						22 394.8	.3	747
	681	Not Reported						22 744.7	.3	747
	681	Not Reported						23 180.0	.3	747
	681	Not Reported						23 208.1	.3	747
	681	Not Reported						23 215.3	.3	747
	681	Not Reported						23 270.	1.	747
	681	Not Reported						23 323.0	.3	747
	681	Not Reported						23 360.4	.3	747
	681	Not Reported						23 647.1	.3	747
	681	Not Reported						23 770.5	.3	747
	681	Not Reported						23 814.4	.3	747
	681	Not Reported						23 994.7	.3	747
	681	Not Reported						24 020.5	.3	747
	681	Not Reported						24 149.3	.3	747
	681	Not Reported						24 293.6	.3	747
	681	Not Reported						24 323.2	.3	747
	681	Not Reported						24 353.3	.3	747
	681	Not Reported						24 357.4	.3	747
	681	Not Reported						24 449.6	.3	747
	681	Not Reported						24 543.3	.3	747
	681	Not Reported						24 581.4	.3	747
	681	Not Reported						24 639.8	.3	747
	681	Not Reported						24 729.2	.3	747
	681	Not Reported						24 809.7	.3	747
	681	Not Reported						25 352.1	.3	747
	681	Not Reported						25 448.9	.3	747
	681	Not Reported						25 516.3	.3	747
	681	Not Reported						25 742.0	.3	747
	681	Not Reported						26 116.4	.3	747
	681	Not Reported						26 329.0	.3	747
	681	Not Reported						26 335.6	.3	747
	681	Not Reported						26 726.	1.	747
	681	Not Reported						26 832.0	.3	747
	681	Not Reported						27 112.0	.3	747
	681	Not Reported						27 218.6	.3	747
	681	Not Reported						27 680.4	.3	747
	681	Not Reported						28 177.7	.3	747
	681	Not Reported						28 414.2	.3	747
	681	Not Reported						28 438.1	.3	747
	681	Not Reported						28 455.1	.3	747
	681	Not Reported						28 945.3	.3	747

Isotopic Species	Id. No.	Rotational Quantum Nos.	Vib. State	Hyperfine				Frequency MHz	Acc. ±MHz	Ref.
				F_1'	F'	F_1	F			
HDC^{12}:C^{12}F$_2^{19}$	682	2, 0, 2← 1, 0, 1	Ground					25 912.7	.3	747
	682	2, 1, 1← 1, 1, 0	Ground					33 723.1	.3	747
	682	2, 1, 2← 1, 1, 1	Ground					24 805.5	.3	747
	682	3, 0, 3← 2, 0, 2	Ground					35 751.8	.3	747
	682	3, 1, 3← 2, 1, 2	Ground					35 512.2	.3	747
	682	3, 1, 2← 3, 1, 3	Ground					24 595.1	.3	747
	682	3, 2, 2← 3, 0, 3	Ground					25 937.9	.3	747
	682	3, 3, 1← 3, 1, 2	Ground					21 056.0	.3	747
	682	4, 2, 3← 4, 0, 4	Ground					35 643.3	.3	747
	682	4, 2, 2← 4, 2, 3	Ground					23 004.8	.3	747
	682	4, 3, 2← 4, 1, 3	Ground					26 815.7	.3	747
	682	4, 4, 1← 4, 2, 2	Ground					26 327.3	.3	747
	682	5, 2, 3← 5, 2, 4	Ground					34 704.2	.3	747
	682	5, 3, 3← 5, 1, 4	Ground					35 643.3	.3	747
	682	5, 3, 2← 5, 3, 3	Ground					20 529.2	.3	747
	682	5, 4, 2← 5, 2, 3	Ground					28 728.4	.3	747
	682	5, 5, 1← 5, 3, 2	Ground					33 792.5	.3	747
	682	6, 4, 3← 6, 2, 4	Ground					35 922.1	.3	747
	682	6, 5, 2← 6, 3, 3	Ground					32 264.5	.3	747
	682	7, 4, 3← 7, 4, 4	Ground					30 804.8	.3	747
D$_2$C^{12}:C^{12}F$_2^{19}$	683	2, 0, 2← 1, 0, 1	Ground					24 778.2	.3	747
	683	2, 1, 2← 1, 1, 1	Ground					23 561.7	.3	747
	683	3, 0, 3← 2, 0, 2	Ground					34 178.2	.3	747
	683	3, 1, 3← 2, 1, 2	Ground					33 857.0	.3	747
	683	3, 1, 2← 3, 1, 3	Ground					23 114.4	.3	747
	683	3, 2, 2← 3, 0, 3	Ground					24 950.7	.3	747
	683	3, 3, 1← 3, 1, 2	Ground					21 587.7	.3	747
	683	4, 2, 4← 4, 1, 4	Ground					33 619.5	.3	747
	683	4, 2, 2← 4, 2, 3	Ground					21 084.9	.3	747
	683	4, 3, 2← 4, 2, 4	Ground					26 203.4	.3	747
	683	5, 2, 3← 5, 2, 4	Ground					32 567.8	.3	747
	683	5, 4, 2← 5, 2, 3	Ground					28 933.9	.3	747

$C_2H_2F_2$ C_{2v} HFC:CHF

Isotopic Species	Pt. Gp.	Id. No.	A MHz		B MHz		C MHz		D_J MHz	D_{JK} MHz	Δ Amu A^2	κ
c-HF^{19}C^{12}:C^{12}HF^{19}	C_{2v}	691	21 103.31	M	5 930.35	M	4 622.27	M	.01082		.1688	
c-HF^{19}C^{12}:C^{13}HF^{19}	C_s	692	20 752.10	M	5 900.17	M	4 586.92	M	.01007		.1702	

Id. No.	μ_a Debye		μ_b Debye		μ_c Debye		eQq Value(MHz) Rel.	eQq Value(MHz) Rel.	eQq Value(MHz) Rel.	ω_a d 1/cm	ω_b d 1/cm	ω_c d 1/cm	ω_d d 1/cm
691	0.	X	2.42	M	0.	X				400 1			

References:

ABC: 944 D_J: 944 Δ: 944 μ: 944 ω: 944

For species 691, excited state: $D_J = 0.01076$ MHz. Ref. 944.

---

cis-1,2-Difluoroethene   Spectral Line Table

Isotopic Species	Id. No.	Rotational Quantum Nos.	Vib. State v_a	Hyperfine F_1'	F'	F_1	F	Frequency MHz	Acc. $\pm$MHz	Ref.
c-HF^{19}C^{12}:C^{12}HF^{19}	691	1, 1, 1← 0, 0, 0	Ground					25 725.58	.05	944
	691	1, 1, 1← 0, 0, 0	1					25 848.69	.05	944
	691	2, 1, 2← 1, 0, 1	Ground					34 970.12	.05	944
	691	2, 1, 2← 1, 0, 1	1					35 083.75	.05	944
	691	2, 1, 1← 2, 0, 2	Ground					17 870.03	.05	944
	691	2, 1, 1← 2, 0, 2	1					18 006.54	.05	944
	691	3, 1, 2← 3, 0, 3	Ground					20 101.45	.05	944
	691	3, 1, 2← 3, 0, 3	1					20 243.53	.05	944
	691	4, 0, 4← 3, 1, 3	1					29 020.76	.05	944
	691	4, 0, 4← 3, 1, 3	Ground					29 158.73	.05	944
	691	4, 1, 3← 4, 0, 4	Ground					23 336.67	.05	944
	691	4, 1, 3← 4, 0, 4	1					23 485.96	.05	944
	691	4, 3, 2← 5, 2, 3	Ground					24 090.66	.05	944
	691	5, 1, 4← 5, 0, 5	Ground					27 750.08	.05	944
	691	5, 1, 4← 5, 0, 5	1					27 908.28	.05	944
	691	6, 1, 5← 6, 0, 6	Ground					33 478.20	.05	944
c-HF^{19}C^{12}:C^{13}HF^{19}	692	1, 1, 1← 0, 0, 0	Ground					25 339.02	.05	944
	692	2, 1, 2← 1, 0, 1	Ground					34 512.86	.05	944
	692	3, 1, 2← 3, 0, 3	Ground					19 808.46	.05	944
	692	4, 0, 4← 3, 1, 3	Ground					29 198.25	.05	944
	692	4, 1, 3← 4, 0, 4	Ground					23 070.58	.05	944
	692	5, 1, 4← 5, 0, 5	Ground					27 524.86	.05	944
	692	6, 1, 5← 6, 0, 6	Ground					33 305.61	.05	944

C₂H₂FN

C_s

CH₂FCN

Isotopic Species	Pt. Gp.	Id. No.	A MHz		B MHz		C MHz		D_J MHz	D_{JK} MHz	Δ Amu A²	κ
C¹²H₂F¹⁹C¹²N¹⁴	C_s	701	36 578.4	M	4 780.8	M	4 339.2	M	.005	−.07		

Id. No.	μ_a Debye		μ_b Debye		μ_c Debye		eQq Value(MHz)	Rel.	eQq Value(MHz)	Rel.	eQq Value(MHz)	Rel.	ω_a 1/cm	d	ω_b 1/cm	d	ω_c 1/cm	d	ω_d 1/cm	d
701	2.61	M	2.23	M	0.	X															

References:

ABC: 961 D_J: 961 D_{JK}: 961 μ: 961

Add. Ref. 942

Monofluoroacetonitrile

Isotopic Species	Id. No.	Rotational Quantum Nos.	Vib. State	Hyperfine				Frequency MHz	Acc. ±MHz	Ref.
				F_1'	F'	F_1	F			
C¹²H₂F¹⁹C¹²N¹⁴	701	2, 1, 1← 2, 0, 2	Ground					32 684.	1.0	961
	701	3, 0, 3← 2, 0, 2	Ground					27 345.	1.0	961
	701	3, 1, 2← 2, 1, 1	Ground					28 022.6	.2	961
	701	3, 1, 3← 2, 1, 2	Ground					26 697.8	.2	961
	701	3, 2, 1← 2, 2, 0	Ground					27 382.	1.0	961
	701	3, 2, 2← 2, 2, 1	Ground					27 365.	1.0	961
	701	3, 1, 2← 3, 0, 3	Ground					33 363.2	.2	961
	701	4, 1, 3← 4, 0, 4	Ground					34 282.8	.2	961
	701	5, 1, 4← 5, 0, 5	Ground					35 457.5	.2	961
	701	6, 1, 5← 5, 2, 4	Ground					36 752.8	.2	961
	701	12, 1,11←11, 2,10	Ground					29 854.	1.0	961

Ethenone, Carbomethene, Keten

C_2H_2O C_{2v} $H_2C{:}CO$

Isotopic Species	Pt. Gp.	Id. No.	A MHz	B MHz	C MHz	D_J MHz	D_{JK} MHz	Δ Amu A^2	κ
$H_2C^{12}{:}C^{12}O^{16}$	C_{2v}	711	280 000. M	10 293.29 M	9 915.87 M	.0036	1.5761		−.9973
$D_2C^{12}{:}C^{12}O^{16}$	C_{2v}	712		9 120.80 M	8 552.66 M	.0034	1.0674		
$HDC^{12}{:}C^{12}O^{16}$	C_s	713		9 647.05 M	9 174.63 M				

Id. No.	μ_a Debye	μ_b Debye	μ_c Debye	eQq Value(MHz) Rel.	eQq Value(MHz) Rel.	eQq Value(MHz) Rel.	ω_a d 1/cm	ω_b d 1/cm	ω_c d 1/cm	ω_d d 1/cm
711	1.414 M	0. X	0. X				443 1	617 1	527 1	
712	1.423 M	0. X	0. X				386 1	528 1	437 1	
713	1.442 M	0. u	0. X				397 1	558 1	506 1	

References:

ABC: 203,364 D_J: 879 D_{JK}: 879 κ: 981 μ: 364 ω: 988

Add. Ref. 180,283,284,412

For excited vibrational states the following constants are given in ref. 364 (all values in MHz):

Species	State	$a_b + a_c$	$a_b - a_c$	D_J	D_{JK}
711 712 713	$v_9 = 1$	57.66 53.27 55.97	18.39	.020	.505
711 713	$v_8 = 1$	10.72 11.08	−27.0	−.026	
711 712 713	$v_7 = 1$	22.39 25.18	−27.88 34.66	.019	.550

Ref. 879 also gives $D_K = 39.0535$ MHz for species 711 and 9.1058 MHz for species 712.

The values reported for D_J and D_{JK} vary significantly from those previously given by Johnson and Strandberg.

Isotopic Species	Id. No.	Rotational Quantum Nos.	Vib. State v_a; v_b; v_c	F_1'	F'	F_1	F	Frequency MHz	Acc. ±MHz	Ref.
$H_2C^{12}:C^{12}O^{16}$	711	1, 0, 1← 0, 0, 0	Ground					20 209.20		364
	711	1, 0, 1← 0, 0, 0	0; 0; 1					20 219.92		988
	711	1, 0, 1← 0, 0, 0	Excited					20 231.59		364
	711	1, 0, 1← 0, 0, 0	1; 0; 0					20 266.86		988
	711	2, 0, 2← 1, 0, 1	Ground					40 417.90		364
	711	2, 0, 2← 1, 0, 1	0; 0; 1					40 440.17		988
	711	2, 0, 2← 1, 0, 1	0; 1; 0					40 462.26		988
	711	2, 0, 2← 1, 0, 1	1; 0; 0					40 532.78		988
	711	2, 1, 1← 1, 1, 0	Ground					40 793.62		364
	711	2, 1, 1← 1, 1, 0	Excited					40 809.98		364
	711	2, 1, 1← 1, 1, 0	1; 0; 0					40 926.95		988
	711	2, 1, 2← 1, 1, 1	Ground					40 038.80		364
	711	2, 1, 2← 1, 1, 1	0; 0; 1					40 087.77		988
	711	2, 1, 2← 1, 1, 1	Excited					40 110.92		364
	711	2, 1, 2← 1, 1, 1	1; 0; 0					40 135.35		988
	711	3, 0, 3← 2, 0, 2	Ground					60 625.68		364
	711	3, 1, 2← 2, 1, 1	Ground					61 190.24		364
	711	3, 1, 3← 2, 1, 2	Ground					60 057.92		364
	711	3, 2, 1← 2, 2, 0	Ground					60 617.30		364
	711	3, 2, 2← 2, 2, 1	Ground					60 615.88		364
	711	6, 1, 5← 6, 1, 6	Ground					7 925.18		364
	711	9, 1, 8← 9, 1, 9	Ground					16 980.97		364
	711	10, 1, 9←10, 1,10	Ground					20 753.90		364
	711	11, 1,10←11, 1,11	Ground					24 903.53		364
	711	12, 1,11←12, 1,12	Ground					29 430.02		364
	711	13, 1,12←13, 1,13	Ground					34 333.14		364
	711	14, 1,13←14, 1,14	Ground					39 612.55		364
	711	27, 2,25←27, 2,26	Ground					9 188.20		364
	711	28, 2,26←28, 2,27	Ground					10 588.88		364
$D_2C^{12}:C^{12}O^{16}$	712	1, 0, 1← 0, 0, 0	Ground					17 673.61		988
	712	1, 0, 1← 0, 0, 0	0; 0; 1					17 692.59		988
	712	1, 0, 1← 0, 0, 0	1; 0; 0					17 695.00		988
	712	1, 0, 1← 0, 0, 0	1; 0; 0					17 727.13		988
	712	1, 0, 1← 0, 0, 0	0; 0; 1					35 383.36		988
	712	2, 0, 2← 1, 0, 1	Ground					35 345.20		988
	712	2, 0, 2← 1, 0, 1	0; 1; 1					35 388.20		988
	712	2, 0, 2← 1, 0, 1	1; 0; 0					35 451.85		988
	712	2, 1, 1← 1, 0, 1	Ground					35 913.83		988
	712	2, 1, 1← 1, 0, 1	1; 0; 0					36 048.80		988
	712	2, 1, 1← 1, 1, 0	0; 0; 1					35 924.40		988
	712	2, 1, 1← 1, 1, 0	0; 1; 0					35 941.43		988
	712	2, 1, 2← 1, 1, 1	Ground					34 777.62		988
	712	2, 1, 2← 1, 1, 1	0; 1; 0					34 835.37		988
	712	2, 1, 2← 1, 1, 1	0; 0; 1					34 842.90		988
	712	2, 1, 2← 1, 1, 1	1; 0; 0					34 857.04		988
	712	5, 1, 4← 5, 1, 5	Ground					8 521.53		364
	712	8, 1, 7← 8, 1, 8	Ground					20 448.71		364
	712	9, 1, 8← 9, 1, 9	Ground					25 558.93		364
	712	10, 1, 9←10, 1,10	Ground					31 235.60		364

Isotopic Species	Id. No.	Rotational Quantum Nos.	Vib. State v_a ; v_b ; v_c	Hyperfine F_1'	F'	F_1	F	Frequency MHz	Acc. ±MHz	Ref.
$D_2C^{12}:C^{12}O^{16}$	712	18, 2,16←18, 2,17	Ground					8 687.06		364
	712	22, 2,20←22, 2,21	Ground					18 737.62		364
	712	23, 2,21←23, 2,22	Ground					22 181.36		364
	712	24, 2,22←24, 2,23	Ground					26 050.17		364
	712	25, 2,23←25, 2,24	Ground					30 368.36		364
	712	26, 2,24←26, 2,25	Ground					35 159.06		364
$HDC^{12}:C^{12}O^{16}$	713	1, 0, 1← 0, 0, 0	Ground					18 821.68		364
	713	1, 0, 1← 0, 0, 0	0; 0; 1					18 832.76		988
	713	1, 0, 1← 0, 0, 0	Excited					18 846.86		364
	713	1, 0, 1← 0, 0, 0	1; 0; 0					18 877.65		988
	713	2, 0, 2← 1, 0, 1	Ground					37 642.41		988
	713	2, 0, 2← 1, 0, 1	0; 0; 1					37 664.81		988
	713	2, 0, 2← 1, 0, 1	0; 1; 0					37 693.05		988
	713	2, 0, 2← 1, 0, 1	1; 0; 0					37 754.59		988
	713	2, 1, 1← 1, 1, 0	0; 0; 1					38 113.22		988
	713	2, 1, 1← 1, 1, 0	Ground					38 114.40		988
	713	2, 1, 1← 1, 1, 0	0; 1; 0					38 144.71		988
	713	2, 1, 1← 1, 1, 0	1; 0; 0					38 254.38		988
	713	2, 1, 2← 1, 1, 1	Ground					37 169.75		988
	713	2, 1, 2← 1, 1, 1	0; 0; 1					37 215.86		988
	713	2, 1, 2← 1, 1, 1	1; 0; 0					37 240.17		988
	713	2, 1, 2← 1, 1, 1	1; 0; 0					37 254.26		988
	713	6, 1, 5← 6, 1, 6	Ground					9 919.95		364
	713	9, 1, 8← 9, 1, 9	Ground					21 254.31		364
	713	10, 1, 9←10, 1,10	Ground					25 975.83		364
	713	21, 2,19←21, 2,20	Ground					7 901.08		364
	713	22, 2,20←22, 2,21	Ground					9 462.69		364
	713	27, 2,25←27, 2,26	Ground					20 855.93		364

Vinyl Bromide, Bromoethylene

C_2H_3Br C_s $CH_2:CHBr$

Isotopic Species	Pt. Gp.	Id. No.	A MHz		B MHz		C MHz		D_J MHz	D_{JK} MHz	Δ Amu A^2	κ
$C^{12}H_2:C^{12}DBr^{79}$	C_s	721			4 103.81	M	3 740.73	M				
$C^{12}H_2:C^{12}DBr^{81}$	C_s	722			4 079.20	M	3 720.22	M				
$C^{12}D_2:C^{12}HBr^{79}$	C_s	723			3 718.93	M	3 432.76	M				
$C^{12}D_2:C^{12}HBr^{81}$	C_s	724			3 696.21	M	3 413.33	M				
$C^{12}D_2:C^{12}DBr^{79}$	C_s	725	36 128.19	M	3 676.93	M	3 337.26	M				
$C^{12}D_2:C^{12}DBr^{81}$	C_s	726	36 062.24	M	3 654.2	M	3 317.98	M				
$C^{12}H_2:C^{12}HBr^{79}$	C_s	727			4 162.67	M	3 862.64	M	.00008			
$C^{12}H_2:C^{12}HBr^{81}$	C_s	728			4 138.34	M	3 841.63	M				
$C^{13}H_2:C^{13}HBr^{79}$	C_s	733			3 959.27	M	3 676.44	M				
$C^{13}H_2:C^{13}HBr^{81}$	C_s	734			3 977.92	M	3 699.97	M				
$C^{13}H_2:C^{12}HBr^{79}$	C_s	735			4 002.14	M	3 721.01	M				
$C^{13}H_2:C^{!2}HBr^{81}$	C_s	736			3 934.85	M	3 655.17	M				
$C^{12}H_2:C^{13}HBr^{79}$	C_s	737			4 114.49	M	3 813.14	M				
$C^{12}H_2:C^{13}HBr^{81}$	C_s	738			4 089.94	M	3 791.89	M				
c-$C^{12}HD:C^{12}HBr^{79}$	C^1	739	45 667.	M	4 021.61	M	3 689.26	M				
t-$C^{12}HD:C^{12}HBr^{79}$	C_1	741	56 868.	M	3 834.91	M	3 578.87	M				
c-$C^{12}HD:C^{12}HBr^{81}$	C_1	742	46 140.	M	3 996.75	M	3 668.11	M				
t-$C^{12}HD:C^{12}HBr^{81}$	C_1	743	55 382.41	M	3 812.10	M	3 558.90	M				
c-$C^{12}HD:C^{12}DBr^{79}$	C_1	744	42 987.68	M	3 787.02	M	3 475.70	M				
t-$C^{12}HD:C^{12}DBr^{79}$	C_1	745	36 511.	M	3 971.22	M	3 577.84	M				
c-$C^{12}HD:C^{12}DBr^{81}$	C_1	746	43 501.	M	3 763.86	M	3 456.11	M				
t-$C^{12}HD:C^{12}DBr^{81}$	C_1	747	36 786.	M	3 946.67	M	3 557.81	M				

Id. No.	μ_a Debye	μ_b Debye	μ_c Debye	eQq Value(MHz)	Rel.	eQq Value(MHz)	Rel.	eQq Value(MHz)	Rel.	ω_a 1/cm	d	ω_b 1/cm	d	ω_c 1/cm	d	ω_d 1/cm	d
721										319	1						
727	1.415 T		0. X	469	aa	−219	bb										
728				393	aa	−181	bb										

References:

ABC: 756,992,998,1001 D_J: 756 μ: 1030   eQq: 259 ω: 998

Add. Ref. 191

Ref. 259 gives the following constants for the excited vibrational state $v_5 = 1$, which a later reference identifies as being v_{10}:

Species No.	B (MHz)	C (MHz)	eQV$_{aa}$ (MHz)
727	4159.7	3857.6	465
728	4135.1	3837.3	

Isotopic Species	Id. No.	Rotational Quantum Nos.	Vib. State v_a	Hyperfine				Frequency MHz	Acc. ±MHz	Ref.
				F_1'	F'	F_1	F			
$C^{12}H_2{:}C^{12}DBr^{79}$	721	2, 0, 2← 1, 0, 1	1		3/2		3/2	15 574.7		1001
	721	2, 0, 2← 1, 0, 1	Ground		3/2		3/2	15 590.2		1001
	721	2, 0, 2← 1, 0, 1	1		5/2		3/2	15 659.9		1001
	721	2, 0, 2← 1, 0, 1	1		7/2		5/2	15 659.9		1001
	721	2, 0, 2← 1, 0, 1	Ground		5/2		3/2	15 675.3		1001
	721	2, 0, 2← 1, 0, 1	Ground		7/2		5/2	15 675.3		1001
	721	2, 0, 2← 1, 0, 1	Ground					15 685.14	.06	992
	721	2, 0, 2← 1, 0, 1	Ground		1/2		1/2	15 686.8		1001
	721	2, 0, 2← 1, 0, 1	Ground		5/2		5/2	15 795.4		1001
	721	2, 1, 1← 1, 1, 0	Ground		3/2		1/2	15 989.0		1001
	721	2, 1, 1← 1, 1, 0	1		7/2		5/2	16 010.1		1001
	721	2, 1, 1← 1, 1, 0	Ground		7/2		5/2	16 022.7		1001
	721	2, 1, 1← 1, 1, 0	Ground					16 052.06	.05	992
	721	2, 1, 1← 1, 1, 0	Ground		5/2		5/2	16 079.8		1001
	721	2, 1, 1← 1, 1, 0	Ground		3/2		3/2	16 103.1		1001
	721	2, 1, 1← 1, 1, 0	Ground		5/2		3/2	16 143.2		1001
	721	2, 1, 2← 1, 1, 1	Ground		1/2		1/2	15 206.2		1001
	721	2, 1, 2← 1, 1, 1	Ground		3/2		1/2	15 270.1		1001
	721	2, 1, 2← 1, 1, 1	1		7/2		5/2	15 281.1		1001
	721	2, 1, 2← 1, 1, 1	Ground		7/2		5/2	15 298.0		1001
	721	2, 1, 2← 1, 1, 1	Ground					15 326.89	.04	992
	721	2, 1, 2← 1, 1, 1	Ground		5/2		5/2	15 361.9		1001
	721	2, 1, 2← 1, 1, 1	Ground		3/2		3/2	15 372.1		1001
	721	2, 1, 2← 1, 1, 1	1		5/2		3/2	15 399.6		1001
	721	2, 1, 2← 1, 1, 1	Ground		5/2		3/2	15 417.1		1001
	721	3, 0, 3← 2, 0, 2	Ground		3/2		3/2	23 426.2		1001
	721	3, 0, 3← 2, 0, 2	1		9/2		7/2	23 493.0		1001
	721	3, 0, 3← 2, 0, 2	1		7/2		5/2	23 493.0		1001
	721	3, 0, 3← 2, 0, 2	Ground		7/2		5/2	23 516.0		1001
	721	3, 0, 3← 2, 0, 2	Ground		9/2		7/2	23 516.0		1001
	721	3, 0, 3← 2, 0, 2	Ground					23 523.62	.08	992
	721	3, 0, 3← 2, 0, 2	1		5/2		3/2	23 524.3		1001
	721	3, 0, 3← 2, 0, 2	1		3/2		1/2	23 524.3		1001
	721	3, 0, 3← 2, 0, 2	Ground		3/2		1/2	23 546.8		1001
	721	3, 0, 3← 2, 0, 2	Ground		5/2		3/2	23 546.8		1001
	721	3, 1, 2← 2, 1, 1	1		3/2		3/2	23 987.6		1001
	721	3, 1, 2← 2, 1, 1	Ground		3/2		3/2	24 005.7		1001
	721	3, 1, 2← 2, 1, 1	1		9/2		7/2	24 044.0		1001
	721	3, 1, 2← 2, 1, 1	1		3/2		1/2	24 044.0		1001
	721	3, 1, 2← 2, 1, 1	Ground		5/2		5/2	24 052.5		1001
	721	3, 1, 2← 2, 1, 1	Ground		3/2		1/2	24 062.2		1001
	721	3, 1, 2← 2, 1, 1	Ground		9/2		7/2	24 062.8		1001
	721	3, 1, 2← 2, 1, 1	1		5/2		3/2	24 073.7		1001
	721	3, 1, 2← 2, 1, 1	1		7/2		5/2	24 073.7		1001
	721	3, 1, 2← 2, 1, 1	Ground					24 075.33	.05	992
	721	3, 1, 2← 2, 1, 1	Ground		7/2		5/2	24 092.5		1001
	721	3, 1, 2← 2, 1, 1	Ground		5/2		3/2	24 092.5		1001
	721	3, 1, 2← 2, 1, 1	Ground		7/2		7/2	24 149.4		1001
	721	3, 1, 3← 2, 1, 2	Ground		3/2		3/2	22 910.9		1001
	721	3, 1, 3← 2, 1, 2	1		5/2		5/2	22 932.0		1001

Isotopic Species	Id. No.	Rotational Quantum Nos.	Vib. State v_a	Hyperfine F_1'	F'	F_1	F	Frequency MHz	Acc. ±MHz	Ref.
$C^{12}H_2{:}C^{12}DBr^{79}$	721	3, 1, 3← 2, 1, 2	1		9/2		7/2	22 945.5		1001
	721	3, 1, 3← 2, 1, 2	Ground		5/2		5/2	22 959.2		1001
	721	3, 1, 3← 2, 1, 2	Ground		9/2		7/2	22 972.8		1001
	721	3, 1, 3← 2, 1, 2	Ground		3/2		1/2	22 974.0		1001
	721	3, 1, 3← 2, 1, 2	Ground					22 985.73	.04	992
	721	3, 1, 3← 2, 1, 2	Ground		7/2		5/2	23 002.5		1001
	721	3, 1, 3← 2, 1, 2	Ground		5/2		3/2	23 004.1		1001
	721	3, 1, 3← 2, 1, 2	Ground		7/2		7/2	23 066.5		1001
	721	3, 2, 1← 2, 2, 0	Ground		3/2		1/2	23 422.25		1001
	721	3, 2, 1← 2, 2, 0	Ground		7/2		7/2	23 508.1		1001
	721	3, 2, 1← 2, 2, 0	Ground		9/2		7/2	23 508.1		1001
	721	3, 2, 1← 2, 2, 0	Ground					23 542.70	.06	992
	721	3, 2, 1← 2, 2, 0	Ground		3/2		3/2	23 542.9		1001
	721	3, 2, 1← 2, 2, 0	Ground		5/2		3/2	23 542.9		1001
	721	3, 2, 1← 2, 2, 0	Ground		7/2		5/2	23 628.5		1001
	721	3, 2, 1← 2, 2, 0	Ground		5/2		5/2	23 628.5		1001
	721	3, 2, 2← 2, 2, 1	Ground		3/2		1/2	23 413.7		1001
	721	3, 2, 2← 2, 2, 1	Ground		7/2		7/2	23 498.6		1001
	721	3, 2, 2← 2, 2, 1	Ground		9/2		7/2	23 498.6		1001
	721	3, 2, 2← 2, 2, 1	Ground					23 532.93	.06	992
	721	3, 2, 2← 2, 2, 1	Ground		3/2		3/2	23 533.0		1001
	721	3, 2, 2← 2, 2, 1	Ground		5/2		3/2	23 533.0		1001
	721	3, 2, 2← 2, 2, 1	Ground		7/2		5/2	23 618.7		1001
	721	3, 2, 2← 2, 2, 1	Ground		5/2		5/2	23 618.7		1001
	721	4, 0, 4← 3, 0, 3	Ground		7/2		7/2	31 305.1		1001
	721	4, 0, 4← 3, 0, 3	1		9/2		7/2	31 327.65		1001
	721	4, 0, 4← 3, 0, 3	1		11/2		9/2	31 327.65		1001
	721	4, 0, 4← 3, 0, 3	Ground		9/2		7/2	31 347.4		1001
	721	4, 0, 4← 3, 0, 3	Ground		11/2		9/2	31 347.4		1001
	721	4, 0, 4← 3, 0, 3	Ground					31 351.06		1001
	721	4, 1, 3← 3, 1, 2	Ground		5/2		5/2	32 009.00		1001
	721	4, 1, 3← 3, 1, 2	1		7/2		7/2	32 040.1		1001
	721	4, 1, 3← 3, 1, 2	1		11/2		9/2	32 066.2		1001
	721	4, 1, 3← 3, 1, 2	Ground		7/2		7/2	32 069.6		1001
	721	4, 1, 3← 3, 1, 2	1		9/2		7/2	32 077.1		1001
	721	4, 1, 3← 3, 1, 2	1		7/2		5/2	32 084.5		1001
	721	4, 1, 3← 3, 1, 2	Ground		11/2		9/2	32 091.6		1001
	721	4, 1, 3← 3, 1, 2	Ground		5/2		3/2	32 095.9		1001
	721	4, 1, 3← 3, 1, 2	Ground					32 097.13		1001
	721	4, 1, 3← 3, 1, 2	Ground		9/2		7/2	32 102.4		1001
	721	4, 1, 3← 3, 1, 2	Ground		7/2		5/2	32 109.55		1001
	721	4, 1, 3← 3, 1, 2	Ground		9/2		9/2	32 188.9		1001
	721	4, 1, 4← 3, 1, 3	1		5/2		5/2	30 515.7		1001
	721	4, 1, 4← 3, 1, 3	Ground		5/2		5/2	30 550.2		1001
	721	4, 1, 4← 3, 1, 3	1		7/2		7/2	30 576.1		1001
	721	4, 1, 4← 3, 1, 3	1		11/2		9/2	30 600.9		1001
	721	4, 1, 4← 3, 1, 3	Ground		7/2		7/2	30 612.55		1001
	721	4, 1, 4← 3, 1, 3	Ground		11/2		9/2	30 637.2		1001
	721	4, 1, 4← 3, 1, 3	Ground		5/2		3/2	30 643.7		1001
	721	4, 1, 4← 3, 1, 3	Ground					30 644.41		1001

Isotopic Species	Id. No.	Rotational Quantum Nos.	Vib. State v_a	Hyperfine				Frequency MHz	Acc. ±MHz	Ref.
				F_1'	F'	F_1	F			
$C^{12}H_2{:}C^{12}DBr^{79}$	721	4, 1, 4← 3, 1, 3	Ground		9/2		7/2	30 649.3		1001
	721	4, 1, 4← 3, 1, 3	Ground		7/2		5/2	30 655.7		1001
	721	4, 1, 4← 3, 1, 3	Ground		9/2		9/2	30 742.8		1001
	721	4, 2, 2← 3, 2, 1	Ground		5/2		3/2	31 365.1		1001
	721	4, 2, 2← 3, 2, 1	Ground		5/2		5/2	31 365.1		1001
	721	4, 2, 2← 3, 2, 1	Ground		11/2		9/2	31 382.2		1001
	721	4, 2, 2← 3, 2, 1	Ground					31 399.58		1001
	721	4, 2, 2← 3, 2, 1	Ground		7/2		5/2	31 413.4		1001
	721	4, 2, 2← 3, 2, 1	Ground		7/2		7/2	31 413.4		1001
	721	4, 2, 2← 3, 2, 1	Ground		9/2		9/2	31 430.1		1001
	721	4, 2, 2← 3, 2, 1	Ground		9/2		7/2	31 430.1		1001
	721	4, 2, 3← 3, 2, 2	1		5/2		3/2	31 316.1		1001
	721	4, 2, 3← 3, 2, 2	1		5/2		5/2	31 316.1		1001
	721	4, 2, 3← 3, 2, 2	1		11/2		9/2	31 334.6		1001
	721	4, 2, 3← 3, 2, 2	Ground		5/2		3/2	31 339.9		1001
	721	4, 2, 3← 3, 2, 2	Ground		5/2		5/2	31 339.9		1001
	721	4, 2, 3← 3, 2, 2	Ground		11/2		9/2	31 357.0		1001
	721	4, 2, 3← 3, 2, 2	1		7/2		5/2	31 368.7		1001
	721	4, 2, 3← 3, 2, 2	1		7/2		7/2	31 368.7		1001
	721	4, 2, 3← 3, 2, 2	Ground					31 374.41		1001
	721	4, 2, 3← 3, 2, 2	Ground		7/2		5/2	31 388.4		1001
	721	4, 2, 3← 3, 2, 2	Ground		7/2		7/2	31 388.4		1001
	721	4, 2, 3← 3, 2, 2	Ground		9/2		7/2	31 404.85		1001
	721	4, 2, 3← 3, 2, 2	Ground		9/2		9/2	31 404.85		1001
	721	4, 3, 2← 3, 3, 1	Ground		5/2		3/2	31 291.5		1001 [1]
	721	4, 3, 2← 3, 3, 1	Ground		9/2		9/2	31 305.1		1001 [1]
	721	4, 3, 2← 3, 3, 1	1		11/2		9/2	31 326.2		1001 [1]
	721	4, 3, 2← 3, 3, 1	Ground		11/2		9/2	31 346.6		1001 [1]
	721	4, 3, 2← 3, 3, 1	1		7/2		5/2	31 374.2		1001 [1]
	721	4, 3, 2← 3, 3, 1	Ground					31 381.76		1001 [1]
	721	4, 3, 2← 3, 3, 1	Ground		7/2		5/2	31 399.7		1001 [1]
	721	4, 3, 2← 3, 3, 1	Ground		9/2		7/2	31 455.2		1001 [1]
	721	4, 3, 2← 3, 3, 1	Ground		7/2		7/2	31 470.1		1001 [1]
	721	Not Reported	Ground		3/2		3/2	15 587.4	.1	903
$C^{12}H_2{:}C^{12}DBr^{81}$	722	2, 0, 2← 1, 0, 1	1		3/2		3/2	15 499.9		1001
	722	2, 0, 2← 1, 0, 1	Ground		3/2		3/2	15 515.2		1001
	722	2, 0, 2← 1, 0, 1	1		7/2		5/2	15 571.2		1001
	722	2, 0, 2← 1, 0, 1	1		5/2		3/2	15 571.2		1001
	722	2, 0, 2← 1, 0, 1	Ground		7/2		5/2	15 586.5		1001
	722	2, 0, 2← 1, 0, 1	Ground		5/2		3/2	15 586.5		1001
	722	2, 0, 2← 1, 0, 1	Ground					15 594.47	.05	992
	722	2, 0, 2← 1, 0, 1	Ground		5/2		5/2	15 686.8		1001
	722	2, 1, 1← 1, 1, 0	Ground		1/2		1/2	15 857.3		1001
	722	2, 1, 1← 1, 1, 0	Ground		3/2		1/2	15 904.5		1001
	722	2, 1, 1← 1, 1, 0	1		7/2		5/2	15 920.2		1001
	722	2, 1, 1← 1, 1, 0	Ground		7/2		5/2	15 932.7		1001
	722	2, 1, 1← 1, 1, 0	Ground					15 957.28	.04	992
	722	2, 1, 1← 1, 1, 0	Ground		5/2		5/2	15 980.4		1001
	722	2, 1, 1← 1, 1, 0	Ground		3/2		3/2	15 999.8		1001

1. All $4_{32} \leftarrow 3_{31}$ assignments are uncertain; they may be $4_{31} \leftarrow 3_{30}$.

Isotopic Species	Id. No.	Rotational Quantum Nos.	Vib. State v_a	Hyperfine F_1'	F'	F_1	F	Frequency MHz	Acc. ±MHz	Ref.
$C^{12}H_2{:}C^{12}DBr^{81}$	722	2, 1, 1← 1, 1, 0	Ground		5/2		3/2	16 033.4		1001
	722	2, 1, 2← 1, 1, 1	Ground		1/2		1/2	15 139.1		1001
	722	2, 1, 2← 1, 1, 1	Ground		3/2		1/2	15 191.9		1001
	722	2, 1, 2← 1, 1, 1	1		7/2		5/2	15 202.0		1001
	722	2, 1, 2← 1, 1, 1	Ground		7/2		5/2	15 216.7		1001
	722	2, 1, 2← 1, 1, 1	Ground					15 240.22	.04	992
	722	2, 1, 2← 1, 1, 1	Ground		5/2		5/2	15 269.3		1001
	722	2, 1, 2← 1, 1, 1	Ground		3/2		3/2	15 278.1		1001
	722	2, 1, 2← 1, 1, 1	Ground		5/2		3/2	15 315.7		1001
	722	3, 0, 3← 2, 0, 2	Ground		3/2		3/2	23 306.7		1001
	722	3, 0, 3← 2, 0, 2	1		7/2		5/2	23 358.4		1001
	722	3, 0, 3← 2, 0, 2	1		9/2		7/2	23 358.4		1001
	722	3, 0, 3← 2, 0, 2	Ground		9/2		7/2	23 381.4		1001
	722	3, 0, 3← 2, 0, 2	Ground		7/2		5/2	23 381.4		1001
	722	3, 0, 3← 2, 0, 2	Ground					23 386.18	.2	992
	722	3, 0, 3← 2, 0, 2	Ground		3/2		1/2	23 407.5		1001
	722	3, 0, 3← 2, 0, 2	Ground		5/2		3/2	23 407.5		1001
	722	3, 0, 3← 2, 0, 2	Ground		7/2		7/2	23 481.7		1001
	722	3, 1, 2← 2, 1, 1	Ground		3/2		3/2	23 875.3		1001
	722	3, 1, 2← 2, 1, 1	1		9/2		7/2	23 904.1		1001
	722	3, 1, 2← 2, 1, 1	1		3/2		1/2	23 904.1		1001
	722	3, 1, 2← 2, 1, 1	Ground		5/2		5/2	23 914.3		1001
	722	3, 1, 2← 2, 1, 1	Ground		3/2		1/2	23 922.9		1001
	722	3, 1, 2← 2, 1, 1	Ground		9/2		7/2	23 922.9		1001
	722	3, 1, 2← 2, 1, 1	1		7/2		5/2	23 929.0		1001
	722	3, 1, 2← 2, 1, 1	1		5/2		3/2	23 929.0		1001
	722	3, 1, 2← 2, 1, 1	Ground					23 933.48	.08	992
	722	3, 1, 2← 2, 1, 1	Ground		5/2		3/2	23 947.8		1001
	722	3, 1, 2← 2, 1, 1	Ground		7/2		5/2	23 947.8		1001
	722	3, 1, 2← 2, 1, 1	Ground		7/2		7/2	23 995.3		1001
	722	3, 1, 3← 2, 1, 2	Ground		3/2		3/2	22 793.6		1001
	722	3, 1, 3← 2, 1, 2	1		9/2		7/2	22 818.3		1001
	722	3, 1, 3← 2, 1, 2	Ground		5/2		5/2	22 834.1		1001
	722	3, 1, 3← 2, 1, 2	1		7/2		5/2	22 843.4		1001
	722	3, 1, 3← 2, 1, 2	Ground		9/2		7/2	22 845.1		1001
	722	3, 1, 3← 2, 1, 2	Ground					22 855.91	.05	992
	722	3, 1, 3← 2, 1, 2	Ground		7/2		5/2	22 870.2		1001
	722	3, 1, 3← 2, 1, 2	Ground		5/2		3/2	22 871.5		1001
	722	3, 1, 3← 2, 1, 2	Ground		7/2		7/2	22 923.4		1001
	722	3, 2, 1← 2, 2, 0	Ground		3/2		1/2	23 306.7		1001
	722	3, 2, 1← 2, 2, 0	Ground		9/2		7/2	23 377.6		1001
	722	3, 2, 1← 2, 2, 0	Ground		7/2		7/2	23 377.6		1001
	722	3, 2, 1← 2, 2, 0	1		3/2		3/2	23 386.6		1001
	722	3, 2, 1← 2, 2, 0	1		5/2		3/2	23 386.6		1001
	722	3, 2, 1← 2, 2, 0	Ground					23 406.34	.3	992
	722	3, 2, 1← 2, 2, 0	Ground		5/2		3/2	23 409.0		1001
	722	3, 2, 1← 2, 2, 0	Ground		3/2		3/2	23 409.0		1001
	722	3, 2, 1← 2, 2, 0	Ground		5/2		5/2	23 478.2		1001
	722	3, 2, 1← 2, 2, 0	Ground		7/2		5/2	23 478.2		1001
	722	3, 2, 2← 2, 2, 1	Ground		3/2		1/2	23 297.2		1001

Isotopic Species	Id. No.	Rotational Quantum Nos.	Vib. State v_a	F_1'	F'	F_1	F	Frequency MHz	Acc. ±MHz	Ref.
$C^{12}H_2:C^{12}DBr^{81}$	722	3, 2, 2← 2, 2, 1	1		7/2		7/2	23 345.6		1001
	722	3, 2, 2← 2, 2, 1	1		9/2		7/2	23 345.6		1001
	722	3, 2, 2← 2, 2, 1	Ground		7/2		7/2	23 368.2		1001
	722	3, 2, 2← 2, 2, 1	Ground		9/2		7/2	23 368.2		1001
	722	3, 2, 2← 2, 2, 1	Ground		3/2		3/2	23 396.9		1001
	722	3, 2, 2← 2, 2, 1	Ground		5/2		3/2	23 396.9		1001
	722	3, 2, 2← 2, 2, 1	Ground					23 397.06	.06	992
	722	3, 2, 2← 2, 2, 1	Ground		7/2		5/2	23 468.6		1001
	722	3, 2, 2← 2, 2, 1	Ground		5/2		5/2	23 468.6		1001
	722	4, 0, 4← 3, 0, 3	Ground		5/2		5/2	31 074.8		1001
	722	4, 1, 3← 3, 1, 2	Ground		5/2		5/2	31 834.4		1001
	722	4, 1, 3← 3, 1, 2	1		7/2		7/2	31 859.3		1001
	722	4, 1, 3← 3, 1, 2	1		11/2		9/2	31 878.0		1001
	722	4, 1, 3← 3, 1, 2	1		5/2		3/2	31 882.0		1001
	722	4, 1, 3← 3, 1, 2	Ground		7/2		7/2	31 884.5		1001
	722	4, 1, 3← 3, 1, 2	1		9/2		7/2	31 887.3		1001
	722	4, 1, 3← 3, 1, 2	1		7/2		5/2	31 892.8		1001
	722	4, 1, 3← 3, 1, 2	Ground		11/2		9/2	31 903.1		1001
	722	4, 1, 3← 3, 1, 2	Ground		5/2		3/2	31 907.05		1001
	722	4, 1, 3← 3, 1, 2	Ground					31 908.00		1001
	722	4, 1, 3← 3, 1, 2	Ground		9/2		7/2	31 912.35		1001
	722	4, 1, 3← 3, 1, 2	Ground		7/2		5/2	31 918.1		1001
	722	4, 1, 3← 3, 1, 2	1		9/2		9/2	31 962.6		1001
	722	4, 1, 3← 3, 1, 2	Ground		9/2		9/2	31 984.65		1001
	722	4, 1, 4← 3, 1, 3	Ground		5/2		5/2	30 392.85		1001
	722	4, 1, 4← 3, 1, 3	1		7/2		7/2	30 408.3		1001
	722	4, 1, 4← 3, 1, 3	1		11/2		9/2	30 429.5		1001
	722	4, 1, 4← 3, 1, 3	1		5/2		3/2	30 434.9		1001
	722	4, 1, 4← 3, 1, 3	1		9/2		7/2	30 441.8		1001
	722	4, 1, 4← 3, 1, 3	Ground		7/2		7/2	30 444.8		1001
	722	4, 1, 4← 3, 1, 3	Ground		11/2		9/2	30 465.6		1001
	722	4, 1, 4← 3, 1, 3	Ground		5/2		3/2	30 470.9		1001
	722	4, 1, 4← 3, 1, 3	Ground					30 471.54		1001
	722	4, 1, 4← 3, 1, 3	Ground		9/2		7/2	30 475.6		1001
	722	4, 1, 4← 3, 1, 3	Ground		7/2		5/2	30 481.0		1001
	722	4, 1, 4← 3, 1, 3	1		9/2		9/2	30 517.9		1001
	722	4, 1, 4← 3, 1, 3	Ground		9/2		9/2	30 553.8		1001
	722	4, 2, 2← 3, 2, 1	Ground					31 218.09		1001
	722	4, 2, 2← 3, 2, 1	Ground		7/2		5/2	31 229.5		1001
	722	4, 2, 2← 3, 2, 1	Ground		7/2		7/2	31 229.5		1001
	722	4, 2, 2← 3, 2, 1	Ground		9/2		7/2	31 243.5		1001
	722	4, 2, 2← 3, 2, 1	Ground		9/2		9/2	31 243.5		1001
	722	4, 3, 2← 3, 3, 1	Ground					31 199.71		1001 [1]
	722	4, 3, 2← 3, 3, 1	1		7/2		7/2	31 252.3		1001 [1]
	722	4, 3, 2← 3, 3, 1	Ground		9/2		7/2	31 262.4		1001 [1]
	722	4, 3, 2← 3, 3, 1	Ground		7/2		7/2	31 275.0		1001 [1]
$C^{12}D_2:C^{12}HBr^{79}$	723	2, 0, 2← 1, 0, 1	Ground		3/2		3/2	14 208.7		1001
	723	2, 0, 2← 1, 0, 1	Ground		5/2		3/2	14 290.2		1001
	723	2, 0, 2← 1, 0, 1	Ground		7/2		5/2	14 290.2		1001

1. All $4_{32} \leftarrow 3_{31}$ assignments are uncertain; they may be $4_{31} \leftarrow 3_{30}$.

Isotopic Species	Id. No.	Rotational Quantum Nos.	Vib. State v_a	F_1'	F'	F_1	F	Frequency MHz	Acc. ±MHz	Ref.
$C^{12}D_2{:}C^{12}HBr^{79}$	723	2, 0, 2← 1, 0, 1	Ground					14 301.72	.02	992
	723	2, 0, 2← 1, 0, 1	Ground		5/2		5/2	14 405.4		1001
	723	2, 1, 1← 1, 1, 0	Ground		1/2		1/2	14 469.1		1001
	723	2, 1, 1← 1, 1, 0	Ground		3/2		1/2	14 525.5		1001
	723	2, 1, 1← 1, 1, 0	Ground		7/2		5/2	14 560.8		1001
	723	2, 1, 1← 1, 1, 0	Ground					14 587.84	.04	992
	723	2, 1, 1← 1, 1, 0	Ground		5/2		5/2	14 612.6		1001
	723	2, 1, 1← 1, 1, 0	Ground		3/2		3/2	14 639.9		1001
	723	2, 1, 1← 1, 1, 0	Ground		5/2		3/2	14 676.3		1001
	723	2, 1, 2← 1, 1, 1	Ground		1/2		1/2	13 901.6		1001
	723	2, 1, 2← 1, 1, 1	Ground		3/2		1/2	13 963.9		1001
	723	2, 1, 2← 1, 1, 1	Ground		7/2		5/2	13 986.4		1001
	723	2, 1, 2← 1, 1, 1	Ground					14 015.74	.05	992
	723	2, 1, 2← 1, 1, 1	Ground		5/2		5/2	14 050.6		1001
	723	2, 1, 2← 1, 1, 1	Ground		3/2		3/2	14 057.3		1001
	723	2, 1, 2← 1, 1, 1	Ground		5/2		3/2	14 102.4		1001
	723	3, 0, 3← 2, 0, 2	Ground		7/2		5/2	21 440.9		1001
	723	3, 0, 3← 2, 0, 2	Ground		9/2		7/2	21 440.9		1001
	723	3, 0, 3← 2, 0, 2	Ground					21 446.15	.2	992
	723	3, 0, 3← 2, 0, 2	Ground		3/2		1/2	21 468.6		1001
	723	3, 0, 3← 2, 0, 2	Ground		5/2		3/2	21 468.6		1001
	723	3, 1, 2← 2, 1, 1	Ground		3/2		3/2	21 816.4		1001
	723	3, 1, 2← 2, 1, 1	1		9/2		7/2	21 855.9		1001
	723	3, 1, 2← 2, 1, 1	Ground		3/2		1/2	21 869.2		1001
	723	3, 1, 2← 2, 1, 1	Ground		9/2		7/2	21 869.2		1001
	723	3, 1, 2← 2, 1, 1	Ground					21 880.88	.05	992
	723	3, 1, 2← 2, 1, 1	1		7/2		5/2	21 884.0		1001
	723	3, 1, 2← 2, 1, 1	1		5/2		3/2	21 884.0		1001
	723	3, 1, 2← 2, 1, 1	Ground		7/2		5/2	21 897.2		1001
	723	3, 1, 2← 2, 1, 1	Ground		5/2		3/2	21 897.2		1001
	723	3, 1, 2← 2, 1, 1	Ground		7/2		7/2	21 949.4		1001
	723	3, 1, 3← 2, 1, 2	Ground		3/2		3/2	20 948.2		1001
	723	3, 1, 3← 2, 1, 2	1		9/2		7/2	20 990.4		1001
	723	3, 1, 3← 2, 1, 2	Ground		5/2		5/2	20 995.2		1001
	723	3, 1, 3← 2, 1, 2	Ground		9/2		7/2	21 009.6		1001
	723	3, 1, 3← 2, 1, 2	1		5/2		3/2	21 020.1		1001
	723	3, 1, 3← 2, 1, 2	Ground					21 021.98	.04	992
	723	3, 1, 3← 2, 1, 2	Ground		7/2		5/2	21 038.0		1001
	723	3, 1, 3← 2, 1, 2	Ground		5/2		3/2	21 040.2		1001
	723	3, 1, 3← 2, 1, 2	Ground		7/2		7/2	21 102.2		1001
	723	3, 2, 1← 2, 2, 0	Ground		9/2		7/2	21 425.0		1001
	723	3, 2, 1← 2, 2, 0	Ground		7/2		7/2	21 425.0		1001
	723	3, 2, 1← 2, 2, 0	Ground					21 458.14	.06	992
	723	3, 2, 1← 2, 2, 0	Ground		5/2		3/2	21 458.9		1001
	723	3, 2, 1← 2, 2, 0	Ground		3/2		3/2	21 458.9		1001
	723	3, 2, 1← 2, 2, 0	Ground		7/2		5/2	21 540.9		1001
	723	3, 2, 1← 2, 2, 0	Ground		5/2		5/2	21 540.9		1001
	723	3, 2, 2← 2, 2, 1	Ground		7/2		7/2	21 420.3		1001
	723	3, 2, 2← 2, 2, 1	Ground		9/2		7/2	21 420.3		1001
	723	3, 2, 2← 2, 2, 1	Ground					21 453.22	.06	992

Isotopic Species	Id. No.	Rotational Quantum Nos.	Vib. State v_a	F'$_1$	F'	F$_1$	F	Frequency MHz	Acc. ±MHz	Ref.
C^{12}D$_2$:C^{12}HBr^{79}	723	3, 2, 2← 2, 2, 1	Ground		3/2		3/2	21 453.6		1001
	723	3, 2, 2← 2, 2, 1	Ground		5/2		3/2	21 453.6		1001
	723	3, 2, 2← 2, 2, 1	Ground		7/2		5/2	21 535.5		1001
	723	3, 2, 2← 2, 2, 1	Ground		5/2		5/2	21 535.5		1001
	723	Not Reported	Ground		1/2		1/2	14 294.7	.2	903
C^{12}D$_2$:C^{12}HBr^{81}	724	2, 0, 2← 1, 0, 1	Ground		3/2		3/2	14 139.6		1001
	724	2, 0, 2← 1, 0, 1	Ground		5/2		3/2	14 207.8		1001
	724	2, 0, 2← 1, 0, 1	Ground		7/2		5/2	14 207.8		1001
	724	2, 0, 2← 1, 0, 1	Ground					14 216.76	.07	992
	724	2, 0, 2← 1, 0, 1	Ground		5/2		5/2	14 304.1		1001
	724	2, 1, 1← 1, 1, 0	Ground		1/2		1/2	14 405.4		1001
	724	2, 1, 1← 1, 1, 0	Ground		3/2		1/2	14 448.3		1001
	724	2, 1, 1← 1, 1, 0	Ground		7/2		5/2	14 477.8		1001
	724	2, 1, 1← 1, 1, 0	Ground					14 501.15	.04	992
	724	2, 1, 1← 1, 1, 0	Ground		5/2		5/2	14 521.2		1001
	724	2, 1, 1← 1, 1, 0	Ground		3/2		3/2	14 543.8		1001
	724	2, 1, 1← 1, 1, 0	Ground		5/2		3/2	14 574.2		1001
	724	2, 1, 2← 1, 1, 1	Ground		1/2		1/2	13 839.4		1001
	724	2, 1, 2← 1, 1, 1	Ground		3/2		1/2	13 892.1		1001
	724	2, 1, 2← 1, 1, 1	Ground		7/2		5/2	13 910.5		1001
	724	2, 1, 2← 1, 1, 1	Ground					13 936.08	.06	992
	724	2, 1, 2← 1, 1, 1	Ground		5/2		5/2	13 963.9		1001
	724	2, 1, 2← 1, 1, 1	Ground		3/2		3/2	13 973.7		1001
	724	2, 1, 2← 1, 1, 1	Ground		5/2		3/2	14 007.2		1001
	724	3, 0, 3← 2, 0, 2	Ground		7/2		5/2	21 315.8		1001
	724	3, 0, 3← 2, 0, 2	Ground		9/2		7/2	21 315.8		1001
	724	3, 0, 3← 2, 0, 2	Ground					21 319.8	.3	992
	724	3, 0, 3← 2, 0, 2	Ground		3/2		1/2	21 339.0		1001
	724	3, 0, 3← 2, 0, 2	Ground		5/2		3/2	21 339		1001
	724	3, 1, 2← 2, 1, 1	1		9/2		7/2	21 726.8		1001
	724	3, 1, 2← 2, 1, 1	Ground		5/2		5/2	21 732.7		1001
	724	3, 1, 2← 2, 1, 1	Ground		9/2		7/2	21 739.9		1001
	724	3, 1, 2← 2, 1, 1	Ground		3/2		1/2	21 739.9		1001
	724	3, 1, 2← 2, 1, 1	Ground					21 749.84	.05	992
	724	3, 1, 2← 2, 1, 1	1		5/2		3/2	21 750.6		1001
	724	3, 1, 2← 2, 1, 1	1		7/2		5/2	21 750.6		1001
	724	3, 1, 2← 2, 1, 1	Ground		5/2		3/2	21 763.7		1001
	724	3, 1, 2← 2, 1, 1	Ground		7/2		5/2	21 763.7		1001
	724	3, 1, 2← 2, 1, 1	Ground		7/2		7/2	21 807.1		1001
	724	3, 1, 3← 2, 1, 2	1		9/2		7/2	20 871.4		1001
	724	3, 1, 3← 2, 1, 2	Ground		5/2		5/2	20 878.4		1001
	724	3, 1, 3← 2, 1, 2	Ground		9/2		7/2	20 890.5		1001
	724	3, 1, 3← 2, 1, 2	Ground					20 900.84	.04	992
	724	3, 1, 3← 2, 1, 2	Ground		7/2		5/2	20 914.3		1001
	724	3, 1, 3← 2, 1, 2	Ground		5/2		3/2	20 916.1		1001
	724	3, 1, 3← 2, 1, 2	Ground		7/2		7/2	20 967.8		1001
	724	3, 2, 1← 2, 2, 0	Ground		3/2		1/2	21 235.9		1001
	724	3, 2, 1← 2, 2, 0	Ground		9/2		7/2	21 304.8		1001
	724	3, 2, 1← 2, 2, 0	Ground		7/2		7/2	21 304.8		1001
	724	3, 2, 1← 2, 2, 0	Ground					21 332.29	.07	992

Isotopic Species	Id. No.	Rotational Quantum Nos.	Vib. State v_a	Hyperfine				Frequency MHz	Acc. ±MHz	Ref.
				F_1'	F'	F_1	F			
$C^{12}D_2{:}C^{12}HBr^{81}$	724	3, 2, 1← 2, 2, 0	Ground		3/2		3/2	21 332.5		1001
	724	3, 2, 1← 2, 2, 0	Ground		5/2		3/2	21 332.5		1001
	724	3, 2, 1← 2, 2, 0	Ground		7/2		5/2	21 401.1		1001
	724	3, 2, 1← 2, 2, 0	Ground		5/2		5/2	21 401.1		1001
	724	3, 2, 2← 2, 2, 1	Ground		3/2		1/2	21 231.7		1001
	724	3, 2, 2← 2, 2, 1	Ground		9/2		7/2	21 299.6		1001
	724	3, 2, 2← 2, 2, 1	Ground		7/2		7/2	21 299.6		1001
	724	3, 2, 2← 2, 2, 1	Ground					21 326.95	.06	992
	724	3, 2, 2← 2, 2, 1	Ground		5/2		3/2	21 327.2		1001
	724	3, 2, 2← 2, 2, 1	Ground		3/2		3/2	21 327.2		1001
	724	3, 2, 2← 2, 2, 1	Ground		7/2		5/2	21 395.1		1001
	724	3, 2, 2← 2, 2, 1	Ground		5/2		5/2	21 395.1		1001
	724	Not Reported	Ground		5/2		5/2	13 969.6	.1	903
$C^{12}D_2{:}C^{12}DBr^{79}$	725	2, 0, 2← 1, 0, 1	Ground		7/2		5/2	14 015.6		1001
	725	2, 0, 2← 1, 0, 1	Ground		5/2		3/2	14 015.6		1001
	725	2, 0, 2← 1, 0, 1	Ground					14 022.12	.05	992
	725	2, 0, 2← 1, 0, 1	Ground		5/2		5/2	14 133.0		1001
	725	2, 1, 1← 1, 1, 0	Ground		1/2		1/2	14 252.2		1001
	725	2, 1, 1← 1, 1, 0	Ground		3/2		1/2	14 305.9		1001
	725	2, 1, 1← 1, 1, 0	Ground		7/2		5/2	14 340.4		1001
	725	2, 1, 1← 1, 1, 0	Ground					14 369.00	.04	992
	725	2, 1, 1← 1, 1, 0	Ground		5/2		5/2	14 394.4		1001
	725	2, 1, 1← 1, 1, 0	Ground		3/2		3/2	14 420.0		1001
	725	2, 1, 1← 1, 1, 0	Ground		5/2		3/2	14 458.0		1001
	725	2, 1, 2← 1, 1, 1	Ground		1/2		1/2	13 572.4		1001
	725	2, 1, 2← 1, 1, 1	Ground		3/2		1/2	13 635.9		1001
	725	2, 1, 2← 1, 1, 1	Ground		7/2		5/2	13 660.5		1001
	725	2, 1, 2← 1, 1, 1	Ground					13 689.58	.04	992
	725	2, 1, 2← 1, 1, 1	Ground		5/2		5/2	13 724.5		1001
	725	2, 1, 2← 1, 1, 1	Ground		3/2		3/2	13 732.7		1001
	725	2, 1, 2← 1, 1, 1	Ground		5/2		3/2	13 777.6		1001
	725	3, 0, 3← 2, 0, 2	Ground		3/2		3/2	20 935.9		1001
	725	3, 0, 3← 2, 0, 2	Ground		5/2		5/2	20 971.0		1001
	725	3, 0, 3← 2, 0, 2	Ground		9/2		7/2	21 026.3		1001
	725	3, 0, 3← 2, 0, 2	Ground		7/2		5/2	21 026.3		1001
	725	3, 0, 3← 2, 0, 2	Ground					21 032.00	.25	992
	725	3, 0, 3← 2, 0, 2	Ground		3/2		1/2	21 055.4		1001
	725	3, 0, 3← 2, 0, 2	Ground		5/2		3/2	21 055.4		1001
	725	3, 0, 3← 2, 0, 2	Ground		7/2		7/2	21 143.3		1001
	725	3, 1, 2← 2, 1, 1	Ground		3/2		3/2	21 477.7		1001
	725	3, 1, 2← 2, 1, 1	1		5/2		5/2	21 516.3		1001
	725	3, 1, 2← 2, 1, 1	1		3/2		1/2	21 525.5		1001
	725	3, 1, 2← 2, 1, 1	1		9/2		7/2	21 525.5		1001
	725	3, 1, 2← 2, 1, 1	Ground		5/2		5/2	21 529.3		1001
	725	3, 1, 2← 2, 1, 1	Ground		9/2		7/2	21 538.4		1001
	725	3, 1, 2← 2, 1, 1	Ground		3/2		1/2	21 538.4		1001
	725	3, 1, 2← 2, 1, 1	Ground					21 548.27	.05	992
	725	3, 1, 2← 2, 1, 1	1		5/2		3/2	21 554.1		1001
	725	3, 1, 2← 2, 1, 1	1		7/2		5/2	21 554.1		1001

Isotopic Species	Id. No.	Rotational Quantum Nos.	Vib. State v_a	F'_1	F'	F_1	F	Frequency MHz	Acc. ±MHz	Ref.
$C^{12}D_2$:$C^{12}DBr^{79}$	725	3, 1, 2← 2, 1, 1	Ground		5/2		3/2	21 567.2		1001
	725	3, 1, 2← 2, 1, 1	Ground		7/2		5/2	21 567.2		1001
	725	3, 1, 2← 2, 1, 1	1		7/2		7/2	21 608.0		1001
	725	3, 1, 2← 2, 1, 1	Ground		7/2		7/2	21 621.4		1001
	725	3, 1, 3← 2, 1, 2	Ground		3/2		3/2	20 456.8		1001
	725	3, 1, 3← 2, 1, 2	1		9/2		7/2	20 499.1		1001
	725	3, 1, 3← 2, 1, 2	Ground		5/2		5/2	20 504.4		1001
	725	3, 1, 3← 2, 1, 2	Ground		9/2		7/2	20 518.4		1001
	725	3, 1, 3← 2, 1, 2	Ground		3/2		1/2	20 519.9		1001
	725	3, 1, 3← 2, 1, 2	1		7/2		5/2	20 528.1		1001
	725	3, 1, 3← 2, 1, 2	1		5/2		3/2	20 530.2		1001
	725	3, 1, 3← 2, 1, 2	Ground					20 531.11	.04	992
	725	3, 1, 3← 2, 1, 2	Ground		7/2		5/2	20 547.4		1001
	725	3, 1, 3← 2, 1, 2	Ground		5/2		3/2	20 549.4		1001
	725	3, 1, 3← 2, 1, 2	Ground		7/2		7/2	20 611.4		1001
	725	3, 2, 1← 2, 2, 0	Ground		3/2		1/2	20 935.9		1001
	725	3, 2, 1← 2, 2, 0	Ground		7/2		7/2	21 019.5		1001
	725	3, 2, 1← 2, 2, 0	Ground		9/2		7/2	21 019.5		1001
	725	3, 2, 1← 2, 2, 0	Ground					21 053.15	.06	992
	725	3, 2, 1← 2, 2, 0	Ground		3/2		3/2	21 053.5		1001
	725	3, 2, 1← 2, 2, 0	Ground		5/2		3/2	21 053.5		1001
	725	3, 2, 1← 2, 2, 0	Ground		7/2		5/2	21 137.0		1001
	725	3, 2, 1← 2, 2, 0	Ground		5/2		5/2	21 137.0		1001
	725	3, 2, 2← 2, 2, 1	Ground		3/2		1/2	20 925.7		1001
	725	3, 2, 2← 2, 2, 1	Ground		7/2		7/2	21 009.4		1001
	725	3, 2, 2← 2, 2, 1	Ground		9/2		7/2	21 009.4		1001
	725	3, 2, 2← 2, 2, 1	Ground					21 043.02	.07	992
	725	3, 2, 2← 2, 2, 1	Ground		5/2		3/2	21 043.2		1001
	725	3, 2, 2← 2, 2, 1	Ground		3/2		3/2	21 043.2		1001
	725	3, 2, 2← 2, 2, 1	Ground		7/2		5/2	21 126.9		1001
	725	3, 2, 2← 2, 2, 1	Ground		5/2		5/2	21 126.9		1001
	725	Not Reported	Ground		3/2		3/2	13 922.7	.1	903
	725	Not Reported	Ground		1/2		1/2	14 021.0	.1	903
	725	Not Reported	Ground		1/2		1/2	14 248.4	.1	903
$C^{12}D_2$:$C^{12}DBr^{81}$	726	2, 0, 2← 1, 0, 1	Ground		3/2		3/2	13 863.9		1001
	726	2, 0, 2← 1, 0, 1	Ground		7/2		5/2	13 933.4		1001
	726	2, 0, 2← 1, 0, 1	Ground		5/2		3/2	13 933.4		1001
	726	2, 0, 2← 1, 0, 1	Ground					13 942.31	.06	992
	726	2, 0, 2← 1, 0, 1	Ground		5/2		5/2	14 031.4		1001
	726	2, 1, 1← 1, 1, 0	Ground		1/2		1/2	14 183.1		1001
	726	2, 1, 1← 1, 1, 0	Ground		3/2		1/2	14 228.0		1001
	726	2, 1, 1← 1, 1, 0	Ground		7/2		5/2	14 256.9		1001
	726	2, 1, 1← 1, 1, 0	Ground					14 280.71	.04	992
	726	2, 1, 1← 1, 1, 0	Ground		5/2		5/2	14 302.1		1001
	726	2, 1, 1← 1, 1, 0	Ground		3/2		3/2	14 323.2		1001
	726	2, 1, 1← 1, 1, 0	Ground		5/2		3/2	14 355.2		1001
	726	2, 1, 2← 1, 1, 1	Ground		1/2		1/2	13 511.6		1001
	726	2, 1, 2← 1, 1, 1	Ground		3/2		1/2	13 564.3		1001
	726	2, 1, 2← 1, 1, 1	Ground		7/2		5/2	13 585.0		1001

Isotopic Species	Id. No.	Rotational Quantum Nos.	Vib. State v_a	F_1'	F'	F_1	F	Frequency MHz	Acc. ±MHz	Ref.
$C^{12}D_2{:}C^{12}DBr^{81}$	726	2, 1, 2← 1, 1, 1	Ground					13 609.29	.05	992
	726	·2, 1, 2← 1, 1, 1	Ground		5/2		5/2	13 638.4		1001
	726	2, 1, 2← 1, 1, 1	Ground		3/2		3/2	13 645.4		1001
	726	2, 1, 2← 1, 1, 1	Ground		5/2		3/2	13 683.0		1001
	726	3, 0, 3← 2, 0, 2	Ground		3/2		3/2	20 828.5		1001
	726	3, 0, 3← 2, 0, 2	Ground		7/2		5/2	20 901.4		1001
	726	3, 0, 3← 2, 0, 2	Ground		9/2		7/2	20 901.4		1001
	726	3, 0, 3← 2, 0, 2	Ground					20 906.26	.2	992
	726	3, 0, 3← 2, 0, 2	Ground		3/2		1/2	20 925.7		1001
	726	3, 0, 3← 2, 0, 2	Ground		5/2		3/2	20 925.7		1001
	726	3, 1, 2← 2, 1, 1	Ground		3/2		3/2	21 362.8		1001
	726	3, 1, 2← 2, 1, 1	1		3/2		1/2	21 395.6		1001
	726	3, 1, 2← 2, 1, 1	1		9/2		7/2	21 395.6		1001
	726	3, 1, 2← 2, 1, 1	Ground		5/2		5/2	21 400.7		1001
	726	3, 1, 2← 2, 1, 1	Ground		9/2		7/2	21 408.5		1001
	726	3, 1, 2← 2, 1, 1	Ground		3/2		1/2	21 408.5		1001
	726	3, 1, 2← 2, 1, 1	1		5/2		3/2	21 419.7		1001
	726	3, 1, 2← 2, 1, 1	1		7/2		5/2	21 419.7		1001
	726	3, 1, 2← 2, 1, 1	Ground					21 420.24	.05	992
	726	3, 1, 2← 2, 1, 1	Ground		7/2		5/2	21 432.6		1001
	726	3, 1, 2← 2, 1, 1	Ground		5/2		3/2	21 432.6		1001
	726	3, 1, 2← 2, 1, 1	Ground		7/2		7/2	21 483.7		1001
	726	3, 1, 3← 2, 1, 2	Ground		3/2		3/2	20 349.3		1001
	726	3, 1, 3← 2, 1, 2	1		9/2		7/2	20 381.3		1001
	726	3, 1, 3← 2, 1, 2	Ground		5/2		5/2	20 388.7		1001
	726	3, 1, 3← 2, 1, 2	Ground		9/2		7/2	20 400.5		1001
	726	3, 1, 3← 2, 1, 2	Ground		3/2		1/2	20 401.7		1001
	726	3, 1, 3← 2, 1, 2	1		7/2		5/2	20 405.6		1001
	726	3, 1, 3← 2, 1, 2	Ground					20 411.14	.05	992
	726	3, 1, 3← 2, 1, 2	Ground		7/2		5/2	20 424.6		1001
	726	3, 1, 3← 2, 1, 2	Ground		5/2		3/2	20 426.4		1001
	726	3, 1, 3← 2, 1, 2	Ground		7/2		7/2	20 478.9		1001
	726	3, 2, 1← 2, 2, 0	Ground		3/2		1/2	20 827.4		1001
	726	3, 2, 1← 2, 2, 0	Ground		9/2		7/2	20 898.6		1001
	726	3, 2, 1← 2, 2, 0	Ground		7/2		7/2	20 898.6		1001
	726	3, 2, 1← 2, 2, 0	Ground					20 926.71	.06	992
	726	3, 2, 1← 2, 2, 0	Ground		5/2		3/2	20 927.0		1001
	726	3, 2, 1← 2, 2, 0	Ground		3/2		3/2	20 927.0		1001
	726	3, 2, 1← 2, 2, 0	Ground		5/2		5/2	20 997.1		1001
	726	3, 2, 1← 2, 2, 0	Ground		7/2		5/2	20 997.1		1001
	726	3, 2, 2← 2, 2, 1	Ground		3/2		1/2	20 819.6		1001
	726	3, 2, 2← 2, 2, 1	Ground		9/2		7/2	20 888.9		1001
	726	3, 2, 2← 2, 2, 1	Ground		7/2		7/2	20 888.9		1001
	726	3, 2, 2← 2, 2, 1	Ground		5/2		3/2	20 917.0		1001
	726	3, 2, 2← 2, 2, 1	Ground		3/2		3/2	20 917.0		1001
	726	3, 2, 2← 2, 2, 1	Ground					20 917.00	.06	992
	726	3, 2, 2← 2, 2, 1	Ground		5/2		5/2	20 987.0		1001
	726	3, 2, 2← 2, 2, 1	Ground		7/2		5/2	20 987.0		1001
	726	Not Reported	Ground		1/2		3/2	13 592.6	.1	903
	726	Not Reported	Ground		3/2		5/2	13 592.6	.1	903

Isotopic Species	Id. No.	Rotational Quantum Nos.	Vib. State v_a	F_1'	F'	F_1	F	Frequency MHz	Acc. ±MHz	Ref.
$C^{12}D_2 : C^{12}DBr^{81}$	726	Not Reported	Ground		3/2		3/2	13 647.7	.2	903
	726	Not Reported	Ground		1/2		1/2	13 938.8	.1	903
$C^{12}H_2 : C^{12}HBr^{79}$	727	1, 0, 1← 0, 0, 0	Ground		5/2		3/2	8 001.82		1001
	727	1, 0, 1← 0, 0, 0	Ground		3/2		3/2	8 119.60		1001
	727	2, 0, 2← 1, 0, 1	Ground		3/2		3/2	15 955.1		1001
	727	2, 0, 2← 1, 0, 1	1		7/2		5/2	16 023.0		1001
	727	2, 0, 2← 1, 0, 1	1		5/2		3/2	16 023.0		1001
	727	2, 0, 2← 1, 0, 1	Ground		5/2		3/2	16 038.6		1001
	727	2, 0, 2← 1, 0, 1	Ground		7/2		5/2	16 038.6		1001
	727	2, 0, 2← 1, 0, 1	Ground		7/2		5/2	16 038.7		937
	727	2, 0, 2← 1, 0, 1	Ground					16 048.91	.06	992
	727	2, 0, 2← 1, 0, 1	Ground		5/2		3/2	16 156.4		1001
	727	2, 0, 2← 1, 0, 1	Ground		5/2		5/2	16 156.5	.2	937
	727	2, 1, 1← 1, 1, 0	Ground		1/2		1/2	16 233.5		1001
	727	2, 1, 1← 1, 1, 0	Ground		3/2		1/2	16 287.6		1001
	727	2, 1, 1← 1, 1, 0	1		7/2		5/2	16 309.1		1001
	727	2, 1, 1← 1, 1, 0	Ground		7/2		5/2	16 322.1		1001
	727	2, 1, 1← 1, 1, 0	Ground					16 350.85	.04	992
	727	2, 1, 1← 1, 1, 0	Ground		5/2		5/2	16 376.7		1001
	727	2, 1, 1← 1, 1, 0	Ground		3/2		3/2	16 401.8		1001
	727	2, 1, 1← 1, 1, 0	Ground		5/2		3/2	16 440.3		1001
	727	2, 1, 2← 1, 1, 1	Ground		1/2		1/2	15 633.5		1001
	727	2, 1, 2← 1, 1, 1	Ground		3/2		1/2	15 696.9		1001
	727	2, 1, 2← 1, 1, 1	1		7/2		5/2	15 702.1		1001
	727	2, 1, 2← 1, 1, 1	Ground		7/2		5/2	15 720.8		937
	727	2, 1, 2← 1, 1, 1	Ground		7/2		5/2	15 720.8		1001
	727	2, 1, 2← 1, 1, 1	Ground					15 750.57	.04	992
	727	2, 1, 2← 1, 1, 1	Ground		5/2		5/2	15 784.8		937
	727	2, 1, 2← 1, 1, 1	Ground		3/2		3/2	15 794.2		1001
	727	2, 1, 2← 1, 1, 1	Ground		5/2		3/2	15 839.2		1001
	727	2, 1, 2← 1, 1, 1	Ground		5/2		3/2	15 839.3		937
	727	3, 0, 3← 2, 0, 2	Ground		3/2		3/2	23 975.0		1001
	727	3, 0, 3← 2, 0, 2	Ground		5/2		5/2	24 001.6	.3	259
	727	3, 0, 3← 2, 0, 2	Ground		5/2		5/2	24 009.30		1001
	727	3, 0, 3← 2, 0, 2	1		9/2		7/2	24 040.9	.3	259 [2]
	727	3, 0, 3← 2, 0, 2	Ground		9/2		7/2	24 063.7		937
	727	3, 0, 3← 2, 0, 2	Ground		7/2		5/2	24 064.00		1001
	727	3, 0, 3← 2, 0, 2	Ground		9/2		7/2	24 064.00		1001
	727	3, 0, 3← 2, 0, 2	1		3/2		1/2	24 069.10		1001
	727	3, 0, 3← 2, 0, 2	1		5/2		3/2	24 069.10		1001
	727	3, 0, 3← 2, 0, 2	Ground					24 069.18	.08	992
	727	3, 0, 3← 2, 0, 2	Ground		3/2		1/2	24 092.4		1001
	727	3, 0, 3← 2, 0, 2	Ground		5/2		3/2	24 092.4		1001
	727	3, 0, 3← 2, 0, 2	Ground		5/2		3/2	24 092.5		937
	727	3, 0, 3← 2, 0, 2	Ground		7/2		7/2	24 181.45		1001
	727	3, 1, 2← 2, 1, 1	Ground		3/2		3/2	24 457.1		1001
	727	3, 1, 2← 2, 1, 1	1		5/2		5/2	24 483.0		1001
	727	3, 1, 2← 2, 1, 1	1		9/2		7/2	24 492.55		1001
	727	3, 1, 2← 2, 1, 1	1		3/2		1/2	24 492.55		1001

2. The author indicates excited mode as v_5, but later measurements on other lines in this same group have been identified as v_{10}.

Isotopic Species	Id. No.	Rotational Quantum Nos.	Vib. State v_a	F_1'	F'	F_1	F	Frequency MHz	Acc. ±MHz	Ref.
$C^{12}H_2{:}C^{12}HBr^{79}$	727	3, 1, 2← 2, 1, 1	Ground		5/2		5/2	24 502.3		1001
	727	3, 1, 2← 2, 1, 1	Ground		3/2		1/2	24 511.95		1001
	727	3, 1, 2← 2, 1, 1	Ground		9/2		7/2	24 511.95		1001
	727	3, 1, 2← 2, 1, 1	1		7/2		5/2	24 521.3	.3	295 [2]
	727	3, 1, 2← 2, 1, 1	1		5/2		3/2	24 521.5		1001
	727	3, 1, 2← 2, 1, 1	1		7/2		5/2	24 521.5		1001
	727	3, 1, 2← 2, 1, 1	Ground					24 524.28	.05	992
	727	3, 1, 2← 2, 1, 1	Ground		7/2		5/2	24 541.1		1001
	727	3, 1, 2← 2, 1, 1	Ground		5/2		3/2	24 541.1		1001
	727	3, 1, 2← 2, 1, 1	Ground		7/2		7/2	24 595.70		1001
	727	3, 1, 3← 2, 1, 2	1		3/2		3/2	23 522.4		1001
	727	3, 1, 3← 2, 1, 2	Ground		3/2		3/2	23 549.8	.2	937
	727	3, 1, 3← 2, 1, 2	1		5/2		5/2	23 570.4		1001
	727	3, 1, 3← 2, 1, 2	1		9/2		7/2	23 584.4		1001
	727	3, 1, 3← 2, 1, 2	1		9/2		7/2	23 585.4	.3	259 [2]
	727	3, 1, 3← 2, 1, 2	Ground		5/2		5/2	23 597.55		1001
	727	3, 1, 3← 2, 1, 2	Ground		9/2		7/2	23 611.6		937
	727	3, 1, 3← 2, 1, 2	Ground		9/2		7/2	23 612.0	.3	259
	727	3, 1, 3← 2, 1, 2	Ground		3/2		1/2	23 613.0		1001
	727	3, 1, 3← 2, 1, 2	1		5/2		3/2	23 615.6		1001
	727	3, 1, 3← 2, 1, 2	Ground					23 624.28	.04	992
	727	3, 1, 3← 2, 1, 2	Ground		7/2		5/2	23 640.65		1001
	727	3, 1, 3← 2, 1, 2	Ground		7/2		5/2	23 640.7		937
	727	3, 1, 3← 2, 1, 2	Ground		5/2		3/2	23 642.50		1001
	727	3, 1, 3← 2, 1, 2	Ground		7/2		7/2	23 704.8		1001
	727	3, 1, 3← 2, 1, 2	Ground		7/2		7/2	23 704.9		937
	727	3, 1, 3← 2, 1, 2	Ground		7/2		7/2	23 795.8	.3	259
	727	3, 2, 1← 2, 2, 0	Ground		3/2		1/2	23 963.6		1001
	727	3, 2, 1← 2, 2, 0	1		7/2		7/2	24 024.05		1001
	727	3, 2, 1← 2, 2, 0	1		9/2		7/2	24 024.05		1001
	727	3, 2, 1← 2, 2, 0	Ground		7/2		7/2	24 047.4		1001
	727	3, 2, 1← 2, 2, 0	Ground		9/2		7/2	24 047.4		1001
	727	3, 2, 1← 2, 2, 0	Ground					24 081.22	.05	992
	727	3, 2, 1← 2, 2, 0	Ground		5/2		3/2	24 081.3		1001
	727	3, 2, 1← 2, 2, 0	Ground		3/2		3/2	24 081.3		1001
	727	3, 2, 1← 2, 2, 0	1		5/2		5/2	24 142.1		1001
	727	3, 2, 1← 2, 2, 0	1		7/2		5/2	24 142.1		1001
	727	3, 2, 1← 2, 2, 0	Ground		5/2		5/2	24 165.2		1001
	727	3, 2, 1← 2, 2, 0	Ground		7/2		5/2	24 165.2		1001
	727	3, 2, 2← 2, 2, 1	1		3/2		1/2	23 935.0		1001
	727	3, 2, 2← 2, 2, 1	Ground		3/2		1/2	23 958.5		1001
	727	3, 2, 2← 2, 2, 1	Ground		9/2		7/2	24 042.1		1001
	727	3, 2, 2← 2, 2, 1	Ground		7/2		7/2	24 042.1		1001
	727	3, 2, 2← 2, 2, 1	1		5/2		3/2	24 052.5		1001
	727	3, 2, 2← 2, 2, 1	1		3/2		3/2	24 052.5		1001
	727	3, 2, 2← 2, 2, 1	Ground		3/2		3/2	24 076.1		1001
	727	3, 2, 2← 2, 2, 1	Ground		5/2		3/2	24 076.1		1001
	727	3, 2, 2← 2, 2, 1	Ground					24 076.44	.05	992
	727	3, 2, 2← 2, 2, 1	Ground		5/2		3/2	24 076.7		937
	727	3, 2, 2← 2, 2, 1	1		7/2		5/2	24 136.9		1001

2. The author indicates excited mode as v_5, but later measurements on other lines in this same group have been identified as v_{10}.

Isotopic Species	Id. No.	Rotational Quantum Nos.	Vib. State v_a	Hyperfine				Frequency MHz	Acc. ±MHz	Ref.
				F_1'	F'	F_1	F			
$C^{12}H_2$:$C^{12}HBr^{79}$	727	3, 2, 2← 2, 2, 1	1		5/2		5/2	24 136.9		1001
	727	3, 2, 2← 2, 2, 1	Ground		7/2		5/2	24 160.0		1001
	727	3, 2, 2← 2, 2, 1	Ground		5/2		5/2	24 160.0		1001
	727	3, 2, 2← 2, 2, 1	Ground		7/2		5/2	24 160.8	.2	937
	727	4, 0, 4← 3, 0, 3	Ground		7/2		5/2	32 098.6		1001
	727	4, 0, 4← 3, 0, 3	Ground		5/2		3/2	32 098.6		1001
	727	4, 1, 3← 3, 1, 2	Ground		5/2		5/2	32 612.0		1001
	727	4, 1, 3← 3, 1, 2	1		11/2		9/2	32 664.9		1001
	727	4, 1, 3← 3, 1, 2	Ground		7/2		7/2	32 670.5		1001
	727	4, 1, 3← 3, 1, 2	1		9/2		7/2	32 676.7		1001
	727	4, 1, 3← 3, 1, 2	1		7/2		5/2	32 682.2		1001
	727	4, 1, 3← 3, 1, 2	Ground		11/2		9/2	32 691.7		1001
	727	4, 1, 3← 3, 1, 2	Ground		5/2		3/2	32 695.9		1001
	727	4, 1, 3← 3, 1, 2	Ground					32 697.20		1001
	727	4, 1, 3← 3, 1, 2	Ground		9/2		7/2	32 702.2		1001
	727	4, 1, 3← 3, 1, 2	Ground		7/2		5/2	32 708.0		1001
	727	4, 1, 3← 3, 1, 2	Ground		9/2		9/2	32 785.9		1001
	727	4, 1, 4← 3, 1, 3	Ground		5/2		5/2	31 403.6		1001
	727	4, 1, 4← 3, 1, 3	1		7/2		7/2	31 428.7		1001
	727	4, 1, 4← 3, 1, 3	1		11/2		9/2	31 453.6		1001
	727	4, 1, 4← 3, 1, 3	1		5/2		3/2	31 460.2		1001
	727	4, 1, 4← 3, 1, 3	Ground		7/2		7/2	31 465.4		1001
	727	4, 1, 4← 3, 1, 3	Ground		11/2		9/2	31 490.0		1001
	727	4, 1, 4← 3, 1, 3	Ground		5/2		3/2	31 496.4		1001
	727	4, 1, 4← 3, 1, 3	Ground					31 497.07		1001
	727	4, 1, 4← 3, 1, 3	Ground		9/2		7/2	31 501.7		1001
	727	4, 1, 4← 3, 1, 3	Ground		7/2		5/2	31 508.25		1001
	727	4, 1, 4← 3, 1, 3	Ground		9/2		9/2	31 594.8		1001
	727	4, 2, 2← 3, 2, 1	1		5/2		3/2	32 048.1		1001
	727	4, 2, 2← 3, 2, 1	1		5/2		5/2	32 048.1		1001
	727	4, 2, 2← 3, 2, 1	Ground		5/2		5/2	32 079.25		1001
	727	4, 2, 2← 3, 2, 1	Ground		5/2		3/2	32 079.25		1001
	727	4, 2, 2← 3, 2, 1	Ground		11/2		9/2	32 095.8		1001
	727	4, 2, 2← 3, 2, 1	Ground					32 112.88		1001
	727	4, 2, 2← 3, 2, 1	Ground		7/2		7/2	32 126.35		1001
	727	4, 2, 2← 3, 2, 1	Ground		7/2		5/2	32 126.35		1001
	727	4, 2, 2← 3, 2, 1	Ground		9/2		9/2	32 142.8		1001
	727	4, 2, 2← 3, 2, 1	Ground		9/2		7/2	32 142.8		1001
	727	4, 2, 3← 3, 2, 2	1		5/2		3/2	32 034.6		1001
	727	4, 2, 3← 3, 2, 2	1		5/2		5/2	32 034.6		1001
	727	4, 2, 3← 3, 2, 2	Ground		5/2		5/2	32 065.4		1001
	727	4, 2, 3← 3, 2, 2	Ground		5/2		3/2	32 065.4		1001
	727	4, 2, 3← 3, 2, 2	Ground		11/2		9/2	32 082.5		1001
	727	4, 2, 3← 3, 2, 2	Ground					32 099.42		1001
	727	4, 2, 3← 3, 2, 2	Ground		7/2		5/2	32 113.0		1001
	727	4, 2, 3← 3, 2, 2	Ground		7/2		7/2	32 113.0		1001
	727	4, 2, 3← 3, 2, 2	Ground		9/2		9/2	32 129.4		1001
	727	4, 2, 3← 3, 2, 2	Ground		9/2		7/2	32 129.4		1001
	727	4, 3, 2← 3, 3, 1	1		9/2		9/2	32 000.4		1001 [1]
	727	4, 3, 2← 3, 3, 1	Ground		5/2		3/2	32 016.6		1001 [1]

1. All $4_{32} \leftarrow 3_{31}$ assignments are uncertain; they may be $4_{31} \leftarrow 3_{30}$.

143

Isotopic Species	Id. No.	Rotational Quantum Nos.	Vib. State v_a	Hyperfine F_1'	F'	F_1	F	Frequency MHz	Acc. ±MHz	Ref.
$C^{12}H_2:C^{12}HBr^{79}$	727	4, 3, 2← 3, 3, 1	Ground		9/2		9/2	32 029.6		1001 [1]
	727	4, 3, 2← 3, 3, 1	l		11/2		9/2	32 039.4		1001 [1]
	727	4, 3, 2← 3, 3, 1	Ground		11/2		9/2	32 070.4		1001 [1]
	727	4, 3, 2← 3, 3, 1	l		7/2		5/2	32 091.5		1001 [1]
	727	4, 3, 2← 3, 3, 1	Ground					32 104.75		1001 [1]
	727	4, 3, 2← 3, 3, 1	Ground		7/2		5/2	32 122.2		1001 [1]
	727	4, 3, 2← 3, 3, 1	l		9/2		7/2	32 145.8		1001 [1]
	727	4, 3, 2← 3, 3, 1	l		7/2		7/2	32 159.7		1001 [1]
	727	4, 3, 2← 3, 3, 1	Ground		5/2		5/2	32 164.2		1001 [1]
	727	4, 3, 2← 3, 3, 1	Ground		9/2		7/2	32 176.65		1001 [1]
	727	4, 3, 2← 3, 3, 1	Ground		7/2		7/2	32 190.7		1001 [1]
$C^{12}H_2:C^{12}HBr^{81}$	728	1, 0, 1← 0, 0, 0	Ground		5/2		3/2	7 960.00		1001
	728	1, 0, 1← 0, 0, 0	Ground		3/2		3/2	8 058.90		1001
	728	2, 0, 2← 1, 0, 1	Ground		3/2		3/2	15 879.6		937
	728	2, 0, 2← 1, 0, 1	Ground		3/2		3/2	15 879.7		1001
	728	2, 0, 2← 1, 0, 1	l		7/2		5/2	15 934.2		1001
	728	2, 0, 2← 1, 0, 1	l		5/2		3/2	15 934.2		1001
	728	2, 0, 2← 1, 0, 1	Excited		7/2		5/2	15 934.3		937
	728	2, 0, 2← 1, 0, 1	Ground		7/2		5/2	15 949.5		1001
	728	2, 0, 2← 1, 0, 1	Ground		5/2		3/2	15 949.5		1001
	728	2, 0, 2← 1, 0, 1	Ground					15 958.13	.06	992
	728	2, 0, 2← 1, 0, 1	Ground		5/2		5/2	16 048.1		1001
	728	2, 1, 1← 1, 1, 0	Ground		1/2		1/2	16 158.4		1001
	728	2, 1, 1← 1, 1, 0	Ground		3/2		1/2	16 203.7		1001
	728	2, 1, 1← 1, 1, 0	Ground		3/2		1/2	16 203.8		937
	728	2, 1, 1← 1, 1, 0	l		7/2		5/2	16 219.8		1001
	728	2, 1, 1← 1, 1, 0	Ground		7/2		5/2	16 232.7		1001
	728	2, 1, 1← 1, 1, 0	Ground					16 256.67	.04	992
	728	2, 1, 1← 1, 1, 0	Ground		5/2		5/2	16 278.3		1001
	728	2, 1, 1← 1, 1, 0	Ground		3/2		3/2	16 299.2		1001
	728	2, 1, 1← 1, 1, 0	Ground		5/2		3/2	16 331.4		1001
	728	2, 1, 2← 1, 1, 1	Ground		1/2		1/2	15 565.1		1001
	728	2, 1, 2← 1, 1, 1	Ground		3/2		1/2	15 618.0		1001
	728	2, 1, 2← 1, 1, 1	Ground		3/2		1/2	15 618.2	.2	937
	728	2, 1, 2← 1, 1, 1	l		7/2		5/2	15 620.1		1001
	728	2, 1, 2← 1, 1, 1	Ground		7/2		5/2	15 638.2		1001
	728	2, 1, 2← 1, 1, 1	Ground					15 663.84	.04	992
	728	2, 1, 2← 1, 1, 1	Ground		5/2		5/2	15 691.3		1001
	728	2, 1, 2← 1, 1, 1	Ground		5/2		5/2	15 691.7		937
	728	2, 1, 2← 1, 1, 1	Ground		3/2		3/2	15 699.5		1001
	728	2, 1, 2← 1, 1, 1	Ground		5/2		3/2	15 737.2		1001
	728	3, 0, 3← 2, 0, 2	Ground		5/2		5/2	23 850	5	259
	728	3, 0, 3← 2, 0, 2	Ground		3/2		3/2	23 854.6		1001
	728	3, 0, 3← 2, 0, 2	Ground		5/2		5/2	23 883.1		1001
	728	3, 0, 3← 2, 0, 2	l		9/2		7/2	23 904.9	.3	259 [2]
	728	3, 0, 3← 2, 0, 2	Ground		9/2		7/2	23 928.9		1001
	728	3, 0, 3← 2, 0, 2	Ground		7/2		5/2	23 928.9		1001
	728	3, 0, 3← 2, 0, 2	Ground		7/2		5/2	23 929.30	.1	259
	728	3, 0, 3← 2, 0, 2	Ground		9/2		7/2	23 929.30	.1	259
	728	3, 0, 3← 2, 0, 2	Ground					23 933.31	.07	992
	728	3, 0, 3← 2, 0, 2	Ground		3/2		1/2	23 952.7		1001
	728	3, 0, 3← 2, 0, 2	Ground		5/2		3/2	23 952.7		1001
	728	3, 0, 3← 2, 0, 2	Ground		3/2		1/2	23 955.14	.1	259
	728	3, 0, 3← 2, 0, 2	Ground		5/2		3/2	23 955.14	.1	259

1. $4_{32} ← 3_{31}$ assignments are uncertain; they may be $4_{31} ← 3_{30}$.
2. The author indicates excited mode as v_5, but later measurements on other lines in this same group have been identified as v_{10}.

Isotopic Species	Id. No.	Rotational Quantum Nos.	Vib. State	v_a	F'_1	F'	F_1	F	Frequency MHz	Acc. ±MHz	Ref.
$C^{12}H_2:C^{12}HBr^{81}$	728	3, 0, 3← 2, 0, 2	Ground			7/2		7/2	24 027.15		1001
	728	3, 1, 2← 2, 1, 1		1		3/2		3/2	24 307.8		1001
	728	3, 1, 2← 2, 1, 1	Ground			3/2		3/2	24 327.1		1001
	728	3, 1, 2← 2, 1, 1		1		5/2		5/2	24 345.6		1001
	728	3, 1, 2← 2, 1, 1		1		3/2		1/2	24 353.60		1001
	728	3, 1, 2← 2, 1, 1		1		9/2		7/2	24 353.60		1001
	728	3, 1, 2← 2, 1, 1		1		7/2		5/2	24 353.9	.3	259 [2]
	728	3, 1, 2← 2, 1, 1		1		9/2		7/2	24 353.9	.3	259 [2]
	728	3, 1, 2← 2, 1, 1	Ground			5/2		5/2	24 365.0		259
	728	3, 1, 2← 2, 1, 1	Ground			9/2		7/2	24 372.90	.1	259
	728	3, 1, 2← 2, 1, 1	Ground			3/2		1/2	24 372.90	.1	259
	728	3, 1, 2← 2, 1, 1		1		5/2		3/2	24 378.0		1001
	728	3, 1, 2← 2, 1, 1		1		7/2		5/2	24 378.0		1001
	728	3, 1, 2← 2, 1, 1	Ground						24 383.26	.05	992
	728	3, 1, 2← 2, 1, 1	Ground			5/2		3/2	24 397.25		1001
	728	3, 1, 2← 2, 1, 1	Ground			7/2		5/2	24 397.25	.3	259
	728	3, 1, 2← 2, 1, 1		1		7/2		7/2	24 423.4		1001
	728	3, 1, 2← 2, 1, 1	Ground			7/2		7/2	24 443.0		1001
	728	3, 1, 3← 2, 1, 2	Ground			3/2		3/2	23 431.00		1001
	728	3, 1, 3← 2, 1, 2		1		9/2		7/2	23 455.00		1001
	728	3, 1, 3← 2, 1, 2	Excited			9/2		7/2	23 455.5		937
	728	3, 1, 3← 2, 1, 2		1		9/2		7/2	23 460.0	.3	259 [2]
	728	3, 1, 3← 2, 1, 2	Ground			5/2		5/2	23 470.9		937
	728	3, 1, 3← 2, 1, 2		1		7/2		5/2	23 480.00		1001
	728	3, 1, 3← 2, 1, 2	Ground			9/2		7/2	23 482.5		1001
	728	3, 1, 3← 2, 1, 2	Ground			3/2		1/2	23 483.90		1001
	728	3, 1, 3← 2, 1, 2	Ground						23 493.15	.04	992
	728	3, 1, 3← 2, 1, 2	Ground			7/2		5/2	23 507.1		1001
	728	3, 1, 3← 2, 1, 2	Ground			5/2		3/2	23 508.45		1001
	728	3, 1, 3← 2, 1, 2		1		7/2		7/2	23 531.40		1001
	728	3, 1, 3← 2, 1, 2	Ground			7/2		7/2	23 560.3		937
	728	3, 2, 1← 2, 2, 0	Ground			3/2		1/2	23 846.9		1001
	728	3, 2, 1← 2, 2, 0		1		9/2		7/2	23 893.5		1001
	728	3, 2, 1← 2, 2, 0		1		7/2		7/2	23 893.5		1001
	728	3, 2, 1← 2, 2, 0	Ground			7/2		7/2	23 560.3		937
	728	3, 2, 1← 2, 2, 0	Ground			7/2		7/2	23 916.6		1001
	728	3, 2, 1← 2, 2, 0		1		3/2		3/2	23 922.1		1001
	728	3, 2, 1← 2, 2, 0		1		5/2		3/2	23 922.1		1001
	728	3, 2, 1← 2, 2, 0	Ground						23 944.87	.05	992
	728	3, 2, 1← 2, 2, 0	Ground			3/2		3/2	23 945.1		1001
	728	3, 2, 1← 2, 2, 0	Ground			5/2		3/2	23 945.1		1001
	728	3, 2, 1← 2, 2, 0		1		5/2		5/2	23 992.4		1001
	728	3, 2, 1← 2, 2, 0		1		7/2		5/2	23 992.4		1001
	728	3, 2, 1← 2, 2, 0	Ground			5/2		5/2	24 015.1		1001
	728	3, 2, 1← 2, 2, 0	Ground			7/2		5/2	24 015.1		1001
	728	3, 2, 2← 2, 2, 1	Ground			3/2		1/2	23 841.6		1001
	728	3, 2, 2← 2, 2, 1	Ground			3/2		1/2	23 841.6	.2	937
	728	3, 2, 2← 2, 2, 1		1		9/2		7/2	23 888.4		1001
	728	3, 2, 2← 2, 2, 1		1		7/2		7/2	23 888.4		1001

2. The author indicates excited mode as v_5, but later measurements on other lines in this same group have been identified as v_{10}.

Isotopic Species	Id. No.	Rotational Quantum Nos.	Vib. State v_a	Hyperfine F_1'	F'	F_1	F	Frequency MHz	Acc. ±MHz	Ref.
$C^{12}H_2{:}C^{12}HBr^{81}$	728	3, 2, 2← 2, 2, 1	Ground		7/2		7/2	23 911.6		1001
	728	3, 2, 2← 2, 2, 1	Ground		9/2		7/2	23 911.6		937
	728	3, 2, 2← 2, 2, 1	Ground		5/2		3/2	23 940.0		1001
	728	3, 2, 2← 2, 2, 1	Ground		3/2		3/2	23 940.0		1001
	728	3, 2, 2← 2, 2, 1	Ground					23 940.06	.05	992
	728	3, 2, 2← 2, 2, 1	Ground		7/2		5/2	24 010.3		1001
	728	3, 2, 2← 2, 2, 1	Ground		5/2		5/2	24 010.3		1001
	728	4, 1, 3← 3, 1, 2	1		5/2		5/2	32 399.0		1001
	728	4, 1, 3← 3, 1, 2	Ground		5/2		5/2	32 438.2		1001
	728	4, 1, 3← 3, 1, 2	1		11/2		9/2	32 470.1		1001
	728	4, 1, 3← 3, 1, 2	1		5/2		3/2	32 478.1		1001
	728	4, 1, 3← 3, 1, 2	Ground		7/2		7/2	32 485.7		1001
	728	4, 1, 3← 3, 1, 2	1		7/2		5/2	32 488.0		1001
	728	4, 1, 3← 3, 1, 2	Ground		11/2		9/2	32 503.8		1001
	728	4, 1, 3← 3, 1, 2	Ground		5/2		3/2	32 508.5		1001
	728	4, 1, 3← 3, 1, 2	Ground					32 509.46		1001
	728	4, 1, 3← 3, 1, 2	Ground		9/2		7/2	32 513.7		1001
	728	4, 1, 3← 3, 1, 2	Ground		7/2		5/2	32 518.3		1001
	728	4, 1, 3← 3, 1, 2	Ground		9/2		9/2	32 583.5		1001
	728	4, 1, 4← 3, 1, 3	Ground		5/2		5/2	31 244.2		1001
	728	4, 1, 4← 3, 1, 3	1		7/2		7/2	31 259.6		1001
	728	4, 1, 4← 3, 1, 3	1		11/2		9/2	31 280.4		1001
	728	4, 1, 4← 3, 1, 3	1		5/2		3/2	31 286.0		1001
	728	4, 1, 4← 3, 1, 3	1		9/2		7/2	31 290.3		1001
	728	4, 1, 4← 3, 1, 3	Ground		7/2		7/2	31 295.8		1001
	728	4, 1, 4← 3, 1, 3	Ground		11/2		9/2	31 316.4		1001
	728	4, 1, 4← 3, 1, 3	Ground		5/2		3/2	31 321.8		1001
	728	4, 1, 4← 3, 1, 3	Ground					31 322.27		1001
	728	4, 1, 4← 3, 1, 3	Ground		9/2		7/2	31 326.2		1001
	728	4, 1, 4← 3, 1, 3	Ground		7/2		5/2	31 331.8		1001
	728	4, 1, 4← 3, 1, 3	Ground		9/2		9/2	31 403.6		1001
	728	4, 2, 2← 3, 2, 1	1		11/2		9/2	31 886.1		1001
	728	4, 2, 2← 3, 2, 1	1		7/2		7/2	31 912.05		1001
	728	4, 2, 2← 3, 2, 1	1		7/2		5/2	31 912.05		1001
	728	4, 2, 2← 3, 2, 1	Ground		11/2		9/2	31 917.0		1001
	728	4, 2, 2← 3, 2, 1	1		9/2		7/2	31 925.65		1001
	728	4, 2, 2← 3, 2, 1	1		9/2		9/2	31 925.65		1001
	728	4, 2, 2← 3, 2, 1	Ground					31 931.36		1001
	728	4, 2, 2← 3, 2, 1	Ground		7/2		5/2	31 942.8		1001
	728	4, 2, 2← 3, 2, 1	Ground		7/2		7/2	31 942.8		1001
	728	4, 2, 2← 3, 2, 1	Ground		9/2		7/2	31 956.2		1001
	728	4, 2, 2← 3, 2, 1	Ground		9/2		9/2	31 956.2		1001
	728	4, 2, 3← 3, 2, 2	1		5/2		5/2	31 859.0		1001
	728	4, 2, 3← 3, 2, 2	1		5/2		3/2	31 859.0		1001
	728	4, 2, 3← 3, 2, 2	1		11/2		9/2	31 872.85		1001
	728	4, 2, 3← 3, 2, 2	Ground		5/2		3/2	31 889.95		1001
	728	4, 2, 3← 3, 2, 2	Ground		5/2		5/2	31 889.95		1001
	728	4, 2, 3← 3, 2, 2	1		7/2		7/2	31 898.8		1001
	728	4, 2, 3← 3, 2, 2	1		7/2		5/2	31 898.8		1001
	728	4, 2, 3← 3, 2, 2	Ground		11/2		9/2	31 903.95		1001

Isotopic Species	Id. No.	Rotational Quantum Nos.	Vib. State v_a	F_1'	F'	F_1	F	Frequency MHz	Acc. ±MHz	Ref.
$C^{12}H_2{:}C^{12}HBr^{81}$	728	4, 2, 3← 3, 2, 2	Ground					31 918.09		1001
	728	4, 2, 3← 3, 2, 2	Ground		7/2		7/2	31 929.55		1001
	728	4, 2, 3← 3, 2, 2	Ground		7/2		5/2	31 929.55		1001
	728	4, 2, 3← 3, 2, 2	Ground		9/2		9/2	31 942.8		1001
	728	4, 2, 3← 3, 2, 2	Ground		9/2		7/2	31 942.8		1001
	728	4, 3, 2← 3, 3, 1	1		5/2		3/2	31 818.6		1001 [1]
	728	4, 3, 2← 3, 3, 1	Ground		5/2		3/2	31 849.5		1001 [1]
	728	4, 3, 2← 3, 3, 1	Ground		9/2		9/2	31 860.3		1001 [1]
	728	4, 3, 2← 3, 3, 1	1		11/2		9/2	31 863.7		1001 [1]
	728	4, 3, 2← 3, 3, 1	Ground		11/2		9/2	31 894.55		1001 [1]
	728	4, 3, 2← 3, 3, 1	1		7/2		5/2	31 907.2		1001 [1]
	728	4, 3, 2← 3, 3, 1	Ground					31 923.19		1001 [1]
	728	4, 3, 2← 3, 3, 1	Ground		7/2		5/2	31 937.95		1001 [1]
	728	4, 3, 2← 3, 3, 1	1		9/2		7/2	31 952.4		1001 [1]
	728	4, 3, 2← 3, 3, 1	Ground		5/2		5/2	31 972.8		1001 [1]
	728	4, 3, 2← 3, 3, 1	Ground		9/2		7/2	31 983.25		1001 [1]
	728	4, 3, 2← 3, 3, 1	Ground		7/2		7/2	31 995.00		1001 [1]
	728	Not Reported	Ground		1/2		1/2	15 565.1		937
	728	Not Reported	Ground		5/2		5/2	15 691.3	.1	903
	728	Not Reported	Ground		1/2		1/2	15 962.9	.2	903
	728	Not Reported	Ground		7/2		5/2	16 232.9		937
	728	Not Reported	Ground		5/2		5/2	16 278.4		937
	728	Not Reported	Ground		3/2		3/2	24 327.2		937
$C^{12}H_2{:}C^{12}DBr^b$	729	3, , ← 2, ,	Ground					21 989.0		756
	729	3, , ← 2, ,	Ground					22 010.5		756
	729	3, , ← 2, ,	Ground					22 779.	5.	755
	729	3, , ← 2, ,	Ground					22 908.0		756
	729	3, , ← 2, ,	Ground					22 917.9	.2	902
	729	3, , ← 2, ,	Ground					22 942.0		756
	729	3, , ← 2, ,	Ground					23 318.0	.1	902
	729	3, , ← 2, ,	Ground					23 337.8	.1	902
	729	3, , ← 2, ,	Ground					23 339.0		756
	729	3, , ← 2, ,	Ground					23 354.9	.2	902
	729	3, , ← 2, ,	Ground					23 389.3	.1	902
	729	3, , ← 2, ,	Ground					23 462.9	.2	902
	729	3, , ← 2, ,	Ground					23 511.53		756
	729	3, , ← 2, ,	Ground					23 526.1	.1	902
	729	3, , ← 2, ,	Ground					23 547.9	.1	902
	729	3, , ← 2, ,	Ground					23 611.7	.1	902
	729	3, , ← 2, ,	Ground					23 624.43		756
	729	3, , ← 2, ,	Ground					23 637.19		756
	729	3, , ← 2, ,	Ground					23 772.6	.2	902
	729	3, , ← 2, ,	Ground					23 797.3	.2	902
	729	3, , ← 2, ,	Ground					23 818.8	.1	902
	729	3, , ← 2, ,	Ground					23 843.8	.1	902
	729	3, , ← 2, ,	Ground					23 860.89		756
	729	3, , ← 2, ,	Ground					23 917.74		756
	729	3, , ← 2, ,	Ground					23 924.35		756
	729	3, , ← 2, ,	Ground					23 937.1		756

1. All $4_{32} \leftarrow 3_{31}$ assignments are uncertain; they may be $4_{31} \leftarrow 3_{30}$.

Isotopic Species	Id. No.	Rotational Quantum Nos.	Vib. State v_a	F_1'	F'	F_1	F	Frequency MHz	Acc. ±MHz	Ref.
$C^{12}H_2{:}C^{12}DBr$[b]	729	3, , ← 2, ,	Ground					23 942.9		756
	729	3, , ← 2, ,	Ground					23 957.7	.1	902
	729	3, , ← 2, ,	Ground					24 103.	5.	755
	729	3, , ← 2, ,	Ground					24 103.39		756
	729	3, , ← 2, ,	Ground					24 139.0		756
	729	3, , ← 2, ,	Ground					24 146.7		756
	729	Not Reported	Ground					15 469.5	.2	903
	729	Not Reported	Ground					15 518.7	.1	903
	729	Not Reported	Ground					15 522.1	.1	903
	729	Not Reported	Ground					15 555.6	.2	903
	729	Not Reported	Ground					15 607.0	.1	903
$C^{12}D_2{:}C^{12}HBr$[b]	731	3, , ← 2, ,	Ground					20 956.6	.3	902
	731	3, , ← 2, ,	Ground					21 242.6		756
	731	3, , ← 2, ,	Ground					21 258.5		756
	731	3, , ← 2, ,	Ground					21 284.4	.2	902
	731	3, , ← 2, ,	Ground					21 285.9		756
	731	3, , ← 2, ,	Ground					21 300.0	.05	756
	731	3, , ← 2, ,	Ground					21 343.8	.1	902
	731	3, , ← 2, ,	Ground					21 696.8	.1	902
	731	3, , ← 2, ,	Ground					21 742.6	.3	902
	731	3, , ← 2, ,	Ground					21 861.0		756
	731	3, , ← 2, ,	Ground					21 871.7	.3	902
	731	3, , ← 2, ,	Ground					21 887.5		756
	731	Not Reported	Ground					14 152.9	.2	903
	731	Not Reported	Ground					14 197.2	.1	903
	731	Not Reported	Ground					14 212.1	.2	903
	731	Not Reported	Ground					14 279.5	.1	903
	731	Not Reported	Ground					14 474.1	.1	903
	731	Not Reported	Ground					14 552.1	.1	903
$C^{12}D_2{:}C^{12}DBr$[b]	732	3, , ← 2, ,	Ground					20 505.83	10.	756
	732	3, , ← 2, ,	Ground					20 535.86	10.	756
	732	3, , ← 2, ,	Ground					20 536.65	10.	756
	732	3, , ← 2, ,	Ground					20 554.	5.	755
	732	3, , ← 2, ,	Ground					20 554.03	10.	756
	732	3, , ← 2, ,	Ground					20 556.	5.	755
	732	3, , ← 2, ,	Ground					20 885.52	10.	756
	732	3, , ← 2, ,	Ground					20 911.19	10.	756
	732	3, , ← 2, ,	Ground					21 110.9	.1	902
	732	3, , ← 2, ,	Ground					21 285.4	.2	902
	732	3, , ← 2, ,	Ground					21 321.8	.2	902
	732	3, , ← 2, ,	Ground					21 333.0	.2	902
	732	3, , ← 2, ,	Ground					21 345.8	.1	902
	732	3, , ← 2, ,	Ground					21 437.0	.1	902
	732	3, , ← 2, ,	Ground					21 450.9	.1	902
	732	3, , ← 2, ,	Ground					21 479.8	.1	902
	732	3, , ← 2, ,	Ground					21 512.2	.1	902
	732	3, , ← 2, ,	Ground					21 521.8	.2	902
	732	3, , ← 2, ,	Ground					21 542.9	.1	902
	732	Not Reported	Ground					13 911.7	.2	903

Isotopic Species	Id. No.	Rotational Quantum Nos.	Vib. State v_a	F_1'	F'	F_1	F	Frequency MHz	Acc. ±MHz	Ref.
$C^{12}D_2$:$C^{12}DBr$[b]	732	Not Reported	Ground					14 004.9	.1	903
	732	Not Reported	Ground					14 297.1	.2	903
	732	Not Reported	Ground					14 331.8	.1	903
	732	Not Reported	Ground					21 571.6	.1	902
$C^{13}H_2$:$C^{13}HBr^{79}$	733	2, 0, 2 ← 1, 0, 1	Excited		7/2		5/2	15 244.5		937
	733	2, 0, 2 ← 1, 0, 1	Ground		7/2		5/2	15 259.0		937
	733	2, 0, 2 ← 1, 0, 1	Ground					15 269.00	.11	992
	733	2, 1, 1 ← 1, 1, 0	Ground		7/2		5/2	15 525.3		937
	733	2, 1, 1 ← 1, 1, 0	Ground					15 553.84	.05	992
	733	2, 1, 1 ← 1, 1, 0	Ground		5/2		5/2	15 580.0		937
	733	2, 1, 1 ← 1, 1, 0	Ground		3/2		3/2	15 604.9		937
	733	2, 1, 1 ← 1, 1, 0	Ground		7/2		5/2	15 643.4		937
	733	2, 1, 2 ← 1, 1, 1	Ground					14 988.18	.05	992
	733	3, 0, 3 ← 2, 0, 2	Ground		9/2		7/2	22 894.8		937
	733	3, 0, 3 ← 2, 0, 2	Ground					22 900.18	.08	992
	733	3, 0, 3 ← 2, 0, 2	Ground		9/2		7/2	23 012.2		937
	733	3, 1, 2 ← 2, 1, 1	Ground		3/2		3/2	23 261.8		937
	733	3, 1, 2 ← 2, 1, 1	Excited		9/2		7/2	23 298.5		937
	733	3, 1, 2 ← 2, 1, 1	Ground		5/2		5/2	23 307.2		937
	733	3, 1, 2 ← 2, 1, 1	Excited		7/2		5/2	23 327.6		937
	733	3, 1, 2 ← 2, 1, 1	Ground					23 328.91	.06	992
	733	3, 1, 2 ← 2, 1, 1	Excited		7/2		7/2	23 382.6	.2	937
	733	3, 1, 3 ← 2, 1, 2	Ground					22 480.64	.04	992
	733	3, 1, 3 ← 2, 1, 2	Ground		7/2		7/2	22 561.0		937
	733	3, 2, 1 ← 2, 2, 0	Ground					22 911.06	.05	992
	733	3, 2, 2 ← 2, 2, 1	Ground					22 906.58	.05	992
	733	Not Reported	Ground		5/2		5/2	15 022.1		937
	733	Not Reported	Ground		3/2		3/2	15 033.6	.2	937
	733	Not Reported	Ground		5/2		3/2	15 076.7		937
	733	Not Reported	Ground		7/2		5/2	22 990.5		937
	733	Not Reported	Ground		7/2		7/2	23 400.4	.2	937
$C^{13}H_2$:$C^{13}HBr^{81}$	734	2, 0, 2 ← 1, 0, 1	Excited		7/2		5/2	15 155.3		937
	734	2, 0, 2 ← 1, 0, 1	Ground					15 178.27	.07	992
	734	2, 1, 1 ← 1, 1, 0	Ground		1/2		1/2	15 361.5		937
	734	2, 1, 1 ← 1, 1, 0	Ground		3/2		1/2	15 406.4		937
	734	2, 1, 1 ← 1, 1, 0	Ground		7/2		5/2	15 435.8		937
	734	2, 1, 1 ← 1, 1, 0	Ground					15 459.59	.04	992
	734	2, 1, 1 ← 1, 1, 0	Ground		5/2		5/2	15 481.0		937
	734	2, 1, 1 ← 1, 1, 0	Ground		3/2		3/2	15 502.3		937
	734	2, 1, 2 ← 1, 1, 1	Ground		1/2		1/2	14 802.5	.1	937
	734	2, 1, 2 ← 1, 1, 1	Ground					14 900.24	.05	992
	734	3, 0, 3 ← 2, 0, 2	Ground					22 764.69	.08	992
	734	3, 1, 2 ← 2, 1, 1	Ground		3/2		3/2	23 131.6		937
	734	3, 1, 2 ← 2, 1, 1	Excited		9/2		7/2	23 159.1		937
	734	3, 1, 2 ← 2, 1, 1	Ground		9/2		7/2	23 177.1		937
	734	3, 1, 2 ← 2, 1, 1	Ground					23 187.27	.08	992
	734	3, 1, 2 ← 2, 1, 1	Ground		5/2		3/2	23 198.2	.3	937
	734	3, 1, 3 ← 2, 1, 2	Ground		3/2		3/2	22 286.9		937
	734	3, 1, 3 ← 2, 1, 2	Ground					22 349.17	.04	992

Isotopic Species	Id. No.	Rotational Quantum Nos.	Vib. State v_a	F_1'	F'	F_1	F	Frequency MHz	Acc. ±MHz	Ref.
$C^{13}H_2{:}C^{13}HBr^{81}$	734	3, 2, 1← 2, 2, 0	Ground		3/2		1/2	22 676.4		937
	734	3, 2, 1← 2, 2, 0	Ground					22 774.41	.05	992
	734	3, 2, 2← 2, 2, 1	Ground		9/2		7/2	22 741.8		937
	734	3, 2, 2← 2, 2, 1	Ground					22 770.11	.06	992
	734	3, 2, 2← 2, 2, 1	Ground		5/2		3/2	22 770.4		937
	734	3, 2, 2← 2, 2, 1	Ground		7/2		5/2	22 840.3		937
	734	Not Reported	Ground		5/2		3/2	15 534.3		937
$C^{13}H_2{:}C^{12}HBr^{79}$	735	2, 0, 2← 1, 0, 1	Ground		7/2		5/2	15 433.9		937
	735	2, 0, 2← 1, 0, 1	Ground					15 443.85	.11	992
	735	2, 1, 1← 1, 1, 0	Ground		7/2		5/2	15 698.2		937
	735	2, 1, 1← 1, 1, 0	Ground					15 726.64	.05	992
	735	2, 1, 1← 1, 1, 0	Ground		3/2		3/2	15 777.7		937
	735	2, 1, 1← 1, 1, 0	Ground		5/2		3/2	15 815.7		937
	735	2, 1, 2← 1, 1, 1	Ground		3/2		1/2	15 111.0		937
	735	2, 1, 2← 1, 1, 1	Ground					15 164.38	.05	992
	735	2, 1, 2← 1, 1, 1	Ground		5/2		5/2	15 199.15		937
	735	2, 1, 2← 1, 1, 1	Ground		5/2		3/2	15 251.8		937
	735	3, 0, 2← 2, 0, 1	Ground		7/2		7/2	23 274.0		937
	735	3, 0, 3← 2, 0, 2	Ground					23 162.57	.10	992
	735	3, 1, 2← 2, 1, 1	Ground		5/2		5/2	23 566.7		937
	735	3, 1, 2← 2, 1, 1	Ground					23 588.28	.06	992
	735	3, 1, 2← 2, 1, 1	Ground		7/2		5/2	23 604.9		937
	735	3, 1, 2← 2, 1, 1	Ground		7/2		7/2	23 658.3		937
	735	3, 1, 3← 2, 1, 2	Ground		5/2		5/2	22 719.0		937
	735	3, 1, 3← 2, 1, 2	Ground					22 745.87	.04	992
	735	3, 1, 3← 2, 1, 2	Ground		7/2		5/2	22 762.0		937
	735	3, 1, 3← 2, 1, 2	Ground		7/2		7/2	22 826.2		937
	735	3, 2, 1← 2, 2, 0	Ground					23 171.03	.06	992
	735	3, 2, 2← 2, 2, 1	Ground		9/2		7/2	23 135.1		937
	735	3, 2, 2← 2, 2, 1	Ground					23 166.53	.06	992
	735	3, 2, 2← 2, 2, 1	Ground		7/2		5/2	23 247.3		937
	735	Not Reported	Ground		3/2		3/2	15 351.0		937
$C^{13}H_2{:}C^{12}HBr^{81}$	736	2, 0, 2← 1, 0, 1	Ground					15 354.34	.06	992
	736	2, 0, 2← 1, 0, 1	Ground		5/2		5/2	15 443.6		937
	736	2, 1, 1← 1, 1, 0	Ground					15 633.59	.05	992
	736	2, 1, 2← 1, 1, 1	Ground					15 077.69	.05	992
	736	3, 0, 3← 2, 0, 2	Ground					23 028.10	.08	992
	736	3, 1, 2← 2, 1, 1	Excited		3/2		3/2	23 374.7	.2	937
	736	3, 1, 2← 2, 1, 1	Excited		9/2		7/2	23 420.7		937
	736	3, 1, 2← 2, 1, 1	Ground		9/2		7/2	23 438.4		937
	736	3, 1, 2← 2, 1, 1	Ground					23 448.69	.08	992
	736	3, 1, 2← 2, 1, 1	Ground		7/2		5/2	23 462.6		937
	736	3, 1, 3← 2, 1, 2	Ground					22 615.89	.04	992
	736	3, 1, 3← 2, 1, 2	Ground		7/2		7/2	22 683.0		937
	736	3, 1, 3← 2, 1, 2	Excited		9/2		7/2	22 708.8		937
	736	3, 2, 1← 2, 2, 0	Ground					23 038.04	.06	992
	736	3, 2, 2← 2, 2, 1	Ground		3/2		1/2	22 936.5		937
	736	3, 2, 2← 2, 2, 1	Ground		9/2		7/2	23 005.9		937
	736	3, 2, 2← 2, 2, 1	Ground					23 034.05	.05	992

Isotopic Species	Id. No.	Rotational Quantum Nos.	Vib. State v_a	Hyperfine F_1'	F'	F_1	F	Frequency MHz	Acc. ±MHz	Ref.
$C^{13}H_2{:}C^{12}HBr^{81}$	736	Not Reported	Ground		1/2		1/2	14 980.8		937
	736	Not Reported	Ground		3/2		1/2	15 031.8		937
	736	Not Reported	Ground		7/2		5/2	15 053.2		937
	736	Not Reported	Ground		3/2		3/2	15 276.4		937
	736	Not Reported	Ground		1/2		1/2	15 536.3		937
	736	Not Reported	Ground		7/2		5/2	15 609.8		937
	736	Not Reported	Ground		5/2		5/2	15 654.8		937
	736	Not Reported	Ground		3/2		3/2	22 553.6		937
	736	Not Reported	Excited		9/2		7/2	22 580.6		937
	736	Not Reported	Ground		9/2		7/2	23 023.5		937
	736	Not Reported	Ground		3/2		3/2	23 393.4		937
$C^{12}H_2{:}C^{13}HBr^{79}$	737	2, 0, 2← 1, 0, 1	Ground		7/2		5/2	15 843.4		937
	737	2, 0, 2← 1, 0, 1	Ground					15 853.47	.11	992
	737	2, 1, 1← 1, 1, 0	Ground					16 156.52	.09	992
	737	2, 1, 2← 1, 1, 1	Ground		3/2		1/2	15 499.7	.2	937
	737	2, 1, 2← 1, 1, 1	Ground		7/2		5/2	15 524.2		937
	737	2, 1, 2← 1, 1, 1	Ground					15 553.83	.06	992
	737	2, 1, 2← 1, 1, 1	Ground		5/2		5/2	15 588.3		937
	737	3, 0, 3← 2, 0, 2	Ground					23 776.36	.09	992
	737	3, 1, 2← 2, 1, 1	Excited		9/2		7/2	24 201.6		937
	737	3, 1, 2← 2, 1, 1	Ground					24 233.53	.06	992
	737	3, 1, 3← 2, 1, 2	Ground		3/2		3/2	23 256.4		937
	737	3, 1, 3← 2, 1, 2	Ground		5/2		5/2	23 302.9		937
	737	3, 1, 3← 2, 1, 2	Ground					23 330.08	.05	992
	737	3, 1, 3← 2, 1, 2	Ground		7/2		5/2	23 347.8	.3	937
	737	3, 1, 3← 2, 1, 2	Ground		7/2		7/2	23 410.1		937
	737	3, 2, 1← 2, 2, 0	Ground					23 788.66	.06	992
	737	3, 2, 2← 2, 2, 1	Ground					23 783.36	.08	992
	737	3, 2, 2← 2, 2, 1	Ground		7/2		5/2	23 868.0		937
	737	Not Reported	Ground		7/2		5/2	15 643.4		937
	737	Not Reported	Ground		5/2		5/2	16 208.3		937
	737	Not Reported	Ground		5/2		3/2	16 247.0		937
$C^{12}H_2{:}C^{13}HBr^{81}$	738	2, 0, 2← 1, 0, 1	Ground		3/2		3/2	15 682.9		937
	738	2, 0, 2← 1, 0, 1	Ground		7/2		5/2	15 753.2		937
	738	2, 0, 2← 1, 0, 1	Ground					15 761.79	.06	992
	738	2, 1, 1← 1, 1, 0	Ground		1/2		1/2	15 962.1		937
	738	2, 1, 1← 1, 1, 0	Ground					16 061.50	.05	992
	738	2, 1, 1← 1, 1, 0	Ground		5/2		5/2	16 083.6		937
	738	2, 1, 2← 1, 1, 1	Ground		1/2		1/2	15 366.9		937
	738	2, 1, 2← 1, 1, 1	Ground		3/2		1/2	15 419.7		937
	738	2, 1, 2← 1, 1, 1	Ground		7/2		5/2	15 440.6		937
	738	2, 1, 2← 1, 1, 1	Ground					15 465.40	.06	992
	738	2, 1, 2← 1, 1, 1	Ground		5/2		3/2	15 540.1		937
	738	3, 0, 3← 2, 0, 2	Ground		3/2		3/2	23 558.9		937
	738	3, 0, 3← 2, 0, 2	Ground		5/2		5/2	23 587.1		937
	738	3, 0, 3← 2, 0, 2	Ground					23 638.68	.10	992
	738	3, 1, 2← 2, 1, 1	Excited		9/2		7/2	24 060.5		937
	738	3, 1, 2← 2, 1, 1	Ground		5/2		5/2	24 071.9		937

151

Isotopic Species	Id. No.	Rotational Quantum Nos.	Vib. State v_a	Hyperfine				Frequency MHz	Acc. ±MHz	Ref.
				F_1'	F'	F_1	F			
$C^{12}H_2:C^{13}HBr^{81}$	738	3, 1, 2← 2, 1, 1	Ground					24 090.58	.06	992
	738	3, 1, 2← 2, 1, 1	Ground		5/2		3/2	24 102.5	.5	937
	738	3, 1, 3← 2, 1, 1	Ground		9/2		7/2	24 080.2		937
	738	3, 1, 3← 2, 1, 2	Ground		5/2		5/2	23 174.5		937
	738	3, 1, 3← 2, 1, 2	Ground		9/2		7/2	23 186.0		937
	738	3, 1, 3← 2, 1, 2	Ground					23 196.72	.04	992
	738	3, 1, 3← 2, 1, 2	Ground		7/2		5/2	23 210.6		937
	738	3, 1, 3← 2, 1, 2	Ground		7/2		7/2	23 264.0		937
	738	3, 2, 1← 2, 2, 0	Ground		5/2		3/2	23 650.7		937
	738	3, 2, 1← 2, 2, 0	Ground					23 650.76	.06	992
	738	3, 2, 2← 2, 2, 1	Ground		3/2		1/2	23 546.5		937
	738	3, 2, 2← 2, 2, 1	Ground		9/2		7/2	23 617.0		937
	738	3, 2, 2← 2, 2, 1	Ground					23 645.38	.06	992
	738	3, 2, 2← 2, 2, 1	Ground		7/2		5/2	23 716.1		937
c-$C^{12}HD:C^{12}HBr^{79}$	739	2, 0, 2← 1, 0, 1	Ground		3/2		3/2	15 328.8	.1	938
	739	2, 0, 2← 1, 0, 1	Ground		7/2		5/2	15 409.8	.5	938
	739	2, 0, 2← 1, 0, 1	Ground		5/2		3/2	15 409.8	.5	938
	739	2, 0, 2← 1, 0, 1	Ground					15 419.91	.06	938
	739	2, 0, 2← 1, 0, 1	Ground		5/2		5/2	15 524.5	.1	938
	739	2, 0, 2← 1, 0, 1	Ground		3/2		1/2	15 534.7	.2	938
	739	2, 1, 1← 1, 1, 0	Ground		3/2		1/2	15 691.5	.1	938
	739	2, 1, 1← 1, 1, 0	Ground		7/2		5/2	15 726.9	.1	938
	739	2, 1, 1← 1, 1, 0	Ground					15 754.71	.05	938
	739	2, 1, 1← 1, 1, 0	Ground		5/2		5/2	15 778.1	.1	938
	739	2, 1, 1← 1, 1, 0	Ground		3/2		3/2	15 805.95	.1	938
	739	2, 1, 1← 1, 1, 0	Ground		5/2		3/2	15 842.4	.2	938
	739	2, 1, 2← 1, 1, 1	Ground		7/2		5/2	15 060.3	.1	938
	739	2, 1, 2← 1, 1, 1	Ground					15 089.11	.06	938
	739	2, 1, 2← 1, 1, 1	Ground		5/2		5/2	15 124.7	.1	938
	739	2, 1, 2← 1, 1, 1	Ground		3/2		3/2	15 130.7	.1	938
	739	2, 1, 2← 1, 1, 1	Ground		5/2		3/2	15 175.8	.1	938
	739	3, 0, 3← 2, 0, 2	Ground		3/2		3/2	23 031.6	.1	938
	739	3, 0, 3← 2, 0, 2	Ground		5/2		5/2	23 064.9	.2	938
	739	3, 0, 3← 2, 0, 2	Ground		9/2		7/2	23 118.8	.1	938
	739	3, 0, 3← 2, 0, 2	Ground		7/2		5/2	23 118.8	.1	938
	739	3, 0, 3← 2, 0, 2	Ground					23 123.76	.06	938
	739	3, 0, 3← 2, 0, 2	Ground		3/2		1/2	23 146.2	.2	938
	739	3, 0, 3← 2, 0, 2	Ground		5/2		3/2	23 146.2	.2	938
	739	3, 0, 3← 2, 0, 2	Ground		7/2		7/2	23 233.3	.1	938
	739	3, 1, 2← 2, 1, 1	1		3/2		3/2	23 549.5	.1	938
	739	3, 1, 2← 2, 1, 1	Ground		3/2		3/2	23 565.7	.2	938
	739	3, 1, 2← 2, 1, 1	1		5/2		5/2	23 593.4	.1	938
	739	3, 1, 2← 2, 1, 1	1		9/2		7/2	23 601.7	.2	938
	739	3, 1, 2← 2, 1, 1	Ground		5/2		5/2	23 609.6	.2	938
	739	3, 1, 2← 2, 1, 1	Ground		9/2		7/2	23 617.9	.2	938
	739	3, 1, 2← 2, 1, 1	1		7/2		5/2	23 629.8	.2	938
	739	3, 1, 2← 2, 1, 1	Ground					23 629.80	.06	938
	739	3, 1, 2← 2, 1, 1	Ground		7/2		5/2	23 646.3	.1	938
	739	3, 1, 2← 2, 1, 1	Ground		7/2		7/2	23 697.6	.1	938

Isotopic Species	Id. No.	Rotational Quantum Nos.	Vib. State v_a	F_1'	F'	F_1	F	Frequency MHz	Acc. ±MHz	Ref.
c-$C^{12}HD$:$C^{12}HBr^{79}$	739	3, 1, 3← 2, 1, 2	Ground		3/2		3/2	22 558.8	.1	938
	739	3, 1, 3← 2, 1, 2	Ground		5/2		5/2	22 605.8	.2	938
	739	3, 1, 3← 2, 1, 2	Ground		9/2		7/2	22 620.2	.1	938
	739	3, 1, 3← 2, 1, 2	Ground					22 632.77	.08	938
	739	3, 1, 3← 2, 1, 2	Ground		7/2		5/2	22 648.6	.2	938
	739	3, 1, 3← 2, 1, 2	Ground		5/2		3/2	22 651.1	.2	938
	739	3, 1, 3← 2, 1, 2	Ground		7/2		7/2	22 712.8	.2	938
	739	3, 2, 1← 2, 2, 0	Ground		3/2		1/2	23 026.1	.2	938
	739	3, 2, 1← 2, 2, 0	Ground		7/2		7/2	23 107.95	.05	938
	739	3, 2, 1← 2, 2, 0	Ground		9/2		7/2	23 107.95	.05	938
	739	3, 2, 1← 2, 2, 0	Ground					23 140.72	.06	938
	739	3, 2, 1← 2, 2, 0	Ground		3/2		3/2	23 140.9	.1	938
	739	3, 2, 1← 2, 2, 0	Ground		5/2		3/2	23 140.9	.1	938
	739	3, 2, 1← 2, 2, 0	Ground		5/2		5/2	23 222.6	.2	938
	739	3, 2, 1← 2, 2, 0	Ground		7/2		5/2	23 222.6	.2	938
	739	3, 2, 2← 2, 2, 1	Ground		3/2		1/2	23 019.1	.2	938
	739	3, 2, 2← 2, 2, 1	Ground		7/2		7/2	23 100.5	.1	938
	739	3, 2, 2← 2, 2, 1	Ground		9/2		7/2	23 100.5	.1	938
	739	3, 2, 2← 2, 2, 1	Ground					23 133.36	.06	938
	739	3, 2, 2← 2, 2, 1	Ground		3/2		3/2	23 133.6	.2	938
	739	3, 2, 2← 2, 2, 1	Ground		5/2		3/2	23 133.6	.2	938
	739	3, 2, 2← 2, 2, 1	Ground		5/2		5/2	23 215.2	.2	938
	739	3, 2, 2← 2, 2, 1	Ground		7/2		5/2	23 215.2	.2	938
	739	4, 0, 4← 3, 0, 3	Ground		7/2		5/2	30 829.0		998
	739	4, 1, 3← 3, 1, 2	Ground		5/2		5/2	31 422.4		998
	739	4, 1, 3← 3, 1, 2	1		11/2		9/2	31 475.9		998
	739	4, 1, 3← 3, 1, 2	Ground		7/2		7/2	31 477.4		998
	739	4, 1, 3← 3, 1, 2	1		5/2		3/2	31 480.7		998
	739	4, 1, 3← 3, 1, 2	1					31 482.30	.25	998
	739	4, 1, 3← 3, 1, 2	1		9/2		7/2	31 487.0		998
	739	4, 1, 3← 3, 1, 2	1		7/2		5/2	31 492.7		998
	739	4, 1, 3← 3, 1, 2	Ground		11/2		9/2	31 497.7		998
	739	4, 1, 3← 3, 1, 2	Ground		5/2		3/2	31 502.2		998
	739	4, 1, 3← 3, 1, 2	Ground					31 503.89	.17	998
	739	4, 1, 3← 3, 1, 2	Ground		9/2		7/2	31 508.6		998
	739	4, 1, 3← 3, 1, 2	Ground		7/2		5/2	31 514.3		998
	739	4, 1, 3← 3, 1, 2	Ground		9/2		9/2	31 588.2		998
	739	4, 1, 4← 3, 1, 3	Ground		5/2		5/2	30 081.8		998
	739	4, 1, 4← 3, 1, 3	1		11/2		9/2	30 137.4		998
	739	4, 1, 4← 3, 1, 3	1		5/2		3/2	30 142.6		998
	739	4, 1, 4← 3, 1, 3	Ground		7/2		7/2	30 142.65		998
	739	4, 1, 4← 3, 1, 3	1					30 144.14	.47	998
	739	4, 1, 4← 3, 1, 3	1		9/2		7/2	30 149.1		998
	739	4, 1, 4← 3, 1, 3	1		7/2		5/2	30 154.6		998
	739	4, 1, 4← 3, 1, 3	Ground		11/2		9/2	30 167.4		998
	739	4, 1, 4← 3, 1, 3	Ground		5/2		3/2	30 173.8		998
	739	4, 1, 4← 3, 1, 3	Ground					30 174.38	.13	998
	739	4, 1, 4← 3, 1, 3	Ground		9/2		7/2	30 179.7		998
	739	4, 1, 4← 3, 1, 3	Ground		7/2		5/2	30 185.3		998
	739	4, 1, 4← 3, 1, 3	1		9/2		9/2	30 239.6		998

Isotopic Species	Id. No.	Rotational Quantum Nos.	Vib. State v_a	F'_1	F'	F_1	F	Frequency MHz	Acc. ±MHz	Ref.
c-$C^{12}HD{:}C^{12}HBr^{79}$	739	4, 1, 4← 3, 1, 3	Ground		9/2		9/2	30 271.1		998
	739	4, 2, 2← 3, 2, 1	Ground		11/2		9/2	30 845.2		998
	739	4, 2, 2← 3, 2, 1	Ground					30 861.81		998
	739	4, 2, 2← 3, 2, 1	Ground		7/2		5/2	30 874.9		998
	739	4, 2, 3← 3, 2, 2	1		5/2		3/2	30 782.6		998
	739	4, 2, 3← 3, 2, 2	1		11/2		9/2	30 798.75		998
	739	4, 2, 3← 3, 2, 2	Ground		5/2		3/2	30 809.2		998
	739	4, 2, 3← 3, 2, 2	1					30 815.49		998
	739	4, 2, 3← 3, 2, 2	Ground		11/2		9/2	30 825.4		998
	739	4, 2, 3← 3, 2, 2	Ground					30 841.92	.11	998
	739	4, 2, 3← 3, 2, 2	Ground		7/2		5/2	30 855.2		998
	739	4, 2, 3← 3, 2, 2	Ground		9/2		7/2	30 870.8		998
	739	4, 3, 2← 3, 3, 1	Ground		5/2		3/2	30 762.6		998
	739	4, 3, 2← 3, 3, 1	Ground		11/2		9/2	30 814.9		998
	739	4, 3, 2← 3, 3, 1	Ground					30 848.47	.19	998
	739	4, 3, 2← 3, 3, 1	Ground		7/2		5/2	30 865.3		998
	739	4, 3, 2← 3, 3, 1	Ground		5/2		5/2	30 906.6		998
	739	4, 3, 2← 3, 3, 1	Ground		9/2		7/2	30 918.5		998
	739	4, 3, 2← 3, 3, 1	Ground		7/2		7/2	30 932.0		998
	739	5, 2, 3← 4, 2, 2	Ground		7/2		5/2	38 576.4		998
	739	5, 2, 3← 4, 2, 2	Ground		13/2		11/2	38 579.4		998
	739	5, 2, 3← 4, 2, 2	Ground					38 589.30	.07	998
	739	5, 2, 3← 4, 2, 2	Ground		9/2		7/2	38 599.3		998
	739	5, 2, 3← 4, 2, 2	Ground		11/2		9/2	38 602.5		998
t-$C^{12}HD{:}C^{12}HBr^{79}$	741	2, 0, 2← 1, 0, 1	Ground		5/2		3/2	14 816.1	.1	938
	741	2, 0, 2← 1, 0, 1	Ground		7/2		5/2	14 816.1	.1	938
	741	2, 0, 2← 1, 0, 1	Ground		1/2		1/2	14 825.9	.1	938
	741	2, 0, 2← 1, 0, 1	Ground					14 826.14	.06	938
	741	2, 0, 2← 1, 0, 1	Ground		5/2		5/2	14 934.1	.1	938
	741	2, 1, 1← 1, 1, 0	Ground		7/2		5/2	15 055.3	1.	938
	741	2, 1, 1← 1, 1, 0	Ground					15 083.70	.07	938
	741	2, 1, 1← 1, 1, 0	Ground		5/2		5/2	15 110.0	.2	938
	741	2, 1, 1← 1, 1, 0	Ground		3/2		3/2	15 135.2	.2	938
	741	2, 1, 1← 1, 1, 0	Ground		5/2		3/2	15 173.7	.1	938
	741	2, 1, 2← 1, 1, 1	Ground		1/2		1/2	14 454.7	.1	938
	741	2, 1, 2← 1, 1, 1	Ground		3/2		1/2	14 517.5	.2	938
	741	2, 1, 2← 1, 1, 1	Ground		7/2		5/2	14 541.2	.1	938
	741	2, 1, 2← 1, 1, 1	Ground					14 571.50	.05	938
	741	2, 1, 2← 1, 1, 1	Ground		5/2		5/2	14 605.5	.2	938
	741	2, 1, 2← 1, 1, 1	Ground		3/2		3/2	14 615.5	.1	938
	741	2, 1, 2← 1, 1, 1	Ground		5/2		3/2	14 660.0	.1	938
	741	3, 0, 3← 2, 0, 2	Ground		9/2		7/2	22 231.0	.2	938
	741	3, 0, 3← 2, 0, 2	Ground		7/2		5/2	22 231.0	.2	938
	741	3, 0, 3← 2, 0, 2	Ground					22 236.53	.1	938
	741	3, 0, 3← 2, 0, 2	Ground		5/2		3/2	22 260.0	.2	938 [3]
	741	3, 0, 3← 2, 0, 2	Ground		7/2		7/2	22 349.0	.2	938
	741	3, 1, 2← 2, 1, 1	Ground		3/2		3/2	22 556.8	.1	938
	741	3, 1, 2← 2, 1, 1	1		5/2		5/2	22 587.3	.1	938
	741	3, 1, 2← 2, 1, 1	1		9/2		7/2	22 596.55	.05	938

3. The author indicates an additional hyperfine assignment for this frequency, but due to a misprint, the quantum numbers violate the selection rule $\Delta F = 0, +1$, and therefore it has been omitted.

Isotopic Species	Id. No.	Rotational Quantum Nos.	Vib. State	v_a	F_1'	F'	F_1	F	Frequency MHz	Acc. ±MHz	Ref.
t-$C^{12}HD$:$C^{12}HBr^{79}$	741	3, 1, 2← 2, 1, 1	Ground			5/2		5/2	22 602.16	.05	938
	741	3, 1, 2← 2, 1, 1	Ground			3/2		1/2	22 610.8	.2	938
	741	3, 1, 2← 2, 1, 1	Ground			9/2		7/2	22 611.8	.1	938
	741	3, 1, 2← 2, 1, 1	Ground						22 624.10	.04	938
	741	3, 1, 2← 2, 1, 1		1		7/2		5/2	22 625.9	.2	938
	741	3, 1, 2← 2, 1, 1	Ground			7/2		5/2	22 640.9	.1	938
	741	3, 1; 2← 2, 1, 1		1		7/2		7/2	22 680.9	.1	938
	741	3, 1, 2← 2, 1, 1	Ground			7/2		7/2	22 695.9	.2	938
	741	3, 1, 3← 2, 1, 2		1		3/2		3/2	21 760.0	.2	938
	741	3, 1, 3← 2, 1, 2	Ground			3/2		3/2	21 781.7	.2	938
	741	3, 1, 3← 2, 1, 2		1		5/2		5/2	21 808.6	.1	938
	741	3, 1, 3← 2, 1, 2		1		9/2		7/2	21 822.8	.2	938
	741	3, 1, 3← 2, 1, 2	Ground			5/2		5/2	21 829.7	.2	938
	741	3, 1, 3← 2, 1, 2	Ground			9/2		7/2	21 844.3	.1	938
	741	3, 1, 3← 2, 1, 2		1		7/2		5/2	21 853.1	.2	938
	741	3, 1, 3← 2, 1, 2	Ground						21 856.51	.08	938
	741	3, 1, 3← 2, 1, 2	Ground			7/2		5/2	21 873.1	.1	938
	741	3, 1, 3← 2, 1, 2	Ground			5/2		3/2	21 874.3	.2	938
	741	3, 1, 3← 2, 1, 2		1		7/2		7/2	21 916.3	.2	938
	741	3, 1, 3← 2, 1, 2	Ground			7/2		7/2	21 937.15	.05	938
	741	3, 2, 1← 2, 2, 0	Ground			3/2		1/2	22 127.6	.1	938
	741	3, 2, 1← 2, 2, 0	Ground			9/2		7/2	22 211.3	.1	938
	741	3, 2, 1← 2, 2, 0	Ground			7/2		7/2	22 211.3	.1	938
	741	3, 2, 1← 2, 2, 0	Ground						22 245.32	.04	938
	741	3, 2, 1← 2, 2, 0	Ground			5/2		3/2	22 245.5	.1	938
	741	3, 2, 1← 2, 2, 0	Ground			3/2		3/2	22 245.5	.1	938
	741	3, 2, 1← 2, 2, 0	Ground			5/2		5/2	22 329.6	.1	938
	741	3, 2, 1← 2, 2, 0	Ground			7/2		5/2	22 329.6	.1	938
	741	3, 2, 2← 2, 2, 1	Ground			3/2		1/2	22 124.0	.2	938
	741	3, 2, 2← 2, 2, 1	Ground			9/2		7/2	22 208.0	.1	938
	741	3, 2, 2← 2, 2, 1	Ground			7/2		7/2	22 208.0	.1	938
	741	3, 2, 2← 2, 2, 1	Ground						22 241.69	.04	938
	741	3, 2, 2← 2, 2, 1	Ground			5/2		3/2	22 241.7	.2	938
	741	3, 2, 2← 2, 2, 1	Ground			3/2		3/2	22 241.7	.2	938
	741	3, 2, 2← 2, 2, 1	Ground			7/2		5/2	22 325.8	.1	938
	741	3, 2, 2← 2, 2, 1	Ground			5/2		5/2	22 325.8	.1	938
	741	4, 0, 4← 3, 0, 3	Ground			7/2		7/2	29 599.2		998
	741	4, 0, 4← 3, 0, 3	Ground			9/2		7/2	29 640.5		998
	741	4, 0, 4← 3, 0, 3	Ground			11/2		9/2	29 640.5		998
	741	4, 0, 4← 3, 0, 3	Ground						29 644.39	.24	998
	741	4, 0, 4← 3, 0, 3	Ground			5/2		3/2	29 654.8		998
	741	4, 0, 4← 3, 0, 3	Ground			7/2		5/2	29 654.8		998
	741	4, 1, 3← 3, 1, 2	Ground			5/2		5/2	30 078.9		998
	741	4, 1, 3← 3, 1, 2		1		11/2		9/2	30 129.6		998
	741	4, 1, 3← 3, 1, 2	Ground			7/2		7/2	30 135.5		998
	741	4, 1, 3← 3, 1, 2		1		5/2		3/2	30 135.5		998
	741	4, 1, 3← 3, 1, 2		1					30 137.14	.42	998
	741	4, 1, 3← 3, 1, 2		1		9/2		7/2	30 142.6		998
	741	4, 1, 3← 3, 1, 2		1		7/2		5/2	30 147.0		998
	741	4, 1, 3← 3, 1, 2	Ground			11/2		9/2	30 157.35		998

Isotopic Species	Id. No.	Rotational Quantum Nos.	Vib. State v_a	F'_1	F'	F_1	F	Frequency MHz	Acc. ±MHz	Ref.
t-C^{12}HD:C^{12}HBr^{79}	741	4, 1, 3← 3, 1, 2	Ground		5/2		3/2	30 162.7		998
	741	4, 1, 3← 3, 1, 2	Ground					30 164.08	.048	998
	741	4, 1, 3← 3, 1, 2	Ground		9/2		7/2	30 169.2		998
	741	4, 1, 3← 3, 1, 2	Ground		7/2		5/2	30 174.8		998
	741	4, 1, 3← 3, 1, 2	Ground		9/2		9/2	30 252.8		998
	741	4, 1, 4← 3, 1, 3	Ground		5/2		5/2	29 046.5		998
	741	4, 1, 4← 3, 1, 3	1		11/2		9/2	29 104.4		998
	741	4, 1, 4← 3, 1, 3	Ground		7/2		7/2	29 108.0		998
	741	4, 1, 4← 3, 1, 3	1		5/2		3/2	29 110.9		998
	741	4, 1, 4← 3, 1, 3	1					29 111.48	.5	998
	741	4, 1, 4← 3, 1, 3	1		9/2		7/2	29 115.9		998
	741	4, 1, 4← 3, 1, 3	1		7/2		5/2	29 122.8		998
	741	4, 1, 4← 3, 1, 3	Ground		11/2		9/2	29 132.8		998
	741	4, 1, 4← 3, 1, 3	Ground		5/2		3/2	29 139.3		998
	741	4, 1, 4← 3, 1, 3	Ground					29 139.88	.04	998
	741	4, 1, 4← 3, 1, 3	Ground		9/2		7/2	29 144.6		998
	741	4, 1, 4← 3, 1, 3	Ground		7/2		5/2	29 151.1		998
	741	4, 1, 4← 3, 1, 3	Ground		9/2		9/2	29 237.6		998
	741	4, 2, 2← 3, 2, 1	Ground		5/2		3/2	29 629.2		998
	741	4, 2, 2← 3, 2, 1	Ground		11/2		9/2	29 646.5		998
	741	4, 2, 2← 3, 2, 1	Ground					29 663.22	.19	998
	741	4, 2, 2← 3, 2, 1	Ground		7/2		5/2	29 676.6		998
	741	4, 2, 2← 3, 2, 1	Ground		9/2		7/2	29 693.25		998
	741	4, 2, 3← 3, 2, 2	1		5/2		3/2	29 596.0		998
	741	4, 2, 3← 3, 2, 2	1		11/2		9/2	29 612.6		998
	741	4, 2, 3← 3, 2, 2	Ground		5/2		3/2	29 620.3		998
	741	4, 2, 3← 3, 2, 2	1					29 629.81	.07	998
	741	4, 2, 3← 3, 2, 2	Ground		11/2		9/2	29 636.9		998
	741	4, 2, 3← 3, 2, 2	1		7/2		5/2	29 643.5		998
	741	4, 2, 3← 3, 2, 2	Ground					29 654.06	.06	998
	741	4, 2, 3← 3, 2, 2	1		9/2		7/2	29 659.8		998
	741	4, 2, 3← 3, 2, 2	Ground		7/2		5/2	29 667.3		998
	741	4, 2, 3← 3, 2, 2	Ground		9/2		7/2	29 684.2		998
	741	4, 3, 2← 3, 3, 1	Ground		5/2		3/2	29 570.4		998
	741	4, 3, 2← 3, 3, 1	Ground		9/2		9/2	29 581.8		998
	741	4, 3, 2← 3, 3, 1	Ground		11/2		9/2	29 622.9		998
	741	4, 3, 2← 3, 3, 1	Ground					29 657.76	.42	998
	741	4, 3, 2← 3, 3, 1	Ground		7/2		5/2	29 675.4		998
	741	4, 3, 2← 3, 3, 1	Ground		9/2		7/2	29 729.8		998
	741	4, 3, 2← 3, 3, 1	Ground		7/2		7/2	29 743.7		998
	741	5, 1, 5← 4, 1, 4	1		13/2		11/2	36 382.6		998
	741	5, 1, 5← 4, 1, 4	1		11/2		9/2	36 388.5		998
	741	5, 1, 5← 4, 1, 4	1		9/2		7/2	36 393.5		998
	741	5, 1, 5← 4, 1, 4	Ground		13/2		11/2	36 418.4		998
	741	5, 1, 5← 4, 1, 4	Ground		11/2		9/2	36 423.9		998
c-C^{12}HD:C^{12}HBr^{81}	742	2, 0, 2← 1, 0, 1	Ground		3/2		3/2	15 254.6	.1	938
	742	2, 0, 2← 1, 0, 1	Ground		5/2		3/2	15 322.6	.1	938
	742	2, 0, 2← 1, 0, 1	Ground		7/2		5/2	15 322.6	.1	938
	742	2, 0, 2← 1, 0, 1	Ground					15 330.96	.05	938
	742	2, 0, 2← 1, 0, 1	Ground		5/2		5/2	15 418.4	.1	938

Isotopic Species	Id. No.	Rotational Quantum Nos.	Vib. State v_a	F_1'	F'	F_1	F	Frequency MHz	Acc. ±MHz	Ref.
c-C^{12}HD:C^{12}HBr^{81}	742	2, 0, 2← 1, 0, 1	Ground		3/2		1/2	15 426.7	.1	938
	742	2, 1, 1← 1, 1, 0	Ground		1/2		1/2	15 566.2	.1	938
	742	2, 1, 1← 1, 1, 0	Ground		3/2		1/2	15 609.0	.1	938
	742	2, 1, 1← 1, 1, 0	Ground		7/2		5/2	15 638.7	.1	938
	742	2, 1, 1← 1, 1, 0	Ground					15 661.85	.04	938
	742	2, 1, 1← 1, 1, 0	Ground		5/2		5/2	15 681.5	.2	938
	742	2, 1, 1← 1, 1, 0	Ground		3/2		3/2	15 704.5	.1	938
	742	2, 1, 1← 1, 1, 0	Ground		5/2		3/2	15 734.7	.1	938
	742	2, 1, 2← 1, 1, 1	Ground		1/2		1/2	14 909.1	.2	938
	742	2, 1, 2← 1, 1, 1	Ground		3/2		1/2	14 962.1	.1	938
	742	2, 1, 2← 1, 1, 1	Ground		7/2		5/2	14 979.9	.1	938
	742	2, 1, 2← 1, 1, 1	Ground					15 004.18	.06	938
	742	2, 1, 2← 1, 1, 1	Ground		5/2		5/2	15 033.5	.1	938
	742	2, 1, 2← 1, 1, 1	Ground		5/2		3/2	15 076.3	.1	938
	742	3, 0, 3← 2, 0, 2	Ground		3/2		3/2	22 913.8	.1	938
	742	3, 0, 3← 2, 0, 2	Ground		5/2		5/2	22 941.4	.1	938
	742	3, 0, 3← 2, 0, 2	Ground		9/2		7/2	22 986.35	.1	938
	742	3, 0, 3← 2, 0, 2	Ground		7/2		5/2	22 986.35	.1	938
	742	3, 0, 3← 2, 0, 2	Ground					22 990.63	.04	938
	742	3, 0, 3← 2, 0, 2	Ground		3/2		1/2	23 009.4	.1	938
	742	3, 0, 3← 2, 0, 2	Ground		5/2		3/2	23 009.4	.1	938
	742	3, 0, 3← 2, 0, 2	Ground		7/2		7/2	23 082.2	1.	938
	742	3, 0, 3← 2, 0, 2	Ground		9/2		7/2	23 082.2	1.	938
	742	3, 1, 2← 2, 1, 1	1		3/2		3/2	23 421.1	.1	938
	742	3, 1, 2← 2, 1, 1	Ground		3/2		3/2	23 437.1	.1	938
	742	3, 1, 2← 2, 1, 1	1		9/2		7/2	23 464.6	.2	938
	742	3, 1, 2← 2, 1, 1	Ground		5/2		5/2	23 473.3	.2	938
	742	3, 1, 2← 2, 1, 1	Ground		9/2		7/2	23 480.75	.1	938
	742	3, 1, 2← 2, 1, 1	1		7/2		5/2	23 488.3	.2	938
	742	3, 1, 2← 2, 1, 1	Ground					23 490.67	.07	938
	742	3, 1, 2← 2, 1, 1	Ground		7/2		5/2	23 504.4	.1	938
	742	3, 1, 2← 2, 1, 1	Ground		7/2		7/2	23 547.7	.2	938
	742	3, 1, 3← 2, 1, 2	Ground		3/2		3/2	22 443.1	.2	938
	742	3, 1, 3← 2, 1, 2	Ground		9/2		7/2	22 494.35	.05	938
	742	3, 1, 3← 2, 1, 2	Ground		3/2		1/2	22 496.1	.1	938
	742	3, 1, 3← 2, 1, 2	Ground					22 504.70	.05	938
	742	3, 1, 3← 2, 1, 2	Ground		7/2		5/2	22 518.0	.1	938
	742	3, 1, 3← 2, 1, 2	Ground		5/2		3/2	22 520.0	.1	938
	742	3, 1, 3← 2, 1, 2	Ground		7/2		7/2	22 571.6	.2	938
	742	3, 2, 1← 2, 2, 0	Ground		3/2		1/2	22 911.7	.2	938
	742	3, 2, 1← 2, 2, 0	Ground		7/2		7/2	22 979.5	.2	938
	742	3, 2, 1← 2, 2, 0	Ground		9/2		7/2	22 979.5	.2	938
	742	3, 2, 1← 2, 2, 0	Ground					23 007.09	.08	938
	742	3, 2, 1← 2, 2, 0	Ground		3/2		3/2	23 007.4	.2	938
	742	3, 2, 1← 2, 2, 0	Ground		5/2		3/2	23 007.4	.2	938
	742	3, 2, 1← 2, 2, 0	Ground		7/2		5/2	23 075.4	.1	938
	742	3, 2, 1← 2, 2, 0	Ground		5/2		5/2	23 075.4	.1	938
	742	3, 2, 2← 2, 2, 1	Ground		3/2		1/2	22 903.9	.1	938
	742	3, 2, 2← 2, 2, 1	Ground		9/2		7/2	22 972.5	.2	938
	742	3, 2, 2← 2, 2, 1	Ground		7/2		7/2	22 972.5	.2	938

Isotopic Species	Id. No.	Rotational Quantum Nos.	Vib. State v_a	F_1'	F'	F_1	F	Frequency MHz	Acc. ±MHz	Ref.
c-C^{12}HD:C^{12}HBr^{81}	742	3, 2, 2← 2, 2, 1	Ground		5/2		3/2	22 999.8	.2	938
	742	3, 2, 2← 2, 2, 1	Ground		3/2		3/2	22 999.8	.2	938
	742	3, 2, 2← 2, 2, 1	Ground					22 999.86	.07	938
	742	3, 2, 2← 2, 2, 1	Ground		5/2		5/2	23 068.4	.1	938
	742	3, 2, 2← 2, 2, 1	Ground		7/2		5/2	23 068.4	.1	938
	742	4, 1, 3← 3, 1, 2	1		7/2		7/2	31 274.8		998
	742	4, 1, 3← 3, 1, 2	1		11/2		9/2	31 291.6		998
	742	4, 1, 3← 3, 1, 2	Ground		7/2		7/2	31 296.1		998
	742	4, 1, 3← 3, 1, 2	1					31 297.15	.12	998
	742	4, 1, 3← 3, 1, 2	1		9/2		7/2	31 301.0		998
	742	4, 1, 3← 3, 1, 2	1		7/2		5/2	31 305.7		998
	742	4, 1, 3← 3, 1, 2	Ground		11/2		9/2	31 313.2		998
	742	4, 1, 3← 3, 1, 2	Ground		5/2		3/2	31 317.1		998
	742	4, 1, 3← 3, 1, 2	Ground					31 318.52	.08	998
	742	4, 1, 3← 3, 1, 2	Ground		9/2		7/2	31 322.5		998
	742	4, 1, 3← 3, 1, 2	Ground		7/2		5/2	31 327.1		998
	742	4, 1, 3← 3, 1, 2	1		9/2		9/2	31 367.3		998
	742	4, 1, 3← 3, 1, 2	Ground		9/2		9/2	31 388.9		998
	742	4, 1, 4← 3, 1, 3	Ground		5/2		5/2	29 926.4		998
	742	4, 1, 4← 3, 1, 3	1		5/2		3/2	29 975.7		998
	742	4, 1, 4← 3, 1, 3	1					29 975.83		998
	742	4, 1, 4← 3, 1, 3	Ground		7/2		7/2	29 976.5		998
	742	4, 1, 4← 3, 1, 3	1		9/2		7/2	29 979.45		998
	742	4, 1, 4← 3, 1, 3	Ground		11/2		9/2	29 998.0		998
	742	4, 1, 4← 3, 1, 3	Ground		5/2		3/2	30 003.3		998
	742	4, 1, 4← 3, 1, 3	Ground					30 003.58	.09	998
	742	4, 1, 4← 3, 1, 3	Ground		9/2		7/2	30 007.4		998
	742	4, 1, 4← 3, 1, 3	Ground		7/2		5/2	30 013.0		998
	742	4, 1, 4← 3, 1, 3	Ground		9/2		9/2	30 084.65		998
	742	4, 2, 2← 3, 2, 1	Ground		5/2		3/2	30 655.8		998
	742	4, 2, 2← 3, 2, 1	Ground		11/2		9/2	30 669.6		998
	742	4, 2, 2← 3, 2, 1	Ground					30 683.39	.06	998
	742	4, 2, 2← 3, 2, 1	Ground		7/2		5/2	30 694.5		998
	742	4, 2, 2← 3, 2, 1	Ground		9/2		7/2	30 707.9		998
	742	4, 2, 3← 3, 2, 2	1		5/2		3/2	30 610.2		998
	742	4, 2, 3← 3, 2, 2	1		11/2		9/2	30 623.7		998
	742	4, 2, 3← 3, 2, 2	Ground		5/2		3/2	30 636.7		998
	742	4, 2, 3← 3, 2, 2	1					30 637.56	.03	998
	742	4, 2, 3← 3, 2, 2	Ground		11/2		9/2	30 649.9		998
	742	4, 2, 3← 3, 2, 2	1		9/2		7/2	30 661.9		998
	742	4, 2, 3← 3, 2, 2	Ground					30 663.99	.08	998
	742	4, 2, 3← 3, 2, 2	Ground		7/2		5/2	30 675.1		998
	742	4, 2, 3← 3, 2, 2	Ground		9/2		7/2	30 688.3		998
	742	4, 3, 2← 3, 3, 1	Ground		5/2		3/2	30 598.9		998
	742	4, 3, 2← 3, 3, 1	Ground		11/2		9/2	30 642.9		998
	742	4, 3, 2← 3, 3, 1	Ground					30 670.60	.15	998
	742	4, 3, 2← 3, 3, 1	Ground		7/2		5/2	30 684.6		998
	742	4, 3, 2← 3, 3, 1	Ground		5/2		5/2	30 718.9		998
	742	4, 3, 2← 3, 3, 1	Ground		9/2		7/2	30 728.9		998

Isotopic Species	Id. No.	Rotational Quantum Nos.	Vib. State	v_a	F_1'	F'	F_1	F	Frequency MHz	Acc. ±MHz	Ref.
t-$C^{12}HD$:$C^{12}HBr^{81}$	743	2, 0, 2← 1, 0, 1	Ground		3/2			3/2	14 662.6	.1	938
	743	2, 0, 2← 1, 0, 1	Ground		7/2			5/2	14 732.4	.1	938
	743	2, 0, 2← 1, 0, 1	Ground		5/2			3/2	14 732.4	.1	938
	743	2, 0, 2← 1, 0, 1	Ground						14 741.07	.05	938
	743	2, 0, 2← 1, 0, 1	Ground			5/2		5/2	14 831.0	.1	938
	743	2, 0, 2← 1, 0, 1	Ground		3/2			1/2	14 839.6	.1	938
	743	2, 0, 2← 1, 0, 1	Ground		3/2			1/2	14 944.5	.1	938
	743	2, 1, 1← 1, 1, 0	Ground		1/2			1/2	14 897.3	.1	938
	743	2, 1, 1← 1, 1, 0	Ground		3/2			1/2	14 942.5	.2	938
	743	2, 1, 1← 1, 1, 0	Ground		7/2			5/2	14 971.4	.1	938
	743	2, 1, 1← 1, 1, 0	Ground						14 995.68	.04	938
	743	2, 1, 1← 1, 1, 0	Ground			5/2		5/2	15 017.7	.1	938
	743	2, 1, 1← 1, 1, 0	Ground			3/2		3/2	15 038.6	.1	938
	743	2, 1, 1← 1, 1, 0	Ground			5/2		3/2	15 070.4	.1	938
	743	2, 1, 2← 1, 1, 1	Ground		1/2			1/2	14 388.4	.2	938
	743	2, 1, 2← 1, 1, 1	Ground		7/2			5/2	14 463.4	.1	938
	743	2, 1, 2← 1, 1, 1	Ground						14 488.07	.06	938
	743	2, 1, 2← 1, 1, 1	Ground			5/2		5/2	14 517.5	.2	938
	743	2, 1, 2← 1, 1, 1	Ground			3/2		3/2	14 525.7	.2	938
	743	2, 1, 2← 1, 1, 1	Ground			5/2		3/2	14 562.7	.1	938
	743	3, 0, 3← 2, 0, 2	Ground		3/2			3/2	22 030.2	.2	938
	743	3, 0, 3← 2, 0, 2	Ground		5/2			5/2	22 058.1	.1	938
	743	3, 0, 3← 2, 0, 2	Ground		7/2			5/2	22 104.1	.2	938
	743	3, 0, 3← 2, 0, 2	Ground		9/2			7/2	22 104.1	.2	938
	743	3, 0, 3← 2, 0, 2	Ground						22 108.79	.06	938
	743	3, 0, 3← 2, 0, 2	Ground		3/2			1/2	22 128.3	.2	938
	743	3, 0, 3← 2, 0, 2	Ground		5/2			3/2	22 128.3	.2	938
	743	3, 0, 3← 2, 0, 2	Ground		7/2			7/2	22 202.7	.1	938
	743	3, 1, 2← 2, 1, 1	Ground		3/2			3/2	22 436.0	.2	938
	743	3, 1, 2← 2, 1, 1		1	9/2			7/2	22 466.4	.1	938
	743	3, 1, 2← 2, 1, 1	Ground		5/2			5/2	22 473.4	.1	938
	743	3, 1, 2← 2, 1, 1	Ground		9/2			7/2	22 481.5	.1	938
	743	3, 1, 2← 2, 1, 1		1	7/2			5/2	22 490.9	.2	938
	743	3, 1, 2← 2, 1, 1	Ground						22 491.90	.06	938
	743	3, 1, 2← 2, 1, 1	Ground		7/2			5/2	22 505.85	.05	938
	743	3, 1, 2← 2, 1, 1	Ground		7/2			7/2	22 551.7	.1	938
	743	3, 1, 3← 2, 1, 2		1	3/2			3/2	21 648.9	.1	938
	743	3, 1, 3← 2, 1, 2	Ground		3/2			3/2	21 670.2	.2	938
	743	3, 1, 3← 2, 1, 2		1	5/2			5/2	21 689.6	.1	938
	743	3, 1, 3← 2, 1, 2		1	9/2			7/2	21 701.13	.2	938
	743	3, 1, 3← 2, 1, 2	Ground		5/2			5/2	21 710.6	.2	938
	743	3, 1, 3← 2, 1, 2	Ground		9/2			7/2	21 722.7	.1	938
	743	3, 1, 3← 2, 1, 2		1	7/2			5/2	21 725.5	.2	938
	743	3, 1, 3← 2, 1, 2		1	5/2			3/2	21 726.7	.2	938
	743	3, 1, 3← 2, 1, 2	Ground						21 733.35	.07	938
	743	3, 1, 3← 2, 1, 2	Ground		7/2			5/2	21 747.6	.1	938
	743	3, 1, 3← 2, 1, 2		1	7/2			7/2	21 779.4	.1	938
	743	3, 1, 3← 2, 1, 2	Ground		7/2			7/2	21 800.6	.1	938
	743	3, 2, 1← 2, 2, 0	Ground		3/2			1/2	22 018.5	.1	938
	743	3, 2, 1← 2, 2, 0	Ground		9/2			7/2	22 088.7	.1	938

Isotopic Species	Id. No.	Rotational Quantum Nos.	Vib. State v_a	Hyperfine F_1'	F'	F_1	F	Frequency MHz	Acc. ±MHz	Ref.
t-$C^{12}HD$:$C^{12}HBr^{81}$	743	3, 2, 1← 2, 2, 0	Ground		7/2		7/2	22 088.7	.1	938
	743	3, 2, 1← 2, 2, 0	Ground					22 117.13	.06	938
	743	3, 2, 1← 2, 2, 0	Ground		3/2		3/2	22 117.2	.2	938
	743	3, 2, 1← 2, 2, 0	Ground		5/2		3/2	22 117.2	.2	938
	743	3, 2, 1← 2, 2, 0	Ground		7/2		5/2	22 187.6	.2	938
	743	3, 2, 1← 2, 2, 0	Ground		5/2		5/2	22 187.6	.2	938
	743	3, 2, 2← 2, 2, 1	Ground		3/2		1/2	22 015.5	.1	938
	743	3, 2, 2← 2, 2, 1	Ground		7/2		7/2	22 085.5	.1	938
	743	3, 2, 2← 2, 2, 1	Ground		9/2		7/2	22 085.5	.1	938
	743	3, 2, 2← 2, 2, 1	Ground					22 113.79	.06	938
	743	3, 2, 2← 2, 2, 1	Ground		5/2		3/2	22 113.9	.2	938
	743	3, 2, 2← 2, 2, 1	Ground		3/2		3/2	22 113.9	.2	938
	743	3, 2, 2← 2, 2, 1	Ground		5/2		5/2	22 184.2	.2	938
	743	3, 2, 2← 2, 2, 1	Ground		7/2		5/2	22 184.2	.2	938
	743	4, 0, 4← 3, 0, 3	Ground		7/2		7/2	29 434.5		998
	743	4, 0, 4← 3, 0, 3	Ground		9/2		7/2	29 470.5		998
	743	4, 0, 4← 3, 0, 3	Ground		11/2		9/2	29 470.5		998
	743	4, 0, 4← 3, 0, 3	Ground					29 473.75	.30	998
	743	4, 0, 4← 3, 0, 3	Ground		7/2		5/2	29 482.5		998
	743	4, 0, 4← 3, 0, 3	Ground		5/2		3/2	29 482.5		998
	743	4, 0, 4← 3, 0, 3	Ground		9/2		9/2	29 569.0		998
	743	4, 1, 3← 3, 1, 2	Ground		5/2		5/2	29 916.2		998
	743	4, 1, 3← 3, 1, 2	1		7/2		7/2	29 944.4		998
	743	4, 1, 3← 3, 1, 2	1		11/2		9/2	29 962.3		998
	743	4, 1, 3← 3, 1, 2	Ground		7/2		7/2	29 964.0		998
	743	4, 1, 3← 3, 1, 2	1		5/2		3/2	29 966.5		998
	743	4, 1, 3← 3, 1, 2	1					29 967.75	.12	998
	743	4, 1, 3← 3, 1, 2	1		9/2		7/2	29 971.9		998
	743	4, 1, 3← 3, 1, 2	Ground		11/2		9/2	29 982.2		998
	743	4, 1, 3← 3, 1, 2	Ground		5/2		3/2	29 986.65		998
	743	4, 1, 3← 3, 1, 2	Ground					29 987.78	.05	998
	743	4, 1, 3← 3, 1, 2	Ground		9/2		7/2	29 991.85		998
	743	4, 1, 3← 3, 1, 2	Ground		7/2		5/2	29 996.6		998
	743	4, 1, 3← 3, 1, 2	Ground		9/2		9/2	30 062.2		998
	743	4, 1, 4← 3, 1, 3	Ground		5/2		5/2	28 896.8		998
	743	4, 1, 4← 3, 1, 3	1		11/2		9/2	28 940.9		998
	743	4, 1, 4← 3, 1, 3	1					28 947.00	.33	998
	743	4, 1, 4← 3, 1, 3	1		9/2		7/2	28 950.6		998
	743	4, 1, 4← 3, 1, 3	1		7/2		5/2	28 957.0		998
	743	4, 1, 4← 3, 1, 3	Ground		11/2		9/2	28 969.2		998
	743	4, 1, 4← 3, 1, 3	Ground		5/2		3/2	28 974.5		998
	743	4, 1, 4← 3, 1, 3	Ground					28 975.00	.06	998
	743	4, 1, 4← 3, 1, 3	Ground		9/2		7/2	28 978.9		998
	743	4, 1, 4← 3, 1, 3	Ground		7/2		5/2	28 984.4		998
	743	4, 1, 4← 3, 1, 3	1		9/2		9/2	29 029.6		998
	743	4, 1, 4← 3, 1, 3	Ground		9/2		9/2	29 056.8		998
	743	4, 2, 2← 3, 2, 1	Ground		5/2		3/2	29 464.2		998
	743	4, 2, 2← 3, 2, 1	Ground		11/2		9/2	29 478.2		998
	743	4, 2, 2← 3, 2, 1	Ground					29 492.59	.11	998
	743	4, 2, 2← 3, 2, 1	Ground		7/2		5/2	29 504.1		998

Isotopic Species	Id. No.	Rotational Quantum Nos.	Vib. State v_a	F_1'	F'	F_1	F	Frequency MHz	Acc. ±MHz	Ref.
-C¹²HD:C¹²HBr⁸¹	743	4, 2, 2← 3, 2, 1	Ground		9/2		7/2	29 517.6		998
	743	4, 2, 3← 3, 2, 2	1		5/2		3/2	29 430.8		998
	743	4, 2, 3← 3, 2, 2	1		11/2		9/2	29 445.3		998
	743	4, 2, 3← 3, 2, 2	Ground		5/2		3/2	29 455.0		998
	743	4, 2, 3← 3, 2, 2	1					29 460.33		998
	743	4, 2, 3← 3, 2, 2	Ground		11/2		9/2	29 469.1		998
	743	4, 2, 3← 3, 2, 2	Ground					29 483.16	.09	998
	743	4, 2, 3← 3, 2, 2	Ground		7/2		5/2	29 494.5		998
	743	4, 2, 3← 3, 2, 2	Ground		9/2		7/2	29 508.05		998
	743	4, 3, 2← 3, 3, 1	Ground		5/2		3/2	29 413.1		998
	743	4, 3, 2← 3, 3, 1	Ground		9/2		9/2	29 424.3		998
	743	4, 3, 2← 3, 3, 1	Ground		11/2		9/2	29 458.2		998
	743	4, 3, 2← 3, 3, 1	Ground					29 487.72	.62	998
	743	4, 3, 2← 3, 3, 1	Ground		7/2		5/2	29 503.0		998
	743	4, 3, 2← 3, 3, 1	Ground		5/2		5/2	29 536.3		998
	743	4, 3, 2← 3, 3, 1	Ground		9/2		7/2	29 547.2		998
	743	4, 3, 2← 3, 3, 1	Ground		7/2		7/2	29 558.3		998
	743	5, 1, 5← 4, 1, 4	1		13/2		11/2	36 177.6		998
	743	5, 1, 5← 4, 1, 4	1		7/2		5/2	36 182.1		998
	743	5, 1, 5← 4, 1, 4	1		11/2		9/2	36 184.2		998
	743	5, 1, 5← 4, 1, 4	Ground		13/2		11/2	36 212.5		998
	743	5, 1, 5← 4, 1, 4	Ground		7/2		5/2	36 217.2		998
-C¹²HD:C¹²DBr⁷⁹	744	2, 0, 2← 1, 0, 1	Ground		3/2		3/2	14 427.8	.1	938
	744	2, 0, 2← 1, 0, 1	Ground		7/2		5/2	14 512.9	.1	938
	744	2, 0, 2← 1, 0, 1	Ground		5/2		3/2	14 512.9	.1	938
	744	2, 0, 2← 1, 0, 1	Ground		1/2		1/2	14 523.2	.1	938
	744	2, 0, 2← 1, 0, 1	Ground					14 523.38	.04	938
	744	2, 0, 2← 1, 0, 1	Ground		5/2		5/2	14 633.0	.2	938
	744	2, 0, 2← 1, 0, 1	Ground		3/2		1/2	14 643.5	.1	938
	744	2, 1, 1← 1, 1, 0	Ground		1/2		1/2	14 717.5	.1	938
	744	2, 1, 1← 1, 1, 0	Ground		3/2		1/2	14 774.0	.2	938
	744	2, 1, 1← 1, 1, 0	Ground		7/2		5/2	14 807.8	.1	938
	744	2, 1, 1← 1, 1, 0	Ground					14 837.07	.04	938
	744	2, 1, 1← 1, 1, 0	Ground		5/2		5/2	14 864.6	.1	938
	744	2, 1, 1← 1, 1, 0	Ground		3/2		3/2	14 888.1	.1	938
	744	2, 1, 1← 1, 1, 0	Ground		5/2		3/2	14 928.0	.1	938
	744	2, 1, 2← 1, 1, 1	Ground		1/2		1/2	14 093.1	.2	938
	744	2, 1, 2← 1, 1, 1	Ground		3/2		1/2	14 157.4	.1	938
	744	2, 1, 2← 1, 1, 1	Ground		7/2		5/2	14 183.9	.1	938
	744	2, 1, 2← 1, 1, 1	Ground					14 213.95	.05	938
	744	2, 1, 2← 1, 1, 1	Ground		5/2		5/2	14 248.0	.1	938
	744	2, 1, 2← 1, 1, 1	Ground		3/2		3/2	14 259.8	.1	938
	744	2, 1, 2← 1, 1, 1	Ground		5/2		3/2	14 304.9	.1	938
	744	3, 0, 3← 2, 0, 2	Ground		3/2		3/2	21 683.5	.2	938
	744	3, 0, 3← 2, 0, 2	Ground		5/2		5/2	21 718.1	.2	938
	744	3, 0, 3← 2, 0, 2	Ground		7/2		5/2	21 774.1	.1	938
	744	3, 0, 3← 2, 0, 2	Ground		9/2		7/2	21 774.1	.1	938
	744	3, 0, 3← 2, 0, 2	Ground					21 779.57	.06	938
	744	3, 0, 3← 2, 0, 2	Ground		5/2		3/2	21 803.1		938

Isotopic Species	Id. No.	Rotational Quantum Nos.	Vib. State v_a	F_1'	F'	F_1	F	Frequency MHz	Acc. ±MHz	Ref.
c-C^{12}HD:C^{12}DBr^{79}	744	3, 0, 3← 2, 0, 2	Ground		3/2		1/2	21 803.1		938
	744	3, 0, 3← 2, 0, 2	Ground		7/2		7/2	21 893.15	.1	938
	744	3, 1, 2← 2, 1, 1	Ground		3/2		3/2	22 184.2	.1	938
	744	3, 1, 2← 2, 1, 1	1		5/2		5/2	22 215.9	.2	938
	744	3, 1, 2← 2, 1, 1	1		9/2		7/2	22 226.4	.1	938
	744	3, 1, 2← 2, 1, 1	Ground		5/2		5/2	22 230.7	.1	938
	744	3, 1, 2← 2, 1, 1	Ground		9/2		7/2	22 240.8	.1	938
	744	3, 1, 2← 2, 1, 1	Ground					22 253.42	.06	938
	744	3, 1, 2← 2, 1, 1	1		7/2		5/2	22 255.9	.2	938
	744	3, 1, 2← 2, 1, 1	Ground		7/2		5/2	22 270.55	.1	938
	744	3, 1, 2← 2, 1, 1	Ground		7/2		7/2	22 327.4	.2	938
	744	3, 1, 3← 2, 1, 2	1		3/2		3/2	21 222.7	.2	938
	744	3, 1, 3← 2, 1, 2	Ground		3/2		3/2	21 244.5	.1	938
	744	3, 1, 3← 2, 1, 2	1		5/2		5/2	21 271.4	.2	938
	744	3, 1, 3← 2, 1, 2	1		9/2		7/2	21 285.1	.1	938
	744	3, 1, 3← 2, 1, 2	Ground		5/2		5/2	21 292.8	.1	938
	744	3, 1, 3← 2, 1, 2	Ground		9/2		7/2	21 306.5	.1	938
	744	3, 1, 3← 2, 1, 2	1		7/2		5/2	21 314.8	.2	938
	744	3, 1, 3← 2, 1, 2	Ground					21 319.36	.05	938
	744	3, 1, 3← 2, 1, 2	Ground		7/2		5/2	21 336.2	.1	938
	744	3, 1, 3← 2, 1, 2	Ground		5/2		3/2	21 337.9	.1	938
	744	3, 1, 3← 2, 1, 2	Ground		7/2		7/2	21 400.0	.2	938
	744	3, 2, 1← 2, 2, 0	Ground		3/2		1/2	21 675.6	.1	938
	744	3, 2, 1← 2, 2, 0	Ground		9/2		7/2	21 760.9	.1	938
	744	3, 2, 1← 2, 2, 0	Ground		7/2		7/2	21 760.9	.1	938
	744	3, 2, 1← 2, 2, 0	Ground					21 795.45	.04	938
	744	3, 2, 1← 2, 2, 0	Ground		5/2		3/2	21 795.6	.1	938
	744	3, 2, 1← 2, 2, 0	Ground		3/2		3/2	21 795.6	.1	938
	744	3, 2, 1← 2, 2, 0	Ground		7/2		5/2	21 881.2	.1	938
	744	3, 2, 1← 2, 2, 0	Ground		5/2		5/2	21 881.2	.1	938
	744	3, 2, 2← 2, 2, 1	Ground		3/2		1/2	21 668.2	.2	938
	744	3, 2, 2← 2, 2, 1	Ground		7/2		7/2	21 753.6	.1	938
	744	3, 2, 2← 2, 2, 1	Ground		9/2		7/2	21 753.6	.1	938
	744	3, 2, 2← 2, 2, 1	Ground					21 788.03	.04	938
	744	3, 2, 2← 2, 2, 1	Ground		5/2		3/2	21 788.2	.1	938
	744	3, 2, 2← 2, 2, 1	Ground		3/2		3/2	21 788.2	.1	938
	744	3, 2, 2← 2, 2, 1	Ground		5/2		5/2	21 873.7	.1	938
	744	3, 2, 2← 2, 2, 1	Ground		7/2		5/2	21 873.7	.1	938
	744	4, 0, 4← 3, 0, 3	Ground		7/2		7/2	28 984.1		998
	744	4, 0, 4← 3, 0, 3	Ground		9/2		7/2	29 027.4		998
	744	4, 0, 4← 3, 0, 3	Ground		11/2		9/2	29 027.4		998
	744	4, 0, 4← 3, 0, 3	Ground					29 031.45	1.53	998
	744	4, 0, 4← 3, 0, 3	Ground		5/2		3/2	29 043.8		998
	744	4, 0, 4← 3, 0, 3	Ground		7/2		5/2	29 043.8		998
	744	4, 0, 4← 3, 0, 3	Ground		9/2		9/2	29 148.0		998
	744	4, 1, 3← 3, 1, 2	Ground		5/2		5/2	29 580.5		998
	744	4, 1, 3← 3, 1, 2	Ground		7/2		7/2	29 639.7		998
	744	4, 1, 3← 3, 1, 2	1		11/2		9/2	29 642.6		998
	744	4, 1, 3← 3, 1, 2	1		5/2		3/2	29 648.0		998
	744	4, 1, 3← 3, 1, 2	1					29 649.35	.16	998

Isotopic Species	Id. No.	Rotational Quantum Nos.	v_a	Vib. State	F_1'	F'	F_1	F	Frequency MHz	Acc. ±MHz	Ref.
c-C^{12}HD:C^{12}DBr^{79}	744	4, 1, 3← 3, 1, 2	1			9/2		7/2	29 654.2		998
	744	4, 1, 3← 3, 1, 2	1			7/2		5/2	29 660.2		998
	744	4, 1, 3← 3, 1, 2		Ground		11/2		9/2	29 662.0		998
	744	4, 1, 3← 3, 1, 2		Ground		5/2		3/2	29 667.2		998
	744	4, 1, 3← 3, 1, 2		Ground					29 668.64	.16	998
	744	4, 1, 3← 3, 1, 2		Ground		9/2		7/2	29 673.4		998
	744	4, 1, 3← 3, 1, 2		Ground		7/2		5/2	29 679.7		998
	744	4, 1, 3← 3, 1, 2		Ground		9/2		9/2	29 760.1		998
	744	4, 1, 4← 3, 1, 3		Ground		5/2		5/2	28 329.3		998
	744	4, 1, 4 ←3, 1, 3	1			11/2		9/2	28 387.4		998
	744	4, 1, 4← 3, 1, 3		Ground		7/2		7/2	28 391.3		998
	744	4, 1, 4← 3, 1, 3	1			5/2		3/2	28 394.1		998
	744	4, 1, 4← 3, 1, 3	1						28 394.63	.04	998
	744	4, 1, 4← 3, 1, 3	1			9/2		7/2	28 399.45		998
	744	4, 1, 4← 3, 1, 3	1			7/2		5/2	28 406.2		998
	744	4, 1, 4← 3, 1, 3		Ground		11/2		9/2	28 416.1		998
	744	4, 1, 4← 3, 1, 3		Ground		5/2		3/2	28 422.4		998
	744	4, 1, 4← 3, 1, 3		Ground					28 423.22	.07	998
	744	4, 1, 4← 3, 1, 3		Ground		9/2		7/2	28 427.9		998
	744	4, 1, 4← 3, 1, 3		Ground		7/2		5/2	28 434.6		998
	744	4, 1, 4← 3, 1, 3		Ground		9/2		9/2	28 521.6		998
	744	4, 2, 2← 3, 2, 1		Ground		5/2		3/2	29 032.5		998
	744	4, 2, 2← 3, 2, 1		Ground		11/2		9/2	29 049.7		998
	744	4, 2, 2← 3, 2, 1		Ground					29 066.90	.11	998
	744	4, 2, 2← 3, 2, 1		Ground		7/2		5/2	29 080.4		998
	744	4, 2, 2← 3, 2, 1		Ground		9/2		7/2	29 097.4		998
	744	4, 2, 3← 3, 2, 2	1			5/2		3/2	28 990.7		998
	744	4, 2, 3← 3, 2, 2	1			11/2		9/2	29 007.4		998
	744	4, 2, 3← 3, 2, 2		Ground		5/2		3/2	29 014.1		998
	744	4, 2, 3← 3, 2, 2	1						29 024.96	.18	998
	744	4, 2, 3← 3, 2, 2		Ground		11/2		9/2	29 031.1		998
	744	4, 2, 3← 3, 2, 2	1			7/2		5/2	29 038.6		998
	744	4, 2, 3← 3, 2, 2		Ground					29 048.49	.07	998
	744	4, 2, 3← 3, 2, 2	1			9/2		7/2	29 056.5		998
	744	4, 2, 3← 3, 2, 2		Ground		7/2		5/2	29 062.1		998
	744	4, 2, 3← 3, 2, 2		Ground		9/2		7/2	29 079.2		998
	744	4, 3, 2← 3, 3, 1		Ground		5/2		3/2	28 964.6		998
	744	4, 3, 2← 3, 3, 1		Ground		9/2		9/2	28 978.7		998
	744	4, 3, 2← 3, 3, 1		Ground		11/2		9/2	29 029.4		998
	744	4, 3, 2← 3, 3, 1		Ground					29 054.80	.28	998
	744	4, 3, 2← 3, 3, 1		Ground		7/2		5/2	29 072.4		998
	744	4, 3, 2← 3, 3, 1		Ground		5/2		5/2	29 115.4		998
	744	4, 3, 2← 3, 3, 1		Ground		9/2		7/2	29 127.7		998
	744	4, 3, 2← 3, 3, 1		Ground		7/2		7/2	29 144.3		998
-C^{12}HD:C^{12}DBr^{79}	745	2, 0, 2← 1, 0, 1		Ground		3/2		3/2	15 001.35	.05	938
	745	2, 0, 2← 1, 0, 1		Ground		5/2		3/2	15 084.1	.1	938
	745	2, 0, 2← 1, 0, 1		Ground		7/2		5/2	15 084.1	.1	938
	745	2, 0, 2← 1, 0, 1		Ground					15 094.31	.05	938
	745	2, 0, 2← 1, 0, 1		Ground		5/2		5/2	15 201.1	.1	938

Bromoethene

Isotopic Species	Id. No.	Rotational Quantum Nos.	Vib. State v_a	F_1'	F'	F_1	F	Frequency MHz	Acc. ±MHz	Ref.
t-$C^{12}HD$:$C^{12}DBr^{79}$	745	2, 0, 2← 1, 0, 1	Ground		3/2		1/2	15 211.2	.1	938
	745	2, 1, 1← 1, 1, 0	Ground		3/2		1/2	15 428.7	.2	938
	745	2, 1, 1← 1, 1, 0	Ground		7/2		5/2	15 463.4	.1	938
	745	2, 1, 1← 1, 1, 0	Ground					15 491.73	.06	938
	745	2, 1, 1← 1, 1, 0	Ground		5/2		5/2	15 517.0	.2	938
	745	2, 1, 1← 1, 1, 0	Ground		3/2		3/2	15 543.1	.1	938
	745	2, 1, 1← 1, 1, 0	Ground		5/2		3/2	15 580.6	.2	938
	745	2, 1, 2← 1, 1, 1	Ground		3/2		1/2	14 652.0	.2	938
	745	2, 1, 2← 1, 1, 1	Ground		7/2		5/2	14 675.8	.1	938
	745	2, 1, 2← 1, 1, 1	Ground					14 704.98	.06	938
	745	2, 1, 2← 1, 1, 1	Ground		5/2		5/2	14 740.0	.2	938
	745	2, 1, 2← 1, 1, 1	Ground		3/2		3/2	14 748.1	.2	938
	745	2, 1, 2← 1, 1, 1	Ground		5/2		3/2	14 793.1	.1	938
	745	3, 0, 3← 2, 0, 2	Ground		3/2		3/2	22 538.2	.1	938
	745	3, 0, 3← 2, 0, 2	Ground		5/2		5/2	22 572.6	.2	938
	745	3, 0, 3← 2, 0, 2	Ground		9/2		7/2	22 626.5	.1	938
	745	3, 0, 3← 2, 0, 2	Ground		7/2		5/2	22 626.5	.1	938
	745	3, 0, 3← 2, 0, 2	Ground					22 631.95	.04	938
	745	3, 0, 3← 2, 0, 2	Ground		3/2		1/2	22 655.2	.1	938
	745	3, 0, 3← 2, 0, 2	Ground		5/2		3/2	22 655.2	.1	938
	745	3, 0, 3← 2, 0, 2	Ground		7/2		7/2	22 743.2	.1	938
	745	3, 1, 2← 2, 1, 1	Ground		7/2		7/2	22 613.1	.1	938
	745	3, 1, 2← 2, 1, 1	Ground		9/2		7/2	22 613.1	.1	938
	745	3, 1, 2← 2, 1, 1	Ground		3/2		3/2	23 167.8	.2	938
	745	3, 1, 2← 2, 1, 1	1		9/2		7/2	23 206.4	.1	938
	745	3, 1, 2← 2, 1, 1	Ground		5/2		5/2	23 213.5	.1	938
	745	3, 1, 2← 2, 1, 1	Ground		3/2		1/2	23 221.3	.1	938
	745	3, 1, 2← 2, 1, 1	Ground		9/2		7/2	23 222.6	.1	938
	745	3, 1, 2← 2, 1, 1	Ground					23 234.53	.04	938
	745	3, 1, 2← 2, 1, 1	1		7/2		5/2	23 235.6	.1	938
	745	3, 1, 2← 2, 1, 1	Ground		7/2		5/2	23 251.2	.1	938
	745	3, 1, 2← 2, 1, 1	1		7/2		7/2	23 288.8	.2	938
	745	3, 1, 2← 2, 1, 1	Ground		7/2		7/2	23 304.8	.1	938
	745	3, 1, 3← 2, 1, 2	1		3/2		3/2	21 955.7	.2	938
	745	3, 1, 3← 2, 1, 2	Ground		3/2		3/2	21 979.8	.1	938
	745	3, 1, 3← 2, 1, 2	1		5/2		5/2	22 003.4	.2	938
	745	3, 1, 3← 2, 1, 2	1		9/2		7/2	22 017.3	.1	938
	745	3, 1, 3← 2, 1, 2	Ground		5/2		5/2	22 027.3	.1	938
	745	3, 1, 3← 2, 1, 2	Ground		9/2		7/2	22 041.3	.1	938
	745	3, 1, 3← 2, 1, 2	Ground		3/2		1/2	22 042.9	.1	938
	745	3, 1, 3← 2, 1, 2	1		7/2		5/2	22 046.3	.1	938
	745							22 054.15	.05	938
	745	3, 1, 3← 2, 1, 2	Ground		7/2		5/2	22 071.4	.1	938
	745	3, 1, 3← 2, 1, 2	Ground		5/2		3/2	22 072.5	.1	938
	745	3, 2, 1← 2, 2, 0	Ground		3/2		1/2	22 544.4	.1	938
	745	3, 2, 1← 2, 2, 0	Ground		9/2		7/2	22 627.7	.1	938
	745	3, 2, 1← 2, 2, 0	Ground		7/2		7/2	22 627.7	.1	938
	745	3, 2, 1← 2, 2, 0	Ground					22 661.15	.04	938
	745	3, 2, 1← 2, 2, 0	Ground		3/2		3/2	22 661.3	.1	938
	745	3, 2, 1← 2, 2, 0	Ground		5/2		3/2	22 661.3	.1	938

Isotopic Species	Id. No.	Rotational Quantum Nos.	Vib. State	v_a	Hyperfine				Frequency MHz	Acc. ±MHz	Ref.
					F'_1	F'	F_1	F			
t-$C^{12}HD$:$C^{12}DBr^{79}$	745	3, 2, 1← 2, 2, 0	Ground			5/2		5/2	22 744.5	.1	938
	745	3, 2, 1← 2, 2, 0	Ground			7/2		5/2	22 744.5	.1	938
	745	3, 2, 2← 2, 2, 1	Ground			3/2		1/2	22 530.3	.1	938
	745	3, 2, 2← 2, 2, 1	Ground						22 646.88	.04	938
	745	3, 2, 2← 2, 2, 1	Ground			3/2		3/2	22 647.1	.1	938
	745	3, 2, 2← 2, 2, 1	Ground			5/2		3/2	22 647.1	.1	938
	745	3, 2, 2← 2, 2, 1	Ground			5/2		5/2	22 730.4	.1	938
	745	3, 2, 2← 2, 2, 1	Ground			7/2		5/2	22 730.4	.1	938
	745	4, 0, 4← 3, 0, 3	Ground			5/2		5/2	30 050.8		998
	745	4, 0, 4← 3, 0, 3	Ground			7/2		7/2	30 115.2		998
	745	4, 0, 4← 3, 0, 3	Ground			11/2		9/2	30 155.6		998
	745	4, 0, 4← 3, 0, 3	Ground			9/2		7/2	30 155.6		998
	745	4, 0, 4← 3, 0, 3	Ground						30 158.51	.91	998
	745	4, 0, 4← 3, 0, 3	Ground			5/2		3/2	30 167.3		998
	745	4, 0, 4← 3, 0, 3	Ground			7/2		5/2	30 167.3		998
	745	4, 1, 3← 3, 1, 2	Ground			5/2		5/2	30 889.6		998
	745	4, 1, 3← 3, 1, 2	Ground			7/2		7/2	30 945.2		998
	745	4, 1, 3← 3, 1, 2		1		5/2		3/2	30 951.2		998
	745	4, 1, 3← 3, 1, 2		1					30 953.09		998
	745	4, 1, 3← 3, 1, 2		1		9/2		7/2	30 957.6		998
	745	4, 1, 3← 3, 1, 2		1		7/2		5/2	30 961.2		998
	745	4, 1, 3← 3, 1, 2	Ground			11/2		9/2	30 966.4		998
	745	4, 1, 3← 3, 1, 2	Ground			5/2		3/2	30 973.1		998
	745	4, 1, 3← 3, 1, 2	Ground						30 973.48	.45	998
	745	4, 1, 3← 3, 1, 2	Ground			9/2		7/2	30 979.2		998
	745	4, 1, 3← 3, 1, 2	Ground			7/2		5/2	30 982.7		998
	745	4, 1, 4← 3, 1, 3	Ground			5/2		5/2	29 307.4		998
	745	4, 1, 4← 3, 1, 3		1		11/2		9/2	29 361.9		998
	745	4, 1, 4← 3, 1, 3		1					29 368.86		998
	745	4, 1, 4← 3, 1, 3	Ground			7/2		7/2	29 369.0		998
	745	4, 1, 4← 3, 1, 3		1		9/2		7/2	29 373.3		998
	745	4, 1, 4← 3, 1, 3		1		7/2		5/2	29 380.1		998
	745	4, 1, 4← 3, 1, 3	Ground			11/2		9/2	29 393.9		998
	745	4, 1, 4← 3, 1, 3	Ground			5/2		3/2	29 400.1		998
	745	4, 1, 4← 3, 1, 3	Ground						29 400.73	.09	998
	745	4, 1, 4← 3, 1, 3	Ground			9/2		7/2	29 405.2		998
	745	4, 1, 4← 3, 1, 3	Ground			7/2		5/2	29 411.9		998
	745	4, 1, 4← 3, 1, 3	Ground			9/2		9/2	29 498.3		998
	745	4, 2, 2← 3, 2, 1	Ground			5/2		3/2	30 194.05		998
	745	4, 2, 2← 3, 2, 1	Ground			11/2		9/2	30 210.8		998
	745	4, 2, 2← 3, 2, 1	Ground						30 227.62	.15	998
	745	4, 2, 2← 3, 2, 1	Ground			7/2		5/2	30 242.1		998
	745	4, 2, 2← 3, 2, 1	Ground			9/2		7/2	30 256.9		998
	745	4, 2, 3← 3, 2, 2		1		5/2		3/2	30 132.0		998
	745	4, 2, 3← 3, 2, 2		1		11/2		9/2	30 148.9		998
	745	4, 2, 3← 3, 2, 2	Ground			5/2		3/2	30 158.8		998
	745	4, 2, 3← 3, 2, 2		1					30 165.20		998
	745	4, 2, 3← 3, 2, 2	Ground			11/2		9/2	30 175.6		998
	745	4, 2, 3← 3, 2, 2		1		7/2		5/2	30 178.7		998
	745	4, 2, 3← 3, 2, 2	Ground						30 192.38	.11	998

Bromoethene

Isotopic Species	Id. No.	Rotational Quantum Nos.	Vib. State v_a	Hyperfine				Frequency MHz	Acc. ±MHz	Ref.
				F_1'	F'	F_1	F			
t-C^{12}HD:C^{12}DBr^{79}	745	4, 2, 3← 3, 2, 2	Ground		7/2		5/2	30 205.85		998
	745	4, 2, 3← 3, 2, 2	Ground		9/2		7/2	30 222.0		998
	745	4, 3, 2← 3, 3, 1	Ground		9/2		9/2	30 128.7		998
	745	4, 3, 2← 3, 3, 1	Ground		11/2		9/2	30 169.0		998
	745	4, 3, 2← 3, 3, 1	Ground					30 202.98	.21	998
	745	4, 3, 2← 3, 3, 1	Ground		7/2		5/2	30 220.5		998
	745	4, 3, 2← 3, 3, 1	Ground		9/2		7/2	30 273.8		998
	745	4, 3, 2← 3, 3, 1	Ground		7/2		7/2	30 289.1		998
	745	5, 1, 4← 4, 1, 3	1		13/2		11/2	38 681.3		998
	745	5, 1, 4← 4, 1, 3	1		7/2		5/2	38 684.5		998
	745	5, 1, 4← 4, 1, 3	1					38 684.81	.6	998
	745	5, 1, 4← 4, 1, 3	1		11/2		9/2	38 686.0		998
	745	5, 1, 4← 4, 1, 3	1		9/2		7/2	38 691.5		998
	745	5, 1, 4← 4, 1, 3	Ground		13/2		11/2	38 707.5		998
	745	5, 1, 4← 4, 1, 3	Ground		7/2		5/2	38 711.6		998
	745	5, 1, 4← 4, 1, 3	Ground					38 711.82	.29	998
	745	5, 1, 4← 4, 1, 3	Ground		11/2		9/2	38 713.4		998
	745	5, 1, 4← 4, 1, 3	Ground		9/2		7/2	38 718.8		998
c-C^{12}HD:C^{12}DBr^{81}	746	2, 0, 2← 1, 0, 1	Ground		3/2		3/2	14 357.9	.1	938
	746	2, 0, 2← 1, 0, 1	Ground		7/2		5/2	14 429.9	.1	938
	746	2, 0, 2← 1, 0, 1	Ground		5/2		3/2	14 429.9	.1	938
	746	2, 0, 2← 1, 0, 1	Ground		1/2		1/2	14 437.7	.1	938
	746	2, 0, 2← 1, 0, 1	Ground					14 438.20	.04	938
	746	2, 0, 2← 1, 0, 1	Ground		5/2		5/2	14 529.6	.1	938
	746	2, 0, 2← 1, 0, 1	Ground		3/2		1/2	14 538.7	.1	938
	746	2, 1, 1← 1, 1, 0	Ground		1/2		1/2	14 647.9	.1	938
	746	2, 1, 1← 1, 1, 0	Ground		3/2		1/2	14 695.05	.1	938
	746	2, 1, 1← 1, 1, 0	Ground		7/2		5/2	14 723.4	.1	938
	746	2, 1, 1← 1, 1, 0	Ground					14 747.95	.04	938
	746	2, 1, 1← 1, 1, 0	Ground		5/2		5/2	14 771.0	.1	938
	746	2, 1, 1← 1, 1, 0	Ground		3/2		3/2	14 790.6	.1	938
	746	2, 1, 1← 1, 1, 0	Ground		5/2		3/2	14 824.1	.1	938
	746	2, 1, 2← 1, 1, 1	Ground		1/2		1/2	14 033.0	.1	938
	746	2, 1, 2← 1, 1, 1	Ground		3/2		1/2	14 085.6	.1	938
	746	2, 1, 2← 1, 1, 1	Ground		7/2		5/2	14 107.3	.1	938
	746	2, 1, 2← 1, 1, 1	Ground					14 132.47	.05	938
	746	2, 1, 2← 1, 1, 1	Ground		5/2		5/2	14 160.6	.2	938
	746	2, 1, 2← 1, 1, 1	Ground		3/2		3/2	14 170.0	.3	938
	746	2, 1, 2← 1, 1, 1	Ground		5/2		3/2	14 208.2	.1	938
	746	2, 1, 2← 1, 1, 1	Ground		5/2		3/2	14 694.5	.1	938
	746	3, 0, 3← 2, 0, 2	Ground		3/2		3/2	21 571.8	.2	938
	746	3, 0, 3← 2, 0, 2	Ground		5/2		5/2	21 600.5	.1	938
	746	3, 0, 3← 2, 0, 2	Ground		7/2		5/2	21 647.4	.1	938
	746	3, 0, 3← 2, 0, 2	Ground		9/2		7/2	21 647.4	.1	938
	746	3, 0, 3← 2, 0, 2	Ground					21 651.91	.04	938
	746	3, 0, 3← 2, 0, 2	Ground		5/2		3/2	21 671.5	.1	938
	746	3, 0, 3← 2, 0, 2	Ground		3/2		1/2	21 671.5	.1	938
	746	3, 0, 3← 2, 0, 2	Ground		7/2		7/2	21 747.4	.2	938
	746	3, 1, 2← 2, 1, 1	1		3/2		3/2	22 048.4	.1	938

Isotopic Species	Id. No.	Rotational Quantum Nos.	Vib. State v_a	F_1'	F'	F_1	F	Frequency MHz	Acc. ±MHz	Ref.
c-$C^{12}HD{:}C^{12}DBr^{81}$	746	3, 1, 2← 2, 1, 1	Ground		3/2		3/2	22 062.1	.2	938
	746	3, 1, 2← 2, 1, 1	1		5/2		5/2	22 086.4	.2	938
	746	3, 1, 2← 2, 1, 1	1		9/2		7/2	22 094.9	.1	938
	746	3, 1, 2← 2, 1, 1	Ground		5/2		5/2	22 100.75	.1	938
	746	3, 1, 2← 2, 1, 1	Ground		9/2		7/2	22 109.5	.1	938
	746	3, 1, 2← 2, 1, 1	1		7/2		5/2	22 120.0	.1	938
	746	3, 1, 2← 2, 1, 1	Ground					22 120.00	.06	938
	746	3, 1, 2← 2, 1, 1	Ground		7/2		5/2	22 134.4	.1	938
	746	3, 1, 2← 2, 1, 1	Ground		7/2		7/2	22 181.8	.1	938
	746	3, 1, 3← 2, 1, 2	Ground		3/2		3/2	21 134.2	.2	938
	746	3, 1, 3← 2, 1, 2	1		5/2		5/2	21 153.1	.2	938
	746	3, 1, 3← 2, 1, 2	1		9/2		7/2	21 164.7	.1	938
	746	3, 1, 3← 2, 1, 2	Ground		5/2		5/2	21 174.6	.2	938
	746	3, 1, 3← 2, 1, 2	Ground		9/2		7/2	21 185.9	.1	938
	746	3, 1, 3← 2, 1, 2	1		7/2		5/2	21 189.5	.1	938
	746	3, 1, 3← 2, 1, 2	1		5/2		3/2	21 191.0	.2	938
	746	3, 1, 3← 2, 1, 2	Ground					21 196.72	.05	938
	746	3, 1, 3← 2, 1, 2	Ground		7/2		5/2	21 210.8	.1	938
	746	3, 1, 3← 2, 1, 2	Ground		5/2		3/2	21 212.3	.1	938
	746	3, 1, 3← 2, 1, 2	Ground		7/2		7/2	21 264.0	.2	938
	746	3, 2, 1← 2, 2, 0	Ground		3/2		1/2	21 567.2	.1	938
	746	3, 2, 1← 2, 2, 0	Ground		7/2		7/2	21 638.5	.1	938
	746	3, 2, 1← 2, 2, 0	Ground		9/2		7/2	21 638.5	.1	938
	746	3, 2, 1← 2, 2, 0	Ground					21 667.30	.04	938
	746	3, 2, 1← 2, 2, 0	Ground		3/2		3/2	21 667.5	.1	938
	746	3, 2, 1← 2, 2, 0	Ground		5/2		3/2	21 667.5	.1	938
	746	3, 2, 1← 2, 2, 0	Ground		5/2		5/2	21 738.8	.1	938
	746	3, 2, 1← 2, 2, 0	Ground		7/2		5/2	21 738.8	.1	938
	746	3, 2, 2← 2, 2, 1	Ground		3/2		1/2	21 560.0	.1	938
	746	3, 2, 2← 2, 2, 1	Ground		7/2		7/2	21 631.4	.1	938
	746	3, 2, 2← 2, 2, 1	Ground		9/2		7/2	21 631.4	.1	938
	746	3, 2, 2← 2, 2, 1	Ground					21 660.16	.04	938
	746	3, 2, 2← 2, 2, 1	Ground		5/2		3/2	21 660.2	.1	938
	746	3, 2, 2← 2, 2, 1	Ground		3/2		3/2	21 660.2	.1	938
	746	3, 2, 2← 2, 2, 1	Ground		5/2		5/2	21 731.8	.1	938
	746	3, 2, 2← 2, 2, 1	Ground		7/2		5/2	21 731.8	.1	938
	746	4, 0, 4← 3, 0, 3	Ground		5/2		5/2	28 767.2		998
	746	4, 0, 4← 3, 0, 3	Ground		7/2		7/2	28 822.7		998
	746	4, 0, 4← 3, 0, 3	Ground					28 862.34		998
	746	4, 1, 3← 3, 1, 2	Ground		5/2		5/2	29 417.5		998
	746	4, 1, 3← 3, 1, 2	1		11/2		9/2	29 466.7		998
	746	4, 1, 3← 3, 1, 2	1		5/2		3/2	29 470.4		998
	746	4, 1, 3← 3, 1, 2	1					29 472.0	.38	998
	746	4, 1, 3← 3, 1, 2	1		9/2		7/2	29 475.9		998
	746	4, 1, 3← 3, 1, 2	1		7/2		5/2	29 481.0		998
	746	4, 1, 3← 3, 1, 2	Ground		11/2		9/2	29 485.4		998
	746	4, 1, 3← 3, 1, 2	Ground		5/2		3/2	29 489.3		998
	746	4, 1, 3← 3, 1, 2	Ground					29 490.94	.13	998
	746	4, 1, 3← 3, 1, 2	Ground		9/2		7/2	29 495.2		998
	746	4, 1, 3← 3, 1, 2	Ground		7/2		5/2	29 500.3		998

Isotopic Species	Id. No.	Rotational Quantum Nos.	Vib. State v_a	F_1'	F'	F_1	F	Frequency MHz	Acc. ±MHz	Ref.
c-$C^{12}HD{:}C^{12}DBr^{81}$	746	4, 1, 3← 3, 1, 2	Ground		9/2		9/2	29 567.4		998
	746	4, 1, 4← 3, 1, 3	Ground		5/2		5/2	28 181.3		998
	746	4, 1, 4← 3, 1, 3	1		11/2		9/2	28 225.6		998
	746	4, 1, 4← 3, 1, 3	1		5/2		3/2	28 231.0		998
	746	4, 1, 4← 3, 1, 3	1					28 231.54	.08	998
	746	4, 1, 4← 3, 1, 3	Ground		7/2		7/2	28 233.2		998
	746	4, 1, 4← 3, 1, 3	1		9/2		7/2	28 235.5		998
	746	4, 1, 4← 3, 1, 3	1		7/2		5/2	28 241.2		998
	746	4, 1, 4← 3, 1, 3	Ground		11/2		9/2	28 253.9		998
	746	4, 1, 4← 3, 1, 3	Ground		5/2		3/2	28 259.3		998
	746	4, 1, 4← 3, 1, 3	Ground					28 259.86	.06	998
	746	4, 1, 4← 3, 1, 3	Ground		9/2		7/2	28 263.7		998
	746	4, 1, 4← 3, 1, 3	Ground		7/2		5/2	28 269.4		998
	746	4, 1, 4← 3, 1, 3	Ground		9/2		9/2	28 342.3		998
	746	4, 2, 2← 3, 2, 1	Ground		5/2		3/2	28 867.1		998
	746	4, 2, 2← 3, 2, 1	Ground		11/2		9/2	28 881.5		998
	746	4, 2, 2← 3, 2, 1	Ground					28 895.88	.08	998
	746	4, 2, 2← 3, 2, 1	Ground		7/2		5/2	28 907.3		998
	746	4, 2, 2← 3, 2, 1	Ground		9/2		7/2	28 921.4		998
	746	4, 2, 3← 3, 2, 2	1		5/2		3/2	28 825.5		998
	746	4, 2, 3← 3, 2, 2	1		11/2		9/2	28 839.8		998
	746	4, 2, 3← 3, 2, 2	Ground		5/2		3/2	28 849.5		998
	746	4, 2, 3← 3, 2, 2	1					28 854.59	.02	998
	746	4, 2, 3← 3, 2, 2	Ground		11/2		9/2	28 863.7		998
	746	4, 2, 3← 3, 2, 2	1		7/2		5/2	28 866.2		998
	746	4, 2, 3← 3, 2, 2	Ground					28 878.13	.09	998
	746	4, 2, 3← 3, 2, 2	Ground		7/2		5/2	28 889.7		998
	746	4, 2, 3← 3, 2, 2	Ground		9/2		7/2	28 903.4		998
	746	4, 3, 2← 3, 3, 1	Ground		5/2		3/2	28 808.8		998
	746	4, 3, 2← 3, 3, 1	Ground		9/2		9/2	28 819.8		998
	746	4, 3, 2← 3, 3, 1	Ground		11/2		9/2	28 854.8		998
	746	4, 3, 2← 3, 3, 1	Ground					28 883.84	.13	998
	746	4, 3, 2← 3, 3, 1	Ground		7/2		5/2	28 898.7		998
	746	4, 3, 2← 3, 3, 1	Ground		5/2		5/2	28 934.5		998
	746	4, 3, 2← 3, 3, 1	Ground		9/2		7/2	28 944.7		998
	746	4, 3, 2← 3, 3, 1	Ground		7/2		7/2	28 956.4		998
t-$C^{12}HD{:}C^{12}DBr^{81}$	747	2, 0, 2← 1, 0, 1	Ground		7/2		5/2	14 996.9	.1	938
	747	2, 0, 2← 1, 0, 1	Ground		5/2		3/2	14 996.9	.1	938
	747	2, 0, 2← 1, 0, 1	Ground					15 005.28	.07	938
	747	2, 0, 2← 1, 0, 1	Ground		5/2		5/2	15 094.8	.2	938
	747	2, 0, 2← 1, 0, 1	Ground		3/2		1/2	15 103.1	.1	938
	747	2, 1, 1← 1, 1, 0	Ground		1/2		1/2	15 300.8	.1	938
	747	2, 1, 1← 1, 1, 0	Ground		3/2		1/2	15 345.5	.1	938
	747	2, 1, 1← 1, 1, 0	Ground		7/2		5/2	15 374.5	.1	938
	747	2, 1, 1← 1, 1, 0	Ground					15 398.28	.05	938
	747	2, 1, 1← 1, 1, 0	Ground		5/2		5/2	15 419.2	.1	938
	747	2, 1, 1← 1, 1, 0	Ground		3/2		3/2	15 440.9	.1	938
	747	2, 1, 1← 1, 1, 0	Ground		5/2		3/2	15 472.4	.2	938
	747	2, 1, 2← 1, 1, 1	Ground		1/2		1/2	14 523.2	.1	938

Isotopic Species	Id. No.	Rotational Quantum Nos.	Vib. State v_a	F_1'	F'	F_1	F	Frequency MHz	Acc. ±MHz	Ref.
t-$C^{12}HD{:}C^{12}DBr^{81}$	747	2, 1, 2← 1, 1, 1	Ground		3/2		1/2	14 576.0	.1	938
	747	2, 1, 2← 1, 1, 1	Ground		7/2		5/2	14 596.2	.1	938
	747	2, 1, 2← 1, 1, 1	Ground					14 620.65	.04	938
	747	2, 1, 2← 1, 1, 1	Ground		5/2		5/2	14 649.4	.2	938
	747	2, 1, 2← 1, 1, 1	Ground		3/2		3/2	14 656.3	.1	938
	747	3, 0, 3← 2, 0, 2	Ground		3/2		3/2	22 420.5	.2	938
	747	3, 0, 3← 2, 0, 2	Ground		5/2		5/2	22 449.4	.1	938
	747	3, 0, 3← 2, 0, 2	Ground		9/2		7/2	22 494.2	.1	938
	747	3, 0, 3← 2, 0, 2	Ground		7/2		5/2	22 494.2	.1	938
	747	3, 0, 3← 2, 0, 2	Ground					22 498.94	.04	938
	747	3, 0, 3← 2, 0, 2	Ground		3/2		1/2	22 518.2	.1	938
	747	3, 0, 3← 2, 0, 2	Ground		5/2		3/2	22 518.2	.1	938
	747	3, 0, 3← 2, 0, 2	Ground		7/2		7/2	22 592.0	.2	938
	747	3, 1, 2← 2, 1, 1	1		3/2		3/2	23 022.7	.2	938
	747	3, 1, 2← 2, 1, 1	Ground		3/2		3/2	23 038.7	.1	938
	747	3, 1, 2← 2, 1, 1	1		9/2		7/2	23 068.2	.2	938
	747	3, 1, 2← 2, 1, 1	Ground		5/2		5/2	23 076.5	.2	938
	747	3, 1, 2← 2, 1, 1	Ground		9/2		7/2	23 084.3	.1	938
	747	3, 1, 2← 2, 1, 1	1		7/2		5/2	23 092.3	.2	938
	747	3, 1, 2← 2, 1, 1	Ground					23 094.27	.05	938
	747	3, 1, 2← 2, 1, 1	Ground		7/2		5/2	23 108.25	.1	938
	747	3, 1, 2← 2, 1, 1	Ground		7/2		7/2	23 153.0	.1	938
	747	3, 1, 3← 2, 1, 2	Ground		3/2		3/2	21 865.5	.1	938
	747	3, 1, 3← 2, 1, 2	Ground		5/2		5/2	21 905.2	.2	938
	747	3, 1, 3← 2, 1, 2	Ground		9/2		7/2	21 917.0	.1	938
	747	3, 1, 3← 2, 1, 2	Ground					21 927.64	.04	938
	747	3, 1, 3← 2, 1, 2	Ground		7/2		5/2	21 941.1	.1	938
	747	3, 1, 3← 2, 1, 2	Ground		5/2		3/2	21 943.0	.1	938
	747	3, 1, 3← 2, 1, 2	Ground		7/2		7/2	21 994.6	.1	938
	747	3, 2, 1← 2, 2, 0	Ground		3/2		1/2	22 429.6	.1	938
	747	3, 2, 1← 2, 2, 0	Ground		9/2		7/2	22 499.25	.1	938
	747	3, 2, 1← 2, 2, 0	Ground		7/2		7/2	22 499.25	.1	938
	747	3, 2, 1← 2, 2, 0	Ground					22 527.26	.04	938
	747	3, 2, 1← 2, 2, 0	Ground		3/2		3/2	22 527.4	.1	938
	747	3, 2, 1← 2, 2, 0	Ground		5/2		3/2	22 527.4	.1	938
	747	3, 2, 1← 2, 2, 0	Ground		5/2		5/2	22 597.0	.1	938
	747	3, 2, 1← 2, 2, 0	Ground		7/2		5/2	22 597.0	.1	938
	747	3, 2, 2← 2, 2, 1	Ground		3/2		1/2	22 416.1	.1	938
	747	3, 2, 2← 2, 2, 1	Ground		7/2		7/2	22 485.7	.1	938
	747	3, 2, 2← 2, 2, 1	Ground		9/2		7/2	22 485.7	.1	938
	747	3, 2, 2← 2, 2, 1	Ground					22 513.63	.04	938
	747	3, 2, 2← 2, 2, 1	Ground		5/2		3/2	22 513.7	.1	938
	747	3, 2, 2← 2, 2, 1	Ground		3/2		3/2	22 513.7	.1	938
	747	3, 2, 2← 2, 2, 1	Ground		7/2		5/2	22 583.3	.1	938
	747	3, 2, 2← 2, 2, 1	Ground		5/2		5/2	22 583.3	.1	938
	747	4, 0, 4← 3, 0, 3	Ground		11/2		9/2	29 979.6		998
	747	4, 0, 4← 3, 0, 3	Ground		9/2		7/2	29 979.6		998
	747	4, 0, 4← 3, 0, 3	Ground					29 982.50		998
	747	4, 0, 4← 3, 0, 3	Ground		5/2		3/2	29 990.7		998
	747	4, 0, 4← 3, 0, 3	Ground		7/2		5/2	29 990.7		998

Isotopic Species	Id. No.	Rotational Quantum Nos.	Vib. State v_a	F_1'	F'	F_1	F	Frequency MHz	Acc. ±MHz	Ref.
t-$C^{12}HD{:}C^{12}DBr^{81}$	747	4, 1, 3← 3, 1, 2	Ground		5/2		5/2	30 717.2		998
	747	4, 1, 3← 3, 1, 2	1		11/2		9/2	30 759.9		998
	747	4, 1, 3← 3, 1, 2	Ground		7/2		7/2	30 763.4		998
	747	4, 1, 3← 3, 1, 2	1		5/2		3/2	30 765.4		998
	747	4, 1, 3← 3, 1, 2	1					30 765.91	.63	998
	747	4, 1, 3← 3, 1, 2	1		9/2		7/2	30 770.5		998
	747	4, 1, 3← 3, 1, 2	1		7/2		5/2	30 773.7		998
	747	4, 1, 3← 3, 1, 2	Ground		11/2		9/2	30 781.4		998
	747	4, 1, 3← 3, 1, 2	Ground		5/2		3/2	30 786.9		998
	747	4, 1, 3← 3, 1, 2	Ground					30 787.25	.24	998
	747	4, 1, 3← 3, 1, 2	Ground		9/2		7/2	30 791.9		998
	747	4, 1, 3← 3, 1, 2	Ground		7/2		5/2	30 795.05		998
	747	4, 1, 3← 3, 1, 2	Ground		9/2		9/2	30 860.8		998
	747	4, 1, 4← 3, 1, 3	Ground		5/2		5/2	29 154.2		998
	747	4, 1, 4← 3, 1, 3	1		5/2		3/2	29 205.5		998
	747	4, 1, 4← 3, 1, 3	1					29 205.71		998
	747	4, 1, 4← 3, 1, 3	1		9/2		7/2	29 209.7		998
	747	4, 1, 4← 3, 1, 3	1		7/2		5/2	29 215.7		998
	747	4, 1, 4← 3, 1, 3	Ground		11/2		9/2	29 226.5		998
	747	4, 1, 4← 3, 1, 3	Ground		5/2		3/2	29 231.7		998
	747	4, 1, 4← 3, 1, 3	Ground					29 232.17	.06	998
	747	4, 1, 4← 3, 1, 3	Ground		9/2		7/2	29 235.9		998
	747	4, 1, 4← 3, 1, 3	Ground		7/2		5/2	29 241.9		998
	747	4, 1, 4← 3, 1, 3	Ground		9/2		9/2	29 313.75		998
	747	4, 2, 2← 3, 2, 1	Ground		5/2		3/2	30 021.1		998
	747	4, 2, 2← 3, 2, 1	Ground		11/2		9/2	30 035.1		998
	747	4, 2, 2← 3, 2, 1	Ground					30 049.17	.07	998
	747	4, 2, 2← 3, 2, 1	Ground		7/2		5/2	30 060.4		998
	747	4, 2, 2← 3, 2, 1	Ground		9/2		7/2	30 074.0		998
	747	4, 2, 3← 3, 2, 2	1		5/2		3/2	29 960.0		998
	747	4, 2, 3← 3, 2, 2	1		11/2		9/2	29 974.3		998
	747	4, 2, 3← 3, 2, 2	Ground		5/2		3/2	29 986.7		998
	747	4, 2, 3← 3, 2, 2	1					29 987.70	.4	998
	747	4, 2, 3← 3, 2, 2	1		7/2		5/2	29 997.9		998
	747	4, 2, 3← 3, 2, 2	Ground		11/2		9/2	30 000.7		998
	747	4, 2, 3← 3, 2, 2	1		9/2		7/2	30 013.1		998
	747	4, 2, 3← 3, 2, 2	Ground					30 014.77	.07	998
	747	4, 2, 3← 3, 2, 2	Ground		7/2		5/2	30 026.0		998
	747	4, 2, 3← 3, 2, 2	Ground		9/2		7/2	30 039.6		998
	747	4, 3, 2← 3, 3, 1	Ground		5/2		3/2	29 952.5		998
	747	4, 3, 2← 3, 3, 1	Ground		9/2		9/2	29 963.8		998
	747	4, 3, 2← 3, 3, 1	Ground		11/2		9/2	29 995.9		998
	747	4, 3, 2← 3, 3, 1	Ground					30 025.14	.23	998
	747	4, 3, 2← 3, 3, 1	Ground		9/2		7/2	30 084.8		998
	747	4, 3, 2← 3, 3, 1	Ground		7/2		7/2	30 096.9		998
	747	5, 1, 4← 4, 1, 3	1		13/2		11/2	38 448.8		998
	747	5, 1, 4← 4, 1, 3	1					38 452.19		998
	747	5, 1, 4← 4, 1, 3	1		7/2		5/2	38 452.8		998
	747	5, 1, 4← 4, 1, 3	1		9/2		7/2	38 457.4		998
	747	5, 1, 4← 4, 1, 3	Ground		13/2		11/2	38 480.3		998

Isotopic Species	Id. No.	Rotational Quantum Nos.	Vib. State v_a	Hyperfine				Frequency MHz	Acc. ±MHz	Ref.
				F_1'	F'	F_1	F			
t-$C^{12}HD$:$C^{12}DBr^{81}$	747	5, 1, 4← 4, 1, 3	Ground		7/2		5/2	38 484.4		998
	747	5, 1, 4← 4, 1, 3	Ground					38 484.56		998
	747	5, 1, 4← 4, 1, 3	Ground		9/2		7/2	38 490.8		998
$C^{12}HD$:$C^{12}HBr^b$	748	Not Reported						14 391.1	.1	938
	748	Not Reported						14 399.0	.1	938
	748	Not Reported						14 409.6	.4	938
	748	Not Reported						14 444.0	.2	938
	748	Not Reported						14 449.5	.2	938
	748	Not Reported						14 526.9	.2	938
	748	Not Reported						14 720.4	.1	938
	748	Not Reported						14 754.7	.1	938
	748	Not Reported						14 849.3	.1	938
	748	Not Reported						14 937.0	.2	938
	748	Not Reported						14 964.2	.1	938
	748	Not Reported						14 966.6	.1	938
	748	Not Reported						14 976.1	.1	938
	748	Not Reported						14 996.1	.1	938
	748	Not Reported						15 028.5	.1	938
	748	Not Reported						15 045.0	.2	938
	748	Not Reported						15 309.4	.1	938
	748	Not Reported						15 396.55	.5	938
	748	Not Reported						15 716.5	.1	938
	748	Not Reported						15 720.9	.2	938
	748	Not Reported						15 723.9	.2	938
	748	Not Reported						21 657.0	.2	938
	748	Not Reported						21 662.0	.2	938
	748	Not Reported						21 693.5	.5	938
	748	Not Reported						21 746.6	.2	938
	748	Not Reported						21 764.4	.2	938
	748	Not Reported						22 149.4	.2	938
	748	Not Reported						22 153.7	.2	938
	748	Not Reported						22 229.8	.2	938
	748	Not Reported						22 470.5	.2	938
	748	Not Reported						22 607.45	.05	938
	748	Not Reported						22 688.5	.1	938
	748	Not Reported						22 703.0	.2	938
	748	Not Reported						22 969.0	.2	938
	748	Not Reported						23 011.5	.3	938
	748	Not Reported						23 049.0	.2	938
	748	Not Reported						23 581.5	.3	938
	748	Not Reported						23 640.8	.2	938
	748	Not Reported	Ground					28 628.4	.2	998
	748	Not Reported	Ground					28 857.8	.1	998
	748	Not Reported	Ground					28 872.8	.1	998
	748	Not Reported	Ground					28 875.1	.1	998
	748	Not Reported	Ground					28 905.4	.1	998
	748	Not Reported	Ground					28 990.6	.3	998
	748	Not Reported	Ground					29 012.9	.1	998
	748	Not Reported	Ground					29 116.2	.1	998

Isotopic Species	Id. No.	Rotational Quantum Nos.	Vib. State v_a	Hyperfine F_1'	F'	F_1	F	Frequency MHz	Acc. ±MHz	Ref.
$C^{12}HD:C^{12}HBr^b$	748	Not Reported	Ground					29 255.9	.1	998
	748	Not Reported	Ground					29 267.0	.1	998
	748	Not Reported	Ground					29 321.3	.1	998
	748	Not Reported	Ground					29 323.2	.1	998
	748	Not Reported	Ground					29 345.6	.1	998
	748	Not Reported	Ground					29 484.6	.1	998
	748	Not Reported	Ground					29 591.9	.1	998
	748	Not Reported	Ground					29 616.0	.1	998
	748	Not Reported	Ground					29 627.0	.1	998
	748	Not Reported	Ground					29 630.1	.1	998
	748	Not Reported	Ground					29 654.8	.1	998
	748	Not Reported	Ground					29 662.4	.1	998
	748	Not Reported	Ground					29 804.6	.1	998
	748	Not Reported	Ground					30 035.65	.1	998
	748	Not Reported	Ground					30 044.7	.1	998
	748	Not Reported	Ground					30 116.9	.1	998
	748	Not Reported	Ground					30 147.0	.1	998
	748	Not Reported	Ground					30 153.4	.1	998
	748	Not Reported	Ground					30 299.6	.2	998
	748	Not Reported	Ground					30 340.15	.1	998
	748	Not Reported	Ground					30 387.3	.1	998
	748	Not Reported	Ground					30 421.6	.1	998
	748	Not Reported	Ground					30 521.8	.1	998
	748	Not Reported	Ground					30 524.5	.1	998
	748	Not Reported	Ground					30 616.1	.1	998
	748	Not Reported	Ground					30 629.6	.1	998
	748	Not Reported	Ground					30 702.05	.05	998
	748	Not Reported	Ground					30 771.9	.1	998
	748	Not Reported	Ground					30 788.6	.1	998
	748	Not Reported	Ground					30 819.3	.1	998
	748	Not Reported	Ground					30 836.2	.1	998
	748	Not Reported	Ground					30 839.3	.1	998
	748	Not Reported	Ground					30 884.5	.1	998
	748	Not Reported	Ground					31 331.7	.1	998
	748	Not Reported	Ground					31 400.6	.1	998
	748	Not Reported	Ground					31 464.9	.1	998
	748	Not Reported	Ground					31 490.0	.1	998
	748	Not Reported	Ground					36 192.0	.2	998
	748	Not Reported	Ground					36 222.1	.1	998
	748	Not Reported	Ground					38 548.2	.2	998
	748	Not Reported	Ground					38 553.4	.2	998
	748	Not Reported	Ground					38 562.6	.1	998
	748	Not Reported	Ground					38 583.2	.1	998
	748	Not Reported	Ground					38 585.5	.1	998
	748	Not Reported	Ground					38 594.6	.1	998
	748	Not Reported	Ground					38 608.6	.1	998
	748	Not Reported	Ground					38 620.3	.1	998
	748	Not Reported	Ground					38 640.0	.1	998
	748	Not Reported	Ground					38 652.7	.2	998

Isotopic Species	Id. No.	Rotational Quantum Nos.	Vib. State v_a	Hyperfine				Frequency MHz	Acc. ±MHz	Ref.
				F_1'	F'	F_1	F			
$C^{12}HD:C^{12}DBr^b$	749	Not Reported						14 095.4	.2	938
	749	Not Reported						14 147.0	.1	938
	749	Not Reported						14 149.0	.2	938
	749	Not Reported						14 417.5	.1	938
	749	Not Reported						14 463.4	.2	938
	749	Not Reported						14 500.9	.1	938
	749	Not Reported						14 517.7	.2	938
	749	Not Reported						14 541.2	.1	938
	749	Not Reported						14 548.2	.2	938
	749	Not Reported						14 580.3	.2	938
	749	Not Reported						14 588.75	.05	938
	749	Not Reported						14 660.0	.1	938
	749	Not Reported						14 705.9	.1	938
	749	Not Reported						14 713.7	.1	938
	749	Not Reported						14 732.4	.1	938
	749	Not Reported						14 761.8	.2	938
	749	Not Reported						14 777.5	.2	938
	749	Not Reported						14 781.3	.2	938
	749	Not Reported						14 798.1	.1	938
	749	Not Reported						14 804.8	.2	938
	749	Not Reported						14 814.3	.2	938
	749	Not Reported						14 816.0	.1	938
	749	Not Reported						14 918.6	.1	938
	749	Not Reported						14 980.1	.1	938
	749	Not Reported						14 983.7	.1	938
	749	Not Reported						15 004.8	.1	938
	749	Not Reported						15 016.9	.1	938
	749	Not Reported						15 045.1	.3	938
	749	Not Reported						15 055.2	.1	938
	749	Not Reported						15 060.4	.1	938
	749	Not Reported						15 070.7	.1	938
	749	Not Reported						15 173.7	.2	938
	749	Not Reported						15 175.9	.2	938
	749	Not Reported						15 305.0	.1	938
	749	Not Reported						15 322.7	.1	938
	749	Not Reported						15 363.9	.1	938
	749	Not Reported						15 410.0	.1	938
	749	Not Reported						15 452.6	.1	938
	749	Not Reported						21 194.0	.2	938
	749	Not Reported						21 197.5	.1	938
	749	Not Reported						21 202.2	.1	938
	749	Not Reported						21 318.1	.2	938
	749	Not Reported						21 347.8	.2	938
	749	Not Reported						21 581.3	.2	938
	749	Not Reported						21 595.6	.1	938
	749	Not Reported						21 596.9	.1	938
	749	Not Reported						21 673.9	.1	938
	749	Not Reported						21 722.5	.1	938
	749	Not Reported						21 739.9	.1	938
	749	Not Reported						21 746.7	.1	938

Isotopic Species	Id. No.	Rotational Quantum Nos.	Vib. State v_a	Hyperfine				Frequency MHz	Acc. ±MHz	Ref.
				F_1'	F'	F_1	F			
$C^{12}HD:C^{12}DBr^b$	749	Not Reported						21 781.6	.1	938
	749	Not Reported						21 805.5	.2	938
	749	Not Reported						21 844.7	.1	938
	749	Not Reported						21 869.1	.2	938
	749	Not Reported						21 874.6	.2	938
	749	Not Reported						21 897.9	.1	938
	749	Not Reported						21 926.8	.2	938
	749	Not Reported						21 950.6	.2	938
	749	Not Reported						21 971.3	.2	938
	749	Not Reported						21 973.7	.2	938
	749	Not Reported						22 007.2	.2	938
	749	Not Reported						22 050.9	.2	938
	749	Not Reported						22 080.3	.2	938
	749	Not Reported						22 082.7	.1	938
	749	Not Reported						22 103.9	.2	938
	749	Not Reported						22 112.3	.1	938
	749	Not Reported						22 137.3	.1	938
	749	Not Reported						22 143.7	.1	938
	749	Not Reported						22 151.0	.2	938
	749	Not Reported						22 187.7	.1	938
	749	Not Reported						22 205.5	.1	938
	749	Not Reported						22 207.7	.1	938
	749	Not Reported						22 243.4	.1	938
	749	Not Reported						22 418.5	.2	938
	749	Not Reported						22 455.1	.2	938
	749	Not Reported						22 473.4	.1	938
	749	Not Reported						22 474.4	.1	938
	749	Not Reported						22 481.4	.1	938
	749	Not Reported						22 482.4	.1	938
	749	Not Reported						22 506.0	.1	938
	749	Not Reported						22 507.7	.2	938
	749	Not Reported						22 577.4	.2	938
	749	Not Reported						22 602.05	.1	938
	749	Not Reported						22 606.4	.1	938
	749	Not Reported						22 611.8	.1	938
	749	Not Reported						22 620.1	.2	938
	749	Not Reported						22 641.0	.1	938
	749	Not Reported						22 648.4	.1	938
	749	Not Reported						22 651.0	.1	938
	749	Not Reported						22 688.4	.1	938
	749	Not Reported						22 834.2	.2	938
	749	Not Reported						23 027.2	.3	938
	749	Not Reported						23 057.6	.2	938
	749	Not Reported						23 145.9	.2	938
	749	Not Reported						23 219.1	.3	938
	749	Not Reported						23 504.5	.1	938
	749	Not Reported	Ground					27 998.4	.1	998
	749	Not Reported	Ground					28 000.5	.1	998
	749	Not Reported	Ground					28 037.5	.2	998
	749	Not Reported	Ground					28 064.0	.1	998

Isotopic Species	Id. No.	Rotational Quantum Nos.	Vib. State v_a	Hyperfine F_1'	F'	F_1	F	Frequency MHz	Acc. ±MHz	Ref.
$C^{12}HD:C^{12}DBr$[b]	749	Not Reported	Ground					28 144.2	.1	998
	749	Not Reported	Ground					28 279.5	.1	998
	749	Not Reported	Ground					28 284.85	.1	998
	749	Not Reported	Ground					28 443.7	.1	998
	749	Not Reported	Ground					28 450.0	.2	998
	749	Not Reported	Ground					28 542.2	.1	998
	749	Not Reported	Ground					28 556.9	.1	998
	749	Not Reported	Ground					28 645.6	.2	998
	749	Not Reported	Ground					28 647.0	.2	998
	749	Not Reported	Ground					28 648.8	.2	998
	749	Not Reported	Ground					28 718.4	.1	998
	749	Not Reported	Ground					28 726.4	.1	998
	749	Not Reported	Ground					28 734.4	.1	998
	749	Not Reported	Ground					28 740.5	.2	998
	749	Not Reported	Ground					28 749.6	.1	998
	749	Not Reported	Ground					28 806.75	.1	998
	749	Not Reported	Ground					28 809.5	.1	998
	749	Not Reported	Ground					28 829.8	.2	998
	749	Not Reported	Ground					28 830.3	.1	998
	749	Not Reported	Ground					28 857.8	.1	998
	749	Not Reported	Ground					28 873.2	.1	998
	749	Not Reported	Ground					28 916.6	.1	998
	749	Not Reported	Ground					28 930.6	.2	998
	749	Not Reported	Ground					28 957.8	.1	998
	749	Not Reported	Ground					28 965.5	.1	998
	749	Not Reported	Ground					28 969.1	.1	998
	749	Not Reported	Ground					28 974.4	.1	998
	749	Not Reported	Ground					29 041.4	.1	998
	749	Not Reported	Ground					29 104.3	.1	998
	749	Not Reported	Ground					29 132.8	.1	998
	749	Not Reported	Ground					29 139.3	.1	998
	749	Not Reported	Ground					29 146.6	.1	998
	749	Not Reported	Ground					29 150.7	.1	998
	749	Not Reported	Ground					29 194.85	.1	998
	749	Not Reported	Ground					29 204.4	.1	998
	749	Not Reported	Ground					29 239.7	.2	998
	749	Not Reported	Ground					29 254.7	.1	998
	749	Not Reported	Ground					29 265.0	.1	998
	749	Not Reported	Ground					29 282.2	.1	998
	749	Not Reported	Ground					29 285.3	.1	998
	749	Not Reported	Ground					29 286.8	.1	998
	749	Not Reported	Ground					29 329.5	.1	998
	749	Not Reported	Ground					29 433.9	.2	998
	749	Not Reported	Ground					29 458.4	.1	998
	749	Not Reported	Ground					29 517.7	.1	998
	749	Not Reported	Ground					29 554.05	.1	998
	749	Not Reported	Ground					29 623.3	.2	998
	749	Not Reported	Ground					29 693.3	.1	998
	749	Not Reported	Ground					29 904.9	.2	998
	749	Not Reported	Ground					29 905.9	.2	998

Isotopic Species	Id. No.	Rotational Quantum Nos.	Vib. State v_a	Hyperfine				Frequency MHz	Acc. ±MHz	Ref.
				F_1'	F'	F_1	F			
$C^{12}HD{:}C^{12}DBr^b$	749	Not Reported	Ground					29 925.4	.1	998
	749	Not Reported	Ground					29 951.2	.1	998
	749	Not Reported	Ground					29 970.1	.1	998
	749	Not Reported	Ground					29 982.2	.1	998
	749	Not Reported	Ground					30 009.0	.1	998
	749	Not Reported	Ground					30 105.4	.2	998
	749	Not Reported	Ground					30 121.6	.1	998
	749	Not Reported	Ground					30 137.3	.1	998
	749	Not Reported	Ground					30 139.4	.2	998
	749	Not Reported	Ground					30 142.1	.1	998
	749	Not Reported	Ground					30 147.1	.1	998
	749	Not Reported	Ground					30 157.5	.1	998
	749	Not Reported	Ground					30 184.5	.1	998
	749	Not Reported	Ground					30 189.8	.2	998
	749	Not Reported	Ground					30 465.7	.1	998
	749	Not Reported	Ground					30 475.5	.1	998
	749	Not Reported	Ground					30 480.9	.1	998
	749	Not Reported	Ground					30 684.9	.1	998
	749	Not Reported	Ground					30 688.4	.1	998
	749	Not Reported	Ground					30 694.5	.1	998
	749	Not Reported	Ground					30 815.2	.1	998
	749	Not Reported	Ground					30 825.5	.1	998
	749	Not Reported	Ground					30 845.4	.1	998
	749	Not Reported	Ground					30 855.4	.1	998
	749	Not Reported	Ground					30 865.0	.2	998
	749	Not Reported	Ground					30 870.8	.1	998
	749	Not Reported	Ground					30 874.8	.1	998
	749	Not Reported	Ground					30 890.8	.1	998
	749	Not Reported	Ground					30 997.5	.1	998
	749	Not Reported	Ground					38 492.2	.2	998
$C^{12}H_2{:}C^{12}HBr^b$	751	3, , ← 2, ,	Ground					23 374.4	.1	902
	751	3, , ← 2, ,	Ground					23 481.6	.2	902
	751	3, , ← 2, ,	Ground					23 585.6	.1	902
	751	3, , ← 2, ,	Ground					23 646.3	.2	902
	751	3, , ← 2, ,	Ground					24 256.0	.3	902
	751	3, , ← 2, ,	Ground					24 280.5	.1	902
	751	3, , ← 2, ,	Ground					24 384.9	.1	902
	751	3, , ← 2, ,	Ground					24 472.6	.2	902
	751	3, , ← 2, ,	Ground					24 528.8	.2	902
	751	Not Reported						16 224.3	.1	903
$C^bH_2{:}C^bHBr^b$	752	3, 0, 2← 2, 0, 2	Ground		5/2		3/2	23 799.7	.2	937
	752	3, 0, 3← 2, 0, 2	Ground		9/2		7/2	22 760.3		937
	752	3, 0, 3← 2, 0, 2	Ground		5/2		3/2	22 783.1		937
	752	3, 0, 3← 2, 0, 2	Ground		5/2		3/2	22 923.4		937
	752	3, 0, 3← 2, 0, 2	Ground		5/2		3/2	23 047.4		937
	752	3, 0, 3← 2, 0, 2	Ground		9/2		7/2	23 157.0		937
	752	3, 0, 3← 2, 0, 2	Ground		9/2		7/2	23 634.0		937
	752	3, 0, 3← 2, 0, 2	Ground		9/2		7/2	23 770.8		937
	752	3, 0, 3← 2, 0, 2	Ground		5/2		3/2	23 952.7		937
	752	3, 0, 3← 2, 0, 2	Ground		7/2		7/2	24 181.5	.2	937

Isotopic Species	Id. No.	Rotational Quantum Nos.	Vib. State v_a	F'_I	F'	F_I	F	Frequency MHz	Acc. ±MHz	Ref.
$C^bH_2:C^bHBr^b$	752	3, 1, 2← 2, 1, 1	Excited		7/2		5/2	24 230.8		937
	752	3, 2, 1← 2, 2, 0	Ground		9/2		7/2	22 745.7		937
	752	3, 2, 1← 2, 2, 0	Ground		5/2		3/2	22 774.7		937
	752	3, 2, 1← 2, 2, 0	Ground		3/2		1/2	22 793.5		937
	752	3, 2, 1← 2, 2, 0	Ground		7/2		5/2	22 844.7		937
	752	3, 2, 1← 2, 2, 0	Ground		9/2		7/2	22 877.1		937
	752	3, 2, 1← 2, 2, 0	Ground		5/2		3/2	22 911.3		937
	752	3, 2, 1← 2, 2, 0	Ground		7/2		5/2	22 995.1		937
	752	3, 2, 1← 2, 2, 0	Ground		9/2		7/2	23 009.8		937
	752	3, 2, 1← 2, 2, 0	Ground		5/2		3/2	23 038.4		937
	752	3, 2, 1← 2, 2, 0	Ground		7/2		5/2	23 107.9		937
	752	3, 2, 1← 2, 2, 0	Ground		9/2		7/2	23 139.35		937
	752	3, 2, 1← 2, 2, 0	Ground		7/2		5/2	23 252.2		937
	752	3, 2, 1← 2, 2, 0	Ground		9/2		7/2	23 622.1		937
	752	3, 2, 1← 2, 2, 0	Ground		7/2		5/2	23 721.3		937
	752	3, 2, 1← 2, 2, 0	Ground		9/2		7/2	23 754.6		937
	752	3, 2, 1← 2, 2, 0	Ground		5/2		3/2	23 788.9		937
	752	3, 2, 1← 2, 2, 0	Ground		7/2		5/2	23 873.3		937
	752	3, 2, 1← 2, 2, 0	Ground		9/2		7/2	23 916.4		937
	752	3, 2, 1← 2, 2, 0	Ground		5/2		3/2	23 945.0		937
	752	3, 2, 1← 2, 2, 0	Ground		7/2		5/2	24 015.3		937
	752	3, 2, 1← 2, 2, 0	Ground		9/2		7/2	24 047.5		937
	752	3, 2, 1← 2, 2, 0	Ground		7/2		5/2	24 165.6		937
	752	3, 2, 2← 2, 2, 1	Ground		5/2		3/2	22 906.7		937
	752	3, 2, 2← 2, 2, 1	Ground		5/2		3/2	23 034.1		937
	752	3, 2, 2← 2, 2, 1	Ground		5/2		3/2	23 940.2		937
	752	3, 2, 2← 2, 2, 1	Ground		7/2		5/2	24 010.3		937
	752	Not Reported	Ground		5/2		5/2	14 928.9		937
	752	Not Reported	Ground					14 941.3		937
	752	Not Reported	Ground					15 118.4		937
	752	Not Reported	Ground		7/2		5/2	15 169.7		937
	752	Not Reported	Ground		3/2		3/2	15 207.9		937
	752	Not Reported	Excited		7/2		5/2	15 331.6		937
	752	Not Reported	Ground		7/2		5/2	15 345.8		937
	752	Not Reported	Ground					15 376.6		937
	752	Not Reported	Ground					15 423.5		937
	752	Not Reported	Ground					15 522.3	.2	937
	752	Not Reported	Ground					15 551.0		937
	752	Not Reported	Ground					15 593.1	.2	937
	752	Not Reported	Ground		3/2		3/2	15 598.2		937
	752	Not Reported	Excited		7/2		5/2	15 620.2	.2	937
	752	Not Reported	Ground					15 631.7	.2	937
	752	Not Reported	Ground					15 686.2	.2	937
	752	Not Reported	Ground		3/2		3/2	15 794.3		937
	752	Not Reported	Ground					15 827.9		937
	752	Not Reported	Ground		7/2		5/2	15 949.5		937
	752	Not Reported	Ground					15 978.2		937
	752	Not Reported	Ground		3/2		1/2	16 008.7		937
	752	Not Reported	Excited		7/2		5/2	16 023.0		937
	752	Not Reported	Ground		7/2		5/2	16 037.5		937

Isotopic Species	Id. No.	Rotational Quantum Nos.	Vib. State v_a	F_1'	F'	F_1	F	Frequency MHz	Acc. ±MHz	Ref.
$C^bH_2{:}C^bHBr^b$	752	Not Reported	Ground		3/2		3/2	16 104.1		937
	752	Not Reported	Ground		3/2		3/2	16 401.8		937
	752	Not Reported	Ground		5/2		3/2	16 440.3		937
	752	Not Reported	Excited		9/2		7/2	22 313.4		937
	752	Not Reported	Ground		5/2		5/2	22 326.9		937
	752	Not Reported	Ground		9/2		7/2	22 338.5		937
	752	Not Reported	Ground		7/2		5/2	22 363.0		937
	752	Not Reported	Ground		5/2		3/2	22 364.4		937
	752	Not Reported	Ground		5/2		5/2	22 454.0		937
	752	Not Reported	Ground		9/2		7/2	22 468.1		937
	752	Not Reported	Excited		7/2		5/2	22 471.7		937
	752	Not Reported	Ground		7/2		5/2	22 497.0		937
	752	Not Reported	Ground		5/2		3/2	22 498.8		937
	752	Not Reported	Ground		5/2		5/2	22 593.5		937
	752	Not Reported	Ground		9/2		7/2	22 605.45		937
	752	Not Reported	Ground		7/2		5/2	22 629.6		937
	752	Not Reported	Ground		5/2		3/2	22 631.1		937
	752	Not Reported	Ground					22 642.3		937
	752	Not Reported	Ground		9/2		7/2	22 733.4		937
	752	Not Reported	Ground		5/2		3/2	22 764.0		937
	752	Not Reported	Ground					23 082.8	.2	937
	752	Not Reported	Ground					23 092.2	.2	937
	752	Not Reported	Ground					23 113.8	.2	937
	752	Not Reported	Ground		5/2		5/2	23 169.1		937
	752	Not Reported	Excited		7/2		5/2	23 183.6		937
	752	Not Reported	Ground		3/2		1/2	23 187.5	.3	937
	752	Not Reported	Ground		7/2		5/2	23 201.6		937
	752	Not Reported	Ground		5/2		3/2	23 212.1		937
	752	Not Reported	Excited		7/2		7/2	23 235.7	.2	937
	752	Not Reported	Ground					23 254.8		937
	752	Not Reported	Ground					23 289.2		937
	752	Not Reported	Ground		9/2		7/2	23 316.6		937
	752	Not Reported	Ground		7/2		5/2	23 345.6		937
	752	Not Reported	Ground		5/2		5/2	23 426.7	.2	937
	752	Not Reported	Ground		9/2		7/2	23 482.6		937
	752	Not Reported	Ground		7/2		5/2	23 507.0		937
	752	Not Reported	Ground		5/2		3/2	23 508.5	.2	937
	752	Not Reported	Ground		9/2		7/2	23 576.1		937
	752	Not Reported	Ground					23 698.3	.2	937
	752	Not Reported	Excited		7/2		5/2	24 085.5		937
	752	Not Reported	Ground		7/2		5/2	24 104.5		937
	752	Not Reported	Excited		7/2		7/2	24 127.0	.2	937
	752	Not Reported	Ground		7/2		7/2	24 150.9		937
	752	Not Reported	Ground		5/2		5/2	24 211.3		937
	752	Not Reported	Ground		9/2		7/2	24 221.0		937
	752	Not Reported	Ground		7/2		5/2	24 250.4		937
	752	Not Reported	Excited		7/2		7/2	24 289.2		937
	752	Not Reported	Ground		7/2		7/2	24 305.7		937
	752	Not Reported	Excited		9/2		7/2	24 353.5		937
	752	Not Reported	Ground		5/2		5/2	24 365.1		937

Isotopic Species	Id. No.	Rotational Quantum Nos.	Vib. State v_a	Hyperfine F_1'	F'	F_1	F	Frequency MHz	Acc. ±MHz	Ref.
$C^bH_2:C^bHBr^b$	752	Not Reported	Ground		9/2		7/2	24 373.0		937
	752	Not Reported	Excited		7/2		5/2	24 377.9		937
	752	Not Reported	Ground		7/2		5/2	24 397.3		937
	752	Not Reported	Ground		7/2		7/2	24 423.4		937
	752	Not Reported	Ground		7/2		7/2	24 442.9		937
	752	Not Reported	Ground					24 472.7		937
	752	Not Reported	Excited		9/2		7/2	24 492.4		937
	752	Not Reported	Ground					24 500.0	.3	937
	752	Not Reported	Ground		5/2		5/2	24 502.4		937
	752	Not Reported	Ground		9/2		7/2	24 512.0		937
	752	Not Reported	Excited		7/2		5/2	24 521.6		937
	752	Not Reported	Ground					24 528.6		937
	752	Not Reported	Ground					24 533.1		937
	752	Not Reported	Ground		7/2		5/2	24 541.1		937
	752	Not Reported	Ground		7/2		7/2	24 595.7		937
$C^bH_2:C^bHBr^{79}$	753	3, 2, 2← 2, 2, 1	Ground		3/2		1/2	22 789.2		937
	753	3, 2, 2← 2, 2, 1	Ground		9/2		7/2	22 872.8		937
	753	3, 2, 2← 2, 2, 1	Ground		3/2		1/2	23 052.2		937
	753	3, 2, 2← 2, 2, 1	Ground		9/2		7/2	23 749.5		937
	753	3, 2, 2← 2, 2, 1	Ground		9/2		7/2	24 042.5		937
	753	Not Reported	Ground		1/2		1/2	14 871.1		937
	753	Not Reported	Ground		3/2		1/2	14 934.5		937
	753	Not Reported	Ground		7/2		5/2	14 958.2		937
	753	Not Reported	Ground		7/2		5/2	15 135.0		937
	753	Not Reported	Ground		3/2		3/2	15 175.7		937
	753	Not Reported	Ground		3/2		1/2	15 490.5		937
	753	Not Reported	Ground		1/2		1/2	15 633.6	.2	937
	753	Not Reported	Ground		3/2		1/2	15 663.4		937
	753	Not Reported	Ground		3/2		1/2	15 696.8	.2	937
	753	Not Reported	Ground		3/2		3/2	15 759.5		937
	753	Not Reported	Ground		3/2		3/2	15 955.0		937
	753	Not Reported	Ground		7/2		5/2	16 128.1		937
	753	Not Reported	Ground		3/2		1/2	16 287.6		937
	753	Not Reported	Ground		7/2		5/2	16 322.1		937
	753	Not Reported	Ground		5/2		5/2	16 376.8		937
	753	Not Reported	Ground		3/2		3/2	22 406.1		937
	753	Not Reported	Excited		9/2		7/2	22 442.6		937
	753	Not Reported	Ground		3/2		3/2	22 671.5		937
	753	Not Reported	Ground		3/2		3/2	24 457.1		937
$C^bH_2:C^bHBr^{81}$	754	2, 0, 2← 1, 0, 1	Ground		3/2		3/2	15 100.1	.2	937
	754	2, 1, 1← 1, 1, 0	Ground		5/2		3/2	16 136.6		937
	754	3, 2, 2← 2, 2, 1	Ground		7/2		5/2	23 103.9		937
	754	Not Reported	Ground		7/2		5/2	14 875.4		937
	754	Not Reported	Ground		3/2		3/2	14 936.7		937
	754	Not Reported	Ground		5/2		3/2	14 974.3		937
	754	Not Reported	Ground		3/2		3/2	15 114.1		937
	754	Not Reported	Ground		5/2		3/2	15 151.6		937
	754	Not Reported	Ground		5/2		5/2	15 268.1		937
	754	Not Reported	Ground		7/2		5/2	15 638.4		937
	754	Not Reported	Ground		3/2		3/2	15 676.1		937
	754	Not Reported	Ground		5/2		3/2	15 707.9		937
	754	Not Reported	Ground		5/2		3/2	15 737.3		937
	754	Not Reported	Ground		5/2		5/2	15 852.1		937
	754	Not Reported	Ground		5/2		5/2	16 048.1		937
	754	Not Reported	Ground		3/2		3/2	16 299.4	.2	937
	754	Not Reported	Ground		5/2		3/2	16 331.4		937

Vinyl Chloride, Chloroethylene

C_2H_3Cl C_s $H_2C{:}CHCl$

Isotopic Species	Pt. Gp.	Id. No.	A MHz	B MHz	C MHz	D_J MHz	D_{JK} MHz	Δ Amu A^2	κ
t-HDC^{12}:C^{12}DCl^{35}	C_s	761		5 818.24 M	5 039.01 M				
c-DHC^{12}:C^{12}DCl^{35}	C_s	762		5 518.05 M	4 903.64 M				
D$_2$C^{12}:C^{12}DCl^{35}	C_s	763		5 379.52 M	4 705.74 M				
t-HDC^{12}:C^{12}DCl^{37}	C_s	764		5 692.06 M	4 942.94 M				
D$_2$C^{12}:C^{12}DCl^{37}	C_s	765		5 261.84 M	4 614.71 M				
H$_2$C^{12}:C^{12}HCl^{35}	C_s	766		6 029.96 M	5 445.29 M				
H$_2$C^{12}:C^{12}HCl^{37}	C_s	767		5 903.56 M	5 341.26 M				
H$_2$C^{12}:C^{13}HCl^{35}	C_s	768		5 999.28 M	5 405.44 M				
H$_2$C^{13}:C^{12}HCl^{35}	C_s	769		5 826.82 M	5 274.51 M				

Id. No.	μ_a Debye	μ_b Debye	μ_c Debye	eQq Value(MHz) Rel.	eQq Value(MHz) Rel.	eQq Value(MHz) Rel.	ω_a d 1/cm	ω_b d 1/cm	ω_c d 1/cm	ω_d d 1/cm
761	1.42 M		0. X							
766				−57.15 aa						
767				−45.19 aa						
768				−57.8 aa						
769				−56.5 aa						

References:

ABC: 910 μ: 910   eQq: 910

Add. Ref. 132,197,679

Isotopic Species	Id. No.	Rotational Quantum Nos.	Vib. State	Hyperfine				Frequency MHz	Acc. ±MHz	Ref.
				F_1'	F'	F_1	F			
t-HDC^{12}:C^{12}DCl^{35}	761	2, 0, 2← 1, 0, 1	Ground		7/2		5/2	21 701.78	.2	910
	761	2, 1, 1← 1, 1, 0	Ground					22 483.25		448
	761	2, 1, 1← 1, 1, 0	Ground		7/2		5/2	22 497.05	.2	910
	761	2, 1, 2← 1, 1, 1	Ground					20 924.41		448
	761	2, 1, 2← 1, 1, 1	Ground		7/2		5/2	20 938.70	.2	910
	761	3, 1, 2← 2, 1, 1	Ground					33 729.57		448
	761	3, 1, 2← 2, 1, 1	Ground					33 733.08		448
c-DHC^{12}:C^{12}DCl^{35}	762	2, 0, 2← 1, 0, 1	Ground		7/2		5/2	20 837.33	.2	910
	762	2, 1, 1← 1, 1, 0	Ground					21 446.65		448
	762	2, 1, 1← 1, 1, 0	Ground		7/2		5/2	21 461.10	.2	910
	762	2, 1, 2← 1, 1, 1	Ground		7/2		5/2	20 232.40	.2	910
	762	3, 1, 3← 2, 1, 2	Ground					30 336.75		448
	762	3, 1, 3← 2, 1, 2	Ground					30 340.50		448
D$_2$C^{12}:C^{12}DCl^{35}	763	2, 0, 2← 1, 0, 1	Ground		7/2		5/2	20 161.10	.2	910
	763	2, 1, 1← 1, 1, 0	Ground					20 833.55		448
	763	2, 1, 1← 1, 1, 0	Ground		7/2		5/2	20 847.59	.2	910
	763	2, 1, 2← 1, 1, 1	Ground					19 485.95		448
	763	2, 1, 2← 1, 1, 1	Ground		7/2		5/2	19 500.15	.2	910
	763	3, 0, 3← 2, 0, 2	Ground					30 211.10		448
	763	3, 0, 3← 2, 0, 2	Ground					30 214.65		448
	763	3, 1, 3← 2, 1, 2	Ground					29 235.95		448
	763	3, 1, 3← 2, 1, 2	Ground					29 239.54		448
	763	3, 2, 1← 2, 2, 0	Ground					30 296.		448
	763	3, 2, 2← 2, 2, 1	Ground					30 260.80		448
t-HDC^{12}:C^{12}DCl^{37}	764	2, 0, 2← 1, 0, 1	Ground		7/2		5/2	21 257.18	.2	910
	764	2, 1, 1← 1, 1, 0	Ground					22 009.90		448
	764	2, 1, 1← 1, 1, 0	Ground		7/2		5/2	22 021.70	.2	910
	764	2, 1, 2← 1, 1, 1	Ground		7/2		5/2	20 523.55	.2	910
D$_2$C^{12}:C^{12}DCl^{37}	765	2, 0, 2← 1, 0, 1	Ground		7/2		5/2	19 744.98	.2	910
	765	2, 1, 2← 1, 1, 1	Ground		7/2		5/2	19 109.40	.2	910
	765	3, 1, 2← 2, 1, 1	Ground					30 593.66		448
	765	3, 1, 2← 2, 1, 1	Ground					30 596.50		448
H$_2$C^{12}:C^{12}HCl^{35}	766	2, 0, 2← 1, 0, 1	Ground		3/2		1/2	22 931.02	.02	910
	766	2, 0, 2← 1, 0, 1	Ground		5/2		5/2	22 932.24	.02	910
	766	2, 0, 2← 1, 0, 1	Ground		5/2		3/2	22 946.55	.02	910
	766	2, 0, 2← 1, 0, 1	Ground		7/2		5/2	22 946.55	.02	910
	766	2, 0, 2← 1, 0, 1	Ground		3/2		3/2	22 956.74	.02	910
	766	2, 1, 1← 1, 1, 0	Ground		5/2		3/2	23 524.30	.02	910
	766	2, 1, 1← 1, 1, 0	Ground		3/2		3/2	23 528.86	.02	910
	766	2, 1, 1← 1, 1, 0	Ground		5/2		5/2	23 532.19	.02	910
	766	2, 1, 1← 1, 1, 0	Ground		7/2		5/2	23 538.58	.02	910
	766	2, 1, 1← 1, 1, 0	Ground		3/2		1/2	23 543.09	.02	910
	766	2, 1, 1← 1, 1, 0	Ground		1/2		1/2	23 549.45	.02	910
	766	2, 1, 2← 1, 1, 1	Ground		5/2		3/2	22 355.06	.02	910
	766	2, 1, 2← 1, 1, 1	Ground		3/2		3/2	22 360.71	.02	910
	766	2, 1, 2← 1, 1, 1	Ground		5/2		5/2	22 361.47	.02	910
	766	2, 1, 2← 1, 1, 1	Ground		7/2		5/2	22 369.38	.02	910

Isotopic Species	Id. No.	Rotational Quantum Nos.	Vib. State	Hyperfine				Frequency MHz	Acc. ±MHz	Ref.
				F_1'	F'	F_1	F			
$H_2C^{12}:C^{12}HCl^{35}$	766	2, 1, 2← 1, 1, 1	Ground		1/2		1/2	22 380.11	.02	910
$H_2C^{12}:C^{12}HCl^{37}$	767	2, 0, 2← 1, 0, 1	Ground		5/2		5/2	22 474.53	.02	910
	767	2, 0, 2← 1, 0, 1	Ground		5/2		3/2	22 485.81	.02	910
	767	2, 0, 2← 1, 0, 1	Ground		7/2		5/2	22 485.81	.02	910
	767	2, 0, 2← 1, 0, 1	Ground		3/2		3/2	22 493.89	.02	910
	767	2, 0, 2← 1, 0, 1	Ground					22 986.0		131
	767	2, 1, 1← 1, 1, 0	Ground		5/2		3/2	23 043.39	.02	910
	767	2, 1, 1← 1, 1, 0	Ground		3/2		3/2	23 047.00	.02	910
	767	2, 1, 1← 1, 1, 0	Ground		5/2		5/2	23 049.57	.02	910
	767	2, 1, 1← 1, 1, 0	Ground		7/2		5/2	23 054.65	.02	910
	767	2, 1, 1← 1, 1, 0	Ground		3/2		1/2	23 058.19	.02	910
	767	2, 1, 1← 1, 1, 0	Ground		1/2		1/2	23 063.29	.02	910
	767	2, 1, 2← 1, 1, 1	Ground		5/2		3/2	21 918.85	.02	910
	767	2, 1, 2← 1, 1, 1	Ground		7/2		5/2	21 930.15	.02	910
	767	2, 1, 2← 1, 1, 1	Ground		1/2		1/2	21 938.65	.02	910
$H_2C^{12}:C^{13}HCl^{35}$	768	2, 0, 2← 1, 0, 1	Ground		7/2		5/2	22 805.33	.10	910
	768	2, 0, 2← 1, 0, 1	Ground		5/2		3/2	22 805.33	.10	910
	768	2, 1, 1← 1, 1, 0	Ground		5/2		3/2	23 392.21	.10	910
	768	2, 1, 1← 1, 1, 0	Ground		7/2		5/2	23 406.72	.10	910
	768	2, 1, 2← 1, 1, 1	Ground		5/2		3/2	22 204.76	.10	910
	768	2, 1, 2← 1, 1, 1	Ground		7/2		5/2	22 219.16	.10	910
$H_2C^{13}:C^{12}HCl^{35}$	769	2, 0, 2← 1, 0, 1	Ground		5/2		3/2	22 199.21	.10	910
	769	2, 0, 2← 1, 0, 1	Ground		7/2		5/2	22 199.21	.10	910
	769	2, 1, 1← 1, 1, 0	Ground		5/2		3/2	22 744.21	.10	910
	769	2, 1, 1← 1, 1, 0	Ground		7/2		5/2	22 758.33	.10	910
$H_2^bC^b:C^bH^bCl^b$	772	Not Reported	Ground					19 190.48		448
	772	Not Reported	Ground					20 368.40		448
	772	Not Reported	Ground					20 375.00		448
	772	Not Reported	Ground					20 411.65		448
	772	Not Reported	Ground					20 560.75		448
	772	Not Reported	Ground					20 795.87		448
	772	Not Reported	Ground					21 228.05		448
	772	Not Reported	Ground					21 428.		131
	772	Not Reported	Ground					21 829.		131
	772	Not Reported	Ground					21 849.		448
	772	Not Reported	Ground					22 086.55		448
	772	Not Reported	Ground					22 092.50		448
	772	Not Reported	Ground					22 870.		131
	772	Not Reported	Ground					23 961.		131
	772	Not Reported	Ground					24 975.		131
	772	Not Reported	Ground					28 120.18		448
	772	Not Reported	Ground					28 122.02		448
	772	Not Reported	Ground					30 599.32		448
	772	Not Reported	Ground					30 770.90		448
	772	Not Reported	Ground					30 773.62		448

Vinyl Fluoride, Fluoroethylene

C_2H_3F C_s $H_2C{:}CHF$

Isotopic Species	Pt. Gp.	Id. No.	A MHz		B MHz		C MHz		D_J MHz	D_{JK} MHz	Δ Amu A^2	κ
$H_2C^{12}{:}C^{12}HF^{19}$	C_s	781	64 582.7	M	10 636.79	M	9 118.19	M			.0160	−.94466
$H_2C^{13}{:}C^{12}HF^{19}$	C_s	782			10 295.26	M	8 859.05	M				
$H_2C^{12}{:}C^{13}HF^{19}$	C_s	783			10 635.02	M	9 082.78	M				
$H_2C^{12}{:}C^{12}DF^{19}$	C_s	784	48 960.	M	10 635.60	M	8 753.27	M			−.1042	−.9064
c-HDC^{12}:C^{12}HF^{19}	C_s	787	53 400.	M	10 278.20	M	8 610.48	M			.0597	−.92553
t-DHC^{12}:C^{12}HF^{19}	C_s	788	62 440.	M	9 668.14	M	8 384.03	M			−.0875	−.95249
c-HDC^{12}:C^{12}DF^{19}	C_s	789	49 250.	M	9 667.07	M	8 077.02	M			.0299	−.92276
t-DHC^{12}:C^{12}DF^{19}	C_s	791	42 700.	M	10 274.57	M	8 272.36	M			.0694	−.88368

Id. No.	μ_a Debye		μ_b Debye		μ_c Debye		eQq Value(MHz)	Rel.	eQq Value(MHz)	Rel.	eQq Value(MHz)	Rel.	ω_a d 1/cm	ω_b d 1/cm	ω_c d 1/cm	ω_d d 1/cm
781	1.280	M	.629	M	0.	X										

References:

ABC: 804,875,948 Δ: 804 κ: 804 μ: 948

Add. Ref. 945

For species 781, excited state $v_{12} = 1$: B = 10632.16 MHz and C = 9106.67 MHz. Ref. 875.

Fluoroethene

Isotopic Species	Id. No.	Rotational Quantum Nos.	Vib. State v_a	F_1'	F'	F_1	F	Frequency MHz	Acc. ±MHz	Ref.
$H_2C^{12}{:}C^{12}HF^{19}$	781	1, 0, 1← 0, 0, 0	1					19 738.83	.05	875
	781	1, 0, 1← 0, 0, 0	Ground					19 755.01	.05	875
	781	2, 0, 2← 1, 0, 1	1					39 445.40	.05	875
	781	2, 0, 2← 1, 0, 1	Ground					39 477.97	.05	875
	781	2, 1, 1← 1, 1, 0	1					41 003.04	.05	875
	781	2, 1, 1← 1, 1, 0	Ground					41 028.57	.05	875
	781	2, 1, 2← 1, 1, 1	1					37 952.07	.05	875
	781	2, 1, 2← 1, 1, 1	Ground					37 991.28	.05	875
	781	3, 1, 2← 3, 1, 3	Ground					9 111.32	.05	948
	781	5, 1, 5← 4, 0, 4	Ground					142 347.30	.05	948
	781	5, 1, 4← 5, 1, 5	Ground					22 765.2	.1	804
	781	5, 1, 4← 5, 1, 5	Ground					22 765.84	.05	948
	781	6, 1, 6← 5, 0, 5	Ground					157 931.82	.05	948
	781	6, 1, 5← 5, 1, 4	Ground					122 813.35	.05	948
	781	6, 1, 6← 5, 1, 5	Ground					113 727.20	.05	948
	781	6, 2, 4← 5, 2, 3	Ground					119 474.65	.05	948
	781	6, 2, 5← 5, 2, 4	Ground					118 383.85	.05	948
	781	6, 3, 3← 5, 3, 2	Ground					118 715.90	.05	948
	781	6, 3, 4← 5, 3, 3	Ground					118 693.10	.05	948
	781	6, 4, 2← 5, 4, 1	Ground					118 644.25	.05	948
	781	6, 4, 3← 5, 4, 2	Ground					118 644.35	.05	948
	781	6, 5, ← 5, 5,	Ground					118 623.85	.05	948
	781	6, 6, 6← 5, 6, 5	Ground					117 431.32	.05	948
	781	8, 0, 8← 7, 0, 7	Ground					155 476.98	.05	948
	781	8, 1, 7← 7, 1, 6	Ground					163 410.32	.05	948

Isotopic Species	Id. No.	Rotational Quantum Nos.	Vib. State v_a	Hyperfine F_1'	F'	F_1	F	Frequency MHz	Acc. ±MHz	Ref.
$H_2C^{12}:C^{12}HF^{19}$	781	8, 1, 8← 7, 1, 7	Ground					151 370.98	.05	948
	781	8, 2, 6← 7, 2, 5	Ground					160 185.15	.05	948
	781	8, 2, 7← 7, 2, 6	Ground					157 641.90	.05	948
	781	8, 3, 5← 7, 3, 4	Ground					158 464.81	.05	948
	781	8, 3, 6← 7, 3, 5	Ground					158 362.00	.05	948
	781	8, 4, 4← 7, 4, 3	Ground					158 264.18	.05	948
	781	8, 4, 5← 7, 4, 4	Ground					158 262.74	.05	948
	781	8, 5, ← 7, 5,	Ground					158 206.80	.05	948
	781	8, 6, ← 7, 6,	Ground					158 184.68	.05	948
	781	8, 7, ← 7, 7,	Ground					158 179.25	.05	948
$H_2C^{13}:C^{12}HF^{19}$	782	1, 0, 1← 0, 0, 0	Ground					19 154.31	.05	875
	782	2, 0, 2← 1, 0, 1	Ground					38 279.95	.05	875
	782	2, 1, 1← 1, 1, 0	Ground					39 744.74	.05	875
	782	2, 1, 2← 1, 1, 1	Ground					36 872.34	.05	875
$H_2C^{12}:C^{13}HF^{19}$	783	1, 0, 1← 0, 0, 0	Ground					19 717.80	.05	875
	783	2, 0, 2← 1, 0, 1	Ground					39 401.13	.05	875
	783	2, 1, 1← 1, 1, 0	Ground					40 987.77	.05	875
	783	2, 1, 2← 1, 1, 1	Ground					37 883.28	.05	875
$H_2C^{12}:C^{12}DF^{19}$	784	1, 0, 1← 0, 0, 0	Ground					19 388.1	.1	804
	784	2, 0, 2← 1, 0, 1	Ground					38 710.6	.3	804
	784	2, 1, 1← 1, 1, 0	Ground					40 660.7	.3	804
	784	2, 1, 2← 1, 1, 1	Ground					36 895.8	.3	804
	784	4, 1, 3← 4, 1, 4	Ground					18 810.9	.1	804
	784	5, 1, 4← 5, 1, 5	Ground					28 189.7	.1	804
$H_2C^{13}:C^{12}DF^{19}$	785	1, 0, 1← 0, 0, 0	Ground					18 806.0	.1	804
$H_2C^{12}:C^{13}DF^{19}$	786	1, 0, 1← 0, 0, 0	Ground					19 358.0	.1	804
$c\text{-}HDC^{12}:C^{12}HF^{19}$	787	1, 0, 1← 0, 0, 0	Ground					18 888.8	.1	804
	787	2, 0, 2← 1, 0, 1	Ground					37 729.8	.3	804
	787	2, 1, 1← 1, 1, 0	Ground					39 445.0	.3	804
	787	2, 1, 2← 1, 1, 1	Ground					36 109.5	.3	804
	787	5, 1, 4← 5, 1, 5	Ground					24 990.6	.1	804
$t\text{-}DHC^{12}:C^{12}HF^{19}$	788	1, 0, 1← 0, 0, 0	Ground					18 052.0	.1	804
	788	2, 0, 2← 1, 0, 1	Ground					36 081.3	.3	804
	788	2, 1, 1← 1, 1, 0	Ground					37 388.6	.3	804
	788	2, 1, 2← 1, 1, 1	Ground					34 820.3	.3	804
	788	5, 1, 4← 5, 1, 5	Ground					19 253.7	.1	804
	788	6, 1, 5← 6, 1, 6	Ground					26 942.9	.1	804
$c\text{-}HDC^{12}:C^{12}DF^{19}$	789	1, 0, 1← 0, 0, 0	Ground					17 743.8	.1	804
	789	2, 0, 2← 1, 0, 1	Ground					35 441.4	.3	804
	789	2, 1, 1← 1, 1, 0	Ground					37 078.4	.3	804
	789	2, 1, 2← 1, 1, 1	Ground					33 898.3	.3	804
	789	5, 1, 4← 5, 1, 5	Ground					23 826.7	.1	804
$t\text{-}DHC^{12}:C^{12}DF^{19}$	791	1, 0, 1← 0, 0, 0	Ground					18 547.0	.1	804
	791	2, 0, 2← 1, 0, 1	Ground					37 003.9	.3	804
	791	2, 1, 1← 1, 1, 0	Ground					39 096.0	.3	804
	791	2, 1, 2← 1, 1, 1	Ground					35 091.6	.3	804
	791	4, 1, 3← 4, 1, 4	Ground					20 004.8	.1	804

Vinyl Iodide, Iodoethylene

C_2H_3I C_s $H_2C:CHI$

Isotopic Species	Pt. Gp.	Id. No.	A MHz	B MHz	C MHz	D_J MHz	D_{JK} MHz	Δ Amu A²	κ
$H_2C^{12}:C^{12}HI^{127}$	C_s	801	44 606.75 M	3 258.77 M	3 066.75 M				−.991079

Id. No.	μ_a Debye	μ_b Debye	μ_c Debye	eQq Value(MHz)	Rel.	eQq Value(MHz)	Rel.	eQq Value(MHz)	Rel.	ω_a d 1/cm	ω_b d 1/cm	ω_c d 1/cm	ω_d d 1/cm
801	1.27 G		0. X	−1656	aa	−765	ab	770	bb				

References:

ABC: 509 κ: 509 μ: 995   eQq: 548

Add. Ref. 380

For species 801, B − C = 192.02 MHz; B + C = 6325.52 MHz; and A − C = 41540 MHz. Ref. 509.

Iodoethene   Spectral Line Table

Isotopic Species	Id. No.	Rotational Quantum Nos.	Vib. State	F_1'	F'	F_1	F	Frequency MHz	Acc. ±MHz	Ref.
$H_2C^{12}:C^{12}HI^{127}$	801	4, 0, 4 ← 3, 0, 3	Ground					25 295.40		509
	801	4, 1, 3 ← 3, 1, 2	Ground		7/2		5/2	25 641.14	.05	548
	801	4, 1, 3 ← 3, 1, 2	Ground		5/2		3/2	25 650.43	.05	548
	801	4, 1, 3 ← 3, 1, 2	Ground		9/2		7/2	25 657.34	.05	548
	801	4, 1, 3 ← 3, 1, 2	Ground		9/2		9/2	25 657.89	.05	548
	801	4, 1, 3 ← 3, 1, 2	Ground					25 684.77		509
	801	4, 1, 3 ← 3, 1, 2	Ground		11/2		9/2	25 685.91	.05	548
	801	4, 1, 3 ← 3, 1, 2	Ground		3/2		1/2	25 688.71	.05	548
	801	4, 1, 3 ← 3, 1, 2	Ground		13/2		11/2	25 709.28	.05	548
	801	4, 1, 3 ← 3, 1, 2	Ground		7/2		7/2	25 752.80	.05	548
	801	4, 1, 3 ← 3, 1, 2	Ground		5/2		5/2	25 788.26	.05	548
	801	4, 1, 3 ← 3, 1, 2	Ground		3/2		3/2	25 793.83	.05	548
	801	4, 1, 4 ← 3, 1, 3	Ground		11/2		11/2	24 683.33	.05	548
	801	4, 1, 4 ← 3, 1, 3	Ground		7/2		5/2	24 869.49	.05	548
	801	4, 1, 4 ← 3, 1, 3	Ground		5/2		3/2	24 877.60	.05	548
	801	4, 1, 4 ← 3, 1, 3	Ground		9/2		7/2	24 888.45	.05	548
	801	4, 1, 4 ← 3, 1, 3	Ground		9/2		9/2	24 889.69	.05	548
	801	4, 1, 4 ← 3, 1, 3	Ground					24 916.68		509
	801	4, 1, 4 ← 3, 1, 3	Ground		3/2		1/2	24 918.05	.05	548
	801	4, 1, 4 ← 3, 1, 3	Ground		11/2		9/2	24 919.85	.05	548
	801	4, 1, 4 ← 3, 1, 3	Ground		13/2		11/2	24 942.47	.05	548
	801	4, 1, 4 ← 3, 1, 3	Ground		7/2		7/2	24 991.87	.05	548
	801	4, 1, 4 ← 3, 1, 3	Ground		5/2		5/2	25 029.18	.05	548
	801	4, 1, 4 ← 3, 1, 3	Ground		3/2		3/2	25 034.06	.05	548

Ethanenitrile, Methyl Cyanide

C_2H_3N C_{3v} CH_3CN

Isotopic Species	Pt. Gp.	Id. No.	A MHz	B MHz		C MHz		D_J MHz	D_{JK} MHz	Δ Amu A^2	κ
$C^{12}H_3C^{12}N^{14}$	C_{3v}	811		9 198.83	M	9 198.83	M	.00381	.1769		
$C^{12}H_2DC^{12}N^{14}$	C_s	812		8 759.18	M	8 608.51	M		.15		−.997
$C^{12}HD_2C^{12}N^{14}$	C_s	813		8 320.06	M	8 164.43	M		.15		−.997
$C^{12}H_3C^{12}N^{15}$	C_{3v}	814		8 921.81	M	8 921.81	M				
$C^{12}H_3C^{13}N^{14}$	C_{3v}	815		9 194.28	M	9 194.28	M				
$C^{12}D_3C^{12}N^{14}$	C_{3v}	816		7 857.93	M	7 857.93	M		.113		
$C^{12}D_3C^{13}N^{14}$	C_{3v}	817		7 848.51	M	7 848.51	M		.110		
$C^{13}H_3C^{12}N^{14}$	C_{3v}	818		8 933.15	M	8 933.15	M				
$C^{13}D_3C^{12}N^{14}$	C_{3v}	819		7 695.19	M	7 695.19	M				
$C^{12}D_3C^{12}N^{15}$	C_{3v}	821		7 619.32	M	7 619.32	M				

Id. No.	μ_a Debye		μ_b Debye		μ_c Debye		eQq Value(MHz)	Rel.	eQq Value(MHz)	Rel.	eQq Value(MHz)	Rel.	ω_a 1/cm	d	ω_b 1/cm	d	ω_c 1/cm	d	ω_d 1/cm	d
811 811	3.92	M	0.	X	0.	X	−4.35	N^{14}					361	2	847	2	1112	1	831	1
816													333	2						

References:

ABC: 189,205,639,973 D_J: 954 D_{JK}: 205,639,954 κ: 639 μ: 434   eQq: 205 ω: 973,1028

Add. Ref. 53,54,154,222,248,302,741,810

All data in MHz:

Species	State	A	B	D_J	D_{JK}	α	q	ζ
811	$v_8 = 1$	158400	9226.444	.0039	.1777	−27.3	17.775	.878
816				.004	.141	−23.01	13.92	.862
819							13.35	
811	$v_8 = 2$		9254.125	.0030	.143			
811	$v_7 = 1$					5.2		
816						−6.02	18.85	
811	$v_4 = 1$					46.3		
816						47.19		
816	$v_3 = 1$					40.13		
References		954	954	954 973	954 973	189 864 973	954 973	954 923

Isotopic Species	Id. No.	Rotational Quantum Nos.	Vib. State v_a^l ; v_b^l ; v_c ; v_d	F_1'	F'	F_1	F	Frequency MHz	Acc. ±MHz	Ref.
$C^{12}H_3C^{12}N^{14}$	811	1, ← 0,	1, 0;0, 0;0;0					18 452.2		864
	811	1, ← 0,	2, 0;0, 0;0;0					18 506.9		864
	811	2, ← 1,	2, 0;0, 0;0;0					37 013.2		864
	811	2, 0← 1, 0	Ground		1		0	36 794.26		205
	811	2, 0← 1, 0	Ground		3		2	36 795.38		205
	811	2, 0← 1, 0	Ground		2		1	36 795.38		205
	811	2, 0← 1, 0	Ground		1		1	36 797.52		205
	811	2, 0← 1, 0	1,±1;0, 0;0;0					36 903.40		864 [1]
	811	2, 1← 1, 1	Ground		2		1	36 793.64		205
	811	2, 1← 1, 1	Ground		2		2	36 794.26		205
	811	2, 1← 1, 1	Ground		3		2	36 794.88		205
	811	2, 1← 1, 1	Ground		1		0	36 796.27		205
	811	2, 1← 1, 1	1,−1;0, 0;0;0					36 870.94		864 [1]
	811	2, 1← 1, 1	1,∓1;0, 0;0;0					36 903.40		864
	811	2, 1← 1, 1	1,+1;0, 0;0;0					36 942.15		864 [1]
	811	3, ← 2,	Ground					55 187.8		864
	811	3, ← 2,	Ground					55 192.8		864
	811	3, ← 2,	2, 0;0, 0;0;0					55 519.4		864
	811	3, 0← 2, 0	1,±1;0, 0;0;0					55 353.0		864
	811	3, 1← 2, 1	1,−1;0, 0;0;0					55 307.0		864
	811	3, 1← 2, 1	1,∓1;0, 0;0;0					55 353.0		864
	811	3, 1← 2, 1	1,+1;0, 0;0;0					55 412.5		864
	811	3, 2← 2, 2	1,∓1;0, 0;0;0					55 344.9		864
	811	3, 2← 2, 2	1,±1;0, 0;0;0					55 359.5		864
	811	5, ← 4,	2, 0;0, 0;0;0					92 510.7	.25	954
	811	5, ← 4,	2, 0;0, 0;0;0					92 519.6	.25	954
	811	5, ← 4,	2, 0;0, 0;0;0					92 524.2	.25	954
	811	5, ← 4,	2, 0;0, 0;0;0					92 525.1	.25	954
	811	5, 0← 4, 0	Ground					91 987.07	.25	954
	811	5, 0← 4, 0	1,∓1;0, 0;0;0					92 261.53	.25	954
	811	5, 1← 4, 1	Ground					91 985.35	.25	954
	811	5, 1← 4, 1	1,±1;0, 0;0;0					92 175.46	.25	954
	811	5, 1← 4, 1	1,∓1;0, 0;0;0					92 256.53	.25	954
	811	5, 1← 4, 1	1,±1;0, 0;0;0					92 353.67	.25	954
	811	5, 1← 4, 1	2, 0;0, 0;0;0					92 538.7	.25	954
	811	5, 2← 4, 2	Ground					91 980.00	.25	954
	811	5, 2← 4, 2	1,∓1;0, 0;0;0					92 247.42	.25	954
	811	5, 2← 4, 2	1,±1;0, 0;0;0					92 264.23	.25	954
	811	5, 2← 4, 2	2, 0;0, 0;0;0					92 533.0	.25	954
	811	5, 3← 4, 3	Ground					91 970.62	.25	954 [2]
	811	5, 3← 4, 3	Ground					91 971.35	.25	954 [3]
	811	5, 3← 4, 3	1,∓1;0, 0;0;0					92 234.88	.25	954
	811	5, 3← 4, 3	1,±1;0, 0;0;0					92 258.59	.25	954
	811	5, 3← 4, 3	2, 0;0, 0;0;0					92 526.9	.25	954
	811	5, 4← 4, 4	Ground					91 957.94	.25	954 [2]
	811	5, 4← 4, 4	Ground					91 959.20	.25	954 [2]
	811	5, 4← 4, 4	1,∓1;0, 0;0;0					92 218.88	.25	954
	811	5, 4← 4, 4	1,±1;0, 0;0;0					92 250.14	.25	954
	811	6, ← 5,	2, 0;0, 0;0;0					110 975.7	.25	954
	811	6, ← 5,	2, 0;0, 0;0;0					110 987.8	.25	954

1. These lines were observed for the excited vibrational state associated with the lowest fundamental vibrational mode at a frequency of 361 cm^{-1}, but hyperfine structure was not resolved.
2. These lines were taken to be hyperfine structure components $F = J + 1 ← F = J$.
3. These lines were taken to be a superposition of the hyperfine structure components $F = J ← F = J - 1$ and $F = J + 2 ← F = J + 1$.

Isotopic Species	Id. No.	Rotational Quantum Nos.	Vib. State $v_a^l ; v_b^l ; v_c ; v_d$	Hyperfine F_1'	F'	F_1	F	Frequency MHz	Acc. ±MHz	Ref.
$C^{12}H_3C^{12}N^{14}$	811	6, · ← 5,	2, 0; 0, 0; 0; 0					111 009.3	.25	954
	811	6, ← 5,	2, 0; 0, 0; 0; 0					111 021.5	.25	954
	811	6, ← 5,	2, 0; 0, 0; 0; 0					111 028.8	.25	954
	811	6, ← 5,	2, 0; 0, 0; 0; 0					111 039.1	.25	954
	811	6, ← 5,	2, 0; 0, 0; 0; 0					111 049.4	.25	954
	811	6, 0← 5, 0	Ground					110 383.47	.25	954
	811	6, 0← 5, 0	1, ∓1; 0, 0; 0; 0					110 712.17	.25	954
	811	6, 0← 5, 0	2, 0; 0, 0; 0; 0					111 047.4	.25	954
	811	6, 1← 5, 1	Ground					110 381.39	.25	954
	811	6, 1← 5, 1	1, ±1; 0, 0; 0; 0					110 609.77	.25	954
	811	6, 1← 5, 1	1, ∓1; 0, 0; 0; 0					110 706.25	.25	954
	811	6, 1← 5, 1	1, ±1; 0, 0; 0; 0					110 823.16	.25	954
	811	6, 1← 5, 1	2, 0; 0, 0; 0; 0					111 046.0	.25	954
	811	6, 2← 5, 2	Ground					110 375.01	.25	954
	811	6, 2← 5, 2	1, ∓1; 0, 0; 0; 0					110 697.07	.25	954
	811	6, 2← 5, 2	1, ±1; 0, 0; 0; 0					110 716.32	.25	954
	811	6, 2← 5, 2	2, 0; 0, 0; 0; 0					111 040.9	.25	954
	811	6, 3← 5, 3	Ground					110 364.02	.25	954 [2]
	811	6, 3← 5, 3	Ground					110 364.52	.25	954 [3]
	811	6, 3← 5, 3	1, ∓1; 0, 0; 0; 0					110 680.36	.25	954
	811	6, 3← 5, 3	1, ±1; 0, 0; 0; 0					110 709.40	.25	954
	811	6, 3← 5, 3	2, 0; 0, 0; 0; 0					111 030.8	.25	954
	811	6, 4← 5, 4	Ground					110 349.00	.25	954 [2]
	811	6, 4← 5, 4	Ground					110 349.68	.25	954 [3]
	811	6, 4← 5, 4	1, ∓1; 0, 0; 0; 0					110 660.88	.25	954
	811	6, 4← 5, 4	1, ±1; 0, 0; 0; 0					110 698.62	.25	954
	811	6, 5← 5, 5	Ground					110 329.70	.25	954
	811	6, 5← 5, 5	Ground					110 330.79	.25	954 [3]
	811	6, 5← 5, 5	1, ∓1; 0, 0; 0; 0					110 637.22	.25	954
	811	6, 5← 5, 5	1, ±1; 0, 0; 0; 0					110 684.52	.25	954
	811	7, 0← 6, 0	Ground					128 779.43	.25	954
	811	7, 1← 6, 1	Ground					128 777.02	.25	954
	811	7, 2← 6, 2	Ground					128 769.52	.25	954
	811	7, 3← 6, 3	Ground					128 757.03	.25	954
	811	7, 4← 6, 4	Ground					128 739.39	.25	954 [2]
	811	7, 4← 6, 4	Ground					128 739.91	.25	954 [3]
	811	7, 5← 6, 5	Ground					128 717.18	.25	954 [2]
	811	7, 5← 6, 5	Ground					128 717.69	.25	954 [3]
	811	7, 6← 6, 6	Ground					128 689.48	.25	954 [2]
	811	7, 6← 6, 6	Ground					128 690.57	.25	954 [3]
	811	8, ← 7,	2, 0; 0, 0; 0; 0					147 910.6	.25	954
	811	8, ← 7,	2, 0; 0, 0; 0; 0					147 921.1	.25	954
	811	8, ← 7,	2, 0; 0, 0; 0; 0					147 939.1	.25	954
	811	8, ← 7,	2, 0; 0, 0; 0; 0					147 973.5	.25	954
	811	8, ← 7,	2, 0; 0, 0; 0; 0					148 019.2	.25	954
	811	8, ← 7,	2, 0; 0, 0; 0; 0					148 032.5	.25	954
	811	8, ← 7,	2, 0; 0, 0; 0; 0					148 035.3	.25	954
	811	8, ← 7,	2, 0; 0, 0; 0; 0					148 052.0	.25	954
	811	8, ← 7,	2, 0; 0, 0; 0; 0					148 064.0	.25	954
	811	8, ← 7,	2, 0; 0, 0; 0; 0					148 082.0	.25	954

2. These lines were taken to be hyperfine structure components $F = J + 1 ← F = J$.
3. These lines were taken to be a superposition of the hyperfine structure components $F = J ← F = J - 1$ and $F = J + 2 ← F = J + 1$.

Isotopic Species	Id. No.	Rotational Quantum Nos.	Vib. State v_a^l ; v_b^l ; v_c ; v_d	Hyperfine F_1'	F'	F_1	F	Frequency MHz	Acc. ±MHz	Ref.
$C^{12}H_3C^{12}N^{14}$	811	8, 0← 7, 0	Ground					147 174.72	.25	954
	811	8, 0← 7, 0	1,∓1; 0, 0; 0; 0					147 611.01	.25	954
	811	8, 0← 7, 0	2, 0; 0, 0; 0; 0					148 060.0	.25	954
	811	8, 1← 7, 1	Ground					147 171.95	.25	954
	811	8, 1← 7, 1	1,±1; 0, 0; 0; 0					147 476.07	.25	954
	811	8, 1← 7, 1	1,∓1; 0, 0; 0; 0					147 603.96	.25	954
	811	8, 1← 7, 1	1,±1; 0, 0; 0; 0					147 760.38	.25	954
	811	8, 1← 7, 1	2, 0; 0, 0; 0; 0					148 058.7	.25	954
	811	8, 2← 7, 2	Ground					147 163.46	.25	954
	811	8, 2← 7, 2	1,∓1; 0, 0; 0; 0					147 589.93	.25	954
	811	8, 2← 7, 2	1,±1; 0, 0; 0; 0					147 620.13	.25	954
	811	8, 2← 7, 2	2, 0; 0, 0; 0; 0					148 050.4	.25	954
	811	8, 3← 7, 3	Ground					147 149.28	.25	954
	811	8, 3← 7, 3	1,∓1; 0, 0; 0; 0					147 569.86	.25	954
	811	8, 3← 7, 3	1,±1; 0, 0; 0; 0					147 609.76	.25	954
	811	8, 3← 7, 3	2, 0; 0, 0; 0; 0					148 038.6	.25	954
	811	8, 4← 7, 4	Ground					147 129.41	.25	954
	811	8, 4← 7, 4	1,∓1; 0, 0; 0; 0					147 543.84	.25	954
	811	8, 4← 7, 4	1,±1; 0, 0; 0; 0					147 595.40	.25	954
	811	8, 4← 7, 4	2, 0; 0, 0; 0; 0					148 022.7	.25	954
	811	8, 5← 7, 5	Ground					147 104.07	.25	954
	811	8, 5← 7, 5	1,∓1; 0, 0; 0; 0					147 512.50	.25	954
	811	8, 5← 7, 5	1,±1; 0, 0; 0; 0					147 575.44	.25	954
	811	8, 5← 7, 5	2, 0; 0, 0; 0; 0					148 004.9	.25	954
	811	8, 6← 7, 6	Ground					147 072.27	.25	954 [2]
	811	8, 6← 7, 6	Ground					147 073.04	.25	954 [3]
	811	8, 6← 7, 6	1,±1; 0, 0; 0; 0					147 550.12	.25	954
	811	8, 6← 7, 6	2, 0; 0, 0; 0; 0					147 977.5	.25	954
	811	8, 7← 7, 7	Ground					147 035.89	.25	954 [3]
	811	8, 7← 7, 7	1,±1; 0, 0; 0; 0					147 519.34	.25	954
	811	8, 7← 7, 7	2, 0; 0, 0; 0; 0					147 947.9	.25	954
	811	9, 0← 8, 0	Ground					165 568.95	.5	954
	811	9, 0← 8, 0	1,∓1; 0, 0; 0; 0					166 059.13	.5	954
	811	9, 1← 8, 1	Ground					165 565.71	.5	954
	811	9, 1← 8, 1	1,±1; 0, 0; 0; 0					165 908.28	.5	954
	811	9,.1← 8, 1	1,∓1; 0, 0; 0; 0					166 051.73	.5	954
	811	9, 1← 8, 1	1,±1; 0, 0; 0; 0					166 228.53	.5	954
	811	9, 2← 8, 2	Ground					165 556.18	.5	954
	811	9, 2← 8, 2	1,∓1; 0, 0; 0; 0					166 036.03	.5	954
	811	9, 2← 8, 2	1,±1; 0, 0; 0; 0					166 071.30	.5	954
	811	9, 3← 8, 3	Ground					165 540.31	.5	954
	811	9, 3← 8, 3	1,∓1; 0, 0; 0; 0					166 013.38	.5	954
	811	9, 3← 8, 3	1,±1; 0, 0; 0; 0					166 059.13	.5	954
	811	9, 4← 8, 4	Ground					165 517.93	.5	954
	811	9, 4← 8, 4	1,∓1; 0, 0; 0; 0					165 983.98	.5	954
	811	9, 4← 8, 4	1,±1; 0, 0; 0; 0					166 042.93	.5	954
	811	9, 5← 8, 5	Ground					165 489.39	.5	954
	811	9, 5← 8, 5	1,∓1; 0, 0; 0; 0					165 948.85	.5	954
	811	9, 5← 8, 5	1,±1; 0, 0; 0; 0					166 020.50	.5	954
	811	9, 6← 8, 6	Ground					165 454.09	.5	954 [2]

2. These lines were taken to be hyperfine structure components $F = J + 1 \leftarrow F = J$.
3. These lines were taken to be a superposition of the hyperfine structure components $F = J \leftarrow F = J - 1$ and $F = J + 2 \leftarrow F = J + 1$.

Isotopic Species	Id. No.	Rotational Quantum Nos.	Vib. State v_a^l ; v_b^l ; v_c ; v_d	F_1'	F'	F_1	F	Frequency MHz	Acc. ±MHz	Ref.
$C^{12}H_3C^{12}N^{14}$	811	9, 6← 8, 6	Ground					165 454.34	.5	954 [3]
	811	9, 6← 8, 6	1,±1; 0, 0; 0; 0					165 992.06	.5	954
	811	9, 7← 8, 7	Ground					165 412.93	.5	954 [3]
	811	9, 7← 8, 7	1,±1; 0, 0; 0; 0					165 957.03	.5	954
	811	9, 8← 8, 8	Ground					165 365.43	.5	954 [3]
	811	10, 0← 9, 0	Ground					183 962.62	.5	954
	811	10, 0← 9, 0	1,±1; 0, 0; 0; 0					184 505.64	.5	954
	811	10, 1← 9, 1	Ground					183 959.08	.5	954
	811	10, 1← 9, 1	1,±1; 0, 0; 0; 0					184 339.70	.5	954
	811	10, 1← 9, 1	1,∓1; 0, 0; 0; 0					184 498.20	.5	954
	811	10, 1← 9, 1	1,±1; 0, 0; 0; 0					184 695.21	.5	954
	811	10, 2← 9, 2	Ground					183 948.49	.5	954
	811	10, 2← 9, 2	1,∓1; 0, 0; 0; 0					184 481.06	.5	954
	811	10, 2← 9, 2	1,±1; 0, 0; 0; 0					184 522.32	.5	954
	811	10, 3← 9, 3	Ground					183 930.79	.5	954
	811	10, 3← 9, 3	1,∓1; 0, 0; 0; 0					184 456.00	.5	954
	811	10, 3← 9, 3	1,±1; 0, 0; 0; 0					184 508.45	.5	954
	811	10, 4← 9, 4	Ground					183 906.05	.5	954
	811	10, 4← 9, 4	1,∓1; 0, 0; 0; 0					184 424.19	.5	954
	811	10, 4← 9, 4	1,±1; 0, 0; 0; 0					184 489.56	.5	954
	811	10, 5← 9, 5	Ground					183 874.21	.5	954
	811	10, 5← 9, 5	1,∓1; 0, 0; 0; 0					184 384.64	.5	954
	811	10, 5← 9, 5	1,±1; 0, 0; 0; 0					184 464.84	.5	954
	811	10, 6← 9, 6	Ground					183 835.34	.5	954 [3]
	811	10, 6← 9, 6	1,±1; 0, 0; 0; 0					184 433.11	.5	954
	811	10, 7← 9, 7	Ground					183 789.31	.5	954 [3]
	811	10, 7← 9, 7	1,±1; 0, 0; 0; 0					184 394.55	.5	954
	811	10, 8← 9, 8	Ground					183 736.28	.5	954 [3]
	811	11, ←10,	2, 0; 0, 0; 0; 0					203 300.	.5	954
	811	11, ←10,	2, 0; 0, 0; 0; 0					203 303.5	.5	954
	811	11, ←10,	2, 0; 0, 0; 0; 0					203 409.5	.5	954
	811	11, ←10,	2, 0; 0, 0; 0; 0					203 413.7	.5	954
	811	11, ←10,	2, 0; 0, 0; 0; 0					203 441.2	.5	954
	811	11, ←10,	2, 0; 0, 0; 0; 0					203 449.5	.5	954
	811	11, ←10,	2, 0; 0, 0; 0; 0					203 485.0	.5	954
	811	11, ←10,	2, 0; 0, 0; 0; 0					203 514.6	.5	954
	811	11, ←10,	2, 0; 0, 0; 0; 0					203 517.7	.5	954
	811	11, ←10,	2, 0; 0, 0; 0; 0					203 529.3	.5	954
	811	11, ←10,	2, 0; 0, 0; 0; 0					203 533.9	.5	954
	811	11, ←10,	2, 0; 0, 0; 0; 0					203 553.8	.5	954
	811	11, ←10,	2, 0; 0, 0; 0; 0					203 564.4	.5	954
	811	11, ←10,	2, 0; 0, 0; 0; 0					203 583.3	.5	954
	811	11, 0←10, 0	Ground					202 355.61	.5	954
	811	11, 0←10, 0	1,∓1; 0, 0; 0; 0					202 950.97	.5	954
	811	11, 0←10, 0	2, 0; 0, 0; 0; 0					203 574.3	.5	954
	811	11, 1←10, 1	Ground					202 351.45	.5	954
	811	11, 1←10, 1	1,±1; 0, 0; 0; 0					202 769.94	.5	954
	811	11, 1←10, 1	1,∓1; 0, 0; 0; 0					202 943.39	.5	954
	811	11, 1←10, 1	1,±1; 0, 0; 0; 0					203 161.23	.5	954
	811	11, 1←10, 1	2, 0; 0, 0; 0; 0					203 572.6	.5	954

---

3. These lines were taken to be a superposition of the hyperfine structure components $F = J \leftarrow F = J - 1$ and $F = J + 2 \leftarrow F = J + 1$.

Isotopic Species	Id. No.	Rotational Quantum Nos.	Vib. State $v_a^l ; v_b^l ; v_c ; v_d$	Hyperfine F_1'	F'	F_1	F	Frequency MHz	Acc. ±MHz	Ref.
$C^{12}H_3C^{12}N^{14}$	811	11, 2←10, 2	Ground					202 340.10	.5	954
	811	11, 2←10, 2	1,∓1; 0, 0; 0; 0					202 924.94	.5	954
	811	11, 2←10, 2	1,±1; 0, 0; 0; 0					202 972.63	.5	954
	811	11, 2←10, 2	2, 0; 0, 0; 0; 0					203 563.1	.5	954
	811	11, 3←10, 3	Ground					202 321.54	.5	954
	811	11, 3←10, 3	1,∓1; 0, 0; 0; 0					202 897.68	.5	954
	811	11, 3←10, 3	1,±1; 0, 0; 0; 0					202 956.31	.5	954
	811	11, 3←10, 3	2, 0; 0, 0; 0; 0					203 544.4	.5	954
	811	11, 4←10, 4	Ground					202 293.78	.5	954
	811	11, 4←10, 4	1,∓1; 0, 0; 0; 0					202 862.38	.5	954
	811	11, 4←10, 4	1,±1; 0, 0; 0; 0					202 935.67	.5	954
	811	11, 5←10, 5	Ground					202 257.87	.5	954
	811	11, 5←10, 5	1,∓1; 0, 0; 0; 0					202 819.06	.5	954
	811	11, 5←10, 5	1,±1; 0, 0; 0; 0					202 907.98	.5	954
	811	11, 6←10, 6	Ground					202 215.87	.5	954 [3]
	811	11, 6←10, 6	1,∓1; 0, 0; 0; 0					202 768.06	.5	954
	811	11, 6←10, 6	1,±1; 0, 0; 0; 0					202 872.91	.5	954
	811	11, 6←10, 6	2, 0; 0, 0; 0; 0					230 461.2	.5	954
	811	11, 7←10, 7	Ground					202 164.93	.5	954 [3]
	811	11, 7←10, 7	1,∓1; 0, 0; 0; 0					202 709.07	.5	954
	811	11, 7←10, 7	1,±1; 0, 0; 0; 0					202 830.05	.5	954
	811	11, 8←10, 8	Ground					202 106.80	.5	954 [3]
	811	11, 8←10, 8	1,∓1; 0, 0; 0; 0					202 642.27	.5	954
	811	11, 8←10, 8	1,±1; 0, 0; 0; 0					202 779.70	.5	954
	811	11, 9←10, 9	Ground					202 040.76	.5	954 [3]
	811	11, 9←10, 9	1,±1; 0, 0; 0; 0					202 721.62	.5	954
	811	11,10←10,10	Ground					201 967.02	.5	954 [3]
	811	11,10←10,10	1,±1; 0, 0; 0; 0					202 655.71	.5	954
	811	12, 0←11, 0	Ground					220 747.24	.5	954
	811	12, 0←11, 0	1,∓1; 0, 0; 0; 0					221 394.15	.5	954
	811	12, 1←11, 1	Ground					220 742.99	.5	954
	811	12, 1←11, 1	1,±1; 0, 0; 0; 0					221 200.23	.5	954
	811	12, 1←11, 1	1,∓1; 0, 0; 0; 0					221 387.30	.5	954
	811	12, 1←11, 1	1,±1; 0, 0; 0; 0					221 625.91	.5	954
	811	12, 2←11, 2	Ground					220 730.27	.5	954
	811	12, 2←11, 2	1,∓1; 0, 0; 0; 0					221 367.67	.5	954
	811	12, 2←11, 2	1,±1; 0, 0; 0; 0					221 422.37	.5	954
	811	12, 3←11, 3	Ground					220 709.08	.5	954
	811	12, 3←11, 3	1,∓1; 0, 0; 0; 0					221 338.22	.5	954
	811	12, 3←11, 3	1,±1; 0, 0; 0; 0					221 403.82	.5	954
	811	12, 4←11, 4	Ground					220 679.32	.5	954
	811	12, 4←11, 4	1,∓1; 0, 0; 0; 0					221 299.88	.5	954
	811	12, 4←11, 4	1,±1; 0, 0; 0; 0					221 380.74	.5	954
	811	12, 5←11, 5	Ground					220 641.12	.5	954
	811	12, 5←11, 5	1,∓1; 0, 0; 0; 0					221 252.93	.5	954
	811	12, 5←11, 5	1,±1; 0, 0; 0; 0					221 350.37	.5	954
	811	12, 6←11, 6	Ground					220 594.50	.5	954 [3]
	811	12, 6←11, 6	1,±1; 0, 0; 0; 0					221 311.95	.5	954
	811	12, 7←11, 7	Ground					220 539.30	.5	954 [3]
	811	12, 7←11, 7	1,±1; 0, 0; 0; 0					221 265.54	.5	954

3. These lines were taken to be a superposition of the hyperfine structure components $F = J \leftarrow F = J - 1$ and $F = J + 2 \leftarrow F = J + 1$.

Isotopic Species	Id. No.	Rotational Quantum Nos.	Vib. State $v_a^l ; v_b^l ; v_c ; v_d$	Hyperfine				Frequency MHz	Acc. ±MHz	Ref.
				F_1'	F'	F_1	F			
$C^{12}H_3C^{12}N^{14}$	811	12, 8←11, 8	Ground					220 476.04	.5	954 [3]
	811	12, 9←11, 9	Ground					220 403.96	.5	954 [3]
$C^{12}H_2DC^{12}N^{14}$	812	2, 0, 2← 1, 0, 1	Ground		2		2	34 734.07	.1	639
	812	2, 0, 2← 1, 0, 1	Ground		1		0	34 734.07	.1	639
	812	2, 0, 2← 1, 0, 1	Ground		3		2	34 735.23	.1	639
	812	2, 0, 2← 1, 0, 1	Ground		2		1	34 735.23	.1	639
	812	2, 0, 2← 1, 0, 1	Ground		1		1	34 737.20	.1	639
	812	2, 1, 1← 1, 1, 0	Ground		2		1	34 884.32	.1	639
	812	2, 1, 1← 1, 1, 0	Ground		3		2	34 885.68	.1	639
	812	2, 1, 1← 1, 1, 0	Ground		1		0	34 886.97	.1	639
	812	2, 1, 2← 1, 1, 1	Ground		2		1	34 583.03	.1	639
	812	2, 1, 2← 1, 1, 1	Ground		3		2	34 584.33	.1	639
	812	2, 1, 2← 1, 1, 1	Ground		1		0	34 585.59	.1	639
$C^{12}HD_2C^{12}N^{14}$	813	2, 0, 2← 1, 0, 1	Ground		1		0	32 967.44	.1	639
	813	2, 0, 2← 1, 0, 1	Ground		2		2	32 967.44	.1	639
	813	2, 0, 2← 1, 0, 1	Ground		3		2	32 968.70	.1	639
	813	2, 0, 2← 1, 0, 1	Ground		2		1	32 968.70	.1	639
	813	2, 0, 2← 1, 0, 1	Ground		1		1	32 970.70	.1	639
	813	2, 1, 1← 1, 1, 0	Ground		2		1	33 122.87	.1	639
	813	2, 1, 1← 1, 1, 0	Ground		3		2	33 124.23	.1	639
	813	2, 1, 1← 1, 1, 0	Ground		1		0	33 125.55	.1	639
	813	2, 1, 2← 1, 1, 1	Ground		2		1	32 811.64	.1	639
	813	2, 1, 2← 1, 1, 1	Ground		3		2	32 812.95	.1	639
	813	2, 1, 2← 1, 1, 1	Ground		1		0	32 814.30	.1	639
$C^{12}H_3C^{13}N^{14}$	815	2, 0← 1, 0	Ground		3		2	36 777.18		205
	815	2, 0← 1, 0	Ground		2		1	36 777.18		205
$C^{12}D_3C^{12}N^{14}$	816	2, ← 1,	0, 0; 0, 0; 0; 1					31 242.98		973
	816	2, ← 1,	0, 0; 0, 0; 1; 0					31 271.22		973
	816	2, ← 1,	Ground					31 431.50		973
	816	2, ← 1,	1,±1; 0, 0; 0; 0					31 492.42		973
	816	2, ← 1,	1, 0; 0, 0; 0; 0					31 520.26		973
	816	2, ← 1,	Excited					31 565.0		973
	816	2, ← 1,	Excited					31 588.6		973
	816	2, ← 1,	4, 0; 0, 0; 0; 0					31 770.		973
	816	2, ← 1,	5, 0; 0, 0; 0; 0					31 852.		973
	816	2, 0← 1, 0	0, 0; 1,±1; 0; 0					31 455.80		973
	816	2, 1← 1, 1	0, 0; 1,±1; 0; 0					31 409.64		973
	816	2, 1← 1, 1	0, 0; 1,∓1; 0; 0					31 468.06		973
	816	2, 1← 1, 1	0, 0; 1,±1; 0; 0					31 485.02		973
	816	2, 1← 1, 1	1,±1; 0, 0; 0; 0					31 493.42		973
	816	2, 1← 1, 1	1,∓1; 0, 0; 0; 0					31 520.92		973
	816	2, 1← 1, 1	1,±1; 0, 0; 0; 0					31 549.10		973
	816	2, 1← 1, 1	2,∓2; 0, 0; 0; 0					31 605.10		973
	816	2, 1← 1, 1	2, 0; 0, 0; 0; 0					31 609.10		973
	816	2, 1← 1, 1	2,±2; 0, 0; 0; 0					31 609.10		973
	816	2, 1← 1, 1	3,±1; 0, 0; 0; 0					31 635.		973
	816	2, 1← 1, 1	3,∓3; 0, 0; 0; 0					31 688.5		973

3. These lines were taken to be a superposition of the hyperfine structure components $F = J \leftarrow F = J - 1$ and $F = J + 2 \leftarrow F = J + 1$.

Isotopic Species	Id. No.	Rotational Quantum Nos.	Vib. State $v_a^l; v_b^l; v_c; v_d$	Hyperfine				Frequency MHz	Acc. ±MHz	Ref.
				F_1'	F'	F_1	F			
C^{12}D$_3$C^{12}N^{14}	816	2, 1← 1, 1	3,∓1; 0, 0; 0; 0					31 694.5		973
	816	2, 1← 1, 1	3,±3; 0, 0; 0; 0					31 694.5		973
	816	2, 1← 1, 1	3,±1; 0, 0; 0; 0					31 747.		973
	816	3, 0← 2, 0	Ground		3		3	47 146.00		205
	816	3, 0← 2, 0	Ground		4		3	47 147.60		205
	816	3, 0← 2, 0	Ground		2		1	47 147.60		205
	816	3, 0← 2, 0	Ground		3		2	47 147.60		205
	816	3, 1← 2, 1	Ground		3		3	47 146.00		205
	816	3, 1← 2, 1	Ground		3		2	47 146.68		205
	816	3, 1← 2, 1	Ground		2		1	47 147.00		205
	816	3, 1← 2, 1	Ground		4		3	47 147.00		205
	816	3, 2← 2, 2	Ground		2		3	45 145.20		205
	816	3, 2← 2, 2	Ground		3		3	45 145.20		205
	816	3, 2← 2, 2	Ground		4		3	45 145.20		205
	816	3, 2← 2, 2	Ground		3		2	47 143.85		205
	816	3, 2← 2, 2	Ground		2		2	47 143.85		205
	816	3, 2← 2, 2	Ground		2		1	47 146.00		205
C^{12}D$_3$C^{13}N^{14}	817	3, 0← 2, 0	Ground		3		3	47 089.43		205
	817	3, 0← 2, 0	Ground		2		1	47 091.05		205
	817	3, 0← 2, 0	Ground		3		2	47 091.05		205
	817	3, 0← 2, 0	Ground		4		3	47 091.05		205
	817	3, 1← 2, 1	Ground		3		3	47 089.43		205
	817	3, 1← 2, 1	Ground		2		1	47 090.41		205
	817	3, 1← 2, 1	Ground		4		3	47 090.41		205
	817	3, 2← 2, 2	Ground		2		2	47 087.39		205
	817	3, 2← 2, 2	Ground		3		2	47 087.39		205
	817	3, 2← 2, 2	Ground		2		3	47 088.69		205
	817	3, 2← 2, 2	Ground		3		3	47 088.69		205
	817	3, 2← 2, 2	Ground		4		3	47 088.69		205
	817	3, 2← 2, 2	Ground		2		1	47 089.43		205
C^{13}D$_3$C^{12}N^{14}	819	2, ← 1,	Ground					30 780.90		973
	819	2, ← 1,	1,±1; 0, 0; 0; 0					30 842.1		973
	819	2, ← 1,	1, 0; 0, 0; 0; 0					30 867.1		973
	819	2, ← 1,	1,±1; 0, 0; 0; 0					30 895.5		973
	819	2, ← 1,	2, 0; 0, 0; 0; 0					30 951.3		973
C^{12}D$_3$C^{12}N^{15}	821	2, ← 1,	Ground					30 477.54		973
	821	2, ← 1,	1, 0; 0, 0; 0; 0					30 563.8		973

C_2H_3N C_{3v} CH_3NC

Isotopic Species	Pt. Gp.	Id. No.	A MHz		B MHz		C MHz		D_J MHz	D_{JK} MHz	Δ Amu A²	κ
$C^{12}H_3N^{14}C^{12}$	C_{3v}	831	160 688.8	M	10 052.88	M	10 052.88	M	.004	.228		
$C^{13}H_3N^{14}C^{12}$	C_{3v}	832			9 771.70	M	9 771.70	M				
$C^{12}H_3N^{14}C^{13}$	C_{3v}	833			9 695.91	M	9 695.91	M				
$C^{12}D_3N^{14}C^{12}$	C_{3v}	834			8 581.88	M	8 581.88	M	.002	.133		
$C^{13}D_3N^{14}C^{12}$	C_{3v}	835			8 410.17	M	8 410.17	M				
$C^{12}D_3N^{14}C^{13}$	C_{3v}	836			8 278.79	M	8 278.79	M		.130		
$C^{12}D_3N^{15}C^{12}$	C_{3v}	837			8 567.63	M	8 567.63	M				
$C^{12}H_2DN^{14}C^{12}$	C_s	838			9 578.20	M	9 397.81	M		.18		
$C^{12}HD_2N^{14}C^{12}$	C_s	839			9 096.72	M	8 910.53	M		.19		

Id. No.	μ_a Debye		μ_b Debye		μ_c Debye		eQq Value(MHz)	Rel.	eQq Value(MHz)	Rel.	eQq Value(MHz)	Rel.	ω_a 1/cm	d	ω_b 1/cm	d	ω_c 1/cm	d	ω_d 1/cm	d
831	3.83	M	0.	X	0.	X	<0.5	N^{14}					263	2						

References:

ABC: 205,724,791,864 D_J: 724 D_{JK}: 205,724 μ: 434   eQq: 205 ω: 864

Add. Ref. 53,54,154,222,302

For species 831, $B_v = 10091.86$ MHz, $D_{JK} = 0.27$ MHz; $q_8 = 27.78$ MHz, $\zeta_8 = 0.93$, $a = 1.08$ where $a = q_8\omega_8/2B_e^2$. Ref. 864.

Isotopic Species	Id. No.	Rotational Quantum Nos.	Vib. State v_a^l	Hyperfine F_1'	F'	F_1	F	Frequency MHz	Acc. ±MHz	Ref.
$C^{12}H_3N^{14}C^{12}$	831	1, 0← 0, 0	Ground					20 105.76	.1	724
	831	2, 0← 1, 0	Ground					40 211.36	.1	724
	831	2, 0← 1, 0	1,±1					40 366.55		864
	831	2, 1← 1, 1	Ground					40 210.46	.1	724
	831	2, 1← 1, 1	1,−1					40 313.37		864 [1]
	831	2, 1← 1, 1	1,+1					40 424.49		864 [1]
	831	2, 1← 2, 1	1,∓1					40 364.07		864
	831	3, ← 2,	Excited					60 470.04	.2	724
	831	3, ← 2,	Excited					60 538.71	.2	724
	831	3, ← 2,	Excited					60 545.19	.2	724
	831	3, ← 2,	Excited					60 547.38	.2	724
	831	3, ← 2,	Excited					60 556.62	.2	724
	831	3, ← 2,	Excited					60 636.63	.2	724
	831	3, 0← 2, 0	Ground					60 316.86	.2	724
	831	3, 1← 2, 1	Ground					60 315.48	.2	724
	831	3, 2← 2, 2	Ground					60 311.40	.2	724
$C^{13}H_3N^{14}C^{12}$	832	1, 0← 0, 0	Ground					19 543.4	.3	724
$C^{12}H_3N^{14}C^{13}$	833	2, 0← 1, 0	Ground					38 783.21		205
	833	2, 1← 1, 1	Ground					38 782.21		205
$C^{12}D_3N^{14}C^{12}$	834	2, 0← 1, 0	Ground					34 327.45	.1	724
	834	2, 1← 1, 1	Ground					34 326.93	.1	724
	834	3, 0← 2, 0	Ground					51 491.04	.15	724
	834	3, 1← 2, 1	Ground					51 490.24	.15	724
	834	3, 2← 2, 2	Ground					51 487.84	.15	724
$C^{13}D_3N^{14}C^{12}$	835	2, 0← 1, 0	Ground					33 640.6	.3	724
$C^{12}D_3N^{14}C^{13}$	836	3, 0← 2, 0	Ground					49 671.19		205
	836	3, 1← 2, 1	Ground					49 670.43		205
	836	3, 2← 2, 2	Ground					49 668.07		205
$C^{12}H_2DN^{14}C^{12}$	838	2, 0, 2← 1, 0, 1	Ground					37 951.67	.1	724
	838	2, 1, 1← 1, 1, 0	Ground					38 131.57	.1	724
	838	2, 1, 2← 1, 1, 1	Ground					37 770.79	.1	724
$C^{12}HD_2N^{14}C^{12}$	839	2, 0, 2← 1, 0, 1	Ground					36 013.96	.1	724
	839	2, 1, 1← 1, 1, 0	Ground					36 199.82	.1	724
	839	2, 1, 2← 1, 1, 1	Ground					35 827.44	.1	724

1. When two l-doubling measured frequencies have the same designations, the larger has been reported with $l=+1$, and the smaller with $l=-1$.

C_2H_4O C_{2v} $H_2C_*OC_*H_2$

Isotopic Species	Pt. Gp.	Id. No.	A MHz		B MHz		C MHz		D_J MHz	D_{JK} MHz	Δ Amu A²	κ
$H_2C^{12}{}_*O^{16}C^{12}{}_*H_2$	C_{2v}	841	25 483.7	M	22 120.9	M	14 098.0	M				.40930
$H_2C^{12}{}_*O^{16}C^{13}{}_*H_2$	C_s	842	25 291.2	M	21 597.4	M	13 825.2	M				.3557
$D_2C^{12}{}_*O^{16}C^{12}{}_*D_2$	C_{2v}	843	20 399.	M	15 457.	M	11 544.	M				−.11615
t-HDC$^{12}{}_*O^{16}C^{12}{}_*$HD	C_2	844	22 945.1	M	18 198.6	M	12 585.5	M				.0838546
c-HDC$^{12}{}_*O^{16}C^{12}{}_*$DH	C_s	845	22 700.3	M	18 318.5	M	12 650.3	M				.128012

Id. No.	μ_a Debye		μ_b Debye		μ_c Debye		eQq Value(MHz)	Rel.	eQq Value(MHz)	Rel.	eQq Value(MHz)	Rel.	ω_a 1/cm	d	ω_b 1/cm	d	ω_c 1/cm	d	ω_d 1/cm	d
841	0.	X	1.88	M	0.	X															

References:

ABC: 263,622   κ: 263,622   μ: 263

Add. Ref. 76,99,126,566,609,725

Isotopic Species	Id. No.	Rotational Quantum Nos.	Vib. State	Hyperfine				Frequency MHz	Acc. ±MHz	Ref.
				F_1'	F'	F_1	F			
$H_2C^{12}{}_*O^{16}C^{12}{}_*H_2$	841	1, 1, 1← 0, 0, 0	Ground					39 581.8	.1	263
	841	2, 1, 1← 2, 0, 2	Ground					24 923.66	.03	263
	841	2, 2, 1← 2, 1, 2	Ground					34 157.1	.5	263
	841	3, 2, 1← 3, 1, 2	Ground					23 610.38	.03	263
	841	3, 3, 0← 3, 2, 1	Ground					23 134.21	.03	263
	841	3, 3, 1← 3, 2, 2	Ground					39 680.	1.	263
	841	4, 2, 2← 4, 1, 3	Ground					41 581.	1.	263
	841	4, 3, 1← 4, 2, 2	Ground					24 834.26	.03	263
	841	4, 4, 0← 4, 3, 1	Ground					34 148.3	.5	263
	841	5, 3, 2← 5, 2, 3	Ground					37 781.	1.	263
	841	5, 4, 1← 5, 3, 2	Ground					29 688.	1.	263
	841	6, 4, 2← 6, 3, 3	Ground					35 791.	1.	263
	841	6, 5, 1← 6, 4, 2	Ground					38 702.	1.	263
	841	7, 5, 2← 7, 4, 3	Ground					37 329.	1.	263
	841	8, 6, 2← 8, 5, 3	Ground					43 398.	1.	263
$H_2C^{12}{}_*O^{16}C^{13}{}_*H_2$	842	1, 1, 1← 0, 0, 0	Ground					39 116.4	.5	263
	842	2, 1, 1← 2, 0, 2	Ground					24 352.2	.3	263
	842	2, 2, 1← 2, 1, 2	Ground					34 398.0	.4	263
	842	3, 2, 1← 3, 1, 2	Ground					23 278.5	.3	263
	842	3, 3, 0← 3, 2, 1	Ground					24 667.9	.3	263
	842	4, 3, 1← 4, 2, 2	Ground					25 246.8	.3	263
$D_2C^{12}{}_*O^{16}C^{12}{}_*D_2$	843	1, 1, 1← 0, 0, 0	Ground					31 943.		263
	843	2, 2, 1← 2, 1, 2	Ground					26 565.		263
	843	3, 1, 2← 3, 0, 3	Ground					24 055.		263
	843	3, 2, 2← 3, 1, 3	Ground					33 285.		263
	843	3, 3, 0← 3, 2, 1	Ground					29 080.		263
	843	3, 3, 1← 3, 2, 2	Ground					35 341.		263
	843	4, 2, 2← 4, 1, 3	Ground					21 664.		263
	843	5, 2, 3← 5, 1, 4	Ground					31 280.		263
	843	5, 3, 2← 5, 2, 3	Ground					24 668.		263
	843	5, 4, 1← 5, 3, 2	Ground					39 592.		263
	843	6, 3, 3← 6, 2, 4	Ground					28 495.		263
	843	6, 4, 2← 6, 3, 3	Ground					34 680.		263
	843	7, 4, 3← 7, 3, 4	Ground					32 296.		263
	843	8, 4, 4← 8, 3, 5	Ground					35 068.		263
t-$HDC^{12}{}_*O^{16}C^{12}{}_*HD$	844	1, 1, 1← 0, 0, 0	Ground					35 528.56	.05	622
	844	2, 1, 1← 2, 0, 2	Ground					18 829.50	.05	622
	844	2, 2, 0← 2, 1, 1	Ground					17 092.43	.05	622
	844	2, 2, 1← 2, 1, 2	Ground					31 072.85	.05	622
	844	3, 1, 2← 3, 0, 3	Ground					33 196.80	.05	622
	844	3, 2, 1← 3, 1, 2	Ground					19 742.81	.05	622
	844	3, 2, 2← 3, 1, 3	Ground					40 992.		622
	844	3, 3, 0← 3, 2, 1	Ground					28 947.30	.05	622
	844	3, 3, 1← 3, 2, 2	Ground					39 254.37	.05	622
	844	4, 2, 2← 4, 1, 3	Ground					29 191.8	.2	622
	844	5, 3, 2← 5, 2, 3	Ground					28 611.8	.2	622
	844	5, 4, 1← 5, 3, 2	Ground					37 879.9	.2	622
	844	6, 3, 3← 6, 2, 4	Ground					39 273.7	.2	622
	844	6, 4, 2← 6, 3, 3	Ground					34 017.5	.2	622

Isotopic Species	Id. No.	Rotational Quantum Nos.	Vib. State	Hyperfine				Frequency MHz	Acc. ±MHz	Ref.
				F_1'	F'	F_1	F			
t-HDC^{12}*O^{16}C^{12}*HD	844	7, 4, 3← 7, 3, 4	Ground					37 234.8	.2	622
c-HDC^{12}*O^{16}C^{12}*DH	845	1, 1, 1← 0, 0, 0	Ground					35 350.54	.05	622
	845	2, 1, 1← 2, 0, 2	Ground					18 741.20	.05	622
	845	2, 2, 0← 2, 1, 1	Ground					16 168.15	.05	622
	845	2, 2, 1← 2, 1, 2	Ground					30 150.10	.05	622
	845	3, 1, 2← 3, 0, 3	Ground					33 290.16	.05	622
	845	3, 2, 1← 3, 1, 2	Ground					19 253.07	.05	622
	845	3, 2, 2← 3, 1, 3	Ground					40 231.5	.2	622
	845	3, 3, 0← 3, 2, 1	Ground					27 008.50	.05	622
	845	3, 3, 1← 3, 2, 2	Ground					37 657.0	.2	622
	845	4, 2, 2← 4, 1, 3	Ground					29 326.2	.2	622
	845	5, 3, 2← 5, 2, 3	Ground					28 097.1	.2	622
	845	5, 4, 1← 5, 3, 2	Ground					35 040.3	.2	622
	845	6, 3, 3← 6, 2, 4	Ground					39 828.8	.2	622
	845	6, 4, 2← 6, 3, 3	Ground					32 116.2	.2	622
	845	7, 4, 3← 7, 3, 4	Ground					36 874.3	.2	622
	845	8, 5, 3← 8, 4, 4	Ground					39 467.2	.2	622
H$_2^b$C^b*O^{16}C^b*H$_2^b$	846	Not Reported	Excited					21 692.	2.	263
	846	Not Reported	Excited					22 097.	5.	263
	846	Not Reported	Excited					22 303.	5.	263
	846	Not Reported	Excited					22 340.	5.	263
	846	Not Reported	Excited					22 695.	5.	263
	846	Not Reported	Excited					23 341.4	.3	263
	846	Not Reported	Excited					23 412.0	.3	263
	846	Not Reported	Excited					23 432.6	.3	263
	846	Not Reported	Excited					23 454.6	.3	263
	846	Not Reported	Excited					23 561.1	.3	263
	846	Not Reported	Excited					23 743.5	.3	263
	846	Not Reported	Excited					23 788.8	.3	263
	846	Not Reported	Excited					23 818.0	.3	263
	846	Not Reported	Excited					23 839.3	.3	263
	846	Not Reported	Excited					24 361.0	.3	263
	846	Not Reported	Excited					24 395.8	.3	263
	846	Not Reported	Excited					24 870.	1.	263
	846	Not Reported	Excited					25 002.2	.3	263
	846	Not Reported	Excited					25 210.8	.3	263
	846	Not Reported	Excited					28 751.0	.3	263
	846	Not Reported	Excited					34 407.9	.2	263
	846	Not Reported	Excited					39 462.	3.	263

C_2H_4S C_{2v} $C_*H_2SC_*H_2$

Isotopic Species	Pt. Gp.	Id. No.	A MHz		B MHz		C MHz		D_J MHz	D_{JK} MHz	Δ Amu A^2	κ
$C^{12}{}_*H_2S^{32}C^{12}{}_*H_2$	C_{2v}	851	21 974.	M	10 824.9	M	8 026.3	M				−.5988
$C^{12}{}_*H_2S^{34}C^{12}{}_*H_2$	C_{2v}	852	21 974.	M	10 551.0	M	7 874.7	M				−.62045
$C^{12}{}_*D_2S^{32}C^{12}{}_*D_2$	C_{2v}	853	15 471.	M	9 197.6	M	6 819.0	M				−.45015

Id. No.	μ_a Debye		μ_b Debye		μ_c Debye		eQq Value(MHz)	Rel.	eQq Value(MHz)	Rel.	eQq Value(MHz)	Rel.	ω_a d 1/cm	ω_b d 1/cm	ω_c d 1/cm	ω_d d 1/cm
851	1.84	M	0.	X	0.	X										

References:

ABC: 263 κ: 263 μ: 263

Ethylene Sulfide   Spectral Line Table

Isotopic Species	Id. No.	Rotational Quantum Nos.	Vib. State	F_1'	F'	F_1	F	Frequency MHz	Acc. ±MHz	Ref.
$C^{12}{}_*H_2S^{32}C^{12}{}_*H_2$	851	1, 0, 1← 0, 0, 0	Ground					18 851.4		263 [1]
	851	2, 1, 1← 1, 1, 0	Ground					40 500.5		263 [1]
	851	2, 1, 2← 1, 1, 1	Ground					34 903.6		263 [1]
	851	3, 1, 3← 3, 1, 2	Ground					16 742.4	.2	263
	851	4, 1, 5← 4, 1, 4	Ground					27 648.4	.2	263
	851	5, 1, 5← 5, 1, 4	Ground					40 672.5	.2	263
	851	6, 2, 5← 6, 2, 4	Ground					22 976.4	.2	263
	851	7, 2, 6← 7, 2, 5	Ground					35 515.3		263 [1]
	851	9, 3, 7← 9, 3, 6	Ground					26 973.2	.2	263
	851	10, 3, 8←10, 3, 7	Ground					40 865.1	.2	263
	851	11, 4, 8←11, 4, 7	Ground					17 716.8	.2	263 [1]
$C^{12}{}_*H_2S^{34}C^{12}{}_*H_2$	852	1, 0, 1← 0, 0, 0	Ground					18 425.8	.2	263
	852	2, 1, 1← 1, 1, 0	Ground					39 527.9		263 [1]
	852	2, 1, 2← 1, 1, 1	Ground					34 174.7	1.5	263
	852	4, 1, 4← 4, 1, 3	Ground					26 477.7	.2	263
	852	5, 1, 5← 5, 1, 4	Ground					39 034.6	.2	263
	852	6, 2, 5← 6, 2, 4	Ground					21 245.8	.2	263
	852	7, 2, 6← 7, 2, 5	Ground					33 091.8	.2	263
$C^{12}{}_*D_2S^{32}C^{12}{}_*D_2$	853	2, 0, 2← 1, 0, 1	Ground					31 474.8	.2	263
	853	2, 1, 1← 1, 1, 0	Ground					34 412.0	.2	263
	853	2, 1, 2← 1, 1, 1	Ground					29 654.6	.2	263
	853	4, 1, 4← 4, 1, 3	Ground					23 214.3	.2	263
	853	5, 1, 5← 5, 1, 4	Ground					33 541.5	.2	263
	853	6, 2, 5← 6, 2, 4	Ground					23 016.3	.2	263
	853	7, 2, 6← 7, 2, 5	Ground					33 942.5	.2	263
$C^{12}{}_*H_2^bS^bC^{12}{}_*H_2^b$	854	Not Reported						16 059.6	.5	263
	854	Not Reported						16 161.8	.5	263
	854	Not Reported						17 020.5	.5	263
	854	Not Reported						17 215.5	.5	263
	854	Not Reported						18 030.	.5	263

1. Some lines as listed in the tables of the reference differ from the listing in its Appendix A. Table listings have been assumed correct.

Isotopic Species	Id. No.	Rotational Quantum Nos.	Vib. State	Hyperfine				Frequency MHz	Acc. ±MHz	Ref.
				F_1'	F'	F_1	F			
$C^{12}_*H_2^bS^bC^{12}_*H_2^b$	854	Not Reported						18 446.6	.5	263
	854	Not Reported						18 489.7	.5	263
	854	Not Reported						21 411.4	.2	263
	854	Not Reported						21 795.6	.2	263
	854	Not Reported						22 168.6	.2	263
	854	Not Reported						22 358.2	.2	263
	854	Not Reported						22 468.6	.2	263
	854	Not Reported						22 857.0	.2	263
	854	Not Reported						23 203.6	.2	263
	854	Not Reported						23 215.2	.2	263
	854	Not Reported						23 560.0	10.	263
	854	Not Reported						23 958.1	.2	263
	854	Not Reported						24 043.3	.2	263
	854	Not Reported						24 101.4	.2	263
	854	Not Reported						24 105.0	10.	263
	854	Not Reported						24 175.0	10.	263
	854	Not Reported						24 183.0	.2	263
	854	Not Reported						24 190.0	10.	263
	854	Not Reported						24 192.0	10.	263
	854	Not Reported						24 230.0	10.	263
	854	Not Reported						24 354.0	.2	263
	854	Not Reported						24 632.7	.2	263
	854	Not Reported						25 247.2	.2	263
	854	Not Reported						25 275.0	10.	263
	854	Not Reported						25 550.0	20.	263
	854	Not Reported						26 035.0	20.	263
	854	Not Reported						27 150.0	20.	263
	854	Not Reported						27 725.0	20.	263
	854	Not Reported						28 218.0	.5	263
	854	Not Reported						29 337.0	.5	263
	854	Not Reported						29 890.0	25.	263
	854	Not Reported						29 924.0	.5	263
	854	Not Reported						30 350.0	25.	263
	854	Not Reported						30 400.0	25.	263
	854	Not Reported						30 439.7	.2	263
	854	Not Reported						30 560.3	.3	263
	854	Not Reported						30 762.0	1.	263
	854	Not Reported						31 139.9	.5	263
	854	Not Reported						31 823.7	.5	263
	854	Not Reported						31 990.0	25.	263
	854	Not Reported						32 000.0	25.	263
	854	Not Reported						32 760.0	20.	263
	854	Not Reported						32 850.0	20.	263
	854	Not Reported						33 027.5	.2	263
	854	Not Reported						33 940.0	15.	263
	854	Not Reported						33 960.0	15.	263
	854	Not Reported						34 243.3	.5	263
	854	Not Reported						35 538.0	.5	263
	854	Not Reported						35 785.0	1.	263
	854	Not Reported						39 900.0	25.	263
	854	Not Reported						39 920.0	25.	263

C$_2$H$_5$N C$_s$ [1] C$_*$H$_2$NHC$_*$H$_2$

Isotopic Species	Pt. Gp.	Id. No.	A MHz		B MHz		C MHz		D$_J$ MHz	D$_{JK}$ MHz	Δ Amu A^2	κ
C$^{12}_*$H$_2$N^{14}HC$^{12}_*$H$_2$	C$_s$	861	22 736.1	M	21 192.3	M	13 383.3	M				.6700
C$^{12}_*$H$_2$N^{14}DC$^{12}_*$H$_2$	C$_s$	862	20 697.05	M			12 816.95	M				.9761

Id. No.	μ$_a$ Debye		μ$_b$ Debye		μ$_c$ Debye		eQq Value(MHz) Rel.	eQq Value(MHz) Rel.	eQq Value(MHz) Rel.	ω$_a$ d 1/cm	ω$_b$ d 1/cm	ω$_c$ d 1/cm	ω$_d$ d 1/cm
861	0.	X	1.67	M	.89	M							

1. References say that the hydrogen atom attached to the nitrogen is not in the plane of the nitrogen and two carbon atoms.

References:

ABC: 444,642   κ: 642   μ: 444

Add. Ref. 440,483,490

Species	(A + C)/2 (MHz)	(A − C)/2 (MHz)	Ref.
861	18059.7	4676.0	642
862	16757	3940.05	642

Isotopic Species	Id. No.	Rotational Quantum Nos.	Vib. State	Hyperfine F_1'	F'	F_1	F	Frequency MHz	Acc. ±MHz	Ref.
$C^{12}{}_*H_2N^{14}HC^{12}{}_*H_2$	861	1, 1, 1← 0, 0, 0	Ground					36 119.4		444
	861	2, 1, 1← 2, 0, 2	Ground					23 632.9	.5	482
	861	2, 1, 1← 2, 0, 2	Ground					23 634.41	.1	490
	861	2, 2, 0← 2, 1, 2	Ground					34 540.	10.	726
	861	2, 2, 1← 2, 1, 2	Ground					28 058.34	.1	490
	861	3, 2, 1← 3, 1, 2	Ground					22 261.2	.5	482
	861	3, 2, 1← 3, 1, 2	Ground					22 262.34	.1	490
	861	3, 3, 0← 3, 2, 2	Ground					35 334.4	.5	490
	861	3, 3, 1← 3, 2, 2	Ground					30 494.62	.1	490
	861	4, 2, 2← 4, 1, 3	Ground					40 785.8		444
	861	4, 3, 1← 4, 2, 2	Ground					21 360.6		444
	861	4, 4, 0← 4, 3, 1	Ground					18 577.5		444
	861	4, 4, 1← 4, 3, 2	Ground					33 784.2		444
	861	5, 4, 1← 5, 3, 2	Ground					21 474.0		444
	861	5, 5, 0← 5, 4, 1	Ground					24 682.8		444
	861	5, 5, 1← 5, 4, 1	Ground					22 671.9	.5	482
	861	5, 5, 1← 5, 4, 2	Ground					37 930.	10.	726
	861	6, 4, 2← 6, 3, 3	Ground					36 374.0	.5	444
	861	6, 5, 1← 6, 4, 2	Ground					23 025.3		444
	861	6, 6, 0← 6, 5, 1	Ground					32 185.8		444
	861	6, 6, 1← 6, 5, 1	Ground					31 055.	10.	726
	861	7, 5, 2← 7, 4, 3	Ground					34 053.3		444
	861	7, 6, 1← 7, 5, 2	Ground					26 314.1		444
	861	7, 6, 1← 7, 5, 2	Ground					26 329.2	.5	482
	861	7, 7, 0← 7, 6, 1	Ground					40 588.6		444
	861	8, 6, 2← 8, 5, 3	Ground					32 526.3		444
	861	8, 7, 1← 8, 6, 2	Ground					31 540.	10.	726
	861	8, 7, 2← 8, 6, 2	Ground					25 732.7	.5	482
	861	9, 7, 2← 9, 6, 3	Ground					32 406.3		444
	861	9, 8, 1← 9, 7, 2	Ground					38 500.	10.	726
	861	10, 8, 2←10, 7, 3	Ground					34 165.1		444
	861	11, 9, 2←11, 8, 3	Ground					38 000.	10.	726
	861	12,10, 3←12, 9, 3	Ground					36 250.	10.	726
$C^{12}{}_*H_2N^{14}DC^{12}{}_*H_2$	862	1, 0, 1← 0, 0, 0	Ground					33 420.	20.	642
	862	2, 1, 1← 2, 1, 2	Ground					23 356.3	.5	642
	862	2, 1, 1← 2, 1, 2	Ground					23 359.5	.5	642
	862	2, 2, 1← 2, 0, 2	Ground					23 638.6	.5	642
	862	2, 2, 1← 2, 0, 2	Ground					23 643.6	.5	642
	862	3, 2, 1← 3, 2, 2	Ground					23 217.1	.5	642
	862	3, 3, 1← 3, 1, 2	Ground					23 784.8	.5	642
	862	3, 3, 1← 3, 1, 2	Ground					23 787.8	.5	642
	862	4, 3, 1← 4, 3, 2	Ground					23 028.2	.5	642
	862	4, 4, 1← 4, 2, 2	Ground					23 983.1	.5	642
	862	4, 4, 1← 4, 2, 2	Ground					23 985.1	.5	642
	862	5, 4, 1← 5, 4, 2	Ground					22 792.3	.5	642
	862	5, 5, 1← 5, 3, 2	Ground					24 237.5	.5	642
	862	5, 5, 1← 5, 3, 2	Ground					24 239.1	.5	642
	862	6, 5, 1← 6, 5, 2	Ground					22 509.5	.5	642
	862	6, 6, 1← 6, 4, 2	Ground					24 553.2	.5	642

Isotopic Species	Id. No.	Rotational Quantum Nos.	Vib. State	F_1'	F'	F_1	F	Frequency MHz	Acc. ±MHz	Ref.
$C^{12}{}_*H_2N^{14}DC^{12}{}_*H_2$	862	6, 6, 1← 6, 4, 2	Ground					24 554.6	.5	642
	862	7, 6, 1← 7, 6, 2	Ground					22 181.1	.5	642
	862	7, 7, 1← 7, 5, 2	Ground					24 936.3	.5	642
	862	7, 7, 1← 7, 5, 2	Ground					24 937.3	.5	642
	862	8, 7, 1← 8, 7, 2	Ground					21 805.9	.5	642
	862	8, 8, 1← 8, 6, 2	Ground					25 393.3	.5	642
	862	9, 8, 1← 9, 8, 2	Ground					21 385.1	.5	642
	862	9, 9, 1← 9, 7, 2	Ground					25 920.	10.	642
	862	10, 9, 1←10, 9, 2	Ground					20 918.5	.5	642
	862	11,10, 1←11,10, 2	Ground					20 410.	10.	642
	862	12,11, 1←12,11, 2	Ground					19 860.	10.	642
	862	13,12, 1←13,12, 2	Ground					19 250.	10.	726
$C^b{}_*H_2^bN^bH^bC^b{}_*H_2^b$	863	Not Reported						20 703.8	.5	726
	863	Not Reported						20 941.3	.5	726
	863	Not Reported						21 035.	10.	726
	863	Not Reported						21 068.	10.	726
	863	Not Reported						21 279.8	.5	726
	863	Not Reported						21 340.	10.	726
	863	Not Reported						21 345.	10.	726
	863	Not Reported						21 355.	10.	726
	863	Not Reported						21 380.	10.	726
	863	Not Reported						21 400.	10.	726
	863	Not Reported						21 435.	10.	726
	863	Not Reported						21 751.	10.	726
	863	Not Reported						21 785.	10.	726
	863	Not Reported						21 945.5	.5	726
	863	Not Reported						21 985.	10.	726
	863	Not Reported						21 991.0	.5	726
	863	Not Reported						22 025.	10.	726
	863	Not Reported						22 080.9	.5	726
	863	Not Reported						22 125.	10.	726
	863	Not Reported						22 164.9	.5	726
	863	Not Reported						22 195.3	.5	726
	863	Not Reported						22 238.8	.5	726
	863	Not Reported						22 245.5	.5	726
	863	Not Reported						22 288.8	.5	726
	863	Not Reported						22 365.1	.5	726
	863	Not Reported						22 405.6	.5	726
	863	Not Reported						22 436.5	.5	726
	863	Not Reported						22 486.1	.5	726
	863	Not Reported						22 650.	10.	726
	863	Not Reported						22 680.	10.	726
	863	Not Reported						22 830.	10.	726
	863	Not Reported						22 904.4	.5	726
	863	Not Reported						22 982.2	.5	726
	863	Not Reported						23 035.0	.5	726
	863	Not Reported						23 223.7	.5	726
	863	Not Reported						23 250.	10.	726
	863	Not Reported						23 280.	10.	726

Isotopic Species	Id. No.	Rotational Quantum Nos.	Vib. State	Hyperfine F_1'	F'	F_1	F	Frequency MHz	Acc. ±MHz	Ref.
$C^b_*H^b_2N^bH^bC^b_*H^b_2$	863	Not Reported						23 340.	10.	726
	863	Not Reported						23 395.	10.	726
	863	Not Reported						23 412.9	.5	726
	863	Not Reported						23 612.9	.5	726
	863	Not Reported						23 665.7	.5	726
	863	Not Reported						23 908.	10.	726
	863	Not Reported						23 923.	10.	726
	863	Not Reported						24 290.	10.	726
	863	Not Reported						24 400.	10.	726
	863	Not Reported						24 900.	10.	726
	863	Not Reported						24 907.9	.5	726
	863	Not Reported						24 985.	10.	726
	863	Not Reported						24 991.9	.5	726
	863	Not Reported						25 020.	10.	726
	863	Not Reported						25 032.4	.5	726
	863	Not Reported						25 042.0	.5	726
	863	Not Reported						25 270.	10.	726
	863	Not Reported						25 370.	10.	726
	863	Not Reported						25 550.8	.5	726
	863	Not Reported						25 627.	10.	726
	863	Not Reported						25 731.4	.5	726
	863	Not Reported						25 878.3	.5	726
	863	Not Reported						26 200.	10.	726
	863	Not Reported						26 256.	10.	726
	863	Not Reported						29 830.	10.	726
	863	Not Reported						30 215.	10.	726
	863	Not Reported						30 765.	10.	726
	863	Not Reported						31 110.	10.	726
	863	Not Reported						31 490.	10.	726
	863	Not Reported						31 570.	10.	726
	863	Not Reported						31 740.	10.	726
	863	Not Reported						31 755.	10.	726
	863	Not Reported						31 970.	10.	726
	863	Not Reported						32 142.	10.	726
	863	Not Reported						32 178.	10.	726
	863	Not Reported						32 220.	10.	726
	863	Not Reported						32 350.	10.	726
	863	Not Reported						32 450.	10.	726
	863	Not Reported						32 695.	10.	726
	863	Not Reported						32 745.	10.	726
	863	Not Reported						32 775.	10.	726
	863	Not Reported						32 850.	10.	726
	863	Not Reported						32 885.	10.	726
	863	Not Reported						32 940.	10.	726
	863	Not Reported						33 070.	10.	726
	863	Not Reported						33 410.	10.	726
	863	Not Reported						33 470.	10.	726
	863	Not Reported						33 580.	10.	726
	863	Not Reported						33 590.	10.	726
	863	Not Reported						33 680.	10.	726

Isotopic Species	Id. No.	Rotational Quantum Nos.	Vib. State	Hyperfine				Frequency MHz	Acc. ±MHz	Ref.
				F_1'	F'	F_1	F			
$C^b_*H_2^bN^bH^bC^b_*H_2^b$	863	Not Reported						33 740.	10.	726
	863	Not Reported						33 880.	10.	726
	863	Not Reported						33 900.	10.	726
	863	Not Reported						33 990.	10.	726
	863	Not Reported						34 120.	10.	726
	863	Not Reported						34 200.	10.	726
	863	Not Reported						34 240.	10.	726
	863	Not Reported						34 280.	10.	726
	863	Not Reported						34 380.	10.	726
	863	Not Reported						34 630.	10.	726
	863	Not Reported						34 680.	10.	726
	863	Not Reported						35 480.	10.	726
	863	Not Reported						35 600.	10.	726
	863	Not Reported						35 750.	10.	726
	863	Not Reported						35 900.	10.	726
	863	Not Reported						36 180.	10.	726
	863	Not Reported						36 350.	10.	726
	863	Not Reported						36 575.	10.	726
	863	Not Reported						36 910.	10.	726
	863	Not Reported						37 080.	10.	726
	863	Not Reported						37 410.	10.	726
	863	Not Reported						38 825.	10.	726
	863	Not Reported						38 930.	10.	726
	863	Not Reported						38 970.	10.	726
	863	Not Reported						39 000.	10.	726
	863	Not Reported						39 040.	10.	726
	863	Not Reported						39 100.	10.	726
	863	Not Reported						39 170.	10.	726
	863	Not Reported						39 300.	10.	726
	863	Not Reported						39 325.	10.	726
	863	Not Reported						39 430.	10.	726
	863	Not Reported						39 500.	10.	726
	863	Not Reported						39 600.	10.	726
	863	Not Reported						39 700.	10.	726
	863	Not Reported						40 000.	10.	726

C$_3$HF$_3$ · · · · · · C$_{3v}$ · · · · · · CF$_3$C:CH

Isotopic Species	Pt. Gp.	Id. No.	A MHz	B MHz		C MHz		D$_J$ MHz	D$_{JK}$ MHz	Δ Amu A^2	κ
C^{12}F$_3^{19}$C^{12}:C^{12}H	C$_{3v}$	871		2 877.93	M	2 877.93	M	.00024	.0063		
C^{12}F$_3^{19}$C^{13}:C^{12}H	C$_{3v}$	872		2 854.99	M	2 854.99	M				
C^{12}F$_3^{19}$C^{12}:C^{13}H	C$_{3v}$	873		2 787.63	M	2 787.63	M				
C^{12}F$_3^{19}$C^{12}:C^{12}D	C$_{3v}$	874		2 696.02	M	2 696.02	M	.00026	.0062		

Id. No.	μ$_a$ Debye	μ$_b$ Debye	μ$_c$ Debye	eQq Value(MHz) Rel.	eQq Value(MHz) Rel.	eQq Value(MHz) Rel.	ω$_a$ d 1/cm	ω$_b$ d 1/cm	ω$_c$ d 1/cm	ω$_d$ d 1/cm
871	2.36 M	0. X	0. X				170 2			

References:

ABC: 315 · · · D$_J$: 256 · · · D$_{JK}$: 256 · · · μ: 315 · · · ω: 315

Add. Ref. 200,313

For state $v_{10} = 1$ of species 871, $D_J = .0003$ MHz, $D_{JK} = .008$ MHz, $q_{10} = 3.6125$ MHz, $B_v = 2883.47$ MHz and $\zeta_{10} = .6$, Ref. 864.

Assuming $a = 1.15$ as in methylacetylene (ref. 315), $\alpha_{10} = -6.51$ MHz, Ref. 256.

3,3,3-Trifluoro-1-Propyne

Spectral Line Table

Isotopic Species	Id. No.	Rotational Quantum Nos.	Vib. State v$_a^l$	F$_1'$	F'	F$_1$	F	Frequency MHz	Acc. ±MHz	Ref.
C^{12}F$_3^{19}$C^{12}:C^{12}H	871	4, ← 3,	Ground					23 023.4	.3	315
	871	4, ← 3,	1, 0					23 067.7	.3	315
	871	4, ← 3,	2, 0					23 111.2	.3	315
	871	4, ← 3,	3, 0					23 153.0	.3	315
	871	4, 1← 3, 1	1,−1					23 053.5	.3	315 [1]
	871	4, 1← 3, 1	1,±1					23 082.4	.3	315 [1]
	871	5, 0← 4, 0	Ground					28 779.31	.10	256
	871	5, 0← 4, 0	1,±1					28 835.26	.20	256
	871	5, 1← 4, 1	Ground					28 779.31	.10	256
	871	5, 1← 4, 1	1,−1					28 816.48	.20	256 [1]
	871	5, 1← 4, 1	1,∓1					28 834.45	.20	256 [1]
	871	5, 1← 4, 1	1,+1					28 852.61	.20	256 [1]
	871	5, 2← 4, 2	Ground					27 779.14	.10	256
	871	5, 2← 4, 2	1,±1					28 833.81	.20	256
	871	5, 2← 4, 2	1,∓1					28 833.81	.20	256
	871	5, 3← 4, 3	Ground					27 778.76	.10	256
	871	5, 3← 4, 3	1,∓1					28 833.22	.20	256
	871	5, 3← 4, 3	1,±1					28 834.20	.20	256
	871	5, 4← 4, 4	Ground					27 778.32	.10	256
	871	5, 4← 4, 4	1,∓1					28 832.	1.	256
	871	5, 4← 4, 4	1,±1					28 834.20	.20	256
	871	6, 0← 5, 0	Ground					34 535.09	.10	256
	871	6, 1← 5, 1	Ground					34 535.09	.10	256
	871	6, 2← 5, 2	Ground					34 534.86	.10	256
	871	6, 3← 5, 3	Ground					34 534.47	.10	256

1. When two *l*-doubling measured frequencies have the same designations, the larger has been reported with $l = +1$, and the smaller with $l = -1$.

206

Isotopic Species	Id. No.	Rotational Quantum Nos.	Vib. State v_a^l	F_1'	F'	F_1	F	Frequency MHz	Acc. ±MHz	Ref.
$C^{12}F_3^{19}C^{12}{:}C^{12}H$	871	6, 4← 5, 4	Ground					34 533.91	.10	256
	871	6, 5← 5, 5	Ground					34 533.23	.10	256
	871	9, 0← 8, 0	Ground					51 802.26	.10	256
	871	9, 0← 8, 0	1,±1					51 906.64	.20	256
	871	9, 1← 8, 1	Ground					51 802.26	.10	256
	871	9, 1← 8, 1	1,−1					51 869.14	.20	256 [1]
	871	9, 1← 8, 1	1,∓1					51 903.42	.20	256
	871	9, 1← 8, 1	1,+1					51 934.48	.20	256 [1]
	871	9, 2← 8, 2	Ground					51 801.90	.10	256
	871	9, 2← 8, 2	1,±1					51 896.86	.20	256
	871	9, 2← 8, 2	1,∓1					51 901.68	.20	256
	871	9, 3← 8, 3	Ground					51 801.32	.10	256
	871	9, 3← 8, 3	1,±1					51 899.44	.20	256
	871	9, 3← 8, 3	1,∓1					51 899.99	.20	256
	871	9, 4← 8, 4	Ground					51 800.54	.10	256
	871	9, 4← 8, 4	1,∓1					51 898.61	.20	256
	871	9, 4← 8, 4	1,±1					51 899.99	.20	256
	871	9, 5← 8, 5	Ground					51 799.56	.10	256
	871	9, 5← 8, 5	1,±1					51 899.44	.20	256
	871	9, 6← 8, 6	Ground					51 798.26	.10	256
	871	9, 6← 8, 6	1,∓1					51 894.95	.20	256
	871	9, 6← 8, 6	1,±1					51 898.61	.20	256
	871	9, 7← 8, 7	Ground					51 796.78	.10	256
	871	9, 7← 8, 7	1,∓1					51 892.88	.20	256
	871	9, 7← 8, 7	1,±1					51 897.63	.20	256
	871	9, 8← 8, 8	Ground					51 795.10	.10	256
	871	9, 8← 8, 8	1,∓1					51 890.62	.20	256
$C^{12}F_3^{19}C^{13}{:}C^{12}H$	872	4, ← 3,	Ground					22 839.9	.3	315
$C^{12}F_3^{19}C^{12}{:}C^{13}H$	873	4, ← 3,	Ground					22 301.0	.3	315
$C^{12}F_3^{19}C^{12}{:}C^{12}D$	874	4, ← 3,	Ground					21 568.2	.3	315
	874	6, 0← 5, 0	Ground					32 352.62	.10	256
	874	6, 1← 5, 1	Ground					32 352.62	.10	256
	874	6, 2← 5, 2	Ground					32 352.36	.10	256
	874	6, 3← 5, 3	Ground					32 352.01	.10	256
	874	6, 4← 5, 4	Ground					32 351.47	.10	256
	874	6, 5← 5, 5	Ground					32 350.82	.10	256
	874	9, 0← 8, 0	Ground					48 528.42	.10	256
	874	9, 1← 8, 1	Ground					48 528.42	.10	256
	874	9, 2← 8, 2	Ground					48 528.08	.10	256
	874	9, 3← 8, 3	Ground					48 527.56	.10	256
	874	9, 4← 8, 4	Ground					48 526.74	.10	256
	874	9, 5← 8, 5	Ground					48 525.74	.10	256
	874	9, 6← 8, 6	Ground					48 524.54	.10	256
	874	9, 7← 8, 7	Ground					48 523.08	.10	256
	874	9, 8← 8, 8	Ground					48 521.44	.10	256

1. When two l-doubling measured frequencies have the same designations, the larger has been reported with $l=+1$, and the smaller with $l=-1$.

C₃HN $C_{\infty v}$ HC:CCN

Isotopic Species	Pt. Gp.	Id. No.	A MHz	B MHz	C MHz	D_J MHz	D_{JK} MHz	Δ Amu A²	κ
$HC^{12}:C^{12}C^{12}N^{14}$	$C_{\infty v}$	881		4549.07 M	4549.07 M				
$HC^{13}:C^{12}C^{12}N^{14}$	$C_{\infty v}$	882		4408.45 M	4408.45 M				
$HC^{12}:C^{13}C^{12}N^{14}$	$C_{\infty v}$	883		4529.84 M	4529.84 M				
$HC^{12}:C^{12}C^{13}N^{14}$	$C_{\infty v}$	884		4530.23 M	4530.23 M				
$HC^{12}:C^{12}C^{12}N^{15}$	$C_{\infty v}$	885		4416.91 M	4416.91 M				
$DC^{12}:C^{12}C^{12}N^{14}$	$C_{\infty v}$	886		4221.60 M	4221.60 M				
$DC^{13}:C^{12}C^{12}N^{14}$	$C_{\infty v}$	887		4107.21 M	4107.21 M				
$DC^{12}:C^{13}C^{12}N^{14}$	$C_{\infty v}$	888		4207.59 M	4207.59 M				
$DC^{12}:C^{12}C^{13}N^{14}$	$C_{\infty v}$	889		4202.54 M	4202.54 M				
$DC^{12}:C^{12}C^{12}N^{15}$	$C_{\infty v}$	891		4100.41 M	4100.41 M				

Id. No.	μ_a Debye	μ_b Debye	μ_c Debye	eQq Value(MHz) Rel.	eQq Value(MHz) Rel.	eQq Value(MHz) Rel.	ω_a d 1/cm	ω_b d 1/cm	ω_c d 1/cm	ω_d d 1/cm
881	3.6 M	0. X	0. X	−4.2 N^{14}						

References:

ABC: 251   μ: 251   eQq: 251

Propiolonitrile

Spectral Line Table

Isotopic Species	Id. No.	Rotational Quantum Nos.	Vib. State	F_1'	F'	F_1	F	Frequency MHz	Acc. ±MHz	Ref.
$HC^{12}:C^{12}C^{12}N^{14}$	881	2← 1	Ground					18 196.6		251
	881	3← 2	Ground		3		3	27 293.09		251
	881	3← 2	Ground		4		3	27 294.47		251
	881	3← 2	Ground		2		2	27 296.29		251
$HC^{13}:C^{12}C^{12}N^{14}$	882	3← 2	Ground					26 450.73		251
$HC^{12}:C^{13}C^{12}N^{14}$	883	3← 2	Ground					27 179.10		251
$HC^{12}:C^{12}C^{13}N^{14}$	884	3← 2	Ground					27 181.45		251
$HC^{12}:C^{12}C^{12}N^{15}$	885	3← 2	Ground					26 501.46		251
$DC^{12}:C^{12}C^{12}N^{14}$	886	3← 2	Ground					25 329.62		251
$DC^{13}:C^{12}C^{12}N^{14}$	887	3← 2	Ground					24 643.29		251
$DC^{12}:C^{13}C^{12}N^{14}$	888	3← 2	Ground					25 245.58		251
$DC^{12}:C^{12}C^{13}N^{14}$	889	3← 2	Ground					25 215.30		251
$DC^{12}:C^{12}C^{12}N^{15}$	891	3← 2	Ground					24 602.45		251

Propanedinitrile, Malonic Dinitrile, Methylene Cyanide

C₃H₂N₂   C$_{2v}$ CH$_2$(CN)$_2$

Isotopic Species	Pt. Gp.	Id. No.	A MHz	B MHz	C MHz	D$_J$ MHz	D$_{JK}$ MHz	Δ Amu A²	κ
C^{12}H$_2$(C^{12}N^{14})$_2$	C$_{2v}$	901	20 882.14 M	2 942.477 M	2 616.774 M			.3379	−.9643364
C^{12}HD(C^{12}N^{14})$_2$	C$_s$	902	18 501.73 M	2 931.189 M	2 584.910 M				−.9564889
C^{12}D$_2$(C^{12}N^{14})$_2$	C$_{2v}$	903	16 634.32 M	2 916.905 M	2 556.710 M				−.9488271
C^{12}H$_2$(C^{12}N^{14})(C^{12}N^{15})	C$_s$	904	20 639.15 M	2 863.585 M	2 550.477 M				−.9653806

Id. No.	μ$_a$ Debye	μ$_b$ Debye	μ$_c$ Debye	eQq Value(MHz) Rel.	eQq Value(MHz) Rel.	eQq Value(MHz) Rel.	ω$_a$ d 1/cm	ω$_b$ d 1/cm	ω$_c$ d 1/cm	ω$_d$ d 1/cm
901	0. X	3.735 M	0. X							

References:
ABC: 906   Δ: 906   κ: 906   μ: 906

Add. Ref. 834

For species 905, A − (B + C)/2 = 18120.8 MHz, (B − C)/2 = 160.3 MHz, Ref. 906.

Isotopic Species	Id. No.	Rotational Quantum Nos.	Vib. State	F_1'	F'	F_1	F	Frequency MHz	Acc. ±MHz	Ref.
$C^{12}H_2(C^{12}N^{14})_2$	901	1, 1, 1← 0, 0, 0	Ground					23 498.82		906
	901	1, 1, 0← 1, 0, 1	Ground					18 265.36		906
	901	2, 1, 2← 1, 0, 1	Ground					28 732.50		906
	901	2, 1, 1← 2, 0, 2	Ground					18 595.54		906
	901	3, 1, 2← 3, 0 ,3	Ground					19 098.79		906
	901	4, 1, 3← 4, 0, 4	Ground					19 785.23		906
	901	4, 2, 3← 5, 1, 4	Ground					24 093.24		906
	901	5, 1, 4← 5, 0, 5	Ground					20 667.93		906
	901	5, 2, 4← 6, 1, 5	Ground					17 595.24		906
	901	6, 0, 6← 5, 1, 5	Ground					17 414.02		906
	901	6, 1, 5← 6, 0, 6	Ground					21 761.79		906
	901	7, 0, 7← 6, 1, 6	Ground					23 738.22		906
	901	7, 1, 6← 7, 0, 7	Ground					23 084.25		906
	901	8, 1, 7← 8, 0, 8	Ground					24 656.01		906
	901	9, 1, 8← 9, 0, 9	Ground					26 497.38		906
	901	10, 1, 9←10, 0,10	Ground					28 626.87		906
	901	Not Reported	Ground					17 670.5	.1	921
	901	Not Reported	Ground					19 045.7	.1	921
	901	Not Reported	Ground					19 145.2	.1	921
	901	Not Reported	Ground					19 318.3	.1	921
	901	Not Reported	Ground					19 324.2	.1	921
	901	Not Reported	Ground					19 397.7	.1	921
	901	Not Reported	Ground					19 466.3	.1	921
	901	Not Reported	Ground					19 713.5	.1	921
	901	Not Reported	Ground					19 847.4	.1	921
	901	Not Reported	Ground					19 862.9	.1	921
	901	Not Reported	Ground					19 902.9	.1	921
	901	Not Reported	Ground					20 027.1	.1	921
	901	Not Reported	Ground					20 067.5	.1	921
	901	Not Reported	Ground					20 093.5	.1	921
	901	Not Reported	Ground					20 135.7	.1	921
	901	Not Reported	Ground					20 142.8	.1	921
	901	Not Reported	Ground					20 179.7	.1	921
	901	Not Reported	Ground					20 183.1	.1	921
	901	Not Reported	Ground					20 275.9	.1	921
	901	Not Reported	Ground					20 396.8	.1	921
	901	Not Reported	Ground					20 524.9	.1	921
	901	Not Reported	Ground					20 532.9	.1	921
	901	Not Reported	Ground					20 602.8	.1	921
	901	Not Reported	Ground					20 691.	.1	921
	901	Not Reported	Ground					20 706.8	.1	921
	901	Not Reported	Ground					20 726.	.1	921
	901	Not Reported	Ground					20 738.	.1	921
	901	Not Reported	Ground					20 855.5	.1	921
	901	Not Reported	Ground					20 930.6	.1	921
	901	Not Reported	Ground					20 943.6	.1	921
	901	Not Reported	Ground					21 055.0	.1	921
	901	Not Reported	Ground					21 080.9	.1	921
	901	Not Reported	Ground					21 186.9	.1	921
	901	Not Reported	Ground					21 260.9	.1	921

Isotopic Species	Id. No.	Rotational Quantum Nos.	Vib. State	Hyperfine				Frequency MHz	Acc. ±MHz	Ref.
				F_1'	F'	F_1	F			
$C^{12}H_2(C^{12}N^{14})_2$	901	Not Reported	Ground					21 271.	.1	921
	901	Not Reported	Ground					21 371.5	.1	921
	901	Not Reported	Ground					21 400.6	.1	921
	901	Not Reported	Ground					21 431.9	.1	921
	901	Not Reported	Ground					21 508.8	.1	921
	901	Not Reported	Ground					21 682.7	.1	921
	901	Not Reported	Ground					21 858.	.1	921
	901	Not Reported	Ground					21 981.2	.1	921
	901	Not Reported	Ground					22 037.0	.1	921
	901	Not Reported	Ground					22 051.2	.1	921
	901	Not Reported	Ground					22 110.5	.1	921
	901	Not Reported	Ground					22 173.6	.1	921
	901	Not Reported	Ground					22 224.1	.1	921
	901	Not Reported	Ground					22 290.3	.1	921
	901	Not Reported	Ground					22 293.8	.1	921
	901	Not Reported	Ground					22 327.0	.1	921
	901	Not Reported	Ground					22 356.5	.1	921
	901	Not Reported	Ground					22 380.5	.1	921
	901	Not Reported	Ground					22 383.9	.1	921
	901	Not Reported	Ground					22 389.9	.1	921
	901	Not Reported	Ground					22 417.3	.1	921
	901	Not Reported	Ground					22 446.6	.1	921
	901	Not Reported	Ground					22 477.9	.1	921
	901	Not Reported	Ground					22 598.5	.1	921
	901	Not Reported	Ground					22 616.	.1	921
	901	Not Reported	Ground					22 652.9	.1	921
	901	Not Reported	Ground					22 663.	.1	921
	901	Not Reported	Ground					22 719.7	.1	921
	901	Not Reported	Ground					22 786.4	.1	921
	901	Not Reported	Ground					22 809.1	.1	921
	901	Not Reported	Ground					22 852.9	.1	921
	901	Not Reported	Ground					22 905.0	.1	921
	901	Not Reported	Ground					22 913.4	.1	921
	901	Not Reported	Ground					22 918.7	.1	921
	901	Not Reported	Ground					22 921.2	.1	921
	901	Not Reported	Ground					22 950.2	.1	921
	901	Not Reported	Ground					22 980.3	.1	921
	901	Not Reported	Ground					23 098.1	.1	921
	901	Not Reported	Ground					23 113.3	.1	921
	901	Not Reported	Ground					23 162.	.1	921
	901	Not Reported	Ground					23 210.	.1	921
	901	Not Reported	Ground					23 224.7	.1	921
	901	Not Reported	Ground					23 248.5	.1	921
	901	Not Reported	Ground					23 284.2	.1	921
	901	Not Reported	Ground					23 318.2	.1	921
	901	Not Reported	Ground					23 349.5	.1	921
	901	Not Reported	Ground					23 379.	.1	921
	901	Not Reported	Ground					23 406.8	.1	921
	901	Not Reported	Ground					23 414.8	.1	921
	901	Not Reported	Ground					23 420.5	.1	921

Isotopic Species	Id. No.	Rotational Quantum Nos.	Vib. State	Hyperfine				Frequency MHz	Acc. ±MHz	Ref.
				F_1'	F'	F_1	F			
$C^{12}H_2(C^{12}N^{14})_2$	901	Not Reported	Ground					23 432.	.1	921
	901	Not Reported	Ground					23 456.	.1	921
	901	Not Reported	Ground					23 463.6	.1	921
	901	Not Reported	Ground					23 563.2	.1	921
	901	Not Reported	Ground					23 577.0	.1	921
	901	Not Reported	Ground					23 591.	.1	921
	901	Not Reported	Ground					23 601.5	.1	921
	901	Not Reported	Ground					23 638.	.1	921
	901	Not Reported	Ground					23 667.9	.1	921
	901	Not Reported	Ground					23 723.0	.1	921
	901	Not Reported	Ground					23 738.2	.1	921
	901	Not Reported	Ground					23 769.5	.1	921
	901	Not Reported	Ground					23 780.5	.1	921
	901	Not Reported	Ground					23 803.5	.1	921
	901	Not Reported	Ground					23 825.5	.1	921
	901	Not Reported	Ground					23 856.2	.1	921
	901	Not Reported	Ground					23 925.5	.1	921
	901	Not Reported	Ground					23 947.6	.1	921
	901	Not Reported	Ground					23 964.	.1	921
	901	Not Reported	Ground					23 985.	.1	921
	901	Not Reported	Ground					24 003.2	.1	921
	901	Not Reported	Ground					24 031.7	.1	921
	901	Not Reported	Ground					24 048.	.1	921
	901	Not Reported	Ground					24 120.8	.1	921
	901	Not Reported	Ground					24 126.4	.1	921
	901	Not Reported	Ground					24 138.8	.1	921
	901	Not Reported	Ground					24 191.0	.1	921
	901	Not Reported	Ground					24 200.0	.1	921
	901	Not Reported	Ground					24 207.6	.1	921
	901	Not Reported	Ground					24 240.8	.1	921
	901	Not Reported	Ground					24 248.5	.1	921
	901	Not Reported	Ground					24 267.9	.1	921
	901	Not Reported	Ground					24 298.2	.1	921
	901	Not Reported	Ground					24 313.3	.1	921
	901	Not Reported	Ground					24 369.4	.1	921
	901	Not Reported	Ground					24 371.6	.1	921
	901	Not Reported	Ground					24 438.0	.1	921
	901	Not Reported	Ground					24 445.2	.1	921
	901	Not Reported	Ground					24 490.9	.1	921
	901	Not Reported	Ground					24 507.3	.1	921
	901	Not Reported	Ground					24 539.2	.1	921
	901	Not Reported	Ground					24 571.3	.1	921
	901	Not Reported	Ground					24 607.5	.1	921
	901	Not Reported	Ground					24 638.7	.1	921
	901	Not Reported	Ground					24 641.2	.1	921
	901	Not Reported	Ground					24 670.0	.1	921
	901	Not Reported	Ground					24 764.3	.1	921
	901	Not Reported	Ground					24 824.0	.1	921
	901	Not Reported	Ground					24 830.1	.1	921
	901	Not Reported	Ground					24 852.9	.1	921

Isotopic Species	Id. No.	Rotational Quantum Nos.	Vib. State	Hyperfine				Frequency MHz	Acc. ±MHz	Ref.
				F_1'	F'	F_1	F			
$C^{12}H_2(C^{12}N^{14})_2$	901	Not Reported	Ground					24 917.5	.1	921
	901	Not Reported	Ground					24 955.5	.1	921
	901	Not Reported	Ground					25 008.8	.1	921
	901	Not Reported	Ground					25 013.1	.1	921
	901	Not Reported	Ground					25 018.4	.1	921
	901	Not Reported	Ground					25 024.1	.1	921
	901	Not Reported	Ground					25 096.8	.1	921
	901	Not Reported	Ground					25 105.0	.1	921
	901	Not Reported	Ground					25 128.4	.1	921
	901	Not Reported	Ground					25 156.6	.1	921
	901	Not Reported	Ground					25 162.1	.1	921
	901	Not Reported	Ground					25 186.7	.1	921
	901	Not Reported	Ground					25 197.3	.1	921
	901	Not Reported	Ground					25 347.8	.1	921
	901	Not Reported	Ground					25 895.4	.1	921
	901	Not Reported	Ground					25 926.1	.1	921
	901	Not Reported	Ground					25 947.9	.1	921
	901	Not Reported	Ground					25 962.7	.1	921
	901	Not Reported	Ground					26 069.9	.1	921
	901	Not Reported	Ground					26 194.3	.1	921
	901	Not Reported	Ground					26 398.4	.1	921
	901	Not Reported	Ground					26 554.5	.1	921
	901	Not Reported	Ground					26 560.6	.1	921
	901	Not Reported	Ground					26 619.6	.1	921
	901	Not Reported	Ground					26 906.3	.1	921
	901	Not Reported	Ground					27 072.7	.1	921
	901	Not Reported	Ground					27 244.4	.1	921
	901	Not Reported	Ground					27 281.0	.1	921
	901	Not Reported	Ground					27 309.6	.1	921
	901	Not Reported	Ground					27 707.3	.1	921
	901	Not Reported	Ground					27 804.7	.1	921
	901	Not Reported	Ground					27 842.7	.1	921
	901	Not Reported	Ground					28 368.5	.1	921
	901	Not Reported	Ground					28 409.8	.1	921
	901	Not Reported	Ground					28 453.7	.1	921
	901	Not Reported	Ground					28 480.5	.1	921
	901	Not Reported	Ground					28 509.6	.1	921
	901	Not Reported	Ground					28 541.7	.1	921
	901	Not Reported	Ground					28 605.5	.1	921
	901	Not Reported	Ground					28 612.1	.1	921
	901	Not Reported	Ground					28 648.3	.1	921
	901	Not Reported	Ground					28 697.7	.1	921
	901	Not Reported	Ground					28 850.	.1	921
	901	Not Reported	Ground					28 892.4	.1	921
	901	Not Reported	Ground					28 910.9	.1	921
	901	Not Reported	Ground					28 961.9	.1	921
	901	Not Reported	Ground					28 967.2	.1	921
	901	Not Reported	Ground					28 976.1	.1	921
	901	Not Reported	Ground					28 986.9	.1	921
	901	Not Reported	Ground					28 989.2	.1	921

Isotopic Species	Id. No.	Rotational Quantum Nos.	Vib. State	Hyperfine F_1'	F'	F_1	F	Frequency MHz	Acc. ±MHz	Ref.
$C^{12}H_2(C^{12}N^{14})_2$	901	Not Reported	Ground					29 011.0	.1	921
	901	Not Reported	Ground					29 039.5	.1	921
	901	Not Reported	Ground					29 091.7	.1	921
	901	Not Reported	Ground					29 155.5	.1	921
	901	Not Reported	Ground					29 167.6	.1	921
	901	Not Reported	Ground					29 179.	.1	921
	901	Not Reported	Ground					29 188.	.1	921
	901	Not Reported	Ground					29 218.	.1	921
	901	Not Reported	Ground					29 238.6	.1	921
	901	Not Reported	Ground					29 257.5	.1	921
	901	Not Reported	Ground					29 277.2	.1	921
	901	Not Reported	Ground					29 290.	.1	921
	901	Not Reported	Ground					29 312.7	.1	921
	901	Not Reported	Ground					29 332.5	.1	921
	901	Not Reported	Ground					29 351.	.1	921
	901	Not Reported	Ground					29 376.7	.1	921
	901	Not Reported	Ground					29 403.3	.1	921
	901	Not Reported	Ground					29 416.5	.1	921
	901	Not Reported	Ground					29 484.8	.1	921
	901	Not Reported	Ground					29 509.3	.1	921
	901	Not Reported	Ground					29 550.6	.1	921
	901	Not Reported	Ground					29 582.3	.1	921
	901	Not Reported	Ground					29 586.0	.1	921
	901	Not Reported	Ground					29 599.3	.1	921
	901	Not Reported	Ground					29 708.9	.1	921
	901	Not Reported	Ground					29 719.	.1	921
	901	Not Reported	Ground					29 783.9	.1	921
	901	Not Reported	Ground					29 815.1	.1	921
	901	Not Reported	Ground					29 990.3	.1	921
$C^{12}HD(C^{12}N^{14})_2$	902	1, 1, 1← 0, 0, 0	Ground					21 086.78		906
	902	1, 1, 0← 1, 0, 1	Ground					15 916.73		906
	902	2, 1, 2← 1, 0, 1	Ground					26 256.39		906
	902	2, 1, 1← 2, 0, 2	Ground					16 268.84		906
	902	3, 1, 2← 3, 0, 3	Ground					16 807.95		906
	902	4, 1, 3← 4, 0, 4	Ground					17 545.41		906
	902	5, 1, 4← 5, 0, 5	Ground					18 498.82		906
	902	6, 1, 5← 6, 0, 6	Ground					19 688.76		906
	902	7, 1, 6← 7, 0, 7	Ground					21 135.78		906
	902	8, 1, 7← 8, 0, 8	Ground					22 864.92		906
	902	9, 1, 8← 9, 0, 9	Ground					24 899.76		906
	902	10, 1, 9←10, 0,10	Ground					27 261.51		906
$C^{12}D_2(C^{12}N^{14})_2$	903	1, 1, 1← 0, 0, 0	Ground					19 191.00		906
	903	1, 1, 0← 1, 0, 1	Ground					14 077.66		906
	903	2, 1, 2← 1, 0, 1	Ground					24 304.47		906
	903	2, 1, 1← 2, 0, 2	Ground					14 444.84		906
	903	3, 1, 2← 3, 0, 3	Ground					15 008.56		906
	903	4, 1, 3← 4, 0, 4	Ground					15 784.72		906
	903	5, 1, 4← 5, 0, 5	Ground					16 791.92		906
	903	6, 1, 5← 6, 0, 6	Ground					18 055.72		906

Isotopic Species	Id. No.	Rotational Quantum Nos.	Vib. State	F_1'	F'	F_1	F	Frequency MHz	Acc. ±MHz	Ref.
$C^{12}D_2(C^{12}N^{14})_2$	903	7, 1, 6← 7, 0, 7	Ground					19 601.55		906
	903	8, 1, 7← 8, 0, 8	Ground					21 456.12		906
	903	9, 1, 8← 9, 0, 9	Ground					23 645.34		906
	903	10, 1, 9←10, 0,10	Ground					26 189.70		906
$C^{12}H_2(C^{12}N^{14})(C^{12}N^{15})$	904	1, 1, 1← 0, 0, 0	Ground					23 189.52		906
	904	1, 1, 0← 1, 0, 1	Ground					18 088.67		906
	904	2, 1, 2← 1, 0, 1	Ground					28 290.63		906
	904	2, 1, 1← 2, 0, 2	Ground					18 405.56		906
	904	3, 1, 2← 3, 0, 3	Ground					18 889.37		906
	904	4, 1, 3← 4, 0, 4	Ground					19 548.66		906
	904	5, 1, 4← 5, 0, 5	Ground					20 394.99		906
	904	6, 1, 5← 6, 0, 6	Ground					21 443.07		906
	904	7, 1, 6← 7, 0, 7	Ground					22 709.58		906
	904	8, 1, 7← 8, 0, 8	Ground					24 213.27		906
	904	9, 1, 8← 9, 0, 9	Ground					25 973.43		906
	904	10, 1, 9←10, 0,10	Ground					28 008.57		906
$C^{12}H_2(C^{12}N^{14})(C^{13}N^{14})$	905	9, 1, 8← 9, 0, 9	Ground					26 367.03		906
	905	10, 1, 9←10, 0,10	Ground					28 456.41		906

---

910 — Propiolaldehyde   Molecular Constant Table
        Propynal, Propargyl Aldehyde

C_3H_2O C_s HC:CCHO

Isotopic Species	Pt. Gp.	Id. No.	A MHz	B MHz	C MHz	D_J MHz	D_{JK} MHz	Δ Amu A²	κ
$HC^{12}:C^{12}C^{12}HO^{16}$	C_s	911	68 026.60 M	4 826.223 M	4 499.612 M	.0020	−.141	.1718	−.98812
$HC^{13}:C^{12}C^{12}HO^{16}$	C_s	912		4 667.399 M	4 360.354 M	.0020	−.141		
$HC^{12}:C^{13}C^{12}HO^{16}$	C_s	913		4 802.704 M	4 478.829 M	.0020	−.141		
$HC^{12}:C^{12}C^{13}HO^{16}$	C_s	914		4 805.335 M	4 473.455 M	.0020	−.141		
$HC^{12}:C^{12}C^{12}HO^{18}$	C_s	915		4 612.594 M	4 304.669 M	.0020	−.141		
$DC^{12}:C^{12}C^{12}HO^{16}$	C_s	916	66 768.43 M	4 463.771 M	4 177.876 M	.0020	−.141	.1785	
$DC^{13}:C^{12}C^{12}HO^{16}$	C_s	917		4 334.699 M	4 064.104 M	.0020	−.141		
$DC^{12}:C^{13}C^{12}HO^{16}$	C_s	918		4 446.560 M	4 162.440 M	.0020	−.141		
$DC^{12}:C^{12}C^{13}HO^{16}$	C_s	919		4 442.680 M	4 152.580 M	.0020			
$HC^{12}:C^{12}C^{12}DO^{16}$	C_s	921	51 764.46 M	4 791.439 M	4 378.764 M	.0020	−.141	.1775	
$HC^{13}:C^{12}C^{12}DO^{16}$	C_s	922		4 631.907 M	4 244.209 M	.0020			
$HC^{12}:C^{13}C^{12}DO^{16}$	C_s	923		4 767.282 M	4 358.333 M	.0020			
$HC^{12}:C^{12}C^{13}DO^{16}$	C_s	924		4 771.445 M	4 355.685 M	.0020			
$HC^{12}:C^{12}C^{12}DO^{18}$	C_s	925		4 586.252 M	4 195.843 M	.0020			
$DC^{12}:C^{12}C^{12}DO^{16}$	C_s	926	51 074.93 M	4 429.099 M	4 069.604 M	.0020	−.141	.1848	

Id. No.	μ_a Debye	μ_b Debye	μ_c Debye	eQq Value(MHz) Rel.		eQq Value(MHz) Rel.		eQq Value(MHz) Rel.		ω_a d 1/cm	ω_b d 1/cm	ω_c d 1/cm	ω_d d 1/cm
911	2.39 M	.60 M	0. X							150 1	230 1		

References:

ABC: 850 D_J: 850 D_{JK}: 850   Δ: 850   κ: 612   μ: 612   ω: 612

For species 911, $v_{11} = 1$, B = 4848.84 MHz, C = 4512.41 MHz; for $v_{12} = 1$, B = 4834.39 MHz, C = 4515.52 MHz. Ref. 612.

Isotopic Species	Id. No.	Rotational Quantum Nos.	Vib. State v_a ; v_b	Hyperfine				Frequency MHz	Acc. ±MHz	Ref.
				F_1'	F'	F_1	F			
$HC^{12} \colon C^{12}C^{12}HO^{16}$	911	2, 0, 2← 1, 0, 1	Ground					18 650.33	.02	850
	911	2, 1, 1← 1, 1, 0	Ground					18 978.78	.02	850
	911	2, 1, 2← 1, 1, 1	Ground					18 325.56	.02	850
	911	2, 1, 2← 3, 0, 3	Ground					34 903.64	.02	850
	911	3, 0, 3← 2, 0, 2	0; 1					27 077.49	.15	612
	911	3, 0, 3← 2, 0, 2	Ground					27 972.13	.02	850
	911	3, 0, 3← 2, 0, 2	1; 0					28 044.06	.15	612
	911	3, 1, 2← 2, 1, 1	Ground					28 467.15	.02	850
	911	3, 1, 2← 2, 1, 1	1; 0					28 527.18	.15	612
	911	3, 1, 2← 2, 1, 1	0; 1					28 587.40	.15	612
	911	3, 1, 3← 2, 1, 2	Ground					27 487.48	.02	850
	911	3, 1, 3← 2, 1, 2	1; 0					27 570.57	.15	612
	911	3, 1, 3← 2, 1, 2	0; 1					27 578.10	.15	612
	911	3, 2, 1← 2, 2, 0	Ground					27 985.53	.02	850
	911	3, 2, 2← 2, 2, 1	Ground					27 980.86	.02	850
	911	3, 1, 3← 4, 0, 4	Ground					25 100.65	.02	850
	911	4, 1, 3← 3, 1, 2	1; 0					37 034.77	.15	612
	911	4, 1, 3← 3, 1, 2	Ground					37 954.95	.10	612
	911	4, 1, 3← 3, 1, 2	0; 1					38 144.42	.15	612
	911	4, 1, 4← 5, 0, 5	Ground					15 146.06	.02	850
	911	9, 0, 9← 8, 1, 8	Ground					26 074.66	.02	850
$HC^{13} \colon C^{12}C^{12}HO^{16}$	912	2, 0, 2← 1, 0, 1	Ground					18 054.28	.02	850
	912	2, 1, 1← 1, 1, 0	Ground					18 363.05	.02	850
	912	2, 1, 2← 1, 1, 1	Ground					17 748.96	.02	850
	912	3, 0, 3← 2, 0, 2	Ground					27 078.64	.02	850
	912	3, 1, 2← 2, 1, 1	Ground					27 543.73	.02	850
	912	3, 1, 3← 2, 1, 2	Ground					26 622.49	.02	850
	912	3, 2, 1← 2, 2, 0	Ground					27 090.94	.02	850
	912	3, 2, 2← 2, 2, 1	Ground					27 086.60	.02	850
$HC^{12} \colon C^{13}C^{12}HO^{16}$	913	2, 0, 2← 1, 0, 1	Ground					18 561.73	.02	850
	913	2, 1, 1← 1, 1, 0	Ground					18 887.44	.02	850
	913	2, 1, 2← 1, 1, 1	Ground					18 239.69	.02	850
	913	3, 0, 3← 2, 0, 2	Ground					27 839.50	.02	850
	913	3, 1, 2← 2, 1, 1	Ground					28 330.10	.02	850
	913	3, 1, 3← 2, 1, 2	Ground					27 358.89	.02	850
	913	3, 2, 1← 2, 2, 0	Ground					27 852.84	.02	850
	913	3, 2, 2← 2, 2, 1	Ground					27 847.98	.02	850
$HC^{12} \colon C^{12}C^{13}HO^{16}$	914	2, 0, 2← 1, 0, 1	Ground					18 556.16	.02	850
	914	2, 1, 1← 1, 1, 0	Ground					18 889.96	.02	850
	914	2, 1, 2← 1, 1, 1	Ground					18 226.20	.02	850
	914	3, 0, 3← 2, 0, 2	Ground					27 831.02	.02	850
	914	3, 1, 2← 2, 1, 1	Ground					28 334.12	.02	850
	914	3, 1, 3← 2, 1, 2	Ground					27 338.45	.02	850
	914	3, 2, 1← 2, 2, 0	Ground					27 844.65	.02	850
	914	3, 2, 2← 2, 2, 1	Ground					27 839.37	.02	850
$HC^{12} \colon C^{12}C^{12}HO^{18}$	915	2, 0, 2← 1, 0, 1	Ground					17 833.28	.05	850
	915	2, 1, 1← 1, 1, 0	Ground					18 142.95	.05	850
	915	2, 1, 2← 1, 1, 1	Ground					17 527.10	.05	850

Isotopic Species	Id. No.	Rotational Quantum Nos.	Vib. State v_a ; v_b	Hyperfine				Frequency MHz	Acc. ±MHz	Ref.
				F_1'	F'	F_1	F			
HC^{12}:C^{12}C^{12}HO^{18}	915	3, 0, 3← 2, 0, 2	Ground					26 747.54	.05	850
	915	3, 1, 2← 2, 1, 1	Ground					27 213.69	.05	850
	915	3, 1, 3← 2, 1, 2	Ground					26 289.67	.05	850
	915	3, 2, 1← 2, 2, 0	Ground					26 759.87	.05	850
	915	3, 2, 2← 2, 2, 1	Ground					26 755.15	.05	850
DC^{12}:C^{12}C^{12}HO^{16}	916	2, 0, 2← 1, 0, 1	Ground					17 282.27	.02	850
	916	2, 1, 1← 1, 1, 0	Ground					17 569.69	.02	850
	916	2, 1, 2← 1, 1, 1	Ground					16 997.90	.02	850
	916	2, 1, 2← 3, 0, 3	Ground					36 099.79	.02	850
	916	3, 0, 3← 2, 0, 2	Ground					25 920.83	.02	850
	916	3, 1, 2← 2, 1, 1	Ground					26 353.89	.02	850
	916	3, 1, 3← 2, 1, 2	Ground					25 496.13	.02	850
	916	3, 2, 1← 2, 2, 0	Ground					25 931.96	.02	850
	916	3, 2, 2← 2, 2, 1	Ground					25 928.19	.02	850
	916	3, 1, 3← 4, 0, 4	Ground					27 039.67	.02	850
	916	4, 1, 4← 5, 0, 5	Ground					17 845.34	.02	850
DC^{13}:C^{12}C^{12}HO^{16}	917	2, 0, 2← 1, 0, 1	Ground					16 796.64	.02	850
	917	2, 1, 1← 1, 1, 0	Ground					17 068.70	.02	850
	917	2, 1, 2← 1, 1, 1	Ground					16 527.51	.02	850
	917	3, 0, 3← 2, 0, 2	Ground					25 192.85	.02	850
	917	3, 1, 2← 2, 1, 1	Ground					25 602.50	.02	850
	917	3, 1, 3← 2, 1, 2	Ground					24 790.45	.02	850
	917	3, 2, 1← 2, 2, 0	Ground					25 202.96	.02	850
	917	3, 2, 2← 2, 2, 1	Ground					25 199.51	.02	850
DC^{12}:C^{13}C^{12}HO^{16}	918	2, 0, 2← 1, 0, 1	Ground					17 216.98	.02	850
	918	2, 1, 1← 1, 1, 0	Ground					17 502.62	.02	850
	918	2, 1, 2← 1, 1, 1	Ground					16 934.38	.02	850
	918	3, 0, 3← 2, 0, 2	Ground					25 822.90	.02	850
	918	3, 1, 2← 2, 1, 1	Ground					26 253.29	.02	850
	918	3, 1, 3← 2, 1, 2	Ground					25 400.78	.02	850
	918	3, 2, 1← 2, 2, 0	Ground					25 833.92	.02	850
	918	3, 2, 2← 2, 2, 1	Ground					25 830.23	.02	850
DC^{12}:C^{12}C^{13}HO^{16}	919	2, 0, 2← 1, 0, 1	Ground					17 189.39	.02	850
	919	2, 1, 1← 1, 1, 0	Ground					17 481.12	.02	850
	919	2, 1, 2← 1, 1, 1	Ground					16 900.92	.02	850
	919	3, 0, 3← 2, 0, 2	Ground					25 781.36	.02	850
	919	3, 1, 2← 2, 1, 1	Ground					26 221.04	.02	850
	919	3, 1, 3← 2, 1, 2	Ground					25 350.41	.02	850
	919	3, 2, 1← 2, 2, 0	Ground					25 792.59	.02	850
	919	3, 2, 2← 2, 2, 1	Ground					25 788.78	.02	850
HC^{12}:C^{12}C^{12}DO^{16}	921	1, 1, 1← 2, 0, 2	Ground					28 635.67	.02	850
	921	2, 0, 2← 1, 0, 1	Ground					18 337.94	.02	850
	921	2, 1, 1← 1, 1, 0	Ground					18 753.58	.02	850
	921	2, 1, 2← 1, 1, 1	Ground					17 928.23	.02	850
	921	2, 1, 2← 3, 0, 3	Ground					19 063.79	.02	850
	921	3, 1, 3← 4, 0, 4	Ground					9 300.33	.02	850
	921	6, 0, 6← 5, 1, 5	Ground					10 763.74	.02	850

Isotopic Species	Id. No.	Rotational Quantum Nos.	Vib. State $v_a ; v_b$	Hyperfine				Frequency MHz	Acc. ±MHz	Ref.
				F_1'	F'	F_1	F			
$HC^{13}:C^{12}C^{12}DO^{16}$	922	2, 1, 1← 1, 1, 0	Ground					18 140.43	.02	850
	922	2, 1, 2← 1, 1, 1	Ground					17 365.03	.02	850
	922	Not Reported						17 265.	5.	622
	922	Not Reported						17 325.	5.	622
	922	Not Reported						17 545.	5.	622
	922	Not Reported						17 615.	5.	622
	922	Not Reported						17 637.	5.	622
	922	Not Reported						17 685.	5.	622
	922	Not Reported						18 008.	5.	622
	922	Not Reported						18 100.	5.	622
	922	Not Reported						18 142.	5.	622
	922	Not Reported						18 215.	5.	622
	922	Not Reported						18 240.	5.	622
	922	Not Reported						18 268.	5.	622
	922	Not Reported						18 301.	5.	622
	922	Not Reported						18 340.	5.	622
	922	Not Reported						18 400.	5.	622
	922	Not Reported						18 441.	5.	622
	922	Not Reported						18 530.	5.	622
	922	Not Reported						18 559.	5.	622
	922	Not Reported						18 837.	5.	622
	922	Not Reported						18 869.	5.	622
	922	Not Reported						19 001.	5.	622
	922	Not Reported						19 164.	5.	622
	922	Not Reported						19 202.	5.	622
	922	Not Reported						19 230.	5.	622
	922	Not Reported						19 282.	5.	622
	922	Not Reported						19 342.	5.	622
	922	Not Reported						19 380.	5.	622
	922	Not Reported						19 647.	5.	622
	922	Not Reported						19 878.	5.	622
	922	Not Reported						19 955.	5.	622
	922	Not Reported						20 098.	5.	622
	922	Not Reported						20 142.	5.	622
	922	Not Reported						20 223.	5.	622
	922	Not Reported						20 254.	5.	622
	922	Not Reported						20 287.	5.	622
	922	Not Reported						20 420.	5.	622
	922	Not Reported						20 447.	5.	622
	922	Not Reported						20 512.	5.	622
	922	Not Reported						20 716.	5.	622
	922	Not Reported						20 805.	5.	622
	922	Not Reported						20 819.	5.	622
	922	Not Reported						20 862.	5.	622
	922	Not Reported						20 979.	5.	622
	922	Not Reported						21 005.	5.	622
	922	Not Reported						21 032.	5.	622
	922	Not Reported						21 105.	5.	622
	922	Not Reported						21 144.	5.	622
	922	Not Reported						21 158.	5.	622

Isotopic Species	Id. No.	Rotational Quantum Nos.	Vib. State v_a ; v_b	Hyperfine				Frequency MHz	Acc. ±MHz	Ref.
				F_1'	F'	F_1	F			
$HC^{13}:C^{12}C^{12}DO^{16}$	922	Not Reported						21 165.	5.	622
	922	Not Reported						21 434.	5.	622
	922	Not Reported						21 461.	5.	622
	922	Not Reported						21 548.	5.	622
	922	Not Reported						21 618.	5.	622
	922	Not Reported						21 642.	5.	622
	922	Not Reported						21 650.	5.	622
	922	Not Reported						21 678.	5.	622
	922	Not Reported						21 707.	5.	622
	922	Not Reported						21 725.	5.	622
	922	Not Reported						21 748.	5.	622
	922	Not Reported						21 755.	5.	622
	922	Not Reported						21 783.	5.	622
	922	Not Reported						21 804.	5.	622
	922	Not Reported						21 807.	5.	622
	922	Not Reported						21 817.	5.	622
	922	Not Reported						21 858.	5.	622
	922	Not Reported						21 878.	5.	622
	922	Not Reported						21 921.	5.	622
	922	Not Reported						21 958.	5.	622
	922	Not Reported						21 970.	5.	622
	922	Not Reported						21 987.	5.	622
	922	Not Reported						21 998.	5.	622
	922	Not Reported						22 057.	5.	622
	922	Not Reported						22 108.	5.	622
	922	Not Reported						22 169.	5.	622
	922	Not Reported						22 191.	5.	622
	922	Not Reported						22 203.	5.	622
	922	Not Reported						22 212.	5.	622
	922	Not Reported						22 234.	5.	622
	922	Not Reported						22 266.	5.	622
	922	Not Reported						22 304.	5.	622
	922	Not Reported						22 309.	5.	622
	922	Not Reported						22 313.	5.	622
	922	Not Reported						22 331.	5.	622
	922	Not Reported						22 359.	5.	622
	922	Not Reported						22 427.	5.	622
	922	Not Reported						22 446.	5.	622
	922	Not Reported						22 451.	5.	622
	922	Not Reported						22 458.	5.	622
	922	Not Reported						22 463.	5.	622
	922	Not Reported						22 467.	5.	622
	922	Not Reported						22 532.	5.	622
	922	Not Reported						22 585.	5.	622
	922	Not Reported						22 635.	5.	622
	922	Not Reported						22 688.	5.	622
	922	Not Reported						22 735.	5.	622
	922	Not Reported						22 747.	5.	622
	922	Not Reported						22 842.	5.	622
	922	Not Reported						22 856.	5.	622

Isotopic Species	Id. No.	Rotational Quantum Nos.	Vib. State v_a ; v_b	Hyperfine				Frequency MHz	Acc. ±MHz	Ref.
				F_1'	F'	F_1	F			
$HC^{13}\!:\!C^{12}C^{12}DO^{16}$	922	Not Reported						22 972.	5.	622
	922	Not Reported						23 053.	5.	622
	922	Not Reported						23 106.	5.	622
	922	Not Reported						23 126.	5.	622
	922	Not Reported						23 179.	5.	622
	922	Not Reported						23 184.	5.	622
	922	Not Reported						23 190.	5.	622
	922	Not Reported						23 219.	5.	622
	922	Not Reported						23 245.	5.	622
	922	Not Reported						23 276.	5.	622
	922	Not Reported						23 339.	5.	622
	922	Not Reported						23 393.	5.	622
	922	Not Reported						23 427.	5.	622
	922	Not Reported						23 448.	5.	622
	922	Not Reported						23 455.	5.	622
	922	Not Reported						23 499.	5.	622
	922	Not Reported						23 529.	5.	622
	922	Not Reported						23 587.	5.	622
	922	Not Reported						23 613.	5.	622
	922	Not Reported						23 656.	5.	622
	922	Not Reported						23 670.	5.	622
	922	Not Reported						23 720.	5.	622
	922	Not Reported						23 767.	5.	622
	922	Not Reported						23 775.	5.	622
	922	Not Reported						23 812.	5.	622
	922	Not Reported						23 843.	5.	622
	922	Not Reported						23 870.	5.	622
	922	Not Reported						23 881.	5.	622
	922	Not Reported						23 898.	5.	622
	922	Not Reported						23 930.	5.	622
	922	Not Reported						23 948.	5.	622
	922	Not Reported						23 968.	5.	622
	922	Not Reported						23 980.	5.	622
	922	Not Reported						23 989.	5.	622
	922	Not Reported						24 028.	5.	622
	922	Not Reported						24 045.2		622
	922	Not Reported						24 089.	5.	622
	922	Not Reported						24 138.	5.	622
	922	Not Reported						24 166.	5.	622
	922	Not Reported						24 226.	5.	622
	922	Not Reported						24 232.	5.	622
	922	Not Reported						24 302.	5.	622
	922	Not Reported						24 338.	5.	622
	922	Not Reported						24 376.5		622
	922	Not Reported						24 447.4		622
	922	Not Reported						24 480.0		622
	922	Not Reported						24 532.	5.	622
	922	Not Reported						24 578.6		622
	922	Not Reported						24 640.1		622
	922	Not Reported						24 707.	5.	622

Isotopic Species	Id. No.	Rotational Quantum Nos.	Vib. State v_a ; v_b	Hyperfine				Frequency MHz	Acc. ±MHz	Ref.
				F_1'	F'	F_1	F			
HC^{13}:$C^{12}C^{12}DO^{16}$	922	Not Reported						24 746.	5.	622
	922	Not Reported						24 772.	5.	622
	922	Not Reported						24 783.	5.	622
	922	Not Reported						24 812.	5.	622
	922	Not Reported						24 847.	5.	622
	922	Not Reported						24 910.	5.	622
	922	Not Reported						24 926.	5.	622
	922	Not Reported						24 963.	5.	622
	922	Not Reported						24 968.	5.	622
	922	Not Reported						24 983.	5.	622
	922	Not Reported						25 025.	5.	622
	922	Not Reported						25 057.	5.	622
	922	Not Reported						25 065.	5.	622
	922	Not Reported						25 087.	5.	622
	922	Not Reported						25 103.	5.	622
	922	Not Reported						25 111.	5.	622
	922	Not Reported						25 203.	5.	622
	922	Not Reported						25 261.	5.	622
	922	Not Reported						25 273.	5.	622
	922	Not Reported						25 402.	5.	622
	922	Not Reported						25 467.	5.	622
	922	Not Reported						25 475.	5.	622
	922	Not Reported						25 487.	5.	622
	922	Not Reported						25 595.	5.	622
	922	Not Reported						25 659.	5.	622
	922	Not Reported						25 722.	5.	622
	922	Not Reported						25 833.	5.	622
	922	Not Reported						25 868.	5.	622
	922	Not Reported						25 927.	5.	622
	922	Not Reported						25 940.	5.	622
	922	Not Reported						25 962.	5.	622
	922	Not Reported						25 969.	5.	622
	922	Not Reported						25 981.	5.	622
	922	Not Reported						26 012.	5.	622
	922	Not Reported						26 062.	5.	622
	922	Not Reported						26 099.	5.	622
	922	Not Reported						26 120.	5.	622
	922	Not Reported						26 133.	5.	622
	922	Not Reported						26 277.	5.	622
	922	Not Reported						26 294.	5.	622
	922	Not Reported						26 357.	5.	622
	922	Not Reported						26 403.	5.	622
	922	Not Reported						26 414.	5.	622
	922	Not Reported						26 470.	5.	622
	922	Not Reported						26 494.	5.	622
	922	Not Reported						26 650.	5.	622
	922	Not Reported						26 725.	5.	622
	922	Not Reported						26 746.	5.	622
	922	Not Reported						26 789.	5.	622
	922	Not Reported						26 827.	5.	622

Isotopic Species	Id. No.	Rotational Quantum Nos.	Vib. State v_a ; v_b	Hyperfine				Frequency MHz	Acc. ±MHz	Ref.
				F_1'	F'	F_1	F			
$HC^{13}:C^{12}C^{12}DO^{16}$	922	Not Reported						26 866.	5.	622
	922	Not Reported						26 905.	5.	622
	922	Not Reported						27 036.	5.	622
	922	Not Reported						27 069.	5.	622
	922	Not Reported						27 083.	5.	622
	922	Not Reported						27 164.	5.	622
	922	Not Reported						27 175.	5.	622
	922	Not Reported						27 188.	5.	622
	922	Not Reported						27 245.	5.	622
	922	Not Reported						27 305.	5.	622
	922	Not Reported						27 333.	5.	622
	922	Not Reported						27 346.	5.	622
	922	Not Reported						27 350.	5.	622
	922	Not Reported						27 401.	5.	622
	922	Not Reported						27 510.	5.	622
	922	Not Reported						27 530.	5.	622
	922	Not Reported						27 555.	5.	622
	922	Not Reported						27 656.	5.	622
	922	Not Reported						27 669.	5.	622
	922	Not Reported						27 712.	5.	622
	922	Not Reported						27 738.	5.	622
	922	Not Reported						27 785.	5.	622
	922	Not Reported						27 832.	5.	622
	922	Not Reported						27 875.	5.	622
	922	Not Reported						27 897.	5.	622
	922	Not Reported						27 933.	5.	622
	922	Not Reported						28 154.	5.	622
	922	Not Reported						28 185.	5.	622
	922	Not Reported						28 234.	5.	622
	922	Not Reported						28 246.	5.	622
	922	Not Reported						28 258.	5.	622
	922	Not Reported						28 282.	5.	622
	922	Not Reported						28 336.	5.	622
	922	Not Reported						28 361.	5.	622
	922	Not Reported						28 371.	5.	622
	922	Not Reported						28 407.	5.	622
	922	Not Reported						28 420.	5.	622
	922	Not Reported						28 438.	5.	622
	922	Not Reported						28 468.	5.	622
	922	Not Reported						28 530.	5.	622
	922	Not Reported						28 578.	5.	622
	922	Not Reported						28 602.	5.	622
	922	Not Reported						28 624.	5.	622
	922	Not Reported						28 702.	5.	622
	922	Not Reported						28 716.	5.	622
	922	Not Reported						28 762.	5.	622
	922	Not Reported						28 825.	5.	622
	922	Not Reported						28 830.	5.	622
	922	Not Reported						28 833.	5.	622
	922	Not Reported						28 850.	5.	622

Isotopic Species	Id. No.	Rotational Quantum Nos.	Vib. State $v_a : v_b$	Hyperfine F_1'	F'	F_1	F	Frequency MHz	Acc. ±MHz	Ref.
HC^{13}:C^{12}C^{12}DO^{16}	922	Not Reported						28 874.	5.	622
	922	Not Reported						28 906.	5.	622
	922	Not Reported						28 924.	5.	622
	922	Not Reported						28 930.7		622
	922	Not Reported						28 950.	5.	622
	922	Not Reported						29 021.	5.	622
	922	Not Reported						29 092.	5.	622
	922	Not Reported						29 198.	5.	622
	922	Not Reported						29 293.	5.	622
	922	Not Reported						29 418.	5.	622
	922	Not Reported						29 451.	5.	622
	922	Not Reported						29 488.	5.	622
	922	Not Reported						29 533.	5.	622
	922	Not Reported						29 785.	5.	622
	922	Not Reported						29 841.	5.	622
	922	Not Reported						29 918.	5.	622
	922	Not Reported						30 075.	5.	622
	922	Not Reported						30 105.	5.	622
	922	Not Reported						30 160.	5.	622
	922	Not Reported						30 190.	5.	622
	922	Not Reported						30 228.	5.	622
	922	Not Reported						30 263.	5.	622
	922	Not Reported						30 469.	5.	622
	922	Not Reported						30 491.	5.	622
	922	Not Reported						30 525.	5.	622
	922	Not Reported						30 674.	5.	622
	922	Not Reported						30 699.	5.	622
	922	Not Reported						30 705.	5.	622
	922	Not Reported						30 800.	5.	622
	922	Not Reported						30 840.	5.	622
	922	Not Reported						30 863.	5.	622
	922	Not Reported						30 903.	5.	622
	922	Not Reported						30 958.	5.	622
	922	Not Reported						30 980.	5.	622
	922	Not Reported						31 015.	5.	622
	922	Not Reported						31 050.	5.	622
	922	Not Reported						31 067.	5.	622
	922	Not Reported						31 109.	5.	622
	922	Not Reported						31 168.	5.	622
	922	Not Reported						31 192.	5.	622
	922	Not Reported						31 407.	5.	622
	922	Not Reported						31 440.	5.	622
	922	Not Reported						31 448.	5.	622
	922	Not Reported						31 465.	5.	622
	922	Not Reported						31 590.	5.	622
	922	Not Reported						31 615.	5.	622
	922	Not Reported						31 650.	5.	622
	922	Not Reported						31 658.	5.	622
	922	Not Reported						31 703.	5.	622
	922	Not Reported						31 717.	5.	622

Isotopic Species	Id. No.	Rotational Quantum Nos.	Vib. State v_a ; v_b	Hyperfine				Frequency MHz	Acc. ±MHz	Ref.
				F_1'	F'	F_1	F			
HC^{13}:C^{12}C^{12}DO^{16}	922	Not Reported						31 751.	5.	622
	922	Not Reported						31 784.	5.	622
	922	Not Reported						31 910.	5.	622
	922	Not Reported						31 921.	5.	622
	922	Not Reported						31 940.	5.	622
	922	Not Reported						31 980.	5.	622
	922	Not Reported						32 004.	5.	622
	922	Not Reported						32 193.	5.	622
	922	Not Reported						32 301.	5.	622
	922	Not Reported						32 492.	5.	622
	922	Not Reported						32 575.	5.	622
	922	Not Reported						32 592.	5.	622
	922	Not Reported						32 633.	5.	622
	922	Not Reported						32 675.	5.	622
	922	Not Reported						32 818.	5.	622
	922	Not Reported						32 886.	5.	622
	922	Not Reported						32 960.	5.	622
	922	Not Reported						32 975.	5.	622
	922	Not Reported						32 992.	5.	622
	922	Not Reported						33 062.	5.	622
	922	Not Reported						33 076.	5.	622
	922	Not Reported						33 145.	5.	622
	922	Not Reported						33 170.	5.	622
	922	Not Reported						33 183.	5.	622
	922	Not Reported						33 205.	5.	622
	922	Not Reported						33 239.	5.	622
	922	Not Reported						33 254.	5.	622
	922	Not Reported						33 301.	5.	622
	922	Not Reported						33 342.	5.	622
	922	Not Reported						33 350.	5.	622
	922	Not Reported						33 435.	5.	622
	922	Not Reported						33 461.	5.	622
	922	Not Reported						33 513.	5.	622
	922	Not Reported						33 529.	5.	622
	922	Not Reported						33 540.	5.	622
	922	Not Reported						33 588.	5.	622
	922	Not Reported						33 620.	5.	622
	922	Not Reported						33 635.	5.	622
	922	Not Reported						33 790.	5.	622
	922	Not Reported						33 808.	5.	622
	922	Not Reported						33 822.	5.	622
	922	Not Reported						33 848.	5.	622
	922	Not Reported						33 910.	5.	622
	922	Not Reported						33 924.	5.	622
	922	Not Reported						33 944.	5.	622
	922	Not Reported						33 998.	5.	622
	922	Not Reported						34 025.	5.	622
	922	Not Reported						34 135.	5.	622
	922	Not Reported						34 158.	5.	622
	922	Not Reported						34 188.	5.	622

Isotopic Species	Id. No.	Rotational Quantum Nos.	Vib. State v_a ; v_b	Hyperfine F_1'	F'	F_1	F	Frequency MHz	Acc. ±MHz	Ref.
HC^{13}:C^{12}C^{12}DO^{16}	922	Not Reported						34 220.	5.	622
	922	Not Reported						34 236.	5.	622
	922	Not Reported						34 305.	5.	622
	922	Not Reported						34 317.	5.	622
	922	Not Reported						34 350.	5.	622
	922	Not Reported						34 407.	5.	622
	922	Not Reported						34 437.	5.	622
	922	Not Reported						34 490.	5.	622
	922	Not Reported						34 515.	5.	622
	922	Not Reported						34 557.	5.	622
	922	Not Reported						34 580.	5.	622
	922	Not Reported						34 760.	5.	622
	922	Not Reported						34 796.	5.	622
	922	Not Reported						34 804.	5.	622
	922	Not Reported						34 833.	5.	622
	922	Not Reported						34 847.	5.	622
	922	Not Reported						34 860.	5.	622
	922	Not Reported						34 920.	5.	622
	922	Not Reported						34 939.	5.	622
	922	Not Reported						34 949.	5.	622
	922	Not Reported						35 032.	5.	622
	922	Not Reported						35 057.	5.	622
	922	Not Reported						35 114.	5.	622
	922	Not Reported						35 134.	5.	622
	922	Not Reported						35 210.	5.	622
	922	Not Reported						35 228.	5.	622
	922	Not Reported						35 267.	5.	622
	922	Not Reported						35 288.	5.	622
	922	Not Reported						35 350.	5.	622
	922	Not Reported						35 355.	5.	622
	922	Not Reported						35 392.	5.	622
	922	Not Reported						35 420.	5.	622
	922	Not Reported						35 435.	5.	622
	922	Not Reported						35 486.	5.	622
	922	Not Reported						35 493.	5.	622
	922	Not Reported						35 518.	5.	622
	922	Not Reported						35 578.	5.	622
	922	Not Reported						35 602.	5.	622
	922	Not Reported						35 720.	5.	622
	922	Not Reported						35 830.	5.	622
	922	Not Reported						35 841.	5.	622
	922	Not Reported						35 872.	5.	622
	922	Not Reported						35 890.	5.	622
	922	Not Reported						35 924.	5.	622
HC^{12}:C^{13}C^{12}DO^{16}	923	2, 1, 1← 1, 1, 0	Ground					18 660.68	.02	850
	923	2, 1, 2← 1, 1, 1	Ground					17 842.78	.02	850
HC^{12}:C^{12}C^{13}DO^{16}	924	2, 1, 1← 1, 1, 0	Ground					18 670.52	.02	850
	924	2, 1, 2← 1, 1, 1	Ground					17 839.00	.02	850
HC^{12}:C^{12}C^{12}DO^{18}	925	2, 0, 2← 1, 0, 1	Ground					17 561.76	.05	850
	925	2, 1, 1← 1, 1, 0	Ground					17 955.10	.05	850
	925	2, 1, 2← 1, 1, 1	Ground					17 174.28	.05	850
DC^{12}:C^{12}C^{12}DO^{16}	926	1, 1, 1← 2, 0, 2	Ground					29 650.85	.02	850
	926	2, 1, 1← 1, 1, 0	Ground					17 357.40	.02	850
	926	2, 1, 2← 1, 1, 1	Ground					16 638.41	.02	850
	926	2, 1, 2← 3, 0, 3	Ground					20 801.15	.02	850
	926	3, 1, 3← 4, 0, 4	Ground					11 782.99	.02	850
	926	7, 0, 7← 6, 1, 6	Ground					16 206.27	.02	850

$C_3H_2O_3$ C_{2v} $HC_*:CHOCOO_*$

Isotopic Species	Pt. Gp.	Id. No.	A MHz	B MHz	C MHz	D_J MHz	D_{JK} MHz	Δ Amu A²	κ
$HC^{12}_*:C^{12}HO^{16}C^{12}O^{16}O^{16}_*$	C_{2v}	931	9 346.79 M	4 188.46 M	2 891.54 M				−.59818

Id. No.	μ_a Debye	μ_b Debye	μ_c Debye	eQq Value(MHz) Rel.	eQq Value(MHz) Rel.	eQq Value(MHz) Rel.	ω_a d 1/cm	ω_b d 1/cm	ω_c d 1/cm	ω_d d 1/cm
931	4.51 M	0. X	0. X							

References:

ABC: 563   κ: 563   μ: 563

Add. Ref. 564

Vinylene Carbonate   Spectral Line Table

Isotopic Species	Id. No.	Rotational Quantum Nos.	Vib. State	F_1'	F'	F_1	F	Frequency MHz	Acc. ±MHz	Ref.
$HC^{12}_*:C^{12}HO^{16}C^{12}O^{16}O^{16}_*$	931	3, 1, 2← 2, 1, 1	Ground					23 037.67	.10	563
	931	3, 1, 3← 2, 1, 2	Ground					19 169.62	.10	563
	931	4, 0, 4← 3, 0, 3	Ground					26 450.13	.10	563
	931	4, 1, 3← 3, 1, 2	Ground					30 409.18	.10	563
	931	4, 1, 4← 3, 1, 3	Ground					25 355.29	.10	563
	931	4, 2, 2← 3, 2, 1	Ground					30 019.22	.10	563
	931	4, 2, 3← 3, 2, 2	Ground					28 151.36	.10	563
	931	5, 1, 5← 4, 1, 4	Ground					31 420.68	.10	563

C_3H_3Br C_{3v} $H_3CC\!:\!CBr$

Isotopic Species	Pt. Gp.	Id. No.	A MHz	B MHz	C MHz	D_J MHz	D_{JK} MHz	Δ Amu A²	κ
$H_3C^{12}C^{12}\!:\!C^{12}Br^{79}$	C_{3v}	941		1561.11 M	1561.11 M		.0114		
$H_3C^{12}C^{12}\!:\!C^{12}Br^{81}$	C_{3v}	942		1550.42 M	1550.42 M		.0111		
$D_3C^{12}C^{12}\!:\!C^{12}Br^{79}$	C_{3v}	943		1375.77 M	1375.77 M	.00003	.0078		
$D_3C^{12}C^{12}\!:\!C^{12}Br^{81}$	C_{3v}	944		1365.94 M	1365.94 M	.00003	.0078		

Id. No.	μ_a Debye	μ_b Debye	μ_c Debye	eQq Value(MHz) Rel.		eQq Value(MHz) Rel.	eQq Value(MHz) Rel.	ω_a d 1/cm	ω_b d 1/cm	ω_c d 1/cm	ω_d d 1/cm
941				647	Br^{79}						
942				539	Br^{81}						
943				640	Br^{79}						
944				535	Br^{81}						

References:

ABC: 392,792 D_J: 792 D_{JK}: 392,792   eQq: 392,792

Add. Ref. 234

1-Bromopropyne

Spectral Line Table

Isotopic Species	Id. No.	Rotational Quantum Nos.	Vib. State	F_1'	F'	F_1	F	Frequency MHz	Acc. ±MHz	Ref.
$H_3C^{12}C^{12}\!:\!C^{12}Br^{79}$	941	9, ← 8,	Ground					28 099.90	.1	392
$H_3C^{12}C^{12}\!:\!C^{12}Br^{81}$	942	9, 0← 8, 0	Ground		21/2		19/2	27 906.60		392
	942	9, 0← 8, 0	Ground		19/2		17/2	27 906.60		392
	942	9, 0← 8, 0	Ground		17/2		15/2	27 909.17		392
	942	9, 1← 8, 1	Ground		21/2		19/2	27 906.09		392
	942	9, 1← 8, 1	Ground		19/2		17/2	27 907.14		392
	942	9, 1← 8, 1	Ground		15/2		13/2	27 908.37		392
	942	9, 1← 8, 1	Ground		17/2		15/2	27 909.17		392
	942	9, 2← 8, 2	Ground		21/2		19/2	27 904.18		392
	942	9, 2← 8, 2	Ground		15/2		13/2	27 905.53		392
	942	9, 2← 8, 2	Ground		19/2		17/2	27 908.37		392
	942	9, 2← 8, 2	Ground		15/2		13/2	27 909.17		392
	942	9, 2← 8, 2	Ground		17/2		15/2	27 910.00		392
	942	9, 3← 8, 3	Ground		15/2		13/2	27 900.86		392
	942	9, 3← 8, 3	Ground		21/2		19/2	27 900.86		392
	942	9, 3← 8, 3	Ground		19/2		17/2	27 910.92		392
	942	9, 3← 8, 3	Ground		17/2		15/2	27 910.92		392
	942	9, 4← 8, 4	Ground		15/2		13/2	27 894.38		392
	942	9, 4← 8, 4	Ground		21/2		19/2	27 896.59		392
	942	9, 4← 8, 4	Ground		17/2		15/2	27 912.39		392
	942	9, 4← 8, 4	Ground		19/2		17/2	27 914.35		392
	942	9, 5← 8, 5	Ground		21/2		19/2	27 890.79		392
	942	9, 5← 8, 5	Ground		17/2		15/2	27 914.35		392
	942	9, 5← 8, 5	Ground		19/2		17/2	27 918.60		392
	942	9, 6← 8, 6	Ground		17/2		15/2	27 916.40		392
	942	9, 7← 8, 7	Ground		17/2		15/2	27 918.60		392

C_3H_3Br C_s $H_2BrCC:CH$

Reference: 867

1. The only literature found for this molecule was the letter by Kikuchi, Hirota, and Morino in J.C.P. *31*, 1139L (1959). Since scant information was given, none is recorded here.

C_3H_3Cl C_s $H_2C:C:CHCl$

Isotopic Species	Pt. Gp.	Id. No.	A MHz	B MHz	C MHz	D_J MHz	D_{JK} MHz	Δ Amu A²	κ
$H_2C^{12}:C^{12}:C^{12}HCl^{35}$	C_s	961		2 850.43 M	2 665.20 M				
$H_2C^{12}:C^{12}:C^{12}HCl^{37}$	C_s	962		2 788.59 M	2 609.74 M				

Id. No.	μ_a Debye	μ_b Debye	μ_c Debye	eQq Value(MHz)	Rel.	eQq Value(MHz)	Rel.	eQq Value(MHz)	Rel.	ω_a d 1/cm	ω_b d 1/cm	ω_c d 1/cm	ω_d d 1/cm
961				−41.5	aa	35.3	bb						
962				−32.8	aa	28.3	bb						

References:

ABC: 905   eQq: 915

No Spectral Lines

Methylchloroacetylene

C_3H_3Cl C_{3v} $CH_3C:CCl$

Isotopic Species	Pt. Gp.	Id. No.	A MHz	B MHz	C MHz	D_J MHz	D_{JK} MHz	Δ Amu A^2	κ
$C^{12}H_3C^{12}:C^{12}Cl^{35}$	C_{3v}	971		2 232.271 M	2 232.271 M		.0215		
$C^{12}H_3C^{12}:C^{12}Cl^{37}$	C_{3v}	972		2 183.242 M	2 183.242 M		.0205		
$C^{12}D_3C^{12}:C^{12}Cl^{35}$	C_{3v}	973		1 978.965 M	1 978.965 M		.0150		
$C^{12}D_3C^{12}:C^{12}Cl^{37}$	C_{3v}	974		1 934.460 M	1 934.460 M		.0144		
$C^{12}H_3C^{13}:C^{12}Cl^{35}$	C_{3v}	975		2 217.656 M	2 217.656 M				
$C^{12}H_3C^{13}:C^{12}Cl^{37}$	C_{3v}	976		2 168.284 M	2 168.284 M				
$C^{13}H_3C^{12}:C^{12}Cl^{35}$	C_{3v}	977		2 164.009 M	2 164.009 M				
$C^{13}H_3C^{12}:C^{12}Cl^{37}$	C_{3v}	978		2 115.865 M	2 115.865 M				
$C^{12}D_3C^{13}:C^{12}Cl^{35}$	C_{3v}	979		1 969.605 M	1 969.605 M				
$C^{13}D_3C^{12}:C^{12}Cl^{35}$	C_{3v}	981		1 929.709 M	1 929.709 M				

Id. No.	μ_a Debye	μ_b Debye	μ_c Debye	eQq Value(MHz)	Rel.	eQq Value(MHz)	Rel.	eQq Value(MHz)	Rel.	ω_a d 1/cm	ω_b d 1/cm	ω_c d 1/cm	ω_d d 1/cm
971				−79.6	Cl^{35}								
972				−62.6	Cl^{37}								
973				−79.6	Cl^{35}								
974				−62.7	Cl^{37}								

References:

ABC: 591 D_{JK}: 591   eQq: 591

1-Chloropropyne

Isotopic Species	Id. No.	Rotational Quantum Nos.	Vib. State	F_1'	F'	F_1	F	Frequency MHz	Acc. ±MHz	Ref.
$C^{12}H_3C^{12}:C^{12}Cl^{35}$	971	6, 0 ← 5, 0	Ground		11/2		9/2	26 786.53	.04	591
	971	6, 0 ← 5, 0	Ground		9/2		7/2	26 786.53	.04	591
	971	6, 0 ← 5, 0	Ground		15/2		13/2	26 787.43	.04	591
	971	6, 0 ← 5, 0	Ground		13/2		11/2	26 787.43	.04	591
	971	6, 3 ← 5, 3	Ground		9/2		7/2	26 787.99	.04	591
	971	6, 4 ← 5, 4	Ground		13/2		11/2	26 777.41	.04	591
	971	6, 4 ← 5, 4	Ground		11/2		9/2	26 780.12	.04	591
	971	6, 4 ← 5, 4	Ground		9/2		7/2	26 789.21	.04	591
	971	6, 5 ← 5, 5	Ground		13/2		11/2	26 771.84	.04	591
	971	6, 5 ← 5, 5	Ground		11/2		9/2	26 776.56	.04	591
	971	6, 5 ← 5, 5	Ground		9/2		7/2	26 790.76	.04	591
$C^{12}H_3C^{12}:C^{12}Cl^{37}$	972	6, 0 ← 5, 0	Ground		9/2		7/2	26 198.31	.04	591
	972	6, 0 ← 5, 0	Ground		11/2		9/2	26 198.31	.04	591
	972	6, 0 ← 5, 0	Ground		13/2		11/2	26 199.01	.04	591
	972	6, 0 ← 5, 0	Ground		15/2		13/2	26 199.01	.04	591
	972	6, 2 ← 5, 2	Ground		13/2		11/2	26 198.64	.04	591
	972	6, 2 ← 5, 2	Ground		15/2		13/2	26 198.64	.04	591
	972	6, 3 ← 5, 3	Ground		13/2		11/2	26 194.19	.04	591
	972	6, 3 ← 5, 3	Ground		11/2		9/2	26 195.06	.04	591
	972	6, 3 ← 5, 3	Ground		9/2		7/2	26 199.09	.04	591
	972	6, 4 ← 5, 4	Ground		13/2		11/2	26 190.49	.04	591

Isotopic Species	Id. No.	Rotational Quantum Nos.	Vib. State	F_1'	F'	F_1	F	Frequency MHz	Acc. ±MHz	Ref.
$C^{12}H_3C^{12}:C^{12}Cl^{37}$	972	6, 4 ← 5, 4	Ground		11/2		9/2	26 192.62	.04	591
	972	6, 4 ← 5, 4	Ground		9/2		7/2	26 199.74	.04	591
	972	6, 5 ← 5, 5	Ground		9/2		7/2	26 200.63	.04	591
	972	6, 5 ← 5, 6	Ground		13/2		11/2	26 185.73	.04	591
	972	6, 5 ← 5, 6	Ground		11/2		9/2	26 189.44	.04	591
$C^{12}D_3C^{12}:C^{12}Cl^{35}$	973	6, 0 ← 5, 0	Ground		9/2		7/2	23 746.85	.04	591
	973	6, 0 ← 5, 0	Ground		11/2		9/2	23 746.85	.04	591
	973	6, 0 ← 5, 0	Ground		13/2		11/2	23 747.77	.04	591
	973	6, 0 ← 5, 0	Ground		15/2		13/2	23 747.77	.04	591
	973	6, 1 ← 5, 1	Ground		11/2		9/2	23 746.52	.04	591
	973	6, 2 ← 5, 2	Ground		13/2		11/2	23 745.59	.04	591
	973	6, 2 ← 5, 2	Ground		11/2		9/2	23 745.59	.04	591
	973	6, 3 ← 5, 3	Ground		13/2		11/2	23 742.85	.04	591
	973	6, 3 ← 5, 3	Ground		11/2		9/2	23 743.97	.04	591
	973	6, 3 ← 5, 3	Ground		9/2		7/2	23 749.07	.04	591
	973	6, 4 ← 5, 4	Ground		11/2		9/2	23 741.70	.04	591
	973	6, 4 ← 5, 4	Ground		9/2		7/2	23 750.78	.04	591
	973	6, 5 ← 5, 5	Ground		13/2		11/2	23 734.10	.04	591
	973	6, 5 ← 5, 5	Ground		9/2		7/2	23 753.03	.04	591
$C^{12}D_3C^{12}:C^{12}Cl^{37}$	974	6, 0 ← 5, 0	Ground		15/2		13/2	23 213.64	.04	591
	974	6, 0 ← 5, 0	Ground		13/2		11/2	23 213.64	.04	591
	974	6, 3 ← 5, 3	Ground		13/2		11/2	23 209.48	.04	591
	974	6, 3 ← 5, 3	Ground		11/2		9/2	23 210.33	.04	591
	974	6, 3 ← 5, 3	Ground		9/2		7/2	23 214.38	.04	591
	974	6, 4 ← 5, 4	Ground		13/2		11/2	23 206.22	.04	591
	974	6, 4 ← 5, 4	Ground		11/2		9/2	23 208.34	.04	591
	974	6, 4 ← 5, 4	Ground		9/2		7/2	23 215.53	.04	591
	974	6, 5 ← 5, 5	Ground		13/2		11/2	23 202.09	.04	591
	974	6, 5 ← 5, 5	Ground		9/2		7/2	23 217.02	.04	591
$C^{12}H_3C^{13}:C^{12}Cl^{35}$	975	6, 0 ← 5, 0	Ground		11/2		9/2	26 611.16	.04	591
	975	6, 0 ← 5, 0	Ground		9/2		7/2	26 611.16	.04	591
	975	6, 0 ← 5, 0	Ground		15/2		13/2	26 612.05	.04	591
	975	6, 0 ← 5, 0	Ground		13/2		11/2	26 612.05	.04	591
$C^{12}H_3C^{13}:C^{12}Cl^{37}$	976	6, 0 ← 5, 0	Ground		13/2		11/2	26 019.53	.04	591
	976	6, 0 ← 5, 0	Ground		15/2		13/2	26 019.53	.04	591
$C^{13}H_3C^{12}:C^{12}Cl^{35}$	977	6, 0 ← 5, 0	Ground		11/2		9/2	25 967.36	.04	591
	977	6, 0 ← 5, 0	Ground		9/2		7/2	25 967.36	.04	591
	977	6, 0 ← 5, 0	Ground		13/2		11/2	25 968.26	.04	591
	977	6, 0 ← 5, 0	Ground		15/2		13/2	25 968.26	.04	591
$C^{13}H_3C^{12}:C^{12}Cl^{37}$	978	6, 0 ← 5, 0	Ground		15/2		13/2	25 390.51	.04	591
	978	6, 0 ← 5, 0	Ground		13/2		11/2	25 390.51	.04	591
$C^{12}D_3C^{13}:C^{12}Cl^{35}$	979	6, 0 ← 5, 0	Ground		13/2		11/2	23 635.45	.04	591
	979	6, 0 ← 5, 0	Ground		15/2		13/2	23 635.45	.04	591
$C^{13}D_3C^{12}:C^{12}Cl^{35}$	981	6, 0 ← 5, 0	Ground		9/2		7/2	23 155.79	.04	591
	981	6, 0 ← 5, 0	Ground		11/2		9/2	23 155.79	.04	591
	981	6, 0 ← 5, 0	Ground		15/2		13/2	23 156.71	.04	591
	981	6, 0 ← 5, 0	Ground		13/2		11/2	23 156.71	.04	591

Propargyl Chloride, Chloromethylacetylene

C₃H₃Cl   C_s   Molecular Constant Table   CH₂ClC⫶CH

Isotopic Species	Pt. Gp.	Id. No.	A MHz		B MHz		C MHz		D_J MHz	D_{JK} MHz	Δ Amu A²	κ
C¹²H₂Cl³⁵C¹²⫶C¹²H	C_s	991	24 299.28	M	3 079.77	M	2 777.73	M	.0021	−.057		−.97193
C¹²H₂Cl³⁷C¹²⫶C¹²H	C_s	992	24 146.48	M	3 013.80	M	• 2 721.98	M				−.97276

Id. No.	μ_a Debye		μ_b Debye		μ_c Debye		eQq Value(MHz)	Rel.	eQq Value(MHz)	Rel.	eQq Value(MHz)	Rel.	ω_a 1/cm	d	ω_b 1/cm	d	ω_c 1/cm	d	ω_d 1/cm	d
991	.99	M	1.36	M	0.	X	−30.4	aa	−7.58	bb										

References:

ABC: 824 D_J: 824 D_{JK}: 824 κ: 824 μ: 824   eQq: 824

3-Chloropropyne   Spectral Line Table

Isotopic Species	Id. No.	Rotational Quantum Nos.	Vib. State	F_1'	F'	F_1	F	Frequency MHz	Acc. ±MHz	Ref.
C¹²H₂Cl³⁵C¹²⫶C¹²H	991	1, 1, 1← 0, 0, 0	Ground					27 077.01		824
	991	3, 0, 3← 2, 0, 2	Ground					17 559.47		824
	991	3, 1, 2← 2, 1, 1	Ground					18 023.22		824
	991	3, 1, 3← 2, 1, 2	Ground					17 117.51		824
	991	4, 0, 4← 3, 0, 3	Ground					23 397.48		824
	991	4, 1, 3← 3, 1, 2	Ground					24 028.26		824
	991	4, 1, 4← 3, 1, 3	Ground					22 819.59		824
	991	4, 2, 2← 3, 2, 1	Ground					23 460.52		824
	991	4, 2, 3← 3, 2, 2	Ground					23 429.2		824
	991	4, 1, 3← 4, 0, 4	Ground					22 921.58		824
	991	5, 0, 5← 4, 0, 4	Ground					29 222.29		824
	991	5, 1, 5← 4, 1, 4	Ground					28 518.24		824
	991	5, 2, 3← 4, 2, 2	Ground					29 345.7		824
	991	5, 2, 4← 4, 2, 3	Ground					29 281.78		824
	991	5, 3, 2← 4, 3, 1	Ground					29 302.15		824
	991	5, 3, 3← 4, 3, 2	Ground					29 302.15		824
	991	5, 1, 4← 5, 0, 5	Ground					23 727.87		824
	991	6, 1, 5← 6, 0, 6	Ground					24 719.31		824
	991	7, 1, 6← 7, 0, 7	Ground					25 911.16		824
	991	8, 0, 8← 7, 1, 7	Ground					29 132.01		824
	991	8, 1, 7← 8, 0, 8	Ground					27 319.22		824
	991	9, 1, 8← 9, 0, 9	Ground					28 958.40		824
	991	12, 1,11←11, 2,10	Ground					17 442.95		824
	991	13, 1,12←12, 2,11	Ground					25 079.7		824
C¹²H₂Cl³⁷C¹²⫶C¹²H	992	1, 1, 1← 0, 0, 0	Ground					26 868.46		824
	992	4, 1, 3← 3, 1, 2	Ground					23 520.91		824
	992	5, 0, 5← 4, 0, 4	Ground					28 617.83		824
	992	5, 1, 4← 4, 1, 3	Ground					29 394.		824
	992	5, 1, 5← 4, 1, 4	Ground					27 936.00		824
	992	5, 2, 4← 4, 2, 3	Ground					28 672.8		824
	992	5, 3, 2← 4, 3, 1	Ground					28 694.4		824
	992	5, 3, 3← 4, 3, 2	Ground					28 694.4		824
	992	6, 1, 5← 6, 0, 6	Ground					24 508.91		824
	992	7, 1, 6← 7, 0, 7	Ground					25 656.94		824
	992	8, 1, 7← 8, 0, 8	Ground					27 011.07		824
	992	9, 1, 8← 9, 0, 9	Ground					28 587.43		824

Propargyl Fluoride, Fluoromethylacetylene

C₃H₃F C_s H₂FCC:CH

Isotopic Species	Pt. Gp.	Id. No.	A MHz	B MHz	C MHz	D_J MHz	D_{JK} MHz	Δ Amu A²	κ
H₂F¹⁹C¹²C¹²:C¹²H	C_s	1001	35 637.79 M	4 608.79 M	4 183.60 M	.003	−.072		

References:

ABC: 969 D_J: 969 D_{JK}: 969

No Spectral Lines

Methyl Iodoacetylene

C₃H₃I C_{3v} H₃CC:CI

Isotopic Species	Pt. Gp.	Id. No.	A MHz	B MHz	C MHz	D_J MHz	D_{JK} MHz	Δ Amu A²	κ
H₃C¹²C¹²:C¹²I¹²⁷	C_{3v}	1011		1 259.02 M	1 259.02 M		.0072		
D₃C¹²C¹²:C¹²I¹²⁷	C_{3v}	1012		1 107.73 M	1 107.73 M		.0053		

Id. No.	μ_a Debye	μ_b Debye	μ_c Debye	eQq Value(MHz) Rel.	eQq Value(MHz) Rel.	eQq Value(MHz) Rel.	ω_a d 1/cm	ω_b d 1/cm	ω_c d 1/cm	ω_d d 1/cm
1011	1.21 L	0. X	0. X	−2230 I¹²⁷						

References:

ABC: 392,1011 D_{JK}: 392,1011   μ: 995   eQq: 392

Isotopic Species	Id. No.	Rotational Quantum Nos.	Vib. State	Hyperfine				Frequency MHz	Acc. ±MHz	Ref.
				F_1'	F'	F_1	F			
$H_3C^{12}C^{12}:C^{12}I^{127}$	1011	13, 0←12, 0	Ground		23/2		21/2	32 729.00		392
	1011	13, 0←12, 0	Ground		25/2		23/2	32 731.19		392
	1011	13, 0←12, 0	Ground		21/2		19/2	32 731.52		392
	1011	13, 0←12, 0	Ground		27/2		25/2	32 735.35		392
	1011	13, 0←12, 0	Ground		31/2		29/2	32 736.69		392
	1011	13, 0←12, 0	Ground		29/2		27/2	32 738.45		392
	1011	13, 1←12, 1	Ground		23/2		21/2	32 729.00		392
	1011	13, 1←12, 1	Ground		25/2		23/2	32 730.52		392
	1011	13, 1←12, 1	Ground		21/2		19/2	32 732.18		392
	1011	13, 1←12, 1	Ground		27/2		25/2	32 734.45		392
	1011	13, 1←12, 1	Ground		31/2		29/2	32 737.22		392
	1011	13, 1←12, 1	Ground		29/2		27/2	32 737.72		392
	1011	13, 2←12, 2	Ground		25/2		23/2	32 728.68		392
	1011	13, 2←12, 2	Ground		23/2		21/2	32 729.00		392
	1011	13, 2←12, 2	Ground		27/2		25/2	32 731.83		392
	1011	13, 2←12, 2	Ground		21/2		19/2	32 734.26		392
	1011	13, 2←12, 2	Ground		29/2		27/2	32 736.05		392
	1011	13, 2←12, 2	Ground		31/2		29/2	32 738.45		392
	1011	13, 3←12, 3	Ground		25/2		23/2	32 725.28		392
	1011	13, 3←12, 3	Ground		21/2		19/2	32 727.32		392
	1011	13, 3←12, 3	Ground		27/2		25/2	32 727.74		392
	1011	13, 3←12, 3	Ground		23/2		21/2	32 728.68		392
	1011	13, 3←12, 3	Ground		29/2		27/2	32 733.05		392
	1011	13, 3←12, 3	Ground		31/2		29/2	32 740.77		392
	1011	13, 4←12, 4	Ground		25/2		23/2	32 720.85		392
	1011	13, 4←12, 4	Ground		27/2		25/2	32 721.36		392
	1011	13, 4←12, 4	Ground		25/2		23/2	32 728.14		392
	1011	13, 4←12, 4	Ground		29/2		27/2	32 729.7		392
	1011	13, 4←12, 4	Ground		21/2		19/2	32 743.89		392
	1011	13, 4←12, 4	Ground		31/2		29/2	32 743.89		392
	1011	13, 5←12, 5	Ground		27/2		25/2	32 713.35		392
	1011	13, 5←12, 5	Ground		25/2		23/2	32 714.74		392
	1011	13, 5←12, 5	Ground		29/2		27/2	32 724.73		392
	1011	13, 5←12, 5	Ground		23/2		21/2	32 727.10		392
	1011	13, 5←12, 5	Ground		31/2		29/2	32 747.90		392
	1011	13, 5←12, 5	Ground		21/2		19/2	32 748.69		392
	1011	13, 6←12, 6	Ground		25/2		23/2	32 707.68		392
	1011	13, 6←12, 6	Ground		29/2		27/2	32 717.22		392
	1011	13, 6←12, 6	Ground		23/2		21/2	32 726.06		392
	1011	13, 6←12, 6	Ground		31/2		29/2	32 752.83		392
	1011	13, 7←12, 7	Ground		23/2		21/2	32 725.28		392
	1011	13, 8←12, 8	Ground		23/2		21/2	32 724.73		392
	1011	13, 9←12, 9	Ground		23/2		21/2	32 723.61		392

C_3H_3N C_s $CH_2{:}CHCN$

Isotopic Species	Pt. Gp.	Id. No.	A MHz	B MHz	C MHz	D_J MHz	D_{JK} MHz	Δ Amu A^2	κ
$C^{12}H_2{:}C^{12}HC^{12}N^{14}$	C_s	1021	49 847.1 M	4 971.125 M	4 513.875 M	.0030	−.098	.1598	−.979477
$C^{13}H_2{:}C^{12}HC^{12}N^{14}$	C_s	1022	49 180. M	4 837.539 M	4 398.194 M			.1598	
$C^{12}H_2{:}C^{13}HC^{12}N^{14}$	C_s	1023	48 645. M	4 948.741 M	4 485.416 M			.1598	
$C^{12}H_2{:}C^{12}HC^{13}N^{14}$	C_s	1024	49 781. M	4 948.434 M	4 494.619 M			.1598	
$C^{12}H_2{:}C^{12}HC^{12}N^{15}$	C_s	1025	49 647. M	4 819.619 M	4 387.054 M			.1598	
$C^{12}H_2{:}C^{12}DC^{12}N^{14}$	C_s	1026	40 194.8 M	4 934.338 M	4 388.398 M		−.040	.1685	

Id. No.	μ_a Debye	μ_b Debye	μ_c Debye	eQq Value(MHz) Rel.	eQq Value(MHz) Rel.	eQq Value(MHz) Rel.	ω_a d 1/cm	ω_b d 1/cm	ω_c d 1/cm	ω_d d 1/cm
1021	3.68 M	1.25 M	0. X	−4.21 CN						

References:

ABC: 851 D_J: 851 D_{JK}: 851 Δ: 851 κ: 573 μ: 573   eQq: 851

Add. Ref. 409

Ref. 573 gives the rotational constant $\delta = 0.0102615$ (species not given). For species 1021, $R_6 = -0.0026$ MHz (for the C^{13} and N^{15} species, $R_6 =$ same value within experimental error). Ref. 851.

Acrylonitrile   Spectral Line Table

Isotopic Species	Id. No.	Rotational Quantum Nos.	Vib. State	F_1'	F'	F_1	F	Frequency MHz	Acc. ±MHz	Ref.
$C^{12}H_2{:}C^{12}HC^{12}N^{14}$	1021	1, 1, 1← 2, 0, 2	Ground		0		1	25 908.70		851
	1021	1, 1, 1← 2, 0, 2	Ground		2		3	25 910.08		851
	1021	1, 1, 1← 2, 0, 2	Ground		2		2	25 911.28		851
	1021	1, 1, 1← 2, 0, 2	Ground		1		2	25 911.78		851
	1021	2, 0, 2← 1, 0, 1	Ground		1		0	18 965.48		851
	1021	2, 0, 2← 1, 0, 1	Ground		2		2	18 965.48		851
	1021	2, 0, 2← 1, 0, 1	Ground		3		2	18 966.61		851
	1021	2, 0, 2← 1, 0, 1	Ground		1		1	18 968.41		851
	1021	2, 1, 1← 1, 1, 0	Ground		2		1	19 426.67		851
	1021	2, 1, 1← 1, 1, 0	Ground		3		2	19 427.80		851
	1021	2, 1, 1← 1, 1, 0	Ground		1		0	19 429.06		851
	1021	2, 1, 2← 1, 1, 1	Ground		2		1	18 512.14		851
	1021	2, 1, 2← 1, 1, 1	Ground		2		2	18 512.68		851
	1021	2, 1, 2← 1, 1, 1	Ground		3		2	18 513.31		851
	1021	2, 1, 2← 1, 1, 1	Ground		1		0	18 514.43		851
	1021	3, 0, 3← 2, 0, 2	Ground					28 440.84		851
	1021	3, 1, 2← 2, 1, 1	Ground					29 139.01		851
	1021	3, 1, 3← 2, 1, 2	Ground					27 767.31		851
	1021	3, 2, 1← 2, 2, 0	Ground					28 470.75		851
	1021	3, 2, 2← 2, 2, 1	Ground					28 457.34		851
	1021	7, 0, 7← 6, 1, 6	Ground					25 699.42		851

Isotopic Species	Id. No.	Rotational Quantum Nos.	Vib. State	F_1'	F'	F_1	F	Frequency MHz	Acc. ±MHz	Ref.
$C^{13}H_2:C^{12}HC^{12}N^{14}$	1022	2, 0, 2← 1, 0, 1	Ground					18 468.17		851
	1022	2, 1, 1← 1, 1, 0	Ground					18 911.15		851
	1022	2, 1, 2← 1, 1, 1	Ground					18 032.46		851
	1022	3, 0, 3← 2, 0, 2	Ground					27 693.9	.15	851
	1022	3, 1, 2← 2, 1, 1	Ground					28 364.5	.15	851
	1022	3, 1, 3← 2, 1, 2	Ground					27 046.5	.15	851
	1022	3, 2, 1← 2, 2, 0	Ground					27 722.0	.15	851
	1022	3, 2, 2← 2, 2, 1	Ground					27 709.5	.15	851
$C^{12}H_2:C^{13}HC^{12}N^{14}$	1023	2, 0, 2← 1, 0, 1	Ground					18 864.60		851
	1023	2, 1, 1← 1, 1, 0	Ground					19 331.98		851
	1023	2, 1, 2← 1; 1, 1	Ground					18 405.33		851
	1023	3, 0, 3← 2, 0, 2	Ground					28 287.5	.15	851
	1023	3, 1, 2← 2, 1, 1	Ground					28 995.49	.15	851
	1023	3, 1, 3← 2, 1, 2	Ground					27 605.5	.15	851
	1023	3, 2, 1← 2, 2, 0	Ground					28 319.0	.15	851
	1023	3, 2, 2← 2, 2, 1	Ground					28 304.8	.15	851
$C^{12}H_2:C^{12}HC^{13}N^{14}$	1024	2, 0, 2← 1, 0, 1	Ground					18 882.63		851
	1024	2, 1, 1← 1, 1, 0	Ground					19 340.26		851
	1024	2, 1, 2← 1, 1, 1	Ground					18 432.63		851
	1024	3, 0, 3← 2, 0, 2	Ground					28 315.2	.15	851
	1024	3, 1, 2← 2, 1, 1	Ground					29 008.07	.15	851
	1024	3, 1, 3← 2, 1, 2	Ground					27 646.6	.15	851
	1024	3, 2, 1← 2, 2, 0	Ground					28 344.7	.15	851
	1024	3, 2, 2← 2, 2, 1	Ground					28 331.6	.15	851
$C^{12}H_2:C^{12}HC^{12}N^{15}$	1025	2, 0, 2← 1, 0, 1	Ground					18 410.18		851
	1025	2, 1, 1← 1, 1, 0	Ground					18 846.25		851
	1025	2, 1, 2← 1, 1, 1	Ground					17 981.12		851
	1025	3, 0, 3← 2, 0, 2	Ground					27 607.3	.15	851
	1025	3, 1, 2← 2, 1, 1	Ground					28 267.2	.15	851
	1025	3, 1, 3← 2, 1, 2	Ground					26 969.6	.15	851
	1025	3, 2, 1← 2, 2, 0	Ground					27 634.4	.15	851
	1025	3, 2, 2← 2, 2, 1	Ground					27 622.4	.15	851
$C^{12}H_2:C^{12}DC^{12}N^{14}$	1026	2, 0, 2← 1, 0, 1	Ground		2		2	18 638.14		851
	1026	2, 0, 2← 1, 0, 1	Ground		1		0	18 638.14		851
	1026	2, 0, 2← 1, 0, 1	Ground		3		2	18 639.17		851
	1026	2, 0, 2← 1, 0, 1	Ground		1		1	18 640.99		851
	1026	2, 1, 1← 1, 1, 0	Ground		2		1	19 190.57		851
	1026	2, 1, 1← 1, 1, 0	Ground		3		2	19 191.72		851
	1026	2, 1, 1← 1, 1, 0	Ground		1		0	19 192.98		851
	1026	2, 1, 2← 1, 1, 1	Ground		2		1	18 098.69		851
	1026	2, 1, 2← 1, 1, 1	Ground		2		2	18 099.18		851
	1026	2, 1, 2← 1, 1, 1	Ground		3		2	18 099.85		851
	1026	2, 1, 2← 1, 1, 1	Ground		1		0	18 100.97		851

Isotopic Species	Pt. Gp.	Id. No.	A MHz		B MHz		C MHz		D_J MHz	D_{JK} MHz	Δ Amu A^2	κ
C$^{12}_*$H$_2$C^{12}H:C$^{12}_*$H	C$_{2v}$	1031	30 063.7	M	21 825.6	M	13 795.7	M				
C$^{12}_*$H$_2$C^{12}D:C$^{12}_*$H	C$_s$	1032	26 898.7	M	20 520.1	M	12 606.1	M				
C$^{12}_*$H$_2$C^{12}D:C$^{12}_*$D	C$_{2v}$	1033	23 179.6	M	20 102.0	M	11 585.4	M				
C$^{12}_*$DHC^{12}H:C$^{12}_*$H	C$_s$	1034	28 794.6	M	19 356.5	M	13 011.6	M				

Id. No.	μ_a Debye		μ_b Debye		μ_c Debye		eQq Value(MHz) Rel.	eQq Value(MHz) Rel.	eQq Value(MHz) Rel.	ω_a d 1/cm	ω_b d 1/cm	ω_c d 1/cm	ω_d d 1/cm
1031	.454	M	0.	X	0.	X							
1032	.433	M	.156	M									
1033	.461	M	0.	X	0.	X							
1034	.466	M	0.	X	.043	M							

References:
ABC: 866 μ: 866

Add. Ref. 343

Cyclopropene   Spectral Line Table

Isotopic Species	Id. No.	Rotational Quantum Nos.	Vib. State	F$_1'$	F'	F$_1$	F	Frequency MHz	Acc. ±MHz	Ref.
C$^{12}_*$H$_2$C^{12}H:C$^{12}_*$H	1031	1, 0, 1← 0, 0, 0	Ground					35 621.3		866
	1031	2, 1, 1← 2, 1, 2	Ground					24 089.6		866
	1031	4, 2, 2← 4, 2, 3	Ground					34 935.4		866
	1031	5, 3, 2← 5, 3, 3	Ground					21 544.2		866
	1031	7, 4, 3← 7, 4, 4	Ground					27 599.8		866
	1031	9, 5, 4← 9, 5, 5	Ground					33 353.4		866
	1031	12, 7, 5←12, 7, 6	Ground					19 784.3		866
C$^{12}_*$H$_2$C^{12}D:C$^{12}_*$H	1032	1, 0, 1← 0, 0, 0	Ground					33 126.1		866
	1032	2, 1, 1← 2, 1, 2	Ground					23 741.6		866
	1032	4, 2, 2← 4, 2, 3	Ground					36 498.7		866
	1032	5, 3, 2← 5, 3, 3	Ground					25 209.6		866
	1032	7, 4, 3← 7, 4, 4	Ground					34 726.6		866
	1032	8, 5, 3← 8, 5, 4	Ground					20 700.9		866
	1032	10, 6, 4←10, 6, 5	Ground					27 649.2		866
	1032	12, 7, 5←12, 7, 6	Ground					35 306.1		866
C$^{12}_*$H$_2$C^{12}D:C$^{12}_*$D	1033	1, 0, 1← 0, 0, 0	Ground					31 687.4		866
	1033	2, 1, 1← 2, 1, 2	Ground					25 549.3		866
	1033	3, 2, 1← 3, 2, 2	Ground					21 429.0		866
	1033	5, 3, 2← 5, 3, 3	Ground					37 894.2		866
	1033	6, 4, 2← 6, 4, 3	Ground					30 819.7		866
	1033	7, 5, 2← 7, 5, 3	Ground					23 124.0		866
	1033	10, 7, 3←10, 7, 4	Ground					28 955.5		866
	1033	11, 8, 3←11, 8, 4	Ground					19 647.9		866
C$^{12}_*$DHC^{12}H:C$^{12}_*$H	1034	1, 0, 1← 0, 0, 0	Ground					32 368.0		866
	1034	2, 1, 1← 2, 1, 2	Ground					19 034.3		866
	1034	3, 1, 2← 3, 1, 3	Ground					37 508.9		866
	1034	4, 2, 2← 4, 2, 3	Ground					24 449.1		866
	1034	6, 3, 3← 6, 3, 4	Ground					27 632.7		866
	1034	8, 4, 4← 8, 4, 5	Ground					29 165.3		866
	1034	10, 5, 5←10, 5, 6	Ground					29 446.2		866
	1034	12, 6, 6←12, 6, 7	Ground					28 790.1		866

Methylacetylene, Propine

C$_3$H$_4$ C$_{3v}$ CH$_3$C:CH

Isotopic Species	Pt. Gp.	Id. No.	A MHz	B MHz		C MHz		D$_J$ MHz	D$_{JK}$ MHz	Δ Amu A^2	κ
C^{12}H$_3$C^{12}:C^{12}H	C$_{3v}$	1041		8 545.84	M	8 545.84	M	.00296	.1629		
C^{12}H$_3$C^{13}:C^{12}H	C$_{3v}$	1042		8 542.28	M	8 542.28	M		.16		
C^{13}H$_3$C^{12}:C^{12}H	C$_{3v}$	1043		8 313.23	M	8 313.23	M		.16		
C^{12}H$_3$C^{12}:C^{13}H	C$_{3v}$	1044		8 290.24	M	8 290.24	M		.13		
C^{12}H$_3$C^{12}:C^{12}D	C$_{3v}$	1045		7 788.14	M	7 788.14	M	.002	.142		
C^{12}D$_3$C^{12}:C^{12}D	C$_{3v}$	1046		6 734.31	M	6 734.31	M	.002	.090		
C^{12}H$_2$DC^{12}:C^{12}H	C$_s$	1047		8 155.67	M	8 025.46	M	.003	.13		−.997
C^{12}HD$_2$C^{12}:C^{12}H	C$_s$	1048		7 765.73	M	7 630.99	M	.002	.13		−.997
C^{12}H$_2$DC^{12}:C^{12}D	C$_s$	1049		7 440.77	M	7 331.96	M	.001	.12		−.997
C^{12}HD$_2$C^{12}:C^{12}D	C$_s$	1051		7 095.09	M	6 982.56	M	.004	.11		−.997
C^{12}D$_3$C^{12}:C^{12}H	C$_{3v}$	1052		7 355.75	M	7 355.75	M	.002	.102		

Id. No.	μ$_a$ Debye		μ$_b$ Debye		μ$_c$ Debye		eQq Value(MHz) Rel.	eQq Value(MHz) Rel.	eQq Value(MHz) Rel.	ω$_a$ d 1/cm	ω$_b$ d 1/cm	ω$_c$ d 1/cm	ω$_d$ d 1/cm
1041	.75	M	0.	X	0.	X				633 2	328 2		

References:

ABC: 247,639   D$_J$: 639,745   D$_{JK}$: 247,639,745   κ: 639   μ: 434   ω: 864

Add. Ref. 200,282,302,426,461,666,687

For species 1041, v$_9$ = 1, B = 3551.1 MHz, q$_9$ = 9.06 MHz, ζ$_9$ = 0.95; v$_{10}$ = 1, B = 8569.75 MHz, q$_{10}$ = 16.76 MHz, ζ$_{10}$ = 1.00. Ref. 864. α$_{10}$ = −23.92 MHz, a = 1.15. Ref. 200.

Isotopic Species	Id. No.	Rotational Quantum Nos.	Vib. State v_a^l ; v_b^l	Hyperfine				Frequency MHz	Acc. ±MHz	Ref.
				F_1'	F'	F_1	F			
$C^{12}H_3C^{12}\vdots C^{12}H$	1041	2, 0← 1, 0	Ground					34 183.37	.10	247
	1041	2, 0← 1, 0	0, 0; 1, ±1					34 278.98		864
	1041	2, 1← 1, 1	Ground					34 182.71	.10	247
	1041	2, 1← 1, 1	0, 0; 1, ±1					34 246.30		864 [1]
	1041	2, 1← 1, 1	0, 0; 1, ∓1					34 277.05		864
	1041	2, 1← 1, 1	0, 0; 1, ±1					34 313.21		864 [1]
	1041	3, 0← 2, 0	Ground					51 274.75	.10	247
	1041	3, 0← 2, 0	1, ±1; 0, 0					51 307.53		864
	1041	3, 0← 2, 0	0, 0; 1, ±1					51 418.75		864
	1041	3, 1← 2, 1	Ground					51 273.76	.10	247
	1041	3, 1← 2, 1	1, ±1; 0, 0					51 280.45		864 [1]
	1041	3, 1← 2, 1	1, ∓1; 0, 0					51 304.05		864
	1041	3, 1← 2, 1	1, ±1; 0, 0					51 334.81		864 [1]
	1041	3, 1← 2, 1	0, 0; 1, ±1					51 369.12		864 [1]
	1041	3, 1← 2, 1	0, 0; 1, ∓1					51 415.35		864
	1041	3, 1← 2, 1	0, 0; 1, ±1					51 469.85		864 [1]
	1041	3, 2← 2, 2	Ground					51 270.86	.10	247
	1041	3, 2← 2, 2	1, ∓1; 0, 0					51 296.33		864
	1041	3, 2← 2, 2	1, ±1; 0, 0					51 305.93		864
	1041	3, 2← 2, 2	0, 0; 1, ∓1					51 410.51		864
	1041	3, 2← 2, 2	0, 0; 1, ±1					51 418.23		864
	1041	5, 0← 4, 0						85 457.29	.20	745
	1041	5, 1← 4, 1						85 455.67	.20	745
	1041	5, 2← 4, 2						85 450.78	.20	745
	1041	5, 3← 4, 3						85 442.61	.20	745
	1041	5, 4← 4, 4						85 431.34	.20	745
	1041	7, 0← 6, 0						119 638.22	.25	745
	1041	7, 1← 6, 1						119 635.97	.25	745
	1041	7, 2← 6, 2						119 629.13	.25	745
	1041	7, 3← 6, 3						119 617.67	.25	745
	1041	7, 4← 6, 4						119 601.62	.25	745
	1041	7, 5← 6, 5						119 581.12	.25	745
	1041	7, 6← 6, 6						119 556.00	.25	745
	1041	8, 0← 7, 0						136 727.93	.30	745
	1041	8, 1← 7, 1						136 725.36	.30	745
	1041	8, 2← 7, 2						136 717.60	.30	745
	1041	8, 3← 7, 3						136 704.48	.30	745
	1041	8, 4← 7, 4						136 686.19	.30	745
	1041	8, 5← 7, 5						136 662.74	.30	745
	1041	8, 6← 7, 6						136 634.03	.30	745
	1041	8, 7← 7, 7						136 600.15	.30	745
	1041	9, 0← 8, 0						153 817.16	.30	745
	1041	9, 1← 8, 1						153 814.27	.30	745
	1041	9, 2← 8, 2						153 805.37	.30	745
	1041	9, 3← 8, 3						153 790.66	.30	745
	1041	9, 4← 8, 4						153 770.15	.30	745
	1041	9, 5← 8, 5						153 743.72	.30	745
	1041	9, 6← 8, 6						153 711.55	.30	745
	1041	10, 0← 9, 0						170 905.66	.35	745
	1041	10, 1← 9, 1						170 902.37	.35	745

1. Certain frequencies with l-doubling have the same designations in Jaseja's article.

Isotopic Species	Id. No.	Rotational Quantum Nos.	Vib. State v_a^l ; v_b^l	Hyperfine F_1'	F'	F_1	F	Frequency MHz	Acc. ±MHz	Ref.
$C^{12}H_3C^{12}{:}C^{12}H$	1041	10, 2← 9, 2						170 892.59	.35	745
	1041	10, 3← 9, 3						170 876.27	.35	745
	1041	10, 4← 9, 4						170 853.50	.35	745
	1041	10, 5← 9, 5						170 824.13	.35	745
	1041	10, 6← 9, 6						170 788.29	.35	745
	1041	10, 7← 9, 7						170 746.05	.35	745
	1041	11, 0←10, 0						187 993.69	.40	745
	1041	11, 1←10, 1						187 990.02	.40	745
	1041	11, 2←10, 2						187 979.34	.40	745
	1041	11, 3←10, 3						187 961.41	.40	745
	1041	11, 4←10, 4						187 936.34	.40	745
	1041	11, 5←10, 5						187 903.96	.40	745
	1041	11, 6←10, 6						187 864.42	.40	745
	1041	11, 7←10, 7						187 817.95	.40	745
	1041	11, 8←10, 8						187 763.96	.40	745
	1041	13, 0←12, 0						222 166.71	.45	745
	1041	13, 1←12, 1						222 162.46	.45	745
	1041	13, 2←12, 2						222 149.80	.45	745
	1041	13, 3←12, 3						222 128.61	.45	745
	1041	13, 4←12, 4						222 099.05	.45	745
	1041	13, 5←12, 5						222 060.95	.45	745
	1041	13, 6←12, 6						222 014.44	.45	745
	1041	13, 7←12, 7						221 959.38	.45	745
	1041	14, 0←13, 0						239 252.14	.50	745
	1041	14, 1←13, 1						239 247.62	.50	745
	1041	14, 2←13, 2						239 233.92	.50	745
	1041	14, 3←13, 3						239 210.93	.50	745
	1041	14, 4←13, 4						239 178.97	.50	745
	1041	14, 5←13, 5						239 138.04	.50	745
	1041	14, 6←13, 6						239 087.81	.50	745
$C^{12}H_3C^{13}{:}C^{12}H$	1042	2, 0← 1, 0	Ground					34 169.13	.14	247
	1042	2, 1← 1, 1	Ground					34 168.47	.14	247
$C^{13}H_3C^{12}{:}C^{12}H$	1043	2, 0← 1, 0	Ground					33 252.88	.10	247
	1043	2, 1← 1, 1	Ground					33 252.22	.10	247
$C^{12}H_3C^{12}{:}C^{13}H$	1044	2, 0← 1, 0	Ground					33 160.94	.10	247
	1044	2, 1← 1, 1	Ground					33 160.35	.10	247
$C^{12}H_3C^{12}{:}C^{12}D$	1045	2, 0← 1, 0	Ground					31 152.56	.10	247
	1045	2, 1← 1, 1	Ground					31 152.00	.10	247
	1045	3, 0← 2, 0	Ground					46 728.72	.1	639
	1045	3, 1← 2, 1	Ground					46 727.86	.1	639
	1045	3, 2← 2, 2	Ground					46 725.32	.1	639
$C^{12}D_3C^{12}{:}C^{12}D$	1046	2, 0← 1, 0	Ground					26 937.24	.10	247
	1046	2, 1← 1, 1	Ground					26 936.87	.10	247
	1046	3, 0← 2, 0	Ground					40 405.75	.1	639
	1046	3, 1← 2, 1	Ground					40 405.21	.1	639
	1046	3, 2← 2, 2	Ground					40 403.60	.1	639
$C^{12}H_2DC^{12}{:}C^{12}H$	1047	1, 0, 1← 0, 0, 0	Ground					16 181.12	.1	639
	1047	2, 0, 2← 1, 0, 1	Ground					32 362.08	.1	639

Isotopic Species	Id. No.	Rotational Quantum Nos.	Vib. State $v_a^1 ; v_b^1$	Hyperfine				Frequency MHz	Acc. ±MHz	Ref.
				F_1'	F'	F_1	F			
$C^{12}H_2DC^{12}:C^{12}H$	1047	2, 1, 1← 1, 1, 0	Ground					32 491.86	.1	639
	1047	2, 1, 2← 1, 1, 1	Ground					32 231.44	.1	639
	1047	3, 0, 3← 2, 0, 2	Ground					48 542.62	.1	639
	1047	3, 1, 2← 2, 1, 1	Ground					48 737.52	.1	639
	1047	3, 1, 3← 2, 1, 2	Ground					48 346.90	.1	639
	1047	3, 2, 1← 2, 2, 0	Ground					48 540.33	.1	639
	1047	3, 2, 2← 2, 2, 1	Ground					48 539.96	.1	639
$C^{12}HD_2C^{12}:C^{12}H$	1048	2, 0, 2← 1, 0, 1	Ground					30 793.13	.1	639
	1048	2, 1, 1← 1, 1, 0	Ground					30 927.55	.1	639
	1048	2, 1, 2← 1, 1, 1	Ground					30 658.07	.1	639
	1048	3, 0, 3← 2, 0, 2	Ground					46 189.01	.1	639
	1048	3, 1, 2← 2, 1, 1	Ground					46 391.00	.1	639
	1048	3, 1, 3← 2, 1, 2	Ground					45 986.74	.1	639
	1048	3, 2, 1← 2, 2, 0	Ground					46 187.46	.1	639
	1048	3, 2, 2← 2, 2, 1	Ground					46 186.84	.1	639
$C^{12}H_2DC^{12}:C^{12}D$	1049	2, 0, 2← 1, 0, 1	Ground					29 545.33	.1	639
	1049	2, 1, 1← 1, 1, 0	Ground					29 653.70	.1	639
	1049	2, 1, 2← 1, 1, 1	Ground					29 436.09	.1	639
	1049	3, 0, 3← 2, 0, 2	Ground					44 317.72	.1	639
	1049	3, 1, 2← 2, 1, 1	Ground					44 480.44	.1	639
	1049	3, 1, 3← 2, 1, 2	Ground					44 154.10	.1	639
	1049	3, 2, 1← 2, 2, 0	Ground					44 315.50	.1	639
	1049	3, 2, 2← 2, 2, 1	Ground					44 315.24	.1	639
$C^{12}HD_2C^{12}:C^{12}D$	1051	2, 0, 2← 1, 0, 1	Ground					28 155.14	.1	639
	1051	2, 1, 1← 1, 1, 0	Ground					28 267.33	.1	639
	1051	2, 1, 2← 1, 1, 1	Ground					28 042.28	.1	639
	1051	3, 0, 3← 2, 0, 2	Ground					42 232.30	.1	639
	1051	3, 1, 2← 2, 1, 1	Ground					42 400.70	.1	639
	1051	3, 1, 3← 2, 1, 2	Ground					42 063.15	.1	639
	1051	3, 2, 1← 2, 2, 0	Ground					42 230.61	.1	639
	1051	3, 2, 2← 2, 2, 1	Ground					42 230.22	.1	639
$C^{12}D_3C^{12}:C^{12}H$	1052	2, 0← 1, 0	Ground					29 422.89	.1	639
	1052	2, 1← 1, 1	Ground					29 422.50	.1	639
	1052	3, 0← 2, 0	Ground					44 134.19	.1	639
	1052	3, 1← 2, 1	Ground					44 133.62	.1	639
	1052	3, 2← 2, 2	Ground					44 131.76	.1	639

$C_3H_4Cl_2$ C_{2v} $C_*H_2CH_2C_*Cl_2$

Isotopic Species	Pt. Gp.	Id. No.	A MHz		B MHz		C MHz		D_J MHz	D_{JK} MHz	Δ Amu A²	κ
$C^{12}_*H_2C^{12}H_2C^{12}_*Cl^{35}_2$	C_{2v}	1061	3 981.81	M	2 919.15	M	1 949.39	M				−.04570
$C^{12}_*H_2C^{12}H_2C^{12}_*Cl^{35}Cl^{37}$	C_s	1062	3 955.25	M	2 849.15	M	1 911.65	M				−.08250
$C^{12}_*D_2C^{12}D_2C^{12}_*Cl^{35}_2$	C_{2v}	1063	3 454.55	M	2 760.81	M	1 833.55	M				.14407
$C^{12}_*D_2C^{12}D_2C^{12}_*Cl^{35}Cl^{37}$	C_s	1064	3 431.18	M	2 697.41	M	1 799.02	M				.10085
$C^{12}_*HDC^{12}H_2C^{12}_*Cl^{35}_2$	C_1	1065	3 835.71	M	2 878.11	M	1 917.95	M				.00133
c-$C^{12}_*HDC^{12}H_2C^{12}_*Cl^{35}Cl^{37}$	C_1	1066										−.04213
t-$C^{12}_*HDC^{12}H_2C^{12}_*Cl^{35}Cl^{37}$	C_1	1067										−.03178
$C^{12}_*H_2C^{13}H_2C^{12}_*Cl^{35}_2$	C_{2v}	1068	3 888.15	M	2 909.60	M	1 930.87	M				.0001

Id. No.	μ_a Debye		μ_b Debye		μ_c Debye		eQq Value(MHz)	Rel.	eQq Value(MHz)	Rel.	eQq Value(MHz)	Rel.	ω_a d 1/cm	ω_b d 1/cm	ω_c d 1/cm	ω_d d 1/cm
1061	0.	X	1.58	M	0.	X	−43.545	aa	4.100	bb	39.445	cc				
1063							−43.45	aa	3.96	bb	39.49	cc				

References:

ABC: 964 κ: 964 μ: 964   eQq: 964

For species 1066 $(A-C)/2 = 966.94$ MHz; for species 1067 $(A-C)/2 = 962.37$ MHz. Ref. 964.

Isotopic Species	Id. No.	Rotational Quantum Nos.	Vib. State	F_1'	F'	F_1	F	Frequency MHz	Acc. ±MHz	Ref.
$C^{12}_*H_2C^{12}H_2C^{12}_*Cl_2^{35}$	1061	2, 0, 2← 1, 1, 1	Ground					8 250.4		964
	1061	2, 1, 2← 1, 0, 1	Ground					9 833.3		964
	1061	2, 2, 1← 1, 1, 0	Ground					13 899.3		964
	1061	3, 1, 3← 2, 0, 2	Ground					13 447.3		964
	1061	4, 1, 3← 4, 0, 4	Ground					8 955.2		964
	1061	4, 2, 3← 4, 1, 4	Ground					9 967.5		964
	1061	4, 3, 2← 4, 2, 3	Ground					8 909.3		964
	1061	4, 4, 0← 4, 3, 1	Ground					9 598.3		964
	1061	5, 3, 3← 5, 2, 4	Ground					10 441.2		964
	1061	5, 4, 2← 5, 3, 3	Ground					10 694.4		964
	1061	6, 2, 4← 6, 1, 5	Ground					11 007.3		964
	1061	6, 4, 3← 6, 3, 4	Ground					11 521.0		964
	1061	7, 3, 4← 7, 2, 5	Ground					9 489.9		964
	1061	7, 5, 2← 7, 4, 3	Ground					10 603.1		964
	1061	8, 4, 4← 8, 3, 5	Ground					9 447.6		964
	1061	8, 5, 3← 8, 4, 4	Ground					9 252.1		964
	1061	9, 5, 4← 9, 4, 5	Ground					8 836.8		964
$C^{12}_*H_2C^{12}H_2C^{12}_*Cl^{35}Cl^{37}$	1062	2, 2, 1← 1, 1, 0	Ground					13 781.8		964
	1062	3, 1, 3← 2, 0, 2	Ground					13 232.1		964
	1062	4, 1, 4← 3, 0, 3	Ground					16 826.1		964
	1062	4, 2, 3← 4, 1, 4	Ground					9 861.9		964
	1062	4, 4, 0← 4, 3, 1	Ground					9 899.9		964
	1062	4, 4, 1← 4, 3, 2	Ground					10 580.7		964
	1062	5, 4, 1← 5, 3, 2	Ground					8 857.4		964
	1062	5, 4, 2← 5, 3, 3	Ground					10 863.7		964
	1062	6, 4, 3← 6, 3, 4	Ground					11 601.2		964
	1062	7, 5, 2← 7, 4, 3	Ground					11 121.9		964
	1062	8, 4, 4← 8, 3, 5	Ground					8 259.2		964
	1062	9, 5, 4← 9, 4, 5	Ground					9 025.6		964
$C^{12}_*D_2C^{12}D_2C^{12}_*Cl_2^{35}$	1063	3, 0, 3← 2, 1, 2	Ground					12 062.6		964
	1063	3, 1, 3← 2, 0, 2	Ground					12 403.3		964
	1063	3, 2, 2← 2, 1, 1	Ground					15 869.4		964
	1063	4, 0, 4← 3, 1, 3	Ground					15 860.2		964
	1063	4, 2, 3← 4, 1, 4	Ground					8 605.0		964
	1063	5, 5, 0← 5, 4, 1	Ground					8 958.4		964
	1063	5, 5, 1← 5, 4, 2	Ground					9 448.0		964
	1063	6, 4, 3← 6, 3, 4	Ground					9 242.3		964
	1063	6, 5, 2← 6, 4, 3	Ground					9 554.8		964
	1063	7, 3, 4← 7, 2, 5	Ground					9 425.7		964
	1063	8, 4, 4← 8, 3, 5	Ground					8 297.0		964
	1063	8, 6, 2← 8, 5, 3	Ground					9 075.4		964
$C^{12}_*D_2C^{12}D_2C^{12}_*Cl^{35}Cl^{37}$	1064	2, 1, 2← 1, 0, 1	Ground					8 831.3		964
	1064	2, 2, 1← 1, 1, 0	Ground					12 096.5		964
	1064	3, 1, 3← 2, 0, 2	Ground					12 202.5		964
	1064	4, 1, 4← 3, 0, 3	Ground					15 665.6		964
	1064	5, 5, 1← 5, 4, 2	Ground					9 743.7		964
	1064	6, 4, 3← 6, 3, 4	Ground					9 275.7		964
	1064	6, 5, 2← 6, 4, 3	Ground					9 806.0		964

Isotopic Species	Id. No.	Rotational Quantum Nos.	Vib. State	F_1'	F'	F_1	F	Frequency MHz	Acc. ±MHz	Ref.
$C^{12}_*D_2C^{12}D_2C^{12}_*Cl^{35}Cl^{37}$	1064	7, 3, 4← 7, 2, 5	Ground					9 064.3		964
	1064	8, 6, 2← 8, 5, 3	Ground					9 689.8		964
	1064	10, 5, 5←10, 4, 6	Ground					9 560.8		964
$C^{12}_*HDC^{12}H_2C^{12}_*Cl^{35}_2$	1065	2, 0, 2← 1, 1, 1	Ground					8 191.0		964
	1065	2, 1, 2← 1, 0, 1	Ground					9 592.8		964
	1065	2, 2, 1← 1, 1, 0	Ground					13 429.5		964
	1065	3, 0, 3← 2, 1, 2	Ground					12 569.5		964
	1065	4, 1, 3← 4, 0, 4	Ground					8 767.0		964
	1065	5, 1, 4← 5, 0, 5	Ground					11 772.1		964
	1065	5, 4, 2← 5, 3, 3	Ground					9 964.8		964
	1065	5, 5, 0← 5, 4, 1	Ground					11 752.0		964
	1065	6, 4, 3← 6, 3, 4	Ground					10 863.8		964
	1065	7, 3, 4← 7, 2, 5	Ground					9 488.8		964
	1065	9, 4, 5← 9, 3, 6	Ground					11 236.2		964
$c\text{-}C^{12}_*HDC^{12}H_2C^{12}_*Cl^{35}Cl^{37}$	1066	5, 1, 4← 5, 0, 5	Ground					11 529.8		964
	1066	6, 4, 3← 6, 3, 4	Ground					10 957.8		964
	1066	6, 5, 1← 6, 4, 2	Ground					11 361.1		964
	1066	9, 4, 5← 9, 3, 6	Ground					10 652.8		964
	1066	10, 5, 5←10, 4, 6	Ground					9 516.9		964
	1066	10, 6, 4←10, 5, 5	Ground					10 455.2		964
$t\text{-}C^{12}_*HDC^{12}H_2C^{12}_*Cl^{35}Cl^{37}$	1067	4, 4, 0← 4, 3, 1	Ground					8 995.9		964
	1067	6, 4, 3← 6, 3, 4	Ground					10 903.5		964
	1067	6, 5, 1← 6, 4, 2	Ground					11 206.5		964
	1067	7, 3, 4← 7, 2, 5	Ground					9 141.0		964
	1067	8, 5, 3← 8, 4, 4	Ground					8 625.4		964
	1067	9, 4, 5← 9, 3, 6	Ground					10 758.8		964
	1067	10, 6, 4←10, 5, 5	Ground					10 270.1		964
$C^{12}_*H_2C^{13}H_2C^{12}_*Cl^{35}_2$	1068	4, 2, 3← 3, 1, 2	Ground					20 853.91		964
	1068	5, 5, 1← 4, 0, 4	Ground					20 728.54		964
	1068	8, 4, 4← 8, 3, 5	Ground					8 487.4		964
	1068	8, 5, 3← 8, 4, 4	Ground					8 485.0		964
	1068	9, 5, 5← 9, 4, 6	Ground					14 958.		964
	1068	9, 6, 4← 9, 5, 5	Ground					14 958.		964
	1068	10, 5, 5←10, 4, 6	Ground					10 060.25		964
	1068	10, 6, 4←10, 5, 5	Ground					10 058.20		964
	1068	11, 5, 6←11, 4, 7	Ground					13 171.		964
	1068	11, 7, 4←11, 6, 5	Ground					13 171.		964

C$_3$H$_4$O$_2$ C$_s$ O$_*$CH$_2$CH$_2$C$_*$O

Isotopic Species	Pt. Gp.	Id. No.	A MHz	B MHz	C MHz	D$_J$ MHz	D$_{JK}$ MHz	Δ Amu A^2	κ
O$^{16}_*$C^{12}H$_2$C^{12}H$_2$C$^{12}_*$O^{16}	C$_s$	1071	12 408.76 M	5 244.39 M	3 869.19 M				−.677923

Id. No.	μ_a Debye	μ_b Debye	μ_c Debye	eQq Value(MHz) Rel.	eQq Value(MHz) Rel.	eQq Value(MHz) Rel.	ω_a d 1/cm	ω_b d 1/cm	ω_c d 1/cm	ω_d d 1/cm
1071	3.67 M	2.00 M	0. X				120 1			

References:

ABC: 694 κ: 694 μ: 694 ω: 694

Add, Ref. 616, 695

Species 1071	A (MHz)	B (MHz)	C (MHz)	κ	Ref.
v = 1	12340.20	5247.05	3876.35	−.6761	694
v = 2	12261.00	5249.75	3883.57	−.6738	694

β-Propiolactone   Spectral Line Table

Isotopic Species	Id. No.	Rotational Quantum Nos.	Vib. State v$_a$	F$_1'$	F'	F$_1$	F	Frequency MHz	Acc. ±MHz	Ref.
O$^{16}_*$C^{12}H$_2$C^{12}H$_2$C$^{12}_*$O^{16}	1071	2, 0, 2← 1, 0, 1	Ground					18 047.49	.1	694
	1071	2, 1, 1← 1, 1, 0	Ground					19 602.54	.1	694
	1071	3, 0, 3← 2, 0, 2	Ground					26 641.64	.1	694
	1071	3, 1, 2← 2, 1, 1	Ground					29 282.90	.1	694
	1071	3, 1, 2← 2, 1, 1	1					29 305.27	.10	694
	1071	3, 1, 2← 2, 1, 1	2					29 327.65	.20	694
	1071	3, 1, 3← 2, 1, 2	Ground					25 172.11	.1	694
	1071	3, 1, 3← 2, 1, 2	1					25 208.13	.10	694
	1071	3, 1, 3← 2, 1, 2	2					25 244.25	.20	694
	1071	3, 2, 1← 2, 2, 0	Ground					28 039.96	.1	694
	1071	3, 2, 2← 2, 2, 1	Ground					27 340.69	.1	694
	1071	3, 2, 2← 2, 2, 1	1					27 370.21	.10	694
	1071	3, 2, 2← 2, 2, 1	2					27 399.95	.20	694
	1071	4, 1, 4← 3, 1, 3	Ground					33 384.4	.2	694

C$_3$H$_5$Cl   C$_s$ C$_*$H$_2$CH$_2$C$_*$HCl

Isotopic Species	Pt. Gp.	Id. No.	A MHz		B MHz		C MHz		D$_J$ MHz	D$_{JK}$ MHz	Δ Amu A^2	κ
C$^{12}_*$H$_2$C^{12}H$_2$C$^{12}_*$HCl^{35}	C$_s$	1081	16 625.	M	3 905.4	M	3 622.5	M				−.9565 [1]
C$^{12}_*$H$_2$C^{12}H$_2$C$^{12}_*$HCl^{37}	C$_s$	1082	16 085.	M	3 810.4	M	3 540.8	M				−.9569 [1]

Id. No.	μ$_a$ Debye	μ$_b$ Debye	μ$_c$ Debye	eQq Value(MHz) Rel.		eQq Value(MHz) Rel.		eQq Value(MHz) Rel.		ω$_a$ d 1/cm	ω$_b$ d 1/cm	ω$_c$ d 1/cm	ω$_d$ d 1/cm
1081	1.78 L		0. X	−55.6	aa	23.5	bb	32.1	cc				

1. The asymmetry parameter, κ, was computed according to $\kappa = (3b_p + 1)/(b_p - 1)$.

References:

ABC: 818   κ: 818   μ: 995   eQq: 818

Add. Ref. 428,511

1-Chlorocyclopropane

Isotopic Species	Id. No.	Rotational Quantum Nos.	Vib. State	F$_1'$	F'	F$_1$	F	Frequency MHz	Acc. ±MHz	Ref.
C$^{12}_*$H$_2$C^{12}H$_2$C$^{12}_*$HCl^{35}	1081	3, 0, 3← 2, 0, 2	Ground		3/2		1/2	22 562.3	.1	818
	1081	3, 0, 3← 2, 0, 2	Ground		5/2		3/2	22 562.3	.1	818
	1081	3, 0, 3← 2, 0, 2	Ground		9/2		7/2	22 565.9	.1	818
	1081	3, 0, 3← 2, 0, 2	Ground		7/2		5/2	22 565.9	.1	818
	1081	3, 1, 2← 2, 1, 1	Ground		7/2		7/2	22 996.5	.1	818
	1081	3, 1, 2← 2, 1, 1	Ground		5/2		3/2	23 003.1	.1	818
	1081	3, 1, 2← 2, 1, 1	Ground		7/2		5/2	23 003.1	.1	818
	1081	3, 1, 2← 2, 1, 1	Ground		3/2		1/2	23 007.0	.1	818
	1081	3, 1, 2← 2, 1, 1	Ground		9/2		7/2	23 007.0	.1	818
	1081	3, 1, 2← 2, 1, 1	Ground		5/2		5/2	23 009.4	.1	818
	1081	3, 1, 2← 2, 1, 1	Ground		3/2		3/2	23 015.6	.1	818
	1081	3, 1, 3← 2, 1, 2	Ground		7/2		7/2	22 148.7	.1	818
	1081	3, 1, 3← 2, 1, 2	Ground		5/2		3/2	22 154.5	.1	818
	1081	3, 1, 3← 2, 1, 2	Ground		7/2		5/2	22 154.5	.1	818
	1081	3, 1, 3← 2, 1, 2	Ground		9/2		7/2	22 157.8	.1	818
	1081	3, 1, 3← 2, 1, 2	Ground		3/2		1/2	22 157.8	.1	818
	1081	3, 1, 3← 2, 1, 2	Ground		5/2		5/2	22 160.4	.1	818
	1081	3, 1, 3← 2, 1, 2	Ground		3/2		3/2	22 164.0	.1	818
	1081	3, 2, 1← 2, 2, 0	Ground		7/2		5/2	22 592.1	.1	818
	1081	3, 2, 1← 2, 2, 0	Ground		5/2		3/2	22 602.4	.1	818
	1081	3, 2, 1← 2, 2, 0	Ground		9/2		7/2	22 606.6	.1	818
	1081	3, 2, 1← 2, 2, 0	Ground		3/2		1/2	22 616.5	.1	818
	1081	3, 2, 2← 2, 2, 1	Ground		7/2		5/2	22 573.5	.1	818
	1081	3, 2, 2← 2, 2, 1	Ground		5/2		3/2	22 583.8	.1	818
	1081	3, 2, 2← 2, 2, 1	Ground		9/2		7/2	22 587.8	.1	818
	1081	3, 2, 2← 2, 2, 1	Ground		3/2		1/2	22 597.6	.1	818
	1081	4, 1, 3← 3, 1, 2	Ground		7/2		5/2	30 667.5	.1	818
	1081	4, 1, 3← 3, 1, 2	Ground		9/2		7/2	30 668.8	.1	818
	1081	4, 1, 4← 3, 1, 3	Ground		9/2		9/2	29 527.1	.1	818
	1081	4, 1, 4← 3, 1, 3	Ground		7/2		5/2	29 535.1	.1	818

Isotopic Species	Id. No.	Rotational Quantum Nos.	Vib. State	F_1'	F'	F_1	F	Frequency MHz	Acc. ±MHz	Ref.
$C^{12}{}_*H_2C^{12}H_2C^{12}{}_*HCl^{35}$	1081	4, 1, 4← 3, 1, 3	Ground		9/2		7/2	29 535.6	.1	818
	1081	4, 1, 4← 3, 1, 3	Ground		5/2		3/2	29 536.4	.1	818
	1081	4, 1, 4← 3, 1, 3	Ground		11/2		9/2	29 536.9	.1	818
	1081	4, 1, 4← 3, 1, 3	Ground		7/2		7/2	29 539.1	.1	818
	1081	4, 1, 4← 3, 1, 3	Ground		5/2		5/2	29 545.4	.1	818
	1081	4, 2, 2← 3, 2, 1	Ground		9/2		7/2	30 151.0	.1	818
	1081	4, 2, 2← 3, 2, 1	Ground		7/2		5/2	30 153.1	.1	818
	1081	4, 2, 2← 3, 2, 1	Ground		11/2		9/2	30 156.9	.1	818
	1081	4, 2, 2← 3, 2, 1	Ground		5/2		3/2	30 158.3	.1	818
	1081	4, 2, 3← 3, 2, 2	Ground		9/2		7/2	30 104.3	.1	818
	1081	4, 2, 3← 3, 2, 2	Ground		7/2		5/2	30 106.2	.1	818
	1081	4, 2, 3← 3, 2, 2	Ground		11/2		9/2	30 109.8	.1	818
	1081	4, 2, 3← 3, 2, 2	Ground		5/2		3/2	30 112.0	.1	818
	1081	4, 3, 1← 3, 3, 0	Ground		7/2		7/2	30 112.0	.1	818
	1081	4, 3, 1← 3, 3, 0	Ground		5/2		5/2	30 113.5	.1	818
	1081	4, 3, 1← 3, 3, 0	Ground		9/2		7/2	30 113.5	.1	818
	1081	4, 3, 1← 3, 3, 0	Ground		7/2		5/2	30 118.8	.1	818
	1081	4, 3, 1← 3, 3, 0	Ground		11/2		9/2	30 124.9	.1	818
	1081	4, 3, 1← 3, 3, 0	Ground		5/2		3/2	30 131.6	.1	818
	1081	4, 3, 1← 3, 3, 0	Ground		9/2		9/2	30 131.6	.1	818
	1081	4, 3, 2← 3, 3, 1	Ground		7/2		7/2	30 112.0	.1	818
	1081	4, 3, 2← 3, 3, 1	Ground		9/2		7/2	30 113.5	.1	818
	1081	4, 3, 2← 3, 3, 1	Ground		5/2		5/2	30 113.5	.1	818
	1081	4, 3, 2← 3, 3, 1	Ground		7/2		5/2	30 118.8	.1	818
	1081	4, 3, 2← 3, 3, 1	Ground		11/2		9/2	30 124.9	.1	818
	1081	4, 3, 2← 3, 3, 1	Ground		9/2		9/2	30 131.6	.1	818
	1081	4, 3, 2← 3, 3, 1	Ground		5/2		3/2	30 131.6	.1	818
$C^{12}{}_*H_2C^{12}H_2C^{12}{}_*HCl^{37}$	1082	3, 0, 3← 2, 0, 2	Ground		9/2		7/2	22 036.2	.1	818
	1082	3, 0, 3← 2, 0, 2	Ground		7/2		5/2	22 036.2	.1	818
	1082	3, 1, 2← 2, 1, 1	Ground		7/2		5/2	22 453.2	.1	818
	1082	3, 1, 2← 2, 1, 1	Ground		5/2		3/2	22 453.2	.1	818
	1082	3, 1, 2← 2, 1, 1	Ground		3/2		1/2	22 456.0	.1	818
	1082	3, 1, 2← 2, 1, 1	Ground		9/2		7/2	22 456.0	.1	818
	1082	3, 1, 2← 2, 1, 1	Ground		5/2		5/2	22 457.7	.1	818
	1082	3, 1, 3← 2, 1, 2	Ground		5/2		3/2	21 644.8	.1	818
	1082	3, 1, 3← 2, 1, 2	Ground		7/2		5/2	21 644.8	.1	818
	1082	3, 1, 3← 2, 1, 2	Ground		3/2		1/2	21 647.5	.1	818
	1082	3, 1, 3← 2, 1, 2	Ground		9/2		7/2	21 647.5	.1	818
	1082	3, 2, 1← 2, 2, 0	Ground		7/2		5/2	22 062.7	.1	818
	1082	3, 2, 1← 2, 2, 0	Ground		5/2		3/2	22 069.5	.1	818
	1082	3, 2, 1← 2, 2, 0	Ground		9/2		7/2	22 072.7	.1	818
	1082	3, 2, 2← 2, 2, 1	Ground		7/2		5/2	22 045.2	.1	818
	1082	3, 2, 2← 2, 2, 1	Ground		5/2		3/2	22 053.1	.1	818
	1082	3, 2, 2← 2, 2, 1	Ground		9/2		7/2	22 056.7	.1	818
	1082	3, 2, 2← 2, 2, 1	Ground		3/2		1/2	22 062.7	.1	818
	1082	4, 0, 4← 3, 0, 3	Ground					29 358.4	.3	818
	1082	4, 1, 3← 3, 1, 2	Ground					29 935.1		600
	1082	4, 1, 4← 3, 1, 3	Ground					28 856.2		600
	1082	4, 2, 2← 3, 2, 1	Ground		9/2		7/2	29 439.9	.1	818
	1082	4, 2, 2← 3, 2, 1	Ground		7/2		5/2	29 441.7	.1	818
	1082	4, 2, 2← 3, 2, 1	Ground		11/2		9/2	29 444.6	.1	818
	1082	4, 2, 2← 3, 2, 1	Ground		5/2		3/2	29 446.2	.1	818
	1082	4, 2, 3← 3, 2, 2	Ground					29 403.8		600

1,3-Epoxypropane, Oxetane

C_3H_6O C_{2v} $C_*H_2CH_2CH_2O_*$

Isotopic Species	Pt. Gp.	Id. No.	A MHz		B MHz		C MHz		D_J MHz	D_{JK} MHz	Δ Amu A²	κ
$C^{12}_*H_2C^{12}H_2C^{12}H_2O^{16}_*$	C_{2v}	1091	12 045.2	M	11 734.0	M	6 730.7	M				
$C^{12}_*D_2C^{12}H_2C^{12}D_2O^{16}_*$	C_{2v}	1092	10 910.53	M	8 976.55	M	5 860.99	M				
$C^{12}_*H_2C^{12}H_2C^{12}H_2O^{18}_*$	C_{2v}	1093	12 044.6	M	11 207.5	M	6 554.6	M				
$C^{12}_*H_2C^{12}HDC^{12}H_2O^{16}_*$	C_s	1094	11 839.53	M	10 781.54	M	6 466.75	M				

Id. No.	μ_a Debye		μ_b Debye		μ_c Debye		eQq Value(MHz) Rel.	eQq Value(MHz) Rel.	eQq Value(MHz) Rel.	ω_a d 1/cm	ω_b d 1/cm	ω_c d 1/cm	ω_d d 1/cm
1091	1.93	M	0.	X	0.	X				100 1			
1094	1.93	M	0.	X	.02	M							

References:

ABC: 896 μ: 896 ω: 896

Add. Ref. 598,862

Ref. 896 gives the following rotational constants for excited states:

Excited States	Species: (all values in MHz)			
	1091	1092	1093	1094
1st	A: 12058.0	10899.39	12054.8	11852.0
	B: 11726.0	8991.74	11201.3	10776.7
	C: 6772.6	5894.41	6594.8	6504.6
2nd	A: 12058.9	10890.54		
	B: 11718.8	8996.45		
	C: 6789.1	5908.07		
3rd	A: 12060.2	10880.5		
	B: 11710.0	9002.0		
	C: 6809.6	5924.3		
4th	A: 12058.0			
	B: 11698.7			
	C: 6827.6			

Ref. 896 also gives, for species 1091, an inversion barrier height = 35 cm^{-1}.

Isotopic Species	Id. No.	Rotational Quantum Nos.	Vib. State v_a	F'_1	F'	F_1	F	Frequency MHz	Acc. ±MHz	Ref.
$C^{12}{}_*H_2C^{12}H_2C^{12}H_2O^{16}{}_*$	1091	1, 0, 1← 0, 0, 0	Ground					18 465.0	.1	896
	1091	1, 0, 1← 0, 0, 0	1					18 498.		896
	1091	2, 0, 2← 1, 0, 1	Ground					32 223.3	.1	896
	1091	2, 0, 2← 1, 0, 1	1					32 358.6	.1	896
	1091	2, 0, 2← 1, 0, 1	2					32 405.	2.	896
	1091	2, 0, 2← 1, 0, 1	3					32 471.3	.1	896
	1091	2, 0, 2← 1, 0, 1	4					32 521.6	.1	896
	1091	2, 1, 1← 1, 1, 0	Ground					41 932.7	.1	896
	1091	2, 1, 1← 1, 1, 0	3					41 940.5	.1	896
	1091	2, 1, 1← 1, 1, 0	2					41 945.7	.1	896
	1091	2, 1, 1← 1, 1, 0	1					41 950.5	.1	896
	1091	2, 1, 2← 1, 1, 1	Ground					31 926.0	.1	896
	1091	2, 1, 2← 1, 1, 1	1					32 043.4	.1	896
	1091	2, 1, 2← 1, 1, 1	2					32 086.0	.2	896
	1091	2, 1, 2← 1, 1, 1	3					32 138.9	.1	896
	1091	2, 1, 2← 1, 1, 1	4					32 182.	2.	896
	1091	3, 3, 1← 3, 1, 2	2					16 417.5	.1	896
	1091	3, 3, 1← 3, 1, 2	1					16 443.8	.1	896
	1091	3, 3, 1← 3, 1, 2	Ground					16 490.2	.1	896
	1091	4, 4, 1← 4, 2, 2	Ground					17 268.2	.1	896
	1091	4, 4, 1← 4, 2, 2	3					17 278.1	.1	896
	1091	4, 4, 1← 4, 2, 2	2					17 281.6	.1	896
	1091	4, 4, 1← 4, 2, 2	1					17 283.2	.1	896
	1091	5, 5, 1← 5, 3, 2	Ground					18 346.5	.1	896
	1091	5, 5, 1← 5, 3, 2	1					18 452.1	.1	896
	1091	5, 5, 1← 5, 3, 2	2					18 490.0	.1	896
	1091	6, 5, 2← 6, 3, 3	1					25 928.		896
	1091	6, 5, 2← 6, 3, 3	Ground					26 081.		896
	1091	7, 5, 2← 7, 5, 3	1					24 240.		896
	1091	7, 5, 2← 7, 5, 3	Ground					24 592.		896
	1091	7, 6, 2← 7, 4, 3	2					26 152.0	.1	896
	1091	7, 6, 2← 7, 4, 3	1					26 215.		896
	1091	7, 6, 2← 7, 4, 3	Ground					26 333.		896
	1091	8, 6, 2← 8, 6, 3	1					23 460.		896
	1091	8, 6, 2← 8, 6, 3	Ground					23 890.		896
	1091	8, 7, 2← 8, 5, 3	3					26 587.1	.1	896
	1091	8, 7, 2← 8, 5, 3	2					26 630.1	.1	896
	1091	8, 7, 2← 8, 5, 3	1					26 665.		896
	1091	8, 7, 2← 8, 5, 3	Ground					26 725.		896
	1091	9, 7, 2← 9, 7, 3	Ground					22 976.		896
	1091	9, 8, 2← 9, 6, 3	Ground					27 302.9	.1	896
	1091	9, 8, 2← 9, 6, 3	1					27 332.3	.1	896
	1091	9, 8, 2← 9, 6, 3	2					27 337.0	.1	896
	1091	9, 8, 2← 9, 6, 3	4					27 341.8	.1	896
	1091	9, 8, 2← 9, 6, 3	3					27 345.0	.1	896
	1091	10, 9, 2←10, 7, 3	Ground					28 125.0	.1	896
	1091	10, 9, 2←10, 7, 3	1					28 279.0	.1	896
	1091	10, 9, 2←10, 7, 3	2					28 342.4	.1	896
	1091	10, 9, 2←10, 7, 3	3					28 421.2	.1	896
	1091	10, 9, 2←10, 7, 3	4					28 485.0	.1	896

Isotopic Species	Id. No.	Rotational Quantum Nos.	Vib. State v_a	Hyperfine F_1'	F'	F_1	F	Frequency MHz	Acc. ±MHz	Ref.
$C^{12}{}_*H_2C^{12}H_2C^{12}H_2O^{16}{}_*$	1091	11,10, 2←11, 8, 3	Ground					29 252.6	.1	896
	1091	11,10, 2←11, 8, 3	2					29 717.5	.1	896
	1091	11,10, 2←11, 8, 3	3					29 893.2	.1	896
	1091	11,10, 2←11, 8, 3	4					30 040.8	.1	955
$C^{12}{}_*D_2C^{12}H_2C^{12}D_2O^{16}{}_*$	1092	2, 1, 2← 1, 0, 1	1					28 582.65	.1	955
	1092	2, 1, 2← 1, 0, 1	2					28 614.75	.1	955
	1092	2, 1, 2← 1, 0, 1	3					28 654.5	.5	955
	1092	2, 1, 2← 1, 0, 1	Ground					29 493.5	.1	955
	1092	2, 2, 1← 1, 1, 0	3					38 565.8	.1	955
	1092	2, 2, 1← 1, 1, 0	2					38 579.7	.1	955
	1092	2, 2, 1← 1, 1, 0	Ground					39 592.57	.1	955
	1092	2, 2, 1← 1, 1, 0	1					39 592.6	.1	955
	1092	3, 0, 3← 2, 1, 2	Ground					38 735.5	.1	955
	1092	3, 0, 3← 2, 1, 2	1					38 915.7	.3	955
	1092	3, 2, 2← 3, 1, 3	1					20 610.72	.1	955
	1092	4, 1, 3← 4, 0, 4	3					26 426.	2.	955
	1092	4, 1, 3← 4, 0, 4	1					26 623.1	.1	955
	1092	4, 1, 3← 4, 0, 4	Ground					28 806.9	.1	955
	1092	4, 2, 3← 4, 1, 4	3					27 327.	2.	955
	1092	4, 2, 3← 4, 1, 4	Ground					27 758.8	.1	955
	1092	4, 3, 2← 4, 2, 3	1					22 097.3	.1	955
	1092	4, 3, 2← 4, 2, 3	Ground					22 291.15	.1	955
	1092	4, 4, 0← 4, 3, 1	Ground					18 674.62	.1	955
	1092	4, 4, 1← 4, 3, 2	Ground					22 740.19	.1	955
	1092	5, 2, 3← 5, 1, 4	1					24 812.0	.1	955
	1092	5, 2, 3← 5, 1, 4	Ground					24 960.27	.1	955
	1092	5, 3, 3← 5, 2, 4	3					27 613.	2.	955
	1092	5, 3, 3← 5, 2, 4	2					27 741.2	.1	955
	1092	5, 3, 3← 5, 2, 4	1					27 847.3	.1	955
	1092	5, 3, 3← 5, 2, 4	Ground					28 065.3	.1	955
	1092	5, 4, 2← 5, 3, 3	2					24 531.6	.1	955
	1092	5, 4, 2← 5, 3, 3	Ground					24 889.50	.1	955
	1092	5, 5, 1← 5, 4, 2	3					27 263.	2.	955
	1092	5, 5, 1← 5, 4, 2	2					27 450.	2.	955
	1092	6, 3, 3← 6, 2, 4	Ground					22 306.55	.1	955
	1092	6, 4, 3← 6, 3, 4	3					28 573.	2.	955
	1092	6, 4, 3← 6, 3, 4	2					28 716.5	.5	955
	1092	6, 4, 3← 6, 3, 4	1					28 836.27	.1	955
	1092	6, 4, 3← 6, 3, 4	Ground					29 080.7	.1	955
	1092	6, 5, 2← 6, 4, 3	3					27 993.	2.	955
	1092	6, 5, 2← 6, 4, 3	Ground					28 624.7	.5	955
	1092	8, 4, 4← 8, 3, 5	Ground					28 818.6	.1	955
	1092	8, 6, 2← 8, 5, 3	Ground					24 785.55	.1	955
	1092	10, 7, 3←10, 6, 4	1					26 885.	5.	955
	1092	10, 7, 3←10, 6, 4	Ground					27 293.4	.1	955
	1092	12, 8, 4←12, 7, 5	Ground					28 778.	2.	955
$C^{12}{}_*H_2C^{12}H_2C^{12}H_2O^{18}{}_*$	1093	2, 0, 2← 1, 0, 1	Ground					31 605.4	.1	955
	1093	2, 0, 2← 1, 0, 1	1					31 730.	3.	955
	1093	2, 1, 2← 1, 1, 1	Ground					30 871.35	.1	955

Isotopic Species	Id. No.	Rotational Quantum Nos.	Vib. State v_a	Hyperfine F_1'	F'	F_1	F	Frequency MHz	Acc. ±MHz	Ref.
$C^{12}{}_*H_2C^{12}H_2C^{12}H_2O^{18}{}_*$	1093	2, 1, 2← 1, 1, 1	1					30 985.7	.1	955
	1093	6, 3, 3← 6, 3, 4	Ground					34 624.3	.1	955
	1093	6, 5, 2← 6, 3, 3	1					27 582.45	.1	955
	1093	6, 5, 2← 6, 3, 3	Ground					27 656.0	.1	955
	1093	7, 6, 2← 7, 4, 3	Ground					29 751.	5.	955
	1093	7, 6, 2← 7, 4, 3	1					29 783.0	.1	955
	1093	8, 5, 4← 8, 5, 3	1					32 002.9	.1	955
	1093	8, 5, 4← 8, 5, 3	Ground					32 440.75	.1	955
	1093	8, 7, 2← 8, 5, 3	Ground					32 973.15	.1	955
	1093	9, 6, 3← 9, 6, 4	1					29 910.5	.1	955
$C^{12}{}_*H_2C^{12}HDC^{12}H_2O^{16}{}_*$	1094	1, 0, 1← 0, 0, 0	Ground					17 248.3	.1	955
	1094	1, 0, 1← 0, 0, 0	1					17 281.33	.1	955
	1094	2, 0, 2← 1, 0, 1	Ground					31 067.9	.1	955
	1094	2, 0, 2← 1, 0, 1	1					31 187.2	.1	955
	1094	2, 1, 1← 1, 1, 0	Ground					38 811.4	.1	955
	1094	2, 1, 1← 1, 1, 0	1					38 834.73	.1	955
	1094	2, 1, 2← 1, 1, 1	Ground					30 181.8	.1	955
	1094	2, 1, 2← 1, 1, 1	1					30 290.5	.1	955
	1094	3, 1, 2← 3, 1, 3	Ground					23 488.5	.1	955
	1094	3, 2, 2← 3, 0, 3	Ground					24 336.87	.1	955
	1094	3, 3, 1← 3, 1, 2	1					18 621.8	.1	955
	1094	3, 3, 1← 3, 1, 2	Ground					18 635.55	.1	955
	1094	4, 1, 3← 4, 1, 4	1					33 267.02	.1	955
	1094	4, 1, 3← 4, 1, 4	Ground					33 522.45	.1	955
	1094	4, 2, 3← 4, 0, 4	1					33 398.61	.1	955
	1094	4, 2, 3← 4, 0, 4	Ground					33 646.22	.1	955
	1094	4, 3, 2← 4, 1, 3	1					24 730.0	.1	955
	1094	4, 3, 2← 4, 1, 3	Ground					24 872.5	.1	955
	1094	5, 2, 3← 5, 2, 4	Ground					33 123.27	.1	955
	1094	5, 3, 3← 5, 1, 4	1					33 367.40	.1	955
	1094	5, 3, 3← 5, 1, 4	Ground					33 610.18	.1	955
	1094	6, 4, 3← 6, 2, 4	1					33 462.65	.1	955
	1094	6, 4, 3← 6, 2, 4	Ground					33 695.12	.1	955
	1094	7, 5, 3← 7, 3, 4	1					33 884.65	.1	955
	1094	7, 5, 3← 7, 3, 4	Ground					34 082.82	.1	955
	1094	8, 6, 3← 8, 4, 4	1					34 926.2	.1	955
	1094	8, 6, 3← 8, 4, 4	Ground					35 051.33	.1	955

s-Trioxane, α-Trioxymethylene

C₃H₆O₃ — $C_3H_6O_3$

C₃ᵥ — C_{3v}

O*CH₂OCH₂OC*H₂ — $O_*CH_2OCH_2OC_*H_2$

Isotopic Species	Pt. Gp.	Id. No.	A MHz	B MHz	C MHz	D_J MHz	D_JK MHz	Δ Amu A²	κ
$O^{16}{}_*C^{12}H_2O^{16}C^{12}H_2O^{16}C^{12}{}_*H_2$	C_{3v}	1101	5 273.6 M	5 273.6 M					
$O^{16}{}_*C^{12}H_2O^{16}C^{12}H_2O^{16}C^{13}{}_*H_2$	C_s	1102	5 225.0 M	5 225.0 M					

Id. No.	μ_a Debye	μ_b Debye	μ_c Debye	eQq Value(MHz) Rel.	eQq Value(MHz) Rel.	eQq Value(MHz) Rel.	ω_a d 1/cm	ω_b d 1/cm	ω_c d 1/cm	ω_d d 1/cm
1101	0. X	0. X	2.08 M							

References:

ABC: 1029   μ: 253

1,3,5-Trioxane

Isotopic Species	Id. No.	Rotational Quantum Nos.	Vib. State	F'_1	F'	F_1	F	Frequency MHz	Acc. ±MHz	Ref.
$O^{16}{}_*C^{12}H_2O^{16}C^{12}H_2O^{16}C^{12}{}_*H_2$	1101	2, ← 1,	Ground					21 094.3		253
$O^{16}{}_*C^{12}H_2O^{16}C^{16}H_2O^{16}C^{13}{}_*H_2$	1102	2, 0, ← 1, 0,	Ground					20 900.		253
	1102	2, 1, ← 1, 1,	Ground					20 804.		253

C₃N₃P — C_3N_3P

C₃ᵥ — C_{3v}

P(CN)₃ — $P(CN)_3$

Isotopic Species	Pt. Gp.	Id. No.	A MHz	B MHz	C MHz	D_J MHz	D_JK MHz	Δ Amu A²	κ
$P^b(C^{12}N^{14})_3$	C_{3v}	1111		2 326. M					

References:

ABC: 446

Phosphorus Tricyanide

Isotopic Species	Id. No.	Rotational Quantum Nos.	Vib. State	F'_1	F'	F_1	F	Frequency MHz	Acc. ±MHz	Ref.
$P^b(C^{12}N^{14})_3$	1111	5, ← 4,	Ground					23 265.		446

Methylcyanoacetylene

C_4H_3N C_{3v} $H_3CC:CCN$

Isotopic Species	Pt. Gp.	Id. No.	A MHz	B MHz		C MHz		D_J MHz	D_{JK} MHz	Δ Amu A²	κ
$H_3C^{12}C^{12}:C^{12}C^{12}N^{14}$	C_{3v}	1121		2 065.73	M	2 065.73	M	.0001	.0198		
$H_3C^{13}C^{12}:C^{12}C^{12}N^{14}$	C_{3v}	1122		2 010.63	M	2 010.63	M				
$H_3C^{12}C^{13}:C^{12}C^{12}N^{14}$	C_{3v}	1123		2 054.77	M	2 054.77	M				
$H_3C^{12}C^{12}:C^{12}C^{13}N^{14}$	C_{3v}	1124		2 048.81	M	2 048.81	M				
$H_3C^{12}C^{12}:C^{12}C^{12}N^{15}$	C_{3v}	1125		2 011.57	M	2 011.57	M				
$D_3C^{12}C^{12}:C^{12}C^{12}N^{14}$	C_{3v}	1126		1 858.15	M	1 858.15	M		.0145		
$D_3C^{13}C^{12}:C^{12}C^{12}N^{14}$	C_{3v}	1127		1 817.75	M	1 817.75	M				
$D_3C^{12}C^{12}:C^{12}C^{13}N^{14}$	C_{3v}	1128		1 841.79	M	1 841.79	M				

Id. No.	μ_a Debye	μ_b Debye	μ_c Debye	eQq Value(MHz)	Rel.	eQq Value(MHz)	Rel.	eQq Value(MHz)	Rel.	ω_a d 1/cm	ω_b d 1/cm	ω_c d 1/cm	ω_d d 1/cm
1121				−4.4	N^{14}								

References:

ABC: 559,1012 D_J: 559 D_{JK}: 559,1009   eQq: 559

Isotopic Species	Id. No.	Rotational Quantum Nos.	Vib. State	Hyperfine				Frequency MHz	Acc. ±MHz	Ref.
				F_1'	F'	F_1	F			
$H_3C^{12}C^{12}:C^{12}C^{12}N^{14}$	1121	6, 0← 5, 0	Ground					24 788.69		1009
	1121	6, 1← 5, 1	Ground					24 788.51		1009
	1121	6, 2← 5, 2	Ground					24 787.85		1009
	1121	6, 3← 5, 3	Ground		6		5	24 786.31		1009
	1121	6, 3← 5, 3	Ground		5		4	24 786.61		1009
	1121	6, 3← 5, 3	Ground		7		6	24 786.61		1009
	1121	6, 4← 5, 4	Ground		6		5	24 784.37		1009
	1121	6, 4← 5, 4	Ground		7		6	24 785.14		1009
	1121	6, 4← 5, 4	Ground		5		4	24 785.14		1009
	1121	9, 0← 8, 0	Ground					37 182.96		1009
	1121	9, 1← 8, 1	Ground					37 182.67		1009
	1121	9, 2← 8, 2	Ground					37 181.60		1009
	1121	9, 3← 8, 3	Ground					37 179.80		1009
	1121	9, 4← 8, 4	Ground					37 177.31		1009
	1121	9, 5← 8, 5	Ground					37 174.24		1009
	1121	9, 6← 8, 6	Ground					37 170.29		1009
	1121	9, 7← 8, 7	Ground					37 165.6		1009
	1121	11, 0←10, 0	Ground					45 445.60		1009
	1121	11, 1←10, 1	Ground					45 445.20		1009
	1121	11, 2←10, 2	Ground					45 443.90		1009
	1121	11, 3←10, 3	Ground					45 441.70		1009
	1121	11, 4←10, 4	Ground					45 438.66		1009
	1121	11, 5←10, 5	Ground					45 434.72		1009
	1121	11, 6←10, 6	Ground					45 429.98		1009
	1121	11, 7←10, 7	Ground					45 424.2		1009
	1121	11, 8←10, 8	Ground					45 417.5		1009
	1121	11, 9←10, 9	Ground					45 410.2		1009
$D_3C^{12}C^{12}:C^{12}C^{12}N^{14}$	1126	10, 1← 9, 1	Ground					37 162.40	.2	1009
	1126	10, 2← 9, 2	Ground					37 161.55	.2	1009
	1126	10, 3← 9, 3	Ground					37 160.12	.2	1009
	1126	10, 4← 9, 4	Ground					37 158.03	.2	1009
	1126	10, 5← 9, 5	Ground					37 155.40	.2	1009
	1126	10, 6← 9, 6	Ground					37 152.26	.2	1009
	1126	10, 0←11, 0	Ground					37 162.66	.2	1009

C₄H₄ → C_4H_4

C_s

$HC:CCH:CH_2$

Isotopic Species	Pt. Gp.	Id. No.	A MHz	B MHz	C MHz	D_J MHz	D_{JK} MHz	Δ Amu A²	κ
$HC^{12}:C^{12}C^{12}H:C^{12}H_2$	C_s	1131		4 744.85 M	4 329.73 M				−.982316

Id. No.	μ_a Debye	μ_b Debye	μ_c Debye	eQq Value(MHz) Rel.	eQq Value(MHz) Rel.	eQq Value(MHz) Rel.	ω_a d 1/cm	ω_b d 1/cm	ω_c d 1/cm	ω_d d 1/cm
1131	.43 M	.09 M	0. X							

References:

ABC: 379   κ: 379   μ: 983

1-Buten-3-yne

Spectral Line Table

Isotopic Species	Id. No.	Rotational Quantum Nos.	Vib. State	F_1'	F'	F_1	F	Frequency MHz	Acc. ±MHz	Ref.
$HC^{12}:C^{12}C^{12}H:C^{12}H_2$	1131	1, , ← 0, ,	Ground					9 074.72	.03	983
	1131	2, 0, 2← 1, 0, $\bar{1}$	Ground					18 146.52		709
	1131	2, 1, 1← 1, 1, 0	Ground					18 564.74		709
	1131	2, 1, 2← 1, 1, 1	Ground					17 734.50		709
	1131	3, 0, 3← 2, 0, 2	Ground					27 212.71		379
	1131	3, 1, 2← 2, 1, 1	Ground					27 845.35	.05	379
	1131	3, 1, 3← 2, 1, 2	Ground					26 600.00	.05	379
	1131	3, 2, 1← 2, 2, 0	Ground					27 237.09	.05	379
	1131	3, 2, 2← 2, 2, 1	Ground					27 226.03	.05	379
	1131	4, 0, 4← 3, 0, 3	Ground					36 270.16	.05	709
	1131	4, 1, 3← 3, 1, 2	Ground					37 123.87		709
	1131	4, 1, 4← 3, 1, 3	Ground					35 463.23		709
	1131	4, 2, 2← 3, 2, 1	Ground					36 327.98		709
	1131	4, 2, 3← 3, 2, 2	Ground					36 298.87		709
	1131	4, 3, 1← 3, 3, 0	Ground					36 310.02		709
	1131	4, 3, 2← 3, 3, 1	Ground					36 310.02		709

C$_4$H$_4$N$_2$ C$_{2v}$ C$_*$HNCHCH:CHN$_*$

Isotopic Species	Pt. Gp.	Id. No.	A MHz	B MHz	C MHz	D$_J$ MHz	D$_{JK}$ MHz	Δ Amu A^2	κ
C$^{12}_*$HN^{14}C^{12}HC^{12}H:C^{12}HN$^{14}_*$	C$_{2v}$	1141	6 276.84 M	6 067.29 M	3 084.34 M				.87
C$^{12}_*$DN^{14}C^{12}DC^{12}H:C^{12}DN$^{14}_*$	C$_{2v}$	1142	5 692.48 M	5 457.33 M	2 785.76 M				

Id. No.	μ$_a$ Debye	μ$_b$ Debye	μ$_c$ Debye	eQq Value(MHz) Rel.	eQq Value(MHz) Rel.	eQq Value(MHz) Rel.	ω$_a$ d 1/cm	ω$_b$ d 1/cm	ω$_c$ d 1/cm	ω$_d$ d 1/cm
1141	0. X	2.42 ·L	0. X							

References:

ABC: 923   κ: 923   μ: 1031

Pyrimidine   Spectral Line Table

Isotopic Species	Id. No.	Rotational Quantum Nos.	Vib. State	F$_1'$	F'	F$_1$	F	Frequency MHz	Acc. ±MHz	Ref.
C$^{12}_*$HN^{14}C^{12}HC^{12}H:C^{12}HN$^{14}_*$	1141	2, 2, 0← 1, 1, 1	Ground					27 680.		923
	1141	2, 2, 1← 1, 1, 0	Ground					21 914.85		923
	1141	3, 0, 3← 2, 1, 2	Ground					21 586.9		923
	1141	3, 1, 3← 2, 0, 2	Ground					21 598.5		923
	1141	3, 1, 2← 2, 2, 1	Ground					27 403.5		923
	1141	3, 2, 2← 2, 1, 1	Ground					28 083.53		923
	1141	4, 0, 4← 3, 1, 3	Ground					27 761.0		923
	1141	4, 1, 4← 3, 0, 3	Ground					27 761.0		923
	1141	5, 2, 3← 5, 1, 4	Ground					21 575.9		923
	1141	5, 2, 4← 5, 1, 5	Ground					27 761.		923
	1141	6, 3, 3← 6, 2, 4	Ground					21 543.		923
	1141	6, 4, 3← 6, 3, 4	Ground					21 571.8		923
	1141	7, 3, 4← 7, 2, 5	Ground					27 741.1		923
	1141	7, 4, 3← 7, 3, 4	Ground					21 486.3		923
	1141	7, 4, 4← 7, 3, 5	Ground					27 741.1		923
	1141	7, 5, 3← 7, 4, 4	Ground					21 554.9		923
	1141	8, 4, 4← 8, 3, 5	Ground					27 714.5		923
	1141	8, 5, 3← 8, 4, 4	Ground					21 391.6		923
	1141	8, 5, 4← 8, 4, 5	Ground					27 719.9		923
	1141	8, 6, 3← 8, 5, 4	Ground					21 539.		923
	1141	9, 5, 4← 9, 4, 5	Ground					27 680.		923
	1141	9, 6, 3← 9, 5, 4	Ground					21 242.8		923
	1141	9, 6, 4← 9, 5, 5	Ground					27 689.7		923
	1141	10, 6, 4←10, 5, 5	Ground					27 625.2		923
	1141	10, 7, 3←10, 6, 4	Ground					21 019.7		923
	1141	10, 7, 4←10, 6, 5	Ground					27 651.		923
	1141	11, 7, 4←11, 6, 5	Ground					27 549.4		923
	1141	11, 8, 3←11, 7, 4	Ground					20 703.6		923
	1141	11, 8, 4←11, 7, 5	Ground					27 603.		923
	1141	11, 9, 3←11, 8, 4	Ground					21 564.		923
	1141	12, 8, 4←12, 7, 5	Ground					27 442.8		923
	1141	12, 9, 3←12, 8, 4	Ground					20 279.1		923
	1141	12,10, 3←12, 9, 4	Ground					21 631.		923

Isotopic Species	Id. No.	Rotational Quantum Nos.	Vib. State	Hyperfine				Frequency MHz	Acc. ±MHz	Ref.
				F_1'	F'	F_1	F			
$C^{12}_*DN^{14}C^{12}DC^{12}H{:}C^{12}DN^{14}_*$	1142	2, 2, 0← 1, 1, 1	Ground					24 986.1		923
	1142	2, 2, 1← 1, 1, 0	Ground					19 863.7		923
	1142	3, 0, 3← 2, 1, 2	Ground					19 495.3		923
	1142	3, 1, 3← 2, 0, 2	Ground					19 510.6		923
	1142	3, 1, 2← 2, 2, 1	Ground					24 657.4		923
	1142	3, 2, 2← 2, 1, 1	Ground					25 434.7		923
	1142	4, 0, 4← 3, 1, 3	Ground					25 071.0		923
	1142	4, 1, 4← 3, 0, 3	Ground					25 071.0		923
	1142	4, 1, 3← 4, 0, 4	Ground					19 495.3		923
	1142	4, 2, 3← 4, 1, 4	Ground					19 495.3		923
	1142	5, 1, 4← 5, 0, 5	Ground					25 071.0		923
	1142	5, 2, 3← 5, 1, 4	Ground					19 470.3		923
	1142	5, 2, 4← 5, 1, 5	Ground					25 071.0		923
	1142	5, 3, 3← 5, 2, 4	Ground					19 487.1		923
	1142	6, 2, 4← 6, 1, 5	Ground					25 058.5		923
	1142	6, 3, 3← 6, 2, 4	Ground					19 422.6		923
	1142	6, 3, 4← 6, 2, 5	Ground					25 058.5		923
	1142	6, 4, 3← 6, 3, 4	Ground					19 470.3		923
	1142	7, 3, 4← 7, 2, 5	Ground					25 038.4		923
	1142	7, 4, 3← 7, 3, 4	Ground					19 336.6		923
	1142	7, 4, 4← 7, 3, 5	Ground					25 038.4		923
	1142	7, 5, 3← 7, 4, 4	Ground					19 454.0		923
	1142	8, 4, 4← 8, 3, 5	Ground					25 009.0		923
	1142	8, 5, 3← 8, 4, 4	Ground					19 193.3		923
	1142	8, 5, 4← 8, 4, 5	Ground					25 009.0		923
	1142	8, 6, 3← 8, 5, 4	Ground					19 443.3		923
	1142	9, 5, 4← 9, 4, 5	Ground					24 969.9		923
	1142	9, 6, 4← 9, 5, 4	Ground					18 968.0		923
	1142	9, 6, 4← 9, 5, 5	Ground					24 969.9		923
	1142	9, 7, 3← 9, 6, 4	Ground					19 448.4		923
	1142	10, 6, 4←10, 5, 5	Ground					24 920.8		923
	1142	10, 7, 3←10, 6, 4	Ground					18 637.0		923
	1142	10, 7, 4←10, 6, 5	Ground					24 920.8		923
	1142	10, 8, 3←10, 7, 4	Ground					19 480.9		923
	1142	11, 7, 4←11, 6, 5	Ground					24 863.4		923
	1142	11, 8, 3←11, 7, 4	Ground					18 182.3		923
	1142	11, 8, 4←11, 7, 5	Ground					24 863.4		923
	1142	11, 9, 3←11, 8, 4	Ground					19 557.3		923
	1142	12, 8, 4←12, 7, 5	Ground					24 801.8		923
	1142	12, 9, 4←12, 8, 5	Ground					24 801.8		923
	1142	12,10, 3←12, 9, 4	Ground					19 693.7		923

C$_4$H$_4$O $\qquad$ C$_{2v}$ $\qquad$ C$_*$H:CHOCH:C$_*$H

Isotopic Species	Pt. Gp.	Id. No.	A MHz		B MHz		C MHz		D$_J$ MHz	D$_{JK}$ MHz	Δ Amu A²	κ
C$^{12}_*$H:C^{12}HO^{16}C^{12}H:C$^{12}_*$H	C$_{2v}$	1151	9 446.96	M	9 246.61	M	4 670.88	M			.0476	.91614
C$^{12}_*$H:C^{12}HO^{18}C^{12}H:C$^{12}_*$H	C$_{2v}$	1152	9 447.66	M	8 841.72	M	4 565.37	M				
C$^{12}_*$H:C^{12}HO^{16}C^{13}H:C$^{12}_*$H	C$_s$	1153	9 295.41	M	9 178.23	M	4 616.25	M				
C$^{12}_*$H:C^{12}HO^{16}C^{12}D:C$^{12}_*$H	C$_s$	1154	9 280.15	M	8 638.48	M	4 472.12	M			.0477	.73316
C$^{12}_*$H:C^{12}HO^{16}C^{12}H:C$^{12}_*$D	C$_s$	1155	9 383.47	M	8 490.28	M	4 455.53	M			.0476	.63748
C$^{12}_*$H:C^{12}DO^{16}C^{12}D:C$^{12}_*$H	C$_{2v}$	1156	9 033.33	M	8 160.52	M	4 285.87	M			.0451	.63233
C$^{13}_*$H:C^{12}HO^{16}C^{12}H:C$^{12}_*$H	C$_s$	1157	9 403.73	M	9 043.68	M	4 608.15	M				

Id. No.	μ$_a$ Debye		μ$_b$ Debye		μ$_c$ Debye		eQq Value(MHz)	Rel.	eQq Value(MHz)	Rel.	eQq Value(MHz)	Rel.	ω$_a$ d 1/cm	ω$_b$ d 1/cm	ω$_c$ d 1/cm	ω$_d$ d 1/cm
1151	.661	M	0.	X	0.	X										

References:

ABC: 956 $\qquad$ Δ: 580 $\qquad$ κ: 580 $\qquad$ μ: 318

Furan $\hfill$ Spectral Line Table

Isotopic Species	Id. No.	Rotational Quantum Nos.	Vib. State	F$_1'$	F'	F$_1$	F	Frequency MHz	Acc. ±MHz	Ref.
C$^{12}_*$H:C^{12}HO^{16}C^{12}H:C$^{12}_*$H	1151	2, 0, 2← 1, 0, 1	Ground					23 453.13		580
	1151	2, 1, 2← 1, 1, 1	Ground					23 259.30		580
	1151	3, 1, 2← 3, 1, 3	Ground					23 352.47		580
	1151	3, 2, 2← 3, 0, 3	Ground					23 384.46		580
	1151	4, 2, 2← 4, 2, 3	Ground					23 305.88		580
	1151	4, 3, 2← 4, 1, 3	Ground					23 402.53		580
	1151	5, 3, 2← 5, 3, 3	Ground					23 213.45		580
	1151	5, 4, 2← 5, 2, 3	Ground					23 440.06		580
	1151	6, 4, 2← 6, 4, 3	Ground					23 055.80		580
	1151	6, 5, 2← 6, 3, 3	Ground					23 507.71		580
	1151	7, 5, 2← 7, 5, 3	Ground					22 810.92		580
	1151	7, 6, 2← 7, 4, 3	Ground					23 619.06		580
	1151	8, 6, 2← 8, 6, 3	Ground					22 458.99		580
	1151	8, 7, 2← 8, 5, 3	Ground					23 790.73		580
	1151	8, 8, 1← 8, 6, 2	Ground					19 011.46		580
	1151	9, 7, 2← 9, 7, 3	Ground					21 984.30		580
	1151	9, 8, 2← 9, 6, 3	Ground					24 043.08		580
	1151	9, 9, 1← 9, 7, 2	Ground					20 624.34		580
	1151	10, 8, 2←10, 8, 3	Ground					21 377.91		580
	1151	10, 9, 2←10, 7, 3	Ground					24 399.77		580
	1151	10,10, 1←10, 8, 2	Ground					22 540.27		580
	1151	11, 9, 2←11, 9, 3	Ground					20 637.71		580
	1151	11,10, 2←11, 8, 3	Ground					24 888.97		580
	1151	11,11, 1←11, 9, 2	Ground					24 767.50		580
	1151	12,10, 2←12,10, 3	Ground					19 767.98		580
	1151	12,11, 2←12, 9, 3	Ground					25 541.64		580

Isotopic Species	Id. No.	Rotational Quantum Nos.	Vib. State	F_1'	F'	F_1	F	Frequency MHz	Acc. ±MHz	Ref.
$C^{12}_*H{:}C^{12}HO^{18}C^{12}H{:}C^{12}_*H$	1152	2, 0, 2← 1, 0, 1	Ground					23 083.9	.1	956
	1152	2, 1, 2← 1, 1, 1	Ground					22 537.9	.1	956
	1152	3, 1, 2← 3, 1, 3	Ground					22 637.2	.1	956
	1152	3, 2, 2← 3, 0, 3	Ground					22 936.8	.1	956
	1152	4, 2, 2← 4, 2, 3	Ground					22 226.4	.1	956
	1152	4, 3, 2← 4, 1, 3	Ground					23 115.1	.1	956
	1152	4, 4, 1← 4, 2, 2	Ground					17 729.3	.2	956
	1152	5, 3, 2← 5, 3, 3	Ground					21 475.8	.1	956
	1152	5, 4, 2← 5, 2, 3	Ground					23 494.4	.1	956
	1152	5, 5, 1← 5, 3, 2	Ground					20 390.3	.1	956
	1152	6, 4, 2← 6, 4, 3	Ground					20 328.7	.1	956
	1152	6, 5, 2← 6, 3, 3	Ground					24 196.7	.1	956
	1152	6, 6, 1← 6, 4, 2	Ground					23 976.1	.1	956
	1152	7, 5, 2← 7, 5, 3	Ground					18 790.4	.1	956
	1152	7, 6, 2← 7, 4, 3	Ground					25 379.1	.1	956
	1152	8, 7, 2← 8, 5, 3	Ground					27 222.6	.1	956
	1152	9, 6, 3← 9, 6, 4	Ground					29 270.9	.1	956
	1152	10, 7, 3←10, 7, 4	Ground					27 663.6	.1	956
	1152	11, 8, 3←11, 8, 4	Ground					25 586.5	.1	956
	1152	12, 9, 3←12, 9, 4	Ground					23 097.4	.1	956
$C^{12}_*H{:}C^{12}HO^{16}C^{13}H{:}C^{12}_*H$	1153	2, 0, 2← 1, 0, 1	Ground					23 141.9	.1	956
	1153	2, 0, 2← 1, 1, 1	Ground					23 024.7	.1	956
	1153	2, 1, 2← 1, 0, 1	Ground					23 144.2	.1	956
	1153	2, 1, 2← 1, 1, 1	Ground					23 027.0	.1	956
	1153	3, 1, 2← 3, 0, 3	Ground					23 093.8	.1	956
	1153	3, 1, 2← 3, 1, 3	Ground					23 093.8	.1	956
	1153	3, 2, 2← 3, 0, 3	Ground					23 104.9	.1	956
	1153	3, 2, 2← 3, 1, 3	Ground					23 104.9	.1	956
	1153	4, 2, 2← 4, 1, 3	Ground					23 077.5	.1	956
	1153	4, 2, 2← 4, 2, 3	Ground					23 077.5	.1	956
	1153	4, 3, 2← 4, 1, 3	Ground					23 110.9	.1	956
	1153	4, 3, 2← 4, 2, 3	Ground					23 110.9	.1	956
	1153	5, 3, 2← 5, 2, 3	Ground					23 046.0	.1	956
	1153	6, 4, 2← 6, 3, 3	Ground					22 991.5	.1	956
	1153	6, 4, 2← 6, 4, 3	Ground					22 989.3	.1	956
	1153	6, 5, 2← 6, 3, 3	Ground					23 146.5	.1	956
	1153	6, 5, 2← 6, 4, 3	Ground					23 144.2	.2	956
	1153	7, 5, 2← 7, 4, 3	Ground					22 906.5	.1	956
	1153	7, 5, 2← 7, 5, 3	Ground					22 900.8	.1	956
	1153	7, 6, 2← 7, 4, 3	Ground					23 183.7	.1	956
	1153	7, 6, 2← 7, 5, 3	Ground					23 178.1	.1	956
	1153	8, 6, 2← 8, 5, 3	Ground					22 782.7	.1	956
	1153	8, 6, 2← 8, 6, 3	Ground					22 770.4	.1	956
	1153	8, 7, 2← 8, 5, 3	Ground					23 240.7	.1	956
	1153	8, 7, 2← 8, 6, 3	Ground					23 228.5	.1	956
	1153	9, 8, 2← 9, 6, 3	Ground					23 323.8	.1	956
	1153	9, 8, 2← 9, 7, 3	Ground					23 299.4	.1	956

Isotopic Species	Id. No.	Rotational Quantum Nos.	Vib. State	F_1'	F'	F_1	F	Frequency MHz	Acc. ±MHz	Ref.
C¹²*H:C¹²HO¹⁶C¹²D:C¹²*H	1154	2, 0, 2← 1, 1, 1	Ground					21 986.4		580
	1154	2, 1, 2← 1, 0, 1	Ground					22 696.3		580
	1154	3, 1, 2← 3, 0, 3	Ground					22 144.0		580
	1154	3, 2, 2← 3, 1, 3	Ground					22 477.6		580
	1154	4, 2, 2← 4, 1, 3	Ground					21 706.9		580
	1154	4, 3, 2← 4, 2, 3	Ground					22 654.8		580
	1154	5, 3, 2← 5, 2, 3	Ground					20 961.2		580
	1154	5, 4, 2← 5, 3, 3	Ground					22 998.2		580
	1154	5, 5, 1← 5, 4, 2	Ground					18 489.2		580
	1154	6, 4, 2← 6, 3, 3	Ground					19 942.6		580
	1154	6, 5, 2← 6, 4, 3	Ground					23 565.7		580
	1154	6, 6, 1← 6, 5, 2	Ground					20 541.8		580
	1154	7, 5, 2← 7, 4, 3	Ground					18 798.6		580
	1154	7, 6, 2← 7, 5, 3	Ground					24 410.9		580
	1154	7, 7, 1← 7, 6, 2	Ground					22 918.6		580
	1154	8, 7, 2← 8, 6, 3	Ground					25 575.1		580
	1154	8, 8, 0← 8, 7, 1	Ground					21 557.6		580
	1154	8, 8, 1← 8, 7, 2	Ground					25 584.7		580
	1154	9, 9, 0← 9, 8, 1	Ground					25 628.6		580
	1154	10, 9, 1←10, 8, 2	Ground					19 666.2		580
	1154	11, 8, 3←11, 7, 4	Ground					25 427.9		580
	1154	11,10, 1←11, 9, 2	Ground					23 481.7		580
	1154	12, 9, 3←12, 8, 4	Ground					23 934.4		580
	1154	12,10, 2←12, 9, 3	Ground					19 150.6		580
C¹²*H:C¹²HO¹⁶C¹²H:C¹²*D	1155	2, 0, 2← 1, 0, 1	Ground					22 617.5		580
	1155	2, 1, 2← 1, 1, 1	Ground					21 856.8		580
	1155	3, 2, 2← 3, 0, 3	Ground					22 497.9		580
	1155	4, 3, 2← 4, 1, 3	Ground					22 906.5		580
	1155	4, 4, 1← 4, 2, 2	Ground					19 939.7		580
	1155	5, 3, 2← 5, 3, 3	Ground					19 572.5		580
	1155	5, 4, 2← 5, 2, 3	Ground					23 790.0		580
	1155	5, 5, 1← 5, 3, 2	Ground					24 401.7		580
	1155	6, 5, 2← 6, 3, 3	Ground					25 436.8		580
	1155	9, 6, 3← 9, 6, 4	Ground					24 757.1		580
	1155	10, 7, 3←10, 7, 4	Ground					21 685.3		580
	1155	11, 8, 3←11, 8, 4	Ground					18 213.4		580
C¹²*H:C¹²DO¹⁶C¹²D:C¹²*H	1156	2, 0, 2← 1, 1, 1	Ground					20 886.6		580
	1156	2, 1, 2← 1, 0, 1	Ground					21 890.9		580
	1156	3, 1, 2← 3, 0, 3	Ground					21 006.7		580
	1156	4, 2, 2← 4, 1, 3	Ground					20 246.5		580
	1156	4, 3, 2← 4, 2, 3	Ground					21 965.1		580
	1156	5, 3, 2← 5, 2, 3	Ground					19 095.2		580
	1156	5, 4, 2← 5, 3, 3	Ground					22 592.6		580
	1156	5, 5, 1← 5, 4, 2	Ground					19 851.7		580
	1156	6, 5, 2← 6, 4, 3	Ground					23 602.5		580
	1156	6, 6, 1← 6, 5, 2	Ground					22 647.1		580

Isotopic Species	Id. No.	Rotational Quantum Nos.	Vib. State	Hyperfine				Frequency MHz	Acc. ±MHz	Ref.
				F_1'	F'	F_1	F			
C^{12}*H:C^{12}DO^{16}C^{12}D:C^{12}*H	1156	7, 6, 2← 7, 5, 3	Ground					25 061.3		580
	1156	7, 7, 0← 7, 6, 1	Ground					22 436.8		580
	1156	7, 7, 1← 7, 6, 2	Ground					25 828.6		580
	1156	9, 6, 3← 9, 5, 4	Ground					24 720.3		580
	1156	9, 8, 1← 9, 7, 2	Ground					21 683.1		580
	1156	10, 7, 3←10, 6, 4	Ground					22 966.5		580
	1156	10, 8, 2←10, 7, 3	Ground					18 205.2		580
	1156	11, 8, 3←11, 7, 4	Ground					21 707.8		580
	1156	11, 9, 2←11, 8, 3	Ground					21 023.5		580
	1156	12, 9, 3←12, 8, 4	Ground					21 360.1		580
	1156	12,10, 2←12, 9, 3	Ground					24 991.9		580
C^{13}*H:C^{12}HO^{16}C^{12}H:C^{12}*H	1157	2, 0, 2← 1, 0, 1	Ground					23 207.2	.1	956
	1157	2, 1, 2← 1, 0, 1	Ground					23 228.5	.1	956
	1157	2, 1, 2← 1, 1, 1	Ground					22 868.1	.1	956
	1157	3, 1, 2← 3, 1, 3	Ground					22 986.7	.1	956
	1157	3, 2, 2← 3, 0, 3	Ground					23 092.2	.1	956
	1157	4, 2, 2← 4, 2, 3	Ground					22 836.55	.1	956
	1157	4, 3, 2← 4, 1, 3	Ground					23 152.8	.1	956
	1157	5, 3, 2← 5, 3, 3	Ground					22 547.1	.1	956
	1157	5, 4, 2← 5, 2, 3	Ground					23 280.0	.1	956
	1157	6, 4, 2← 6, 4, 3	Ground					22 069.0	.1	956
	1157	6, 5, 2← 6, 3, 3	Ground					23 511.9	.1	956
	1157	6, 6, 1← 6, 4, 2	Ground					19 133.7	.1	956
	1157	7, 5, 2← 7, 5, 3	Ground					21 366.1	.1	956
	1157	7, 6, 2← 7, 4, 3	Ground					23 898.3	.1	956
	1157	7, 6, 2← 7, 5, 3	Ground					23 740.3	.1	956
	1157	8, 6, 2← 8, 6, 3	Ground					20 423.6	.1	956
	1157	8, 7, 2← 8, 5, 3	Ground					24 501.3	.1	956
	1157	8, 7, 2← 8, 6, 3	Ground					24 159.2	.1	956
	1157	8, 8, 1← 8, 6, 2	Ground					24 338.1	.1	956
	1157	9, 7, 2← 9, 7, 3	Ground					19 248.2	.1	956
	1157	9, 8, 2← 9, 6, 3	Ground					25 394.7	.1	956
	1157	9, 8, 2← 9, 7, 3	Ground					24 726.7	.1	956
	1157	10, 8, 2←10, 7, 3	Ground					19 063.6	.1	956
	1157	10, 8, 2←10, 8, 3	Ground					17 863.7	.1	956
	1157	10, 9, 2←10, 8, 3	Ground					25 461.6	.1	956
	1157	10,10, 1←10, 9, 2	Ground					24 174.7	.1	956
	1157	11, 9, 2←11, 8, 3	Ground					18 311.2	.1	956
	1157	11,10, 2←11, 8, 3	Ground					28 384.5	.1	956
	1157	11,10, 2←11, 9, 3	Ground					26 378.0	.1	956
	1157	12, 9, 3←12, 8, 4	Ground					29 605.5	.1	956
	1157	12, 9, 3←12, 9, 4	Ground					29 323.1	.1	956

C_4H_4S C_{2v} $S_*HC:CHCH:C_*H$

Isotopic Species	Pt. Gp.	Id. No.	A MHz		B MHz		C MHz		D_J MHz	D_{JK} MHz	Δ Amu A²	κ
$S^{32}_*HC^{12}:C^{12}HC^{12}H:C^{12}_*H$	C_{2v}	1161	8 041.77	M	5 418.12	M	3 235.77	M			.0652	−.09182
$S^{32}_*DC^{12}:C^{12}HC^{12}H:C^{12}_*H$	C_s	1162	7 437.32	M	5 413.61	M	3 131.82	M			.0640	.05994
$S^{32}_*HC^{12}:C^{12}DC^{12}H:C^{12}_*H$	C_s	1163	7 856.13	M	5 138.14	M	3 105.23	M			.0633	−.14420
$S^{32}_*HC^{12}:C^{12}DC^{12}D:C^{12}_*H$	C_s	1164	7 616.99	M	4 914.50	M	2 985.99	M			.0669	−.16713
$S^{32}_*DC^{12}:C^{12}DC^{12}D:C^{12}_*D$	C_{2v}	1165	6 587.67	M	4 905.66	M	2 810.88	M			.0584	.10929
$S^{34}_*HC^{12}:C^{12}HC^{12}H:C^{12}_*H$	C_{2v}	1166	8 042.29	M	5 274.23	M	3 183.70	M			.0789	−.13945
$S^{32}_*HC^{13}:C^{12}HC^{12}H:C^{12}_*H$	C_s	1167	7 852.89	M	5 418.34	M	3 024.81	M			.0661	−.04757
$S^{32}_*C^{12}H:C^{13}HC^{12}H:C^{12}_*H$	C_s	1168	7 981.43	M	5 319.23	M	3 190.63	M			.0656	−.11138

Id. No.	μ_a Debye	μ_b Debye	μ_c Debye	eQq Value(MHz) Rel.	eQq Value(MHz) Rel.	eQq Value(MHz) Rel.	ω_a d 1/cm	ω_b d 1/cm	ω_c d 1/cm	ω_d d 1/cm
1161	.6 M	0. X	0. X							

References:
ABC: 654,935 Δ: 654,935 κ: 654,935 μ: 654

Thiophene

Spectral Line Table

Isotopic Species	Id. No.	Rotational Quantum Nos.	Vib. State	F_1'	F'	F_1	F	Frequency MHz	Acc. ±MHz	Ref.
$S^{32}_*HC^{12}:C^{12}HC^{12}H:C^{12}_*H$	1161	2, 1, 1← 1, 1, 0	Ground					19 490.2	.1	654
	1161	3, 0, 3← 2, 0, 2	Ground					23 043.8	.1	654
	1161	3, 1, 2← 2, 1, 1	Ground					28 488.6	.1	654
	1161	3, 1, 3← 2, 1, 2	Ground					22 202.3	.1	654
	1161	3, 2, 1← 2, 2, 0	Ground					28 879.1	.1	654
	1161	3, 2, 2← 2, 2, 1	Ground					25 961.8	.1	654
	1161	6, 2, 5← 6, 2, 4	Ground					24 377.0	.1	654
	1161	7, 3, 5← 7, 3, 4	Ground					19 089.0	.1	654
	1161	9, 4, 6← 9, 4, 5	Ground					21 547.5	.1	654
	1161	11, 5, 7←11, 5, 6	Ground					23 652.9	.1	654
$S^{32}_*DC^{12}:C^{12}HC^{12}H:C^{12}_*H$	1162	2, 1, 1← 1, 1, 0	Ground					19 372.4	.1	654
	1162	3, 0, 3← 2, 0, 2	Ground					22 227.9	.1	654
	1162	3, 1, 2← 2, 1, 1	Ground					28 056.1	.1	654
	1162	3, 1, 3← 2, 1, 2	Ground					21 617.0	.1	654
	1162	3, 2, 1← 2, 2, 0	Ground					29 044.7	.1	654
	1162	3, 2, 2← 2, 2, 1	Ground					25 636.3	.1	654
	1162	6, 2, 5← 6, 2, 4	Ground					25 528.1	.1	654
	1162	7, 3, 5← 7, 3, 4	Ground					22 034.3	.1	654
	1162	9, 4, 6← 9, 4, 5	Ground					26 289.8	.1	654
	1162	10, 5, 6←10, 5, 5	Ground					20 514.0	.1	654
	1162	12, 6, 7←12, 6, 6	Ground					23 976.1	.1	654
$S^{32}_*HC^{12}:C^{12}DC^{12}H:C^{12}_*H$	1163	2, 1, 1← 1, 1, 0	Ground					18 519.8	.1	654
	1163	3, 0, 3← 2, 0, 2	Ground					22 134.5	.1	654
	1163	3, 1, 2← 2, 1, 1	Ground					27 144.7	.1	654
	1163	3, 1, 3← 2, 1, 2	Ground					21 254.5	.1	654
	1163	3, 2, 1← 2, 2, 0	Ground					27 325.5	.1	654
	1163	3, 2, 2← 2, 2, 1	Ground					24 730.6	.1	654
	1163	4, 1, 4← 4, 1, 3	Ground					19 012.5	.1	654
	1163	6, 2, 5← 6, 2, 4	Ground					22 549.0	.1	654
	1163	8, 3, 6← 8, 3, 5	Ground					25 470.2	.1	654
	1163	9, 4, 6← 9, 4, 5	Ground					18 569.8	.1	654
	1163	10, 4, 7←10, 4, 6	Ground					27 851.2	.1	654
	1163	11, 5, 7←11, 5, 6	Ground					19 749.1	.1	654

Isotopic Species	Id. No.	Rotational Quantum Nos.	Vib. State	F_1'	F'	F_1	F	Frequency MHz	Acc. ±MHz	Ref.
$S^{32}*HC^{12}:C^{12}DC^{12}D:C^{12}*H$	1164	2, 1, 1← 1, 1, 0	Ground					17 729.6	.1	654
	1164	3, 0, 3← 2, 0, 2	Ground					21 292.2	.1	654
	1164	3, 1, 2← 2, 1, 1	Ground					26 016.3	.1	654
	1164	3, 1, 3← 2, 1, 2	Ground					20 415.8	.1	654
	1164	3, 2, 1← 2, 2, 0	Ground					26 110.8	.1	654
	1164	3, 2, 2← 2, 2, 1	Ground					23 701.5	.1	654
	1164	4, 1, 4← 4, 1, 3	Ground					18 109.1	.1	654
	1164	6, 2, 5← 6, 2, 4	Ground					21 299.7	.1	654
	1164	8, 3, 6← 8, 3, 5	Ground					23 827.0	.1	654
	1164	10, 4, 7←10, 4, 6	Ground					25 769.8	.1	654
	1164	11, 5, 7←11, 5, 6	Ground					17 731.0	.1	654
	1164	12, 5, 8←12, 5, 7	Ground					27 185.0	.1	654
$S^{32}*DC^{12}:C^{12}DC^{12}D:C^{12}*D$	1165	2, 1, 1← 1, 1, 0	Ground					17 527.9	.1	654
	1165	3, 0, 3← 2, 0, 2	Ground					19 925.3	.1	654
	1165	3, 1, 2← 2, 1, 1	Ground					25 297.2	.1	654
	1165	3, 1, 3← 2, 1, 2	Ground					19 433.8	.1	654
	1165	3, 2, 1← 2, 2, 0	Ground					26 373.5	.1	654
	1165	3, 2, 2← 2, 2, 1	Ground					23 149.6	.1	654
	1165	6, 2, 5← 6, 2, 4	Ground					23 332.0	.1	654
	1165	7, 3, 5← 7, 3, 4	Ground					20 651.9	.1	654
	1165	9, 4, 6← 9, 4, 5	Ground					24 899.3	.1	654
	1165	10, 5, 6←10, 5, 5	Ground					20 293.9	.1	654
	1165	12, 6, 7←12, 6, 6	Ground					24 048.8	.1	654
$S^{34}*HC^{12}:C^{12}HC^{12}H:C^{12}*H$	1166	2, 1, 1← 1, 1, 0	Ground					19 003.9	3.	654
	1166	2, 1, 1← 1, 1, 0	Ground					19 006.4	.05	935
	1166	3, 2, 1← 2, 2, 0	Ground					28 054.2	.05	935
	1166	3, 2, 2← 2, 2, 1	Ground					25 373.4	3.	654
$S^{32}*HC^{13}:C^{12}HC^{12}H:C^{12}*H$	1167	2, 0, 2← 1, 0, 1	Ground					16 275.2	.05	935
	1167	2, 1, 1← 1, 0, 1	Ground					19 459.8	.05	935
	1167	2, 1, 2← 1, 1, 1	Ground					15 032.8	.05	935
	1167	3, 0, 3← 2, 0, 2	Ground					22 802.8	.05	935
	1167	3, 1, 2← 2, 1, 1	Ground					28 373.6	.05	935
	1167	3, 1, 3← 2, 1, 2	Ground					22 031.6	.05	935
	1167	3, 2, 1← 2, 2, 0	Ground					28 935.8	.05	935
	1167	3, 2, 2← 2, 2, 1	Ground					25 869.0	3.	654
	1167	4, 1, 3← 4, 1, 4	Ground					20 321.3	.05	935
	1167	6, 2, 4← 6, 2, 5	Ground					24 803.0	.05	935
	1167	7, 3, 4← 7, 3, 5	Ground					20 047.7	.05	935
	1167	8, 3, 5← 8, 3, 6	Ground					28 931.2	.05	935
	1167	9, 4, 5← 9, 4, 6	Ground					23 093.6	.05	935
	1167	11, 5, 6←11, 5, 7	Ground					25 894.3	.05	935
	1167	12, 6, 6←12, 6, 7	Ground					17 858.5	.05	935
$S^{32}*C^{12}H:C^{13}HC^{12}H:C^{12}*H$	1168	2, 0, 2← 1, 0, 1	Ground					16 157.7	.05	935
	1168	2, 1, 1← 1, 1, 0	Ground					19 148.3	.05	935
	1168	2, 1, 2← 1, 1, 1	Ground					14 891.3	.05	935
	1168	3, 0, 3← 2, 0, 2	Ground					22 730.9	.05	935
	1168	3, 1, 2← 2, 1, 1	Ground					28 018.5	.05	935
	1168	3, 1, 3← 2, 1, 2	Ground					21 873.0	.05	935
	1168	3, 2, 1← 2, 2, 0	Ground					28 328.0	.05	935
	1168	3, 2, 2← 2, 2, 1	Ground					25 529.2	3.	654
	1168	5, 1, 4← 5, 1, 5	Ground					27 113.5	.05	935
	1168	6, 2, 4← 6, 2, 5	Ground					23 724.5	.05	935
	1168	7, 3, 4← 7, 3, 5	Ground					18 302.0	.05	935
	1168	8, 3, 5← 8, 3, 6	Ground					27 132.7	.05	935
	1168	9, 4, 5← 9, 4, 6	Ground					20 448.7	.05	935
	1168	12, 6, 6←12, 6, 7	Ground					13 882.7	.05	935

C_4H_5N C_s $C_*H_2CH_2C_*HCN$

Isotopic Species	Pt. Gp.	Id. No.	A MHz	B MHz	C MHz	D_J MHz	D_{JK} MHz	Δ Amu A²	κ
$C^{12}_*H_2C^{12}H_2C^{12}_*HC^{12}N^{14}$	C_s	1171	15 917. M	3 465.06 M	3 286.22 M				−.972
$c\text{-}C^{12}_*H_2C^{12}HDC^{12}_*HC^{12}N^{14}$	C_s	1172	14 543. M	3 419.34 M	3 229.15 M				−.966
$t\text{-}C^{12}_*H_2C^{12}DHC^{12}_*HC^{12}N^{14}$	C_s	1173	15 367. M	3 359.13 M	3 161.60 M				−.967

Id. No.	μ_a Debye	μ_b Debye	μ_c Debye	eQq Value(MHz) Rel.	eQq Value(MHz) Rel.	eQq Value(MHz) Rel.	ω_a d 1/cm	ω_b d 1/cm	ω_c d 1/cm	ω_d d 1/cm
1171	3.78 L		0. X	−71.7 Cl³⁵						

References:

ABC: 818 κ: 818 μ: 995   eQq: 818

Cyclopropanecarbonitrile   Spectral Line Table

Isotopic Species	Id. No.	Rotational Quantum Nos.	Vib. State	F_1'	F'	F_1	F	Frequency MHz	Acc. ±MHz	Ref.
$C^{12}_*H_2C^{12}H_2C^{12}_*HC^{12}N^{14}$	1171	3, 0, 3← 2, 0, 2	Ground					20 246.27	.05	818
	1171	3, 1, 2← 2, 1, 1	Ground					20 521.10	.05	818
	1171	3, 1, 3← 2, 1, 2	Ground					19 984.59	.05	818
	1171	3, 2, 1← 2, 2, 0	Ground					20 261.17	.05	818
	1171	3, 2, 2← 2, 2, 1	Ground					20 253.82	.05	818
	1171	4, 0, 4← 3, 0, 3	Ground					26 985.78	.05	818
	1171	4, 1, 3← 3, 1, 2	Ground					27 358.91	.05	818
	1171	4, 1, 4← 3, 1, 3	Ground					26 643.54	.05	818
	1171	4, 2, 2← 3, 2, 1	Ground					27 022.79	.05	818
	1171	4, 2, 3← 3, 2, 2	Ground					27 003.61	.05	818
	1171	4, 3, 1← 3, 3, 0	Ground					27 008.78	.05	818
	1171	4, 3, 2← 3, 3, 1	Ground					27 008.78	.05	818
$c\text{-}\bar{C}^{12}_*H_2C^{12}HDC^{12}_*HC^{12}N^{14}$	1172	4, 0, 4← 3, 0, 3	Ground					26 569.75	.1	818
	1172	4, 1, 3← 3, 1, 2	Ground					26 969.51	.1	818
	1172	4, 1, 4← 3, 1, 3	Ground					26 208.77	.1	818
	1172	4, 2, 2← 3, 2, 1	Ground					26 616.19	.1	818
	1172	4, 2, 3← 3, 2, 2	Ground					26 591.95	.1	818
	1172	4, 3, 1← 3, 3, 0	Ground					26 598.95	.1	818
	1172	4, 3, 2← 3, 3, 1	Ground					26 598.95	.1	818
$t\text{-}C^{12}_*H_2C^{12}DHC^{12}_*HC^{12}N^{14}$	1173	4, 0, 4← 3, 0, 3	Ground					26 058.50	.1	818
	1173	4, 1, 3← 3, 1, 2	Ground					26 473.41	.1	818
	1173	4, 1, 4← 3, 1, 3	Ground					25 683.31	.1	818
	1173	4, 2, 2← 3, 2, 1	Ground					26 105.89	.1	818
	1173	4, 2, 3← 3, 2, 2	Ground					26 080.94	.1	818
	1173	4, 3, 1← 3, 3, 0	Ground					26 087.74	.1	818
	1173	4, 3, 2← 3, 3, 1	Ground					26 087.74	.1	818

C_4H_5N C_{2v} $C_*H:CHCH:CHN_*H$

Isotopic Species	Pt. Gp.	Id. No.	A^1 MHz		B^1 MHz		C^1 MHz		D_J MHz	D_{JK} MHz	Δ Amu A^2	κ
$C^{12}_*H:C^{12}HC^{12}H:C^{12}HN^{14}_*H$	C_{2v}	1181	9 130.53	M	9 001.30	M	4 532.09	M			.0157	.94380
$C^{12}_*H:C^{12}HC^{12}H:C^{12}HN^{14}_*D$	C_{2v}	1182	9 130.55	M	8 340.60	M	4 358.56	M			.0080	.66892
$C^{12}_*D:C^{12}HC^{12}H:C^{12}HN^{14}_*H$	C_s	1183	9 018.42	M	8 361.90	M	4 338.22	M			.0176	.71945
$C^{12}_*H:C^{12}DC^{12}H:C^{12}HN^{14}_*H$	C_s	1184	9 087.97	M	8 271.52	M	4 330.22	M			.0015	.65679
$C^{12}_*D:C^{12}DC^{12}D:C^{12}DN^{14}_*H$	C_{2v}	1185	7 886.03	M	7 429.62	M	3 825.15	M			.0123	.77522
$C^{12}_*D:C^{12}DC^{12}D:C^{12}DN^{14}_*D$	C_{2v}	1186	7 429.67	M	7 360.26	M	3 697.27	M			.0049	.96281

Id. No.	μ_a Debye		μ_b Debye		μ_c Debye		eQq Value(MHz)	Rel.	eQq Value(MHz)	Rel.	eQq Value(MHz)	Rel.	ω_a d 1/cm	ω_b d 1/cm	ω_c d 1/cm	ω_d d 1/cm
1181	1.80	^{2}M	0.	X	0.	X										

1. These values are averages from three different methods of calculation by the same authors.

2. Normal, 1D, 3D, and totally deuterated isotopic species have "a" type transitions: 2D and 2,3,4,5-D species have "b" type transitions.

References:

ABC: 653 Δ: 653 κ: 653 μ: 653

Add. Ref. 408

Pyrrole   Spectral Line Table

Isotopic Species	Id. No.	Rotational Quantum Nos.	Vib. State	F_1'	F'	F_1	F	Frequency MHz	Acc. ±MHz	Ref.
$C^{12}_*H:C^{12}HC^{12}H:C^{12}HN^{14}_*H$	1181	2, 0, 2← 1, 0, 1	Ground					22 724.1		653
	1181	2, 1, 2← 1, 1, 1	Ground					22 597.5		653
	1181	3, 1, 2← 3, 1, 3	Ground					22 657.5		653
	1181	3, 2, 2← 3, 0, 3	Ground					22 670.9		653
	1181	4, 2, 2← 4, 2, 3	Ground					22 638.6		653
	1181	4, 3, 2← 4, 1, 3	Ground					22 679.6		653
	1181	5, 3, 2← 5, 3, 3	Ground					22 597.5		653
	1181	5, 4, 2← 5, 2, 3	Ground					22 694.1		653
	1181	6, 4, 2← 6, 4, 3	Ground					22 527.6		653
	1181	6, 5, 2← 6, 3, 3	Ground					22 723.0		653
	1181	7, 5, 2← 7, 5, 3	Ground					22 418.8		653
	1181	7, 6, 2← 7, 4, 3	Ground					22 769.1		653
	1181	8, 6, 2← 8, 6, 3	Ground					22 258.8		653
	1181	8, 7, 2← 8, 5, 3	Ground					22 840.5		653
	1181	9, 7, 2← 9, 7, 3	Ground					22 037.0		653
	1181	9, 8, 2← 9, 6, 3	Ground					22 944.6		653
	1181	10, 8, 2←10, 8, 3	Ground					21 742.3		653
	1181	10, 9, 2←10, 7, 3	Ground					23 090.0		653
	1181	10,10, 1←10, 8, 2	Ground					18 537.8		653
	1181	11, 9, 2←11, 9, 3	Ground					21 368.2		653
	1181	11,10, 2←11, 8, 3	Ground					23 289.0		653
	1181	11,11, 1←11, 9, 2	Ground					19 806.8		653
	1181	12,10, 2←12,10, 3	Ground					20 909.7		653
	1181	12,11, 2←12, 9, 3	Ground					23 553.9		653
	1181	12,12, 1←12,10, 2	Ground					21 269.9		653

Isotopic Species	Id. No.	Rotational Quantum Nos.	Vib. State	Hyperfine F_1'	F'	F_1	F	Frequency MHz	Acc. ±MHz	Ref.
C¹²*H:C¹²HC¹²H:C¹²HN¹⁴*D	1182	2, 0, 2← 1, 0, 1	Ground					22 099.9		653
	1182	2, 1, 2← 1, 1, 1	Ground					21 416.3		653
	1182	3, 1, 2← 3, 1, 3	Ground					21 427.8		653
	1182	3, 2, 2← 3, 0, 3	Ground					21 957.1		653
	1182	4, 2, 2← 4, 2, 3	Ground					20 735.		653
	1182	4, 3, 2← 4, 1, 3	Ground					22 281.6		653
	1182	4, 4, 1← 4, 2, 2	Ground					18 730.5		653
	1182	5, 3, 2← 5, 3, 3	Ground					19 540.4		653
	1182	5, 4, 2← 5, 2, 3	Ground					22 979.6		653
	1182	6, 4, 2← 6, 4, 3	Ground					17 834.1		653
	1182	6, 5, 2← 6, 3, 3	Ground					24 281.6		653
	1182	9, 6, 3← 9, 6, 4	Ground					25 360.5		653
	1182	10, 7, 3←10, 7, 4	Ground					22 743.1		653
	1182	11, 8, 3←11, 8, 4	Ground					19 681.4		653
C¹²*D:C¹²HC¹²H:C¹²HN¹⁴*H	1183	2, 0, 2← 1, 1, 1	Ground					21 302.6		653
	1183	2, 1, 2← 1, 0, 1	Ground					22 033.1		653
	1183	3, 1, 2← 3, 0, 3	Ground					21 444.6		653
	1183	3, 2, 2← 3, 1, 3	Ground					21 804.0		653
	1183	4, 2, 2← 4, 1, 3	Ground					20 977.8		653
	1183	5, 3, 2← 5, 2, 3	Ground					20 194.		653
	1183	5, 5, 1← 5, 4, 2	Ground					18 208.1		653
	1183	6, 4, 2← 6, 3, 3	Ground					19 146.1		653
	1183	6, 5, 2← 6, 4, 3	Ground					22 967.7		653
	1183	6, 6, 1← 6, 5, 2	Ground					20 310.1		653
	1183	7, 5, 2← 7, 4, 3	Ground					18 005.6		653
	1183	7, 6, 2← 7, 5, 3	Ground					23 867.2		653
	1183	7, 7, 0← 7, 6, 1	Ground					17 845.0		653
	1183	7, 7, 1← 7, 6, 2	Ground					22 738.3		653
	1183	8, 7, 2← 8, 6, 3	Ground					25 099.3		653
	1183	8, 8, 0← 8, 7, 1	Ground					21 822.2		653
	1183	8, 8, 1← 8, 7, 2	Ground					25 454.5		653
	1183	9, 9, 0← 9, 8, 1	Ground					25 888.6		653
	1183	10, 7, 3←10, 6, 4	Ground					25 765.1		653
	1183	10, 9, 1←10, 8, 2	Ground					20 057.1		653
	1183	11, 8, 3←11, 7, 4	Ground					24 184.2		653
	1183	11,10, 1←11, 9, 2	Ground					24 001.9		653
	1183	12, 9, 3←12, 8, 4	Ground					22 784.7		653
	1183	12,10, 2←12, 9, 3	Ground					19 245.1		653
C¹²*H:C¹²DC¹²H:C¹²HN¹⁴*H	1184	2, 0, 2← 1, 0, 1	Ground					21 964.5		653
	1184	2, 1, 2← 1, 1, 1	Ground					21 262.1		653
	1184	3, 2, 2← 3, 0, 3	Ground					21 825.		653
	1184	4, 2, 2← 4, 2, 3	Ground					20 519.1		653
	1184	4, 3, 2← 4, 1, 3	Ground					22 179.		653
	1184	5, 3, 2← 5, 3, 3	Ground					19 256.		653
	1184	5, 5, 1← 5, 3, 2	Ground					22 901.1		653
	1184	6, 4, 2← 6, 4, 3	Ground					17 476.1		653
	1184	9, 6, 3← 9, 6, 4	Ground					24 762.6		653
	1184	10, 7, 3←10, 7, 4	Ground					22 010.1		653
	1184	11, 8, 3←11, 8, 4	Ground					18 835.7		653
	1184	Not Reported	Ground					21 965.		653

Isotopic Species	Id. No.	Rotational Quantum Nos.	Vib. State	F_1'	F'	F_1	F	Frequency MHz	Acc. ±MHz	Ref.
$C^{12}_*D{:}C^{12}DC^{12}D{:}C^{12}DN^{14}_*H$	1185	2, 0, 2← 1, 1, 1	Ground					18 864.4		653
	1185	2, 1, 2← 1, 0, 1	Ground					19 361.5		653
	1185	3, 1, 2← 3, 0, 3	Ground					18 989.6		653
	1185	3, 2, 2← 3, 1, 3	Ground					19 189.4		653
	1185	4, 2, 2← 4, 1, 3	Ground					18 719.2		653
	1185	4, 3, 2← 4, 2, 3	Ground					19 294.1		653
	1185	5, 3, 2← 5, 2, 3	Ground					18 245.0		653
	1185	5, 4, 2← 5, 3, 3	Ground					19 501.5		653
	1185	6, 5, 2← 6, 4, 3	Ground					19 848.7		653
	1185	7, 6, 2← 7, 5, 3	Ground					20 372.6		653
	1185	7, 7, 1← 7, 6, 2	Ground					18 202.2		653
	1185	8, 7, 2← 8, 6, 3	Ground					21 104.3		653
	1185	8, 8, 1← 8, 7, 2	Ground					20 117.5		653
	1185	9, 6, 3← 9, 5, 4	Ground					25 135.0		653
	1185	9, 8, 2← 9, 7, 3	Ground					22 068.0		653
	1185	9, 9, 0← 9, 8, 1	Ground					18 975.3		653
	1185	9, 9, 1← 9, 8, 2	Ground					22 229.1		653
	1185	10, 7, 3←10, 6, 4	Ground					24 213.0		653
	1185	10, 9, 2←10, 8, 3	Ground					23 277.8		653
	1185	10,10, 0←10, 9, 1	Ground					22 139.2		653
	1185	10,10, 1←10, 9, 2	Ground					24 507.5		653
	1185	11, 8, 3←11, 7, 4	Ground					23 081.1		653
	1185	11,10, 2←11, 9, 3	Ground					24 737.3		653
	1185	11,11, 0←11,10, 1	Ground					25 278.2		653
	1185	12,11, 1←12,10, 2	Ground					19 778.1		653
$C^{12}_*D{:}C^{12}DC^{12}D{:}C^{12}DN^{14}_*D$	1186	2, 0, 2← 1, 0, 1	Ground					18 520.4		653
	1186	2, 1, 2← 1, 1, 1	Ground					18 452.2		653
	1186	3, 1, 2← 3, 1, 3	Ground					18 484.5		653
	1186	3, 2, 2← 3, 0, 3	Ground					18 489.3		653
	1186	4, 2, 2← 4, 2, 3	Ground					18 477.6		653
	1186	4, 3, 2← 4, 1, 3	Ground					18 491.8		653
	1186	5, 3, 2← 5, 3, 3	Ground					18 462.7		653
	1186	5, 4, 2← 5, 2, 3	Ground					18 497.3		653
	1186	6, 4, 2← 6, 4, 3	Ground					18 434.9		653
	1186	6, 5, 2← 6, 3, 3	Ground					18 507.2		653
	1186	7, 6, 2← 7, 4, 3	Ground					18 523.1		653
	1186	8, 6, 2← 8, 6, 3	Ground					18 339.0		653
	1186	8, 7, 2← 8, 5, 3	Ground					18 548.1		653
	1186	9, 7, 2← 9, 7, 3	Ground					18 255.7		653
	1186	9, 8, 2← 9, 6, 3	Ground					18 584.2		653
	1186	10, 8, 2←10, 8, 3	Ground					18 143.6		653
	1186	10, 9, 2←10, 7, 3	Ground					18 634.7		653
	1186	11, 9, 2←11, 9, 3	Ground					17 997.6		653
	1186	11,10, 2←11, 8, 3	Ground					18 703.1		653

C₅H₄

C₃ᵥ

H₃CC:CC:CH

Isotopic Species	Pt. Gp.	Id. No.	A MHz	B MHz	C MHz	D_J MHz	D_{JK} MHz	Δ Amu A²	κ
$H_3C^{12}C^{12}:C^{12}C^{12}:C^{12}H$	C_{3v}	1191		2 035.741 M	2 035.741 M	.00007	.01984		
$H_3C^{12}C^{12}:C^{12}C^{12}:C^{12}D$	C_{3v}	1192		1 929.772 M	1 929.772 M	.00006	.01830		
$D_3C^{12}C^{12}:C^{12}C^{12}:C^{12}H$	C_{3v}	1193		1 834.856 M	1 834.856 M	.0001	.01454		
$D_3C^{12}C^{12}:C^{12}C^{12}:C^{12}D$	C_{3v}	1194		1 742.215 M	1 742.215 M	.0001	.01354		

1. The vibrational frequency for the excited vibrational mode is not known, but it is assumed to be the low frequency bending mode.

References:

ABC: 610 D_J: 610 D_{JK}: 610

Add. Ref. 439

The following additional values were given in ref. 610 for the indicated species:

Species	B_v (MHz)	D_{JK} (MHz)	q (MHz)	ζ (MHz)	X (MHz)	α (MHz)	1.15 ω/a
1191	2040.14	.02000	2.104	0.9	.00015	−4.40	151 cm⁻¹
1192	1933.86	.0187	1.956	0.92	.0002	−4.09	146 cm⁻¹
1193	1838.69	.0146	1.804	0.9	.00023	−3.84	143 cm⁻¹
1194	1745.80	.0140	1.684	0.9	.00023	−3.58	138 cm⁻¹

Penta-1,3-Diyne

Spectral Line Table

Isotopic Species	Id. No.	Rotational Quantum Nos.	Vib. State	F_1'	F'	F_1	F	Frequency MHz	Acc. ±MHz	Ref.
$H_3C^{12}C^{12}:C^{12}C^{12}:C^{12}H$	1191	5, 0← 4, 0	Ground					20 357.38	.1	610
	1191	5, 0← 4, 0	Excited					20 401.24	.1	610
	1191	5, 1← 4, 1	Ground					20 357.38	.1	610
	1191	5, 1← 4, 1	Excited					20 390.87	.1	610
	1191	5, 1← 4, 1	Excited					20 400.64	.1	610
	1191	5, 1← 4, 1	Excited					20 411.95	.1	610
	1191	5, 2← 4, 2	Ground					20 356.56	.1	610
	1191	5, 2← 4, 2	Excited					20 399.69	.1	610
	1191	5, 2← 4, 2	Excited					20 401.24	.1	610
	1191	5, 3← 4, 3	Ground					20 355.55	.1	610
	1191	5, 3← 4, 3	Excited					20 400.64	.1	610
	1191	5, 4← 4, 4	Ground					20 354.18	.1	610
	1191	5, 4← 4, 4	Excited					20 399.69	.1	610
	1191	6, 0← 5, 0	Ground					24 428.82	.1	610
	1191	6, 0← 5, 0	Excited					24 481.52	.1	610
	1191	6, 1← 5, 1	Ground					24 428.60	.1	610
	1191	6, 1← 5, 1	Excited					24 469.11	.1	610
	1191	6, 1← 5, 1	Excited					24 480.78	.1	610
	1191	6, 1← 5, 1	Excited					24 494.38	.1	610
	1191	6, 2← 5, 2	Ground					24 427.85	.1	610
	1191	6, 2← 5, 2	Excited					24 479.62	.1	610
	1191	6, 2← 5, 2	Excited					24 481.52	.1	610
	1191	6, 3← 5, 3	Ground					24 426.69	.1	610
	1191	6, 3← 5, 3	Excited					24 480.78	.1	610
	1191	6, 4← 5, 4	Ground					24 425.03	.1	610

Isotopic Species	Id. No.	Rotational Quantum Nos.	Vib. State	F_1'	F'	F_1	F	Frequency MHz	Acc. ±MHz	Ref.
$H_3C^{12}C^{12}\mathord{:}C^{12}C^{12}\mathord{:}C^{12}H$	1191	6, 4← 5, 4	Excited					24 475.95	.1	610
	1191	6, 4← 5, 4	Excited					24 479.62	.1	610
	1191	6, 5← 5, 5	Ground					24 422.83	.1	610
	1191	9, 0← 8, 0	Ground					36 643.08	.1	610
	1191	9, 0← 8, 0	Excited					36 722.30	.1	610
	1191	9, 1← 8, 1	Ground					36 642.77	.1	610
	1191	9, 1← 8, 1	Excited					36 703.62	.1	610
	1191	9, 1← 8, 1	Excited					36 721.20	.1	610
	1191	9, 1← 8, 1	Excited					36 741.46	.1	610
	1191	9, 2← 8, 2	Ground					36 641.70	.1	610
	1191	9, 2← 8, 2	Excited					36 719.38	.1	610
	1191	9, 2← 8, 2	Excited					36 722.30	.1	610
	1191	9, 3← 8, 3	Ground					36 639.90	.1	610
	1191	9, 3← 8, 3	Excited					36 716.90	.1	610
	1191	9, 3← 8, 3	Excited					36 721.20	.1	610
	1191	9, 4← 8, 4	Ground					36 637.49	.1	610
	1191	9, 4← 8, 4	Excited					36 713.83	.1	610
	1191	9, 4← 8, 4	Excited					36 719.38	.1	610
	1191	9, 5← 8, 4	Excited					36 710.05	.1	610
	1191	9, 5← 8, 5	Ground					36 634.20	.1	610
	1191	9, 5← 8, 5	Excited					36 716.90	.1	610
	1191	9, 6← 8, 6	Ground					36 630.24	.1	610
	1191	10, 0← 9, 0	Ground					40 714.56	.1	610
	1191	10, 1← 9, 1	Ground					40 714.14	.1	610
	1191	10, 2← 9, 2	Ground					40 712.96	.1	610
	1191	10, 3← 9, 3	Ground					40 710.96	.1	610
	1191	10, 4← 9, 4	Ground					40 708.20	.1	610
	1191	10, 5← 9, 5	Ground					40 704.62	.1	610
	1191	10, 6← 9, 6	Ground					40 700.28	.1	610
	1191	10, 7← 9, 7	Ground					40 695.10	.1	610
	1191	11, 0←10, 0	Ground					44 785.92	.1	610
	1191	11, 1←10, 1	Ground					44 785.48	.1	610
	1191	11, 2←10, 2	Ground					44 784.16	.1	610
	1191	11, 3←10, 3	Ground					44 782.02	.1	610
	1191	11, 4←10, 4	Ground					44 778.98	.1	610
	1191	11, 5←10, 5	Ground					44 775.04	.1	610
	1191	11, 6←10, 6	Ground					44 770.20	.1	610
	1191	11, 7←10, 7	Ground					44 764.52	.1	610
	1191	11, 9←10, 9	Ground					44 750.52	.1	610
	1191	Not Reported						36 707.00	.1	610
$H_3C^{12}C^{12}\mathord{:}C^{12}C^{12}\mathord{:}C^{12}D$	1192	5, 0← 4, 0	Ground					19 297.70	.1	610
	1192	5, 0← 4, 0	Excited					19 338.58	.1	610
	1192	5, 1← 4, 1	Ground					19 297.52	.1	610
	1192	5, 1← 4, 1	Excited					19 328.94	.1	610
	1192	5, 1← 4, 1	Excited					19 338.02	.1	610
	1192	5, 1← 4, 1	Excited					19 348.50	.1	610
	1192	5, 2← 4, 2	Ground					19 296.98	.1	610
	1192	5, 2← 4, 2	Excited					19 337.14	.1	610
	1192	5, 2← 4, 2	Excited					19 338.58	.1	610
	1192	5, 3← 4, 3	Ground					19 296.08	.1	610
	1192	5, 3← 4, 3	Excited					19 335.80	.1	610
	1192	5, 3← 4, 3	Excited					19 338.02	.1	610
	1192	5, 4← 4, 4	Excited					19 337.14	.1	610
	1192	6, 0← 5, 0	Ground					23 157.21	.1	610
	1192	6, 0← 5, 0	Excited					23 206.23	.1	610
	1192	6, 1← 5, 1	Ground					23 156.99	.1	610
	1192	6, 1← 5, 1	Excited					23 194.67	.1	610
	1192	6, 1← 5, 1	Excited					23 205.54	.1	610
	1192	6, 1← 5, 1	Excited					23 218.17	.1	610
	1192	6, 2← 5, 2	Ground					23 156.34	.1	610

Isotopic Species	Id. No.	Rotational Quantum Nos.	Vib. State	F_1'	F'	F_1	F	Frequency MHz	Acc. ±MHz	Ref.
$H_3C^{12}C^{12}:C^{12}C^{12}:C^{12}D$	1192	6, 2← 5, 2	Excited					23 204.48	.1	610
	1192	6, 2← 5, 2	Excited					23 206.23	.1	610
	1192	6, 3← 5, 3	Ground					23 155.25	.1	610
	1192	6, 3← 5, 3	Excited					23 202.96	.1	610
	1192	6, 3← 5, 3	Excited					23 205.54	.1	610
	1192	6, 4← 5, 4	Ground					23 153.71	.1	610
	1192	6, 4← 5, 4	Excited					23 204.48	.1	610
	1192	6, 5← 5, 5	Ground					23 151.67	.1	610
	1192	11, 0←10, 0	Ground					42 454.66	.1	610
	1192	11, 1←10, 1	Ground					42 454.25	.1	610
	1192	11, 2←10, 2	Ground					42 453.05	.1	610
	1192	11, 3←10, 3	Ground					42 451.03	.1	610
	1192	11, 4←10, 4	Ground					42 448.24	.1	610
	1192	11, 5←10, 5	Ground					42 444.61	.1	610
	1192	11, 6←10, 6	Ground					42 440.20	.1	610
	1192	11, 7←10, 7	Ground					42 434.89	.1	610
	1192	12, 0←11, 0	Ground					46 314.11	.1	610
	1192	12, 0←11, 0	Excited					46 411.82	.1	610
	1192	12, 1←11, 1	Ground					46 313.66	.1	610
	1192	12, 1←11, 1	Excited					46 389.10	.1	610
	1192	12, 1←11, 1	Excited					46 410.76	.1	610
	1192	12, 1←11, 1	Excited					46 436.04	.1	610
	1192	12, 2←11, 2	Ground					46 312.36	.1	610
	1192	12, 2←11, 2	Excited					46 408.54	.1	610
	1192	12, 2←11, 2	Excited					46 412.42	.1	610
	1192	12, 3←11, 3	Ground					46 310.16	.1	610
	1192	12, 3←11, 3	Excited					46 405.40	.1	610
	1192	12, 3←11, 3	Excited					46 410.76	.1	610
	1192	12, 4←11, 4	Ground					46 307.07	.1	610
	1192	12, 4←11, 4	Excited					46 408.54	.1	610
	1192	12, 5←11, 5	Ground					46 303.13	.1	610
	1192	12, 5←11, 5	Excited					46 396.90	.1	610
	1192	12, 5←11, 5	Excited					46 405.40	.1	610
	1192	12, 6←11, 6	Ground					46 298.32	.1	610
	1192	12, 6←11, 6	Excited					46 390.82	.1	610
	1192	12, 7←11, 7	Excited					46 396.24	.1	610
	1192	12, 8←11, 8	Excited					46 389.94	.1	610
	1192	Not Reported						46 386.78	.1	610
	1192	Not Reported						46 387.80	.1	610
$D_3C^{12}C^{12}:C^{12}C^{12}:C^{12}H$	1193	7, 0← 6, 0	Ground					25 687.84	.1	610
	1193	7, 1← 6, 1	Ground					25 687.66	.1	610
	1193	7, 2← 6, 2	Ground					25 687.03	.1	610
	1193	7, 3← 6, 3	Ground					25 686.05	.1	610
	1193	7, 4← 6, 4	Ground					25 684.57	.1	610
	1193	9, 0← 8, 0	Ground					33 027.09	.1	610
	1193	9, 0← 8, 0	Excited					33 096.06	.1	610
	1193	9, 1← 8, 1	Ground					33 026.86	.1	610
	1193	9, 1← 8, 1	Excited					33 080.20	.1	610
	1193	9, 1← 8, 1	Excited					33 095.39	.1	610
	1193	9, 1← 8, 1	Excited					33 112.67	.1	610
	1193	9, 2← 8, 2	Ground					33 026.08	.1	610
	1193	9, 2← 8, 2	Excited					33 094.08	.1	610
	1193	9, 2← 8, 2	Excited					33 096.27	.1	610
	1193	9, 3← 8, 3	Ground					33 024.77	.1	610
	1193	9, 3← 8, 3	Excited					33 092.23	.1	610
	1193	9, 3← 8, 3	Excited					33 095.39	.1	610
	1193	9, 4← 8, 4	Ground					33 022.94	.1	610
	1193	9, 4← 8, 4	Excited					33 089.98	.1	610
	1193	9, 4← 8, 4	Excited					33 094.08	.1	610

Isotopic Species	Id. No.	Rotational Quantum Nos.	Vib. State	Hyperfine				Frequency MHz	Acc. ±MHz	Ref.
				F_1'	F'	F_1	F			
$D_3C^{12}C^{12}:C^{12}C^{12}:C^{12}H$	1193	9, 5← 8, 5	Ground					33 020.60	.1	610
	1193	9, 5← 8, 5	Excited					33 087.11	.1	610
	1193	9, 5← 8, 5	Excited					33 092.23	.1	610
	1193	9, 6← 8, 6	Ground					33 017.74	.1	610
	1193	9, 7← 8, 7	Ground					33 014.30	.1	610
	1193	9, 8← 8, 8	Ground					33 010.35	.1	610
	1193	9, 8← 8, 8	Excited					33 083.30	.1	610
	1193	12, 0←11, 0	Ground					44 035.80	.1	610
	1193	12, 1←11, 1	Ground					44 035.48	.1	610
	1193	12, 2←11, 2	Ground					44 034.42	.1	610
	1193	12, 3←11, 3	Ground					44 032.64	.1	610
	1193	12, 4←11, 4	Ground					44 030.18	.1	610
	1193	12, 5←11, 5	Ground					44 027.08	.1	610
	1193	12, 6←11, 6	Ground					44 023.24	.1	610
$D_3C^{12}C^{12}:C^{12}C^{12}:C^{12}D$	1194	7, 0← 6, 0	Ground					24 390.85	.1	610
	1194	7, 1← 6, 1	Ground					24 390.68	.1	610
	1194	7, 2← 6, 2	Ground					24 390.12	.1	610
	1194	7, 3← 6, 3	Ground					24 389.17	.1	610
	1194	7, 4← 6, 4	Ground					24 387.83	.1	610
	1194	7, 5← 6, 5	Ground					24 386.10	.1	610
	1194	10, 0← 9, 0	Ground					34 843.89	.1	610
	1194	10, 0← 9, 0	Excited					34 915.35	.1	610
	1194	10, 1← 9, 1	Ground					34 843.65	.1	610
	1194	10, 1← 9, 1	Excited					34 898.93	.1	610
	1194	10, 1← 9, 1	Excited					34 914.63	.1	610
	1194	10, 1← 9, 1	Excited					34 932.59	.1	610
	1194	10, 2← 9, 2	Ground					34 842.86	.1	610
	1194	10, 2← 9, 2	Excited					34 913.28	.1	610
	1194	10, 2← 9, 2	Excited					34 915.70	.1	610
	1194	10, 3← 9, 3	Ground					34 841.53	.1	610
	1194	10, 3← 9, 3	Excited					34 911.46	.1	610
	1194	10, 3← 9, 3	Excited					34 914.63	.1	610
	1194	10, 4← 9, 4	Ground					34 839.63	.1	610
	1194	10, 4← 9, 4	Excited					34 909.06	.1	610
	1194	10, 4← 9, 4	Excited					34 913.28	.1	610
	1194	10, 5← 9, 5	Ground					34 837.21	.1	610
	1194	10, 6← 9, 6	Ground					34 834.26	.1	610
	1194	10, 7← 9, 7	Ground					34 830.70	.1	610
	1194	10, 8← 9, 8	Ground					34 826.67	.1	610
	1194	10, 9← 9, 9	Ground					34 822.04	.1	610
	1194	13, 0←12, 0	Ground					45 296.66	.1	610
	1194	13, 1←12, 1	Ground					45 296.40	.1	610
	1194	13, 2←12, 2	Ground					45 295.30	.1	610
	1194	13, 3←12, 3	Ground					45 293.56	.1	610
	1194	13, 4←12, 4	Ground					45 291.10	.1	610
	1194	13, 5←12, 5	Ground					45 287.88	.1	610
	1194	13, 6←12, 6	Ground					45 284.10	.1	610
	1194	Not Reported						34 900.65	.1	610
	1194	Not Reported						34 901.48	.1	610

1200 — Pyridine

Molecular Constant Table

C_5H_5N C_{2v} $C_{*}HCH:CHCH:CHN_{*}$

Isotopic Species	Pt. Gp.	Id. No.	A MHz	B MHz	C MHz	D_J MHz	D_{JK} MHz	Δ Amu A²	κ
$C^{12}{}_{*}HC^{12}H:C^{12}HC^{12}H:C^{12}HN^{14}{}_{*}$	C_{2v}	1201	6 039.13 M	5 804.70 M	2 959.25 M	−.0036	−.0019	.032	.84777
$C^{12}{}_{*}DC^{12}H:C^{12}HC^{12}H:C^{12}HN^{14}{}_{*}$	C_s	1202	5 900.80 M	5 558.47 M	2 861.76 M			.061	.77471
$C^{12}{}_{*}HC^{12}D:C^{12}HC^{12}H:C^{12}HN^{14}{}_{*}$	C_s	1203	5 889.12 M	5 554.96 M	2 858.02 M			.035	.77951
$C^{12}{}_{*}HC^{12}H:C^{12}DC^{12}H:C^{12}HN^{14}{}_{*}$	C_{2v}	1204	6 038.90 M	5 419.93 M	2 855.78 M			.035	.61109
$C^{13}{}_{*}HC^{12}H:C^{12}HC^{12}H:C^{12}HN^{14}{}_{*}$	C_s	1205	5 962.90 M	5 758.70 M	2 928.94 M			.0336	.86539
$C^{12}{}_{*}HC^{13}H:C^{12}HC^{12}H:C^{12}HN^{14}{}_{*}$	C_s	1206	5 956.33 M	5 755.75 M	2 926.57 M			.0350	.86759

Id. No.	μ_a Debye	μ_b Debye	μ_c Debye	eQq Value(MHz) Rel.	eQq Value(MHz) Rel.	eQq Value(MHz) Rel.	ω_a d 1/cm	ω_b d 1/cm	ω_c d 1/cm	ω_d d 1/cm
1201	2.26 M	0. X	0. X							

References:

ABC: 496,805 D_J: 542 D_{JK}: 542 Δ: 496,805 κ: 496,805 μ: 805

Add. Ref. 414,457,581,657

For species 1201, $D_K = 0.00587$ MHz. Ref. 542.

Pyridine

Spectral Line Table

Isotopic Species	Id. No.	Rotational Quantum Nos.	Vib. State	Hyperfine F_1'	F'	F_1	F	Frequency MHz	Acc. ±MHz	Ref.
$C^{12}{}_{*}HC^{12}H:C^{12}HC^{12}H:C^{12}HN^{14}{}_{*}$	1201	2, 1, 1← 1, 1, 0	Ground					20 374.2		512
	1201	2, 2, 0← 1, 0, 1	Ground					26 783.		512
	1201	3, 0, 3← 2, 0, 2	Ground					20 722.5		495
	1201	3, 1, 2← 2, 1, 1	Ground					26 926.3	.1	542
	1201	3, 1, 3← 2, 1, 2	Ground					20 709.0		495
	1201	3, 2, 1← 2, 2, 0	Ground					31 862.31		512
	1201	3, 2, 2← 2, 2, 1	Ground					26 292.62	.10	512
	1201	3, 3, 0← 2, 1, 1	Ground					38 917.12		512
	1201	4, 0, 4← 3, 0, 3	Ground					26 634.8	.1	542
	1201	4, 1, 3← 3, 1, 2	Ground					32 594.		512
	1201	4, 1, 4← 3, 1, 3	Ground					26 634.8	.1	542
	1201	4, 1, 3← 3, 3, 0	Ground					20 594.0		495
	1201	4, 2, 2← 3, 2, 1	Ground					38 988.91		512
	1201	4, 2, 3← 3, 2, 2	Ground					32 531.		512
	1201	4, 3, 2← 3, 3, 1	Ground					37 788.10		512
	1201	5, 1, 4← 5, 1, 5	Ground					26 636.9	.1	542
	1201	5, 2, 4← 5, 0, 5	Ground					26 636.9	.1	542
	1201	5, 2, 3← 5, 2, 4	Ground					20 691.0		495
	1201	6, 2, 4← 6, 2, 5	Ground					26 623.7	.1	542
	1201	6, 3, 4← 6, 1, 5	Ground					26 624.4	.1	542
	1201	6, 3, 3← 6, 3, 4	Ground					20 648.5		495
	1201	7, 2, 5← 6, 4, 2	Ground					21 399.5		495
	1201	7, 3, 5← 6, 5, 2	Ground					20 481.5		495
	1201	7, 3, 4← 7, 3, 5	Ground					26 602.6	.1	542
	1201	7, 4, 4← 7, 2, 5	Ground					26 605.2	.1	542

Isotopic Species	Id. No.	Rotational Quantum Nos.	Vib. State	Hyperfine				Frequency MHz	Acc. ±MHz	Ref.
				F_1'	F'	F_1	F			
$C^{12}_*HC^{12}H:C^{12}HC^{12}H:C^{12}HN^{14}_*$	1201	8, 4, 4← 8, 4, 5	Ground					26 569.7	.1	542
	1201	8, 5, 4← 8, 3, 5	Ground					26 577.8	.1	542
	1201	8, 5, 3← 8, 5, 4	Ground					20 430.0		495
	1201	8, 6, 3← 8, 4, 4	Ground					20 666.8		495
	1201	9, 5, 4← 9, 5, 5	Ground					26 520.2	.1	542
	1201	9, 6, 4← 9, 4, 5	Ground					26 540.6	.1	542
	1201	9, 6, 3← 9, 6, 4	Ground					20 211.25		495
	1201	9, 7, 3← 9, 5, 4	Ground					20 678.5		495
	1201	10, 6, 4←10, 6, 5	Ground					26 446.7	.1	542
	1201	10, 7, 4←10, 5, 5	Ground					26 494.1	.1	542
	1201	11, 7, 4←11, 7, 5	Ground					26 338.8	.1	542
	1201	11, 8, 4←11, 6, 5	Ground					26 439.9	.1	542
	1201	11, 8, 3←11, 8, 4	Ground					19 405.4		495
	1201	12, 8, 4←12, 8, 5	Ground					26 182.4	.1	542
	1201	12, 9, 4←12, 7, 5	Ground					26 382.3	.1	542
	1201	12, 9, 3←12, 9, 4	Ground					18 760.9		495
	1201	13, 9, 4←13, 9, 5	Ground					25 958.0	.1	542
	1201	13,10, 4←13, 8, 5	Ground					26 328.9	.1	542
	1201	14,10, 4←14,10, 5	Ground					25 640.1	.1	542
	1201	14,11, 4←14, 9, 5	Ground					26 292.3	.1	542
	1201	15,11, 4←15,11, 5	Ground					25 198.9	.1	542
	1201	15,12, 4←15,10, 5	Ground					26 290.2	.1	542
	1201	16,12, 4←16,12, 5	Ground					24 602.0	.1	542
	1201	16,13, 4←16,11, 5	Ground					26 347.5	.1	542
	1201	16,14, 3←16,12, 4	Ground					23 781.4	.1	542
	1201	17,13, 4←17,13, 5	Ground					23 820.8	.1	542
	1201	17,14, 4←17,12, 5	Ground					26 495.8	.1	542
	1201	17,15, 3←17,13, 4	Ground					25 209.6	.1	542
	1201	18,14, 4←18,14, 5	Ground					22 834.8	.1	542
	1201	18,15, 4←18,13, 5	Ground					26 775.7	.1	542
	1201	18,16, 3←18,14, 4	Ground					27 033.0	.1	542
$C^{12}_*DC^{12}H:C^{12}HC^{12}H:C^{12}HN^{14}_*$	1202	2, 1, 1← 1, 1, 0	Ground					19 537.4	.1	496
	1202	2, 2, 0← 1, 0, 1	Ground					25 989.0	.1	496
	1202	3, 0, 3← 2, 0, 2	Ground					20 050.4	.1	496
	1202	3, 1, 2← 2, 1, 1	Ground					26 135.7	.1	496
	1202	3, 1, 3← 2, 1, 2	Ground					20 019.5	.1	496
	1202	3, 2, 2← 2, 2, 1	Ground					25 260.5	.1	496
	1202	6, 3, 3← 6, 3, 4	Ground					19 843.5	.1	496
	1202	7, 4, 3← 7, 4, 4	Ground					19 631.7	.1	496
	1202	9, 6, 3← 9, 6, 4	Ground					18 704.5	.1	496
	1202	9, 7, 3← 9, 5, 4	Ground					20 160.8	.1	496
	1202	10, 8, 3←10, 6, 4	Ground					20 439.5	.1	496
	1202	11, 9, 3←11, 7, 4	Ground					20 943.0	.1	496
$C^{12}_*HC^{12}D:C^{12}HC^{12}H:C^{12}HN^{14}_*$	1203	2, 1, 1← 1, 1, 0	Ground					19 523.6	.1	496
	1203	2, 2, 0← 1, 0, 1	Ground					25 949.0	.1	496
	1203	3, 0, 3← 2, 0, 2	Ground					20 022.3	.1	496
	1203	3, 1, 2← 2, 1, 1	Ground					26 098.0	.1	496
	1203	3, 1, 3← 2, 1, 2	Ground					19 993.5	.1	496

Isotopic Species	Id. No.	Rotational Quantum Nos.	Vib. State	F_1'	F'	F_1	F	Frequency MHz	Acc. ±MHz	Ref.
$C^{12}*HC^{12}D:C^{12}HC^{12}H:C^{12}HN^{14}*$	1203	3, 2, 2← 2, 2, 1	Ground					25 238.6	.1	496
	1203	6, 3, 3← 6, 3, 4	Ground					19 828.8	.1	496
	1203	7, 4, 3← 7, 4, 4	Ground					19 628.8	.1	496
	1203	9, 6, 3← 9, 6, 4	Ground					18 751.5	.1	496
	1203	9, 7, 3← 9, 5, 4	Ground					20 116.7	.1	496
	1203	10, 8, 3←10, 6, 4	Ground					20 372.0	.1	496
	1203	11, 9, 3←11, 7, 4	Ground					20 838.4	.1	496
$C^{12}*HC^{12}H:C^{12}DC^{12}H:C^{12}HN^{14}*$	1204	2, 1, 1← 1, 1, 0	Ground					19 116.5	.1	496
	1204	2, 2, 0← 1, 0, 1	Ground					26 199.3	.1	496
	1204	3, 0, 3← 2, 0, 2	Ground					20 039.6	.1	496
	1204	3, 1, 2← 2, 1, 1	Ground					26 204.2	.1	496
	1204	3, 1, 3← 2, 1, 2	Ground					19 950.7	.1	496
	1204	3, 2, 2← 2, 2, 1	Ground					24 827.8	.1	496
	1204	6, 3, 3← 6, 3, 4	Ground					19 186.0	.1	496
	1204	7, 4, 3← 7, 4, 4	Ground					18 322.2	.1	496
	1204	9, 7, 3← 9, 5, 4	Ground					21 847.6	.1	496
	1204	10, 8, 3←10, 6, 4	Ground					23 734.0	.1	496
	1204	11, 9, 3←11, 7, 4	Ground					26 634.1	.1	496
$C^{13}*HC^{12}H:C^{12}HC^{12}H:C^{12}HN^{14}*$	1205	2, 1, 1← 1, 1, 0	Ground					20 205.5		805
	1205	3, 0, 3← 2, 0, 2	Ground					20 509.3		805
	1205	3, 1, 2← 2, 1, 1	Ground					26 622.8		805
	1205	3, 1, 3← 2, 1, 2	Ground					20 499.0		805
	1205	3, 2, 2← 2, 2, 1	Ground					26 063.3		805
	1205	10, 7, 3←10, 7, 4	Ground					19 898.3		805
	1205	10, 8, 3←10, 6, 4	Ground					20 479.9		805
	1205	11, 8, 3←11, 8, 4	Ground					19 549.9		805
	1205	11, 9, 3←11, 7, 4	Ground					20 543.3		805
	1205	12, 9, 3←12, 9, 4	Ground					19 068.4		805
	1205	12,10, 3←12, 8, 4	Ground					20 669.9		805
	1205	12,11, 2←12, 9, 3	Ground					18 454.0		805
$C^{12}*HC^{13}H:C^{12}HC^{12}H:C^{12}HN^{14}*$	1206	2, 1, 1← 1, 1, 0	Ground					20 194.4		805
	1206	3, 0, 3← 2, 0, 2	Ground					20 492.4		805
	1206	3, 1, 3← 2, 1, 2	Ground					20 482.3		805
	1206	3, 2, 2← 2, 2, 1	Ground					26 047.6		805
	1206	10, 8, 3←10, 6, 4	Ground					20 460.5		805
	1206	11, 8, 3←11, 8, 4	Ground					19 573.1		805
	1206	11, 9, 3←11, 7, 4	Ground					20 518.9		805
	1206	12, 9, 3←12, 9, 4	Ground					19 110.5		805
	1206	12,10, 3←12, 8, 4	Ground					20 636.8		805
	1206	12,11, 2←12, 9, 3	Ground					18 305.9		805

1210 — Nickel Cyclopentadienyl Nitrosyl

C_5H_5NNiO C_{5v} C_5H_5NNiO

Isotopic Species	Pt. Gp.	Id. No.	A MHz	B MHz	C MHz	D_J MHz	D_{JK} MHz	Δ Amu A²	κ
$C_5^{12}H_5N^{14}Ni^{58}O^{16}$	C_{5v}	1211		1 259.25 M		.00005			
$C_5^{12}H_5N^{14}Ni^{60}O^{16}$	C_{5v}	1212		1 258.71 M					

References:

ABC: 815 D_J: 815

No Spectral Lines

C₅H₅Tl ... C₅H₅Tl

Isotopic Species	Pt. Gp.	Id. No.	A MHz	B MHz	C MHz	D_J MHz	D_{JK} MHz	Δ Amu A²	κ
$C_5^{12}H_5Tl^{203}$	C_{5v}	1221		1 467.98 ¹M					
$C_5^{12}H_5Tl^{205}$	C_{5v}	1222		1 465.10 M					

1. It is not certain that the value given for the rotational constant B refers to the ground state. It refers to the state showing the strongest absorption spectra of the molecules in excited vibrational states.

References:

ABC: 884

No Spectral Lines

1230 − 1,3-Cyclopentadiene

C₅H₆ ... C₂ᵥ ... $C_*H_2CH:CHCH:C_*H$

Isotopic Species	Pt. Gp.	Id. No.	A MHz	B MHz	C MHz	D_J MHz	D_{JK} MHz	Δ Amu A²	κ
$C^{12}_*H_2C^{12}H:C^{12}HC^{12}H:C^{12}_*H$	C_{2v}	1231	8 426.09 M	8 225.54 M	4 271.54 M			3.1056	.90346

Id. No.	μ_a Debye	μ_b Debye	μ_c Debye	eQq Value(MHz)	Rel.	eQq Value(MHz)	Rel.	eQq Value(MHz)	Rel.	ω_a d 1/cm	ω_b d 1/cm	ω_c d 1/cm	ω_d d 1/cm
1231	0. X	.416 M	0. X										

References:

ABC: 696   Δ: 696   κ: 696   μ: 696

1,3-Cyclopentadiene ... Spectral Line Table

Isotopic Species	Id. No.	Rotational Quantum Nos.	Vib. State	F_1'	F'	F_1	F	Frequency MHz	Acc. ±MHz	Ref.
$C^{12}_*H_2C^{12}H:C^{12}HC^{12}H:C^{12}_*H$	1231	2, 0, 2← 1, 1, 1	Ground					21 032.72	.1	696
	1231	2, 1, 2← 1, 0, 1	Ground					21 240.70	.1	696
	1231	2, 2, 1← 1, 1, 0	Ground					29 549.8	.1	696
	1231	3, 0, 3← 2, 1, 2	Ground					29 678.4	.1	696
	1231	3, 1, 3← 2, 0, 2	Ground					29 686.2	.1	696
	1231	3, 1, 2← 3, 0, 3	Ground					20 239.74	.1	696
	1231	3, 2, 2← 3, 1, 3	Ground					20 276.76	.1	696
	1231	4, 2, 2← 4, 1, 3	Ground					20 187.1	.1	696
	1231	4, 3, 2← 4, 2, 3	Ground					20 296.7	.1	696
	1231	5, 3, 2← 5, 2, 3	Ground					20 085.22	.1	696
	1231	5, 4, 2← 5, 3, 3	Ground					20 336.7	.1	696
	1231	6, 4, 2← 6, 3, 3	Ground					19 914.7	.1	696
	1231	6, 5, 2← 6, 4, 3	Ground					20 406.4	.1	696
	1231	7, 5, 2← 7, 4, 3	Ground					19 658.9	.1	696
	1231	12,12, 1←12,11, 2	Ground					20 291.3	.1	696

C_5H_8O C_s $C_*H_2CH_2CH_2CH_2C_*O$

Isotopic Species	Pt. Gp.	Id. No.	A MHz	B MHz	C MHz	D_J MHz	D_{JK} MHz	Δ Amu A²	κ
$C^{12}{}_*H_2C^{12}H_2C^{12}H_2C^{12}H_2C^{12}{}_*O^{16}$	C_s	1241	6 624.53 M	3 351.69 M	2 410.35 M				−.55325

Id. No.	μ_a Debye	μ_b Debye	μ_c Debye	eQq Value(MHz) Rel.	eQq Value(MHz) Rel.	eQq Value(MHz) Rel.	ω_a d 1/cm	ω_b d 1/cm	ω_c d 1/cm	ω_d d 1/cm
1241	3.30 M	0. X								

References:

ABC: 869   κ: 869   μ: 869

Add. Ref. 588,825

Cyclopentanone   Spectral Line Table

Isotopic Species	Id. No.	Rotational Quantum Nos.	Vib. State	F_1'	F'	F_1	F	Frequency MHz	Acc. ±MHz	Ref.
$C^{12}{}_*H_2C^{12}H_2C^{12}H_2C^{12}H_2C^{12}{}_*O^{16}$	1241	2, 0, 2← 1, 0, 1	Ground					11 349.1		869
	1241	2, 1, 1← 1, 1, 0	Ground					12 465.9		869
	1241	2, 1, 2← 1, 1, 1	Ground					10 582.3		869
	1241	3, 0, 3← 2, 0, 2	Ground					16 621.8		869
	1241	3, 1, 2← 2, 1, 1	Ground					18 576.2		869
	1241	3, 1, 3← 2, 1, 2	Ground					15 773.9		869
	1241	3, 2, 1← 2, 2, 0	Ground					17 950.7		869
	1241	3, 2, 2← 2, 2, 1	Ground					17 286.7		869
	1241	4, 0, 4← 3, 0, 3	Ground					21 568.5		869
	1241	4, 1, 3← 3, 1, 2	Ground					24 510.9		869
	1241	4, 1, 4← 3, 1, 3	Ground					20 868.2		869
	1241	4, 2, 2← 3, 2, 1	Ground					24 387.2		869
	1241	4, 2, 3← 3, 2, 2	Ground					22 910.0		869
	1241	4, 3, 1← 3, 3, 0	Ground					23 469.59		588
	1241	4, 3, 2← 3, 3, 1	Ground					23 346.0		869
	1241	14, 4,10←14, 4,11	Ground					23 551.	5.	514
	1241	19, 6,13←19, 6,14	Ground					20 893.	5.	514
	1241	22, 7,15←22, 7,16	Ground					22 206.	5.	514
	1241	25, 8,17←25, 8,18	Ground					23 256.	5.	514
	1241	28, 9,19←28, 9,20	Ground					24 089.	5.	514
	1241	31,10,21←31,10,22	Ground					24 723.	5.	514
	1241	34,11,23←34,11,24	Ground					25 172.	5.	514
	1241	37,12,25←37,12,26	Ground					25 460.	5.	514
	1241	40,13,27←40,13,28	Ground					25 594.	5.	514
	1241	43,14,29←43,14,30	Ground					25 589.	5.	514
	1241	46,15,31←46,15,32	Ground					25 464.	5.	514
	1241	49,16,33←49,16,34	Ground					25 234.	5.	514
	1241	52,17,35←52,17,36	Ground					24 899.	5.	514
	1241	55,18,37←55,18,38	Ground					24 484.	5.	514

C$_5$H$_8$O   C$_s$ C$_*$HOCHCH$_2$CH$_2$C$_*$H$_2$

Isotopic Species	Pt. Gp.	Id. No.	A MHz		B MHz		C MHz		D$_J$ MHz	D$_{JK}$ MHz	Δ Amu A²	κ
C$^{12}_*$HO^{16}C^{12}HC^{12}H$_2$C^{12}H$_2$C$^{12}_*$H$_2$	C$_s$	1251	5 708.6	M	4 540.4	M	3 248.6	M				.0502

1. Intensity relations obtained by Erlandsson indicate that the a and c components of the dipole moment are approximately equal.

References:

ABC: 597   κ: 597

Add. Ref. 596

Cyclopentene Oxide

Isotopic Species	Id. No.	Rotational Quantum Nos.	Vib. State	F$_1'$	F'	F$_1$	F	Frequency MHz	Acc. ±MHz	Ref.
C$^{12}_*$HO^{16}C^{12}HC^{12}H$_2$C^{12}H$_2$C$^{12}_*$H$_2$	1251	3, 0, 3← 2, 0, 2	Ground					21 446.	5.	597
	1251	3, 1, 2← 2, 1, 1	Ground					24 751.	5.	597
	1251	3, 1, 3← 2, 1, 2	Ground					21 094.	5.	597
	1251	3, 2, 1← 2, 2, 0	Ground					25 286.	5.	597
	1251	3, 2, 2← 2, 2, 1	Ground					23 365.	5.	597
	1251	10, 5, 6←10, 3, 7	Ground					22 140.	5.	597
	1251	10, 8, 2←10, 7, 4	Ground					24 221.	5.	597
	1251	10, 8, 3←10, 7, 3	Ground					23 717.	5.	597
	1251	11, 4, 7←11, 4, 8	Ground					24 470.	5.	597
	1251	11, 5, 7←11, 3, 8	Ground					25 304.	5.	597
	1251	11, 6, 6←11, 4, 7	Ground					22 775.	5.	597
	1251	11, 7, 4←11, 6, 6	Ground					23 028.	5.	597
	1251	11, 8, 3←11, 7, 5	Ground					23 900.	5.	597
	1251	11, 8, 4←11, 7, 4	Ground					22 070.	5.	597
	1251	12, 5, 7←12, 5, 8	Ground					22 568.	5.	597
	1251	12, 6, 7←12, 4, 8	Ground					25 133.	5.	597
	1251	12, 7, 6←12, 5, 7	Ground					24 968.	5.	597
	1251	12, 7, 6←12, 8, 4	Ground					24 262.	5.	597

C₆H₅Br C_{2v} C₆H₅Br

Isotopic Species	Pt. Gp.	Id. No.	A MHz	B MHz	C MHz	D_J MHz	D_{JK} MHz	Δ Amu A²	κ
$C_6^{12}H_5Br^b$	C_{2v}	1261							

Id. No.	μ_a Debye	μ_b Debye	μ_c Debye	eQq Value (MHz) Rel.	eQq Value (MHz) Rel.	eQq Value (MHz) Rel.	ω_a d 1/cm	ω_b d 1/cm	ω_c d 1/cm	ω_d d 1/cm
1261	1.70 ¹G	0. X	0. X							

1. There is some variation in other references from the value recorded for the dipole moment.

References:

μ: 1030

Bromobenzene   Spectral Line Table

Isotopic Species	Id. No.	Rotational Quantum Nos.	Vib. State	Hyperfine F_1'	F'	F_1	F	Frequency MHz	Acc. ±MHz	Ref.
$C_6^{12}H_5Br^b$	1261	Not Reported						22 050.		117 ¹
	1261	Not Reported						23 690.	5..	117
	1261	Not Reported						23 742.	5.	117

1. Roughly estimated.

Phenyl Chloride

C$_6$H$_5$Cl   C$_{2v}$ C$_6$H$_5$Cl

Isotopic Species	Pt. Gp.	Id. No.	A MHz		B MHz		C MHz		D$_J$ MHz	D$_{JK}$ MHz	Δ Amu A^2	κ
C$_6^{12}$H$_5$Cl^{35}	C$_{2v}$	1271	5 679.97	M	1 576.87	M	1 233.61	M				−.8456
C$_6^{12}$H$_5$Cl^{37}	C$_{2v}$	1272	5 666.7	M	1 532.0	M	1 206.3	M				−.8537

Id. No.	μ$_a$ Debye		μ$_b$ Debye		μ$_c$ Debye		eQq Value(MHz)	Rel.	eQq Value(MHz)	Rel.	eQq Value(MHz)	Rel.	ω$_a$ 1/cm	d	ω$_b$ 1/cm	d	ω$_c$ 1/cm	d	ω$_d$ 1/cm	d
1271	1.70	G	0.	X	0.	X	−66.4	[1]aa	30	bb										

1. This is an average of values given for several different transitions.

References:

ABC: 516,779   κ: 516,779   μ: 1030   eQq: 779

Add. Ref. 486,596,688

Chlorobenzene

Isotopic Species	Id. No.	Rotational Quantum Nos.	Vib. State	F$_1'$	F'	F$_1$	F	Frequency MHz	Acc. ±MHz	Ref.
C$_6^{12}$H$_5$Cl^{35}	1271	5, 0, 5← 4, 0, 4	Ground					13 660.4		779
	1271	5, 1, 4← 4, 1, 3	Ground					14 811.6		779
	1271	5, 1, 5← 4, 1, 4	Ground					13 112.6		779
	1271	5, 2, 3← 4, 2, 2	Ground					14 398.3		779
	1271	5, 2, 3← 4, 2, 2	Ground					14 401.8		779
	1271	5, 2, 4← 4, 2, 3	Ground					14 005.4		779
	1271	5, 2, 4← 4, 2, 3	Ground					14 008.3		779
	1271	6, 0, 6← 5, 0, 5	Ground					16 210.3		779
	1271	6, 1, 5← 5, 1, 4	Ground					17 705.8		779
	1271	6, 1, 6← 5, 1, 5	Ground					15 689.2		779
	1271	6, 2, 4← 5, 2, 3	Ground					17 421.0		779
	1271	6, 2, 4← 5, 2, 3	Ground					17 423.0		779
	1271	6, 2, 5← 5, 2, 4	Ground					16 769.3		779
	1271	6, 2, 5← 5, 2, 4	Ground					16 771.2		779
	1271	6, 3, 3← 5, 3, 2	Ground					16 996.5		779
	1271	6, 3, 3← 5, 3, 2	Ground					17 001.3		779
	1271	6, 3, 4← 5, 3, 3	Ground					16 954.9		779
	1271	6, 3, 4← 5, 3, 3	Ground					16 959.3		779
	1271	7, 0, 7← 6, 0, 6	Ground					18 702.2		779
	1271	7, 1, 6← 6, 1, 5	Ground					20 554.7		779
	1271	7, 1, 7← 6, 1, 6	Ground					18 246.7		779
	1271	7, 2, 5← 6, 2, 4	Ground					20 482.4		779
	1271	7, 2, 6← 6, 2, 5	Ground					19 513.4		779
	1271	7, 3, 4← 6, 3, 3	Ground					19 893.5		779
	1271	7, 3, 4← 6, 3, 3	Ground					19 895.8		779
	1271	7, 3, 5← 6, 3, 4	Ground					19 797.9		779
	1271	7, 3, 5← 6, 3, 4	Ground					19 800.7		779
	1271	8, 0, 8← 7, 0, 7	Ground					21 157.0		779
	1271	8, 1, 7← 7, 1, 6	Ground					23 339.	5.0	516
	1271	8, 1, 7← 7, 1, 6	Ground		13/2		11/2	23 866.4	.1	556

Isotopic Species	Id. No.	Rotational Quantum Nos.	Vib. State	F_1'	F'	F_1	F	Frequency MHz	Acc. ±MHz	Ref.
$C_6^{12}H_5Cl^{35}$	1271	8, 1, 7← 7, 1, 6	Ground		15/2		13/2	23 866.4	.1	556
	1271	8, 1, 7← 7, 1, 6	Ground		19/2		17/2	23 866.8	.1	556
	1271	8, 1, 7← 7, 1, 6	Ground		17/2		15/2	23 866.8	.1	556
	1271	8, 1, 8← 7, 1, 7	Ground					20 785.9		779
	1271	8, 2, 6← 7, 2, 5	Ground					23 554.7		779
	1271	8, 2, 6← 7, 2, 5	Ground		15/2		13/2	24 156.1	.1	556
	1271	8, 2, 6← 7, 2, 5	Ground		17/2		15/2	24 156.1	.1	556
	1271	8, 2, 6← 7, 2, 5	Ground		19/2		17/2	24 156.9	.1	556
	1271	8, 2, 6← 7, 2, 5	Ground		13/2		11/2	24 156.9	.1	556
	1271	8, 2, 7← 7, 2, 6	Ground					22 233.8		779
	1271	8, 3, 5← 7, 3, 4	Ground					22 829.2		779
	1271	8, 3, 5← 7, 3, 4	Ground		17/2		15/2	23 390.7	.1	556
	1271	8, 3, 5← 7, 3, 4	Ground		15/2		13/2	23 390.7	.1	556
	1271	8, 3, 5← 7, 3, 4	Ground		13/2		11/2	23 392.6	.1	556
	1271	8, 3, 5← 7, 3, 4	Ground		19/2		17/2	23 392.6	.1	556
	1271	8, 3, 6← 7, 3, 5	Ground					22 644.0		779
	1271	8, 4, 4← 7, 4, 3	Ground					22 628.	5.0	516
	1271	8, 4, 5← 7, 4, 4	Ground					22 628.	5.0	516
	1271	8, 5, 3← 7, 5, 2	Ground					22 578.	5.0	516
	1271	8, 5, 4← 7, 5, 3	Ground					22 578.	5.0	516
	1271	8, 6, 2← 7, 6, 1	Ground					22 560.	5.0	516
	1271	8, 6, 3← 7, 6, 2	Ground					22 560.	5.0	516
	1271	9, 0, 9← 8, 0, 8	Ground					23 584.	5.0	516
	1271	9, 1, 8← 8, 1, 7	Ground					26 063.	5.0	516
	1271	9, 1, 8← 8, 1, 7	Ground		15/2		13/2	26 631.1	.1	556
	1271	9, 1, 8← 8, 1, 7	Ground		17/2		15/2	26 631.1	.1	556
	1271	9, 1, 8← 8, 1, 7	Ground		19/2		17/2	26 631.1	.1	556
	1271	9, 1, 8← 8, 1, 7	Ground		21/2		19/2	26 631.1	.1	556
	1271	9, 1, 9← 8, 1, 8	Ground					23 308.6		779
	1271	9, 1, 9← 8, 1, 8	Ground		17/2		15/2	23 775.6	.1	556
	1271	9, 1, 9← 8, 1, 8	Ground		15/2		13/2	23 775.6	.1	556
	1271	9, 1, 9← 8, 1, 8	Ground		21/2		19/2	23 775.6	.1	556
	1271	9, 1, 9← 8, 1, 8	Ground		19/2		17/2	23 775.6	.1	556
	1271	9, 2, 7← 8, 2, 6	Ground					26 616.	5.0	516
	1271	9, 2, 8← 8, 2, 7	Ground					24 928.7		779
	1271	9, 2, 8← 8, 2, 7	Ground		17/2		15/2	25 472.2	.1	556
	1271	9, 2, 8← 8, 2, 7	Ground		19/2		17/2	25 472.2	.1	556
	1271	9, 2, 8← 8, 2, 7	Ground		21/2		19/2	25 472.7	.1	556
	1271	9, 2, 8← 8, 2, 7	Ground		15/2		13/2	25 472.7	.1	556
	1271	9, 3, 6← 8, 3, 5	Ground					25 810.9		779
	1271	9, 3, 6← 8, 3, 5	Ground		19/2		17/2	26 460.2	.1	556
	1271	9, 3, 6← 8, 3, 5	Ground		17/2		15/2	26 460.2	.1	556
	1271	9, 3, 6← 8, 3, 5	Ground		21/2		19/2	26 461.4	.1	556
	1271	9, 3, 6← 8, 3, 5	Ground		15/2		13/2	26 461.4	.1	556
	1271	9, 3, 7← 8, 3, 6	Ground					25 483.6		779
	1271	9, 3, 7← 8, 3, 6	Ground		17/2		15/2	26 078.0	.1	556
	1271	9, 3, 7← 8, 3, 6	Ground		19/2		17/2	26 078.0	.1	556
	1271	9, 3, 7← 8, 3, 6	Ground		21/2		19/2	26 079.2	.1	556
	1271	9, 3, 7← 8, 3, 6	Ground		15/2		13/2	26 079.2	.1	556
	1271	9, 4, 5← 8, 4, 4	Ground					25 502.0		779

Isotopic Species	Id. No.	Rotational Quantum Nos.	Vib. State	F_1'	F'	F_1	F	Frequency MHz	Acc. ±MHz	Ref.
$C_6^{12}H_5Cl^{35}$	1271	9, 4, 5← 8, 4, 4	Ground					25 502.0		779
	1271	9, 4, 5← 8, 4, 4	Ground					25 504.2		779
	1271	9, 4, 5← 8, 4, 4	Ground					25 504.2		779
	1271	9, 4, 5← 8, 4, 4	Ground		17/2		15/2	26 110.3	.1	556
	1271	9, 4, 5← 8, 4, 4	Ground		19/2		17/2	26 110.3	.1	556
	1271	9, 4, 5← 8, 4, 4	Ground		15/2		13/2	26 112.3	.1	556
	1271	9, 4, 5← 8, 4, 4	Ground		21/2		19/2	26 112.3	.1	556
	1271	9, 4, 6← 8, 4, 5	Ground					25 483.6		779
	1271	9, 4, 6← 8, 4, 5	Ground					25 483.6		779
	1271	9, 4, 6← 8, 4, 5	Ground					25 485.9		779
	1271	9, 4, 6← 8, 4, 5	Ground					25 485.9		779
	1271	9, 4, 6← 8, 4, 5	Ground		19/2		17/2	26 086.9	.1	556
	1271	9, 4, 6← 8, 4, 5	Ground		17/2		15/2	26 086.9	.1	556
	1271	9, 4, 6← 8, 4, 5	Ground		21/2		19/2	26 088.9	.1	556
	1271	9, 4, 6← 8, 4, 5	Ground		15/2		13/2	26 088.9	.1	556
	1271	9, 5, 4← 8, 5, 3	Ground					25 427.	5.0	516
	1271	9, 5, 4← 8, 5, 3	Ground		17/2		15/2	26 027.1	.1	556
	1271	9, 5, 4← 8, 5, 3	Ground		19/2		17/2	26 027.1	.1	556
	1271	9, 5, 4← 8, 5, 3	Ground		21/2		19/2	26 030.2	.1	556
	1271	9, 5, 4← 8, 5. 3	Ground		15/2		13/2	26 030.2	.1	556
	1271	9, 5, 5← 8, 5, 4	Ground					25 427.	5.0	516
	1271	9, 5, 5← 8, 5, 4	Ground		17/2		15/2	26 026.4	.1	556
	1271	9, 5, 5← 8, 5, 4	Ground		19/2		17/2	26 026.4	.1	556
	1271	9, 5, 5← 8, 5, 4	Ground		15/2		13/2	26 029.5	.1	556
	1271	9, 5, 5← 8, 5, 4	Ground		21/2		19/2	26 029.5	.1	556
	1271	9, 6, 3← 8, 6, 2	Ground					25 395.	5.0	516
	1271	9, 6, 4← 8, 6, 3	Ground					25 395.	5.0	516
	1271	9, 6, 4← 8, 6, 3	Ground		19/2		17/2	25 988.7	.1	556
	1271	9, 6, 4← 8, 6, 3	Ground		17/2		15/2	25 989.6	.1	556
	1271	9, 6, 4← 8, 6, 3	Ground		21/2		19/2	25 993.3	.1	556
	1271	9, 6, 4← 8, 6, 3	Ground		15/2		13/2	25 994.2	.1	556
	1271	9, 7, 2← 8, 7, 1	Ground					25 375.	5.0	516
	1271	9, 7, 3← 8, 7, 2	Ground					25 375.	5.0	516
	1271	9, 8, 1← 8, 8, 0	Ground					25 359.	5.0	516
	1271	9, 8, 2← 8, 8, 1	Ground					25 359.	5.0	516
	1271	Not Reported	Ground					23 775.4	.1	556
	1271	Not Reported	Ground					26 630.9	.1	556
$C_6^{12}H_5Cl^{37}$	1272	9, 2, 8← 8, 2, 7	Ground					24 318.	5.0	516
	1272	9, 3, 6← 8, 3, 5	Ground					25 114.	5.0	516
	1272	9, 3, 7← 8, 3, 6	Ground					24 827.	5.0	516
	1272	9, 4, 5← 8, 4, 4	Ground					24 834.	5.0	516
	1272	9, 4, 6← 8, 4, 5	Ground					24 827.	5.0	516
	1272	9, 5, 4← 8, 5, 3	Ground					24 776.	5.0	516
	1272	9, 5, 5← 8, 5, 4	Ground					24 776.	5.0	516
	1272	9, 6, 3← 8, 6, 2	Ground					24 746.	5.0	516
	1272	9, 6, 4← 8, 6, 3	Ground					24 746.	5.0	516

C₆H₅F   C$_{2v}$ C₆H₅F

Isotopic Species	Pt. Gp.	Id. No.	A MHz	B MHz	C MHz	D$_J$ MHz	D$_{JK}$ MHz	Δ Amu A²	κ
C$_6^{12}$H$_5$F^{19}	C$_{2v}$	1281	5 663.54 M	2 570.64 M	1 767.94 M				−.58789
3d-C$_6^{12}$H$_5$F^{19}	C$_s$	1282	5 394.27 M	2 529.99 M	1 722.07 M				−.55998
4d-C$_6^{12}$H$_5$F^{19}	C$_{2v}$	1283	5 663.64 M	2 459.72 M	1 714.75 M				−.62269
2,4,6d$_3$-C$_6^{12}$H$_5$F^{19}	C$_{2v}$	1284	5 134.71 M	2 445.03 M	1 656.19 M				−.54645

Id. No.	μ_a Debye	μ_b Debye	μ_c Debye	eQq Value(MHz) Rel.	eQq Value(MHz) Rel.	eQq Value(MHz) Rel.	ω_a d 1/cm	ω_b d 1/cm	ω_c d 1/cm	ω_d d 1/cm
1281	1.66 M	0. X	0. X							

References:

ABC: 737   κ: 737   μ: 870

Add. Ref. 543

Fluorobenzene   Spectral Line Table

Isotopic Species	Id. No.	Rotational Quantum Nos.	Vib. State	F$_1'$	F'	F$_1$	F	Frequency MHz	Acc. ±MHz	Ref.
C$_6^{12}$H$_5$F^{19}	1281	3, 0, 3← 2, 0, 2	Ground					12 492.5		870
	1281	3, 1, 2← 2, 1, 1	Ground					14 125.8		870
	1281	3, 1, 3← 2, 1, 2	Ground					11 731.9		870
	1281	3, 2, 1← 2, 2, 0	Ground					13 539.0		870
	1281	3, 2, 2← 2, 2, 1	Ground					13 015.7		870
	1281	4, 0, 4← 3, 0, 3	Ground					16 172.3	.1	737
	1281	4, 1, 3← 3, 1, 2	Ground					18 637.4	.1	737
	1281	4, 1, 4← 3, 1, 3	Ground					15 514.1	.1	737
	1281	4, 2, 2← 3, 2, 1	Ground					18 427.2	.1	737
	1281	4, 2, 3← 3, 2, 2	Ground					17 247.0	.1	737
	1281	4, 3, 1← 3, 3, 0	Ground					17 676.6	.1	737
	1281	4, 3, 2← 3, 3, 1	Ground					17 588.9	.1	737
	1281	5, 0, 5← 4, 0, 4	Ground					19 678.2	.1	737
	1281	5, 1, 4← 4, 1, 3	Ground					22 941.0	.1	737
	1281	5, 1, 5← 4, 1, 4	Ground					19 219.8	.1	737
	1281	5, 2, 3← 4, 2, 2	Ground					23 393.8	.1	737
	1281	5, 2, 4← 4, 2, 3	Ground					21 389.2	.1	737
	1281	5, 3, 2← 4, 3, 1	Ground					22 322.1	.1	737
	1281	5, 3, 3← 4, 3, 2	Ground					22 028.8	.1	737
	1281	5, 4, 1← 4, 4, 0	Ground					22 006.1	.1	737
	1281	5, 4, 2← 4, 4, 1	Ground					21 997.3	.1	737
	1281	6, 0, 6← 5, 0, 5	Ground					23 134.6	.1	737
	1281	6, 1, 5← 5, 1, 4	Ground					26 960.2	.1	737
	1281	6, 1, 6← 5, 1, 5	Ground					22 863.3	.1	737
	1281	6, 2, 5← 5, 2, 4	Ground					25 427.3	.1	737
	1281	6, 3, 3← 5, 3, 2	Ground					27 163.3	.1	737
	1281	6, 3, 4← 5, 3, 3	Ground					26 445.2	.1	737
	1281	6, 4, 2← 5, 4, 1	Ground					26 528.9	.1	737
	1281	6, 4, 3← 5, 4, 2	Ground					26 483.2	.1	737
	1281	6, 5, 1← 5, 5, 0	Ground					26 378.8	.1	737

Isotopic Species	Id. No.	Rotational Quantum Nos.	Vib. State	Hyperfine				Frequency MHz	Acc. ±MHz	Ref.
				F_1'	F'	F_1	F			
$C_6^{12}H_5F^{19}$	1281	6, 5, 2← 5, 5, 1	Ground					26 378.8	.1	737
	1281	7, 0, 7← 6, 0, 6	Ground					26 605.3	.1	737
	1281	7, 1, 7← 6, 1, 6	Ground					26 460.7	.1	737
	1281	7, 1, 6← 7, 1, 7	Ground					19 953.0	.1	737
	1281	7, 2, 6← 7, 0, 7	Ground					22 465.2	.1	737
	1281	8, 1, 7← 8, 1, 8	Ground					24 072.3	.1	737
	1281	8, 3, 6← 8, 1, 7	Ground					24 300.5	.1	737
	1281	9, 2, 7← 9, 2, 8	Ground					19 137.0	.1	737
	1281	10, 2, 8←10, 2, 9	Ground					23 839.2	.1	737
	1281	12, 3, 9←12, 3,10	Ground					21 846.3		595
	1281	12, 3,10←12, 3,11	Ground					21 848.	5.0	430
	1281	13, 4, 9←13, 4,10	Ground					13 246.4		595
	1281	15, 4,11←15, 4,12	Ground					23 976.8		595
	1281	16, 5,11←16, 5,12	Ground					14 036.9		595
	1281	17, 5,12←17, 5,13	Ground					19 535.1		595
	1281	18, 5,13←18, 5,14	Ground					25 596.4		595
	1281	19, 6,13←19, 6,14	Ground					14 475.8		595
	1281	20, 6,14←20, 6,15	Ground					20 280.9		595
	1281	21, 6,15←21, 6,16	Ground					26 758.		595
	1281	22, 7,15←22, 7,16	Ground					14 619.2		595
	1281	23, 7,16←23, 7,17	Ground					20 662.1		595
	1281	25, 8,17←25, 8,18	Ground					14 516.3		595
	1281	26, 8,18←26, 8,19	Ground					20 725.5		595
	1281	28, 9,19←28, 9,20	Ground					14 211.7		595
	1281	29, 9,20←29, 9,21	Ground					20 514.5		595
	1281	31,10,21←31,10,22	Ground					13 746.6		595
	1281	32,10,22←32,10,23	Ground					20 070.3		595
	1281	34,11,23←34,11,24	Ground					13 156.6		595
	1281	35,11,24←35,11,25	Ground					19 432.8		595
	1281	39,12,27←39,12,28	Ground					26 280.	5.	430
	1281	42,13,29←42,13,30	Ground					25 315.0		595
	1281	45,14,31←45,14,32	Ground					24 189.7		595
	1281	48,15,33←48,15,34	Ground					22 943.9		595
	1281	51,16,35←51,16,36	Ground					21 608.	5.	430
	1281	54,17,37←54,17,38	Ground					20 219.	5.	430
	1281	57,18,39←57,18,40	Ground					18 806.	5.	430
	1281	Not Reported						17 724.	5.	430
	1281	Not Reported						18 370.	5.	430
	1281	Not Reported						19 238.	5.	430
	1281	Not Reported						20 063.	5.	430
	1281	Not Reported						20 882.	5.	430
	1281	Not Reported						21 551.	5.	430
	1281	Not Reported						21 889.	5.	430
	1281	Not Reported						22 228.	5.0	431
	1281	Not Reported						23 642.	5.	430
	1281	Not Reported						23 941.	5.	430
	1281	Not Reported						24 895.	5.	430
	1281	Not Reported						25 403.	5.	430
	1281	Not Reported						25 758.	5.	430
	1281	Not Reported						26 174.	5.	430

Isotopic Species	Id. No.	Rotational Quantum Nos.	Vib. State	F_1'	F'	F_1	F	Frequency MHz	Acc. ±MHz	Ref.
3d-$C_6^{12}H_5F^{19}$	1282	4, 1, 3← 3, 1, 2	Ground					18 271.6	.1	737
	1282	4, 2, 2← 3, 2, 1	Ground					18 144.5	.1	737
	1282	5, 0, 5← 4, 0, 4	Ground					19 153.2	.1	737
	1282	5, 1, 4← 4, 1, 3	Ground					22 447.5	.1	737
	1282	5, 1, 5← 4, 1, 4	Ground					18 745.2	.1	737
	1282	5, 2, 3← 4, 2, 2	Ground					23 026.5	.1	737
	1282	5, 2, 4← 4, 2, 3	Ground					20 931.8	.1	737
	1282	5, 3, 3← 4, 3, 2	Ground					21 614.6	.1	737
	1282	5, 4, 1← 4, 4, 0	Ground					21 601.5	.1	737
	1282	5, 4, 2← 4, 4, 1	Ground					21 588.9	.1	737
	1282	6, 0, 6← 5, 0, 5	Ground					22 517.0	.1	737
	1282	6, 1, 5← 5, 1, 4	Ground					26 316.0	.1	737
	1282	6, 1, 6← 5, 1, 5	Ground					22 286.5	.1	737
	1282	6, 2, 4← 5, 2, 3	Ground					27 798.1	.1	737
	1282	6, 2, 5← 5, 2, 4	Ground					24 860.7	.1	737
	1282	6, 3, 3← 5, 3, 2	Ground					26 754.7	.1	737
	1282	6, 3, 4← 5, 3, 3	Ground					25 938.4	.1	737
	1282	6, 4, 2← 5, 4, 1	Ground					26 055.9	.1	737
	1282	6, 4, 3← 5, 4, 2	Ground					25 998.1	.1	737
	1282	6, 5, 1← 5, 5, 0	Ground					25 888.4	.1	737
	1282	6, 5, 2← 5, 5, 1	Ground					25 888.4	.1	737
	1282	8, 3, 6← 8, 1, 7	Ground					23 141.2	.1	737
	1282	9, 2, 7← 9, 2, 8	Ground					19 440.8	.1	737
	1282	9, 3, 7← 9, 1, 8	Ground					25 020.2	.1	737
	1282	10, 2, 8←10, 2, 9	Ground					23 984.4	.1	737
	1282	12, 3, 9←12, 3,10	Ground					22 663.2	.1	737
4d-$C_6^{12}H_5F^{19}$	1283	4, 1, 3← 3, 1, 2	Ground					17 920.4	.1	737
	1283	4, 2, 2← 3, 2, 1	Ground					17 625.8	.1	737
	1283	5, 0, 5← 4, 0, 4	Ground					19 103.1	.1	737
	1283	5, 1, 4← 4, 1, 3	Ground					22 107.4	.1	737
	1283	5, 1, 5← 4, 1, 4	Ground					18 608.4	.1	737
	1283	5, 2, 3← 4, 2, 2	Ground					22 378.0	.1	737
	1283	5, 2, 4← 4, 2, 3	Ground					20 616.2	.1	737
	1283	5, 3, 2← 4, 3, 1	Ground					21 391.0	.1	737
	1283	5, 3, 3← 4, 3, 2	Ground					21 165.0	.1	737
	1283	5, 4, 1← 4, 4, 0	Ground					21 134.6	.1	737
	1283	5, 4, 2← 4, 4, 1	Ground					21 128.5	.1	737
	1283	6, 0, 6← 5, 0, 5	Ground					22 461.6	.1	737
	1283	6, 1, 5← 5, 1, 4	Ground					26 055.5	.1	737
	1283	6, 1, 6← 5, 1, 5	Ground					22 150.4	.1	737
	1283	6, 2, 4← 5, 2, 3	Ground					27 088.5	.1	737
	1283	6, 2, 5← 5, 2, 4	Ground					24 535.6	.1	737
	1283	6, 3, 3← 5, 3, 2	Ground					25 980.0	.1	737
	1283	6, 3, 4← 5, 3, 3	Ground					25 416.4	.1	737
	1283	6, 4, 2← 5, 4, 1	Ground					25 461.1	.1	737
	1283	6, 4, 3← 5, 4, 2	Ground					25 429.9	.1	737
	1283	6, 5, 1← 5, 5, 0	Ground					25 338.6	.1	737
	1283	6, 5, 2← 5, 5, 1	Ground					25 338.6	.1	737
	1283	7, 0, 7← 6, 0, 6	Ground					25 823.5	.1	737
	1283	7, 1, 7← 6, 1, 6	Ground					25 647.9	.1	737
	1283	7, 1, 6← 7, 1, 7	Ground					18 858.0	.1	737
	1283	8, 1, 7← 8, 1, 8	Ground					22 908.8	.1	737
	1283	9, 1, 8← 9, 1, 9	Ground					26 818.2	.1	737
	1283	10, 2, 8←10, 2, 9	Ground					22 018.5	.1	737
	1283	11, 2, 9←11, 2,10	Ground					26 548.8	.1	737
	1283	12, 3, 9←12, 3,10	Ground					19 285.1	.1	737

Isotopic Species	Id. No.	Rotational Quantum Nos.	Vib. State	Hyperfine				Frequency MHz	Acc. ±MHz	Ref.
				F_1'	F'	F_1	F			
$2,4,6d_3\text{-}C_6^{12}H_5F^{19}$	1284	5, 0, 5← 4, 0, 4	Ground					18 413.5	.1	737
	1284	5, 1, 4← 4, 1, 3	Ground					21 631.8	.1	737
	1284	5, 1, 5← 4, 1, 4	Ground					18 038.6	.1	737
	1284	5, 2, 3← 4, 2, 2	Ground					22 253.4	.1	737
	1284	5, 2, 4← 4, 2, 3	Ground					20 175.5	.1	737
	1284	5, 3, 2← 4, 3, 1	Ground					21 209.8	.1	737
	1284	5, 3, 3← 4, 3, 2	Ground					20 859.0	.1	737
	1284	6, 0, 6← 5, 0, 5	Ground					21 648.5	.1	737
	1284	6, 1, 5← 5, 1, 4	Ground					25 330.1	.1	737
	1284	6, 1, 6← 5, 1, 5	Ground					21 441.2	.1	737
	1284	6, 2, 4← 5, 2, 3	Ground					26 847.2	.1	737
	1284	6, 2, 5← 5, 2, 4	Ground					23 950.9	.1	737
	1284	6, 3, 3← 5, 3, 2	Ground					25 869.5	.1	737
	1284	6, 3, 4← 5, 3, 3	Ground					25 027.3	.1	737
	1284	6, 4, 2← 5, 4, 1	Ground					25 159.0	.1	737
	1284	6, 4, 3← 5, 4, 2	Ground					25 096.7	.1	737
	1284	6, 5, 1← 5, 5, 0	Ground					24 987.5	.1	737
	1284	6, 5, 2← 5, 5, 1	Ground					24 987.5	.1	737
	1284	7, 0, 7← 6, 0, 6	Ground					24 905.9	.1	737
	1284	7, 1, 7← 6, 1, 6	Ground					24 802.5	.1	737
	1284	10, 2, 8←10, 2, 9	Ground					23 386.2	.1	737
	1284	10, 3, 8←10, 1, 9	Ground					26 405.5	.1	737
	1284	11, 3, 8←11, 3, 9	Ground					17 523.2	.1	737
	1284	12, 3, 9←12, 3,10·	Ground					22 391.7	.1	737

C$_7$H$_5$N

C$_{2v}$

C$_6$H$_5$CN

Isotopic Species	Pt. Gp.	Id. No.	A MHz		B MHz		C MHz		D$_J$ MHz	D$_{JK}$ MHz	Δ Amu A^2	κ
C$_6^{12}$H$_5$C^{12}N^{14}	C$_{2v}$	1291	5 656.7	M	1 546.84	M	1 214.41	M				−.850
C$_6^{12}$H$_5$C^{12}N^{15}	C$_{2v}$	1292	5 655.7	M	1 502.13	M	1 186.67	M				
C$_6^{12}$H$_5$C^{13}N^{14}	C$_{2v}$	1293	5 655.5	M	1 528.63	M	1 203.14	M				
2d-C$_6^{12}$H$_5$C^{12}N^{14}	C$_s$	1294	5 381.1	M	1 546.14	M	1 200.70	M				
3d-C$_6^{12}$H$_5$C^{12}N^{14}	C$_s$	1295	5 383.9	M	1 526.28	M	1 188.94	M				
4d-C$_6^{12}$H$_5$C^{12}N^{14}	C$_{2v}$	1296	5 653.8	M	1 496.60	M	1 183.23	M				
1C^{13}-C$_6^{12}$H$_5$C^{12}N^{14}	C$_{2v}$	1297	5 655.0	M	1 545.55	M	1 213.61	M				
2C^{13}-C$_6^{12}$H$_5$C^{12}N^{14}	C$_s$	1298	5 564.2	M	1 546.82	M	1 210.10	M				
3C^{13}-C$_6^{12}$H$_5$C^{12}N^{14}	C$_s$	1299	5 565.2	M	1 535.73	M	1 203.39	M				
4C^{13}-C$_6^{12}$H$_5$C^{12}N^{14}	C$_{2v}$	1301	5 654.1	M	1 523.65	M	1 200.09	M				

Id. No.	μ$_a$ Debye		μ$_b$ Debye		μ$_c$ Debye		eQq Value(MHz)	Rel.	eQq Value(MHz)	Rel.	eQq Value(MHz)	Rel.	ω$_a$ d 1/cm	ω$_b$ d 1/cm	ω$_c$ d 1/cm	ω$_d$ d 1/cm
1291	4.14	M	0.	X	0.	X										

References:

ABC: 955   κ: 539   μ: 539

Benzonitrile

Spectral Line Table

Isotopic Species	Id. No.	Rotational Quantum Nos.	Vib. State	F$_1'$	F'	F$_1$	F	Frequency MHz	Acc. ±MHz	Ref.
C$_6^{12}$H$_5$C^{12}N^{14}	1291	5, 0, 5← 4, 0, 4	Ground					13 437.5	.2	955
	1291	5, 1, 4← 4, 1, 3	Ground					14 545.3	.2	955
	1291	5, 1, 5← 4, 1, 4	Ground					12 897.8	.2	955
	1291	5, 2, 3← 4, 2, 2	Ground					14 132.1	.2	955
	1291	5, 2, 4← 4, 2, 3	Ground					13 763.2	.2	955
	1291	6, 0, 6← 5, 0, 5	Ground					15 951.9	.2	955
	1291	6, 1, 5← 5, 1, 4	Ground					17 389.9	.2	955
	1291	6, 1, 6← 5, 1, 5	Ground					15 434.1	.2	955
	1291	6, 2, 4← 5, 2, 3	Ground					17 095.8	.2	955
	1291	6, 2, 5← 5, 2, 4	Ground					16 480.6	.2	955
	1291	6, 3, 3← 5, 3, 2	Ground					16 694.3	.2	955
	1291	6, 5, 1← 5, 5, 0	Ground					16 616.4	.2	955
	1291	6, 5, 2← 5, 5, 1	Ground					16 616.4	.2	955
	1291	7, 0, 7← 6, 0, 6	Ground					18 410.0	.2	955
	1291	7, 1, 6← 6, 1, 5	Ground					20 193.7		540
	1291	7, 1, 7← 6, 1, 6	Ground					17 952.5	.2	955
	1291	7, 2, 5← 6, 2, 4	Ground					20 096.3		540
	1291	7, 2, 6← 6, 2, 5	Ground					19 179.5		540
	1291	7, 3, 4← 6, 3, 3	Ground					19 534.8		540
	1291	7, 3, 5← 6, 3, 4	Ground					19 449.2		540
	1291	7, 4, 3← 6, 4, 2	Ground					19 426.0		540
	1291	7, 4, 4← 6, 4, 3	Ground					19 426.0		540
	1291	7, 5, 2← 6, 5, 1	Ground					19 398.1		540
	1291	7, 5, 3← 6, 5, 2	Ground					19 398.1		540
	1291	7, 6, 1← 6, 6, 0	Ground					19 383.5		540

Isotopic Species	Id. No.	Rotational Quantum Nos.	Vib. State	F_1'	F'	F_1	F	Frequency MHz	Acc. ±MHz	Ref.
$C_6^{12}H_5C^{12}N^{14}$	1291	7, 6, 2← 6, 6, 1	Ground					19 383.5		540
	1291	8, 0, 8← 7, 0, 7	Ground					20 828.3	.2	955
	1291	8, 1, 7← 7, 1, 6	Ground					22 936.		515
	1291	8, 1, 7← 7, 1, 6	Ground					22 943.7		540
	1291	8, 1, 8← 7, 1, 7	Ground					20 453.0	.2	955
	1291	8, 2, 6← 7, 2, 5	Ground					23 111.1		540
	1291	8, 2, 7← 7, 2, 6	Ground					21 856.4		540
	1291	8, 3, 5← 7, 3, 4	Ground					22 406.		515
	1291	8, 3, 6← 7, 3, 5	Ground					22 239.		515
	1291	8, 4, 4← 7, 4, 3	Ground					22 231.		515
	1291	8, 4, 5← 7, 4, 4	Ground					22 220.		515
	1291	8, 5, 3← 7, 5, 2	Ground					22 184.7		540
	1291	8, 5, 4← 7, 5, 3	Ground					22 184.7		540
	1291	8, 6, 2← 7, 6, 1	Ground					22 163.1		540
	1291	8, 6, 3← 7, 6, 2	Ground					22 163.1		540
	1291	8, 7, 1← 7, 7, 0	Ground					22 144.		515
	1291	8, 7, 2← 7, 7, 1	Ground					22 144.		515
	1291	9, 0, 9← 8, 0, 8	Ground					23 227.8	.2	955
	1291	9, 1, 8← 8, 1, 7	Ground					25 628.		515
	1291	9, 1, 9← 8, 1, 8	Ground					22 937.9		540
	1291	9, 2, 8← 8, 2, 7	Ground					24 509.3		540
	1291	9, 3, 6← 8, 3, 5	Ground					25 328.		515
	1291	9, 3, 7← 8, 3, 6	Ground					25 030.		515
	1291	9, 4, 6← 8, 3, 6	Ground					25 032.7		540
	1291	9, 4, 5← 8, 4, 4	Ground					25 044.		515
	1291	9, 4, 6← 8, 4, 5	Ground					25 030.		515
	1291	9, 5, 4← 8, 5, 3	Ground					24 972.		515
	1291	9, 5, 4← 8, 5, 3	Ground					24 978.6		540
	1291	9, 5, 5← 8, 5, 4	Ground					24 972.		515
	1291	9, 5, 5← 8, 5, 4	Ground					24 978.6		540
	1291	9, 6, 3← 8, 6, 2	Ground					24 938.		515
	1291	9, 6, 3← 8, 6, 2	Ground					24 947.5	.2	955
	1291	9, 6, 4← 8, 6, 3	Ground					24 938.		515
	1291	9, 6, 4← 8, 6, 3	Ground					24 947.5	.2	955
	1291	9, 7, 2← 8, 7, 1	Ground					24 929.0	.2	955
	1291	9, 7, 3← 8, 7, 2	Ground					24 929.0	.2	955
	1291	9, 8, 1← 8, 8, 0	Ground					24 912.		515
	1291	9, 8, 1← 8, 8, 0	Ground					24 917.1	.2	955
	1291	9, 8, 2← 8, 8, 1	Ground					24 912.		515
	1291	9, 8, 2← 8, 8, 1	Ground					24 917.1	.2	955
	1291	10, 0,10← 9, 0, 9	Ground					25 625.		515
$C_6^{12}H_5C^{12}N^{15}$	1292	5, 0, 5← 4, 0, 4	Ground					13 113.0	.2	955
	1292	5, 1, 4← 4, 1, 3	Ground					14 150.7	.2	955
	1292	5, 1, 5← 4, 1, 4	Ground					12 586.0	.2	955
	1292	5, 2, 3← 4, 2, 2	Ground					13 736.7	.2	955
	1292	5, 2, 4← 4, 2, 3	Ground					13 405.7	.2	955
	1292	6, 0, 6← 5, 0, 5	Ground					15 577.6	.2	955
	1292	6, 1, 5← 5, 1, 4	Ground					16 924.1	.2	955
	1292	6, 1, 6← 5, 1, 5	Ground					15 064.0	.2	955
	1292	6, 2, 4← 5, 2, 3	Ground					16 609.9	.2	955
	1292	6, 2, 5← 5, 2, 4	Ground					16 055.4	.2	955

Isotopic Species	Id. No.	Rotational Quantum Nos.	Vib. State	Hyperfine				Frequency MHz	Acc. ±MHz	Ref.
				F_1'	F'	F_1	F			
$C_6^{12}H_5C^{12}N^{15}$	1292	8, 1, 8← 7, 1, 7	Ground					19 970.6	.2	955
	1292	9, 0, 9← 8, 0, 8	Ground					22 705.9	.2	955
	1292	9, 7, 3← 8, 7, 2	Ground					24 268.2	.2	955
$C_6^{12}H_5C^{13}N^{14}$	1293	5, 0, 5← 4, 0, 4	Ground					13 305.6	.2	955
	1293	5, 1, 5← 4, 1, 4	Ground					12 770.9	.2	955
	1293	5, 2, 3← 4, 2, 2	Ground					13 970.7	.2	955
	1293	5, 2, 4← 4, 2, 3	Ground					13 617.8	.2	955
	1293	6, 0, 6← 5, 0, 5	Ground					15 799.9	.2	955
	1293	6, 1, 5← 5, 1, 4	Ground					17 200.2	.2	955
	1293	6, 1, 6← 5, 1, 5	Ground					15 283.5	.2	955
	1293	6, 2, 4← 5, 2, 3	Ground					16 897.4	.2	955
	1293	6, 2, 5← 5, 2, 4	Ground					16 307.4	.2	955
	1293	8, 0, 8← 7, 0, 7	Ground					20 637.2	.2	955
	1293	9, 0, 9← 8, 0, 8	Ground					23 015.9	.2	955
$2d\text{-}C_6^{12}H_5C^{12}N^{14}$	1294	5, 0, 5← 4, 0, 4	Ground					13 313.7	.2	955
	1294	5, 1, 5← 4, 1, 4	Ground					12 782.1	.2	955
	1294	5, 2, 3← 4, 2, 2	Ground					14 104.5	.2	955
	1294	5, 2, 4← 4, 2, 3	Ground					13 684.6	.2	955
	1294	6, 0, 6← 5, 0, 5	Ground					15 786.3	.2	955
	1294	6, 1, 5← 5, 1, 4	Ground					17 313.0	.2	955
	1294	6, 1, 6← 5, 1, 5	Ground					15 290.0	.2	955
	1294	6, 2, 4← 5, 2, 3	Ground					17 075.4	.2	955
	1294	6, 2, 5← 5, 2, 4	Ground					16 381.1	.2	955
	1294	8, 0, 8← 7, 0, 7	Ground					20 582.0	.2	955
	1294	9, 0, 9← 8, 0, 8	Ground					22 949.7	.2	955
$3d\text{-}C_6^{12}H_5C^{12}N^{14}$	1295	5, 0, 5← 4, 0, 4	Ground					13 175.7	.2	955
	1295	5, 1, 4← 4, 1, 3	Ground					14 317.8	.2	955
	1295	5, 1, 5← 4, 1, 4	Ground					12 648.5	.2	955
	1295	5, 2, 3← 4, 2, 2	Ground					13 929.0	.2	955
	1295	5, 2, 4← 4, 2, 3	Ground					13 529.1	.2	955
	1295	6, 0, 6← 5, 0, 5	Ground					15 627.4	.2	955
	1295	6, 1, 5← 5, 1, 4	Ground					17 110.3	.2	955
	1295	6, 1, 6← 5, 1, 5	Ground					15 132.1	.2	955
	1295	6, 2, 4← 5, 2, 3	Ground					16 859.0	.2	955
	1295	6, 2, 5← 5, 2, 4	Ground					16 196.3	.2	955
	1295	8, 0, 8← 7, 0, 7	Ground					20 383.8	.2	955
	1295	9, 0, 9← 8, 0, 8	Ground					22 729.0	.2	955
$4d\text{-}C_6^{12}H_5C^{12}N^{14}$	1296	5, 0, 5← 4, 0, 4	Ground					13 072.4	.2	955
	1296	5, 1, 4← 4, 1, 3	Ground					14 101.6	.2	955
	1296	5, 1, 5← 4, 1, 4	Ground					12 547.0	.2	955
	1296	5, 2, 3← 4, 2, 2	Ground					13 687.7	.2	955
	1296	5, 2, 4← 4, 2, 3	Ground					13 361.3	.2	955
	1296	6, 0, 6← 5, 0, 5	Ground					15 530.8	.2	955
	1296	6, 1, 5← 5, 1, 4	Ground					16 866.1	.2	955
	1296	6, 1, 6← 5, 1, 5	Ground					15 017.9	.2	955
	1296	6, 2, 4← 5, 2, 3	Ground					16 550.2	.2	955
	1296	6, 2, 5← 5, 2, 4	Ground					16 002.2	.2	955
	1296	8, 0, 8← 7, 0, 7	Ground					20 298.7	.2	955
	1296	9, 0, 9← 8, 0, 8	Ground					22 641.0	.2	955

Isotopic Species	Id. No.	Rotational Quantum Nos.	Vib. State	Hyperfine				Frequency MHz	Acc. ±MHz	Ref.
				F_1'	F'	F_1	F			
$1C^{13}\text{-}C_6^{12}H_5C^{12}N^{14}$	1297	5, 0, 5← 4, 0, 4	Ground					13 428.4	.1	955
	1297	5, 1, 5← 4, 1, 4	Ground					12 888.8	.2	955
	1297	5, 2, 3← 4, 2, 2	Ground					14 120.6	.2	955
	1297	5, 2, 4← 4, 2, 3	Ground					13 753.1	.1	955
	1297	6, 0, 6← 5, 0, 5	Ground					15 941.3	.1	955
	1297	6, 1, 5← 5, 1, 4	Ground					17 376.0	.1	955
	1297	6, 1, 6← 5, 1, 5	Ground					15 423.6	.1	955
	1297	6, 2, 4← 5, 2, 3	Ground					17 081.3	.2	955
	1297	6, 2, 5← 5, 2, 4	Ground					16 468.3	.1	955
	1297	8, 0, 8← 7, 0, 7	Ground					20 814.5	.1	955
	1297	8, 1, 8← 7, 1, 7	Ground					20 439.0	.1	955
	1297	9, 0, 9← 8, 0, 8	Ground					23 212.6	.1	955
$2C^{13}\text{-}C_6^{12}H_5C^{12}N^{14}$	1298	5, 0, 5← 4, 0, 4	Ground					13 399.6	.1	955
	1298	5, 1, 4← 4, 1, 3	Ground					14 529.1	.2	955
	1298	5, 1, 5← 4, 1, 4	Ground					12 861.7	.3	955
	1298	5, 2, 3← 4, 2, 2	Ground					14 124.8	.1	955
	1298	5, 2, 4← 4, 2, 3	Ground					13 739.3	.1	955
	1298	6, 0, 6← 5, 0, 5	Ground					15 900.6	.2	955
	1298	6, 1, 5← 5, 1, 4	Ground					17 367.4	.1	955
	1298	6, 1, 6← 5, 1, 5	Ground					15 389.6	.1	955
	1298	6, 2, 4← 5, 2, 3	Ground					17 090.5	.1	955
	1298	6, 2, 5← 5, 2, 4	Ground					16 450.5	.2	955
	1298	8, 0, 8← 7, 0, 7	Ground					20 750.9	.2	955
	1298	8, 1, 8← 7, 1, 7	Ground					20 389.0	.2	955
	1298	9, 0, 9← 8, 0, 8	Ground					23 140.0	.1	955
$3C^{13}\text{-}C_6^{12}H_5C^{12}N^{14}$	1299	5, 0, 5← 4, 0, 4	Ground					13 320.8	.1	955
	1299	5, 1, 4← 4, 1, 3	Ground					14 432.3	.2	955
	1299	5, 1, 5← 4, 1, 4	Ground					12 786.1	.2	955
	1299	5, 2, 3← 4, 2, 2	Ground					14 026.5	.1	955
	1299	5, 2, 4← 4, 2, 3	Ground					13 652.0	.2	955
	1299	6, 0, 6← 5, 0, 5	Ground					15 809.9	.1	955
	1299	6, 1, 5← 5, 1, 4	Ground					17 252.9	.1	955
	1299	6, 1, 6← 5, 1, 5	Ground					15 299.3	.2	955
	1299	6, 2, 4← 5, 2, 3	Ground					16 970.1	.1	955
	1299	6, 2, 5← 5, 2, 4	Ground					16 346.0	.2	955
	1299	8, 0, 8← 7, 0, 7	Ground					20 637.5	.1	955
	1299	8, 1, 8← 7, 1, 7	Ground					20 271.9	.2	955
	1299	9, 0, 9← 8, 0, 8	Ground					23 013.9	.2	955
$4C^{13}\text{-}C_6^{12}H_5C^{12}N^{14}$	1301	5, 0, 5← 4, 0, 4	Ground					13 269.7	.2	955
	1301	5, 1, 4← 4, 1, 3	Ground					14 340.7	.2	955
	1301	5, 1, 5← 4, 1, 4	Ground					12 736.6	.3	955
	1301	5, 2, 4← 4, 2, 3	Ground					13 578.1	.1	955
	1301	6, 0, 6← 5, 0, 5	Ground					15 758.6	.1	955
	1301	6, 1, 6← 5, 1, 5	Ground					15 242.5	.2	955
	1301	6, 2, 4← 5, 2, 3	Ground					16 843.6	.2	955
	1301	8, 0, 8← 7, 0, 7	Ground					20 585.1	.2	955

$C_7H_{13}N$ C_{3v} $C_7H_{13}N$

Isotopic Species	Pt. Gp.	Id. No.	A MHz	B MHz	C MHz	D_J MHz	D_{JK} MHz	Δ Amu A^2	κ
$C_7^{12}H_{13}N^{14}$	C_{3v}	1311		2431.4 M		.004	.015		

References:

ABC: 908 D_J: 908 D_{JK}: 908

Quinuclidine   Spectral Line Table

Isotopic Species	Id. No.	Rotational Quantum Nos.	Vib. State	F_1'	F'	F_1	F	Frequency MHz	Acc. ±MHz	Ref.
$C_7^{12}H_{13}N^{14}$	1311	5, ← 4,	Ground					24 311.5		908
	1311	5, ← 4,	Excited					24 325.7		908
	1311	5, ← 4,	Excited					24 349.6		908
	1311	5, ← 4,	Excited					24 391.4		908
	1311	6, ← 5,	Excited					29 121.0		908
	1311	6, ← 5,	Excited					29 160.6		908
	1311	6, ← 5,	Ground					29 173.5		908
	1311	6, ← 5,	Excited					29 187.6		908
	1311	6, ← 5,	Excited					29 211.1		908
	1311	6, ← 5,	Excited					29 251.2		908
	1311	6, ← 5,	Excited					29 294.1		908
	1311	7, ← 6,	Excited					34 019.9		908
	1311	7, ← 6,	Ground					34 033.7		908
	1311	7, ← 6,	Excited					34 048.2		908
	1311	7, ← 6,	Excited					34 072.4		908
	1311	7, ← 6,	Excited					34 113.0		908

1320 — Cyclopentadienyl Manganese Tricarbonyl   Molecular Constant Table

$C_8H_5MnO_3$ C_s $C_5H_5Mn(CO)_3$

Isotopic Species	Pt. Gp.	Id. No.	A MHz	B MHz	C MHz	D_J MHz	D_{JK} MHz	Δ Amu A^2	κ
$C_5^{12}H_5Mn^{55}(C^{12}O^{16})_3$	C_s	1321		826.5 M					

References:

ABC: 884

No Spectral Lines

C₈H₁₃Br $\hspace{4cm}$ C$_{3v}$ $\hspace{4cm}$ C₈H₁₃Br

Isotopic Species	Pt. Gp.	Id. No.	A MHz	B MHz	C MHz	D$_J$ MHz	D$_{JK}$ MHz	Δ Amu A²	κ
C$_8^{12}$H$_{13}$Br^{79}	C$_{3v}$	1331		725.9 M					
C$_8^{12}$H$_{13}$Br^{81}	C$_{3v}$	1332		718.55 M					

References:

ABC: 464

Add. Ref. 382

1-Bromo-Bicyclo(2,2,2)-Octane $\hspace{6cm}$ Spectral Line Table

Isotopic Species	Id. No.	Rotational Quantum Nos.	Vib. State	F$_1'$	F'	F$_1$	F	Frequency MHz	Acc. ±MHz	Ref.
C$_8^{12}$H$_{13}$Br^{79}	1331	6, ← 5,	Excited					8 716.5		464
	1331	16, ←15,	Excited					23 231.		464
	1331	17, ←16,	Excited					24 691.		464
	1331	18, ←17,	Excited					26 146.		464
C$_8^{12}$H$_{13}$Br^{81}	1332	6, ← 5,	Excited					8 627.5		464
	1332	16, ←15,	Excited					22 997.		464
	1332	17, ←16,	Excited					24 441.		464
	1332	18, ←17,	Excited					25 882.		464

C₈H₁₃Cl $\hspace{4cm}$ C$_{3v}$ $\hspace{4cm}$ C₈H₁₃Cl

Isotopic Species	Pt. Gp.	Id. No.	A MHz	B MHz	C MHz	D$_J$ MHz	D$_{JK}$ MHz	Δ Amu A²	κ
C$_8^{12}$H$_{13}$Cl^{35}	C$_{3v}$	1341		1 090.90 M					
C$_8^{12}$H$_{13}$Cl^{37}	C$_{3v}$	1342		1 065.91 M					

References:

ABC: 464

1-Chloro-Bicyclo(2,2,2)-Octane $\hspace{6cm}$ Spectral Line Table

Isotopic Species	Id. No.	Rotational Quantum Nos.	Vib. State	F$_1'$	F'	F$_1$	F	Frequency MHz	Acc. ±MHz	Ref.
C$_8^{12}$H$_{13}$Cl^{35}	1341	11, ←10,	Ground					23 999.8		464
	1341	12, ←11,	Ground					26 182.0		464
C$_8^{12}$H$_{13}$Cl^{37}	1342	11, ←10,	Ground					23 449.6		464
	1342	12, ←11,	Ground					25 582.7		464

$C_9H_6CrO_3$ $C_{3v}{}^1$ $C_6H_6Cr(CO)_3$

Isotopic Species	Pt. Gp.	Id. No.	A MHz	B MHz	C MHz	D_J MHz	D_{JK} MHz	Δ Amu A²	κ
$C_6^{12}H_6Cr^{52}(C^{12}O^{16})_3$	C_{3v}	1351		729.8 M					

1. So far as microwave spectra show, the symmetry might be merely C_{3v}, the symmetry of the tricarbonyl group.

References:

ABC: 884

No Spectral Lines

———————————————

$ClFO_3$ C_{3v} ClO_3F

Isotopic Species	Pt. Gp.	Id. No.	A MHz	B MHz	C MHz	D_J MHz	D_{JK} MHz	Δ Amu A²	κ
$Cl^{35}O_3^{16}F^{19}$	C_{3v}	1361		5 260.66 F					
$Cl^{37}O_3^{16}F^{19}$	C_{3v}	1362		5 230.68 F					

Id. No.	μ_a Debye	μ_b Debye	μ_c Debye	eQq Value(MHz) Rel.	eQq Value(MHz) Rel.	eQq Value(MHz) Rel.	ω_a d 1/cm	ω_b d 1/cm	ω_c d 1/cm	ω_d d 1/cm
1361	.023 ²M									

1. No lines were detected in the 19500-34000 MHz region, nor for the $J = 4 \leftarrow 3$ $K = 3 \leftarrow 3$, $K = 3 \leftarrow 3$ transition near 42000 MHz.

2. Since little information about this molecule is available, the value for the dipole moment has been entered under μ_a purely for convenience in recording.

References:

ABC: 701 μ: 767

Add. Ref. 697

No Spectral Lines

ClF$_3$ C$_{2v}$ ClF$_3$

Isotopic Species	Pt. Gp.	Id. No.	A MHz		B MHz		C MHz		D$_J$ MHz	D$_{JK}$ MHz	Δ Amu A²	κ
Cl^{35}F$_3^{19}$	C$_{2v}$	1371	13747.7	M	4611.72	M	3448.79	M			.125	−.77413
Cl^{37}F$_3^{19}$	C$_{2v}$	1372	13653.2	M	4611.90	M	3442.81	M			.148	−.77095

Id. No.	μ_a Debye		μ_b Debye		μ_c Debye		eQq Value(MHz) Rel.	eQq Value(MHz) Rel.	eQq Value(MHz) Rel.	ω_a d 1/cm	ω_b d 1/cm	ω_c d 1/cm	ω_d d 1/cm
1371	0.	X	.554	G	0.	X	149.8						
1372							118.1						

References:

ABC: 477   Δ: 477   κ: 477   μ: 373   eQq: 477

Add. Ref. 397

Species	χ$_{aa}$ (MHz)	χ$_{bb}$ (MHz)	Ref.
1371	−81.25	−64.67	477
1372	−65.41	−51.11	477

Isotopic Species	Id. No.	Rotational Quantum Nos.	Vib. State	F_1'	F'	F_1	F	Frequency MHz	Acc. ±MHz	Ref.
$Cl^{35}F_3^{19}$	1371	2, 1, 2← 1, 0, 1	Ground		5/2		5/2	24 063.80	.15	477
	1371	2, 1, 2← 1, 0, 1	Ground		3/2		1/2	24 073.44	.15	477
	1371	2, 1, 2← 1, 0, 1	Ground		5/2		3/2	24 084.00	.15	477
	1371	2, 1, 2← 1, 0, 1	Ground		7/2		5/2	24 100.42	.15	477
	1371	2, 1, 2← 1, 0, 1	Ground		3/2		3/2	24 110.32	.15	477
	1371	2, 1, 2← 1, 0, 1	Ground		1/2		1/2	24 110.32	.15	477
	1371	5, 1, 4← 5, 0, 5	Ground		7/2		7/2	20 078.22	.15	477
	1371	5, 1, 4← 5, 0, 5	Ground		13/2		13/2	20 990.64	.15	477
	1371	5, 1, 4← 5, 0, 5	Ground		9/2		9/2	21 021.11	.15	477
	1371	5, 1, 4← 5, 0, 5	Ground		11/2		11/2	21 032.67	.15	477
	1371	5, 2, 3← 5, 1, 4	Ground		7/2		7/2	24 308.65	.15	477
	1371	5, 2, 3← 5, 1, 4	Ground		13/2		13/2	24 309.12	.15	477
	1371	5, 2, 3← 5, 1, 4	Ground		9/2		9/2	24 310.05	.15	477
	1371	5, 2, 3← 5, 1, 4	Ground		11/2		11/2	24 310.43	.15	477
	1371	6, 2, 4← 6, 1, 5	Ground		7/2		7/2	24 320.70	.15	477
	1371	6, 2, 4← 6, 1, 5	Ground		15/2		15/2	24 321.86	.15	477
	1371	6, 2, 4← 6, 1, 5	Ground		9/2		9/2	24 326.60	.15	477
	1371	6, 2, 4← 6, 1, 5	Ground		13/2		13/2	24 328.15	.15	477
	1371	7, 2, 5← 7, 1, 6	Ground		11/2		11/2	25 365.84	.15	477
	1371	7, 2, 5← 7, 1, 6	Ground		17/2		17/2	25 368.11	.15	477
	1371	7, 2, 5← 7, 1, 6	Ground		13/2		13/2	25 376.85	.15	477
	1371	7, 2, 5← 7, 1, 6	Ground		15/2		15/2	25 379.38	.15	477
$Cl^{37}F_3^{19}$	1372	2, 1, 2← 1, 0, 1	Ground		5/2		5/2	23 957.65	.15	477
	1372	2, 1, 2← 1, 0, 1	Ground		3/2		1/2	23 965.23	.15	477
	1372	2, 1, 2← 1, 0, 1	Ground		5/2		3/2	23 973.87	.15	477
	1372	2, 1, 2← 1, 0, 1	Ground		7/2		5/2	23 986.75	.15	477
	1372	2, 1, 2← 1, 0, 1	Ground		1/2		1/2	23 994.55	.15	477
	1372	2, 1, 2← 1, 0, 1	Ground		3/2		3/2	23 994.55	.15	477
	1372	5, 1, 4← 5, 0, 5	Ground		7/2		7/2	20 981.36	.15	477
	1372	5, 1, 4← 5, 0, 5	Ground		9/2		9/2	21 013.87	.15	477
	1372	5, 1, 4← 5, 0, 5	Ground		11/2		11/2	21 023.95	.15	477
	1372	5, 2, 3← 5, 1, 4	Ground		13/2		13/2	24 060.00	.15	477
	1372	5, 2, 3← 5, 1, 4	Ground		11/2		11/2	24 062.09	.15	477
	1372	6, 2, 4← 6, 1, 5	Ground		15/2		15/2	24 111.82	.15	477
	1372	6, 2, 4← 6, 1, 5	Ground		9/2		9/2	24 117.35	.15	477
	1372	6, 2, 4← 6, 1, 5	Ground		13/2		13/2	24 118.20	.15	477
	1372	7, 2, 5← 7, 1, 6	Ground		11/2		11/2	25 213.46	.15	477
	1372	7, 2, 5← 7, 1, 6	Ground		17/2		17/2	25 215.40	.15	477
	1372	7, 2, 5← 7, 1, 6	Ground		13/2		13/2	25 222.97	.15	477
	1372	7, 2, 5← 7, 1, 6	Ground		15/2		15/2	25 224.62	.15	477

ClO$_2$ C$_{2v}$ ClO$_2$

Isotopic Species	Pt. Gp.	Id. No.	A MHz		B MHz		C MHz		D$_J$ MHz	D$_{JK}$ MHz	Δ Amu A²	κ
Cl^{35}O$_2^{16}$	C$_{2v}$	1381	52 077.95	M	9 952.42	M	8 333.21	M				
Cl^{37}O$_2^{16}$	C$_{2v}$	1382	50 733.98	M	9 952.91	M	8 298.38	M				
Cl^{35}O^{16}O^{18}	C$_s$	1383	50 580.9	M	9 379.6	M	7 891.7	M				

Id. No.	μ_a Debye		μ_b Debye		μ_c Debye		eQq Value(MHz)	Rel.	eQq Value(MHz)	Rel.	eQq Value(MHz)	Rel.	ω_a 1/cm	d	ω_b 1/cm	d	ω_c 1/cm	d	ω_d 1/cm	d
1381	0.	X	1.785	[1]M	0.	X															

1. The value of the dipole moment obtained by the Stark effect differs considerably from that previously reported by infrared techniques.

References:

ABC: 960 μ: 984

Add. Ref. 85,736

Isotopic Species	Id. No.	Rotational Quantum Nos.	Vib. State	Hyperfine				Frequency MHz	Acc. ±MHz	Ref.
				F_1'	F'	F_1	F			
$Cl^{35}O_2^{16}$	1381	1, 1, 0← 1, 0, 1	Ground		0		1	43 112.78	.2	936
	1381	1, 1, 0← 1, 0, 1	Ground		1		1	43 124.95	.2	936
	1381	1, 1, 0← 1, 0, 1	Ground		2		1	43 140.30	.2	936
	1381	1, 1, 0← 1, 0, 1	Ground		2		2	43 274.56	.2	936
	1381	1, 1, 0← 1, 0, 1	Ground		3		2	43 285.5	.2	936
	1381	1, 1, 0← 1, 0, 1	Ground		2		3	43 338.16	.2	936
	1381	1, 1, 0← 1, 0, 1	Ground		3		3	43 348.60	.2	936
	1381	1, 1, 0← 1, 0, 1	Ground		1		2	43 398.08	.2	936
	1381	1, 1, 0← 1, 0, 1	Ground		2		2	43 413.60	.2	936
	1381	1, 1, 0← 1, 0, 1	Ground		0		1	43 416.10	.2	936
	1381	1, 1, 0← 1, 0, 1	Ground		3		2	43 423.60	.2	936
	1381	1, 1, 0← 1, 0, 1	Ground		1		1	43 428.72	.2	936
	1381	1, 1, 0← 1, 0, 1	Ground		2		1	43 444.10	.2	936
	1381	1, 1, 0← 1, 0, 1	Ground					43 745.0		980
	1381	1, 1, 0← 1, 0, 1	Ground		1		1	44 242.08	.2	936
	1381	1, 1, 0← 1, 0, 1	Ground		1		2	44 376.00	.2	936
	1381	1, 1, 0← 1, 0, 1	Ground		2		1	44 429.74	.2	936
	1381	1, 1, 0← 1, 0, 1	Ground		1		2	44 515.28	.2	936
	1381	1, 1, 0← 1, 0, 1	Ground		1		1	44 545.80	.2	936
	1381	1, 1, 0← 1, 0, 1	Ground		2		2	44 564.10	.2	936
	1381	1, 1, 0← 1, 0, 1	Ground		2		3	44 627.20	.2	936
	1381	1, 1, 0← 1, 0, 1	Ground		2		2	44 703.00	.2	936
	1381	3, 0, 3← 2, 1, 2	Ground		1		0	13 852.05	.2	936
	1381	3, 0, 3← 2, 1, 2	Ground		2		1	13 858.45	.2	936
	1381	3, 0, 3← 2, 1, 2	Ground		3		2	13 894.5	.2	936
	1381	3, 0, 3← 2, 1, 2	Ground		4		3	13 953.6	.2	936
	1381	3, 0, 3← 2, 1, 2	Ground					14 121.9		980
	1381	3, 0, 3← 2, 1, 2	Ground		5		4	14 231.6	.2	936
	1381	3, 0, 3← 2, 1, 2	Ground		4		3	14 246.7	.2	936
	1381	3, 0, 3← 2, 1, 2	Ground		3		2	14 256.4	.2	936
	1381	3, 0, 3← 2, 1, 2	Ground		2		1	14 262.5	.2	936
	1381	3, 1, 2← 3, 0, 3	Ground		5		5	47 697.2	.2	936
	1381	3, 1, 2← 3, 0, 3	Ground		4		4	47 739.8	.2	936
	1381	3, 1, 2← 3, 0, 3	Ground		3		3	47 764.6	.2	936
	1381	3, 1, 2← 3, 0, 3	Ground		2		2	47 778.2	.2	936
	1381	3, 1, 2← 3, 0, 3	Ground					47 992.2		980
	1381	3, 1, 2← 3, 0, 3	Ground		1		1	48 232.10	.2	936
	1381	3, 1, 2← 3, 0, 3	Ground		2		2	48 276.60	.2	936
	1381	3, 1, 2← 3, 0, 3	Ground		3		3	48 336.0	.2	936
	1381	3, 1, 2← 3, 0, 3	Ground		4		4	48 398.20	.2	936
	1381	4, 0, 4← 3, 1, 3	Ground					34 181.9		1005
	1381	4, 0, 4← 3, 1, 3	Ground					34 189.6		1005
	1381	4, 0, 4← 3, 1, 3	Ground					34 251.6		1005
	1381	4, 0, 4← 3, 1, 3	Ground					34 257.2		1005
	1381	4, 0, 4← 3, 1, 3	Ground					34 372.7		1005
	1381	4, 0, 4← 3, 1, 3	Ground					34 390.8		1005
	1381	4, 0, 4← 3, 1, 3	Ground					34 398.6		1005
	1381	4, 0, 4← 3, 1, 3	Ground					34 429.2		1005
	1381	4, 0, 4← 3, 1, 3	Ground					34 437.0		1005
	1381	4, 0, 4← 3, 1, 3	Ground					34 463.4		1005

Isotopic Species	Id. No.	Rotational Quantum Nos.	Vib. State	Hyperfine				Frequency MHz	Acc. ±MHz	Ref.
				F_1'	F'	F_1	F			
$Cl^{35}O_2^{16}$	1381	4, 0, 4 ← 3, 1, 3	Ground					34 468.4		1005
	1381	4, 0, 4 ← 3, 1, 3	Ground					34 488.3		1005
	1381	4, 0, 4 ← 3, 1, 3	Ground					37 167.6		1005
	1381	4, 0, 4 ← 3, 1, 3	Ground					37 182.2		1005
	1381	4, 0, 4 ← 3, 1, 3	Ground					37 193.1		1005
	1381	4, 0, 4 ← 3, 1, 3	Ground					37 196.1		1005
	1381	4, 0, 4 ← 3, 1, 3	Ground					37 208.8		1005
	1381	4, 0, 4 ← 3, 1, 3	Ground					37 210.3		1005
	1381	4, 0, 4 ← 3, 1, 3	Ground					37 221.8		1005
	1381	4, 0, 4 ← 3, 1, 3	Ground					37 230.0		1005
	1381	4, 0, 4 ← 3, 1, 3	Ground					37 253.6		1005
	1381	4, 0, 4 ← 3, 1, 3	Ground					37 268.0		1005
	1381	4, 0, 4 ← 3, 1, 3	Ground					37 279.3		1005
	1381	4, 0, 4 ← 3, 1, 3	Ground					37 289.2		1005
	1381	4, 2, 3 ← 5, 1, 4	Ground		3		4	25 261.0	.2	936
	1381	4, 2, 3 ← 5, 1, 4	Ground					25 265.80	.05	957
	1381	4, 2, 3 ← 5, 1, 4	Ground		4		5	25 274.6	.2	936
	1381	4, 2, 3 ← 5, 1, 4	Ground		5		6	25 290.1	.2	936
	1381	4, 2, 3 ← 5, 1, 4	Ground		6		7	25 306.7	.2	936
	1381	4, 2, 3 ← 5, 1, 4	Ground					25 403.65	.05	957
	1381	4, 2, 3 ← 5, 1, 4	Ground					25 424.78	.05	957
	1381	4, 2, 3 ← 5, 1, 4	Ground					25 451.	.05	957
	1381	4, 2, 3 ← 5, 1, 4	Ground					25 518.7		980
	1381	4, 2, 3 ← 5, 1, 4	Ground		5		6	25 798.5	.2	936
	1381	4, 2, 3 ← 5, 1, 4	Ground		4		5	25 828.1	.2	936
	1381	4, 2, 3 ← 5, 1, 4	Ground		3		4	25 852.7	.2	936
	1381	4, 2, 3 ← 5, 1, 4	Ground		2		3	25 870.1	.2	936
	1381	5, 0, 5 ← 4, 1, 4	Ground		7/2		5/2	55 055.8		980 [1]
	1381	5, 0, 5 ← 4, 1, 4	Ground		9/2		7/2	55 075.2		980 [1]
	1381	5, 0, 5 ← 4, 1, 4	Ground		13/2		11/2	55 090.6		980 [1]
	1381	5, 0, 5 ← 4, 1, 4	Ground		11/2		9/2	55 104.6		980 [1]
	1381	5, 0, 5 ← 4, 1, 4	Ground					55 112.7		980
	1381	5, 0, 5 ← 4, 1, 4	Ground		11/2		9/2	55 113.0		980 [1]
	1381	5, 0, 5 ← 4, 1, 4	Ground		9/2		7/2	55 126.8		980 [1]
	1381	5, 0, 5 ← 4, 1, 4	Ground		7/2		5/2	55 135.8		980 [1]
	1381	5, 0, 5 ← 4, 1, 4	Ground		13/2		11/2	55 138.8		980 [1]
	1381	5, 1, 4 ← 5, 0, 5	Ground		13/2		11/2	56 011.6		980 [1]
	1381	5, 1, 4 ← 5, 0, 5	Ground		11/2		9/2	56 053.8		980 [1]
	1381	5, 1, 4 ← 5, 0, 5	Ground		9/2		7/2	56 085.4		980 [1]
	1381	5, 1, 4 ← 5, 0, 5	Ground		7/2		5/2	56 103.6		980 [1]
	1381	5, 1, 4 ← 5, 0, 5	Ground					56 330.8		980
	1381	5, 1, 4 ← 5, 0, 5	Ground		7/2		5/2	56 576.9		980 [1]
	1381	5, 1, 4 ← 5, 0, 5	Ground		9/2		7/2	56 623.6		980 [1]
	1381	5, 1, 4 ← 5, 0, 5	Ground		11/2		9/2	56 674.6		980 [1]
	1381	5, 1, 4 ← 5, 0, 5	Ground		13/2		11/2	56 723.0		980 [1]
	1381	5, 2, 3 ← 6, 1, 5	Ground					37 153.7	.05	957
	1381	5, 2, 3 ← 6, 1, 5	Ground					37 166.7	.05	957
	1381	5, 2, 3 ← 6, 1, 5	Ground					37 181.1	.05	957
	1381	5, 2, 3 ← 6, 1, 5	Ground					37 192.2	.05	957
	1381	5, 2, 3 ← 6, 1, 5	Ground					37 195.3	.05	957

1. Spectral lines for these species from reference 980 have hyperfine quantum numbers computed from the following relations:

$$\text{for } J = N + 1/2, \; F = N + 2, \, N + 1, \, N, \, N - 1$$

$$J = N - 1/2, \; F = N - 2, \, N - 1, \, N, \, N + 1$$

Isotopic Species	Id. No.	Rotational Quantum Nos.	Vib. State	F_1'	F'	F_1	F	Frequency MHz	Acc. ±MHz	Ref.
$Cl^{35}O_2^{16}$	1381	5, 2, 3← 6, 1, 5	Ground					37 207.9	.05	957
	1381	5, 2, 3← 6, 1, 5	Ground					37 220.5	.05	957
	1381	5, 2, 3← 6, 1, 5	Ground					37 221.5	.05	957
	1381	5, 2, 3← 6, 1, 5	Ground					37 229.5	.05	957
	1381	5, 2, 3← 6, 1, 5	Ground					37 253.4	.05	957
	1381	5, 2, 3← 6, 1, 5	Ground					37 267.0	.05	957
	1381	5, 2, 3← 6, 1, 5	Ground					37 278.4	.05	957
	1381	5, 2, 3← 6, 1, 5	Ground					37 288.4	.05	957
	1381	7, 1, 6← 6, 2, 5	Ground					20 688.0	.2	936
	1381	7, 1, 6← 6, 2, 5	Ground					20 713.3	.2	936
	1381	7, 1, 6← 6, 2, 5	Ground					20 741.8	.2	936
	1381	7, 1, 6← 6, 2, 5	Ground					20 770.6	.2	936
	1381	7, 1, 6← 6, 2, 5	Ground					20 821.7		980
	1381	7, 1, 6← 6, 2, 5	Ground					20 857.2	.2	936
	1381	7, 1, 6← 6, 2, 5	Ground					20 881.2	.2	936
	1381	7, 1, 6← 6, 2, 5	Ground					20 900.6	.2	936
	1381	7, 1, 6← 6, 2, 5	Ground					20 915.9	.2	936
	1381	9, 3, 6←10, 2, 9	Ground					35 399.6	.05	957
	1381	9, 3, 6←10, 2, 9	Ground					35 418.5	.05	957
	1381	9, 3, 6←10, 2, 9	Ground					35 418.7	.05	957
	1381	9, 3, 6←10, 2, 9	Ground					35 447.1	.05	957
	1381	9, 3, 6←10, 2, 9	Ground					35 456.1	.05	957
	1381	9, 3, 6←10, 2, 9	Ground					35 457.9	.05	957
	1381	9, 3, 6←10, 2, 9	Ground					35 459.6	.05	957
	1381	9, 3, 6←10, 2, 9	Ground					35 475.3	.05	957
	1381	9, 3, 6←10, 2, 9	Ground					35 495.7	.05	957
	1381	9, 3, 6←10, 2, 9	Ground					35 545.3	.05	957
	1381	9, 3, 6←10, 2, 9	Ground					35 589.1	.05	957
	1381	11, 2, 9←10, 3, 8	Ground		19/2		17/2	7 438.2		980 [1]
	1381	11, 2, 9←10, 3, 8	Ground		21/2		19/2	7 457.1		980 [1]
	1381	11, 2, 9←10, 3, 8	Ground		23/2		21/2	7 476.2		980 [1]
	1381	11, 2, 9←10, 3, 8	Ground		25/2		23/2	7 494.9		980 [1]
	1381	11, 2, 9←10, 3, 8	Ground					7 569.1		980
	1381	11, 2, 9←10, 3, 8	Ground					7 627.3		980
	1381	11, 2, 9←10, 3, 8	Ground		23/2		21/2	7 660.0		980 [1]
	1381	11, 2, 9←10, 3, 8	Ground		21/2		19/2	7 672.3		980 [1]
	1381	11, 3, 8←12, 2,11	Ground		27/2		25/2	7 449.0		980 [1]
	1381	11, 3, 8←12, 2,11	Ground		25/2		23/2	7 469.2		980 [1]
	1381	11, 3, 8←12, 2,11	Ground		23/2		21/2	7 483.6		980 [1]
	1381	11, 3, 8←12, 2,11	Ground		21/2		19/2	7 493.8		980 [1]
	1381	11, 3, 8←12, 2,11	Ground					7 840.4		980
	1381	11, 3, 8←12, 2,11	Ground		21/2		19/2	8 214.3		980 [1]
	1381	11, 3, 8←12, 2,11	Ground		23/2		21/2	8 235.3		980 [1]
	1381	11, 3, 8←12, 2,11	Ground		25/2		23/2	8 253.7		980 [1]
	1381	11, 3, 8←12, 2,11	Ground		27/2		25/2	8 268.6		980 [1]
	1381	13, 2,11←14, 1,14	Ground		29/2		31/2	12 742.7		980 [1]
	1381	13, 2,11←14, 1,14	Ground		27/2		29/2	12 822.4		980 [1]
	1381	13, 2,11←14, 1,14	Ground		25/2		27/2	12 883.5		980 [1]
	1381	13, 2,11←14, 1,14	Ground		23/2		25/2	12 929.4		980 [1]
	1381	13, 2,11←14, 1,14	Ground					13 720.7		980

1. Spectral lines for these species from reference 980 have hyperfine quantum numbers computed from the following relations:

for $J = N + 1/2$, $F = N + 2, N + 1, N, N - 1$

$J = N - 1/2$, $F = N - 2, N - 1, N, N + 1$

Isotopic Species	Id. No.	Rotational Quantum Nos.	Vib. State	F_1'	F'	F_1	F	Frequency MHz	Acc. ±MHz	Ref.
$Cl^{35}O_2^{16}$	1381	13, 2,11←14, 1,14	Ground		23/2		25/2	14 549.5		980 [1]
	1381	13, 2,11←14, 1,14	Ground		25/2		27/2	14 630.9		980 [1]
	1381	13, 2,11←14, 1,14	Ground		27/2		29/2	14 706.4		980 [1]
	1381	13, 2,11←14, 1,14	Ground		29/2		31/2	14 770.3		980 [1]
	1381	14, 2,13←13, 3,10	Ground		31/2		29/2	16 522.4		980 [1]
	1381	14, 2,13←13, 3,10	Ground		29/2		27/2	16 542.1		980 [1]
	1381	14, 2,13←13, 3,10	Ground		27/2		25/2	16 566.7		980 [1]
	1381	14, 2,13←13, 3,10	Ground		25/2		23/2	16 592.7		980 [1]
	1381	14, 2,13←13, 3,10	Ground					16 997.1		980
	1381	14, 2,13←13, 3,10	Ground		25/2		23/2	17 379.6		980 [1]
	1381	14, 2,13←13, 3,10	Ground		27/2		25/2	17 393.8		980 [1]
	1381	14, 2,13←13, 3,10	Ground		29/2		27/2	17 413.2		980 [1]
	1381	14, 2,13←13, 3,10	Ground		31/2		29/2	17 438.7		980 [1]
	1381	14, 4,11←15, 3,12	Ground		25/2		27/2	10 558.4		980 [1]
	1381	14, 4,11←15, 3,12	Ground		27/2		29/2	10 565.2		980 [1]
	1381	14, 4,11←15, 3,12	Ground		29/2		31/2	10 572.8		980 [1]
	1381	14, 4,11←15, 3,12	Ground		31/2		33/2	10 581.5		980 [1]
	1381	14, 4,11←15, 3,12	Ground					10 716.9		980
	1381	14, 4,11←15, 3,12	Ground		31/2		33/2	10 868.7		980 [1]
	1381	14, 4,11←15, 3,12	Ground		29/2		31/2	10 877.8		980 [1]
	1381	14, 4,11←15, 3,12	Ground		27/2		29/2	10 887.1		980 [1]
	1381	14, 4,11←15, 3,12	Ground		25/2		27/2	10 896.2		980 [1]
	1381	16, 2,15←15, 3,12	Ground		35/2		33/2	34 372.7		980 [1]
	1381	16, 2,15←15, 3,12	Ground		33/2		31/2	34 398.6		980 [1]
	1381	16, 2,15←15, 3,12	Ground		31/2		29/2	34 429.2		980 [1]
	1381	16, 2,15←15, 3,12	Ground		29/2		27/2	34 463.4		980 [1]
	1381	16, 2,15←15, 3,12	Ground					34 941.3		980
	1381	16, 2,15←15, 3,12	Ground		29/2		27/2	35 399.6		980 [1]
	1381	16, 2,15←15, 3,12	Ground		31/2		29/2	35 418.8		980 [1]
	1381	16, 2,15←15, 3,12	Ground		33/2		31/2	35 444.0		980 [1]
	1381	16, 2,15←15, 3,12	Ground		35/2		33/2	35 477.4		980 [1]
	1381	17, 3,14←16, 4,13	Ground		31/2		29/2	37 167.6		980 [1]
	1381	17, 3,14←16, 4,13	Ground		33/2		31/2	37 182.2		980 [1]
	1381	17, 3,14←16, 4,13	Ground		35/2		33/2	37 196.1		980 [1]
	1381	17, 3,14←16, 4,13	Ground		37/2		35/2	37 208.8		980 [1]
	1381	17, 3,14←16, 4,13	Ground					37 235.6		980
	1381	17, 3,14←16, 4,13	Ground		37/2		35/2	37 253.6		980 [1]
	1381	17, 3,14←16, 4,13	Ground		35/2		33/2	37 268.0		980 [1]
	1381	17, 3,14←16, 4,13	Ground		33/2		31/2	37 279.3		980 [1]
	1381	17, 3,14←16, 4,13	Ground		31/2		29/2	37 289.2		980 [1]
	1381	18, 3,16←17, 4,13	Ground		39/2		37/2	25 745.7		980 [1]
	1381	18, 3,16←17, 4,13	Ground		37/2		35/2	25 751.1		980 [1]
	1381	18, 3,16←17, 4,13	Ground		35/2		33/2	25 758.5		980 [1]
	1381	18, 3,16←17, 4,13	Ground		33/2		31/2	25 767.4		980 [1]
	1381	18, 3,16←17, 4,13	Ground					26 060.2		980
	1381	18, 3,16←17, 4,13	Ground		33/2		31/2	26 337.6		980 [1]
	1381	18, 3,16←17, 4,13	Ground		35/2		33/2	26 341.9		980 [1]
	1381	18, 3,16←17, 4,13	Ground		37/2		35/2	26 348.2		980 [1]
	1381	18, 3,16←17, 4,13	Ground		39/2		37/2	26 356.8		980 [1]
	1381	18, 5,14←19, 4,15	Ground		33/2		35/2	26 239.4		980 [1]

1. Spectral lines for these species from reference 980 have hyperfine quantum numbers computed from the following relations:

$$\text{for} \quad J = N + 1/2, \quad F = N + 2, N + 1, N, N - 1$$

$$J = N - 1/2, \quad F = N - 2, N - 1, N, N + 1$$

Isotopic Species	Id. No.	Rotational Quantum Nos.	Vib. State	Hyperfine				Frequency MHz	Acc. ±MHz	Ref.
				F_1'	F'	F_1	F			
$Cl^{35}O_2^{16}$	1381	18, 5,14←19, 4,15	Ground		35/2		37/2	26 242.6		980 [1]
	1381	18, 5,14←19, 4,15	Ground		37/2		39/2	26 246.1		980 [1]
	1381	18, 5,14←19, 4,15	Ground		39/2		41/2	26 249.4		980 [1]
	1381	18, 5,14←19, 4,15	Ground					26 445.5		980
	1381	18, 5,14←19, 4,15	Ground		39/2		41/2	26 657.1		980 [1]
	1381	18, 5,14←19, 4,15	Ground		37/2		39/2	26 661.3		980 [1]
	1381	18, 5,14←19, 4,15	Ground		35/2		37/2	26 665.1		980 [1]
	1381	18, 5,14←19, 4,15	Ground		33/2		35/2	26 668.4		980 [1]
	1381	21, 4,17←20, 5,16	Ground		39/2		37/2	17 325.0		980 [1]
	1381	21, 4,17←20, 5,16	Ground		41/2		39/2	17 331.6		980 [1]
	1381	21, 4,17←20, 5,16	Ground		43/2		41/2	17 338.2		980 [1]
	1381	21, 4,17←20, 5,16	Ground		45/2		43/2	17 344.7		980 [1]
	1381	21, 4,17←20, 5,16	Ground					17 462.6		980
	1381	21, 4,17←20, 5,16	Ground		45/2		43/2	17 577.4		980 [1]
	1381	21, 4,17←20, 5,16	Ground		43/2		41/2	17 583.8		980 [1]
	1381	21, 4,17←20, 5,16	Ground		41/2		39/2	17 589.5		980 [1]
	1381	21, 4,17←20, 5,16	Ground		39/2		37/2	17 594.7		980 [1]
	1381	28, 7,22←29, 6,23	Ground		53/2		55/2	10 767.2		980 [1]
	1381	28, 7,22←29, 6,23	Ground		55/2		57/2	10 768.9		980 [1]
	1381	28, 7,22←29, 6,23	Ground		57/2		59/2	10 770.7		980 [1]
	1381	28, 7,22←29, 6,23	Ground		59/2		61/2	10 771.9		980 [1]
	1381	28, 7,22←29, 6,23	Ground					10 976.7		980
	1381	28, 7,22←29, 6,23	Ground		59/2		61/2	11 179.3		980 [1]
	1381	28, 7,22←29, 6,23	Ground		57/2		59/2	11 181.2		980 [1]
	1381	28, 7,22←29, 6,23	Ground		55/2		57/2	11 183.0		980 [1]
	1381	28, 7,22←29, 6,23	Ground		53/2		55/2	11 184.5		980 [1]
$Cl^{37}O_2^{16}$	1382	1, 1, 0← 1, 0, 1	Ground				2	41 856.53	.2	936
	1382	1, 1, 0← 1, 0, 1	Ground				2	41 968.48	.2	936
	1382	1, 1, 0← 1, 0, 1	Ground			3	2	41 977.60	.2	936
	1382	1, 1, 0← 1, 0, 1	Ground			2	3	42 047.54	.2	936
	1382	1, 1, 0← 1, 0, 1	Ground			3	3	42 055.84	.2	936
	1382	1, 1, 0← 1, 0, 1	Ground			1	2	42 096.40	.2	936
	1382	1, 1, 0← 1, 0, 1	Ground			2	2	42 109.20	.2	936
	1382	1, 1, 0← 1, 0, 1	Ground			0	1	42 112.40	.2	936
	1382	1, 1, 0← 1, 0, 1	Ground			3	2	42 117.52	.2	936
	1382	1, 1, 0← 1, 0, 1	Ground			2	1	42 135.12	.2	936
	1382	1, 1, 0← 1, 0, 1	Ground					42 435.8		980
	1382	1, 1, 0← 1, 0, 1	Ground			1	1	42 948.50	.2	936
	1382	1, 1, 0← 1, 0, 1	Ground			1	2	43 060.24	.2	936
	1382	1, 1, 0← 1, 0, 1	Ground			2	1	43 104.18	.2	936
	1382	1, 1, 0← 1, 0, 1	Ground			1	2	43 201.54	.2	936
	1382	1, 1, 0← 1, 0, 1	Ground			2	2	43 215.84	.2	936
	1382	1, 1, 0← 1, 0, 1	Ground			1	1	43 227.26	.2	936
	1382	1, 1, 0← 1, 0, 1	Ground			2	3	43 294.8	.2	936
	1382	3, 0, 3← 2, 1, 2	Ground			1	0	15 131.8	.2	936
	1382	3, 0, 3← 2, 1, 2	Ground			2	1	15 136.9	.2	936
	1382	3, 0, 3← 2, 1, 2	Ground			3	2	15 166.8	.2	936
	1382	3, 0, 3← 2, 1, 2	Ground			4	3	15 215.3	.2	936
	1382	3, 0, 3← 2, 1, 2	Ground					15 381.0		980

. Spectral lines for these species from reference 980 have hyperfine quantum numbers computed from the following relations:

$$\text{for} \quad J = N + 1/2, \quad F = N + 2, N + 1, N, N - 1$$

$$J = N - 1/2, \quad F = N - 2, N - 1, N, N + 1$$

Isotopic Species	Id. No.	Rotational Quantum Nos.	Vib. State	F₁'	F'	F₁	F	Frequency MHz	Acc. ±MHz	Ref.
$Cl^{37}O_2^{16}$	1382	3, 0, 3← 2, 1, 2	Ground		5		4	15 486.3	.2	936
	1382	3, 0, 3← 2, 1, 2	Ground		4		3	15 499.2	.2	936
	1382	3, 0, 3← 2, 1, 2	Ground		3		2	15 507.4	.2	936
	1382	3, 0, 3← 2, 1, 2	Ground		2		1	15 512.4	.2	936
	1382	3, 1, 2← 3, 0, 3	Ground		5		5	46 503.6	.2	936
	1382	3, 1, 2← 3, 0, 3	Ground		4		4	46 539.0	.2	936
	1382	3, 1, 2← 3, 0, 3	Ground		3		3	46 560.0	.2	936
	1382	3, 1, 2← 3, 0, 3	Ground		2		2	46 571.0	.2	936
	1382	3, 1, 2← 3, 0, 3	Ground					46 786.6		980
	1382	3, 1, 2← 3, 0, 3	Ground		1		1	47 037.0	.2	936
	1382	3, 1, 2← 3, 0, 3	Ground		2		2	47 074.4	.2	936
	1382	3, 1, 2← 3, 0, 3	Ground		3		3	47 122.6	.2	936
	1382	3, 1, 2← 3, 0, 3	Ground		4		4	47 175.7	.2	936
	1382	4, 0, 4← 3, 1, 3	Ground					35 399.6		1005
	1382	4, 0, 4← 3, 1, 3	Ground					35 418.8		1005
	1382	4, 0, 4← 3, 1, 3	Ground					35 444.0		1005
	1382	4, 0, 4← 3, 1, 3	Ground					35 477.4		1005
	1382	4, 0, 4← 3, 1, 3	Ground					35 496.4		1005
	1382	4, 0, 4← 3, 1, 3	Ground					35 545.2		1005
	1382	4, 0, 4← 3, 1, 3	Ground					35 589.7		1005
	1382	4, 2, 3← 5, 1, 4	Ground		3		4	21 226.3	.2	936
	1382	4, 2, 3← 5, 1, 4	Ground		4		5	21 237.5	.2	936
	1382	4, 2, 3← 5, 1, 4	Ground		5		6	21 250.4	.2	936
	1382	4, 2, 3← 5, 1, 4	Ground		6		7	21 264.3	.2	936
	1382	4, 2, 3← 5, 1, 4	Ground					21 468.6		980
	1382	4, 2, 3← 5, 1, 4	Ground		5		6	21 739.2	.2	936
	1382	4, 2, 3← 5, 1, 4	Ground		4		5	21 764.2	.2	936
	1382	4, 2, 3← 5, 1, 4	Ground		3		4	21 784.1	.2	936
	1382	4, 2, 3← 5, 1, 4	Ground		2		3	21 797.7	.2	936
	1382	7, 1, 6← 6, 2, 5	Ground					24 856.0	.2	936
	1382	7, 1, 6← 6, 2, 5	Ground					24 876.8	.2	936
	1382	7, 1, 6← 6, 2, 5	Ground					24 898.5	.2	936
	1382	7, 1, 6← 6, 2, 5	Ground					24 924.6	.2	936
	1382	7, 1, 6← 6, 2, 5	Ground					24 972.7		980
	1382	7, 1, 6← 6, 2, 5	Ground					25 006.3	.2	936
	1382	7, 1, 6← 6, 2, 5	Ground					25 026.1	.2	936
	1382	7, 1, 6← 6, 2, 5	Ground					25 041.0	.2	936
	1382	7, 1, 6← 6, 2, 5	Ground					25 054.7	.2	936
	1382	11, 2, 9←10, 3, 8	Ground		19/2		17/2	15 002.8		980 [1]
	1382	11, 2, 9←10, 3, 8	Ground		21/2		19/2	15 019.1		980 [1]
	1382	11, 2, 9←10, 3, 8	Ground		23/2		21/2	15 035.1		980 [1]
	1382	11, 2, 9←10, 3, 8	Ground		25/2		23/2	15 051.1		980 [1]
	1382	11, 2, 9←10, 3, 8	Ground					15 114.1		980
	1382	11, 2, 9←10, 3, 8	Ground		25/2		23/2	15 161.9		980 [1]
	1382	11, 2, 9←10, 3, 8	Ground		23/2		21/2	15 177.1		980 [1]
	1382	11, 2, 9←10, 3, 8	Ground		21/2		19/2	15 189.9		980 [1]
	1382	11, 2, 9←10, 3, 8	Ground		19/2		17/2	15 200.4		980 [1]
	1382	14, 2,13←13, 3,10	Ground		31/2		29/2	20 643.6		980 [1]
	1382	14, 2,13←13, 3,10	Ground		29/2		27/2	20 661.4		980 [1]
	1382	14, 2,13←13, 3,10	Ground		27/2		25/2	20 682.2		980 [1]

1. Spectral lines for these species from reference 980 have hyperfine quantum numbers computed from the following relations:

$$\text{for} \quad J = N + 1/2, \qquad F = N+2, N+1, N, N-1$$

$$J = N - 1/2, \qquad F = N-2, N-1, N, N+1$$

Isotopic Species	Id. No.	Rotational Quantum Nos.	Vib. State	F_1'	F'	F_1	F	Frequency MHz	Acc. ±MHz	Ref.
$Cl^{37}O_2^{16}$	1382	14, 2,13←13, 3,10	Ground	25/2		23/2		20 705.5		980 [1]
	1382	14, 2,13←13, 3,10	Ground					21 119.5		980
	1382	14, 2,13←13, 3,10	Ground	25/2		23/2		21 509.9		980 [1]
	1382	14, 2,13←13, 3,10	Ground	27/2		25/2		21 522.6		980 [1]
	1382	14, 2,13←13, 3,10	Ground	29/2		27/2		21 539.8		980 [1]
	1382	14, 2,13←13, 3,10	Ground	31/2		29/2		21 561.8		980 [1]
$Cl^{35}O^{16}O^{18}$	1383	3, 0← 3, 2	Ground		1		0	11 633.8		960
	1383	3, 0← 3, 2	Ground		2		1	11 640.1		960
	1383	3, 0← 3, 2	Ground		3		2	11 677.5		960
	1383	3, 0← 3, 2	Ground		4		3	11 736.2		960
	1383	3, 0← 3, 2	Ground		5		4	12 011.4		960
	1383	3, 0← 3, 2	Ground		4		3	12 026.6		960
	1383	3, 0← 3, 2	Ground		3		2	12 036.7		960
	1383	3, 0← 3, 2	Ground		2		1	12 042.7		960
	1383	4, 2← 3, 5	Ground		3		4	28 308.2		960
	1383	4, 2← 3, 5	Ground		4		5	28 322.1		960
	1383	4, 2← 3, 5	Ground		5		6	28 337.4		960
	1383	4, 2← 3, 5	Ground		6		7	28 354.2		960
	1383	4, 2← 3, 5	Ground		5		6	28 843.6		960
	1383	4, 2← 3, 5	Ground		4		5	28 873.7		960
	1383	4, 2← 3, 5	Ground		3		4	28 897.9		960
	1383	4, 2← 3, 5	Ground		2		3	28 915.8		960
	1383	7, 1← 6, 6	Ground		5		4	14 855.6		960
	1383	7, 1← 6, 6	Ground		6		5	14 881.2		960
	1383	7, 1← 6, 6	Ground		7		6	14 910.0		960
	1383	7, 1← 6, 6	Ground		8		7	14 939.1		960
	1383	7, 1← 6, 6	Ground		9		8	15 035.		960
	1383	7, 1← 6, 6	Ground		8		7	15 059.2		960
	1383	7, 1← 6, 6	Ground		7		6	15 078.4		960
	1383	7, 1← 6, 6	Ground		6		5	15 093.6		960

1. Spectral lines for these species from reference 980 have hyperfine quantum numbers computed from the following relations:

$$\text{for} \quad J = N + 1/2, \quad F = N + 2, N + 1, N, N - 1$$

$$J = N - 1/2, \quad F = N - 2, N - 1, N, N + 1$$

ClF$_2$OP $\qquad$ C$_s$ $\qquad$ POF$_2$Cl

Isotopic Species	Pt. Gp.	Id. No.	A MHz		B MHz		C MHz		D$_J$ MHz	D$_{JK}$ MHz	Δ Amu A²	κ
P^{31}O^{16}F$_2^{19}$Cl^{35}	C$_s$	1391	4 912.	M	2 987.	M	3 055.	M				−.928520
P^{31}O^{16}F$_2^{19}$Cl^{37}	C$_s$	1392										−.934584

Id. No.	μ$_a$ Debye		μ$_b$ Debye	μ$_c$ Debye		eQq Value(MHz)	Rel.	eQq Value(MHz)	Rel.	eQq Value(MHz)	Rel.	ω$_a$ 1/cm	d	ω$_b$ 1/cm	d	ω$_c$ 1/cm	d	ω$_d$ 1/cm	d
1391	.44	M		0.	X	−58.4													
1392						−46.2													

References:

ABC: 860 $\qquad$ κ: 860 $\qquad$ μ: 860 $\qquad$ eQq: 860

Species	D$_{JK}$ (MHz)	D$_K$ (MHz)	A − (B + C)/2(MHz)	Ref.
1391	.00214	−.00143	1890.626	860
1392	.00196	−.00188	1962.93	860

Isotopic Species	Id. No.	Rotational Quantum Nos.	Vib. State	F_1'	F'	F_1	F	Frequency MHz	Acc. ±MHz	Ref.
$P^{31}O^{16}F_2^{19}Cl^{35}$	1391	5, 5, 0← 5, 4, 1	Ground		9/2		9/2	17 015.432		860
	1391	5, 5, 0← 5, 4, 1	Ground		11/2		11/2	17 019.765		860
	1391	6, 5, 1← 6, 4, 2	Ground		11/2		11/2	17 013.200		860
	1391	6, 5, 1← 6, 4, 2	Ground		13/2		13/2	17 015.432		860
	1391	6, 6, 0← 6, 5, 1	Ground		11/2		11/2	20 796.85		860
	1391	6, 6, 0← 6, 5, 1	Ground		13/2		13/2	20 799.30		860
	1391	7, 5, 2← 7, 4, 3	Ground		13/2		13/2	17 009.238		860
	1391	7, 5, 2← 7, 4, 3	Ground		15/2		15/2	17 011.164		860
	1391	7, 6, 1← 7, 5, 2	Ground		13/2		13/2	20 794.37		860
	1391	7, 6, 1← 7, 5, 2	Ground		15/2		15/2	20 795.96		860
	1391	8, 6, 2← 8, 5, 3	Ground		15/2		15/2	20 791.11		860
	1391	8, 6, 2← 8, 5, 3	Ground		17/2		17/2	20 792.26		860
	1391	9, 5, 4← 9, 4, 5	Ground		15/2		15/2	16 993.723		860
	1391	9, 5, 4← 9, 4, 5	Ground		21/2		21/2	16 994.217		860
	1391	9, 5, 4← 9, 4, 5	Ground		17/2		17/2	16 998.066		860
	1391	9, 5, 4← 9, 4, 5	Ground		19/2		19/2	16 998.648		860
	1391	9, 5, 5← 9, 4, 6	Ground		15/2		15/2	16 994.217		860
	1391	9, 5, 5← 9, 4, 6	Ground		21/2		21/2	16 994.723		860
	1391	9, 5, 5← 9, 4, 6	Ground		17/2		17/2	16 998.648		860
	1391	9, 6, 3← 9, 5, 4	Ground		21/2		21/2	20 782.84		860
	1391	9, 6, 3← 9, 5, 4	Ground		17/2		17/2	20 787.31		860
	1391	9, 6, 3← 9, 5, 4	Ground		19/2		19/2	20 787.96		860
	1391	10, 5, 5←10, 4, 6	Ground		17/2		17/2	16 985.587		860
	1391	10, 5, 5←10, 4, 6	Ground		23/2		23/2	16 985.935		860
	1391	10, 5, 5←10, 4, 6	Ground		19/2		19/2	16 989.107		860
	1391	10, 5, 5←10, 4, 6	Ground		21/2		21/2	16 989.580		860
	1391	10, 5, 6←10, 4, 7	Ground		17/2		17/2	16 986.850		860
	1391	10, 5, 6←10, 4, 7	Ground		23/2		23/2	16 987.192		860
	1391	10, 5, 6←10, 4, 7	Ground		19/2		19/2	16 990.473		860
	1391	10, 5, 6←10, 4, 7	Ground		21/2		21/2	16 990.883		860
	1391	11, 5, 6←11, 4, 7	Ground		19/2		19/2	16 974.173		860
	1391	11, 5, 6←11, 4, 7	Ground		25/2		25/2	16 974.578		860
	1391	11, 5, 6←11, 4, 7	Ground		23/2		23/2	16 977.300		860
	1391	11, 5, 6←11, 4, 7	Ground		21/2		21/2	16 977.300		860
	1391	11, 5, 7←11, 4, 8	Ground		23/2		23/2	16 977.300		860
	1391	11, 5, 7←11, 4, 8	Ground		21/2		21/2	16 977.300		860
	1391	11, 5, 7←11, 4, 8	Ground		19/2		19/2	16 980.100		860
	1391	11, 5, 7←11, 4, 8	Ground		25/2		25/2	16 980.100		860
	1391	11, 6, 5←11, 5, 6	Ground		19/2		19/2	20 772.50		860
	1391	11, 6, 5←11, 5, 6	Ground		25/2		25/2	20 772.87		860
	1391	11, 6, 5←11, 5, 6	Ground		21/2		21/2	20 776.10		860
	1391	11, 6, 5←11, 5, 6	Ground		23/2		23/2	20 776.55		860
	1391	11, 6, 6←11, 5, 7	Ground		19/2		19/2	20 772.50		860
	1391	11, 6, 6←11, 5, 7	Ground		25/2		25/2	20 772.87		860
	1391	11, 6, 6←11, 5, 7	Ground		21/2		21/2	20 776.10		860
	1391	11, 6, 6←11, 5, 7	Ground		23/2		23/2	20 776.55		860
	1391	11,10, 1←11, 9, 2	Ground		21/2		21/2	35 917.00		860
	1391	11,10, 1←11, 9, 2	Ground		23/2		23/2	35 917.82		860
	1391	12, 5, 7←12, 4, 8	Ground		21/2		21/2	16 959.161		860
	1391	12, 5, 7←12, 4, 8	Ground		27/2		27/2	16 959.161		860

Isotopic Species	Id. No.	Rotational Quantum Nos.	Vib. State	Hyperfine				Frequency MHz	Acc. ±MHz	Ref.
				F_1'	F'	F_1	F			
$P^{31}O^{16}F_2^{19}Cl^{35}$	1391	12, 5, 7←12, 4, 8	Ground		25/2		25/2	16 961.692		860
	1391	12, 5, 7←12, 4, 8	Ground		23/2		23/2	16 961.692		860
	1391	12, 5, 8←12, 4, 9	Ground		27/2		27/2	16 964.803		860
	1391	12, 5, 8←12, 4, 9	Ground		21/2		21/2	16 964.803		860
	1391	12, 5, 8←12, 4, 9	Ground		23/2		23/2	16 967.329		860
	1391	12, 5, 8←12, 4, 9	Ground		25/2		25/2	16 967.329		860
	1391	12, 6, 6←12, 5, 7	Ground		21/2		21/2	20 765.28		860
	1391	12, 6, 6←12, 5, 7	Ground		27/2		27/2	20 765.28		860
	1391	12, 6, 6←12, 5, 7	Ground		25/2		25/2	20 768.50		860
	1391	12, 6, 6←12, 5, 7	Ground		23/2		23/2	20 768.50		860
	1391	12, 6, 7←12, 5, 8	Ground		27/2		27/2	20 765.28		860
	1391	12, 6, 7←12, 5, 8	Ground		21/2		21/2	20 765.28		860
	1391	12, 6, 7←12, 5, 8	Ground		25/2		25/2	20 768.50		860
	1391	12, 6, 7←12, 5, 8	Ground		23/2		23/2	20 768.50		860
	1391	12,10, 2←12, 9, 3	Ground		27/2		27/2	35 909.08		860
	1391	12,10, 2←12, 9, 3	Ground		21/2		21/2	35 909.76		860
	1391	12,10, 2←12, 9, 3	Ground		23/2		23/2	35 914.45		860
	1391	12,10, 2←12, 9, 3	Ground		25/2		25/2	35 914.97		860
	1391	13, 5, 8←13, 4, 9	Ground		23/2		23/2	16 938.654		860
	1391	13, 5, 8←13, 4, 9	Ground		29/2		29/2	16 938.915		860
	1391	13, 5, 8←13, 4, 9	Ground		25/2		25/2	16 940.860		860
	1391	13, 5, 8←13, 4, 9	Ground		27/2		27/2	16 941.040		860
	1391	13, 5, 9←13, 4,10	Ground		23/2		23/2	16 949.266		860
	1391	13, 5, 9←13, 4,10	Ground		29/2		29/2	16 949.451		860
	1391	13, 5, 9←13, 4,10	Ground		25/2		25/2	16 951.413		860
	1391	13, 5, 9←13, 4,10	Ground		27/2		27/2	16 951.602		860
	1391	13, 6, 7←13, 5, 8	Ground		23/2		23/2	20 755.88		860
	1391	13, 6, 7←13, 5, 8	Ground		29/2		29/2	20 755.88		860
	1391	13, 6, 7←13, 5, 8	Ground		27/2		27/2	20 758.31		860
	1391	13, 6, 7←13, 5, 8	Ground		25/2		25/2	20 758.31		860
	1391	13, 6, 8←13, 5, 9	Ground		23/2		23/2	20 755.88		860
	1391	13, 6, 8←13, 5, 9	Ground		29/2		29/2	20 755.88		860
	1391	13, 6, 8←13, 5, 9	Ground		27/2		27/2	20 758.31		860
	1391	13, 6, 8←13, 5, 9	Ground		25/2		25/2	20 758.31		860
	1391	13,10, 3←13, 9, 4	Ground		23/2		23/2	35 906.92		860
	1391	13,10, 3←13, 9, 4	Ground		29/2		29/2	35 907.40		860
	1391	13,10, 3←13, 9, 4	Ground		25/2		25/2	35 911.44		860
	1391	13,10, 3←13, 9, 4	Ground		27/2		27/2	35 911.93		860
	1391	14, 5, 9←14, 4,10	Ground		25/2		25/2	16 912.140		860
	1391	14, 5, 9←14, 4,10	Ground		31/2		31/2	16 912.340		860
	1391	14, 5, 9←14, 4,10	Ground		27/2		27/2	16 914.020		860
	1391	14, 5, 9←14, 4,10	Ground		29/2		29/2	16 914.222		860
	1391	14, 5,10←14, 4,11	Ground		25/2		25/2	16 931.000		860
	1391	14, 5,10←14, 4,11	Ground		31/2		31/2	16 931.168		860
	1391	14, 5,10←14, 4,11	Ground		29/2		29/2	16 932.868		860
	1391	14, 5,10←14, 4,11	Ground		27/2		27/2	16 933.016		860
	1391	14, 6, 8←14, 5, 9	Ground		25/2		25/2	20 744.17		860
	1391	14, 6, 8←14, 5, 9	Ground		31/2		31/2	20 744.17		860
	1391	14, 6, 8←14, 5, 9	Ground		27/2		27/2	20 746.36		860
	1391	14, 6, 8←14, 5, 9	Ground		29/2		29/2	20 746.36		860

Isotopic Species	Id. No.	Rotational Quantum Nos.	Vib. State	Hyperfine				Frequency MHz	Acc. ±MHz	Ref.
				F_1'	F'	F_1	F			
$P^{31}O^{16}F_2^{19}Cl^{35}$	1391	14, 6, 9←14, 5,10	Ground		25/2		25/2	20 744.17		860
	1391	14, 6, 9←14, 5,10	Ground		31/2		31/2	20 744.17		860
	1391	14, 6, 9←14, 5,10	Ground		27/2		27/2	20 746.36		860
	1391	14, 6, 9←14, 5,10	Ground		29/2		29/2	20 746.36		860
	1391	14,10, 4←14, 9, 5	Ground		25/2		25/2	35 904.14		860
	1391	14,10, 4←14, 9, 5	Ground		31/2		31/2	35 904.84		860
	1391	14,10, 4←14, 9, 5	Ground		27/2		27/2	35 908.02		860
	1391	14,10, 4←14, 9, 5	Ground		29/2		29/2	35 908.44		860
	1391	15, 5,10←15, 4,11	Ground		27/2		27/2	16 877.773		860
	1391	15, 5,10←15, 4,11	Ground		33/2		33/2	16 877.868		860
	1391	15, 5,10←15, 4,11	Ground		29/2		29/2	16 879.408		860
	1391	15, 5,10←15, 4,11	Ground		31/2		31/2	16 879.560		860
	1391	15, 5,11←15, 4,12	Ground		27/2		27/2	16 909.88		860
	1391	15, 5,11←15, 4,12	Ground		33/2		33/2	16 910.044		860
	1391	15, 5,11←15, 4,12	Ground		29/2		29/2	16 911.524		860
	1391	15, 5,11←15, 4,12	Ground		31/2		31/2	16 911.660		860
	1391	15,10, 5←15, 9, 6	Ground		27/2		27/2	35 900.79		860
	1391	15,10, 5←15, 9, 6	Ground		33/2		33/2	35 901.03		860
	1391	15,10, 5←15, 9, 6	Ground		29/2		29/2	35 904.14		860
	1391	15,10, 5←15, 9, 6	Ground		31/2		31/2	35 904.84		860
	1391	16, 5,11←16, 4,12	Ground		29/2		29/2	16 833.360		860
	1391	16, 5,11←16, 4,12	Ground		31/2		31/2	16 834.835		860
	1391	16, 5,11←16, 4,12	Ground		33/2		33/2	16 834.975		860
	1391	16, 5,12←16, 4,13	Ground		35/2		35/2	16 886.162		860
	1391	16, 5,12←16, 4,13	Ground		33/2		33/2	16 887.64		860
	1391	16,10, 6←16, 9, 7	Ground		29/2		29/2	35 896.84		860
	1391	16,10, 6←16, 9, 7	Ground		35/2		35/2	35 896.98		860
	1391	16,10, 6←16, 9, 7	Ground		31/2		31/2	35 899.73		860
	1391	16,10, 6←16, 9, 7	Ground		33/2		33/2	35 900.04		860
	1391	17, 5,12←17, 4,13	Ground		37/2		37/2	16 776.724		860
	1391	17, 5,12←17, 4,13	Ground		31/2		31/2	16 776.724		860
	1391	17, 5,12←17, 4,13	Ground		35/2		35/2	16 778.048		860
	1391	17, 5,12←17, 4,13	Ground		33/2		33/2	16 778.048		860
	1391	17, 5,13←17, 4,14	Ground		31/2		31/2	16 860.600		860
	1391	17, 5,13←17, 4,14	Ground		37/2		37/2	16 860.600		860
	1391	17, 5,13←17, 4,14	Ground		33/2		33/2	16 861.904		860
	1391	17, 5,13←17, 4,14	Ground		35/2		35/2	16 861.904		860
	1391	17,10, 7←17, 9, 8	Ground		31/2		31/2	35 891.94		860
	1391	17,10, 7←17, 9, 8	Ground		37/2		37/2	35 892.06		860
	1391	17,10, 7←17, 9, 8	Ground		33/2		33/2	35 894.58		860
	1391	17,10, 7←17, 9, 8	Ground		35/2		35/2	35 894.82		860
	1391	18, 5,13←18, 4,14	Ground		33/2		33/2	16 704.418		860
	1391	18, 5,13←18, 4,14	Ground		39/2		39/2	16 704.418		860
	1391	18, 5,13←18, 4,14	Ground		35/2		35/2	16 705.627		860
	1391	18, 5,13←18, 4,14	Ground		37/2		37/2	16 705.627		860
	1391	18, 5,14←18, 4,15	Ground		39/2		39/2	16 833.536		860
	1391	18, 5,14←18, 4,15	Ground		33/2		33/2	16 833.536		860
	1391	18, 5,14←18, 4,15	Ground		35/2		35/2	16 834.744		860
	1391	18, 5,14←18, 4,15	Ground		37/2		37/2	16 834.835		860
	1391	18,10, 8←18, 9, 9	Ground		39/2		39/2	35 886.40		860

Isotopic Species	Id. No.	Rotational Quantum Nos.	Vib. State	F_1'	F'	F_1	F	Frequency MHz	Acc. ±MHz	Ref.
$P^{31}O^{16}F_2^{19}Cl^{35}$	1391	18,10, 8←18, 9, 9	Ground		33/2		33/2	35 886.40		860
	1391	18,10, 8←18, 9, 9	Ground		37/2		37/2	35 888.78		860
	1391	18,10, 8←18, 9, 9	Ground		35/2		35/2	35 888.78		860
	1391	19,10, 9←19, 9,10	Ground		41/2		41/2	35 880.25		860
	1391	19,10, 9←19, 9,10	Ground		35/2		35/2	35 880.25		860
	1391	19,10, 9←19, 9,10	Ground		39/2		39/2	35 882.35		860
	1391	19,10, 9←19, 9,10	Ground		37/2		37/2	35 882.35		860
	1391	20,10,10←20, 9,11	Ground		43/2		43/2	35 872.46		860
	1391	20,10,10←20, 9,11	Ground		37/2		37/2	35 872.46		860
	1391	20,10,10←20, 9,11	Ground		41/2		41/2	35 874.50		860
	1391	20,10,10←20, 9,11	Ground		39/2		39/2	35 874.50		860
$P^{31}O^{16}F_2^{19}Cl^{37}$	1392	5, 5, 0← 5, 4, 1	Ground		9/2		9/2	17 666.660		860
	1392	5, 5, 0← 5, 4, 1	Ground		11/2		11/2	17 669.590		860
	1392	6, 5, 1← 6, 4, 2	Ground		11/2		11/2	17 664.240		860
	1392	6, 5, 1← 6, 4, 2	Ground		13/2		13/2	17 666.026		860
	1392	7, 5, 2← 7, 4, 3	Ground		13/2		13/2	17 661.152		860
	1392	7, 5, 2← 7, 4, 3	Ground		15/2		15/2	17 662.260		860
	1392	13, 5, 8←13, 4, 9	Ground		29/2		29/2	17 600.829		860
	1392	13, 5, 8←13, 4, 9	Ground		23/2		23/2	17 600.829		860
	1392	13, 5, 8←13, 4, 9	Ground		25/2		25/2	17 602.509		860
	1392	13, 5, 8←13, 4, 9	Ground		27/2		27/2	17 602.509		860
	1392	13, 5, 9←13, 4,10	Ground		29/2		29/2	17 608.525		860
	1392	13, 5, 9←13, 4,10	Ground		23/2		23/2	17 608.525		860
	1392	13, 5, 9←13, 4,10	Ground		27/2		27/2	17 610.259		860
	1392	13, 5, 9←13, 4,10	Ground		25/2		25/2	17 610.259		860
	1392	14, 5, 9←14, 4,10	Ground		31/2		31/2	17 578.382		860
	1392	14, 5, 9←14, 4,10	Ground		25/2		25/2	17 578.382		860
	1392	14, 5, 9←14, 4,10	Ground		27/2		27/2	17 579.894		860
	1392	14, 5, 9←14, 4,10	Ground		29/2		29/2	17 579.894		860
	1392	14, 5,10←14, 4,11	Ground		31/2		31/2	17 592.120		860
	1392	14, 5,10←14, 4,11	Ground		25/2		25/2	17 592.120		860
	1392	14, 5,10←14, 4,11	Ground		29/2		29/2	17 593.592		860
	1392	14, 5,10←14, 4,11	Ground		27/2		27/2	17 593.592		860
	1392	15, 5,10←15, 4,11	Ground		27/2		27/2	17 549.551		860
	1392	15, 5,10←15, 4,11	Ground		33/2		33/2	17 549.551		860
	1392	15, 5,10←15, 4,11	Ground		29/2		29/2	17 551.220		860
	1392	15, 5,10←15, 4,11	Ground		31/2		31/2	17 551.220		860
	1392	15, 5,11←15, 4,12	Ground		27/2		27/2	17 572.927		860
	1392	15, 5,11←15, 4,12	Ground		33/2		33/2	17 572.927		860
	1392	15, 5,11←15, 4,12	Ground		31/2		31/2	17 574.270		860
	1392	15, 5,11←15, 4,12	Ground		29/2		29/2	17 574.270		860
	1392	16, 5,11←16, 4,12	Ground		35/2		35/2	17 512.654		860
	1392	16, 5,11←16, 4,12	Ground		29/2		29/2	17 512.654		860
	1392	16, 5,11←16, 4,12	Ground		31/2		31/2	17 513.800		860
	1392	16, 5,11←16, 4,12	Ground		33/2		33/2	17 513.800		860
	1392	16, 5,12←16, 4,13	Ground		35/2		35/2	17 551.220		860
	1392	16, 5,12←16, 4,13	Ground		29/2		29/2	17 551.220		860
	1392	16, 5,12←16, 4,13	Ground		31/2		31/2	17 552.406		860
	1392	16, 5,12←16, 4,13	Ground		33/2		33/2	17 552.406		860

Isotopic Species	Id. No.	Rotational Quantum Nos.	Vib. State	Hyperfine				Frequency MHz	Acc. ±MHz	Ref.
				F_1'	F'	F_1	F			
$P^{31}O^{16}F_2^{19}Cl^{37}$	1392	18,10, 8←18, 9, 9	Ground		39/2		39/2	37 264.82		860
	1392	18,10, 8←18, 9, 9	Ground		33/2		33/2	37 264.82		860
	1392	18,10, 8←18, 9, 9	Ground		35/2		35/2	37 267.50		860
	1392	18,10, 8←18, 9, 9	Ground		37/2		37/2	37 267.50		860
	1392	19,10, 9←19, 9,10	Ground		35/2		35/2	37 259.94		860
	1392	19,10, 9←19, 9,10	Ground		41/2		41/2	37 259.94		860
	1392	19,10, 9←19, 9,10	Ground		37/2		37/2	37 261.64		860
	1392	19,10, 9←19, 9,10	Ground		39/2		39/2	37 261.64		860
	1392	20,10,10←20, 9,11	Ground		43/2		43/2	37 251.74		860
	1392	20,10,10←20, 9,11	Ground		37/2		37/2	37 251.74		860
	1392	20,10,10←20, 9,11	Ground		39/2		39/2	37 253.48		860
	1392	20,10,10←20, 9,11	Ground		41/2		41/2	37 253.48		860
	1392	22,10,12←22, 9,13	Ground		47/2		47/2	37 235.84		860
	1392	22,10,12←22, 9,13	Ground		41/2		41/2	37 235.84		860
	1392	22,10,12←22, 9,13	Ground		45/2		45/2	37 237.16		860
	1392	22,10,12←22, 9,13	Ground		43/2		43/2	37 237.16		860
	1392	23,10,13←23, 9,14	Ground		43/2		43/2	37 225.28		860
	1392	23,10,13←23, 9,14	Ground		49/2		49/2	37 225.28		860
	1392	23,10,13←23, 9,14	Ground		47/2		47/2	37 226.46		860
	1392	23,10,13←23, 9,14	Ground		45/2		45/2	37 226.46		860
	1392	24,10,14←24, 9,15	Ground					37 215.60		860
	1392	25,10,15←25, 9,16	Ground					37 204.66		860
	1392	26,10,16←26, 9,17	Ground					37 191.23		860
	1392	27,10,17←27, 9,18	Ground					37 177.86		860
	1392	28,10,18←28, 9,19	Ground					37 158.88		860
	1392	29,10,19←29, 9,20	Ground					37 141.78		860
	1392	30,10,20←30, 9,21	Ground					37 123.80		860

ClF₃Ge   C₃ᵥ   GeF₃Cl

Isotopic Species	Pt. Gp.	Id. No.	A MHz	B MHz	C MHz	D_J MHz	D_{JK} MHz	Δ Amu A²	κ
Ge⁷⁰F₃¹⁹Cl³⁵	C₃ᵥ	1401		2 168.52 M	2 168.52 M	.0006			
Ge⁷⁰F₃¹⁹Cl³⁷	C₃ᵥ	1402		2 108.13 M	2 108.13 M				
Ge⁷²F₃¹⁹Cl³⁵	C₃ᵥ	1403		2 167.53 M	2 167.53 M				
Ge⁷²F₃¹⁹Cl³⁷	C₃ᵥ	1404		2 107.04 M	2 107.04 M				
Ge⁷⁴F₃¹⁹Cl³⁵	C₃ᵥ	1405		2 166.60 M	2 166.60 M				
Ge⁷⁴F₃¹⁹Cl³⁷	C₃ᵥ	1406		2 105.98 M	2 105.98 M				

References:

ABC: 255 D_J: 255

Chlorotrifluorogermane   Spectral Line Table

Isotopic Species	Id. No.	Rotational Quantum Nos.	Vib. State	F_1'	F'	F_1	F	Frequency MHz	Acc. ±MHz	Ref.
Ge⁷⁰F₃¹⁹Cl³⁵	1401	7, ← 6,	Ground					30 358.62	.30	255
	1401	8, ← 7,	Ground					34 694.71	.40	255
	1401	9, ← 8,	Ground					39 031.91	.30	255
Ge⁷⁰F₃¹⁹Cl³⁷	1402	7, ← 6,	Ground					29 512.96	.30	255
	1402	8, ← 7,	Ground					33 728.15	.80	255
	1402	9, ← 8,	Ground					37 945.47	.90	255
Ge⁷²F₃¹⁹Cl³⁵	1403	7, ← 6,	Ground					30 344.56	.30	255
	1403	8, ← 7,	Ground					34 679.32	.30	255
	1403	9, ← 8,	Ground					39 013.81	.20	255
Ge⁷²F₃¹⁹Cl³⁷	1404	7, ← 6,	Ground					29 497.57	.30	255
	1404	8, ← 7,	Ground					33 711.21	.30	255
	1404	9, ← 8,	Ground					37 925.87	.90	255
Ge⁷⁴F₃¹⁹Cl³⁵	1405	7, ← 6,	Ground					30 332.58	1.00	255
	1405	8, ← 7,	Ground					34 664.55	.30	255
	1405	9, ← 8,	Ground					38 996.78	.20	255
Ge⁷⁴F₃¹⁹Cl³⁷	1406	7, ← 6,	Ground					29 482.88	.30	255
	1406	8, ← 7,	Ground					33 694.43	.30	255
	1406	9, ← 8,	Ground					37 905.91	.30	255

ClF$_3$Si   C$_{3v}$ SiF$_3$Cl

Isotopic Species	Pt. Gp.	Id. No.	A MHz	B MHz	C MHz	D$_J$ MHz	D$_{JK}$ MHz	Δ Amu A^2	κ
Si^{28}F$_3^{19}$Cl^{35}	C$_{3v}$	1411		2 477.79 M	2 477.79 M		.0018		
Si^{28}F$_3^{19}$Cl^{37}	C$_{3v}$	1412		2 413.06 M	2 413.06 M				

Id. No.	μ$_a$ Debye	μ$_b$ Debye	μ$_c$ Debye	eQq Value(MHz) Rel.		eQq Value(MHz) Rel.		eQq Value(MHz) Rel.		ω$_a$ d 1/cm	ω$_b$ d 1/cm	ω$_c$ d 1/cm	ω$_d$ d 1/cm
1411				−43.	Cl^{35}								

References:

ABC: 311   D$_{JK}$: 311   eQq: 311
Add. Ref. 233,493

Chlorotrifluorosilane   Spectral Line Table

Isotopic Species	Id. No.	Rotational Quantum Nos.	Vib. State	F$_1'$	F'	F$_1$	F	Frequency MHz	Acc. ±MHz	Ref.
Si^{28}F$_3^{19}$Cl^{35}	1411	10, 0← 9, 0	Ground		21/2		19/2	49 555.72		311
	1411	10, 0← 9, 0	Ground		19/2		17/2	49 555.72		311
	1411	10, 0← 9, 0	Ground		23/2		21/2	49 555.72		311
	1411	10, 0← 9, 0	Ground		17/2		15/2	49 555.72		311
	1411	10, 1← 9, 1	Ground		17/2		15/2	49 555.72		311
	1411	10, 1← 9, 1	Ground		23/2		21/2	49 555.72		311
	1411	10, 1← 9, 1	Ground		21/2		19/2	49 555.72		311
	1411	10, 1← 9, 1	Ground		19/2		17/2	49 555.72		311
	1411	10, 2← 9, 2	Ground		19/2		17/2	49 555.72		311
	1411	10, 2← 9, 2	Ground		21/2		19/2	49 555.72		311
	1411	10, 2← 9, 2	Ground		23/2		21/2	49 555.72		311
	1411	10, 2← 9, 2	Ground		17/2		15/2	49 555.72		311
	1411	10, 3← 9, 3	Ground		21/2		19/2	49 555.20		311
	1411	10, 3← 9, 3	Ground		19/2		17/2	49 555.20		311
	1411	10, 3← 9, 3	Ground		17/2		15/2	49 555.72		311
	1411	10, 3← 9, 3	Ground		23/2		21/2	49 555.72		311
	1411	10, 4← 9, 4	Ground		19/2		17/2	49 554.70		311
	1411	10, 4← 9, 4	Ground		21/2		19/2	49 554.70		311
	1411	10, 4← 9, 4	Ground		23/2		21/2	49 555.72		311
	1411	10, 4← 9, 4	Ground		17/2		15/2	49 555.72		311
	1411	10, 5← 9, 5	Ground		19/2		17/2	49 554.10		311
	1411	10, 5← 9, 5	Ground		21/2		19/2	49 554.10		311
	1411	10, 5← 9, 5	Ground		17/2		15/2	49 555.72		311
	1411	10, 5← 9, 5	Ground		23/2		21/2	49 555.72		311
	1411	10, 6← 9, 6	Ground		21/2		19/2	49 553.12		311
	1411	10, 6← 9, 6	Ground		19/2		17/2	49 553.54		311
	1411	10, 6← 9, 6	Ground		23/2		21/2	49 555.72		311
	1411	10, 6← 9, 6	Ground		17/2		15/2	49 555.72		311
	1411	10, 7← 9, 7	Ground		21/2		19/2	49 552.10		311
	1411	10, 7← 9, 7	Ground		19/2		17/2	49 552.70		311
	1411	10, 7← 9, 7	Ground		23/2		21/2	49 555.20		311
	1411	10, 7← 9, 7	Ground		17/2		15/2	49 555.72		311
	1411	10, 8← 9, 8	Ground		23/2		21/2	49 555.20		311
	1411	10, 8← 9, 8	Ground		17/2		15/2	49 555.72		311
	1411	10, 9← 9, 9	Ground		17/2		15/2	49 555.72		311
Si^{28}F$_3^{19}$Cl^{37}	1412	8, ← 7,	Ground					38 608.96	.1	311

ClF$_5$S   C$_{4v}$ SF$_5$Cl

Isotopic Species	Pt. Gp.	Id. No.	A MHz	B MHz	C MHz	D$_J$ MHz	D$_{JK}$ MHz	Δ Amu A^2	κ
S^{32}F$_5^{19}$Cl^{35}	C$_{4v}$	1421		1 824.560 M			.0026		
S^{32}F$_5^{19}$Cl	C$_{4v}$	1422		1 783.524 M					
S^{34}F$_5^{19}$Cl^{35}	C$_{4v}$	1423		1 823.857 M					
S^{34}F$_5^{19}$Cl^{37}	C$_{4v}$	1424		1 782.70 M					

Id. No.	μ$_a$ Debye	μ$_b$ Debye	μ$_c$ Debye	eQq Value(MHz) Rel.	eQq Value(MHz) Rel.	eQq Value(MHz) Rel.	ω$_a$ d 1/cm	ω$_b$ d 1/cm	ω$_c$ d 1/cm	ω$_d$ d 1/cm
1421				−81.5			270 2			

References:

ABC: 909   D$_{JK}$: 909   eQq: 909   ω: 909

Isotopic Species	Id. No.	Rotational Quantum Nos.	Vib. State v_a	Hyperfine F_1'	F'	F_1	F	Frequency MHz	Acc. ±MHz	Ref.
$S^{32}F_5^{19}Cl^{35}$	1421	6, ← 5,	Ground					21 894.39	.5	909
	1421	7, ← 6,	Ground					25 543.85	.5	909
	1421	8, ← 7,	1					29 163.27		909
	1421	8, 0← 7, 0	Ground		15/2		13/2	29 192.96	.5	909
	1421	8, 0← 7, 0	Ground		17/2		15/2	29 192.96	.5	909
	1421	8, 0← 7, 0	Ground		13/2		11/2	29 192.96	.5	909
	1421	8, 0← 7, 0	Ground		19/2		17/2	29 192.96	.5	909
	1421	8, 1← 7, 1	Ground		17/2		15/2	29 192.96	.5	909
	1421	8, 1← 7, 1	Ground		13/2		11/2	29 192.96	.5	909
	1421	8, 1← 7, 1	Ground		15/2		13/2	29 192.96	.5	909
	1421	8, 1← 7, 1	Ground		19/2		17/2	29 192.96	.5	909
	1421	8, 2← 7, 2	Ground		15/2		13/2	29 192.96	.5	909
	1421	8, 2← 7, 2	Ground		17/2		15/2	29 192.96	.5	909
	1421	8, 2← 7, 2	Ground		19/2		17/2	29 192.96	.5	909
	1421	8, 2← 7, 2	Ground		13/2		11/2	29 192.96	.5	909
	1421	8, 3← 7, 3	Ground		19/2		17/2	29 192.96	.5	909
	1421	8, 3← 7, 3	Ground		15/2		13/2	29 192.96	.5	909
	1421	8, 3← 7, 3	Ground		17/2		15/2	29 192.96	.5	909
	1421	8, 3← 7, 3	Ground		13/2		11/2	29 192.96	.5	909
	1421	8, 4← 7, 4	Ground		15/2		13/2	29 189.65	.5	909
	1421	8, 4← 7, 4	Ground		17/2		15/2	29 189.65	.5	909
	1421	8, 4← 7, 4	Ground		19/2		17/2	29 192.96	.5	909
	1421	8, 4← 7, 4	Ground		13/2		11/2	29 195.41	.5	909
	1421	8, 5← 7, 5	Ground		17/2		15/2	29 188.28	.5	909
	1421	8, 5← 7, 5	Ground		15/2		13/2	29 189.65	.5	909
	1421	8, 5← 7, 5	Ground		13/2		11/2	29 195.41	.5	909
	1421	8, 5← 7, 5	Ground		19/2		17/2	29 195.41	.5	909
	1421	8, 6← 7, 6	Ground		17/2		15/2	29 186.38	.5	909
	1421	8, 6← 7, 6	Ground		15/2		13/2	29 188.28	.5	909
	1421	8, 6← 7, 6	Ground		19/2		17/2	29 195.41	.5	909
	1421	8, 6← 7, 6	Ground		13/2		11/2	29 196.82	.5	909
	1421	8, 7← 7, 7	Ground		17/2		15/2	29 184.14	.5	909
	1421	8, 7← 7, 7	Ground		15/2		13/2	29 186.38	.5	909
	1421	8, 7← 7, 7	Ground		19/2		17/2	29 195.41	.5	909
	1421	8, 7← 7, 7	Ground		13/2		11/2	29 198.53	.5	909
	1421	10, ← 9,	Ground					36 490.59	.5	909
$S^{32}F_5^{19}Cl^{37}$	1422	6, ← 5,	Ground					21 401.	.5	909
	1422	7, ← 6,	Ground					24 969.50	.5	909
	1422	8, ← 7,	Ground					28 536.32	.5	909
	1422	8, ← 7,	2					29 134.05		909
	1422	10, ← 9,	Ground					35 670.59	.5	909
$S^{34}F_5^{19}Cl^{35}$	1423	6, ← 5,	Ground					21 886.64	.5	909
	1423	7, ← 6,	Ground					25 533.97	.5	909
	1423	8, ← 7,	Ground					29 181.73	.5	909
$S^{34}F_5^{19}Cl^{37}$	1424	8, ← 7,	Ground					28 522.04	.5	909
	1424	10, ← 9,	Ground					35 654.02	.5	909

ClGeH$_3$ C$_{3v}$ GeH$_3$Cl

Isotopic Species	Pt. Gp.	Id. No.	A MHz	B MHz		C MHz		D$_J$ MHz	D$_{JK}$ MHz	Δ Amu A^2	κ
Ge^{70}H$_3$Cl^{35}	C$_{3v}$	1431		4 401.71	M	4 401.71	M				
Ge^{74}H$_3$Cl^{35}	C$_{3v}$	1432		4 333.91	M	4 333.91	M				
Ge^{74}H$_3$Cl^{37}	C$_{3v}$	1433		4 177.90	M	4 177.90	M				
Ge^{76}H$_3$Cl^{37}	C$_{3v}$	1434		4 146.5	M	4 146.5	M				

Id. No.	μ_a Debye		μ_b Debye		μ_c Debye		eQq Value(MHz)	Rel.	eQq Value(MHz)	Rel.	eQq Value(MHz)	Rel.	ω_a 1/cm	d	ω_b 1/cm	d	ω_c 1/cm	d	ω_d 1/cm	d
1431	2.124	M	0.	X	0.	X														

References:

ABC: 127 μ: 375

Add. Ref. 269

For the Ge^{73}H$_3$Cl$^{35, 37}$ species, eQq (Ge^{73}) = − 95 MHz, eQq (Cl^{35}) = − 46 MHz, and eQq ($\bar{C}$l^{37}) = − 36 MHz. Lines have been observed, but they are obscured by the complexity of the hyperfine structure. Ref. 172.

Chlorogermane   Spectral Line Table

Isotopic Species	Id. No.	Rotational Quantum Nos.	Vib. State	F$_1'$	F'	F$_1$	F	Frequency MHz	Acc. ±MHz	Ref.
Ge^{70}H$_3$Cl^{35}	1431	3, ← 2,	Ground					26 410.26	.5	375
Ge^{74}H$_3$Cl^{35}	1432	3, ← 2,	Ground					26 003.46	.5	375
Ge^{74}H$_3$Cl^{37}	1433	3, ← 2,	Ground					25 067.4	.5	375
Ge^{76}H$_3$Cl^{37}	1434	3, ← 2,	Ground					24 879.0	.5	375

ClH₃Si

C₃ᵥ

SiH₃Cl

Isotopic Species	Pt. Gp.	Id. No.	A MHz	B MHz		C MHz		D_J MHz	D_{JK} MHz	Δ Amu A²	κ
Si²⁸H₃Cl³⁵	C₃ᵥ	1441		6 673.8	M	6 673.8	M				
Si³⁰H₃Cl³⁵	C₃ᵥ	1442		6 485.8	M	6 485.8	M				
Si²⁸H₃Cl³⁷	C₃ᵥ	1443		6 512.4	M	6 512.4	M				
Si²⁸D₃Cl³⁵	C₃ᵥ	1444		5 917.7	M	5 917.7	M				
Si²⁹D₃Cl³⁵	C₃ᵥ	1445		5 850.6	M	5 850.6	M				
Si³⁰D₃Cl³⁵	C₃ᵥ	1446		5 787.0	M	5 787.0	M				
Si²⁸D₃Cl³⁷	C₃ᵥ	1447		5 772.8	M	5 772.8	M				

Id. No.	μ_a Debye		μ_b Debye		μ_c Debye		eQq Value(MHz)	Rel.	eQq Value(MHz)	Rel.	eQq Value(MHz)	Rel.	ω_a 1/cm	d	ω_b 1/cm	d	ω_c 1/cm	d	ω_d 1/cm	d
1441	1.303	M	0.	X	0.	X	−40.0	Cl³⁵												
1443							−30.8	Cl³⁷												
1444							−39.4	Cl³⁵												

References:

ABC: 413 μ: 375   eQq: 98,413

Add. Ref. 128,269,270,342,493

Chlorosilane

Spectral Line Table

Isotopic Species	Id. No.	Rotational Quantum Nos.	Vib. State	F_1'	F'	F_1	F	Frequency MHz	Acc. ±MHz	Ref.
Si²⁸H₃Cl³⁵	1441	2, ← 1,	Ground					26 695.24		98
Si³⁰H₃Cl³⁵	1442	2, ← 1,	Ground					25 943.2	.1	375
Si²⁸H₃Cl³⁷	1443	2, ← 1,	Ground					26 049.60		98
Si²⁸D₃Cl³⁵	1444	2, ← 1,	Ground					23 670.8	.2	494
Si²⁹D₃Cl³⁵	1445	2, ← 1,	Ground					23 402.6	.3	494
Si³⁰D₃Cl³⁵	1446	2, ← 1,	Ground					23 147.9	.3	494
Si²⁸D₃Cl³⁷	1447	2, ← 1,	Ground					23 091.4	.2	494

ClNO C_s NOCl

Isotopic Species	Pt. Gp.	Id. No.	A MHz		B MHz		C MHz		D_J MHz	D_{JK} MHz	Δ Amu A²	κ
$N^{14}O^{16}Cl^{35}$	C_s	1451	85 420.	M	5 737.50	M	5 376.39	M				−.99094
$N^{14}O^{16}Cl^{37}$	C_s	1452	85 400.	M	5 600.88	M	5 256.17	M				−.99143
$N^{14}O^{18}Cl^{35}$	C_s	1453	82 580.	M	5 439.31	M	5 103.17	M				

Id. No.	μ_a Debye		μ_b Debye	μ_c Debye		eQq Value(MHz)	Rel.	eQq Value(MHz)	Rel.	eQq Value(MHz)	Rel.	ω_a d 1/cm	ω_b d 1/cm	ω_c d 1/cm	ω_d d 1/cm
1451	1.86	M		0.	X	−48.7	aa	29.4	bb	19.3	cc				
1452						−38.4	aa	23.2	bb	15.2	cc				

References:

ABC: 946 κ: 307 μ: 845   eQq: 946

Add. Ref. 183,225,226,309

Nitrosyl Chloride   Spectral Line Table

Isotopic Species	Id. No.	Rotational Quantum Nos.	Vib. State	F_1'	F'	F_1	F	Frequency MHz	Acc. ±MHz	Ref.
$N^{14}O^{16}Cl^{35}$	1451	1, 0, 1← 0, 0, 0	Ground		3/2		3/2	11 104.15		946
	1451	1, 0, 1← 0, 0, 0	Ground		5/2		3/2	11 116.33		946
	1451	1, 0, 1← 0, 0, 0	Ground		1/2		3/2	11 126.05		946
	1451	2, 0, 2← 1, 0, 1	Excited		3/2		1/2	22 106.31	.20	307
	1451	2, 0, 2← 1, 0, 1	Excited		5/2		5/2	22 106.31	.20	307
	1451	2, 0, 2← 1, 0, 1	Excited		5/2		3/2	22 118.93	.07	307
	1451	2, 0, 2← 1, 0, 1	Excited		7/2		5/2	22 118.93	.07	307
	1451	2, 0, 2← 1, 0, 1	Excited		3/2		3/2	22 126.81	.20	307
	1451	2, 0, 2← 1, 0, 1	Ground		3/2		1/2	22 215.08	.10	307
	1451	2, 0, 2← 1, 0, 1	Ground		5/2		5/2	22 215.08	.10	307
	1451	2, 0, 2← 1, 0, 1	Ground		7/2		5/2	22 227.37	.03	307
	1451	2, 0, 2← 1, 0, 1	Ground		5/2		3/2	22 227.37	.03	307
	1451	2, 0, 2← 1, 0, 1	Ground		3/2		3/2	22 236.45	.08	307
	1451	2, 1, 1← 1, 1, 0	Excited		5/2		3/2	22 471.6	.3	307
	1451	2, 1, 1← 1, 1, 0	Excited		3/2		3/2	22 476.2	.3	307
	1451	2, 1, 1← 1, 1, 0	Excited		5/2		5/2	22 476.2	.3	307
	1451	2, 1, 1← 1, 1, 0	Excited		7/2		5/2	22 483.7	.3	307
	1451	2, 1, 1← 1, 1, 0	Excited		3/2		1/2	22 483.7	.3	307
	1451	2, 1, 1← 1, 1, 0	Ground		5/2		3/2	22 580.47	.03	307
	1451	2, 1, 1← 1, 1, 0	Ground		3/2		3/2	22 585.47	.13	307
	1451	2, 1, 1← 1, 1, 0	Ground		5/2		5/2	22 585.47	.13	307
	1451	2, 1, 1← 1, 1, 0	Ground		3/2		1/2	22 592.95	.07	307
	1451	2, 1, 1← 1, 1, 0	Ground		7/2		5/2	22 592.95	.07	307
	1451	2, 1, 1← 1, 1, 0	Ground		1/2		1/2	22 601.72	.14	307
	1451	2, 1, 2← 1, 1, 1	Ground		5/2		3/2	21 857.42	.10	307
	1451	2, 1, 2← 1, 1, 1	Ground		3/2		3/2	21 860.70	.10	307
	1451	2, 1, 2← 1, 1, 1	Ground		5/2		5/2	21 864.59	.15	307
	1451	2, 1, 2← 1, 1, 1	Ground		7/2		5/2	21 869.38	.08	307
	1451	2, 1, 2← 1, 1, 1	Ground		3/2		1/2	21 873.93	.10	307
	1451	2, 1, 2← 1, 1, 1	Ground		1/2		1/2	21 879.07	.10	307

Isotopic Species	Id. No.	Rotational Quantum Nos.	Vib. State	Hyperfine				Frequency MHz	Acc. ±MHz	Ref.
				F_1'	F'	F_1	F			
$N^{14}O^{16}Cl^{37}$	1452	1, 0, 1← 0, 0, 0	Ground		3/2		3/2	10 849.36		946
	1452	1, 0, 1← 0, 0, 0	Ground		5/2		3/2	10 858.96		946
	1452	1, 0, 1← 0, 0, 0	Ground		1/2		3/2	10 866.65		946
	1452	2, 0, 2← 1, 0, 1	Ground		5/2		5/2	21 703.77	.08	307
	1452	2, 0, 2← 1, 0, 1	Ground		3/2		1/2	21 703.77	.08	307
	1452	2, 0, 2← 1, 0, 1	Ground		7/2		5/2	21 713.25	.11	307
	1452	2, 0, 2← 1, 0, 1	Ground		5/2		3/2	21 713.25	.11	307
	1452	2, 0, 2← 1, 0, 1	Ground		3/2		3/2	21 719.68	.20	307
	1452	2, 1; 1← 1, 1, 0	Ground		5/2		3/2	22 052.07	.10	307
	1452	2, 1, 1← 1, 1, 0	Ground		3/2		3/2	22 056.03	.10	307
	1452	2, 1, 1← 1, 1, 0	Ground		5/2		5/2	22 056.03	.10	307
	1452	2, 1, 1← 1, 1, 0	Ground		7/2		5/2	22 062.26	.10	307
	1452	2, 1, 1← 1, 1, 0	Ground		3/2		1/2	22 068.11	.18	307
	1452	2, 1, 1← 1, 1, 0	Ground		1/2		1/2	22 071.33	.16	307
	1452	2, 1, 2← 1, 1, 1	Ground		5/2		3/2	21 362.64	.30	307
	1452	2, 1, 2← 1, 1, 1	Ground		3/2		3/2	21 364.7	.6	307
	1452	2, 1, 2← 1, 1, 1	Ground		5/2		5/2	21 367.47	.20	307
	1452	2, 1, 2← 1, 1, 1	Ground		7/2		5/2	21 371.53	.15	307
$N^{14}O^{18}Cl^{35}$	1453	1, 0, 1← 0, 0, 0	Ground		3/2		3/2	10 532.72		946
	1453	1, 0, 1← 0, 0, 0	Ground		5/2		3/2	10 544.88		946
	1453	2, 2, 0← 3, 2, 1	Ground		5/2		5/2	31 623.50		946
	1453	2, 2, 0← 3, 2, 1	Ground		3/2		5/2	31 623.50		946
	1453	2, 2, 0← 3, 2, 1	Ground		5/2		3/2	31 632.28		946
	1453	2, 2, 0← 3, 2, 1	Ground		3/2		3/2	31 632.28		946
	1453	2, 2, 0← 3, 2, 1	Ground		7/2		5/2	31 632.28		946
	1453	3, 0, 3← 2, 0, 2	Ground		7/2		7/2	31 611.34		946
	1453	3, 0, 3← 2, 0, 2	Ground		3/2		1/2	31 620.48		946
	1453	3, 0, 3← 2, 0, 2	Ground		3/2		5/2	31 620.48		946
	1453	3, 0, 3← 2, 0, 2	Ground		7/2		5/2	31 623.50		946
	1453	3, 0, 3← 2, 0, 2	Ground		9/2		7/2	31 623.50		946
	1453	3, 0, 3← 2, 0, 2	Ground		5/2		5/2	31 628.97		946
	1453	3, 0, 3← 2, 0, 2	Ground		3/2		3/2	31 632.28		946
	1453	3, 1, 2← 2, 1, 1	Ground		7/2		5/2	32 129.17		946
	1453	3, 1, 2← 2, 1, 1	Ground		5/2		3/2	32 129.17		946
	1453	3, 1, 2← 2, 1, 1	Ground		9/2		7/2	32 132.36		946
	1453	3, 1, 3← 2, 1, 2	Ground		5/2		3/2	31 121.14		946
	1453	3, 1, 3← 2, 1, 2	Ground		7/2		5/2	31 121.14		946
	1453	3, 1, 3← 2, 1, 2	Ground		3/2		1/2	31 123.80		946
	1453	3, 1, 3← 2, 1, 2	Ground		9/2		7/2	31 123.80		946
	1453	3, 1, 3← 2, 1, 2	Ground		5/2		5/2	31 123.80		946

ClNO₂ ... C_{2v} ... NO₂Cl

Isotopic Species	Pt. Gp.	Id. No.	A MHz		B MHz		C MHz		D_J MHz	D_{JK} MHz	Δ Amu A²	κ
$N^{14}O_2^{16}Cl^{35}$	C_{2v}	1461	13 250.	M	5 173.77	M	3 721.13	M			.208	−.6951
$N^{14}O_2^{16}Cl^{37}$	C_{2v}	1462	13 250.	M	5 018.97	M	3 640.35	M				−.7131

Id. No.	μ_a Debye		μ_b Debye		μ_c Debye		eQq Value(MHz)	Rel.	eQq Value(MHz)	Rel.	eQq Value(MHz)	Rel.	ω_a d 1/cm	ω_b d 1/cm	ω_c d 1/cm	ω_d d 1/cm
1461	.53	M	0.	X	0.	X	−94.70	aa	52.4	bb	42.3	cc				
1462							−74.58	aa	41.3	bb	33.3	cc				

References:

ABC: 848,849 ... Δ: 979 ... κ: 848 ... μ: 848 ... eQq: 848

Add. Ref. 664

Nitryl Chloride

Spectral Line Table

Isotopic Species	Id. No.	Rotational Quantum Nos.	Vib. State	Hyperfine F_1'	F'	F_1	F	Frequency MHz	Acc. ±MHz	Ref.
$N^{14}O_2^{16}Cl^{35}$	1461	1, 0, 1← 0, 0, 0	Ground		3/2		3/2	8 876.28		832
	1461	1, 0, 1← 0, 0, 0	Ground		5/2		3/2	8 899.90		832
	1461	1, 0, 1← 0, 0, 0	Ground		1/2		3/2	8 918.70		832
	1461	2, 0, 2← 1, 0, 1	Ground		3/2		1/2	17 587.70		832
	1461	2, 0, 2← 1, 0, 1	Ground		5/2		5/2	17 589.88		832
	1461	2, 0, 2← 1, 0, 1	Ground		1/2		1/2	17 610.80		832
	1461	2, 0, 2← 1, 0, 1	Ground		7/2		5/2	17 612.98		832
	1461	2, 0, 2← 1, 0, 1	Ground		5/2		3/2	17 613.52		832
	1461	2, 0, 2← 1, 0, 1	Ground		3/2		3/2	17 629.90		832
	1461	2, 0, 2← 1, 0, 1	Ground		1/2		3/2	17 653.20		832
	1461	3, 0, 3← 2, 0, 2	Ground		7/2		7/2	25 964.70		832
	1461	3, 0, 3← 2, 0, 2	Ground		3/2		1/2	25 981.33		848
	1461	3, 0, 3← 2, 0, 2	Ground		5/2		3/2	25 981.95		848
	1461	3, 0, 3← 2, 0, 2	Ground		9/2		7/2	25 987.43		848
	1461	3, 0, 3← 2, 0, 2	Ground		7/2		5/2	25 988.01		848
	1461	3, 0, 3← 2, 0, 2	Ground		5/2		5/2	25 998.57		848
	1461	3, 0, 3← 2, 0, 2	Ground		3/2		3/2	26 004.59		848
	1461	3, 2, 1← 2, 2, 0	Excited					27 110.		798
	1461	3, 2, 1← 2, 2, 0	Excited					27 145.		798
	1461	3, 2, 1← 2, 2, 0	Ground		7/2		5/2	27 365.61		848
	1461	3, 2, 1← 2, 2, 0	Ground		5/2		5/2	27 366.1		848
	1461	3, 2, 1← 2, 2, 0	Ground		5/2		3/2	27 382.73		848
	1461	3, 2, 1← 2, 2, 0	Ground		3/2		3/2	27 383.8		848
	1461	3, 2, 1← 2, 2, 0	Ground		7/2		7/2	27 388.8		848
	1461	3, 2, 1← 2, 2, 0	Ground		9/2		7/2	27 389.88		848
	1461	3, 2, 1← 2, 2, 0	Ground		3/2		1/2	27 407.00		848
	1461	3, 2, 2← 2, 2, 1	Excited					26 425.		798
	1461	3, 2, 2← 2, 2, 1	Excited					26 450.		798
	1461	3, 2, 2← 2, 2, 1	Ground		5/2		5/2	26 667.77		848
	1461	3, 2, 2← 2, 2, 1	Ground		3/2		3/2	26 684.68		848

Isotopic Species	Id. No.	Rotational Quantum Nos.	Vib. State	Hyperfine				Frequency MHz	Acc. ±MHz	Ref.
				F_1'	F'	F_1	F			
$N^{14}O_2^{16}Cl^{35}$	1461	3, 2, 2← 2, 2, 1	Ground		7/2		7/2	26 691.44		848
	1461	3, 2, 2← 2, 2, 1	Ground		1/2		1/2	26 708.33		848
	1461	4, 0, 4← 3, 0, 3	Ground		9/2		9/2	33 911.19		848
	1461	4, 0, 4← 3, 0, 3	Ground		5/2		3/2	33 930.27		848
	1461	4, 0, 4← 3, 0, 3	Ground		7/2		5/2	33 930.97		848
	1461	4, 0, 4← 3, 0, 3	Ground		11/2		9/2	33 933.17		848
	1461	4, 0, 4← 3, 0, 3	Ground		9/2		7/2	33 933.83		848
	1461	4, 0, 4← 3, 0, 3	Ground		7/2		7/2	33 941.4		848
	1461	4, 0, 4← 3, 0, 3	Ground		5/2		5/2	33 953.0		848
	1461	4, 2, 2← 3, 2, 1	Excited					36 740.		798 [1]
	1461	4, 2, 2← 3, 2, 1	Ground		9/2		7/2	37 079.11		848
	1461	4, 2, 2← 3, 2, 1	Ground		7/2		5/2	37 082.50		848
	1461	4, 2, 2← 3, 2, 1	Ground		11/2		9/2	37 089.15		848
	1461	4, 2, 2← 3, 2, 1	Ground		5/2		3/2	37 092.52		848
	1461	4, 2, 3← 3, 2, 2	Excited					35 116.8		798
	1461	4, 2, 3← 3, 2, 2	Excited					35 120.0		798
	1461	4, 2, 3← 3, 2, 2	Excited					35 126.2		798
	1461	4, 2, 3← 3, 2, 2	Excited					35 129.4		798
	1461	4, 2, 3← 3, 2, 2	Ground		9/2		9/2	35 433.63		848
	1461	4, 2, 3← 3, 2, 2	Ground		7/2		7/2	35 436.91		848
$N^{14}O_2^{16}Cl^{35}$	1461	4, 2, 3← 3, 2, 2	Ground		11/2		11/2	35 443.08		848
	1461	4, 2, 3← 3, 2, 2	Ground		5/2		5/2	35 446.36		848
	1461	6, 2, 5← 6, 2, 4	Ground		13/2		13/2	9 973.90		832
	1461	6, 2, 5← 6, 2, 4	Ground		11/2		11/2	9 973.90		832
	1461	6, 2, 5← 6, 2, 4	Ground		9/2		9/2	9 976.40		832
	1461	6, 2, 5← 6, 2, 4	Ground		15/2		15/2	9 976.40		832
	1461	Not Reported						17 627.90		621
$N^{14}O_2^{16}Cl^{37}$	1462	1, 0, 1← 0, 0, 0	Ground		3/2		3/2	8 644.60		832
	1462	1, 0, 1← 0, 0, 0	Ground		5/2		3/2	8 663.20		832
	1462	1, 0, 1← 0, 0, 0	Ground		1/2		3/2	8 678.10		832
	1462	2, 0, 2← 1, 0, 1	Ground		5/2		3/2	17 161.32		832
	1462	2, 0, 2← 1, 0, 1	Ground		7/2		5/2	17 161.32		832
	1462	3, 0, 3← 2, 0, 2	Ground		7/2		7/2	25 338.08		848
	1462	3, 0, 3← 2, 0, 2	Ground		3/2		1/2	25 351.20		848
	1462	3, 0, 3← 2, 0, 2	Ground		5/2		3/2	25 351.64		848
	1462	3, 0, 3← 2, 0, 2	Ground		9/2		7/2	25 355.97		848
	1462	3, 0, 3← 2, 0, 2	Ground		7/2		5/2	25 356.38		848
	1462	3, 0, 3← 2, 0, 2	Ground		5/2		5/2	25 364.68		848
	1462	3, 0, 3← 2, 0, 2	Ground		3/2		3/2	25 369.52		848
	1462	3, 2, 1← 2, 2, 0	Excited					26 350.		798
	1462	3, 2, 1← 2, 2, 0	Excited					26 375.		798
	1462	3, 2, 1← 2, 2, 0	Ground		7/2		5/2	26 587.14		848
	1462	3, 2, 1← 2, 2, 0	Ground		5/2		5/2	26 587.5		848
	1462	3, 2, 1← 2, 2, 0	Ground		5/2		3/2	26 600.60		848
	1462	3, 2, 1← 2, 2, 0	Ground		3/2		3/2	26 601.4		848
	1462	3, 2, 1← 2, 2, 0	Ground		9/2		7/2	26 606.22		848
	1462	3, 2, 1← 2, 2, 0	Ground		3/2		1/2	26 619.69		848
	1462	3, 2, 2← 2, 2, 1	Excited					25 730.		798
	1462	3, 2, 2← 2, 2, 1	Excited					25 750.		798
	1462	3, 2, 2← 2, 2, 1	Ground		5/2		5/2	25 964.60		848
	1462	3, 2, 2← 2, 2, 1	Ground		3/2		3/2	25 977.92		848
	1462	3, 2, 2← 2, 2, 1	Ground		7/2		7/2	25 983.23		848

---

1. These lines actually consist of four lines each.

Isotopic Species	Id. No.	Rotational Quantum Nos.	Vib. State	Hyperfine F_1'	F'	F_1	F	Frequency MHz	Acc. ±MHz	Ref.
$N^{14}O_2^{16}Cl^{37}$	1462	3, 2, 2← 2, 2, 1	Ground		1/2		1/2	25 996.61		848
	1462	4, 0, 4← 3, 0, 3	Ground		5/2		3/2	33 156.97		848
	1462	4, 0, 4← 3, 0, 3	Ground		7/2		5/2	33 157.49		848
	1462	4, 0, 4← 3, 0, 3	Ground		11/2		9/2	33 159.29		848
	1462	4, 0, 4← 3, 0, 3	Ground		9/2		7/2	33 159.75		848
	1462	4, 2, 2← 3, 2, 1	Excited					35 679.0		798
	1462	4, 2, 2← 3, 2, 1	Excited					35 681.7		798
	1462	4, 2, 2← 3, 2, 1	Excited					35 686.9		798
	1462	4, 2, 2← 3, 2, 1	Excited					35 689.5		798
	1462	4, 2, 2← 3, 2, 1	Ground		9/2		7/2	35 985.36		848
	1462	4, 2, 2← 3, 2, 1	Ground		7/2		5/2	35 988.03		848
	1462	4, 2, 2← 3, 2, 1	Ground		11/2		9/2	35 993.27		848
	1462	4, 2, 2← 3, 2, 1	Ground		5/2		3/2	35 995.90		848
	1462	4, 2, 3← 3, 2, 2	Excited					34 210.		798 [1]
	1462	4, 2, 3← 3, 2, 2	Ground		9/2		9/2	34 508.11		848
	1462	4, 2, 3← 3, 2, 2	Ground		7/2		7/2	34 510.72		848
	1462	4, 2, 3← 3, 2, 2	Ground		11/2		11/2	34 515.53		848
	1462	4, 2, 3← 3, 2, 2	Ground		5/2		5/2	34 518.11		848
	1462	5, 0, 5← 4, 0, 4	Ground		7/2		5/2	40 599.43		848
	1462	5, 0, 5← 4, 0, 4	Ground		9/2		7/2	40 599.89		848
	1462	5, 0, 5← 4, 0, 4	Ground		13/2		11/2	40 600.75		848
	1462	5, 0, 5← 4, 0, 4	Ground		11/2		9/2	40 601.22		848
$N^bO_2^bCl^b$	1463	Not Reported						8 621.50		634
	1463	Not Reported						8 622.20		634
	1463	Not Reported						9 002.52		634
	1463	Not Reported						9 003.58		634
	1463	Not Reported						9 368.20		634
	1463	Not Reported						10 540.90		634
	1463	Not Reported						10 591.50		634
	1463	Not Reported						10 931.70		634
	1463	Not Reported						17 075.74		634
	1463	Not Reported						17 077.00		634
	1463	Not Reported						18 300.		634
	1463	Not Reported						19 202.52		634
	1463	Not Reported						19 203.68		634
	1463	Not Reported						20 800.		634
	1463	Not Reported						21 300.		634
	1463	Not Reported						22 000.		634
	1463	Not Reported						23 376.		634
	1463	Not Reported						24 113.		634
	1463	Not Reported						24 791.		634
	1463	Not Reported						24 971.		634
	1463	Not Reported						25 095.		634
	1463	Not Reported						25 327.		634
	1463	Not Reported						25 640.		634
	1463	Not Reported						25 658.		634

1. These lines actually consist of four lines each.

ClO₃Re $\quad\quad$ C$_{3v}$ $\quad\quad$ ReO₃Cl

Isotopic Species	Pt. Gp.	Id. No.	A MHz	B MHz		C MHz		D$_J$ MHz	D$_{JK}$ MHz	Δ Amu A²	κ
Re^{185}O$_3^{16}$Cl^{35}	C$_{3v}$	1471		2 094.20	M	2 094.20	M				
Re^{185}O$_3^{16}$Cl^{37}	C$_{3v}$	1472		2 025.02	M	2 025.02	M				
Re^{187}O$_3^{16}$Cl^{35}	C$_{3v}$	1473		2 093.58	M	2 093.58	M				
Re^{187}O$_3^{16}$Cl^{37}	C$_{3v}$	1474		2 024.36	M	2 024.36	M				

Id. No.	μ$_a$ Debye	μ$_b$ Debye	μ$_c$ Debye	eQq Value(MHz)	Rel.	eQq Value(MHz)	Rel.	eQq Value(MHz)	Rel.	ω$_a$ d 1/cm	ω$_b$ d 1/cm	ω$_c$ d 1/cm	ω$_d$ d 1/cm
1471				270		−34	Cl^{35}						
1473				253									

References:

ABC: 332,531 $\quad\quad$ eQq: 531

Add. Ref. 254

Rhenium Trioxychloride $\quad\quad$ Spectral Line Table

Isotopic Species	Id. No.	Rotational Quantum Nos.	Vib. State	F$_1'$	F'	F$_1$	F	Frequency MHz	Acc. ±MHz	Ref.
Re^{185}O$_3^{16}$Cl^{35}	1471	6, 0← 5, 0	Ground	17/2		15/2		25 129.42	.08	531
	1471	6, 0← 5, 0	Ground	13/2	8	11/2	9	25 130.80	.2	531
	1471	6, 0← 5, 0	Ground	13/2	5	11/2	6	25 130.80	.2	531
	1471	6, 0← 5, 0	Ground	11/2	6	9/2	5	25 132.56	.1	531
	1471	6, 0← 5, 0	Ground	11/2	5	9/2	4	25 132.56	.1	531
	1471	6, 0← 5, 0	Ground	11/2	4	9/2	3	25 133.38	.1	531
	1471	6, 0← 5, 0	Ground	7/2	5	5/2	4	25 133.38	.1	531
	1471	6, 0← 5, 0	Ground	11/2	7	9/2	6	25 133.38	.1	531
	1471	6, 0← 5, 0	Ground	9/2	4	7/2	3	25 133.38	.1	531
	1471	6, 0← 5, 0	Ground	9/2	5	7/2	4	25 133.38	.1	531
	1471	6, 0← 5, 0	Ground	9/2	3	7/2	2	25 134.59	.07	531
	1471	6, 0← 5, 0	Ground	9/2	6	7/2	5	25 134.59	.07	531
	1471	6, 3← 5, 3	Ground	17/2	7	15/2	6	25 124.35	.06	531
	1471	6, 3← 5, 3	Ground	17/2	10	15/2	9	25 124.35	.06	531
	1471	6, 3← 5, 3	Ground	11/2	5	9/2	4	25 134.24	.1	531
	1471	6, 3← 5, 3	Ground	15/2	7	13/2	6	25 134.24	.1	531
	1471	6, 3← 5, 3	Ground	11/2	6	9/2	5	25 134.24	.1	531
	1471	6, 3← 5, 3	Ground	15/2	8	13/2	7	25 134.24	.1	531
	1471	6, 3← 5, 3	Ground	11/2	6	9/2	5	25 135.82	.06	531
	1471	6, 3← 5, 3	Ground	11/2	7	9/2	6	25 135.82	.06	531
	1471	6, 3← 5, 3	Ground	15/2	6	13/2	5	25 135.82	.06	531
	1471	6, 3← 5, 3	Ground	15/2	9	13/2	8	25 135.82	.06	531
	1471	6, 3← 5, 3	Ground	13/2	7	11/2	6	25 137.40	.1	531
	1471	6, 3← 5, 3	Ground	13/2	6	11/2	5	25 137.40	.1	531
	1471	6, 3← 5, 3	Ground	13/2	8	11/2	7	25 138.80	.07	531
	1471	6, 3← 5, 3	Ground	13/2	5	11/2	4	25 138.80	.07	531
	1471	12, ←11,						50 261.4		332

Isotopic Species	Id. No.	Rotational Quantum Nos.	Vib. State	F_1'	F'	F_1	F	Frequency MHz	Acc. ±MHz	Ref.
$Re^{185}O_3^{16}Cl^{37}$	1472	3, ← 2,	Ground					12 550.		332
	1472	5, ← 4,	Ground					20 950.		332
	1472	12, ←11,	Ground					48 600.5		332
$Re^{187}O_3^{16}Cl^{35}$	1473	6, 0← 5, 0	Ground	17/2		15/2		25 122.08	.05	531
	1473	6, 0← 5, 0	Ground	13/2	5	11/2	6	25 123.45	.20	531
	1473	6, 0← 5, 0	Ground	13/2	8	11/2	9	25 123.45	.20	531
	1473	6, 0← 5, 0	Ground	11/2	5	9/2	4	25 125.00	.1	531
	1473	6, 0← 5, 0	Ground	11/2	6	9/2	5	25 125.00	.1	531
	1473	6, 0← 5, 0	Ground	11/2	4	9/2	3	25 125.75	.1	531
	1473	6, 0← 5, 0	Ground	11/2	7	9/2	6	25 125.75	.1	531
	1473	6, 0← 5, 0	Ground	9/2	5	7/2	4	25 125.75	.1	531
	1473	6, 0← 5, 0	Ground	9/2	4	7/2	3	25 125.75	.1	531
	1473	6, 0← 5, 0	Ground	7/2	5	5/2	4	25 125.75	.1	531
	1473	6, 0← 5, 0	Ground	9/2	3	7/2	2	25 127.03	.06	531
	1473	6, 0← 5, 0	Ground	9/2	6	7/2	5	25 127.03	.06	531
	1473	6, 3← 5, 3	Ground	17/2	9	15/2	8	25 115.33	.1	531
	1473	6, 3← 5, 3	Ground	17/2	8	15/2	7	25 115.33	.1	531
	1473	6, 3← 5, 3	Ground	17/2	10	15/2	9	25 117.36	.06	531
	1473	6, 3← 5, 3	Ground	17/2	7	15/2	6	25 117.66	.04	531
	1473	6, 3← 5, 3	Ground	9/2	4	7/2	3	25 120.10	.1	531
	1473	6, 3← 5, 3	Ground	9/2	5	7/2	4	25 120.10	.1	531
	1473	6, 3← 5, 3	Ground	15/2	8	13/2	7	25 126.52	.1	531
	1473	6, 3← 5, 3	Ground	15/2	7	13/2	6	25 126.52	.1	531
	1473	6, 3← 5, 3	Ground	11/2	6	9/2	5	25 126.52	.1	531
	1473	6, 3← 5, 3	Ground	11/2	5	9/2	4	25 126.52	.1	531
	1473	6, 3← 5, 3	Ground	15/2	6	13/2	5	25 128.15	.05	531
	1473	6, 3← 5, 3	Ground	11/2	7	9/2	6	25 128.15	.05	531
	1473	6, 3← 5, 3	Ground	15/2	9	13/2	8	25 128.15	.05	531
	1473	6, 3← 5, 3	Ground	11/2	6	9/2	5	25 128.15	.05	531
	1473	6, 3← 5, 3	Ground	13/2	7	11/2	6	25 129.47	.1	531
	1473	6, 3← 5, 3	Ground	13/2	6	11/2	5	25 129.47	.1	531
	1473	6, 3← 5, 3	Ground	13/2	5	11/2	4	25 130.93	.06	531
	1473	6, 3← 5, 3	Ground	13/2	8	11/2	7	25 130.93	.06	531
	1473	12, ←11,						50 246.1		332
$Re^{187}O_3^{16}Cl^{37}$	1474	12, ←11,						48 584.6		332
$Re^{b}O_3^{16}Cl^{35}$	1475	6, ← 5,						25 120.		332
$Re^{b}O_3^{16}Cl^{37}$	1476	6, ← 5,						24 290.		332

Cl_2O

C_{2v}

Cl_2O

Isotopic Species	Pt. Gp.	Id. No.	A MHz	B MHz	C MHz	D_J MHz	D_{JK} MHz	Δ Amu A^2	κ
$Cl_2^{35}O^{16}$	C_{2v}	1481	42 044. M	3 682. M	3 380. M				

Id. No.	μ_a Debye	μ_b Debye	μ_c Debye	eQq Value(MHz) Rel.	eQq Value(MHz) Rel.	eQq Value(MHz) Rel.	ω_a d 1/cm	ω_b d 1/cm	ω_c d 1/cm	ω_d d 1/cm
1481	0. X	1.69 T	0. X	−74 aa	66 bb	8 cc				

References:

ABC: 863 μ: 1032   eQq: 863

Chlorine Monoxide

Isotopic Species	Id. No.	Rotational Quantum Nos.	Vib. State	F_1'	F'	F_1	F	Frequency MHz	Acc. ±MHz	Ref.
Cl_2O^{16}	1484	1, 1, 0← 1, 0, 1	Ground					38 635.78		966
	1484	1, 1, 0← 1, 0, 1	Ground	5/2	4	5/2	4	38 649.45		966
	1484	1, 1, 0← 1, 0, 1	Ground	3/2	3	5/2	2	38 656.30		966
	1484	1, 1, 0← 1, 0, 1	Ground					38 659.44		966
	1484	1, 1, 0← 1, 0, 1	Ground	5/2	2	3/2	3	38 661.00		966
	1484	1, 1, 0← 1, 0, 1	Ground		1		2	38 661.00		966
	1484	1, 1, 0← 1, 0, 1	Ground	5/2	2	5/2	2	38 663.02		966
	1484	1, 1, 0← 1, 0, 1	Ground		1		2	38 663.02		966
	1484	1, 1, 0← 1, 0, 1	Ground		0		1	38 675.09		966
	1484	1, 1, 0← 1, 0, 1	Ground	5/2	2	3/2	2	38 675.09		966
	1484	1, 1, 0← 1, 0, 1	Ground	3/2	3	5/2	4	38 675.09		966
	1484	1, 1, 0← 1, 0, 1	Ground	5/2	4	3/2	3	38 678.09		966
	1484	1, 1, 0← 1, 0, 1	Ground	3/2	2	5/2	2	38 678.09		966
	1484	1, 1, 0← 1, 0, 1	Ground		1		0	38 678.09		966
	1484	1, 1, 0← 1, 0, 1	Ground	3/2	2	3/2	2	38 686.52		966
	1484	1, 1, 0← 1, 0, 1	Ground	3/2	3	3/2	2	38 694.43		966
	1484	1, 1, 0← 1, 0, 1	Ground	3/2	2	3/2	3	38 694.43		966
	1484	1, 1, 0← 1, 0, 1	Ground	3/2	3	3/2	3	38 694.43		966
	1484	1, 1, 1← 2, 0, 2	Ground					24 233.95		966
	1484	2, 1, 1← 2, 0, 2	Ground	3/2	3	7/2	2	38 947.12		966
	1484	2, 1, 1← 2, 0, 2	Ground	3/2	1	7/2	2	38 947.12		966
	1484	2, 1, 1← 2, 0, 2	Ground	7/2	2	7/2	2	38 948.98		966
	1484	2, 1, 1← 2, 0, 2	Ground	5/2	2	3/2	3	38 964.53		966
	1484	2, 1, 1← 2, 0, 2	Ground	5/2	2	3/2	1	38 964.53		966
	1484	2, 1, 1← 2, 0, 2	Ground	3/2	1	3/2	1	38 966.29		966
	1484	2, 1, 1← 2, 0, 2	Ground	3/2	0	3/2	1	38 966.29		966
	1484	2, 1, 1← 2, 0, 2	Ground	3/2	1	3/2	0	38 966.29		966
	1484	2, 1, 1← 2, 0, 2	Ground	3/2	3	7/2	4	38 966.29		966
	1484	2, 1, 1← 2, 0, 2	Ground	7/2	4	3/2	3	38 966.29		966
	1484	2, 1, 1← 2, 0, 2	Ground	3/2	3	3/2	3	38 966.29		966
	1484	2, 1, 1← 2, 0, 2	Ground	7/2	4	7/2	4	38 966.29		966
	1484	2, 1, 1← 2, 0, 2	Ground	3/2	0	3/2	0	38 966.29		966
	1484	2, 1, 1← 2, 0, 2	Ground	7/2	2	3/2	1	38 966.31		966
	1484	2, 1, 1← 2, 0, 2	Ground	7/2	2	3/2	3	38 966.31		966

Isotopic Species	Id. No.	Rotational Quantum Nos.	Vib. State	F_1'	F'	F_1	F	Frequency MHz	Acc. ±MHz	Ref.
$Cl_2^b O^{16}$	1484	2, 1, 1← 2, 0, 2	Ground	5/2	2	5/2	2	38 983.65		966
	1484	2, 1, 1← 2, 0, 2	Ground	3/2	3	5/2	2	38 985.48		966
	1484	2, 1, 1← 2, 0, 2	Ground	3/2	1	5/2	2	38 985.48		966
	1484	3, 1, 2← 3, 0, 3	Ground	3/2	2	3/2	1	39 406.50		966
	1484	3, 1, 2← 3, 0, 3	Ground	9/2	4	3/2	3	39 406.50		966
	1484	3, 1, 2← 3, 0, 3	Ground	7/2	5	9/2	6	39 408.03		966
	1484	3, 1, 2← 3, 0, 3	Ground	3/2	2	3/2	3	39 409.46		966
	1484	3, 1, 2← 3, 0, 3	Ground	3/2	0	3/2	1	39 412.14		966
	1484	3, 1, 2← 3, 0, 3	Ground	3/2	3	3/2	3	39 412.14		966
	1484	3, 1, 2← 3, 0, 3	Ground	9/2	6	9/2	6	39 416.36		966
	1484	3, 1, 2← 3, 0, 3	Ground	5/2	3	9/2	4	39 418.40		966
	1484	3, 1, 2← 3, 0, 3	Ground	7/2	5	9/2	4	39 418.40		966
	1484	3, 1, 2← 3, 0, 3	Ground	9/2	4	9/2	4	39 422.61		966
	1484	3, 1, 2← 3, 0, 3	Ground	5/2	3	3/2	2	39 424.22		966
	1484	3, 1, 2← 3, 0, 3	Ground	3/2	2	3/2	2	39 424.22		966
	1484	3, 1, 2← 3, 0, 3	Ground	7/2	4	5/2	3	39 425.92		966
	1484	3, 1, 2← 3, 0, 3	Ground	3/2	1	3/2	2	39 425.92		966
	1484	3, 1, 2← 3, 0, 3	Ground	3/2	3	3/2	2	39 425.92		966
	1484	3, 1, 2← 3, 0, 3	Ground	3/2	3	9/2	4	39 428.66		966
	1484	3, 1, 2← 3, 0, 3	Ground	7/2	4	7/2	5	39 428.66		966
	1484	3, 1, 2← 3, 0, 3	Ground	5/2	3	5/2	3	39 430.00		966
	1484	3, 1, 2← 3, 0, 3	Ground	7/2	5	7/2	5	39 432.12		966
	1484	3, 1, 2← 3, 0, 3	Ground	7/2	2	7/2	2	39 432.78		966
	1484	3, 1, 2← 3, 0, 3	Ground	9/2	4	7/2	5	39 437.30		966
	1484	3, 1, 2← 3, 0, 3	Ground	3/2	2	5/2	3	39 437.30		966
	1484	3, 1, 2← 3, 0, 3	Ground	7/2	4	7/2	4	39 438.70		966
	1484	3, 1, 2← 3, 0, 3	Ground	7/2	5	7/2	4	39 440.42		966
	1484	3, 1, 2← 3, 0, 3	Ground	9/2	6	7/2	5	39 440.42		966
	1484	3, 1, 2← 3, 0, 3	Ground	5/2	3	7/2	4	39 443.04		966
	1484	4, 1, 3← 4, 0, 4	Ground	7/2	3	11/2	4	40 022.64		966
	1484	4, 1, 3← 4, 0, 4	Ground	7/2	5	11/2	4	40 022.64		966
	1484	4, 1, 3← 4, 0, 4	Ground	11/2	4	11/2	4	40 030.56		966
	1484	4, 1, 3← 4, 0, 4	Ground	9/2	4	7/2	3	40 032.76		966
	1484	4, 1, 3← 4, 0, 4	Ground	9/2	4	7/2	5	40 032.76		966
	1484	4, 1, 3← 4, 0, 4	Ground	7/2	2	7/2	2	40 040.96		966
	1484	4, 1, 3← 4, 0, 4	Ground	7/2	2	7/2	3	40 040.96		966
	1484	4, 1, 3← 4, 0, 4	Ground	7/2	3	7/2	2	40 040.96		966
	1484	4, 1, 3← 4, 0, 4	Ground	7/2	5	11/2	6	40 040.96		966
	1484	4, 1, 3← 4, 0, 4	Ground	7/2	5	7/2	5	40 040.96		966
	1484	4, 1, 3← 4, 0, 4	Ground	11/2	6	11/2	6	40 040.96		966
	1484	4, 1, 3← 4, 0, 4	Ground	11/2	6	7/2	5	40 040.96		966
	1484	4, 1, 3← 4, 0, 4	Ground	7/2	3	7/2	3	40 040.98		966
	1484	4, 1, 3← 4, 0, 4	Ground	11/2	4	7/2	5	40 049.20		966
	1484	4, 1, 3← 4, 0, 4	Ground	11/2	4	7/2	3	40 049.20		966
	1484	4, 1, 3← 4, 0, 4	Ground	9/2	4	9/2	4	40 051.18		966
	1484	4, 1, 3← 4, 0, 4	Ground	7/2	3	9/2	4	40 059.51		966
	1484	4, 1, 3← 4, 0, 4	Ground	7/2	5	9/2	4	40 059.51		966
	1484	9, 0, 9← 8, 1, 8	Ground					29 980.5		966
	1484	9, 0, 9← 8, 1, 8	Ground					29 982.10		966
	1484	9, 0, 9← 8, 1, 8	Ground					29 983.12		966

Isotopic Species	Id. No.	Rotational Quantum Nos.	Vib. State	Hyperfine F₁′	F′	F₁	F	Frequency MHz	Acc. ±MHz	Ref.
Cl₂ᵇO¹⁶	1484	9, 0, 9← 8, 1, 8	Ground					29 987.0		966
	1484	9, 0, 9← 8, 1, 8	Ground					29 987.5		966
	1484	9, 0, 9← 8, 1, 8	Ground					29 989.38		966
	1484	9, 0, 9← 8, 1, 8	Ground					29 991.62		966
	1484	9, 0, 9← 8, 1, 8	Ground					29 992.70		966
	1484	9, 0, 9← 8, 1, 8	Ground					29 984.05		966
	1484	9, 2, 8←10, 1, 9	Ground					36 739.68		966
	1484	10, 0,10← 9, 1, 9	Ground					38 165.30		966
	1484	10, 2, 9←11, 1,10	Ground					28 072.16		966
	1484	10, 2, 9←11, 1,10	Ground					28 073.35		966
	1484	10, 2, 9←11, 1,10	Ground					28 076.76		966
	1484	10, 2, 9←11, 1,10	Ground					28 078.35		966
	1484	10, 2, 9←11, 1,10	Ground					28 078.56		966
	1484	10, 2, 9←11, 1,10	Ground					28 080.28		966
	1484	10, 2, 9←11, 1,10	Ground					28 083.75		966
	1484	10, 2, 9←11, 1,10	Ground					28 084.74		966
	1484	12, 2,10←13, 1,13	Ground					39 446.28		966
	1484	13, 2,11←14, 1,14	Ground					35 190.95		966
	1484	13, 2,11←14, 1,14	Ground					35 200.66		966
	1484	13, 2,11←14, 1,14	Ground					35 202.77		966
	1484	13, 2,11←14, 1,14	Ground					35 204.82		966
	1484	13, 2,11←14, 1,14	Ground					35 212.77		966
	1484	17, 1,16←16, 2,15	Ground					26 575.2		966
	1484	17, 1,16←16, 2,15	Ground					26 577.6		966
	1484	17, 1,16←16, 2,15	Ground					26 580.9		966
	1484	17, 1,16←16, 2,15	Ground					26 584.0		966
	1484	17, 1,16←16, 2,15	Ground					26 587.2		966
	1484	17, 1,16←16, 2,15	Ground					26 589.5		966
	1484	18, 1,17←17, 2,16	Ground					36 100.46		966
	1484	22, 3,19←23, 2,22	Ground					36 746.02		966
	1484	24, 3,21←25, 2,24	Ground					25 402.94		966

1490 − Trichlorogermane
       Germanium Chloroform   Molecular Constant Table

Cl₃GeH   C₃ᵥ   GeHCl₃

Isotopic Species	Pt. Gp.	Id. No.	A MHz	B MHz	C MHz	D_J MHz	D_JK MHz	Δ Amu A²	κ
Ge⁷⁰HCl₃³⁵	C₃ᵥ	1491	2 172.75 M	2 172.75 M	1 231.22 M	< .002	< .004	·	
Ge⁷²HCl₃³⁵	C₃ᵥ	1492	2 169.26 M	2 169.26 M	1 231.22 M				
Ge⁷⁴HCl₃³⁵	C₃ᵥ	1493	2 165.84 M	2 165.84 M	1 231.22 M				
Ge⁷⁰HCl₃³⁷	C₃ᵥ	1494	2 063.74 M	2 063.74 M	1 164.66 M				
Ge⁷²HCl₃³⁷	C₃ᵥ	1495	2 060.43 M	2 060.43 M	1 164.66 M				
Ge⁷⁴HCl₃³⁷	C₃ᵥ	1496	2 057.20 M	2 057.20 M	1 164.66 M				

References:

ABC: 484   D_J: 484   D_JK: 484

Add. Ref. 485

Isotopic Species	Id. No.	Rotational Quantum Nos.	Vib. State	F_1'	F'	F_1	F	Frequency MHz	Acc. ±MHz	Ref.
$Ge^{70}HCl_3^{35}$	1491	5, ← 4,	Ground					21 728.34	.07	484
	1491	5, ← 4,	Excited					21 733.38	.07	484
	1491	6, ← 5,	Ground					26 073.24	.07	484
	1491	6, ← 5,	Excited					26 080.10	.07	484
	1491	7, ← 6,	Ground					30 418.73	.07	484
	1491	7, ← 6,	Excited					30 426.84	.07	484
	1491	8, ← 7,	Ground					34 763.48	.07	484
	1491	8, ← 7,	Excited					34 771.98	.07	484
	1491	8, ← 7,	Excited					34 777.01	.07	484
	1491	9, ← 8,	Ground					39 109.31	.07	484
	1491	9, ← 8,	Excited					39 118.91	.07	484
$Ge^{72}HCl_3^{35}$	1492	5, ← 4,	Ground					21 693.05	.07	484
	1492	5, ← 4,	Excited					21 696.76	.07	484
	1492	6, ← 5,	Ground					26 031.05	.07	484
	1492	6, ← 5,	Excited					26 035.67	.07	484
	1492	7, ← 6,	Ground					30 370.06	.07	484
	1492	7, ← 6,	Excited					30 376.20	.07	484
	1492	8, ← 7,	Ground					34 707.67	.07	484
	1492	8, ← 7,	Excited					34 715.29	.07	484
	1492	8, ← 7,	Excited					34 719.42	.07	484
	1492	9, ← 8,	Ground					39 047.07	.07	484
	1492	9, ← 8,	Excited					39 055.31	.07	484
$Ge^{74}HCl_3^{35}$	1493	5, ← 4,	Ground					21 659.15	.07	484
	1493	5, ← 4,	Excited					21 662.45	.07	484
	1493	6, ← 5,	Ground					25 990.93	.07	484
	1493	6, ← 5,	Excited					25 995.03	.07	484
	1493	7, ← 6,	Ground					30 321.85	.07	484
	1493	7, ← 6,	Excited					30 327.36	.07	484
	1493	7, ← 6,	Excited					30 332.33	.07	484
	1493	8, ← 7,	Ground					34 652.66	.07	484
	1493	8, ← 7,	Excited					34 659.03	.07	484
	1493	8, ← 7,	Excited					34 664.83	.07	484
	1493	9, ← 8,	Ground					38 984.86	.07	484
	1493	9, ← 8,	Excited					38 992.19	.07	484
$Ge^{70}HCl_3^{37}$	1494	9, ← 8,	Ground					37 147.40	.07	484
$Ge^{72}HCl_3^{37}$	1495	8, ← 7,	Ground					32 967.08	.07	484
	1495	9, ← 8,	Ground					37 087.62	.07	484
	1495	10, ← 9,	Ground					41 208.73	.07	484
$Ge^{74}HCl_3^{37}$	1496	8, ← 7,	Ground					32 915.41	.07	484
	1496	9, ← 8,	Ground					37 029.43	.07	484
	1496	10, ← 9,	Ground					41 144.02	.07	484

Cl₃HSi $\qquad$ C_{3v} $\qquad$ SiHCl₃

Isotopic Species	Pt. Gp.	Id. No.	A MHz	B MHz	C MHz	D_J MHz	D_{JK} MHz	Δ Amu A²	κ
Si²⁸HCl₃³⁵	C₃ᵥ	1501	2 472.489 M	2 472.489 M			< .010		
Si²⁸HCl₃³⁷	C₃ᵥ	1504	2 346.071 M	2 346.071 M					

Id. No.	μ_a Debye	μ_b Debye	μ_c Debye	eQq Value(MHz) Rel.	eQq Value(MHz) Rel.	eQq Value(MHz) Rel.	ω_a d 1/cm	ω_b d 1/cm	ω_c d 1/cm	ω_d d 1/cm
1501	0. X	0. X	.858 G							

References:

ABC: 462 $\qquad$ D_{JK}: 462 $\qquad$ μ: 1030

Add. Ref. 378

Trichlorosilane $\hfill$ Spectral Line Table

Isotopic Species	Id. No.	Rotational Quantum Nos.	Vib. State	F_1'	F'	F_1	F	Frequency MHz	Acc. ±MHz	Ref.
Si²⁸HCl₃³⁵	1501	6, ← 5,	Ground					29 669.87	.5	462
	1501	7, ← 6,	Ground					34 614.34	.5	462
	1501	8, ← 7,	Ground					39 560.	10.	462
	1501	9, ← 8,	Ground					44 500.	10.	462
Si²⁸HCl₂³⁵Cl³⁷	1502	6, , ← 5, ,	Ground					29 140.	20.	462
	1502	7, , ← 6, ,	Ground					34 000.	20.	462
	1502	8, , ← 7, ,	Ground					38 834.	1.	462
	1502	9, , ← 8, ,	Ground					43 690.	10.	462
Si²⁸HCl³⁵Cl₂³⁷	1503	8, , ← 7, ,	Ground					38 180.	10.	462
Si²⁸HCl₃³⁷	1504	6, ← 5,	Ground					28 152.85	.5	462
	1504	7, ← 6,	Ground					32 845.02	.5	462

Cl$_3$OP $\qquad$ C$_{3v}$ $\qquad$ POCl$_3$

Isotopic Species	Pt. Gp.	Id. No.	A MHz	B MHz	C MHz	D$_J$ MHz	D$_{JK}$ MHz	Δ Amu A^2	κ
P^{31}O^{16}Cl$_3^{35}$	C$_{3v}$	1511		2 015.20 M	2 015.20 M				
P^{31}O^{16}Cl$_3^{37}$	C$_{3v}$	1512		1 932.38 M	1 932.38 M				

Id. No.	μ$_a$ Debye	μ$_b$ Debye	μ$_c$ Debye	eQq Value(MHz) Rel.		eQq Value(MHz) Rel.		eQq Value(MHz) Rel.		ω$_a$ 1/cm	d	ω$_b$ 1/cm	d	ω$_c$ 1/cm	d	ω$_d$ 1/cm	d
1511	2.39 L	0. X	0. X														

References:

ABC: 411 $\qquad$ μ: 995

Phosphoryl Chloride $\hfill$ Spectral Line Table

Isotopic Species	Id. No.	Rotational Quantum Nos.	Vib. State	F$_1'$	F'	F$_1$	F	Frequency MHz	Acc. ±MHz	Ref.
P^{31}O^{16}Cl$_3^{35}$	1511	7, ← 6,	Ground					28 212.8	.5	411
	1511	8, ← 7,	Ground					32 242.9	.5	411
	1511	9, ← 8,	Ground					36 273.5	.5	411
P^{31}O^{16}Cl$_3^{37}$	1512	7, ← 6,	Ground					27 052.0	.5	411
	1512	8, ← 7,	Ground					30 918.4	.5	411
	1512	9, ← 8,	Ground					34 783.0	.5	411
	1512	10, ← 9,	Ground					38 648.6	.5	411

Cl_3P C_{3v} PCl_3

Isotopic Species	Pt. Gp.	Id. No.	A MHz	B MHz		C MHz		D_J MHz	D_{JK} MHz	Δ Amu A²	κ
$P^{31}Cl_3^{35}$	C_{3v}	1521		2617.1	M	2617.1	M				
$P^{31}Cl_3^{37}$	C_{3v}	1524		2487.5	M	2487.5	M				

Id. No.	μ_a Debye		μ_b Debye		μ_c Debye		eQq Value(MHz)	Rel.	eQq Value(MHz)	Rel.	eQq Value(MHz)	Rel.	ω_a 1/cm	d	ω_b 1/cm	d	ω_c 1/cm	d	ω_d 1/cm	d
1521	.80	M	0.	X	0.	X														

References:

ABC: 206 μ: 1029

Add. Ref. 207

Phosphorus Trichloride Spectral Line Table

Isotopic Species	Id. No.	Rotational Quantum Nos.	Vib. State	F_1'	F'	F_1	F	Frequency MHz	Acc. ±MHz	Ref.
$P^{31}Cl_3^{35}$	1521	5, ← 4,	Excited					26 152.	10.	206
	1521	5, ← 4,	Ground					26 171.	1.	206
	1521	5, ← 4,	Excited					26 190.	10.	206
	1521	5, ← 4,	Excited					26 210.	10.	206
$P^{31}Cl_2^{35}Cl^{37}$	1522	5, 1, 4← 4, 0, 4	Ground					25 725.	10.	206
	1522	5, 2, 3← 4, 1, 3	Ground					25 716.	10.	206
	1522	5, 2, 4← 4, 1, 4	Ground					25 725.	10.	206
	1522	5, 3, 2← 4, 2, 2	Ground					25 649.	10.	206
	1522	5, 3, 3← 4, 2, 3	Ground					25 725.	10.	206
	1522	5, 4, 1← 4, 3, 1	Ground					25 552.	1.	206
	1522	5, 4, 2← 4, 3, 2	Ground					25 748.	10.	206
	1522	5, 5, 1← 4, 4, 1	Ground					25 971.	1.	206
$P^{31}Cl^{35}Cl_2^{37}$	1523	5, , ← 4, ,	Ground					25 306.	10.	206
$P^{31}Cl_3^{37}$	1524	5, ← 4,	Ground					24 875.	1.	206

Cl₃PS   C$_{3v}$ SPCl₃

Isotopic Species	Pt. Gp.	Id. No.	A MHz	B MHz	C MHz	D$_J$ MHz	D$_{JK}$ MHz	Δ Amu A²	κ
S^{32}P^{31}Cl$_3^{35}$	C$_{3v}$	1531		1 402.64 M	1 402.64 M				
S^{32}P^{31}Cl$_2^{35}$Cl^{37}	C$_s$	1532	1 470. M	1 397. M	1 375. M				−.52
S^{32}P^{31}Cl^{35}Cl$_2^{37}$	C$_s$	1533							−.41
S^{32}P^{31}Cl$_3^{37}$	C$_{3v}$	1534		1 355.72 M	1 355.72 M				
S^{34}P^{31}Cl$_3^{35}$	C$_{3v}$	1535		1 370.13 M	1 370.13 M				

Id. No.	μ$_a$ Debye	μ$_b$ Debye	μ$_c$ Debye	eQq Value(MHz) Rel.	eQq Value(MHz) Rel.	eQq Value(MHz) Rel.	ω$_a$ d 1/cm	ω$_b$ d 1/cm	ω$_c$ d 1/cm	ω$_d$ d 1/cm
1531	1.41 [1]L	0. X	0. X							

1. For molecules with identical chlorine nuclei, the axis of least moment of inertia (a axis) is the figure axis; hence for these molecules, $\mu_b = \mu_c = 0$. C. P. Smyth et al. found the dipole moment of an isotopic mixture to be 1.41 Debye, using classical techniques. In the absence of Stark effect data, we have taken this value for μ_a. However, according to Itoh, "For SPCl$_2^{35}$Cl^{37}, $\mu_b^2 = .02$D, μ_a^2 and $\mu_c^2 = 0$" which seems inconsistent with the above.

References:

ABC: 411,613   κ: 641   μ: 613,1032

Add. Ref. 536

Thiophosphoryl Chloride   Spectral Line Table

Isotopic Species	Id. No.	Rotational Quantum Nos.	Vib. State	F$_1'$	F'	F$_1$	F	Frequency MHz	Acc. ±MHz	Ref.
S^{32}P^{31}Cl$_3^{35}$	1531	10, ← 9,	Ground					28 053.1	.5	411
	1531	11, ←10,	Ground					30 857.9	.5	411
	1531	12, ←11,	Ground					33 662.9	.5	411
S^{32}P^{31}Cl$_3^{37}$	1534	11, ←10,	Ground					29 825.5	1.0	411
	1534	12, ←11,	Ground					32 537.4	1.0	411
S^{34}P^{31}Cl$_3^{35}$	1535	11, ←10,	Ground					30 143.2	1.0	411
	1535	12, ←11,	Ground					32 882.4	1.0	411
S^{b}P^{b}Cl$_3^b$	1536	Not Reported						2 629.		641 [1]
	1536	Not Reported						2 647.		641 [1]
	1536	Not Reported						2 689.		641 [1]
	1536	Not Reported						2 710.		641 [1]
	1536	Not Reported						2 760.		641 [1]
	1536	Not Reported						2 805.		641 [1]

1. For an evaluation of the unassigned lines, see the 1955 article by Tsukada.

Cl₃Sb   C₃ᵥ   SbCl₃

Isotopic Species	Pt. Gp.	Id. No.	A MHz	B MHz	C MHz	D_J MHz	D_{JK} MHz	Δ Amu A²	κ
$Sb^{121}Cl_3^{35}$	C_{3v}	1541	1 753.9 M	1 753.9 M					
$Sb^{123}Cl_3^{35}$	C_{3v}	1542	1 750.7 M	1 750.7 M					

Id. No.	μ_a Debye	μ_b Debye	μ_c Debye	eQq Value(MHz) Rel.	eQq Value(MHz) Rel.	eQq Value(MHz) Rel.	ω_a d 1/cm	ω_b d 1/cm	ω_c d 1/cm	ω_d d 1/cm
1541	0. X	0. X	3.9 M							

References:

ABC: 446   μ: 534

Antimony Trichloride   Spectral Line Table

Isotopic Species	Id. No.	Rotational Quantum Nos.	Vib. State	F_1'	F'	F_1	F	Frequency MHz	Acc. ±MHz	Ref.
$Sb^{121}Cl_3^{35}$	1541	7, ← 6,	Ground					24 554.	10.0	446
$Sb^{123}Cl_3^{35}$	1542	7, ← 6,	Ground					24 510.	10.0	446

FH$_3$Si

C$_{3v}$

SiH$_3$F

Isotopic Species	Pt. Gp.	Id. No.	A MHz	B MHz		C MHz		D$_J$ MHz	D$_{JK}$ MHz	Δ Amu A²	κ
Si^{28}H$_3$F^{19}	C$_{3v}$	1551		14 327.9	M	14 327.9	M				
Si^{29}H$_3$F^{19}	C$_{3v}$	1552		14 196.7	M	14 196.7	M				
Si^{30}H$_3$F^{19}	C$_{3v}$	1553		14 072.6	M	14 072.6	M				
Si^{28}D$_3$F^{19}	C$_{3v}$	1554		12 253.5	M	12 253.5	M				
Si^{29}D$_3$F^{19}	C$_{3v}$	1555		12 176.1	M	12 176.1	M				
Si^{30}D$_3$F^{19}	C$_{3v}$	1556		12 102.2	M	12 102.2	M				

Id. No.	μ_a Debye	μ_b Debye	μ_c Debye	eQq Value(MHz)	Rel.	eQq Value(MHz)	Rel.	eQq Value(MHz)	Rel.	ω_a d 1/cm	ω_b d 1/cm	ω_c d 1/cm	ω_d d 1/cm
1551	1.268 M	0. X	0. X										

References:

ABC: 413 μ: 231

Add. Ref. **489, 493**

Fluorosilane

Spectral Line Table

Isotopic Species	Id. No.	Rotational Quantum Nos.	Vib. State	Hyperfine				Frequency MHz	Acc. ±MHz	Ref.
				F$_1'$	F'	F$_1$	F			
Si^{28}H$_3$F^{19}	1551	1, ← 0,	Ground					28 655.8	.1	413
Si^{29}H$_3$F^{19}	1552	1, ← 0,	Ground					28 393.4	.2	413
Si^{30}H$_3$F^{19}	1553	1, ← 0,	Ground					28 145.2	.2	413
Si^{28}D$_3$F^{19}	1554	1, ← 0,	Ground					24 507.0	.1	413
Si^{29}D$_3$F^{19}	1555	1, ← 0,	Ground					24 352.2	.1	413
Si^{30}D$_3$F^{19}	1556	1, ← 0,	Ground					24 204.5	.2	413

FMnO$_3$ C$_{3v}$ MnO$_3$F

Isotopic Species	Pt. Gp.	Id. No.	A MHz	B MHz	C MHz	D$_J$ MHz	D$_{JK}$ MHz	Δ Amu A^2	κ
Mn^{55}O$_3^{16}$F^{19}	C$_{3v}$	1561		4 129.106 M	4 129.106 M		.01		
Mn^{55}O$_2^{16}$O^{18}F^{19}	C$_s$	1562	4 488.81 M	4 098.088 M	3 963.552 M				

Id. No.	μ$_a$ Debye	μ$_b$ Debye	μ$_c$ Debye	eQq Value(MHz) Rel.	eQq Value(MHz) Rel.	eQq Value(MHz) Rel.	ω$_a$ d 1/cm	ω$_b$ d 1/cm	ω$_c$ d 1/cm	ω$_d$ d 1/cm
1561	1.5 M	0. X	0. X	16.8			350 2			

References:

ABC: 531   D$_{JK}$: 531   μ: 531   eQq: 531   ω: 531

Add. Ref. 360

For unidentified species of molecule 1560, ref. 531 gives in MHz:

Mode	v$_3$	v$_4$	v$_5$	v$_6$
α	7.77	14.38	−12.80	5.87
q		5.90	16.20	9.81

Manganese Trioxyfluoride   Spectral Line Table

Isotopic Species	Id. No.	Rotational Quantum Nos.	Vib. State v$_a^l$	F$_1'$	F'	F$_1$	F	Frequency MHz	Acc. ±MHz	Ref.
Mn^{55}O$_3^{16}$F^{19}	1561	3, 0 ← 2, 0	Ground		3/2		5/2	24 771.56	.1	531
	1561	3, 0 ← 2, 0	Ground		1/2		3/2	24 772.39	.1	531
	1561	3, 0 ← 2, 0	Ground		5/2		5/2	24 773.505	.05	531
	1561	3, 0 ← 2, 0	Ground		5/2		3/2	24 773.85	.15	531
	1561	3, 0 ← 2, 0	Ground		9/2		7/2	24 774.250	.05	531
	1561	3, 0 ← 2, 0	Ground		11/2		9/2	24 774.445	.05	531
	1561	3, 0 ← 2, 0	Ground		7/2		5/2	24 775.125	.05	531
	1561	3, 0 ← 2, 0	Ground		5/2		3/2	24 775.949	.05	531
	1561	3, 0 ← 2, 0	Ground		9/2		9/2	24 777.526	.05	531
	1561	3, 1 ← 2, 1	1, ±1		1/2		1/2	24 768.00	.05	531
	1561	3, 1 ← 2, 1	1, ±1		3/2		3/2	24 768.00	.05	531
	1561	3, 1 ← 2, 1	1, ±1		11/2		9/2	24 768.38	.05	531
	1561	3, 1 ← 2, 1	1, ±1		9/2		7/2	24 769.14	.05	531
	1561	3, 1 ← 2, 1	1, ±1		3/2		1/2	24 769.14	.05	531
	1561	3, 1 ← 2, 1	1, ±1		7/2		7/2	24 769.14	.05	531
	1561	3, 1 ← 2, 1	1, ±1		7/2		5/2	24 769.55	.05	531
	1561	3, 1 ← 2, 1	1, ±1		5/2		3/2	24 769.55	.05	531
	1561	3, 1 ← 2, 1	1, ±1		9/2		9/2	24 770.68	.05	531
	1561	3, 2 ← 2, 2	1, 0		9/2		9/2	24 738.20	.05	531
	1561	3, 2 ← 2, 2	1, 0		11/2		9/2	24 738.20	.05	531
	1561	3, 2 ← 2, 2	1, 0		1/2		3/2	24 738.20	.05	531
	1561	3, 2 ← 2, 2	1, 0		3/2		3/2	24 738.20	.05	531
	1561	3, 2 ← 2, 2	1, 0		5/2		3/2	24 738.20	.05	531
	1561	3, 2 ← 2, 2	1, 0		7/2		9/2	24 738.20	.05	531
	1561	3, 2 ← 2, 2	1, 0		5/2		5/2	24 740.60	.05	531

Isotopic Species	Id. No.	Rotational Quantum Nos.	Vib. State v_a^1	Hyperfine F_1'	F'	F_1	F	Frequency MHz	Acc. ±MHz	Ref.
$Mn^{55}O_3^{16}F^{19}$	1561	3, 2← 2, 2	1, 0		7/2		5/2	24 740.60	.05	531
	1561	3, 2← 2, 2	1, 0		3/2		5/2	24 740.60	.05	531
	1561	3, 2← 2, 2	1, 0		7/2		7/2	24 741.49	.05	531
	1561	3, 2← 2, 2	1, 0		9/2		7/2	24 741.49	.05	531
	1561	3, 2← 2, 2	1, 0		5/2		7/2	24 741.49	.05	531
	1561	3, 2← 2, 2	1, 0		1/2		1/2	24 755.95	.15	531
	1561	3, 2← 2, 2	1, 0		3/2		1/2	24 755.95	.15	531
$Mn^{55}O_2^{16}O^{18}F^{19}$	1562	3, 0, 3← 2, 0, 2	Ground		1/2		1/2	24 181.70	.150	531
	1562	3, 0, 3← 2, 0, 2	Ground		3/2		1/2	24 181.70	.150	531
	1562	3, 1, 2← 2, 1, 1	Ground		11/2		9/2	24 365.64	.030	531
	1562	3, 1, 2← 2, 1, 1	Ground		7/2		7/2	24 366.331	.060	531
	1562	3, 1, 2← 2, 1, 1	Ground		9/2		7/2	24 366.331	.060	531
	1562	3, 1, 2← 2, 1, 1	Ground		5/2		3/2	24 366.80	.050	531
	1562	3, 1, 2← 2, 1, 1	Ground		7/2		5/2	24 366.80	.050	531
	1562	3, 1, 2← 2, 1, 1	Ground		9/2		9/2	24 367.89	.030	531
	1562	3, 1, 3← 2, 1, 2	Ground		11/2		9/2	23 966.310	.040	531
	1562	3, 1, 3← 2, 1, 2	Ground		7/2		5/2	23 967.106	.040	531
	1562	3, 1, 3← 2, 1, 2	Ground		7/2		7/2	23 967.106	.040	531
	1562	3, 1, 3← 2, 1, 2	Ground		9/2		7/2	23 967.106	.040	531
	1562	3, 1, 3← 2, 1, 2	Ground		5/2		3/2	23 967.106	.040	531
	1562	3, 2, 1← 2, 2, 0	Ground		11/2		9/2	24 292.35	.08	531
	1562	3, 2, 1← 2, 2, 0	Ground		3/2		3/2	24 292.35	.08	531
	1562	3, 2, 1← 2, 2, 0	Ground		9/2		9/2	24 292.35	.08	531
	1562	3, 2, 1← 2, 2, 0	Ground		7/2		9/2	24 292.35	.08	531
	1562	3, 2, 1← 2, 2, 0	Ground		5/2		3/2	24 292.35	.08	531
	1562	3, 2, 1← 2, 2, 0	Ground		1/2		3/2	24 292.35	.08	531
	1562	3, 2, 1← 2, 2, 0	Ground		7/2		7/2	24 295.45	.1	531
	1562	3, 2, 1← 2, 2, 0	Ground		9/2		7/2	24 295.45	.1	531
	1562	3, 2, 1← 2, 2, 0	Ground		5/2		7/2	24 295.45	.1	531
	1562	3, 2, 2← 2, 2, 1	Ground		9/2		9/2	24 183.70	.05	531
	1562	3, 2, 2← 2, 2, 1	Ground		3/2		3/2	24 183.70	.05	531
	1562	3, 2, 2← 2, 2, 1	Ground		5/2		3/2	24 183.70	.05	531
	1562	3, 2, 2← 2, 2, 1	Ground		1/2		3/2	24 183.70	.05	531
	1562	3, 2, 2← 2, 2, 1	Ground		7/2		9/2	24 183.70	.05	531
	1562	3, 2, 2← 2, 2, 1	Ground		11/2		9/2	24 183.70	.05	531
	1562	3, 2, 2← 2, 2, 1	Ground		5/2		7/2	24 186.85	.1	531
	1562	3, 2, 2← 2, 2, 1	Ground		7/2		7/2	24 186.85	.1	531
	1562	3, 2, 2← 2, 2, 1	Ground		9/2		7/2	24 186.85	.1	531

Isotopic Species	Pt. Gp.	Id. No.	A MHz	B MHz	C MHz	D_J MHz	D_{JK} MHz	Δ Amu A²	κ
$N^{14}O^{16}F^{19}$	C_s	1571	95 191.73 M	11 843.91 M	10 508.45 M				−.968460

Id. No.	μ_a Debye	μ_b Debye	μ_c Debye	eQq Value(MHz) Rel.	eQq Value(MHz) Rel.	eQq Value(MHz) Rel.	ω_a d 1/cm	ω_b d 1/cm	ω_c d 1/cm	ω_d d 1/cm
1571	1.70 M	.62 M	0. X							

References:

ABC: 293 κ: 293 μ: 292

No Spectral Lines

FNO$_2$ $\qquad$ C$_{2v}$ $\qquad$ NO$_2$F

Isotopic Species	Pt. Gp.	Id. No.	A MHz	B MHz	C MHz	D$_J$ MHz	D$_{JK}$ MHz	Δ Amu A^2	κ
N^{14}O$_2^{16}$F^{19}	C$_{2v}$	1581	13 199.99 M	11 444.43 M	6 119.08 M				

Id. No.	μ$_a$ Debye	μ$_b$ Debye	μ$_c$ Debye	eQq Value(MHz)	Rel.	eQq Value(MHz)	Rel.	eQq Value(MHz)	Rel.	ω$_a$ d 1/cm	ω$_b$ d 1/cm	ω$_c$ d 1/cm	ω$_d$ d 1/cm
1581	.47 M	0. X	0. X	.7	aa	1.5	bb	−2.2	cc				

References:

ABC: 398 $\qquad$ μ: 398 $\qquad$ eQq: 398

Add. Ref. 668

Nitryl Fluoride $\hfill$ Spectral Line Table

Isotopic Species	Id. No.	Rotational Quantum Nos.	Vib. State	F$_1'$	F'	F$_1$	F	Frequency MHz	Acc. ±MHz	Ref.
N^{14}O$_2^{16}$F^{19}	1581	1, 0, 1← 0, 0, 0	Ground					17 566.2		1004
	1581	2, 0, 2← 1, 0, 1	Ground					31 192.3		1004
	1581	2, 2, 1← 2, 0, 2	Ground					21 613.7		1004
	1581	3, 2, 2← 3, 0, 3	Ground					31 282.1		1004
	1581	4, 2, 3← 4, 2, 2	Ground					27 409.7		1004
	1581	5, 4, 1← 5, 4, 2	Ground					7 623.1		1004
	1581	6, 4, 3← 6, 4, 2	Ground					20 155.1		1004
	1581	8, 6, 2← 8, 6, 3	Ground					11 120.0		1004

FO$_3$Re $\qquad$ C$_{3v}$ $\qquad$ ReO$_3$F

Isotopic Species	Pt. Gp.	Id. No.	A MHz	B MHz	C MHz	D$_J$ MHz	D$_{JK}$ MHz	Δ Amu A^2	κ
Re^{185}O$_3^{16}$F^{19}	C$_{3v}$	1591		3 566.801 M	3 566.801 M				
Re^{187}O$_3^{16}$F^{19}	C$_{3v}$	1592		3 566.751 M	3 566.751 M	.00036	.0024		
Re^{b}O$_2^{16}$O^{18}F^{19}	C$_s$	1593	3 983.98 M	3 542.24 M	3 426.33 M				.5843

Id. No.	μ_a Debye	μ_b Debye	μ_c Debye	eQq Value(MHz) Rel.	eQq Value(MHz) Rel.	eQq Value(MHz) Rel.	ω_a d 1/cm	ω_b d 1/cm	ω_c d 1/cm	ω_d d 1/cm
1591 1592	.85 M	0. X	0. X	−48.4 Re^{187}						

References:

ABC: 873 $\qquad$ D$_J$: 873 $\qquad$ D$_{JK}$: 873 $\qquad$ κ: 873 $\qquad$ μ: 873 $\qquad$ eQq: 873

Add. Ref. 541,829

For species 1592, the following parameters for various vibrational states have been reported:

State:	eQq (MHz)	α^B (MHz)	q$_l$ (MHz)
v$_3$ = 1:	−27.0	12.3	
v$_3$ = 2:	−17.0		
v$_3$ = 1, v$_6$ = 1:	−37.0		
v$_3$ = 2, v$_6$ = 1:	−25.7		
v$_5$ = 1:	−34.9	−10.91	16.31
v$_6$ = 1:	−58.2	2.52	5.00

B$_0^{185}$ − B$_0^{187}$ ≈ .050 MHz, eQq (Re^{185}) ≈ 1.07 eQq (Re^{187}). Ref. 873.

Rhenium Trioxyfluoride $\qquad$ Spectral Line Table

Isotopic Species	Id. No.	Rotational Quantum Nos.	Vib. State	F$_1'$	F'	F$_1$	F	Frequency MHz	Acc. ±MHz	Ref.
Re^{185}O$_3^{16}$F^{19}	1591	4, 3← 3, 3	Ground					28 537.27		873
Re^{187}O$_3^{16}$F^{19}	1592	4, 3← 3, 3	Ground					28 529.35		873
Re^{b}O$_2^{16}$O^{18}F^{19}	1593	4, 0, 4← 3, 0, 3	Ground					27 702.56	.20	873
	1593	4, 1, 3← 3, 1, 2	Ground					28 059.01	.20	873
	1593	4, 1, 4← 3, 1, 3	Ground					27 608.25	.20	873
	1593	4, 2, 2← 3, 2, 1	Ground					28 030.30	.20	873
	1593	4, 2, 3← 3, 2, 2	Ground					27 858.61	.20	873
Re^{b}O$_3^{16}$F^{19}	1594	3, ← 2,	Ground					21 400.	50.	873

F₂H₂Si   C₂ᵥ   SiH₂F₂

Isotopic Species	Pt. Gp.	Id. No.	A MHz	B MHz	C MHz	D_J MHz	D_{JK} MHz	Δ Amu A²	κ
$Si^{28}H_2F_2^{19}$	C_{2v}	1601	24690.70 M	7801.90 M	6377.09 M				−.844398
$Si^{29}H_2F_2^{19}$	C_{2v}	1602	24403.00 M		6357.58 M				−.839923
$Si^{28}D_2F_2^{19}$	C_{2v}	1603	18884.68 M	7447.42 M	6126.38 M				−.792913
$Si^{29}D_2F_2^{19}$	C_{2v}	1604	18739.91 M		6110.89 M				−.788348
$Si^{30}D_2F_2^{19}$	C_{2v}	1605	18606.18 M		6096.52 M				−.784024

Id. No.	μ_a Debye	μ_b Debye	μ_c Debye	eQq Value(MHz) Rel.	eQq Value(MHz) Rel.	eQq Value(MHz) Rel.	ω_a d 1/cm	ω_b d 1/cm	ω_c d 1/cm	ω_d d 1/cm
1601	0. X	1.54 M	0. X				322 1			
1603	0. X	1.53 M	0. X				322 1			

References:

ABC: 763   κ: 763   μ: 763   ω: 763

For $v_4 = 1$, ref. 763 gives:

Species	A (MHz)	B (MHz)	C (MHz)	κ
1601	24933.53	7799.72	6364.17	−.845384
1603	19012.95	7443.37	6113.65	−.793831

Isotopic Species	Id. No.	Rotational Quantum Nos.	v_a	Vib. State	F_1'	F'	F_1	F	Frequency MHz	Acc. ±MHz	Ref.
$Si^{28}H_2F_2^{19}$	1601	1, 1, 1← 0, 0, 0		Ground					31 067.80	.10	763
	1601	1, 1, 1← 0, 0, 0	1						31 297.65	.10	763
	1601	1, 1, 0← 1, 0, 1		Ground					18 313.54	.10	763
	1601	1, 1, 0← 1, 0, 1	1	Ground					18 569.32	.10	763
	1601	2, 1, 1← 2, 0, 2		Ground					19 824.84	.10	763
	1601	2, 1, 1← 2, 0, 2	1						20 091.49	.10	763
	1601	3, 0, 3← 2, 1, 2	1						26 363.39	.10	763
	1601	3, 0, 3← 2, 1, 2		Ground					26 642.93	.10	763
	1601	3, 1, 2← 3, 0, 3		Ground					22 249.96	.10	763
	1601	3, 1, 2← 3, 0, 3	1						22 533.09	.10	763
	1601	4, 1, 3← 4, 0, 4		Ground					25 761.92	.10	763
	1601	4, 1, 3← 4, 0, 4	1						26 067.51	.10	763
	1601	5, 1, 4← 4, 2, 3		Ground					28 190.00	.10	763
	1601	5, 1, 4← 5, 0, 5		Ground					30 548.92	.10	763
	1601	5, 1, 4← 5, 0, 5	1						30 884.00	.10	763
$Si^{29}H_2F_2^{19}$	1602	3, 1, 2← 3, 0, 3		Ground					22 046.65	.10	763
	1602	4, 1, 3← 4, 0, 4		Ground					25 625.45	.10	763
	1602	5, 1, 4← 5, 0, 5		Ground					30 509.25	.10	763
$Si^{28}D_2F_2^{19}$	1603	1, 1, 1← 0, 0, 0		Ground					25 011.17	.10	763
	1603	1, 1, 1← 0, 0, 0	1						25 126.60	.10	763
	1603	1, 1, 0← 1, 0, 1		Ground					12 758.30	.10	763
	1603	2, 0, 2← 1, 1, 1		Ground					15 602.20	.10	763
	1603	2, 1, 2← 1, 0, 1		Ground					37 263.80	.20	763
	1603	2, 1, 2← 1, 0, 1	1						37 353.90	.20	763
	1603	2, 1, 1← 2, 0, 2		Ground					14 187.40	.10	763
	1603	3, 0, 3← 2, 1, 2	1						29 894.05	.10	763
	1603	3, 0, 3← 2, 1, 2		Ground					30 069.60	.10	763
	1603	3, 1, 2← 3, 0, 3		Ground					16 525.65	.10	763
	1603	3, 1, 2← 3, 0, 3	1						16 689.30	.10	763
	1603	3, 2, 1← 3, 1, 2		Ground					32 935.18	.10	763
	1603	3, 2, 1← 3, 1, 2	1						33 320.42	.10	763
	1603	4, 1, 3← 3, 2, 2		Ground					24 302.40	.10	763
	1603	4, 1, 3← 4, 0, 4		Ground					19 972.85	.10	763
	1603	4, 1, 3← 4, 0, 4	1						20 155.80	.10	763
	1603	4, 2, 2← 4, 1, 3		Ground					31 481.59	.10	763
	1603	4, 2, 2← 4, 1, 3	1						31 852.40	.10	763
	1603	5, 1, 4← 5, 0, 5		Ground					24 712.75	.10	763
	1603	5, 1, 4← 5, 0, 5	1						24 921.75	.10	763
	1603	5, 2, 3← 5, 1, 4		Ground					30 441.65	.10	763
	1603	5, 2, 3← 5, 1, 4	1						30 796.10	.10	763
	1603	6, 1, 5← 6, 0, 6		Ground					30 828.35	.10	763
	1603	6, 1, 5← 6, 0, 6	1						31 072.20	.10	763
	1603	6, 2, 4← 6, 1, 5		Ground					30 186.00	.10	763
	1603	6, 2, 4← 6, 1, 5	1						30 524.85	.10	763
	1603	7, 2, 5← 7, 1, 6		Ground					31 024.20	.10	763
	1603	8, 2, 6← 8, 1, 7		Ground					33 195.45	.10	763
$Si^{29}D_2F_2^{19}$	1604	4, 1, 3← 4, 0, 4		Ground					19 955.10	.10	763
	1604	4, 2, 3← 4, 1, 4		Ground					31 054.52	.10	763
	1604	5, 1, 4← 5, 0, 5		Ground					24 774.80	.10	763
	1604	5, 2, 3← 5, 1, 4		Ground					30 049.00	.10	763
	1604	6, 2, 4← 6, 1, 5		Ground					29 859.10	.10	763
$Si^{30}D_2F_2^{19}$	1605	4, 1, 3← 4, 0, 4		Ground					19 940.69	.10	763
	1605	5, 1, 4← 5, 0, 5		Ground					24 836.28	.10	763

F₂O   C₂ᵥ   OF₂

Isotopic Species	Pt. Gp.	Id. No.	A MHz	B MHz	C MHz	D_J MHz	D_JK MHz	Δ Amu A²	κ
O¹⁶F₂¹⁹	C₂ᵥ	1611	58 782.63 M	10 896.43 M	9 167.412 M				

Id. No.	μ_a Debye	μ_b Debye	μ_c Debye	eQq Value(MHz) Rel.	eQq Value(MHz) Rel.	eQq Value(MHz) Rel.	ω_a d 1/cm	ω_b d 1/cm	ω_c d 1/cm	ω_d d 1/cm
1611	0. X	.297 M	0. X							

References:

ABC: 996 μ: 949

Oxygen Difluoride   Spectral Line Table

Isotopic Species	Id. No.	Rotational Quantum Nos.	Vib. State	F'₁	F'	F₁	F	Frequency MHz	Acc. ±MHz	Ref.
O¹⁶F₂¹⁹	1611	1, 1, 0← 1, 0, 1	Ground					49 613.55	.10	996
	1611	2, 1, 1← 2, 0, 2	Ground					51 388.95	.10	996
	1611	3, 0, 3← 2, 1, 2	Ground					13 804.03	.10	996
	1611	3, 1, 2← 3, 0, 3	Ground					54 137.05	.10	996
	1611	3, 2, 2← 4, 1, 3	Ground					57 457.10	.10	996
	1611	4, 0, 4← 3, 1, 3	Ground					36 028.73	.10	996
	1611	4, 1, 3← 4, 0, 4	Ground					57 958.30	.10	996
	1611	4, 2, 2← 5, 1, 5	Ground					59 846.20	.10	996
	1611	4, 2, 3← 5, 1, 4	Ground					33 251.80	.10	996
	1611	5, 0, 5← 4, 1, 4	Ground					58 725.50	.10	996
	1611	5, 2, 4← 6, 1, 5	Ground					8 299.51	.10	996
	1611	6, 2, 4← 7, 1, 7	Ground					34 044.45	.10	996
	1611	7, 1, 6← 6, 2, 5	Ground					17 354.71	.10	996
	1611	7, 2, 5← 8, 1, 8	Ground					23 842.38	.10	996
	1611	8, 2, 6← 9, 1, 9	Ground					15 771.44	.10	996
	1611	8, 3, 6← 9, 2, 7	Ground					52 302.40	.10	996
	1611	9, 2, 7←10, 1,10	Ground					10 057.72	.10	996
	1611	9, 3, 6←10, 2, 9	Ground					48 919.17	.10	996
	1611	9, 3, 7←10, 2, 8	Ground					27 495.80	.10	996
	1611	10, 3, 7←11, 2,10	Ground					31 774.50	.10	996
	1611	11, 3, 8←12, 2,11	Ground					15 624.82	.10	996
	1611	12, 2,10←11, 3, 9	Ground					25 527.21	.10	996
	1611	12, 2,10←13, 1,13	Ground					8 610.91	.10	996
	1611	13, 2,11←12, 3,10	Ground					53 708.85	.10	996
	1611	13, 2,11←14, 1, 4	Ground					13 575.51	.01	997
	1611	13, 2,11←14, 1,14	Ground					13 575.01	.01	997
	1611	13, 2,11←14, 1,14	Ground					13 575.28	.01	997
	1611	13, 4, 9←14, 3,12	Ground					58 183.80	.10	996
	1611	13, 4,10←14, 3,11	Ground					49 186.21	.10	996
	1611	14, 2,13←13, 3,10	Ground					12 782.59	.10	996
	1611	14, 2,12←15, 1,15	Ground					21 224.06	.10	996
	1611	14, 4,10←15, 3,13	Ground					38 408.88	.10	996
	1611	14, 4,11←15, 3,12	Ground					25 155.14	.10	996
	1611	15, 2,14←14, 3,11	Ground					24 533.39	.10	996
	1611	15, 2,13←16, 1,16	Ground					31 461.07	.10	996

Isotopic Species	Id. No.	Rotational Quantum Nos.	Vib. State	Hyperfine				Frequency MHz	Acc. ±MHz	Ref.
				F_1'	F'	F_1	F			
$O^{16}F_2^{19}$	1611	15, 4,11←16, 3,14	Ground					19 009.18	.10	996
	1611	16, 2,15←15, 3,12	Ground					34 291.10	.10	996
	1611	17, 3,14←16, 4,13	Ground					26 192.22	.10	996
	1611	17, 2,15←18, 1,18	Ground					59 137.55	.10	996
	1611	18, 3,15←17, 4,14	Ground					53 739.10	.10	996
	1611	18, 3,16←17, 4,13	Ground					18 095.67	.10	996
	1611	18, 5,13←19, 4,16	Ground					50 393.23	.10	996
	1611	19, 3,17←18, 4,14	Ground					35 455.35	.10	996
	1611	19, 5,14←20, 4,17	Ground					29 533.85	.10	996
	1611	19, 5,15←20, 4,16	Ground					22 378.53	.10	996
	1611	20, 3,18←19, 4,15	Ground					51 747.90	.10	996
	1611	20, 5,15←21, 4,18	Ground					8 782.91	.10	996
	1611	22, 2,21←21, 3,18	Ground					38 675.10	.10	996
	1611	22, 4,18←21, 5,17	Ground					26 359.08	.10	996
	1611	22, 4,19←21, 5,16	Ground					11 778.86	.10	996
	1611	23, 2,22←22, 3,19	Ground					29 365.40	.10	996
	1611	23, 4,19←22, 5,18	Ground					52 223.40	.10	996
	1611	23, 4,20←22, 5,17	Ground					32 050.75	.10	996
	1611	24, 2,23←23, 3,20	Ground					17 257.70	.01	997
	1611	24, 2,23←23, 3,20	Ground					17 257.96	.01	997
	1611	24, 2,23←23, 3,20	Ground					17 258.26	.01	997
	1611	24, 4,21←23, 5,18	Ground					51 911.88	.10	996
	1611	24, 6,18←25, 5,21	Ground					22 039.31	.10	996
	1611	24, 6,19←25, 5,20	Ground					18 474.80	.10	996
	1611	25, 3,22←26, 2,25	Ground					14 720.26	.01	997
	1611	25, 3,22←26, 2,25	Ground					14 720.63	.01	997
	1611	25, 3,22←26, 2,25	Ground					14 720.93	.01	997
	1611	27, 5,22←26, 6,21	Ground					28 286.80	.10	996
	1611	27, 5,23←26, 6,20	Ground					20 910.24	.10	996
	1611	27, 7,20←28, 6,23	Ground					58 772.25	.10	996
	1611	27, 7,21←28, 6,22	Ground					58 013.45	.10	996
	1611	28, 5,23←27, 6,22	Ground					52 676.00	.10	996
	1611	28, 7,21←29, 6,24	Ground					37 184.65	.10	996
	1611	28, 7,22←29, 6,23	Ground					36 045.54	.10	996
	1611	29, 7,22←30, 6,25	Ground					15 480.54	.10	996
	1611	29, 7,23←30, 6,24	Ground					13 797.53	.10	996
	1611	31, 6,25←30, 7,24	Ground					8 779.55	.10	996
	1611	32, 6,26←31, 7,25	Ground					31 744.90	.10	996
	1611	32, 6,27←31, 7,24	Ground					28 231.06	.10	996
	1611	32, 8,24←33, 7,27	Ground					52 988.75	.10	996
	1611	32, 8,25←33, 7,26	Ground					52 643.25	.10	996
	1611	33, 6,27←32, 7,26	Ground					55 170.63	.10	996
	1611	33, 6,28←32, 7,25	Ground					50 198.50	.10	996
	1611	33, 8,25←34, 7,28	Ground					31 325.95	.10	996
	1611	33, 8,26←34, 7,27	Ground					30 808.75	.10	996
	1611	34, 8,26←35, 7,29	Ground					9 533.43	.10	996
	1611	34, 8,27←35, 7,28	Ground					8 769.86	.10	996
	1611	36, 7,29←35, 8,28	Ground					13 500.68	.10	996
	1611	36, 7,30←35, 8,27	Ground					12 387.79	.10	996
	1611	36, 4,32←37, 3,35	Ground					29 473.73	.10	996
	1611	37, 7,30←36, 8,29	Ground					36 036.82	.10	996
	1611	37, 7,31←36, 8,28	Ground					34 434.95	.10	996
	1611	38, 7,31←37, 8,30	Ground					58 879.50	.10	996
	1611	38, 7,32←37, 8,29	Ground					56 600.15	.10	996
	1611	38, 9,29←39, 8,32	Ground					25 796.85	.10	996
	1611	38, 9,30←39, 8,31	Ground					25 569.29	.10	996

F$_2$OS

C$_s$

SOF$_2$

Isotopic Species	Pt. Gp.	Id. No.	A MHz	B MHz	C MHz	D$_J$ MHz	D$_{JK}$ MHz	Δ Amu A²	κ
S^{32}O^{16}F$_2^{19}$	C$_s$	1621	.8614.75 M	8356.98 M	4952.96 M				.859213
S^{32}O^{18}F$_2^{19}$	C$_s$	1622	8582.33 M	7843.37 M	4777.90 M				.611526

Id. No.	μ$_a$ Debye	μ$_b$ Debye	μ$_c$ Debye	eQq Value(MHz) Rel.	eQq Value(MHz) Rel.	eQq Value(MHz) Rel.	ω$_a$ d 1/cm	ω$_b$ d 1/cm	ω$_c$ d 1/cm	ω$_d$ d 1/cm
1621	1.618 M	0. X								

References:

ABC: 517   κ: 517   μ: 517

Add. Ref. 433

Thionyl Fluoride

Spectral Line Table

Isotopic Species	Id. No.	Rotational Quantum Nos.	Vib. State	F$_1'$	F'	F$_1$	F	Frequency MHz	Acc. ±MHz	Ref.
S^{32}O^{16}F$_2^{19}$	1621	1, 1, 0← 0, 0, 0	Ground					16 971.79	.1	517
	1621	2, 0, 2← 1, 0, 1	Ground					23 459.49	.1	517
	1621	2, 1, 1← 1, 0, 1	Ground					33 685.69	.1	517
	1621	2, 1, 1← 1, 1, 0	Ground					30 023.90	.1	517
	1621	2, 1, 2← 1, 1, 1	Ground					23 216.06	.1	517
	1621	2, 2, 0← 1, 1, 0	Ground					33 957.43	.1	517
	1621	2, 2, 1← 1, 1, 1	Ground					34 201.22	.1	517
	1621	3, 0, 3← 2, 0, 2	Ground					33 255.43	.1	517
	1621	3, 1, 3← 2, 1, 2	Ground					33 242.36	.1	517
	1621	4, 1, 3← 4, 1, 4	Ground					24 703.35	.1	517
	1621	4, 2, 2← 4, 2, 3	Ground					17 500.91	.1	517
	1621	5, 2, 3← 5, 2, 4	Ground					24 682.55	.1	517
	1621	5, 3, 2← 5, 3, 3	Ground					17 304.89	.1	517
	1621	6, 3, 3← 6, 3, 4	Ground					24 642.79	.1	517
	1621	6, 4, 2← 6, 4, 3	Ground					16 980.58	.1	517
	1621	7, 4, 3← 7, 4, 4	Ground					24 553.93	.1	517
	1621	8, 5, 3← 8, 5, 4	Ground					24 419.36	.1	517
	1621	9, 6, 3← 9, 6, 4	Ground					24 205.50	.1	517
	1621	10, 7, 3←10, 7, 4	Ground					23 879.57	.1	517
	1621	11, 8, 3←11, 8, 4	Ground					23 409.65	.1	517
	1621	12, 9, 3←12, 9, 4	Ground					22 765.11	.1	517
	1621	12,12, 1←12,11, 1						18 053.13	.1	675
	1621	13,10, 3←13,10, 4	Ground					21 924.71	.1	517
	1621	14,11, 3←14,11, 4	Ground					20 881.41	.1	517
	1621	14,14, 1←14,13, 1						22 687.47	.1	675
	1621	15,15, 1←15,14, 1						24 959.04	.1	675
	1621	Not Reported						17 253.25	.1	675
	1621	Not Reported						17 653.47	.1	675
	1621	Not Reported						17 798.42	.1	675
	1621	Not Reported						17 951.97	.1	675

Isotopic Species	Id. No.	Rotational Quantum Nos.	Vib. State	Hyperfine				Frequency MHz	Acc. ±MHz	Ref.
				F_1'	F'	F_1	F			
$S^{32}O^{16}F_2^{19}$	1621	Not Reported						17 964.73	.1	675
	1621	Not Reported						18 027.08	.1	675
	1621	Not Reported						18 208.41	.1	675
	1621	Not Reported						18 232.98	.1	675
	1621	Not Reported						18 401.14	.1	675
	1621	Not Reported						18 434.40	.1	675
	1621	Not Reported						18 569.62	.1	675
	1621	Not Reported						18 588.99	.1	675
	1621	Not Reported						18 608.96	.1	675
	1621	Not Reported						18 609.60	.1	675
	1621	Not Reported						21 711.98	.1	675
	1621	Not Reported						22 030.52	.1	675
	1621	Not Reported						22 034.25	.1	675
	1621	Not Reported						22 063.28	.1	675
	1621	Not Reported						22 730.80	.1	675
	1621	Not Reported						23 166.07	.1	675
	1621	Not Reported						23 315.91	.1	675
	1621	Not Reported						23 389.66	.1	675
	1621	Not Reported						23 444.36	.1	675
	1621	Not Reported						23 506.65	.1	675
	1621	Not Reported						23 601.78	.1	675
	1621	Not Reported						23 611.04	.1	675
	1621	Not Reported						23 625.64	.1	675
	1621	Not Reported						23 787.68	.1	675
	1621	Not Reported						23 814.64	.1	675
	1621	Not Reported						23 855.24	.1	675
	1621	Not Reported						23 872.54	.1	675
	1621	Not Reported						23 958.29	.1	675
	1621	Not Reported						24 253.17	.1	675
	1621	Not Reported						24 278.55	.1	675
	1621	Not Reported						24 389.65	.1	675
	1621	Not Reported						24 416.70	.1	675
	1621	Not Reported						24 591.78	.1	675
	1621	Not Reported						24 646.04	.1	675
	1621	Not Reported						24 656.31	.1	675
	1621	Not Reported						24 694.31	.1	675
	1621	Not Reported						24 776.46	.1	675
	1621	Not Reported						24 832.14	.1	675
	1621	Not Reported						24 898.42	.1	675
	1621	Not Reported						24 990.58	.1	675
	1621	Not Reported						25 078.51	.1	675
	1621	Not Reported						25 268.61	.1	675
	1621	Not Reported						29 042.10	.1	675
	1621	Not Reported						29 052.39	.1	675
	1621	Not Reported						29 087.73	.1	675
	1621	Not Reported						29 106.51	.1	675
	1621	Not Reported						29 801.69	.1	675
	1621	Not Reported						29 807.10	.1	675
	1621	Not Reported						29 878.60	.1	675
	1621	Not Reported						30 370.65	.1	675
	1621	Not Reported						31 092.46	.1	675
	1621	Not Reported						31 182.88	.1	675
	1621	Not Reported						31 288.35	.1	675
	1621	Not Reported						31 418.02	.1	675
	1621	Not Reported						31 446.05	.1	675

Isotopic Species	Id. No.	Rotational Quantum Nos.	Vib. State	Hyperfine F_1'	F'	F_1	F	Frequency MHz	Acc. ±MHz	Ref.
$S^{32}O^{16}F_2^{19}$	1621	Not Reported						31 479.26	.1	675
	1621	Not Reported						31 483.57	.1	675
	1621	Not Reported						32 997.06	.1	675
	1621	Not Reported						33 002.16	.1	675
	1621	Not Reported						33 078.20	.1	675
	1621	Not Reported						33 144.27	.1	675
	1621	Not Reported						33 153.04	.1	675
	1621	Not Reported						33 158.40	.1	675
	1621	Not Reported						33 222.20	.1	675
	1621	Not Reported						33 336.80	.1	675
	1621	Not Reported						33 341.66	.1	675
	1621	Not Reported						33 430.28	.1	675
	1621	Not Reported						33 470.86	.1	675
	1621	Not Reported						33 612.42	.1	675
	1621	Not Reported						33 673.74	.1	675
	1621	Not Reported						33 726.11	.1	675
	1621	Not Reported						33 737.34	.1	675
	1621	Not Reported						33 740.12	.1	675
	1621	Not Reported						33 760.50	.1	675
	1621	Not Reported						33 779.84	.1	675
	1621	Not Reported						33 804.39	.1	675
	1621	Not Reported						33 890.15	.1	675
	1621	Not Reported						33 896.54	.1	675
	1621	Not Reported						33 936.10	.1	675
	1621	Not Reported						33 974.94	.1	675
	1621	Not Reported						34 013.42	.1	675
	1621	Not Reported						34 017.44	.1	675
	1621	Not Reported						34 097.96	.1	675
	1621	Not Reported						34 117.04	.1	675
	1621	Not Reported						34 136.22	.1	675
	1621	Not Reported						34 241.32	.1	675
	1621	Not Reported						34 313.21	.1	675
	1621	Not Reported						34 339.55	.1	675
	1621	Not Reported						34 379.87	.1	675
	1621	Not Reported						34 384.88	.1	675
	1621	Not Reported						34 426.34	.1	675
	1621	Not Reported						34 598.00	.1	675
	1621	Not Reported						34 637.09	.1	675
	1621	Not Reported						35 169.52	.1	675
	1621	Not Reported						35 242.02	.1	675
	1621	Not Reported						35 574.23	.1	675
$S^{32}O^{18}F_2^{19}$	1622	2, 1, 1← 1, 0, 1	Ground					32 112.44	.1	517
	1622	2, 2, 0← 1, 1, 0	Ground					32 969.48	.1	517
	1622	2, 2, 1← 1, 1, 1	Ground					33 590.51	.1	517

F_2O_2S C_{2v} SO_2F_2

Isotopic Species	Pt. Gp.	Id. No.	A MHz		B MHz		C MHz		D_J MHz	D_{JK} MHz	Δ Amu A^2	κ
$S^{32}O_2^{16}F_2^{19}$	C_{2v}	1631	5 134.26	M	5 073.04	M	5 057.04	M				
$S^{34}O_2^{16}F_2^{19}$	C_{2v}	1632	5 133.74	M	5 070.00	M	5 054.07	M				

Id. No.	μ_a Debye		μ_b Debye		μ_c Debye		eQq Value(MHz)	Rel.	eQq Value(MHz)	Rel.	eQq Value(MHz)	Rel.	ω_a d 1/cm	ω_b d 1/cm	ω_c d 1/cm	ω_d d 1/cm
1631	1.110	M	0.	X	0.	X										

References:

ABC: 764 μ: 764

Add. Ref. 345,346,698

Reference 764 has listed for species 1631 the following rotational constants for excited vibrational states (assignments not completely determined):

     for $v_4 = 1$ or $v_5 = 1$, $B_e = 5057.27$ MHz, $C_e = 5055.83$ MHz;

     for $v_5 = 1$ or $v_4 = 1$, $A_e = 4923.7$ MHz, $B_e = 5067.03$ MHz, $C_e = 5050.38$ MHz;

     for $v_7 = 1$ or $v_9 = 1$, $B_e = (B + C)/2 = 5065.47$ MHz.

Isotopic Species	Id. No.	Rotational Quantum Nos.	Vib. State	F_1'	F'	F_1	F	Frequency MHz	Acc. ±MHz	Ref.
$S^{32}O_2^{16}F_2^{19}$	1631	2, 0, 2 ← 1, 0, 1	Excited					20 226.11	.10	764 [1]
	1631	2, 0, 2 ← 1, 0, 1	Ground					20 257.49	.10	764
	1631	2, 0, 2 ← 1, 0, 1	Excited					20 259.98	.10	764 [3]
	1631	2, 1, 1 ← 1, 0, 1	Excited					20 218.15	.10	764 [2]
	1631	2, 1, 1 ← 1, 1, 0	Excited					20 227.69	.10	764 [1]
	1631	2, 1, 1 ← 1, 1, 0	Ground					20 276.20	.10	764
	1631	2, 1, 2 ← 1, 1, 1	Excited					20 224.70	.10	764 [1]
	1631	2, 1, 2 ← 1, 1, 1	Ground					20 244.21	.10	764
	1631	2, 2, 0 ← 1, 1, 0	Excited					20 236.36	.10	764 [2]
	1631	2, 2, 1 ← 1, 1, 1	Excited					20 251.58	.10	764 [2]
	1631	3, 0, 3 ← 2, 0, 2	Excited					30 339.09	.10	764 [1]
	1631	3, 0, 3 ← 2, 0, 2	Ground					30 379.74	.10	764
	1631	3, 0, 3 ← 2, 0, 2	Excited					30 385.22	.10	764 [3]
	1631	3, 1, 2 ← 2, 1, 1	Excited					30 341.39	.10	764 [1]
	1631	3, 1, 2 ← 2, 1, 1	Ground					30 412.33	.10	764
	1631	3, 1, 3 ← 2, 1, 2	Excited					30 337.21	.10	764 [1]
	1631	3, 1, 3 ← 2, 1, 2	Ground					30 364.62	.10	764
	1631	3, 2, 1 ← 2, 1, 1	Excited					30 328.21	.10	764 [2]
	1631	3, 2, 1 ← 2, 2, 0	Ground					30 400.64	.10	764
	1631	3, 2, 2 ← 2, 2, 1	Ground					30 390.31	.10	764
	1631	3, 3, 0 ← 2, 2, 0	Excited					30 358.29	.10	764 [2]
	1631	3, 3, 1 ← 2, 2, 1	Excited					30 378.13	.10	764 [2]
$S^{34}O_2^{16}F_2^{19}$	1632	2, 0, 2 ← 1, 0, 1	Ground					20 245.		764 [4]
	1632	2, 1, 1 ← 1, 1, 0	Ground					20 264.11	.10	764
	1632	2, 1, 2 ← 1, 1, 1	Ground					20 232.31	.10	764
	1632	3, 0, 3 ← 2, 0, 2	Ground					30 362.12	.10	764
	1632	3, 1, 2 ← 2, 1, 1	Ground					30 394.30	.10	764
	1632	3, 1, 3 ← 2, 1, 2	Ground					30 346.69	.10	764

1. Relative intensities of these lines yield a vibrational frequency of 380 ± 35 cm^{-1}. Therefore these lines have been tentatively assigned to the $\omega_4 = 388 \pm 15$ cm^{-1} bending mode or to the $\omega_5 = 388$ cm^{-1} torsional mode.
2. These lines have been tentatively assigned to the $\omega_5 = 388$ cm^{-1} torsional mode or to the $\omega_4 = 388 \pm 15$ cm^{-1} bending mode.
3. Relative intensities give a vibrational frequency of 550 ± 50 cm^{-1}, making tentative assignment possible either to the $\omega_9 = 539$ cm^{-1} or $\omega_7 = 553$ cm^{-1} vibrational modes.
4. Measurements inaccurate because of overlapping by strong line at 20 244.21 MHz.

F$_3$HSi   C$_{3v}$ SiF$_3$H

Isotopic Species	Pt. Gp.	Id. No.	A MHz	B MHz	C MHz	D$_J$ MHz	D$_{JK}$ MHz	Δ Amu A^2	κ
Si^{28}F$_3^{19}$H	C$_{3v}$	1641	7 208.049 M	7 208.049 M		.00756	−.0125		
Si^{28}F$_3^{19}$D	C$_{3v}$	1642	6 890.08 M	6 890.08 M		.004			
Si^{29}F$_3^{19}$H	C$_{3v}$	1643	7 195.70 M	7 195.70 M					
Si^{29}F$_3^{19}$D	C$_{3v}$	1644	6 880.15 M	6 880.15 M					
Si^{30}F$_3^{19}$H	C$_{3v}$	1645	7 183.74 M	7 183.74 M					
Si^{30}F$_3^{19}$D	C$_{3v}$	1646	6 870.53 M	6 870.53 M					

Id. No.	μ_a Debye	μ_b Debye	μ_c Debye	eQq Value(MHz) Rel.	eQq Value(MHz) Rel.	eQq Value(MHz) Rel.	ω_a d 1/cm	ω_b d 1/cm	ω_c d 1/cm	ω_d d 1/cm.
1641	0. X	0. X	1.26 M							

References:

ABC: 527,745   D$_J$: 527,745   D$_{JK}$: 745 μ: 434

Add. Ref. 233,493

Isotopic Species	Id. No.	Rotational Quantum Nos.	Vib. State	F_1'	F'	F_1	F	Frequency MHz	Acc. ±MHz	Ref.
Si^{28}F$_3^{19}$H	1641	2, ← 1,	Ground					28 831.90	.1	311
	1641	3, ← 2,	Ground					43 247.49	.1	527
	1641	5, ← 4,	Ground					72 076.8	1.	527
	1641	6, 0← 5, 0	Ground					86 490.06	.20	745
	1641	6, 1← 5, 1	Ground					86 490.06	.20	745
	1641	6, 2← 5, 2	Ground					86 490.67	.20	745
	1641	6, 3← 5, 3	Ground					86 491.38	.20	745
	1641	6, 4← 5, 4	Ground					86 492.42	.20	745
	1641	6, 5← 5, 5	Ground					86 493.80	.20	745
	1641	10, 0← 9, 0	Ground					144 130.75	.30	745
	1641	10, 1← 9, 1	Ground					144 131.02	.30	745
	1641	10, 2← 9, 2	Ground					144 131.75	.30	745
	1641	10, 3← 9, 3	Ground					144 132.99	.30	745
	1641	10, 4← 9, 4	Ground					144 134.75	.30	745
	1641	10, 5← 9, 5	Ground					144 137.00	.30	745
	1641	10, 6← 9, 6	Ground					144 139.76	.30	745
	1641	10, 7← 9, 7	Ground					144 143.00	.30	745
	1641	10, 8← 9, 8	Ground					144 146.72	.30	745
	1641	10, 9← 9, 9	Ground					144 151.04	.30	745
	1641	13, 0←12, 0	Ground					187 342.87	.40	745
	1641	13, 1←12, 1	Ground					187 343.20	.40	745
	1641	13, 2←12, 2	Ground					187 344.13	.40	745
	1641	13, 3←12, 3	Ground					187 345.74	.40	745
	1641	13, 4←12, 4	Ground					187 348.03	.40	745
	1641	13, 5←12, 5	Ground					187 350.98	.40	745
	1641	13, 6←12, 6	Ground					187 354.52	.40	745
	1641	13, 7←12, 7	Ground					187 358.70	.40	745
	1641	13, 8←12, 8	Ground					187 363.55	.40	745
	1641	13, 9←12, 9	Ground					187 369.10	.40	745
	1641	13,10←12,10	Ground					187 375.26	.40	745
	1641	13,11←12,11	Ground					187 382.00	.40	745
	1641	13,12←12,12	Ground					187 389.49	.40	745
Si^{28}F$_3^{19}$D	1642	2, ← 1,	Ground					27 560.17	.1	527
	1642	3, ← 2,	Ground					41 340.00	.1	527
	1642	4, ← 3,	Ground					55 119.4	1.	527
Si^{29}F$_3^{19}$H	1643	2, ← 1,	Ground					28 782.65	.1	311
Si^{29}F$_3^{19}$D	1644	3, ← 2,	Ground					41 280.46	.1	527
Si^{30}F$_3^{19}$H	1645	2, ← 1,	Ground					28 734.80	.1	311
Si^{30}F$_3^{19}$D	1646	3, ← 2,	Ground					41 222.73	.2	527

$_3$N C_{3v} NF$_3$

Isotopic Species	Pt. Gp.	Id. No.	A MHz	B MHz	C MHz	D_J MHz	D_{JK} MHz	Δ Amu A²	κ
$^{14}F_3^{19}$	C_{3v}	1651	10 680.96 M	10 680.96 M		.0095	−.022		
$^{15}F_3^{19}$	C_{3v}	1652	10 629.35 M	10 629.35 M					

Id. No.	μ_a Debye	μ_b Debye	μ_c Debye	eQq Value(MHz) Rel.	eQq Value(MHz) Rel.	eQq Value(MHz) Rel.	ω_a d 1/cm	ω_b d 1/cm	ω_c d 1/cm	ω_d d 1/cm
1651	0. X	0. X	.234 M	−7.09 N^{14}						

References:

BC: 235 D_J: 897 D_{JK}: 897 μ: 434   eQq: 897

Add. Ref. 53,692,703,704

Isotopic Species	Id. No.	Rotational Quantum Nos.	Vib. State	F_1'	F'	F_1	F	Frequency MHz	Acc. ±MHz	Ref.
$^{14}F_3^{19}$	1651	2, ← 1,	Ground					42 723.84	.10	235
	1651	2, 0← 1, 0	Ground		2		2	42 721.73		235
	1651	2, 0← 1, 0	Ground		1		0	42 722.16		235
	1651	2, 0← 1, 0	Ground		2		1	42 723.94		235
	1651	2, 0← 1, 0	Ground		3		2	42 723.94		235
	1651	2, 0← 1, 0	Ground		1		1	42 727.39		235
	1651	2, 1← 1, 1	Ground		2		1	42 722.16		235
	1651	2, 1← 1, 1	Ground		2		2	42 723.28		235
	1651	2, 1← 1, 1	Ground		3		2	42 724.36		235
	1651	2, 1← 1, 1	Ground		1		0	42 726.60		235
	1651	5, 0← 4, 0	Ground					106 805.93	.20	282
	1651	5, 2← 4, 2	Ground					106 804.54	.50	282
	1651	5, 3← 4, 3	Ground					106 803.62	.20	282
$^{15}F_3^{19}$	1652	2, ← 1,	Ground					42 517.38	.10	235

F$_3$NS   C$_{3v}$ NSF$_3$

Isotopic Species	Pt. Gp.	Id. No.	A MHz		B MHz		C MHz	D$_J$ MHz	D$_{JK}$ MHz	Δ Amu A^2	κ
N^{14}S^{32}F$_3^{19}$	C$_{3v}$	1661	4636.24	M	4636.24	M					
N^{14}S^{33}F$_3^{19}$	C$_{3v}$	1662	4633.24	M	4633.24	M					
N^{14}S^{34}F$_3^{19}$	C$_{3v}$	1663	4630.31	M	4630.31	M					
N^{15}S^{32}F$_3^{19}$	C$_{3v}$	1664	4520.20	M	4520.20	M					

Id. No.	μ_a Debye		μ_b Debye		μ_c Debye		eQq Value(MHz)	Rel.	eQq Value(MHz)	Rel.	eQq Value(MHz)	Rel.	ω_a d 1/cm	ω_b d 1/cm	ω_c d 1/cm	ω_d d 1/cm
1661	0.	X	0.	X	1.91	M	1.19	N^{14}								

References:

ABC: 971 μ: 971   eQq: 971

Trifluorosulfur Nitride   Spectral Line Table

Isotopic Species	Id. No.	Rotational Quantum Nos.	Vib. State	F$_1'$	F'	F$_1$	F	Frequency MHz	Acc. ±MHz	Ref.
N^{14}S^{32}F$_3^{19}$	1661	1, ← 0,	Ground					9272.56		971
	1661	1, 0← 0, 0	Ground		0		1	9271.90		971
	1661	1, 0← 0, 0	Ground		2		1	9272.42		971
	1661	1, 0← 0, 0	Ground		1		1	9272.79		971
	1661	2, ← 1,	Ground					18545.10		971
	1661	3, ← 2,	Ground					27817.67		971
	1661	6, ← 5,	Ground					55633.94		971
N^{14}S^{33}F$_3^{19}$	1662	3, ← 2,	Ground					27799.71		971
	1662	6, ← 5,	Ground					55598.35		971
N^{14}S^{34}F$_3^{19}$	1663	2, ← 1,	Ground					18521.34		971
	1663	3, ← 2,	Ground					27782.10		971
	1663	6, ← 5,	Ground					55563.27		971
N^{15}S^{32}F$_3^{19}$	1664	3, ← 2,	Ground					27121.49		971
	1664	6, ← 5,	Ground					54241.76		971

F$_3$OP C$_{3v}$ POF$_3$

Isotopic Species	Pt. Gp.	Id. No.	A MHz	B MHz	C MHz	D$_J$ MHz	D$_{JK}$ MHz	Δ Amu A²	κ
P^{31}O^{16}F$_3^{19}$	C$_{3v}$	1671		4 594.262 M	4 594.262 M	.00102	.00128		
P^{31}O^{18}F$_3^{19}$	C$_{3v}$	1672		4 395.27 M	4 395.27 M				

Id. No.	μ$_a$ Debye	μ$_b$ Debye	μ$_c$ Debye	eQq Value(MHz) Rel.	eQq Value(MHz) Rel.	eQq Value(MHz) Rel.	ω$_a$ d 1/cm	ω$_b$ d 1/cm	ω$_c$ d 1/cm	ω$_d$ d 1/cm
1671	1.77 M	0. X	0. X							

References:

ABC: 411,745 D$_J$: 745 D$_{JK}$: 745 μ: 434

Add. Ref. 228,666

Phosphoryl Fluoride Spectral Line Table

Isotopic Species	Id. No.	Rotational Quantum Nos.	Vib. State	F$_1'$	F'	F$_1$	F	Frequency MHz	Acc. ±MHz	Ref.
P^{31}O^{16}F$_3^{19}$	1671	2, ← 1,	Ground					18 377.03	.05	356
	1671	3, ← 2,	Excited					27 539.39	.05	356
	1671	3, ← 2,	Ground					27 565.42	.05	356
	1671	4, ← 3,	Ground					36 753.83	.05	411
	1671	10, 0← 9, 0	Ground					91 881.16	.20	745
	1671	10, 1← 9, 1	Ground					91 881.16	.20	745
	1671	10, 2← 9, 2	Ground					91 881.16	.20	745
	1671	10, 3← 9, 3	Ground					91 881.00	.20	745
	1671	10, 4← 9, 4	Ground					91 880.83	.20	745
	1671	10, 5← 9, 5	Ground					91 880.56	.20	745
	1671	10, 6← 9, 6	Ground					91 880.27	.20	745
	1671	10, 7← 9, 7	Ground					91 879.92	.20	745
	1671	10, 8← 9, 8	Ground					91 879.53	.20	745
	1671	10, 9← 9, 9	Ground					91 879.08	.20	745
	1671	11, 0←10, 0	Ground					101 068.35	.20	282
	1671	11, 3←10, 3	Ground					101 068.04	.20	282
	1671	11, 5←10, 5	Ground					101 067.62	.40	282
	1671	11, 6←10, 6	Ground					101 067.33	.20	282
	1671	11, 7←10, 7	Ground					101 066.91	.20	282
	1671	11, 8←10, 8	Ground					101 066.52	.20	282
	1671	11, 9←10, 9	Ground					101 066.06	.20	282
	1671	13, ←12,	Ground					119 441.32	.45	282
	1671	14, ←13,	Ground					128 626.60	.20	282
	1671	16, 0←15, 0	Ground					146 999.65	.30	745
	1671	16, 1←15, 1	Ground					146 999.65	.30	745
	1671	16, 2←15, 2	Ground					146 999.65	.30	745
	1671	16, 3←15, 3	Ground					146 999.36	.30	745
	1671	16, 4←15, 4	Ground					146 999.07	.30	745
	1671	16, 5←15, 5	Ground					146 998.68	.30	745
	1671	16, 6←15, 6	Ground					146 998.02	.30	745

Isotopic Species	Id. No.	Rotational Quantum Nos.	Vib. State	F_1'	F'	F_1	F	Frequency MHz	Acc. ±MHz	Ref.
$P^{31}O^{16}F_3^{19}$	1671	16, 7←15, 7	Ground					146 997.65	.30	745
	1671	16, 8←15, 8	Ground					146 997.06	.30	745
	1671	16, 9←15, 9	Ground					146 996.38	.30	745
	1671	16,10←15,10	Ground					146 995.62	.30	745
	1671	16,11←15,11	Ground					146 994.77	.30	745
	1671	16,12←15,12	Ground					146 993.83	.30	745
	1671	16,13←15,13	Ground					146 992.79	.30	745
	1671	16,14←15,14	Ground					146 991.67	.30	745
	1671	16,15←15,15	Ground					146 990.43	.30	745
	1671	21, 0←20, 0	Ground					192 921.13	.40	745
	1671	21, 1←20, 1	Ground					192 921.13	.40	745
	1671	21, 2←20, 2	Ground					192 921.13	.40	745
	1671	21, 3←20, 3	Ground					192 920.72	.40	745
	1671	21, 4←20, 4	Ground					192 920.35	.40	745
	1671	21, 5←20, 5	Ground					192 919.93	.40	745
	1671	21, 6←20, 6	Ground					192 919.30	.40	745
	1671	21, 7←20, 7	Ground					192 918.60	.40	745
	1671	21, 8←20, 8	Ground					192 917.78	.40	745
	1671	21, 9←20, 9	Ground					192 916.89	.40	745
	1671	21,10←20,10	Ground					192 915.83	.40	745
$P^{31}O^{16}F_3^{19}$	1671	21,11←20,11	Ground					192 914.71	.40	745
	1671	21,12←20,12	Ground					192 913.48	.40	745
	1671	21,13←20,13	Ground					192 912.14	.40	745
	1671	21,14←20,14	Ground					192 910.67	.40	745
	1671	21,15←20,15	Ground					192 909.10	.40	745
	1671	21,16←20,16	Ground					192 907.43	.40	745
	1671	21,17←20,17	Ground					192 905.66	.40	745
	1671	21,18←20,18	Ground					192 903.75	.40	745
	1671	21,19←20,19	Ground					192 901.76	.40	745
	1671	21,20←20,20	Ground					192 899.61	.40	745
	1671	26, 0←25, 0	Ground					238 829.94	.50	745
	1671	26, 1←25, 1	Ground					238 829.94	.50	745
	1671	26, 2←25, 2	Ground					238 829.94	.50	745
	1671	26, 3←25, 3	Ground					238 829.43	.50	745
	1671	26, 4←25, 4	Ground					238 828.98	.50	745
	1671	26, 5←25, 5	Ground					238 828.39	.50	745
	1671	26, 6←25, 6	Ground					238 827.58	.50	745
	1671	26, 7←25, 7	Ground					238 826.80	.50	745
	1671	26, 8←25, 8	Ground					238 825.78	.50	745
	1671	26, 9←25, 9	Ground					238 824.54	.50	745
	1671	26,10←25,10	Ground					238 823.32	.50	745
	1671	26,11←25,11	Ground					238 821.94	.50	745
	1671	26,12←25,12	Ground					238 820.35	.50	745
	1671	26,13←25,13	Ground					238 818.73	.50	745
	1671	26,14←25,14	Ground					238 816.96	.50	745
	1671	26,15←25,15	Ground					238 815.05	.50	745
	1671	26,16←25,16	Ground					238 812.98	.50	745
	1671	26,17←25,17	Ground					238 810.71	.50	745
	1671	26,18←25,18	Ground					238 808.33	.50	745
	1671	26,20←25,20	Ground					238 803.16	.50	745
	1671	26,21←25,21	Ground					238 800.41	.50	745
	1671	26,22←25,22	Ground					238 797.54	.50	745
	1671	26,23←25,23	Ground					238 794.51	.50	745
	1671	26,24←25,24	Ground					238 791.31	.50	745
$P^{31}O^{18}F_3^{19}$	1672	3, ← 2,	Ground					26 371.7	.06	411
	1672	3, ← 2,	Ground					26 391.61	.10	356
	1672	4, ← 3,	Ground					35 162.0	.5	411

350

F₃P — F_3P

Isotopic Species	Pt. Gp.	Id. No.	A MHz	B MHz	C MHz	D_J MHz	D_{JK} MHz	Δ Amu A²	κ
$P^{31}F_3^{19}$	C_{3v}	1681	7 819.90 M	7 819.900 M		.0075	−.0117		

C_{3v} — PF_3

Id. No.	μ_a Debye	μ_b Debye	μ_c Debye	eQq Value(MHz) Rel.	eQq Value(MHz) Rel.	eQq Value(MHz) Rel.	ω_a 1/cm d	ω_b 1/cm d	ω_c 1/cm d	ω_d 1/cm d
1681	0. X	0. X	1.03 M				892 1	487 1	860 2	344 2

References:

ABC: 129 D_J: 282 D_{JK}: 282 μ: 434 ω: 1028

Add. Ref. 130,666

Phosphorus Trifluoride   Spectral Line Table

Isotopic Species	Id. No.	Rotational Quantum Nos.	Vib. State v_a ; v_b ; v_c ; v_d	F_1'	F'	F_1	F	Frequency MHz	Acc. ±MHz	Ref.
$P^{31}F_3^{19}$	1681	2, ← 1,	Ground					31 279.60	.10	129
	1681	3, ← 2,	Excited					46 940.	10.	129
	1681	3, ← 2,	Excited					47 010.	10.	129
	1681	3, ← 2,	Excited					47 033.	10.	129
	1681	3, ← 2,	Excited					47 040.	10.	129
	1681	3, 0← 2, 0	Ground					46 918.82	.18	129
	1681	3, 1← 2, 1	Ground					46 918.82	.18	129
	1681	3, 2← 2, 2	Ground					46 919.02	.18	129
	1681	8, 0← 7, 0	Ground					125 103.89	.20	282
	1681	8, 3← 7, 3	Ground					125 105.60	.20	282
	1681	8, 4← 7, 4	Ground					125 106.85	.25	282
	1681	8, 5← 7, 5	Ground					125 108.60	.20	282
	1681	8, 6← 7, 6	Ground					125 110.64	.30	282
	1681	Not Reported	1; 0; 0; 0					31 130.	.10	446
	1681	Not Reported	0; 2; 0; 0					31 194.	5.0	446
	1681	Not Reported	0; 0; 0; 1					31 224.	3.0	446
	1681	Not Reported	0; 1; 0; 0					31 237.	3.0	446
	1681	Not Reported	0; 0; 0; 1					31 294.	2.0	446
	1681	Not Reported	0; 0; 0; 2					31 307.	3.0	446
	1681	Not Reported	0; 0; 1; 0					31 335.	3.0	446
	1681	Not Reported	0; 0; 0; 1					31 359.	3.0	446

F$_3$PS   C$_{3v}$ SPF$_3$

Isotopic Species	Pt. Gp.	Id. No.	A MHz	B MHz	C MHz	D$_J$ MHz	D$_{JK}$ MHz	Δ Amu A^2	κ
S^{32}P^{31}F$_3^{19}$	C$_{3v}$	1691		2 657.63 M	2 657.63 M	.0003	.0018		
S^{33}P^{31}F$_3^{19}$	C$_{3v}$	1692		2 614.73 M	2 614.73 M				
S^{34}P^{31}F$_3^{19}$	C$_{3v}$	1693		2 579.77 M	2 579.77 M				

Id. No.	μ$_a$ Debye	μ$_b$ Debye	μ$_c$ Debye	eQq Value(MHz) Rel.	eQq Value(MHz) Rel.	eQq Value(MHz) Rel.	ω$_a$ d 1/cm	ω$_b$ d 1/cm	ω$_c$ d 1/cm	ω$_d$ d 1/cm
1691	.633 M	0. X	0. X							

References:

ABC: 411   D$_J$: 282   D$_{JK}$: 282   μ: 356

Thiophosphoryl Fluoride   Spectral Line Table

Isotopic Species	Id. No.	Rotational Quantum Nos.	Vib. State	F$_1'$	F'	F$_1$	F	Frequency MHz	Acc. ±MHz	Ref.
S^{32}P^{31}F$_3^{19}$	1691	4, ← 3,	Ground					21 260.95	.05	356
	1691	5, ← 4,	Excited					26 531.5	1.	356
	1691	5, ← 4,	Excited					26 553.58	.08	356
	1691	5, ← 4,	Ground					26 576.36	.03	356
	1691	5, ← 4,	Excited					26 595.5	.1	356
	1691	6, 0← 5, 0	Ground					31 891.62	.05	411
	1691	6, 1← 5, 1	Ground					31 891.62	.05	411
	1691	6, 2← 5, 2	Ground					31 891.62	.05	411
	1691	6, 3← 5, 3	Ground					31 891.45	.05	411
	1691	6, 4← 5, 4	Ground					31 891.27	.05	411
	1691	6, 5← 5, 5	Ground					31 891.13	.05	411
	1691	7, 0← 6, 0	Ground					37 206.77	.05	411
	1691	7, 1← 6, 1	Ground					37 206.77	.05	411
	1691	7, 2← 6, 2	Ground					37 206.67	.05	411
	1691	7, 3← 6, 3	Ground					37 206.55	.05	411
	1691	7, 4← 6, 4	Ground					37 206.36	.05	411
	1691	7, 5← 6, 5	Ground					37 206.13	.05	411
	1691	7, 6← 6, 6	Ground					37 205.84	.05	411
	1691	23, 0←22, 0						122 237.90	.30	282
	1691	23, 5←22, 5						122 235.80	.30	282 [1]
	1691	23, 6←22, 6						122 235.80	.30	282 [1]
	1691	23, 9←22, 9						122 231.30	.30	282
	1691	23,12←22,12						122 225.50	.30	282
	1691	23,15←22,15						122 219.00	.30	282
S^{33}P^{31}F$_3^{19}$	1692	6, ← 5,	Ground					31 412.7	.3	411
S^{34}P^{31}F$_3^{19}$	1693	5, ← 4,	Excited					25 775.3	.5	356
	1693	5, ← 4,	Ground					25 797.87	.03	356
	1693	5, ← 4,	Excited					25 818.0	2.	356
	1693	7, 0← 6, 0	Ground					36 116.72	.05	411
	1693	7, 1← 6, 1	Ground					36 116.72	.05	411
	1693	7, 2← 6, 2	Ground					36 116.72	.05	411
	1693	7, 3← 6, 3	Ground					36 116.47	.05	411
	1693	7, 6← 6, 6	Ground					36 115.78	.05	411

1. The article by Johnson, Trambarulo and Gordy indicated some uncertainty in the value of K, so this frequency value has been given with both K=5 and K=6.

HN$_3$ C$_s$ HN$_3$

Isotopic Species	Pt. Gp.	Id. No.	A MHz		B MHz		C MHz		D$_J$ MHz	D$_{JK}$ MHz	Δ Amu A^2	κ
HN$_3^{14}$	C$_s$	1701	618 050.	M	12 034.14	M	11 781.48	M	.00491		.083	−.99916
DN$_3^{14}$	C$_s$	1702	352 643.	M	11 350.22	M	10 965.49	M	.00421		.129	−.99774
HN$_2^{14}$N^{15}	C$_s$	1703	616 868.	M	11 641.76	M	11 405.08	M	.00453		.082	
HN^{15}N$_2^{14}$	C$_s$	1704	619 916.	M	11 667.54	M	11 427.86	M	.00471		.089	
HN^{14}N^{15}N^{14}	C$_s$	1705										

Id. No.	μ$_a$ Debye	μ$_b$ Debye	μ$_c$ Debye	eQq Value(MHz)	Rel.	eQq Value(MHz)	Rel.	eQq Value(MHz)	Rel.	ω$_a$ d 1/cm	ω$_b$ d 1/cm	ω$_c$ d 1/cm	ω$_d$ d 1/cm
1701	.847 M		0. X	−4.67									
1703				−1.35	N^{15}								
1704				4.85	N^{15}								
1705				<.7	N^{15}								

References:

ABC: 1003   D$_J$: 1003   Δ: 1003   κ: 1000   μ: 178   eQq: 387,991

Add. Ref. 308

Species	1701	1702	1703	1704	1705
B$_0$ + C$_0$ (MHz)	23815.7	22316.1	23048.2	23096.7	23814

Isotopic Species	Id. No.	Rotational Quantum Nos.	Vib. State	Hyperfine				Frequency MHz	Acc. ±MHz	Ref.
				F_1'	F'	F_1	F			
HN_3^{14}	1701	1, 0, 1← 0, 0, 0	Ground	0	1	1	1	23 813.28	.05	991
	1701	1, 0, 1← 0, 0, 0	Ground	2	2	1	2	23 815.19	.05	991
	1701	1, 0, 1← 0, 0, 0	Ground	2	3	1	3	23 815.56	.05	991
	1701	1, 0, 1← 0, 0, 0	Ground					23 815.7		178
	1701	1, 0, 1← 0, 0, 0	Ground	1	0	1	0	23 816.56	.05	991
	1701	1, 0, 1← 0, 0, 0	Ground	1	2	1	2	23 816.87	.05	991
	1701	1, 0, 1← 0, 0, 0	Ground	1	1	1	1	23 817.17	.05	991
	1701	3, 0, 3← 2, 0, 2	Ground					66 945.50	.25	1000
	1701	3, 1, 2← 2, 1, 1	Ground					67 520.75	.25	1000
	1701	3, 1, 3← 2, 1, 2	Ground					66 366.68	.25	1000
	1701	3, 2, 1← 2, 2, 0	Ground		2		1	66 935.94	.25	1000
	1701	3, 2, 1← 2, 2, 0	Ground		4		3	66 937.27	.25	1000
	1701	3, 2, 1← 2, 2, 0	Ground		3		2	66 938.71	.25	1000
	1701	3, 2, 2← 2, 2, 1	Ground		4		3	66 935.94	.25	1000
	1701	3, 2, 2← 2, 2, 1	Ground		3		2	66 937.27	.25	1000
	1701	4, 0, 4← 3, 0, 3	Ground					89 258.55	.25	1000
	1701	4, 0, 4← 3, 0, 3	Ground					95 260.50	.25	1000
	1701	4, 1, 3← 3, 1, 2	Ground					90 026.99	.25	1000
	1701	4, 1, 3← 3, 1, 2	Ground					95 760.15	.25	1000
	1701	4, 1, 4← 3, 1, 3	Ground					88 488.03	.25	1000
	1701	4, 1, 4← 3, 1, 3	Ground					94 749.42	.25	1000
	1701	4, 2, 2← 3, 2, 1	Ground		3		2	89 250.42	.25	1000
	1701	4, 2, 2← 3, 2, 1	Ground		5		4	89 250.42	.25	1000
	1701	4, 2, 2← 3, 2, 1	Ground		4		3	89 251.21	.25	1000
	1701	4, 2, 2← 3, 2, 1	Ground		5		4	95 236.17	.25	1000
	1701	4, 2, 2← 3, 2, 1	Ground		3		2	95 236.17	.25	1000
	1701	4, 2, 2← 3, 2, 1	Ground		4		3	95 237.03	.25	1000
	1701	4, 2, 3← 3, 2, 2	Ground		5		4	89 247.11	.25	1000
	1701	4, 2, 3← 3, 2, 2	Ground		3		2	89 247.11	.25	1000
	1701	4, 2, 3← 3, 2, 2	Ground		4		3	89 247.84	.25	1000
	1701	4, 2, 3← 3, 2, 2	Ground		3		2	95 235.45	.25	1000
	1701	4, 2, 3← 3, 2, 2	Ground		5		4	95 235.45	.25	1000
	1701	4, 2, 3← 3, 2, 2	Ground		4		3	95 236.17	.25	1000
	1701	4, 3, 1← 3, 3, 0	Ground		5		4	89 229.74	.25	1000
	1701	4, 3, 1← 3, 3, 0	Ground		4		3	89 231.45	.25	1000
	1701	4, 3, 2← 3, 3, 1	Ground		3		2	89 229.74	.25	1000
	1701	4, 3, 2← 3, 3, 1	Ground		3		2	95 203.10	.25	1000
	1701	4, 3, 2← 3, 3, 1	Ground		5		4	95 203.10	.25	1000
	1701	4, 4, 1← 3, 3, 0	Ground		4		3	95 205.12	.25	1000
	1701	5, 0, 5← 4, 0, 4	Ground					111 569.92	.25	1000
	1701	5, 0, 5← 4, 0, 4	Ground					119 074.12	.25	1000
	1701	5, 1, 4← 4, 1, 3	Ground					112 532.50	.25	1000
	1701	5, 1, 4← 4, 1, 3	Ground					119 698.95	.25	1000
	1701	5, 1, 5← 4, 1, 4	Ground					110 608.76	.25	1000
	1701	5, 1, 5← 4, 1, 4	Ground					118 435.82	.25	1000
	1701	5, 2, 3← 4, 2, 2	Ground		6		5	111 564.52	.25	1000
	1701	5, 2, 3← 4, 2, 2	Ground		4		3	111 564.52	.25	1000
	1701	5, 2, 3← 4, 2, 2	Ground		5		4	111 564.81	.25	1000
	1701	5, 2, 3← 4, 2, 2	Ground					119 044.95	.25	1000
	1701	5, 2, 4← 4, 2, 3	Ground		6		5	111 557.82	.25	1000

Isotopic Species	Id. No.	Rotational Quantum Nos.	Vib. State	F_1'	F'	F_1	F	Frequency MHz	Acc. ±MHz	Ref.
HN_3^{14}	1701	5, 2, 4 ← 4, 2, 3	Ground		4		3	111 557.82	.25	1000
	1701	5, 2, 4 ← 4, 2, 3	Ground		5		4	111 558.09	.25	1000
	1701	5, 2, 4 ← 4, 2, 3	Ground					119 043.41	.25	1000
	1701	5, 3, 2 ← 4, 3, 1	Ground		6		5	111 536.98	.25	1000
	1701	5, 3, 2 ← 4, 3, 1	Ground		5		4	111 537.86	.25	1000
	1701	5, 3, 2 ← 4, 3, 1	Ground		6		5	119 003.46	.25	1000
	1701	5, 3, 2 ← 4, 3, 1	Ground		5		4	119 004.33	.25	1000
	1701	5, 3, 3 ← 4, 3, 2	Ground		4		3	111 536.98	.25	1000
	1701	5, 3, 3 ← 4, 3, 2	Ground		4		3	119 003.46	.25	1000
	1701	5, 4, 1 ← 4, 4, 0	Ground		6		5	111 506.13	.25	1000
	1701	5, 4, 1 ← 4, 4, 0	Ground		5		4	111 506.13	.25	1000
	1701	5, 4, 1 ← 4, 4, 0	Ground					118 946.42	.25	1000
	1701	5, 4, 2 ← 4, 4, 1	Ground		4		3	111 504.67	.25	1000
	1701	5, 4, 2 ← 4, 4, 1	Ground					118 946.42	.25	1000
	1701	6, 0, 6 ← 5, 0, 5	Ground					133 879.14	.25	1000
	1701	6, 0, 6 ← 5, 0, 5	Ground					142 886.92	.25	1000
	1701	6, 1, 5 ← 5, 1, 4	Ground					135 036.98	.25	1000
	1701	6, 1, 5 ← 5, 1, 4	Ground					143 637.49	.25	1000
	1701	6, 1, 6 ← 5, 1, 5	Ground					132 728.67	.25	1000
	1701	6, 1, 6 ← 5, 1, 5	Ground					142 121.48	.25	1000
	1701	6, 2, 4 ← 5, 2, 3	Ground					133 879.14	.25	1000
	1701	6, 2, 4 ← 5, 2, 3	Ground					142 853.84	.25	1000
	1701	6, 2, 5 ← 5, 2, 4	Ground					133 867.80	.25	1000
	1701	6, 2, 5 ← 5, 2, 4	Ground					142 851.03	.25	1000
	1701	6, 3, 3 ← 5, 3, 2	Ground					133 844.21	.25	1000
	1701	6, 3, 3 ← 5, 3, 2	Ground					142 803.67	.25	1000
	1701	6, 3, 4 ← 5, 3, 3	Ground					133 844.21	.25	1000
	1701	6, 3, 4 ← 5, 3, 3	Ground					142 803.67	.25	1000
	1701	6, 4, 2 ← 5, 4, 1	Ground					133 805.16	.25	1000
	1701	6, 4, 2 ← 5, 4, 1	Ground		7		6	142 734.66	.25	1000
	1701	6, 4, 2 ← 5, 4, 1	Ground		6		5	142 735.63	.25	1000
	1701	6, 4, 3 ← 5, 4, 2	Ground					133 805.16	.25	1000
	1701	6, 4, 3 ← 5, 4, 2	Ground		5		4	142 734.66	.25	1000
	1701	6, 5, 1 ← 5, 5, 0	Ground					133 756.98	.25	1000
	1701	6, 5, 1 ← 5, 5, 0	Ground					142 644.26	.25	1000
	1701	6, 5, 2 ← 5, 5, 1	Ground					133 756.98	.25	1000
	1701	6, 5, 2 ← 5, 5, 1	Ground					142 644.26	.25	1000
	1701	7, 0, 7 ← 6, 0, 6	Ground					166 698.40	.25	1000
	1701	7, 1, 6 ← 6, 1, 5	Ground					167 575.07	.25	1000
	1701	7, 1, 7 ← 6, 1, 6	Ground					165 806.33	.25	1000
	1701	7, 2, 5 ← 6, 2, 4	Ground					166 662.05	.25	1000
	1701	7, 2, 6 ← 6, 2, 5	Ground					166 657.66	.25	1000
	1701	7, 3, 4 ← 6, 3, 3	Ground					166 602.85	.25	1000
	1701	7, 3, 5 ← 6, 3, 4	Ground					166 602.85	.25	1000
	1701	7, 4, 3 ← 6, 4, 2	Ground					166 522.79	.25	1000
	1701	7, 4, 4 ← 6, 4, 3	Ground					166 522.79	.25	1000
	1701	7, 5, 2 ← 6, 5, 1	Ground					166 416.76	.25	1000
	1701	7, 5, 3 ← 6, 5, 2	Ground					166 416.76	.25	1000
	1701	8, 0, 8 ← 7, 0, 7	Ground					178 489.31	.25	1000
	1701	8, 1, 7 ← 7, 1, 6	Ground					180 042.27	.25	1000

Isotopic Species	Id. No.	Rotational Quantum Nos.	Vib. State	Hyperfine				Frequency MHz	Acc. ±MHz	Ref.
				F_1'	F'	F_1	F			
HN_3^{14}	1701	8, 1, 8← 7, 1, 7	Ground					176 964.18	.25	1000
	1701	8, 2, 6← 7, 2, 5	Ground					178 512.13	.25	1000
	1701	8, 2, 7← 7, 2, 6	Ground					178 484.73	.25	1000
	1701	8, 3, 5← 7, 3, 4	Ground					178 456.64	.25	1000
	1701	8, 3, 6← 7, 3, 5	Ground					178 456.64	.25	1000
	1701	8, 4, 4← 7, 4, 3	Ground					178 404.21	.25	1000
	1701	8, 4, 5← 7, 4, 4	Ground					178 404.21	.25	1000
	1701	8, 5, 3← 7, 5, 2	Ground					178 337.49	.25	1000
	1701	8, 5, 4← 7, 5, 3	Ground					178 337.49	.25	1000
	1701	9, 0, 9← 8, 0, 8	Ground					214 316.89	.15	1003
	1701	9, 1, 8← 8, 1, 7	Ground					215 446.80	.15	1003
	1701	9, 1, 9← 8, 1, 8	Ground					213 173.13	.15	1003
	1701	9, 2, 7← 8, 2, 6	Ground					214 277.22	.15	1003
	1701	9, 2, 8← 8, 2, 7	Ground					214 267.72	.15	1003
	1701	9, 3, 6← 8, 3, 5	Ground					214 198.59	.15	1003
	1701	9, 3, 7← 8, 3, 6	Ground					214 198.59	.15	1003
	1701	9, 4, 5← 8, 4, 4	Ground					214 095.93	.15	1003
	1701	9, 4, 6← 8, 4, 5	Ground					214 095.93	.15	1003
	1701	9, 5, 4← 8, 5, 3	Ground					213 959.72	.15	1003
	1701	9, 5, 5← 8, 5, 4	Ground					213 959.72	.15	1003
	1701	9, 6, 3← 8, 6, 2	Ground					213 788.90	.15	1003
	1701	9, 6, 4← 8, 6, 3	Ground					213 788.90	.15	1003
DN_3^{14}	1702	1, 0, 1← 0, 0, 0	Ground					22 316.1		178
$HN_2^{14}N^{15}$	1703	1, 0, 1← 0, 0, 0	Ground		0		1	23 044.40	.05	991
	1703	1, 0, 1← 0, 0, 0	Ground		2		1	23 046.55	.05	991
	1703	1, 0, 1← 0, 0, 0	Ground		1		1	23 048.03	.05	991
	1703	1, 0, 1← 0, 0, 0	Ground					23 048.2		178
	1703	4, 0, 4← 3, 0, 3	Ground					92 185.47	.15	1003
	1703	4, 1, 3← 3, 1, 2	Ground					92 653.45	.15	1003
	1703	4, 1, 4← 3, 1, 3	Ground					91 706.67	.15	1003
	1703	4, 2, 2← 3, 2, 1	Ground		3		2	92 162.50	.15	1003
	1703	4, 2, 2← 3, 2, 1	Ground		5		4	92 162.50	.15	1003
	1703	4, 2, 2← 3, 2, 1	Ground		2		3	92 162.50	.15	1003
	1703	4, 2, 2← 3, 2, 1	Ground		4		3	92 162.50	.15	1003
	1703	4, 2, 2← 3, 2, 1	Ground		4		3	92 163.45	.15	1003
	1703	4, 2, 3← 3, 2, 2	Ground		5		4	92 161.78	.15	1003
	1703	4, 2, 3← 3, 2, 2	Ground		3		2	92 161.78	.15	1003
	1703	4, 2, 3← 3, 2, 2	Ground		3		4	92 162.50	.15	1003
	1703	4, 3, 1← 3, 3, 0	Ground		5		4	92 131.38	.15	1003
	1703	4, 3, 1← 3, 3, 0	Ground		4		5	92 133.13	.15	1003
	1703	4, 3, 1← 3, 3, 0	Ground		4		3	92 133.13	.15	1003
	1703	4, 3, 2← 3, 3, 1	Ground		3		2	92 131.38	.15	1003
	1703	5, 0, 5← 4, 0, 4	Ground					115 230.58	.15	1003
	1703	5, 1, 4← 4, 1, 3	Ground					115 815.77	.15	1003
	1703	5, 1, 5← 4, 1, 4	Ground					114 632.32	.15	1003
	1703	5, 2, 3← 4, 2, 2	Ground		4		3	115 203.02	.15	1003
	1703	5, 2, 3← 4, 2, 2	Ground		6		5	115 203.02	.15	1003
	1703	5, 2, 3← 4, 2, 2	Ground		5		4	115 203.44	.15	1003

Isotopic Species	Id. No.	Rotational Quantum Nos.	Vib. State	Hyperfine				Frequency MHz	Acc. ±MHz	Ref.
				F_1'	F'	F_1	F			
$HN_2^{14}N^{15}$	1703	5, 2, 4← 4, 2, 3	Ground		6		5	115 201.66	.15	1003
	1703	5, 2, 4← 4, 2, 3	Ground		4		3	115 201.66	.15	1003
	1703	5, 2, 4← 4, 2, 3	Ground		5		4	115 202.04	.15	1003
	1703	5, 3, 2← 4, 3, 1	Ground		5		4	115 164.96	.15	1003
	1703	5, 3, 3← 4, 3, 2	Ground		4		3	115 164.13	.15	1003
	1703	5, 3, 3← 4, 3, 2	Ground		6		5	115 164.13	.15	1003
	1703	6, 0, 6← 5, 0, 5	Ground					138 274.77	.15	1003
	1703	6, 1, 5← 5, 1, 4	Ground					138 977.54	.15	1003
	1703	6, 1, 6← 5, 1, 5	Ground					137 557.30	.15	1003
	1703	6, 2, 4← 5, 2, 3	Ground					138 243.25	.15	1003
	1703	6, 2, 5← 5, 2, 4	Ground					138 240.77	.15	1003
	1703	6, 3, 3← 5, 3, 2	Ground					138 195.85	.15	1003
	1703	6, 3, 4← 5, 3, 3	Ground					138 195.85	.15	1003
	1703	7, 0, 7← 6, 0, 6	Ground					161 317.87	.15	1003
	1703	7, 1, 6← 6, 1, 5	Ground					162 138.74	.15	1003
	1703	7, 1, 7← 6, 1, 6	Ground					160 482.02	.15	1003
	1703	7, 2, 5← 6, 2, 4	Ground					161 282.99	.15	1003
	1703	7, 2, 6← 6, 2, 5	Ground					161 279.15	.15	1003
	1703	7, 3, 4← 6, 3, 3	Ground					161 227.43	.15	1003
	1703	7, 3, 5← 6, 3, 4	Ground					161 227.43	.15	1003
	1703	7, 4, 3← 6, 4, 2	Ground					161 152.23	.15	1003
	1703	7, 4, 4← 6, 4, 3	Ground					161 152.23	.15	1003
	1703	8, 0, 8← 7, 0, 7	Ground					184 359.48	.15	1003
	1703	8, 1, 7← 7, 1, 6	Ground					185 298.79	.15	1003
	1703	8, 1, 8← 7, 1, 7	Ground					183 405.55	.15	1003
	1703	8, 2, 6← 7, 2, 5	Ground					184 322.43	.15	1003
	1703	8, 2, 7← 7, 2, 6	Ground					184 316.65	.15	1003
	1703	8, 3, 5← 7, 3, 4	Ground					184 257.81	.15	1003
	1703	8, 3, 6← 7, 3, 5	Ground					184 257.81	.15	1003
	1703	8, 4, 4← 7, 4, 3	Ground					184 171.91	.15	1003
	1703	8, 4, 5← 7, 4, 4	Ground					184 171.91	.15	1003
	1703	9, 0, 9← 8, 0, 8	Ground					207 400.07	.15	1003
	1703	9, 1, 8← 8, 1, 7	Ground					208 457.72	.15	1003
	1703	9, 1, 9← 8, 1, 8	Ground					206 327.94	.15	1003
	1703	9, 2, 7← 8, 2, 6	Ground					207 361.21	.15	1003
	1703	9, 2, 8← 8, 2, 7	Ground					207 352.88	.15	1003
	1703	9, 3, 6← 8, 3, 5	Ground					207 287.36	.15	1003
	1703	9, 3, 7← 8, 3, 6	Ground					207 287.36	.15	1003
$HN^{15}N_2^{14}$	1704	1, 0, 1← 0, 0, 0	Ground		1		1	23 095.02	.05	991
	1704	1, 0, 1← 0, 0, 0	Ground		2		1	23 095.46	.05	991
	1704	1, 0, 1← 0, 0, 0	Ground		0		1	23 096.05	.05	991
	1704	1, 0, 1← 0, 0, 0	Ground					23 096.7		178
	1704	4, 0, 4← 3, 0, 3	Ground					92 379.68	.15	1003
	1704	4, 1, 3← 3, 1, 2	Ground					92 853.53	.15	1003
	1704	4, 1, 4← 3, 1, 3	Ground					91 894.81	.15	1003
	1704	4, 2, 2← 3, 2, 1	Ground					92 356.46	.15	1003
	1704	4, 2, 3← 3, 2, 2	Ground					92 355.67	.15	1003
	1704	4, 3, 1← 3, 3, 0	Ground					92 325.35	.15	1003
	1704	4, 3, 2← 3, 3, 1	Ground					92 325.35	.15	1003

Isotopic Species	Id. No.	Rotational Quantum Nos.	Vib. State	Hyperfine				Frequency MHz	Acc. ±MHz	Ref.
				F_1'	F'	F_1	F			
$HN^{15}N_2^{14}$	1704	5, 0, 5← 4, 0, 4	Ground					115 473.18	.15	1003
	1704	5, 1, 4← 4, 1, 3	Ground					116 065.79	.15	1003
	1704	5, 1, 5← 4, 1, 4	Ground					114 867.30	.15	1003
	1704	5, 2, 3← 4, 2, 2	Ground					115 445.25	.15	1003
	1704	5, 2, 4← 4, 2, 3	Ground					115 443.81	.15	1003
	1704	5, 3, 2← 4, 3, 1	Ground					115 405.63	.15	1003
	1704	5, 3, 3← 4, 3, 2	Ground					115 405.63	.15	1003
	1704	6, 0, 6← 5, 0, 5	Ground					138 565.84	.15	1003
	1704	6, 1, 5← 5, 1, 4	Ground					139 277.82	.15	1003
	1704	6, 1, 6← 5, 1, 5	Ground					137 839.42	.15	1003
	1704	6, 2, 4← 5, 2, 3	Ground					138 533.67	.15	1003
	1704	6, 2, 5← 5, 2, 4	Ground					138 531.21	.15	1003
	1704	6, 3, 3← 5, 3, 2	Ground					138 485.62	.15	1003
	1704	6, 3, 4← 5, 3, 3	Ground					138 485.62	.15	1003
	1704	6, 4, 2← 5, 4, 1	Ground					138 420.28	.15	1003
	1704	6, 4, 3← 5, 4, 2	Ground					138 420.28	.15	1003
	1704	7, 0, 7← 6, 0, 6	Ground					161 657.34	.15	1003
	1704	7, 1, 6← 6, 1, 5	Ground					162 488.56	.15	1003
	1704	7, 1, 7← 6, 1, 6	Ground					160 810.95	.15	1003
	1704	7, 2, 5← 6, 2, 4	Ground					161 621.98	.15	1003
	1704	7, 2, 6← 6, 2, 5	Ground					161 617.96	.15	1003
	1704	7, 3, 4← 6, 3, 3	Ground					161 565.21	.15	1003
	1704	7, 3, 5← 6, 3, 4	Ground					161 565.21	.15	1003
	1704	8, 0, 8← 7, 0, 7	Ground					184 747.53	.15	1003
	1704	8, 1, 7← 7, 1, 6	Ground					185 698.56	.15	1003
	1704	8, 1, 8← 7, 1, 7	Ground					183 781.39	.15	1003
	1704	8, 2, 6← 7, 2, 5	Ground					184 709.52	.15	1003
	1704	8, 2, 7← 7, 2, 6	Ground					184 703.68	.15	1003
	1704	8, 3, 5← 7, 3, 4	Ground					184 643.88	.15	1003
	1704	8, 3, 6← 7, 3, 5	Ground					184 643.88	.15	1003
	1704	8, 4, 4← 7, 4, 3	Ground					184 556.52	.15	1003
	1704	8, 4, 5← 7, 4, 4	Ground					184 556.52	.15	1003
	1704	9, 0, 9← 8, 0, 8	Ground					207 835.99	.15	1003
	1704	9, 1, 8← 8, 1, 7	Ground					208 907.54	.15	1003
	1704	9, 1, 9← 8, 1, 8	Ground					206 750.75	.15	1003
	1704	9, 2, 7← 8, 2, 6	Ground					207 797.01	.15	1003
	1704	9, 2, 8← 8, 2, 7	Ground					207 788.33	.15	1003
	1704	9, 3, 6← 8, 3, 5	Ground					207 721.44	.15	1003
	1704	9, 3, 7← 8, 3, 6	Ground					207 721.44	.15	1003
$HN^{14}N^{15}N^{14}$	1705	1, 0, 1← 0, 0, 0	Ground					23 814.		178

H_2O C_{2v} H_2O

Isotopic Species	Pt. Gp.	Id. No.	A MHz		B MHz		C MHz		D_J MHz	D_{JK} MHz	Δ Amu A²	κ
H_2O^{16}	C_{2v}	1711	819 332.9	F	436 797.7	F	284 503.1	F				
HDO^{16}	C_s	1712	703 960.9	M	273 595.7	M	192 871.4	M	9.1	36.8		−.684
D_2O^{16}	C_{2v}	1713	461 490.	M	217 740.	M	145 460.	M				
HDO^{17}	C_s	1714										

Id. No.	μ_a Debye		μ_b Debye		μ_c Debye		eQq Value(MHz)	Rel.	eQq Value(MHz)	Rel.	eQq Value(MHz)	Rel.	ω_a d 1/cm	ω_b d 1/cm	ω_c d 1/cm	ω_d d 1/cm
1711	0.	X	1.84	M	0.	X										
1712			1.85	M	0.	X	.313	OD								
1713	0.	X	1.87	M	0.	X	.313	OD								
1714							−8.13									

References:

ABC: 423,555,1029 D_J: 555 D_{JK}: 555 κ: 555 μ: 167,333   eQq: 789,920

Add. Ref. 5,24,48,49,65,93,108,109,139,212,279,362,396,417,424,427,435,468,469,487,497,518,557,558,627,633,643,650,672,678,717,739

787,788,801,842,843,893

For species 1712, the following centrifugal distortion parameters are given in ref. 555: $D_K = 287$ MHz, $\delta_J = 3.333$ MHz, $R_5 = 7.877$ MHz, $R_6 = -0.572$ MHz, $R_7^{(x)} = 3.12$ MHz, $R_8^{(y)} = -8.20$ MHz, $R_9^{(y)} = 50.0$ MHz.

Water   Spectral Line Table

Isotopic Species	Id. No.	Rotational Quantum Nos.	Vib. State	F_1'	F'	F_1	F	Frequency MHz	Acc. ±MHz	Ref.
H_2O^{16}	1711	3, 1, 3← 2, 2, 0	Ground					183 311.30	.30	533
	1711	6, 1, 6← 5, 2, 3	Ground					22 235.22	.05	80
HDO^{16}	1712	1, 1, 0← 1, 1, 1	Ground					80 578.15		673
	1712	2, 1, 1← 2, 1, 2	Ground					241 561.3		673
	1712	2, 2, 1← 2, 2, 0	Ground	1	3/2	2	5/2	10 278.0796	.001	1002 [1]
	1712	2, 2, 1← 2, 2, 0	Ground	3	5/2	3	7/2	10 278.1365	.001	1002 [1]
	1712	2, 2, 1← 2, 2, 0	Ground	1	1/2	1	3/2	10 278.1681	.001	1002 [1]
	1712	2, 2, 1← 2, 2, 0	Ground	3	5/2	1	3/2	10 278.2255	.001	1002 [1]
	1712	2, 2, 1← 2, 2, 0	Ground					10 278.2455	.001	1002 [1]
	1712	2, 2, 1← 2, 2, 0	Ground	1	3/2	3	5/2	10 278.2643	.001	1002 [1]
	1712	2, 2, 1← 2, 2, 0	Ground	1	3/2	1	1/2	10 278.3234	.001	1002 [1]
	1712	2, 2, 1← 2, 2, 0	Ground	3	7/2	3	5/2	10 278.3554	.001	1002 [1]
	1712	2, 2, 1← 2, 2, 0	Ground	2	5/2	1	3/2	10 278.4126	.001	1002 [1]
	1712	3, 2, 1← 3, 2, 2	Ground					50 236.30		673
	1712	3, 3, 0← 3, 3, 1	Ground	3	5/2	2	3/2	824.4754	.002	985 [1]
	1712	3, 3, 0← 3, 3, 1	Ground	4	9/2	4	7/2	824.5074	.002	985 [1]
	1712	3, 3, 0← 3, 3, 1	Ground	3	5/2	4	7/2	824.5247	.002	985 [1]

---

1. See references given for accuracy of frequency differences.

Isotopic Species	Id. No.	Rotational Quantum Nos.	Vib. State	Hyperfine F_1'	F'	F_1	F	Frequency MHz	Acc. ±MHz	Ref.
HDO[16]	1712	3, 3, 0← 3, 3, 1	Ground	3	7/2	3	5/2	824.5488	.002	985 [1]
	1712	3, 3, 0← 3, 3, 1	Ground	3	7/2	4	9/2	824.5685	.002	985 [1]
	1712	3, 3, 0← 3, 3, 1	Ground	2	5/2	4	7/2	824.6042	.002	985 [1]
	1712	3, 3, 0← 3, 3, 1	Ground					824.6706	.002	985 [1]
	1712	3, 3, 0← 3, 3, 1	Ground	2	5/2	4	7/2	824.7419	.002	985 [1]
	1712	3, 3, 0← 3, 3, 1	Ground	3	7/2	4	9/2	824.7730	.002	985 [1]
	1712	3, 3, 0← 3, 3, 1	Ground	3	7/2	3	5/2	824.7904	.002	985 [1]
	1712	3, 3, 0← 3, 3, 1	Ground	3	5/2	4	7/2	824.8136	.002	985 [1]
	1712	3, 3, 0← 3, 3, 1	Ground	4	9/2	4	7/2	824.8341	.002	985 [1]
	1712	3, 3, 0← 3, 3, 1	Ground	3	5/2	2	3/2	824.8637	.002	985 [1]
	1712	3, 2, 1← 4, 1, 4	Ground					20 460.40		167
	1712	4, 2, 2← 4, 2, 3	Ground					143 727.2		673
	1712	4, 3, 1← 4, 3, 2	Ground					5 702.78		555
	1712	5, 0, 5← 4, 2, 2	Ground					2 887.4	.1	415
	1712	5, 1, 5← 4, 2, 2	Ground					120 778.2		673
	1712	5, 3, 2← 5, 3, 3	Ground					22 307.67	.05	110
	1712	5, 4, 1← 5, 4, 2	Ground	5	11/2	5	9/2	486.266	.002	985 [1]
	1712	5, 4, 1← 5, 4, 2	Ground	5	11/2	4	9/2	486.450	.002	985 [1]
	1712	5, 4, 1← 5, 4, 2	Ground	5	11/2	6	13/2	486.487	.002	985 [1]
	1712	5, 4, 1← 5, 4, 2	Ground					486.528	.002	985 [1]
	1712	5, 4, 1← 5, 4, 2	Ground	5	11/2	6	13/2	486.569	.002	985 [1]
	1712	5, 4, 1← 5, 4, 2	Ground	5	11/2	4	9/2	486.606	.002	985 [1]
	1712	6, 1, 6← 5, 2, 3	Ground					138 530.4		673
	1712	6, 4, 2← 6, 4, 3	Ground					2 394.56	.05	415
	1712	7, 1, 7← 6, 2, 4	Ground					26 880.38	.05	555
	1712	7, 4, 3← 7, 4, 4	Ground					8 577.7	.1	555
	1712	8, 4, 4← 8, 4, 5	Ground					24 884.77	.05	555
	1712	9, 5, 4← 9, 5, 5	Ground					3 044.71	.10	415
	1712	10, 5, 5←10, 5, 6	Ground					8 836.95	.1	555
	1712	11, 5, 6←11, 5, 7	Ground					22 581.1	.2	555
	1712	12, 6, 6←12, 6, 7	Ground					2 961.	1.	415
D₂O[16]	1713	3, 1, 3← 2, 2, 0	Ground	7/2	3	5/2	2	10 919.301	.001	777
	1713	3, 1, 3← 2, 2, 0	Ground	9/2	5	7/2	4	10 919.357	.001	777
	1713	3, 1, 3← 2, 2, 0	Ground	5/2	4	3/2	3	10 919.521	.001	777
	1713	3, 1, 3← 2, 2, 0	Ground	9/2	3	7/2	2	10 919.603	.001	777
	1713	4, 2, 3← 3, 3, 0	Ground					43 414.57		673
	1713	4, 4, 0← 5, 3, 3	Ground					55 482.32		673
	1713	4, 4, 1← 5, 3, 2	Ground					10 947.13	.05	555
	1713	6, 1, 6← 5, 2, 3	Ground					90 916.8		673
HDO[17]	1714	2, 2, 0← 2, 2, 1	Ground					10 374.56		789
	1714	3, , ← 2, ,	Ground					23 374.4	.05	902
	1714	3, , ← 2, ,	Ground					23 481.6	.05	902
	1714	3, , ← 2, ,	Ground					23 585.6	.05	902
	1714	3, , ← 2, ,	Ground					23 646.3	.05	902
	1714	3, , ← 2, ,	Ground					24 256.0	.05	902
	1714	3, , ← 2, ,	Ground					24 280.5	.05	902
	1714	3, , ← 2, ,	Ground					24 384.9	.5	902
	1714	3, , ← 2, ,	Ground					24 472.6	.5	902
	1714	3, , ← 2, ,	Ground					24 528.8	.5	902

1. See references given for accuracy of frequency differences.

H₂S C_{2v} H₂S

Isotopic Species	Pt. Gp.	Id. No.	A MHz		B MHz		C MHz		D_J MHz	D_{JK} MHz	Δ Amu A²	κ
H_2S^b	C_{2v}	1721	309 985.4	F	270 832.5	F	141 681.9	F				
D_2S^b	C_{2v}	1722	164 196.4	F	140 722.6	F	73 868.87	F				
H_2S^{32}	C_{2v}	1724	316 304.	M	276 512.	M	147 536.	M				
H_2S^{33}	C_{2v}	1725	315 735.	M	276 512.	M	147 412.	M				
H_2S^{34}	C_{2v}	1726	315 201.	M	276 512.	M	147 296.	M				
HDS^{32}	C_s	1727	290 259.1	M	145 219.5	M	94 314.84	M		27.380		−.47767
D_2S^{32}	C_{2v}	1731										.451

Id. No.	μ_a Debye		μ_b Debye		μ_c Debye		eQq Value(MHz)	Rel.	eQq Value(MHz)	Rel.	eQq Value(MHz)	Rel.	ω_a d 1/cm	ω_b d 1/cm	ω_c d 1/cm	ω_d d 1/cm
1721	.897	G			0.	X											
1725							−32	aa	−8	bb	40	cc				
1727	1.02	M	0.	X	0.	X										

References:

ABC: 278,425,677,846 D_{JK}: 278   κ: 278,846   μ: 278,995   eQq: 425

Add. Ref. 49,137,201,277,301,677,847

For species 1727: $(A-C)/2 = 97924.2$ MHz, $D_K = -4.9166$ MHz, Ref. 278.

No Spectral Lines for Species 1721-2-3

Isotopic Species	Id. No.	Rotational Quantum Nos.	Vib. State	Hyperfine				Frequency MHz	Acc. ±MHz	Ref.
				F_1'	F'	F_1	F			
H_2S^{32}	1724	1, 1, 0← 1, 0, 1	Ground					168 762.51	.35	425
	1724	2, 2, 0← 2, 1, 1	Ground					216 710.42	.45	425
H_2S^{33}	1725	1, 1, 0← 1, 0, 1	Ground					168 322.63	.35	425
H_2S^{34}	1726	1, 1, 0← 1, 0, 1	Ground					167 910.57	.35	425
HDS^{32}	1727	1, 1, 0← 1, 1, 1	Ground					51 073.27	.05	278
	1727	2, 2, 0← 2, 2, 1	Ground					11 283.83	.05	278
	1727	3, 2, 1← 3, 2, 2	Ground					53 200.93	.05	278
	1727	4, 3, 1← 4, 3, 2	Ground					10 861.07	.05	278
	1727	5, 3, 2← 5, 3, 3	Ground					40 929.20	.05	278
	1727	6, 4, 2← 6, 4, 3	Ground					7 936.74	.05	278
	1727	7, 4, 3← 7, 4, 4	Ground					27 566.31	.05	278
	1727	8, 4, 4← 8, 4, 5	Ground					75 551.73	.05	278
	1727	9, 5, 4← 9, 5, 5	Ground					17 212.61	.05	278
	1727	10, 5, 5←10, 5, 6	Ground					47 905.36	.05	278
	1727	11, 6, 5←11, 6, 6	Ground					10 235.81	.05	278
	1727	12, 6, 6←12, 6, 7	Ground					28 842.84	.05	278
HDS^{33}	1728	2, 2, 0← 2, 2, 1	Ground		3/2		1/2	11 251.28		584
	1728	2, 2, 0← 2, 2, 1	Ground		5/2		7/2	11 252.85		584
	1728	2, 2, 0← 2, 2, 1	Ground		5/2		3/2	11 254.82		584
	1728	2, 2, 0← 2, 2, 1	Ground		1/2		1/2	11 257.16		584
	1728	2, 2, 0← 2, 2, 1	Ground		7/2		7/2	11 258.55		584
	1728	2, 2, 0← 2, 2, 1	Ground		3/2		3/2	11 259.09		584
	1728	2, 2, 0← 2, 2, 1	Ground		5/2		5/2	11 260.52		584
	1728	2, 2, 0← 2, 2, 1	Ground		3/2		5/2	11 264.78		584
	1728	2, 2, 0← 2, 2, 1	Ground		7/2		5/2	11 266.35		584
	1728	4, 3, 1← 4, 3, 2	Ground		5/2		5/2	10 830.54		584
	1728	4, 3, 1← 4, 3, 2	Ground		11/2		11/2	10 830.83		584
	1728	4, 3, 1← 4, 3, 2	Ground		7/2		7/2	10 831.37		584
	1728	4, 3, 1← 4, 3, 2	Ground		5/2		5/2	10 831.63		584
HDS^{34}	1729	1, 1, 0← 1, 1, 1	Ground					50 912.27		278
	1729	2, 2, 0← 2, 2, 1	Ground					11 235.45		278
	1729	3, 2, 1← 3, 2, 2	Ground					52 979.67		278
	1729	4, 3, 1← 4, 3, 2	Ground					10 802.36		278
	1729	7, 4, 3← 7, 4, 4	Ground					27 392.00		278

H₂Se   C₂ᵥ   H₂Se

Isotopic Species	Pt. Gp.	Id. No.	A MHz		B MHz		C MHz		D_J MHz	D_JK MHz	Δ Amu A²	κ
H₂Se⁷⁶	C₂ᵥ	1741	245 381.	M	231 778.	M	117 139.	M				
H₂Se⁷⁷	C₂ᵥ	1742	245 299.	M	231 777.	M	117 120.	M				
H₂Se⁷⁸	C₂ᵥ	1743	245 229.	M	231 791.	M	117 107.	M				
H₂Se⁸⁰	C₂ᵥ	1744	245 060.	M	231 772.	M	117 063.	M				
H₂Se⁸²	C₂ᵥ	1745	244 913.	M	231 772.	M	117 029.	M				
D₂Se⁷⁶	C₂ᵥ	1746	125 946.6	M	115 906.4	M	59 614.5	M				
D₂Se⁷⁷	C₂ᵥ	1747	125 864.1	M	115 906.2	M	59 596.0	M				
D₂Se⁷⁸	C₂ᵥ	1748	125 784.0	M	115 906.4	M	59 577.9	M				
D₂Se⁸⁰	C₂ᵥ	1749	125 629.5	M	115 906.1	M	59 542.9	M				
D₂Se⁸²	C₂ᵥ	1751	125 482.7	M	115 906.5	M	59 509.7	M				
HDSe⁷⁷	Cₛ	1755								.00022		−.47889
HDSe⁷⁸	Cₛ	1756								.00021		−.47906
HDSe⁸⁰	Cₛ	1757								.00022		−.47926
HDSe⁸²	Cₛ	1758								.00022		−.47959
HDSe⁷⁶	Cₛ	1759								.00023		−.47882

Id. No.	μ_a Debye	μ_b Debye	μ_c Debye	eQq Value(MHz) Rel.	eQq Value(MHz) Rel.	eQq Value(MHz) Rel.	ω_a d 1/cm	ω_b d 1/cm	ω_c d 1/cm	ω_d d 1/cm
1749		.24 M								
1755		.62 M								
1756		.62 M								
1757		.62 M								

References:

ABC: 888,978   D_JK: 888   κ: 888   μ: 691, 888,

Add. Ref. 49,796,877

Species	1755	1756	1757	1758	1759	Ref.
D_K (MHz)	−.000357	−.000358	−.000338	−.000372	−.000371	888
R_5 (MHz)	−5870	−5750	−6000	−5430	−5710	888
R_6 (MHz)	275000	288000	272000	299000	308000	888
δ_J (MHz)	41000	36400	38500	31000	35600	888

Isotopic Species	Id. No.	Rotational Quantum Nos.	Vib. State	F_1'	F'	F_1	F	Frequency MHz	Acc. ±MHz	Ref.
H_2Se^{76}	1741	1, 1, 0← 1, 0, 1	Ground					128 219.10		691
	1741	2, 2, 0← 2, 1, 1	Ground					142 783.02		691
	1741	3, 3, 0← 3, 2, 1	Ground					166 488.14		691
H_2Se^{77}	1742	1, 1, 0← 1, 0, 1	Ground					128 155.40		691
	1742	2, 2, 0← 2, 1, 1	Ground					142 623.48		691
	1742	3, 3, 0← 3, 2, 1	Ground					166 163.20		691
H_2Se^{78}	1743	1, 1, 0← 1, 0, 1	Ground					128 098.70		691
	1743	2, 2, 0← 2, 1, 1	Ground					142 469.58		691
	1743	3, 3, 0← 3, 2, 1	Ground					165 847.57		691
H_2Se^{80}	1744	1, 1, 0← 1, 0, 1	Ground					127 973.40		691
	1744	2, 2, 0← 2, 1, 1	Ground					142 171.86		691
	1744	3, 3, 0← 3, 2, 1	Ground					165 240.46		691
H_2Se^{82}	1745	1, 1, 0← 1, 0, 1	Ground					127 860.35		691
	1745	2, 2, 0← 2, 1, 1	Ground					141 889.02		691
	1745	3, 3, 0← 3, 2, 1	Ground					164 663.10		691
D_2Se^{76}	1746	1, 1, 1← 0, 0, 0	Ground					185 549.92		691
	1746	1, 1, 0← 1, 0, 1	Ground					66 330.78		691
	1746	2, 1, 1← 2, 0, 2	Ground					170 071.51		691
	1746	2, 2, 0← 2, 1, 1	Ground					77 542.71		691
	1746	3, 3, 0← 3, 2, 1	Ground					96 423.40		691
D_2Se^{77}	1747	1, 1, 1← 0, 0, 0	Ground					185 448.96		691
	1747	1, 1, 0← 1, 0, 1	Ground					66 266.82		691
	1747	2, 1, 1← 2, 0, 2	Ground					170 107.35		691
	1747	2, 2, 0← 2, 1, 1	Ground					77 377.10		691
	1747	3, 3, 0← 3, 2, 1	Ground					96 073.00		691
D_2Se^{78}	1748	1, 1, 1← 0, 0, 0	Ground					185 350.80		691
	1748	1, 1, 0← 1, 0, 1	Ground					66 204.87		691
	1748	2, 1, 1← 2, 0, 2	Ground					170 142.70		691
	1748	2, 2, 0← 2, 1, 1	Ground					77 216.40		691
	1748	3, 3, 0← 3, 2, 1	Ground					95 733.92		691
D_2Se^{80}	1749	1, 1, 1← 0, 0, 0	Ground					185 161.28		691
	1749	1, 1, 0← 1, 0, 1	Ground					66 085.41		691
	1749	2, 1, 1← 2, 0, 2	Ground					170 211.30		691
	1749	2, 2, 0← 2, 1, 1	Ground					76 907.04		691
	1749	3, 3, 0← 3, 2, 1	Ground					95 082.56		691
D_2Se^{82}	1751	1, 1, 1← 0, 0, 0	Ground					184 981.28		691
	1751	1, 1, 0← 1, 0, 1	Ground					65 971.71		691
	1751	2, 1, 1← 2, 0, 2	Ground					170 277.03		691
	1751	2, 2, 0← 2, 1, 1	Ground					76 612.14		691
	1751	3, 3, 0← 3, 2, 1	Ground					94 462.12		691
D_2Se^b	1754	Not Reported	Ground					152 808.24		691
	1754	Not Reported	Ground					152 955.12		691
	1754	Not Reported	Ground					153 126.12		691
	1754	Not Reported	Ground					153 214.14		691
	1754	Not Reported	Ground					153 312.56		691
	1754	Not Reported	Ground					153 842.64		691
	1754	Not Reported	Ground					153 877.80		691
	1754	Not Reported	Ground					153 920.46		691

Isotopic Species	Id. No.	Rotational Quantum Nos.	Vib. State	Hyperfine F_1'	F'	F_1	F	Frequency MHz	Acc. ±MHz	Ref.
D₂Se[b]	1754	Not Reported	Ground					154 001.70		691
	1754	Not Reported	Ground					154 080.18		691
	1754	Not Reported	Ground					181 658.80		691
	1754	Not Reported	Ground					182 226.76		691
	1754	Not Reported	Ground					182 279.28		691
	1754	Not Reported	Ground					184 266.88		691
	1754	Not Reported	Ground					184 655.68		691
	1754	Not Reported						185 062.80		691
HDSe[77]	1755	1, 1, 1← 1, 1, 0	Ground					41 549.0		888
	1755	2, 2, 1← 2, 2, 0	Ground					9 155.85	.1	797
	1755	3, 2, 2← 3, 2, 1	Ground					43 183.1		888
	1755	4, 3, 2← 4, 3, 1	Ground					8 793.95	.1	797
	1755	5, 3, 3← 5, 3, 2	Ground					33 157.4		888
	1755	7, 4, 4← 7, 4, 3	Ground					22 302.3		888
	1755	9, 5, 5← 9, 5, 4	Ground					13 918.3	.1	797
	1755	11, 6, 6←11, 6, 5	Ground					8 280.0		888
HDSe[78]	1756	1, 1, 1← 1, 1, 0	Ground					41 534.1		888
	1756	2, 2, 1← 2, 2, 0	Ground					9 149.65	.1	797
	1756	3, 2, 2← 3, 2, 1	Ground					43 156.8		888
	1756	4, 3, 2← 4, 3, 1	Ground					8 786.05	.1	797
	1756	5, 3, 3← 5, 3, 2	Ground					33 129.7		888
	1756	7, 4, 4← 7, 4, 3	Ground					22 277.6		888
	1756	9, 5, 5← 9, 5, 4	Ground					13 899.3	.1	797
	1756	10, 5, 6←10, 5, 5	Ground					38 724.0		888
	1756	11, 6, 6←11, 6, 5	Ground					8 266.6		888
HDSe[80]	1757	1, 1, 1← 1, 1, 0	Ground					41 504.5		888
	1757	2, 2, 1← 2, 2, 0	Ground					9 138.55	.1	797
	1757	3, 2, 2← 3, 2, 1	Ground					43 106.6		888
	1757	4, 3, 2← 4, 3, 1	Ground					8 771.05	.1	797
	1757	5, 3, 3← 5, 3, 2	Ground					33 075.4		888
	1757	7, 4, 4← 7, 4, 3	Ground					22 230.0		888
	1757	9, 5, 5← 9, 5, 4	Ground					13 862.65	.1	797
	1757	10, 5, 6←10, 5, 5	Ground					38 629.0		888
	1757	11, 6, 6←11, 6, 5	Ground					8 240.6		888
	1757	12, 6, 7←12, 6, 6	Ground					23 249.8		888
HDSe[82]	1758	1, 1, 1← 1, 1, 0	Ground					41 476.1		888
	1758	2, 2, 1← 2, 2, 0	Ground					9 127.75	.1	797
	1758	3, 2, 2← 3, 2, 1	Ground					43 058.4		888
	1758	4, 3, 2← 4, 3, 1	Ground					8 756.7	.1	797
	1758	5, 3, 3← 5, 3, 2	Ground					33 023.0		888
	1758	7, 4, 4← 7, 4, 3	Ground					22 184.4		888
	1758	9, 5, 5← 9, 5, 4	Ground					13 827.7	.1	797
	1758	11, 6, 6←11, 6, 5	Ground					8 215.8		888
HDSe[76]	1759	1, 1, 1← 1, 1, 0	Ground					41 565.4		888
	1759	2, 2, 1← 2, 2, 0	Ground					9 161.50	.1	797
	1759	3, 2, 2← 3, 2, 1	Ground					43 209.8		888
	1759	4, 3, 2← 4, 3, 1	Ground					8 801.85	.1	797
	1759	5, 3, 3← 5, 3, 2	Ground					33 188.2		888
	1759	7, 4, 4← 7, 4, 3	Ground					22 327.7		888
	1759	9, 5, 5← 9, 5, 4	Ground					13 937.8	.1	797
	1759	11, 6, 6←11, 6, 5	Ground					8 293.6		888

Iodosilicane

H₃ISi

C₃ᵥ

SiH₃I

Isotopic Species	Pt. Gp.	Id. No.	A MHz	B MHz	C MHz	D_J MHz	D_{JK} MHz	Δ Amu A²	κ
Si²⁸H₃I¹²⁷	C₃ᵥ	1761		3 215.52 M	3 215.52 M	.0013	.0202		

Id. No.	μ_a Debye	μ_b Debye	μ_c Debye	eQq Value(MHz) Rel.		eQq Value(MHz) Rel.		eQq Value(MHz) Rel.		ω_a d 1/cm	ω_b d 1/cm	ω_c d 1/cm	ω_d d 1/cm
1761				−1244.8	I¹²⁷								

References:

ABC: 720 D_J: 720 D_{JK}: 720   eQq: 720

Add. Ref. 493

Isotopic Species	Id. No.	Rotational Quantum Nos.	Vib. State	F_1'	F'	F_1	F	Frequency MHz	Acc. ±MHz	Ref.
Si²⁸H₃I¹²⁷	1761	4, 0← 3, 0	Ground		7/2		5/2	25 685.10	.1	720
	1761	4, 0← 3, 0	Ground		9/2		7/2	25 713.59	.1	720
	1761	4, 0← 3, 0	Ground		13/2		11/2	25 733.43	.1	720
	1761	4, 0← 3, 0	Ground		11/2		9/2	25 740.94	.1	720
	1761	4, 1← 3, 1	Ground		7/2		5/2	25 689.74	.1	720
	1761	4, 1← 3, 1	Ground		9/2		7/2	25 702.85	.1	720
	1761	4, 1← 3, 1	Ground		11/2		9/2	25 725.23	.1	720
	1761	4, 1← 3, 1	Ground		13/2		11/2	25 742.37	.1	720
	1761	4, 2← 3, 2	Ground		9/2		7/2	25 670.18	.1	720
	1761	4, 2← 3, 2	Ground		11/2		9/2	25 679.38	.1	720
	1761	4, 2← 3, 2	Ground		7/2		5/2	25 704.16	.1	720
	1761	4, 2← 3, 2	Ground		5/2		3/2	25 758.20	.1	720
	1761	4, 2← 3, 2	Ground		13/2		11/2	25 768.92	.1	720
	1761	4, 3← 3, 3	Ground		11/2		9/2	25 605.80	.1	720
	1761	4, 3← 3, 3	Ground		9/2		7/2	25 614.02	.1	720
	1761	4, 3← 3, 3	Ground		13/2		11/2	25 812.69	.1	720
	1761	4, 3← 3, 3	Ground		5/2		3/2	25 864.27	.1	720
	1761	5, 0← 4, 0	Ground		7/2		5/2	32 127.83	.1	720
	1761	5, 0← 4, 0	Ground		9/2		7/2	32 134.64	.1	720
	1761	5, 0← 4, 0	Ground		11/2		9/2	32 151.21	.1	720
	1761	5, 0← 4, 0	Ground		15/2		13/2	32 161.12	.1	720
	1761	5, 0← 4, 0	Ground		13/2		11/2	32 166.04	.1	720
	1761	5, 1← 4, 1	Ground		7/2		5/2	32 134.64	.1	720
	1761	5, 1← 4, 1	Ground		11/2		9/2	32 144.78	.1	720
	1761	5, 1← 4, 1	Ground		13/2		11/2	32 159.40	.1	720
	1761	5, 1← 4, 1	Ground		15/2		13/2	32 166.04	.1	720
	1761	5, 2← 4, 2	Ground		11/2		9/2	32 125.17	.1	720
	1761	5, 2← 4, 2	Ground		9/2		7/2	32 132.25	.1	720
	1761	5, 2← 4, 2	Ground		13/2		11/2	32 138.83	.1	720
	1761	5, 2← 4, 2	Ground		15/2		13/2	32 180.53	.1	720

Isotopic Species	Id. No.	Rotational Quantum Nos.	Vib. State	Hyperfine				Frequency MHz	Acc. ±MHz	Ref.
				F_1'	F'	F_1	F			
$Si^{28}H_3I^{127}$	1761	5, 3← 4, 3	Ground		11/2		9/2	32 092.41	.1	720
	1761	5, 3← 4, 3	Ground		13/2		11/2	32 105.13	.1	720
	1761	5, 3← 4, 3	Ground		9/2		7/2	32 129.89	.1	720
	1761	5, 3← 4, 3	Ground		15/2		13/2	32 204.60	.1	720
	1761	5, 4← 4, 4	Ground		11/2		9/2	32 046.03	.1	720
	1761	5, 4← 4, 4	Ground		13/2		11/2	32 059.38	.1	720
	1761	5, 4← 4, 4	Ground		15/2		13/2	32 237.95	.1	720
	1761	6, 0← 5, 0	Ground		9/2		7/2	38 567.79	.1	720
	1761	6, 0← 5, 0	Ground		11/2		9/2	38 572.70	.1	720
	1761	6, 0← 5, 0	Ground		7/2		5/2	38 575.61	.1	720
	1761	6, 0← 5, 0	Ground		13/2		11/2	38 584.21	.1	720
	1761	6, 0← 5, 0	Ground		17/2		15/2	38 590.02	.1	720
	1761	6, 0← 5, 0	Ground		15/2		13/2	38 593.85	.1	720
	1761	6, 1← 5, 1	Ground		13/2		11/2	38 580.18	.1	720
	1761	6, 1← 5, 1	Ground		7/2		5/2	38 581.91	.1	720
	1761	6, 1← 5, 1	Ground		15/2		13/2	38 590.02	.1	720
	1761	6, 1← 5, 1	Ground		17/2		15/2	38 592.81	.1	720
	1761	6, 2← 5, 2	Ground		13/2		11/2	38 567.79	.1	720
	1761	6, 2← 5, 2	Ground		11/2		9/2	38 567.79	.1	720
	1761	6, 2← 5, 2	Ground		15/2		13/2	38 579.11	.1	720
	1761	6, 2← 5, 2	Ground		9/2		7/2	38 579.11	.1	720
	1761	6, 2← 5, 2	Ground		17/2		15/2	38 601.31	.1	720
	1761	6, 2← 5, 2	Ground		7/2		5/2	38 601.31	.1	720
	1761	6, 3← 5, 3	Ground		13/2		11/2	38 547.50	.1	720
	1761	6, 3← 5, 3	Ground		11/2		9/2	38 561.14	.1	720
	1761	6, 3← 5, 3	Ground		15/2		13/2	38 561.14	.1	720
	1761	6, 3← 5, 3	Ground		9/2		7/2	38 592.81	.1	720
	1761	6, 3← 5, 3	Ground		17/2		15/2	38 615.65	.1	720

H$_3$N   C$_{3v}$ NH$_3$

Isotopic Species	Pt. Gp.	Id. No.	A MHz		B MHz		C MHz	D$_J$ MHz	D$_{JK}$ MHz	Δ Amu A²	κ
N^{14}H$_3$	C$_{3v}$	1771	189 000.	M	298 000.	M		22.680	−41.19		
N^{14}H$_2$D	C$_s$	1773									−.315
N^{14}D$_2$H	C$_s$	1774									−.1385
N^{14}D$_3$	C$_{3v}$	1775						5.670	−10.68		

Id. No.	μ$_a$ Debye	μ$_b$ Debye	μ$_c$ Debye	eQq Value(MHz)	Rel.	eQq Value(MHz)	Rel.	eQq Value(MHz)	Rel.	ω$_a$ d 1/cm	ω$_b$ d 1/cm	ω$_c$ d 1/cm	ω$_d$ d 1/cm
1771	1.468 M			−4.084	N^{14}					950 1			
1773				−4.10	N^{14}								
1774				−4.10	N^{14}								
1775				−4.080	N^{14}								

References:

ABC: 749,1029   D$_J$: 879   D$_{JK}$: 879   κ: 330   μ: 258   eQq: 330,525,823   ω: 1029

Add. Ref. 1,2,4,11,12,13,14,16,17,18,20,23,27,28,36,37,38,44,45,46,50,51,52,56,59,66,67,84,87,88,90,91,92,94,95,101,107,114,122,138,140

148,149,150,151,152,153,179,184,185,214,215,246,260,265,274,280,281,287,290,299,300,303,304,305,310,320,324,325,328,329

337,338,362,365,369,388,399,400,418,419,420,452,453,465,520,524,525,532,567,629,632,640,661,677,681,683,687,693,705,714

722,742,748,751,757,758,760,761,769,770,790,793,794,811,812,821,840,878,879,898,927,1015,1016,1018,1019,1023,1024,1025,1027

Ref. 330 gives: for species 1773, (A − C)/2 = 74350 MHz, κ = − .315; for species 1774, (A − C)/2 = 55200 MHz, κ = − .1385.

For species 1775, ref. 823 gives eQq(D) = 200 MHz.

For species 1774, some inversion splittings are also to be found in ref. 330.

Isotopic Species	Id. No.	Rotational Quantum Nos.	Vib. State v_a	Hyperfine				Frequency MHz	Acc. ±MHz	Ref.
				F_1'	F'	F_1	F			
$N^{14}H_3$	1771	1, 1← 1, 1	Ground					23 694.49	.02	35
	1771	2, 1← 2, 1	Ground					23 098.79	.02	35
	1771	2, 2← 2, 2	Ground					23 722.63	.02	35
	1771	3, 1← 3, 1	Ground					22 234.53	.02	35
	1771	3, 2← 3, 2	Ground					22 834.17	.02	35
	1771	3, 3← 3, 3	Ground					23 870.129		771 [1]
	1771	3, 3← 3, 3	Ground					23 870.129		662 [1]
	1771	3, 3← 3, 3	Ground					23 870.130		841 [1]
	1771	3, 3← 3, 3	Ground					23 870.131		782 [1]
	1771	4, 1← 4, 1	Ground					21 134.29	.02	35
	1771	4, 2← 4, 2	Ground					21 703.36	.02	35
	1771	4, 3← 4, 3	Ground					22 688.29	.02	35
	1771	4, 4← 4, 4	Ground					24 139.41	.02	35
	1771	5, 1← 5, 1	Ground					19 838.26	.02	623
	1771	5, 1← 5, 1	Ground					19 838.41	.02	623
	1771	5, 2← 5, 2	Ground					20 371.46	.02	35
	1771	5, 3← 5, 3	Ground					21 285.27	.02	35
	1771	5, 4← 5, 4	Ground					22 653.00	.02	35
	1771	5, 5← 5, 5	Ground					24 532.98	.02	35
	1771	6, 1← 6, 1	Ground					18 391.46	.04	623
	1771	6, 1← 6, 1	Ground					18 391.65	.04	623
	1771	6, 2← 6, 2	Ground					18 884.76	.04	623
	1771	6, 3← 6, 3	Ground					19 757.40	.04	623
	1771	6, 4← 6, 4	Ground					20 994.61	.02	35
	1771	6, 5← 6, 5	Ground					22 732.43	.02	35
	1771	6, 6← 6, 6	Ground					25 056.02	.02	35
	1771	7, 1← 7, 1	Ground					16 840.95	.04	623
	1771	7, 1← 7, 1	Ground					16 841.16	.04	623
	1771	7, 2← 7, 2	Ground					17 291.54	.04	623
	1771	7, 3← 7, 3	Ground					18 017.42	.04	623
	1771	7, 4← 7, 4	Ground					19 218.36	.04	623
	1771	7, 5← 7, 5	Ground					20 804.83	.02	35
	1771	7, 6← 7, 6	Ground					22 924.94	.02	35
	1771	7, 7← 7, 7	Ground					25 715.17	.02	35
	1771	8, 1← 8, 1	Ground					15 233.12	.04	623
	1771	8, 1← 8, 1	Ground					15 233.36	.04	623
	1771	8, 2← 8, 2	Ground					15 639.84	.06	623
	1771	8, 3← 8, 3	Ground					16 455.13	.04	623
	1771	8, 4← 8, 4	Ground					17 378.14	.04	623
	1771	8, 5← 8, 5	Ground					18 808.56	.04	623
	1771	8, 6← 8, 6	Ground					20 719.21	.02	35
	1771	8, 7← 8, 7	Ground					23 232.24	.02	35
	1771	8, 8← 8, 8	Ground					26 518.91	.10	100
	1771	9, 1← 9, 1	Ground					13 612.08	.04	623
	1771	9, 1← 9, 1	Ground					13 612.36	.04	623
	1771	9, 2← 9, 2	Ground					13 974.54	.08	623
	1771	9, 3← 9, 3	Ground					14 376.56	.06	623
	1771	9, 4← 9, 4	Ground					15 523.96	.06	623
	1771	9, 5← 9, 5	Ground					16 798.22	.04	623
	1771	9, 6← 9, 6	Ground					18 499.28	.04	623

1. These lines were obtained with an accuracy of .00005 MHz. Correct values are: 23 870.12931, 23 870.12942, 23 870.13005, and 23 870.13105 MHz.

Isotopic Species	Id. No.	Rotational Quantum Nos.	Vib. State v_a	Hyperfine				Frequency MHz	Acc. ±MHz	Ref.
				F_1'	F'	F_1	F			
$N^{14}H_3$	1771	9, 7← 9, 7	Ground					20 735.44	.02	35
	1771	9, 8← 9, 8	Ground					23 657.48	.02	35
	1771	9, 9← 9, 9	Ground					27 478.00	.10	100
	1771	10, 1←10, 1	Ground					12 017.02	.020	752
	1771	10, 1←10, 1	Ground					12 017.30	.020	752
	1771	10, 2←10, 2	Ground					12 336.48	.05	752
	1771	10, 3←10, 3	Ground					13 296.37	.06	623
	1771	10, 4←10, 4	Ground					13 700.96	.06	623
	1771	10, 5←10, 5	Ground					14 822.70	.06	623
	1771	10, 6←10, 6	Ground					16 319.38	.04	623
	1771	10, 7←10, 7	Ground					18 285.55	.04	623
	1771	10, 8←10, 8	Ground					20 852.51	.02	35
	1771	10, 9←10, 9	Ground					24 205.29	.02	35
	1771	10,10←10,10	Ground					28 604.73	.10	100
	1771	11, 1←11, 1	Ground					10 481.73	.025	752
	1771	11, 1←11, 1	Ground					10 482.02	.025	752
	1771	11, 2←11, 2	Ground					10 759.82	.05	752
	1771	11, 3←11, 3	Ground					10 536.30	.10	752
	1771	11, 4←11, 4	Ground					11 947.14	.06	623
	1771	11, 5←11, 5	Ground					12 923.10	.06	623
	1771	11, 6←11, 6	Ground					14 224.74	.06	623
	1771	11, 7←11, 7	Ground					15 933.32	.06	623
	1771	11, 8←11, 8	Ground					18 162.54	.04	623
	1771	11, 9←11, 9	Ground					21 070.70	.02	35
	1771	11,10←11,10	Ground					24 881.90	.02	35
	1771	11,11←11,11	Ground					29 914.66	.10	100
	1771	12, 1←12, 1	Ground					9 032.81	.025	752
	1771	12, 1←12, 1	Ground					9 033.13	.025	752
	1771	12, 2←12, 2	Ground					9 272.10	.10	752
	1771	12, 3←12, 3	Ground					10 836.10	.05	752
	1771	12, 4←12, 4	Ground					10 293.46	.10	752
	1771	12, 5←12, 5	Ground					11 132.70	.05	752
	1771	12, 6←12, 6	Ground					12 251.46	.06	623
	1771	12, 7←12, 7	Ground					13 719.51	.08	623
	1771	12, 8←12, 8	Ground					15 632.88	.08	623
	1771	12, 9←12, 9	Ground					18 127.32	.04	623
	1771	12,10←12,10	Ground					21 391.55	.05	57
	1771	12,11←12,11	Ground					25 695.23	.02	35
	1771	12,12←12,12	Ground					31 424.97	.10	100
	1771	13, 2←13, 2	Ground					7 894.37	.05	752
	1771	13, 4←13, 4	Ground					8 762.87	.05	752
	1771	13, 5←13, 5	Ground					9 476.06	.10	752
	1771	13, 6←13, 6	Ground					10 426.76	.10	752
	1771	13, 7←13, 7	Ground					11 673.16	.05	752
	1771	13, 8←13, 8	Ground					13 297.49	.10	623
	1771	13, 9←13, 9	Ground					15 412.52	.06	623
	1771	13,10←13,10	Ground					18 177.99	.06	623
	1771	13,12←13,12	Ground					26 655.00	.10	100
	1771	13,13←13,13	Ground					33 156.95	.10	100
	1771	14, 3←14, 3	Ground					9 670.78	.10	752

Isotopic Species	Id. No.	Rotational Quantum Nos.	Vib. State v_a	Hyperfine				Frequency MHz	Acc. ±MHz	Ref.
				F_1'	F'	F_1	F			
$N^{14}H_3$	1771	14, 4←14, 4	Ground					7 370.05	.05	752
	1771	14, 6←14, 6	Ground					8 766.96	.05	752
	1771	14, 7←14, 7	Ground					9 814.30	.05	752
	1771	14, 8←14, 8	Ground					11 177.38	.05	752
	1771	14, 9←14, 9	Ground					12 951.32	.06	623
	1771	14,10←14,10	Ground					15 268.24	.06	623
	1771	14,11←14,11	Ground					18 313.80	.06	623
	1771	14,11←14,11	Ground					21 818.1	.1	160
	1771	14,12←14,12	Ground					22 355.	.02	1020
	1771	14,13←14,13	Ground					27 772.52	.10	100
	1771	14,14←14,14	Ground					35 134.44	.10	100
	1771	15, 6←15, 6	Ground					7 285.80	.05	752
	1771	15, 7←15, 7	Ground					8 152.68	.10	752
	1771	15, 8←15, 8	Ground					9 283.65	.10	752
	1771	15, 9←15, 9	Ground					10 754.56	.05	752
	1771	15,10←15,10	Ground					12 674.12	.05	752
	1771	15,11←15,11	Ground					15 195.98	.10	623
	1771	15,12←15,12	Ground					18 535.16	.08	623
	1771	15,13←15,13	Ground					23 004.	.02	1020
	1771	15,14←15,14	Ground					29 061.14	.10	100
	1771	15,15←15,15	Ground					37 385.18	.10	100
	1771	16, 8←16, 8	Ground					7 617.90	.15	752
	1771	16, 9←16, 9	Ground					8 823.90	.10	752
	1771	16,10←16,10	Ground					10 397.12	.10	752
	1771	16,11←16,11	Ground					12 461.04	.10	752
	1771	16,12←16,12	Ground					15 193.54	.10	623
	1771	16,13←16,13	Ground					18 842.76	.10	623
	1771	16,14←16,14	Ground					23 777.4	.1	160
	1771	16,16←16,16	Ground					39 941.54	.30	100
	1771	17,12←17,12	Ground					12 308.40	.10	752
	1771	17,15←17,15	Ground					24 680.1	.1	160
	1771	Not Reported	Ground					2 408.		291
	1771	Not Reported	Ground					2 431.		291
	1771	Not Reported	Ground					2 480.		291
	1771	Not Reported	Ground					2 599.		291
	1771	Not Reported	Ground					2 614.		291
	1771	Not Reported	Ground					2 652.		291
	1771	Not Reported	Ground					2 668.		291
	1771	Not Reported	Ground					2 699.		291
	1771	Not Reported	Ground					2 746.		291
	1771	Not Reported	Ground					2 786.		291
	1771	Not Reported	Ground					2 800.		291
	1771	Not Reported	Ground					2 900.		291
	1771	Not Reported	Ground					2 939.		291
	1771	Not Reported	Ground					2 978.		291
	1771	Not Reported	Ground					3 010.		291
	1771	Not Reported	Ground					3 187.		291
	1771	Not Reported	Ground					3 261.		291
	1771	Not Reported	Ground					4 161.		291
	1771	Not Reported	Ground					4 199.		291

Isotopic Species	Id. No.	Rotational Quantum Nos.	Vib. State v_a	Hyperfine				Frequency MHz	Acc. ±MHz	Ref.
				F_1'	F'	F_1	F			
$N^{14}H_3$	1771	Not Reported	Ground					4 216.		291
	1771	Not Reported	Ground					4 219.		291
	1771	Not Reported	Ground					4 241.		291
	1771	Not Reported	Ground					4 282.		291
	1771	Not Reported	Ground					4 407.		291
	1771	Not Reported	Ground					4 410.		291
	1771	Not Reported	Ground					4 511.		291
	1771	Not Reported	Ground					4 721.		291
	1771	Not Reported	Ground					4 850.		291
	1771	Not Reported	Ground					4 859.		291
	1771	Not Reported	Ground					4 907.		291
	1771	Not Reported	Ground					4 915.		291
	1771	Not Reported	Ground					4 938.		291
	1771	Not Reported	Ground					4 948.		291
	1771	Not Reported	Ground					4 956.		291
	1771	Not Reported	Ground					5 025.		291
	1771	Not Reported	Ground					5 030.		291
	1771	Not Reported	Ground					5 122.		291
	1771	Not Reported	Ground					5 124.		291
	1771	Not Reported	Ground					5 192.		291
	1771	Not Reported	Ground					5 213.		291
	1771	Not Reported	Ground					5 230.		291
	1771	Not Reported	Ground					5 236.		291
	1771	Not Reported	Ground					5 364.		291
	1771	Not Reported	Ground					5 368.		291
	1771	Not Reported	Ground					5 392.		291
	1771	Not Reported	Ground					5 415.		291
	1771	Not Reported	Ground					5 495.		291
	1771	Not Reported	Ground					5 549.		291
	1771	Not Reported	Ground					5 574.		291
	1771	Not Reported	Ground					5 689.		291
	1771	Not Reported	Ground					5 726.		291
	1771	Not Reported	Ground					6 463.		291
	1771	Not Reported	Ground					6 598.		291
	1771	Not Reported	Ground					6 975.		291
	1771	Not Reported	Ground					7 104.		291
	1771	Not Reported	Ground					8 278.		291
	1771	Not Reported	Ground					8 283.		291
	1771	Not Reported	Ground					8 778.		291
	1771	Not Reported	Ground					8 903.		291
	1771	Not Reported	Ground					8 922.		291
	1771	Not Reported	Ground					9 014.		291
	1771	Not Reported	Ground					9 521.		291
	1771	Not Reported	Ground					9 636.		291
	1771	Not Reported	Ground					9 829.		291
	1771	Not Reported	Ground					9 967.		291
	1771	Not Reported	Ground					10 091.		291
	1771	Not Reported	Ground					10 660.		291
	1771	Not Reported	Ground					10 844.		291
	1771	Not Reported	Ground					11 400.		291

Isotopic Species	Id. No.	Rotational Quantum Nos.	Vib. State v_a	Hyperfine				Frequency MHz	Acc. ±MHz	Ref.
				F_1'	F'	F_1	F			
$N^{14}H_3$	1771	Not Reported	Ground					11 975.		291
	1771	Not Reported	Ground					11 983.		291
	1771	Not Reported	Ground					12 147.		291
	1771	Not Reported	Ground					12 150.		291
	1771	Not Reported	Ground					12 392.		291
	1771	Not Reported	Ground					12 444.		291
	1771	Not Reported	Ground					12 620.		291
	1771	Not Reported	Ground					12 778.		291
	1771	Not Reported	Ground					13 065.		291
	1771	Not Reported	Ground					13 119.		291
	1771	Not Reported	Ground					13 175.		291
	1771	Not Reported	Ground					13 210.		291
	1771	Not Reported	Ground					13 316.		291
	1771	Not Reported	Ground					13 488.		291
	1771	Not Reported	Ground					13 626.		291
	1771	Not Reported	Ground					13 657.		291
	1771	Not Reported	Ground					13 923.		291
	1771	Not Reported	Ground					14 067.		291
	1771	Not Reported	Ground					14 102.		291
	1771	Not Reported	Ground					14 566.		291
	1771	Not Reported	Ground					15 004.		291
	1771	Not Reported	Ground					15 132.		291
	1771	Not Reported	Ground					15 772.		291
	1771	Not Reported	Ground					16 493.		291
	1771	Not Reported	Ground					16 497.		291
$N^{15}H_3$	1772	1, 1← 1, 1	Ground					22 624.96	.02	35
	1772	2, 1← 2, 1	Ground					22 044.28	.02	35
	1772	2, 2← 2, 2	Ground					22 649.85	.02	35
	1772	3, 1← 3, 1	Ground					21 202.30	.02	35
	1772	3, 2← 3, 2	Ground					21 783.98	.02	35
	1772	3, 3← 3, 3	Ground					22 789.41	.02	35
	1772	4, 1← 4, 1	Ground					20 131.53	.12	623
	1772	4, 2← 4, 2	Ground					20 682.87	.02	35
	1772	4, 3← 4, 3	Ground					21 637.91	.02	35
	1772	4, 4← 4, 4	Ground					23 046.10	.02	35
	1772	5, 2← 5, 2	Ground					19 387.53	.06	623
	1772	5, 3← 5, 3	Ground					20 272.04	.02	35
	1772	5, 4← 5, 4	Ground					21 597.86	.02	35
	1772	5, 5← 5, 5	Ground					23 421.99	.02	35
	1772	6, 2← 6, 2	Ground					17 943.45	.10	623
	1772	6, 3← 6, 3	Ground					18 788.25	.06	623
	1772	6, 4← 6, 4	Ground					19 984.75	.06	623
	1772	6, 5← 6, 5	Ground					21 667.93	.02	35
	1772	6, 6← 6, 6	Ground					23 922.32	.02	35
	1772	7, 3← 7, 3	Ground					17 097.38	.06	623
	1772	7, 4← 7, 4	Ground					18 259.15	.08	623
	1772	7, 5← 7, 5	Ground					19 793.45	.06	623
	1772	7, 6← 7, 6	Ground					21 846.41	.02	35
	1772	7, 7← 7, 7	Ground					24 553.42	.02	35
	1772	8, 3← 8, 3	Ground					15 589.75	.06	623

Isotopic Species	Id. No.	Rotational Quantum Nos.	Vib. State v_a	Hyperfine				Frequency MHz	Acc. ±MHz	Ref.
				F_1'	F'	F_1	F			
$N^{15}H_3$	1772	8, 4← 8, 4	Ground					16 475.23	.06	623
	1772	8, 5← 8, 5	Ground					17 855.57	.08	623
	1772	8, 6← 8, 6	Ground					19 701.99	.06	623
	1772	8, 7← 8, 7	Ground					22 134.89	.02	35
	1772	8, 8← 8, 8	Ground					25 323.51	.02	35
	1772	9, 5← 9, 5	Ground					15 908.53	.10	623
	1772	9, 6← 9, 6	Ground					17 548.42	.08	623
	1772	9, 7← 9, 7	Ground					19 708.24	.04	623
	1772	9, 8← 9, 8	Ground					22 536.26	.02	35
	1772	9, 9← 9, 9	Ground					26 243.0	.5	160
	1772	10, 8←10, 8	Ground					19 810.86	.06	623
	1772	10, 9←10, 9	Ground					23 054.97	.02	35
	1772	11, 9←11, 9	Ground					20 010.05	.08	623
$N^{14}H_2D$	1773	2, 1, 2← 2, 0, 2	Ground					49 962.85	.10	330
	1773	2, 1, 2← 2, 0, 2	Ground					74 155.73	.15	330
	1773	3, 1, 3← 3, 0, 3	Ground					18 807.74	.05	330
	1773	3, 1, 3← 3, 0, 3	Ground					43 042.48	.10	330
	1773	4, 1, 4← 4, 0, 4	Ground					25 023.88	.05	330
	1773	5, 1, 5← 5, 0, 5	Ground					7 562.06	.05	330
	1773	5, 1, 5← 5, 0, 5	Ground					17 052.74	.05	330
	1773	6, 1, 6← 6, 0, 6	Ground					10 842.62	.05	330
	1773	6, 1, 6← 6, 0, 6	Ground					14 104.32	.05	330
	1773	7, 1, 7← 7, 0, 7	Ground					12 154.57	.05	330
	1773	7, 2, 6← 7, 1, 6	Ground					29 186.99	.05	330
	1773	8, 1, 8← 8, 0, 8	Ground					12 784.10	.05	330
	1773	8, 1, 8← 8, 0, 8	Ground					13 119.94	.05	330
	1773	8, 2, 7← 8, 1, 7	Ground					18 254.38	.05	330
	1773	9, 1, 9← 9, 0, 9	Ground					13 217.78	.05	330
	1773	9, 1, 9← 9, 0, 9	Ground					13 320.46	.05	330
	1773	9, 3, 7← 9, 2, 7	Ground					33 909.34	.05	330
	1773	10, 1,10←10, 0,10	Ground					13 626.80	.05	330
	1773	10, 2, 9←10, 1, 9	Ground					12 399.24	.05	330
	1773	11, 1,11←11, 0,11	Ground					14 069.73	.05	330
$N^{14}D_2H$	1774	1, 1, 1← 1, 0, 1	Ground					57 674.76	.15	330
	1774	1, 1, 1← 1, 0, 1	Ground					67 841.52	.15	330
	1774	2, 1, 2← 2, 0, 2	Ground					28 560.90	.05	330
	1774	2, 1, 2← 2, 0, 2	Ground					38 739.13	.10	330
	1774	3, 1, 3← 3, 0, 3	Ground					8 283.92	.05	330
	1774	3, 1, 3← 3, 0, 3	Ground					18 481.91	.05	330
	1774	4, 1, 4← 4, 0, 4	Ground					9 517.55	.05	330
	1774	5, 1, 5← 5, 0, 5	Ground					6 461.09	.05	330
	1774	5, 2, 4← 5, 1, 4	Ground					28 677.86	.05	330
	1774	5, 2, 4← 5, 1, 4	Ground					38 326.84	.10	330
	1774	6, 1, 6← 6, 0, 6	Ground					4 860.20	.05	330
	1774	6, 1, 6← 6, 0, 6	Ground					5 581.08	.05	330
	1774	6, 2, 5← 6, 1, 5	Ground					7 801.38	.05	330
	1774	6, 2, 5← 6, 1, 5	Ground					17 392.56	.05	330
	1774	7, 1, 7← 7, 0, 7	Ground					5 197.56	.05	330
	1774	7, 1, 7← 7, 0, 7	Ground					5 392.07	.05	330

Isotopic Species	Id. No.	Rotational Quantum Nos.	Vib. State v_a	Hyperfine F_1'	F'	F_1	F	Frequency MHz	Acc. $\pm$MHz	Ref.
$N^{14}D_2H$	1774	8, 1, 8← 8, 0, 8	Ground					5 364.03	.05	330
	1774	8, 1, 8← 8, 0, 8	Ground					5 414.15	.05	330
	1774	8, 3, 6← 8, 2, 6	Ground					20 608.77	.05	330
	1774	8, 3, 6← 8, 2, 6	Ground					29 319.47	.05	330
	1774	9, 1, 9← 9, 0, 9	Ground					5 494.98	.05	330
	1774	9, 1, 9← 9, 0, 9	Ground					5 507.75	.05	330
	1774	9, 3, 7← 9, 2, 7	Ground					13 923.95	.05	330
	1774	10, 1,10←10, 0,10	Ground					5 631.97	.05	330
	1774	10, 1,10←10, 0,10	Ground					5 635.18	.05	330
	1774	11, 1,11←11, 0,11	Ground					5 786.44	.05	330
	1774	11, 1,11←11, 0,11	Ground					5 787.16	.05	330
	1774	12, 1,12←12, 0,12	Ground					5 962.30	.20	330
	1774	13, 1,13←13, 0,13	Ground					6 161.86	.20	330
	1774	14, 1,14←14, 0,14	Ground					6 387.23	.20	330
	1774	Not Reported	Ground					3 470.		291
	1774	Not Reported	Ground					3 865.		291
	1774	Not Reported	Ground					4 086.		291
	1774	Not Reported	Ground					6 105.		291
	1774	Not Reported	Ground					6 641.		291
	1774	Not Reported	Ground					6 922.		291
	1774	Not Reported	Ground					7 238.		291
	1774	Not Reported	Ground					7 388.		291
$N^{14}D_3$	1775	1, 0← 0, 0	Ground		2		1	306 735.0		749
	1775	1, 0← 0, 0	Ground		0		1	306 735.0		749
	1775	1, 0← 0, 0	Ground		1		1	306 735.0		749
	1775	1, 0← 0, 0	Ground		1		1	309 908.24		749
	1775	1, 0← 0, 0	Ground		2		1	309 909.54		749
	1775	1, 0← 0, 0	Ground		0		1	309 911.41		749
	1775	1, 1← 1, 1	Ground					1 587.451		1014
	1775	1, 1← 1, 1	Ground					1 588.388		1014
	1775	1, 1← 1, 1	Ground					1 588.987		1014
	1775	2, 1← 2, 1	Ground					1 568.34		823
	1775	2, 2← 2, 2	Ground					1 589.645		1014
	1775	2, 2← 2, 2	Ground					1 590.402		1014
	1775	2, 2← 2, 2	Ground					1 591.687		1014
	1775	2, 2← 2, 2	Ground					1 593.017		1014
	1775	2, 2← 2, 2	Ground					1 593.755		1014
	1775	3, 1← 3, 1	Ground					1 537.81		823
	1775	3, 2← 3, 2	Ground					1 560.78		823
	1775	3, 3← 3, 3	Ground					1 597.395		1014
	1775	3, 3← 3, 3	Ground					1 598.007		1014
	1775	3, 3← 3, 3	Ground					1 599.64		823
	1775	3, 3← 3, 3	Ground					1 599.699		1014
	1775	3, 3← 3, 3	Ground					1 601.369		1014
	1775	3, 3← 3, 3	Ground					1 601.987		1014
	1775	4, 3← 4, 3	Ground					1 557.545		1014
	1775	4, 3← 4, 3	Ground					1 557.702		1014
	1775	4, 3← 4, 3	Ground					1 558.178		1014
	1775	4, 3← 4, 3	Ground					1 558.61		823

Isotopic Species	Id. No.	Rotational Quantum Nos.	Vib. State v_a	Hyperfine F_1'	F'	F_1	F	Frequency MHz	Acc. ±MHz	Ref.
$N^{14}D_3$	1775	4, 3← 4, 3	Ground					1 558.664		1014
	1775	4, 3← 4, 3	Ground					1 558.805		1014
	1775	4, 4← 4, 4	Ground					1 610.550		1014
	1775	4, 4← 4, 4	Ground					1 611.060		1014
	1775	4, 4← 4, 4	Ground					1 613.000		1014
	1775	5, 3← 5, 3	Ground					1 507.50		823
	1775	5, 3← 5, 3	Ground					1 509.22		823
	1775	5, 4← 5, 4	Ground					1 561.15		823
	1775	5, 5← 5, 5	Ground					1 631.82	.05	466
	1775	6, 5← 6, 5	Ground					1 569.05		823
	1775	6, 6← 6, 6	Ground					1 656.18	.05	466
	1775	7, 6← 7, 6	Ground					1 582.22		823
	1775	7, 7← 7, 7	Ground					1 686.46	.05	466
	1775	8, 7← 8, 7	Ground					1 600.58		823
	1775	8, 8← 8, 8	Ground					1 722.85	.05	466
	1775	9, 7← 9, 7	Ground					1 509.65		823
	1775	9, 9← 9, 9	Ground					1 765.80	.05	466
	1775	10,10←10,10	Ground					1 815.37	.05	466
	1775	11, 9←11, 9	Ground					1 540.09		823
	1775	11,11←11,11	Ground					1 872.43	.05	466
	1775	12,10←12,10	Ground					1 563.09		823
	1775	12,12←12,12	Ground					1 937.31	.05	466
	1775	13,13←13,13	Ground					2 010.57	.05	466
	1775	14,14←14,14	Ground					2 092.32	.05	466
	1775	15,15←15,15	Ground					2 183.		466
	1775	16,16←16,16	Ground					2 285.		466
	1775	17,17←17,17	Ground					2 403.		466
	1775	18,18←18,18	Ground					2 540.		466
	1775	Not Reported	Ground					2 094.		291
	1775	Not Reported	Ground					2 186.		291
	1775	Not Reported	Ground					2 290.		291
	1775	Not Reported	Ground					2 533.		291
	1775	Not Reported	1					117 000.		383

Phosphorus Trihydride, Hydrogen Phosphide

H₃P   C$_{3v}$ PH₃

Isotopic Species	Pt. Gp.	Id. No.	A MHz		B MHz		C MHz	D$_J$ MHz	D$_{JK}$ MHz	Δ Amu A²	κ
P³¹H₃	C$_{3v}$	1781	133 478.3	M	133 478.3	M		3.275	−3.759		
P³¹H₂D	C$_s$	1782									−.741384
P³¹HD₂	C$_s$	1783									−2.40671 [1]
P³¹D₃	C$_{3v}$	1784	69 470.41	M	69 470.41	M		.812	−.902		

Id. No.	μ$_a$ Debye		μ$_b$ Debye		μ$_c$ Debye		eQq Value(MHz) Rel.	eQq Value(MHz) Rel.	eQq Value(MHz) Rel.	ω$_a$ d 1/cm	ω$_b$ d 1/cm	ω$_c$ d 1/cm	ω$_d$ d 1/cm
1781	0.	X	0.	X	.578	M							
1782	0.	X			.579	M							
1783	0.	X			.565	M							
1784	0.	X	0.	X	.578	M							

1. This value is denoted by the authors as κ′ and represents the quantity: $[2A - (B + C)]/(C - B)$.

References:

ABC: 506   D$_J$: 879   D$_{JK}$: 879   κ: 476   μ: 476,813

Add. Ref. 289,338,407,505,549,786,847

For species 1781, D$_K$ = 3.545 MHz; for species 1782, (A − C)/2 = 23292.6 MHz; for species 1783, (A − C)/2 = 8533.81 MHz; for species 1784, D$_K$ = 0.857 MHz.

Ref. 476

Phosphine

Isotopic Species	Id. No.	Rotational Quantum Nos.	Vib. State	F$_1'$	F'	F$_1$	F	Frequency MHz	Acc. ±MHz	Ref.
P³¹H₃	1781	1, 0 ← 0, 0	Ground					266 944.69	.55	813
P³¹H₂D	1782	2, 1, 1 ← 2, 1, 2	Ground					18 070.96	.05	476
	1782	3, 1, 3 ← 3, 0, 3	Ground					28 158.53	.05	476
	1782	4, 1, 4 ← 4, 0, 4	Ground					20 815.38	.05	476
	1782	Not Reported	Ground					20 754.57	.05	476
	1782	Not Reported	Ground					22 821.90	.05	476
P³¹HD₂	1783	1, 1, 0 ← 1, 0, 1	Ground					29 073.21	.05	476
	1783	3, 2, 2 ← 3, 1, 2	Ground					19 415.19	.05	476
	1783	5, 3, 3 ← 5, 2, 3	Ground					24 079.48	.05	476
P³¹D₃	1784	1, 0 ← 0, 0	Ground					138 938.14	.30	813
	1784	2, 0 ← 1, 0	Ground					277 857.08	.60	813
P³¹H₃ᵇ	1785	Not Reported						28 759.35	.05	476
	1785	Not Reported						29 833.95	.05	476
	1785	Not Reported						30 531.33	.05	476

Antimony Trihydride

H₃Sb   C₃ᵥ   SbH₃

Isotopic Species	Pt. Gp.	Id. No.	A MHz		B MHz		C MHz		D_J MHz	D_JK MHz	Δ Amu A²	κ
Sb¹²¹H₃	C₃ᵥ	1791	88 031.92	M	88 031.92	M	83 941.90	N				
Sb¹²³H₃	C₃ᵥ	1792	88 015.54	M	88 015.54	M						
Sb¹²¹D₃	C₃ᵥ	1793	44 693.29	M	44 693.29	M						
Sb¹²³D₃	C₃ᵥ	1794	44 677.13	M	44 677.13	M						
Sb¹²¹H₂D	Cₛ	1795										−.9530
Sb¹²³H₂D	Cₛ	1796										−.9530

Id. No.	μ_a Debye	μ_b Debye	μ_c Debye	eQq Value(MHz) Rel.	eQq Value(MHz) Rel.	eQq Value(MHz) Rel.	ω_a d 1/cm	ω_b d 1/cm	ω_c d 1/cm	ω_d d 1/cm
1791				458.7						
1792				586.0						
1793				465.4						
1794				592.8						
1795	0. X	0. u	.116 M	455						
1796	0. X	0. u	.116 M	575						

References:

ABC: 438,614   κ: 614   μ: 213   eQq: 213,614

Stibine   Spectral Line Table

Isotopic Species	Id. No.	Rotational Quantum Nos.	Vib. State	F_I′	F′	F_I	F	Frequency MHz	Acc. ±MHz	Ref.
Sb¹²¹H₃	1791	1, ← 0,	Ground		3/2		5/2	176 004.64	.5	614
	1791	1, ← 0,	Ground		7/2		5/2	176 047.74	.5	614
	1791	1, ← 0,	Ground		5/2		5/2	176 142.92	.5	614
Sb¹²³H₃	1792	1, ← 0,	Ground		5/2		7/2	175 973.49	.5	614
	1792	1, ← 0,	Ground		9/2		7/2	176 008.39	.5	614
	1792	1, ← 0,	Ground		7/2		7/2	176 120.60	.5	614
Sb¹²¹D₃	1793	1, ← 0,	Ground		3/2		5/2	89 322.65	.3	614
	1793	1, ← 0,	Ground		7/2		5/2	89 365.05	.3	614
	1793	1, ← 0,	Ground		5/2		5/2	89 462.46	.3	614
Sb¹²³D₃	1794	1, ← 0,	Ground		5/2		7/2	89 291.93	.3	614
	1794	1, ← 0,	Ground		9/2		7/2	89 326.39	.3	614
	1794	1, ← 0,	Ground		7/2		7/2	89 440.38	.3	614
Sb¹²¹H₂D	1795	1, 1, 1← 1, 0, 1	Ground		5/2		3/2	28 105.28		213
	1795	1, 1, 1← 1, 0, 1	Ground		5/2		7/2	28 108.55		213
	1795	1, 1, 1← 1, 0, 1	Ground		7/2		7/2	28 158.14		213
	1795	1, 1, 1← 1, 0, 1	Ground		7/2		5/2	28 168.04		213
	1795	1, 1, 1← 1, 0, 1	Ground		3/2		5/2	28 187.74		213
Sb¹²³H₂D	1796	1, 1, 1← 1, 0, 1	Ground		7/2		5/2	28 102.66		213
	1796	1, 1, 1← 1, 0, 1	Ground		7/2		9/2	28 105.80		213
	1796	1, 1, 1← 1, 0, 1	Ground		9/2		9/2	28 162.78		213
	1796	1, 1, 1← 1, 0, 1	Ground		9/2		7/2	28 174.48		213
	1796	1, 1, 1← 1, 0, 1	Ground		5/2		5/2	28 175.07		213
	1796	1, 1, 1← 1, 0, 1	Ground		5/2		7/2	28 190.15		213
SbᵇH₂D	1797	2, 1, 2← 2, 0, 2	Ground					26 780.		213

NO₂ NO₂

Isotopic Species	Pt. Gp.	Id. No.	A MHz		B MHz		C MHz		D$_J$ MHz	D$_{JK}$ MHz	Δ Amu A²	κ
N^{14}O$_2^{16}$	C$_{2v}$	1801	239 868.7	M	13 000.12	M	12 303.45	M			.09438	
N^{15}O$_2^{16}$	C$_{2v}$	1802	228 756.1	M	13 003.06	M	12 274.73	M			.09693	
N^{14}O^{16}O^{18}	C$_s$	1803	235 802.5	M	12 264.92	M	11 632.71	M			.09621	

Id. No.	μ$_a$ Debye		μ$_b$ Debye		μ$_c$ Debye		eQq Value(MHz) Rel.	eQq Value(MHz) Rel.	eQq Value(MHz) Rel.	ω$_a$ d 1/cm	ω$_b$ d 1/cm	ω$_c$ d 1/cm	ω$_d$ d 1/cm
1801	0.	X	.29	M	0.	X							
1802	0.	X	.294	M	0.	X							

References:

ABC: 999   Δ: 999   μ: 678,990

Add. Ref. 177,187,220,295,334,582,617,809,871

Isotopic Species	Id. No.	Rotational Quantum Nos.	Vib. State	F_1'	F'	F_1	F	Frequency MHz	Acc. ±MHz	Ref.
$N^{14}O_2^{16}$	1801	7, 1, 7← 8, 0, 8	Ground					14 929.90		659
	1801	7, 1, 7← 8, 0, 8	Ground					14 961.00		659
	1801	7, 1, 7← 8, 0, 8	Ground					15 025.37		659
	1801	7, 1, 7← 8, 0, 8	Ground					15 136.42		659
	1801	7, 1, 7← 8, 0, 8	Ground					15 242.90		659
	1801	7, 1, 7← 8, 0, 8	Ground					15 342.75		659
	1801	7, 1, 7← 8, 0, 8	Ground					15 447.25		659
	1801	7, 1, 7← 8, 0, 8	Ground					15 539.32		659
	1801	7, 1, 7← 8, 0, 8	Ground					15 624.90		659
	1801	7, 1, 7← 8, 0, 8	Ground					15 653.98		659
	1801	10, 0,10← 9, 1, 9	Ground					40 357.96		658
	1801	10, 0,10← 9, 1, 9	Ground					40 467.44		658
	1801	10, 0,10← 9, 1, 9	Ground					40 661.38		658
	1801	10, 0,10← 9, 1, 9	Ground					40 671.06		658
	1801	10, 0,10← 9, 1, 9	Ground					40 703.20		658
	1801	10, 0,10← 9, 1, 9	Ground					40 931.18		658
	1801	10, 0,10← 9, 1, 9	Ground					40 964.38		658
	1801	10, 0,10← 9, 1, 9	Ground					40 993.38		658
	1801	10, 0,10← 9, 1, 9	Ground					41 167.52		658
	1801	10, 0,10← 9, 1, 9	Ground					41 277.92		658
	1801	21, 2,20←22, 1,21	Ground					39 066.70		658
	1801	21, 2,20←22, 1,21	Ground					39 097.80		658
	1801	21, 2,20←22, 1,21	Ground					39 142.46		658
	1801	21, 2,20←22, 1,21	Ground					39 192.94		658
	1801	21, 2,20←22, 1,21	Ground					39 235.98		658
	1801	21, 2,20←22, 1,21	Ground					39 247.28		658
	1801	24, 1,23←23, 2,22	Ground					26 484.		658
	1801	24, 1,23←23, 2,22	Ground					26 563.25		658
	1801	24, 1,23←23, 2,22	Ground					26 569.21		658
	1801	24, 1,23←23, 2,22	Ground					26 577.02		658
	1801	24, 1,23←23, 2,22	Ground					26 603.65		658
	1801	24, 1,23←23, 2,22	Ground					26 619.38		658
	1801	24, 1,23←23, 2,22	Ground					26 633.83		658
	1801	24, 1,23←23, 2,22	Ground					26 647.17		658
	1801	24, 1,23←23, 2,22	Ground					26 681.4		658
	1801	24, 1,23←23, 2,22	Ground					26 695.08		659
	1801	39, 3,37←40, 2,38	Ground					16 008.35		659
	1801	39, 3,37←40, 2,38	Ground					16 014.05		659
	1801	39, 3,37←40, 2,38	Ground					16 019.90		659
	1801	39, 3,37←40, 2,38	Ground					16 023.65		659
	1801	39, 3,37←40, 2,38	Ground					16 025.95		659
	1801	39, 3,37←40, 2,38	Ground					16 031.85		659
$N^{15}O_2^{16}$	1802	3, 1, 3← 4, 0, 4	Ground					111 969.09		999
	1802	3, 1, 3← 4, 0, 4	Ground					112 040.39		999
	1802	3, 1, 3← 4, 0, 4	Ground					113 362.39		999
	1802	3, 1, 3← 4, 0, 4	Ground					113 409.41		999
	1802	5, 1, 5← 6, 0, 6	Ground					58 583.23		999
	1802	5, 1, 5← 6, 0, 6	Ground					58 650.08		999
	1802	5, 1, 5← 6, 0, 6	Ground					59 411.47		999

Isotopic Species	Id. No.	Rotational Quantum Nos.	Vib. State	Hyperfine				Frequency MHz	Acc. ±MHz	Ref.
				F_1'	F'	F_1	F			
$N^{15}O_2^{16}$	1802	5, 1, 5← 6, 0, 6	Ground					59 457.01		999
	1802	7, 1, 7← 8, 0, 8	Ground					3 660.75	.24	990
	1802	7, 1, 7← 8, 0, 8	Ground					3 759.69	.01	990
	1802	7, 1, 7← 8, 0, 8	Ground					3 900.57	.10	990
	1802	7, 1, 7← 8, 0, 8	Ground					4 082.16	.15	990
	1802	7, 1, 7← 8, 0, 8	Ground					4 243.94	.02	990
	1802	7, 1, 7← 8, 0, 8	Ground					4 321.59	.06	990
	1802	10, 0,10← 9, 1, 9	Ground					52 109.4		999
	1802	10, 0,10← 9, 1, 9	Ground					52 110.4		658
	1802	10, 0,10← 9, 1, 9	Ground					52 149.8		658
	1802	10, 0,10← 9, 1, 9	Ground					52 398.4		658
	1802	10, 0,10← 9, 1, 9	Ground					52 419.2		658
	1802	12, 0,12←11, 1,11	Ground					109 559.23		999
	1802	12, 0,12←11, 1,11	Ground					109 576.98		999
	1802	12, 0,12←11, 1,11	Ground					109 743.01		999
	1802	12, 0,12←11, 1,11	Ground					109 743.83		999
	1802	21, 2,20←22, 1,21	Ground					2 746.51	.30	990
	1802	21, 2,20←22, 1,21	Ground					2 767.35	.30	990
	1802	21, 2,20←22, 1,21	Ground					2 861.82	.20	990
	1802	21, 2,20←22, 1,21	Ground					2 902.11	.30	990
	1802	Not Reported						112 408.63		999
$N^{14}O^{16}O^{18}$	1803	7, 1, 7← 8, 0, 8	Ground					23 674.33		999
	1803	7, 1, 7← 8, 0, 8	Ground					23 701.45		999
	1803	7, 1, 7← 8, 0, 8	Ground					23 765.13		999
	1803	7, 1, 7← 8, 0, 8	Ground					23 883.98		999
	1803	7, 1, 7← 8, 0, 8	Ground					23 992.66		999
	1803	7, 1, 7← 8, 0, 8	Ground					24 065.1		999
	1803	7, 1, 7← 8, 0, 8	Ground					24 171.7		999
	1803	7, 1, 7← 8, 0, 8	Ground					24 271.2		999
	1803	7, 1, 7← 8, 0, 8	Ground					24 356.14		999
	1803	7, 1, 7← 8, 0, 8	Ground					24 381.5		999
	1803	10, 0,10← 9, 1, 9	Ground					28 446.98		999
	1803	10, 0,10← 9, 1, 9	Ground					28 626.13		999
	1803	10, 0,10← 9, 1, 9	Ground					28 638.28		999
	1803	10, 0,10← 9, 1, 9	Ground					28 674.0		999
	1803	10, 0,10← 9, 1, 9	Ground					28 886.61		999
	1803	10, 0,10← 9, 1, 9	Ground					28 923.86		999
	1803	10, 0,10← 9, 1, 9	Ground					28 954.85		999
	1803	10, 0,10← 9, 1, 9	Ground					29 114.35		999
	1803	10, 0,10← 9, 1, 9	Ground					29 218.7		999
	1803	25, 1,24←24, 2,23	Ground					25 025.0		999
	1803	25, 1,24←24, 2,23	Ground					25 033.4		999
	1803	25, 1,24←24, 2,23	Ground					25 035.8		999
	1803	25, 1,24←24, 2,23	Ground					25 056.8		999
	1803	25, 1,24←24, 2,23	Ground					25 068.18		999
	1803	25, 1,24←24, 2,23	Ground					25 079.3		999

N₂O $C_{\infty v}$ N₂O

Isotopic Species	Pt. Gp.	Id. No.	A MHz	B MHz		C MHz		D_J MHz	D_{JK} MHz	Δ Amu A²	κ
$N_2^{14}O^{16}$	$C_{\infty v}$	1811		12 561.64	M	12 561.64	M	.00536			
$N^{14}N^{15}O^{16}$	$C_{\infty v}$	1812		12 560.78	M	12 560.78	M				
$N^{15}N^{14}O^{16}$	$C_{\infty v}$	1813		12 137.30	M	12 137.30	M				
$N_2^{15}O^{16}$	$C_{\infty v}$	1814		12 137.39	M	12 137.39	M				

Id. No.	μ_a Debye		μ_b Debye		μ_c Debye		eQq Value(MHz)	Rel.	eQq Value(MHz)	Rel.	eQq Value(MHz)	Rel.	ω_a d 1/cm	ω_b d 1/cm	ω_c d 1/cm	ω_d d 1/cm
1811							−1.03	¹N¹⁴					589 2			
1813	.166	M	0.	X	0.	X	−.27	N¹⁴								

1. The value obtained for the nuclear quadrupole coupling constant was for N¹⁴ in the end position.

References:

ABC: 402,1029 D_J: 663   μ: 236   eQq: 31,104   ω: 1028

Add. Ref. 138,140,236,281,282,507,626,678,680,887,1016

Nitrous Oxide   Spectral Line Table

Isotopic Species	Id. No.	Rotational Quantum Nos.	Vib. State v_a^l	F_1'	F′	F_1	F	Frequency MHz	Acc. ±MHz	Ref.
$N_2^{14}O^{16}$	1811	1← 0	Ground					25 123.25		124
	1811	2← 1	Ground	2	1	2	1	50 245.63	.15	402
	1811	2← 1	Ground	1	1	0	1	50 245.63	.15	402
	1811	2← 1	Ground	1	2	0	1	50 245.63	.15	402
	1811	2← 1	Ground	2	1	2	2	50 245.63	.15	402
	1811	2← 1	Ground	1	0	0	1	50 245.63	.15	402
	1811	2← 1	Ground	2	3	2	2	50 245.63	.15	402
	1811	2← 1	Ground	2	3	2	3	50 245.63	.15	402
	1811	2← 1	Ground	2	2	2	3	50 245.63	.15	402
	1811	2← 1	Ground	2	2	2	1	50 245.63	.15	402
	1811	2← 1	Ground	2	2	2	2	50 245.63	.15	402
	1811	2← 1	Ground	3	3	2	2	50 246.03	.10	402
	1811	2← 1	Ground	1	2	2	3	50 246.03	.10	402
	1811	2← 1	Ground	3	2	2	2	50 246.03	.10	402
	1811	2← 1	Ground	2	2	1	2	50 246.03	.10	402
	1811	2← 1	Ground	3	2	2	1	50 246.03	.10	402
	1811	2← 1	Ground	1	.2	2	2	50 246.03	.10	402
	1811	2← 1	Ground	1	2	2	1	50 246.03	.10	402
	1811	2← 1	Ground	2	1	1	0	50 246.03	.10	402
	1811	2← 1	Ground	2	1	1	2	50 246.03	.10	402
	1811	2← 1	Ground	3	4	2	3	50 246.03	.10	402
	1811	2← 1	Ground	2	1	1	1	50 246.03	.10	402
	1811	2← 1	Ground	3	2	2	3	50 246.03	.10	402
	1811	2← 1	Ground	3	3	2	3	50 246.03	.10	402
	1811	2← 1	Ground	2	3	1	2	50 246.03	.10	402

Isotopic Species	Id. No.	Rotational Quantum Nos.	Vib. State v_a^l	Hyperfine				Frequency MHz	Acc. ±MHz	Ref.
				F_1'	F'	F_1	F			
$N_2^{14}O^{16}$	1811	2← 1	Ground	2	2	1	1	50 246.03	.10	402
	1811	2← 1	Ground	1	1	2	2	50 246.03	.10	402
	1811	2← 1	Ground	1	0	2	1	50 246.03	.10	402
	1811	2← 1	Ground	1	1	2	1	50 246.03	.10	402
	1811	2← 1	Ground	1	0	1	1	50 246.53	.15	402
	1811	2← 1	Ground	1	1	1	2	50 246.53	.15	402
	1811	2← 1	Ground	1	2	1	2	50 246.53	.15	402
	1811	2← 1	Ground	1	2	1	1	50 246.53	.15	402
	1811	2← 1	Ground	1	1	1	1	50 246.53	.15	402
	1811	2← 1	Ground	1	1	1	0	50 246.53	.15	402
	1811	4← 3	Ground					100 491.74	.20	663
	1811	4← 3	1,−1					100 531.65		663
	1811	4← 3	1,+1					100 721.58		663
	1811	5← 4	Ground					125 613.73	.25	663
	1811	5← 4	1,−1					125 663.69		663
	1811	5← 4	1,+1					125 900.99		663
	1811	6← 5	Ground					150 735.13	.30	663
	1811	6← 5	1,−1					150 794.98		663
	1811	6← 5	1,+1					151 079.83		663
	1811	7← 6	Ground					175 855.72	.35	663
	1811	8← 7	Ground					200 975.26	.40	663
	1811	9← 8	Ground					226 093.81	.45	663
	1811	10← 9	Ground					251 211.33	.50	663
	1811	11←10	Ground					276 327.50	.55	663
	1811	12←11	Ground					301 442.38	.60	663
$N^{14}N^{15}O^{16}$	1812	1← 0	Ground					25 121.55		124
$N^{15}N^{14}O^{16}$	1813	1← 0	Ground		1		1	24 274.53		236
	1813	1← 0	Ground		2		1	24 274.61		236
	1813	1← 0	Ground		0		1	24 274.73		236
$N_2^{15}O^{16}$	1814	1← 0	Ground					24 274.78		124

OS$_2$ C$_s$ S$_2$O

Isotopic Species	Pt. Gp.	Id. No.	A MHz	B MHz	C MHz	D$_J$ MHz	D$_{JK}$ MHz	Δ Amu A^2	κ
S$_2^{32}$O^{16}	C$_s$	1821	41 914.40 M	5 059.09 M	4 507.14 M				

Id. No.	μ$_a$ Debye	μ$_b$ Debye	μ$_c$ Debye	eQq Value(MHz) Rel.	eQq Value(MHz) Rel.	eQq Value(MHz) Rel.	ω$_a$ d 1/cm	ω$_b$ d 1/cm	ω$_c$ d 1/cm	ω$_d$ d 1/cm
1821	.875 M	1.18 M	0. X				370 1			

References:

ABC: 874   μ: 874   ω: 874

Add. Ref. 831

For species 1821, first excited state, A = 42478.35 MHz, B = 5059.76 MHz, C = 4500.8 MHz; for species 1822, A$_0$ − C$_0$ = 37356.00 MHz, B$_0$ − C$_0$ = 521.97 MHz. Ref. 874

Disulfur Monoxide   Spectral Line Table

Isotopic Species	Id. No.	Rotational Quantum Nos.	Vib. State v$_a$	F$_1'$	F'	F$_1$	F	Frequency MHz	Acc. ±MHz	Ref.
S$_2^{32}$O^{16}	1821	1, 1, 1← 0, 0, 0	Ground					46 421.6		874
	1821	1, 1, 0← 1, 0, 1	Ground					37 407.2		874
	1821	2, 0, 2← 1, 0, 1	Ground					19 126.4		874
	1821	2, 1, 1← 1, 1, 0	Ground					19 684.3		874
	1821	2, 1, 2← 1, 1, 1	Ground					18 580.2		874
	1821	2, 1, 1← 2, 0, 2	Ground					37 965.3		874
	1821	2, 1, 1← 2, 0, 2	1					38 542.7		874
	1821	3, 0, 3← 2, 0, 2	Ground					28 674.3		874
	1821	3, 1, 2← 2, 1, 1	Ground					29 522.8		874
	1821	3, 1, 3← 2, 1, 2	Ground					27 867.0		874
	1821	3, 2, 1← 2, 2, 0	Ground					28 723.4		874
	1821	3, 2, 2← 2, 2, 1	Ground					28 699.3		874
	1821	3, 1, 2← 3, 0, 3	Ground					38 814.0		874
	1821	3, 1, 2← 3, 0, 3	1					39 402.2		874
	1821	4, 0, 4← 3, 0, 3	1					38 180.0		874
	1821	4, 0, 4← 3, 0, 3	Ground					38 203.0		874
	1821	4, 1, 3← 3, 1, 2	1					39 347.6		874
	1821	4, 1, 3← 3, 1, 2	Ground					39 356.2		874
	1821	4, 1, 4← 3, 1, 3	1					37 112.3		874
	1821	4, 1, 4← 3, 1, 3	Ground					37 149.0		874
	1821	4, 2, 2← 3, 2, 1	Ground					38 322.3		874
	1821	4, 2, 3← 3, 2, 2	Ground					38 260.9		874
	1821	4, 3, 1← 3, 3, 0	Ground					38 279.1		874
	1821	4, 3, 2← 3, 3, 1	Ground					38 279.1		874
	1821	4, 1, 3← 4, 0, 4	Ground					39 966.8		874
	1821	5, 1, 4← 5, 0, 5	Ground					41 442.2		874
S^{34}S^{32}O^{16}	1822	3, 1, 2← 3, 0, 3	Ground					38 685.0		874
	1822	4, 1, 3← 4, 0, 4	Ground					39 772.9		874

O₂S   C$_{2v}$ SO$_2$

Isotopic Species	Pt. Gp.	Id. No.	A MHz		B MHz		C MHz		D$_J$ MHz	D$_{JK}$ MHz	Δ Amu A²	κ
S^{32}O$_2^{16}$	C$_{2v}$	1831	60 778.79	M	10 318.10	M	8 799.96	M				
S^{33}O$_2^{16}$	C$_{2v}$	1832	59 856.49	M	10 318.20	M	8 780.23	M				
S^{34}O$_2^{16}$	C$_{2v}$	1833	58 991.21	M	10 318.40	M	8 761.41	M				

Id. No.	μ_a Debye		μ_b Debye		μ_c Debye		eQq Value(MHz)	Rel.	eQq Value(MHz)	Rel.	eQq Value(MHz)	Rel.	ω_a d 1/cm	ω_b d 1/cm	ω_c d 1/cm	ω_d d 1/cm
1831	1.59	M	0.	X	0.	X										
1832							−1.7	aa	25.71	bb						

References:

ABC: 535,839 μ: 262   eQq: 585

Add. Ref. 32,33,88,90,281,352,594,678,892

Isotopic Species	Id. No.	Rotational Quantum Nos.	Vib. State	Hyperfine F_1'	F'	F_1	F	Frequency MHz	Acc. ±MHz	Ref.
$S^{32}O_2^{16}$	1831	1, 1, 1← 0, 0, 0	Ground					69 576.06	.18	262
	1831	2, 1, 1← 2, 0, 2	Ground					53 529.16	.16	262
	1831	3, 1, 3← 2, 0, 2	Ground					104 033.53		892
	1831	4, 0, 4← 3, 1, 3	Ground					29 321.22	.03	262
	1831	4, 1, 3← 4, 0, 4	Ground					59 225.00	.07	262
	1831	5, 2, 4← 6, 1, 5	Ground					23 414.33	.03	262
	1831	6, 0, 6← 5, 1, 5	Ground					72 758.28		892
	1831	6, 1, 5← 6, 0, 6	Ground					68 972.10		892
	1831	8, 0, 8← 7, 1, 7	Ground					116 980.60		892
	1831	8, 1, 7← 7, 2, 6	Ground					25 392.797	.014	262
	1831	8, 1, 7← 8, 0, 8	Ground					83 687.88		892
	1831	8, 2, 6← 9, 1, 9	Ground					24 083.39	.1	317
	1831	10, 1, 9←10, 0,10	Ground					104 029.43		892
	1831	10, 2, 8←10, 1, 9	Ground					129 514.86		892
	1831	12, 1,11←12, 0,12	Ground					131 014.86		892
	1831	12, 2,10←12, 1,11	Ground					128 605.18		892
	1831	12, 3, 9←13, 2,12	Ground					20 335.47	.1	317
	1831	17, 2,16←16, 3,13	Ground					28 858.11	.1	317
	1831	21, 5,17←22, 4,18	Ground					24 039.50	.1	317
	1831	24, 4,20←23, 5,19	Ground					22 482.51	.1	317
	1831	25, 4,22←24, 5,19	Ground					26 777.20	.1	317
	1831	35, 6,30←34, 7,27	Ground					25 049.13	.1	317
	1831	Not Reported						20 460.05	.1	317
	1831	Not Reported						22 220.32	.1	317
	1831	Not Reported						22 733.83	.1	317
	1831	Not Reported						22 904.95	.1	317
	1831	Not Reported						22 928.45	.1	317
	1831	Not Reported						23 034.83	.1	317
	1831	Not Reported						23 733.03	.1	317
	1831	Not Reported						24 319.67	.1	317
	1831	Not Reported						25 170.97	.1	317
$S^{33}O_2^{16}$	1832	1, 1, 1← 2, 0, 2	Ground	1/2			3/2	11 368.002	.015	839
	1832	1, 1, 1← 2, 0, 2	Ground	1/2			1/2	11 368.002	.015	839
	1832	1, 1, 1← 2, 0, 2	Ground	5/2			3/2	11 373.286	.015	839
	1832	1, 1, 1← 2, 0, 2	Ground	5/2			5/2	11 373.286	.015	839
	1832	1, 1, 1← 2, 0, 2	Ground	5/2			7/2	11 373.286	.015	839
	1832	1, 1, 1← 2, 0, 2	Ground	3/2			3/2	11 379.825	.015	839
	1832	1, 1, 1← 2, 0, 2	Ground	3/2			1/2	11 379.825	.015	839
	1832	1, 1, 1← 2, 0, 2	Ground	3/2			5/2	11 379.825	.015	839
	1832	4, 0, 4← 3, 1, 3	Ground	9/2			7/2	30 190.37		585
	1832	4, 0, 4← 3, 1, 3	Ground	7/2			5/2	30 193.01		585
	1832	4, 0, 4← 3, 1, 3	Ground	11/2			9/2	30 195.91		585
	1832	4, 0, 4← 3, 1, 3	Ground	5/2			3/2	30 198.56		585
	1832	5, 2, 4← 6, 1, 5	Ground	7/2			9/2	20 602.07		585
	1832	5, 2, 4← 6, 1, 5	Ground	13/2			15/2	20 603.55		585
	1832	5, 2, 4← 6, 1, 5	Ground	9/2			11/2	20 608.13		585
$S^{34}O_2^{16}$	1833	1, 1, 1← 2, 0, 2	Ground					10 547.91	.02	839
	1833	4, 0, 4← 3, 1, 3	Ground					31 011.19	.05	500
	1833	5, 2, 4← 6, 1, 5	Ground					17 970.42	.05	500
	1833	8, 1, 7← 7, 2, 6	Ground					30 975.39	.02	839
	1833	8, 2, 6← 9, 1, 9	Ground					20 699.30	.02	839
	1833	10, 2, 8←11, 1,11	Ground					9 650.63	.02	839

O$_3$ C$_{2v}$ O$_3$

Isotopic Species	Pt. Gp.	Id. No.	A MHz		B MHz		C MHz		D$_J$ MHz	D$_{JK}$ MHz	Δ Amu A²	κ
O$_3^{16}$	C$_{2v}$	1841	106 536.1	M	13 349.12	M	11 834.45	M			.09576	−.968012
O^{16}O^{16}O^{18}	C$_s$	1842	104 569.4	M	12 590.4	M	11 214.6	M			.09094	
O^{16}O^{18}O^{16}	C$_{2v}$	1843	98 645.96	M	13 352.51	M	11 731.78	M			.09997	
O^{16}O^{18}O^{18}	C$_s$	1844	96 676.8	M	12 591.4	M	11 115.6	M			.10118	
O$_3^{18}$	C$_{2v}$	1845	94 768.2	M	11 886.5	M	10 536.9	M			.11286	

Id. No.	μ_a Debye		μ_b Debye		μ_c Debye		eQq Value(MHz)	Rel.	eQq Value(MHz)	Rel.	eQq Value(MHz)	Rel.	ω_a d 1/cm	ω_b d 1/cm	ω_c d 1/cm	ω_d d 1/cm
1841	0.	X	.58	M	0.	X										

References:

ABC: 689,716,861   Δ: 689   κ: 861   μ: 689

Add. Ref. 359,403,442,481,574,604,605,606,607,678,710,901

Isotopic Species	Id. No.	Rotational Quantum Nos.	Vib. State	Hyperfine F_1'	F'	F_1	F	Frequency MHz	Acc. ±MHz	Ref.
O_3^{16}	1841	1, 1, 1← 0, 0, 0	Ground					118 364.3	.5	480
	1841	1, .1, 1← 2, 0, 2	Ground					42 832.62	.07	480
	1841	2, 1, 1← 2, 0, 2	Ground					96 228.84	.18	480
	1841	4, 0, 4← 3, 1, 3	Ground					11 073.		689
	1841	4, 1, 3← 4, 0, 4	Ground					101 736.83	.14	480
	1841	10, 1, 9← 9, 2, 8	Ground					10 226.		689
	1841	12, 2,10←13, 1,13	Ground					43 654.		689
	1841	14, 2,12←15, 1,15	Ground					30 181.		689
	1841	15, 3,13←16, 2,14	Ground					30 052.		689
	1841	16, 2,14←17, 1,17	Ground					25 649.		689
	1841	18, 2,16←17, 3,15	Ground					37 832.		689
	1841	18, 2,16←19, 1,19	Ground					30 525.		689
	1841	18, 3,15←19, 2,18	Ground					23 861.		689
	1841	20, 3,17←21, 2,20	Ground					9 201.		689
	1841	23, 2,22←22, 3,19	Ground					36 023.		689
	1841	23, 4,20←24, 3,21	Ground					14 866.		689
	1841	24, 4,20←25, 3,23	Ground					28 960.		689
	1841	27, 3,25←26, 4,22	Ground					16 163.		689
	1841	38, 6,32←39, 5,35	Ground					25 511.		689
	1841	41, 5,37←40, 6,34	Ground					27 862.		689
	1841	45, 7,39←46, 6,40	Ground					25 300.		689
$O^{16}O^{16}O^{18}$	1842	2, 1, 2← 3, 0, 3	Ground					19 263.1		689
	1842	5, 0, 5← 4, 1, 4	Ground					32 743.		689
	1842	8, 2, 7← 9, 1, 8	Ground					33 537.		689
	1842	11, 1,10←10, 2, 9	Ground					27 608.		689
	1842	14, 2,12←15, 1,15	Ground					33 191.		689
	1842	15, 2,13←16, 1,16	Ground					28 915.		689
	1842	16, 2,14←17, 1,17	Ground					26 388.		689
	1842	16, 3,14←17, 2,15	Ground					20 076.0		689
	1842	17, 2,15←18, 1,18	Ground					26 040.		689
	1842	18, 2,16←19, 1,19	Ground					27 862.		689
	1842	19, 2,17←20, 1,20	Ground					32 054.		689
	1842	19, 3,16←20, 2,19	Ground					21 086.0		689
	1842	20, 2,18←21, 1,20	Ground					37 979.		689
	1842	23, 2,22←22, 3,19	Ground					22 918.		689
	1842	24, 2,23←23, 3,20	Ground					34 814.		689
	1842	25, 4,21←26, 3,24	Ground					31 288.		689
$O^{16}O^{18}O^{16}$	1843	1, 1, 1← 2, 0, 2	Ground					35 143.		689
	1843	4, 0, 4← 3, 1, 3	Ground					18 768.		689
	1843	7, 2, 6← 8, 1, 7	Ground					29 227.		689
	1843	10, 1, 9← 9, 2, 8	Ground					35 004.		689
	1843	10, 1, 9← 9, 2, 8	Ground					35 280.		716
	1843	12, 2,10←13, 1,13	Ground					30 877.		689
	1843	12, 2,10←13, 1,13	Ground					30 914.		716
	1843	14, 2,12←15, 1,15	Ground					22 205.4		689
	1843	16, 2,14←17, 1,17	Ground					23 425.		689
	1843	17, 2,16←16, 3,13	Ground					29 889.		689
	1843	18, 2,16←19, 1, 9	Ground					34 662.		689

Isotopic Species	Id. No.	Rotational Quantum Nos.	Vib. State	Hyperfine				Frequency MHz	Acc. ±MHz	Ref.
				F_1'	F'	F_1	F			
$O^{16}O^{18}O^{16}$	1843	23, 3,21←22, 4,18	Ground					26 334.		689
	1843	25, 3,23←24, 4,20	Ground					18 916.		689
	1843	25, 3,23←24, 4,20	Ground					19 076.		716
$O^{16}O^{18}O^{18}$	1844	1, 1, 1← 2, 0, 2	Ground					36 688.		689
	1844	2, 1, 2← 3, 0, 3	Ground					11 575.0		689
	1844	5, 0, 5← 4, 1, 4	Ground					40 476.		689
	1844	7, 2, 6← 8, 1, 7	Ground					38 859.		689
	1844	10, 1, 9← 9, 2, 8	Ground					21 684.0		689
	1844	12, 2,10←13, 1,13	Ground					34 726.		689
	1844	13, 2,11←14, 1,14	Ground					28 384.		689
	1844	14, 2,12←15, 1,15	Ground					24 139.		689
	1844	15, 2,13←16, 1,16	Ground					22 227.		689
	1844	16, 2,14←17, 1,17	Ground					22 527.		689
	1844	17, 2,15←16, 3,14	Ground					25 257.		689
	1844	17, 2,15←18, 1,18	Ground					25 196.		689
	1844	19, 2,17←20, 1,20	Ground					37 347.		689
	1844	21, 2,20←20, 3,17	Ground					20 861.2		689
O_3^{18}	1845	1, 1, 1← 2, 0, 2	Ground					38 054.		689
	1845	12, 2,10←13, 1,13	Ground					39 439.		689
	1845	14, 2,12←15, 1,15	Ground					26 690.		689
	1845	15, 3,13←16, 2,14	Ground					26 334.		689
	1845	17, 1,17←16, 2,14	Ground					22 866.		689
	1845	18, 2,16←17, 3,15	Ground					34 102.		689
	1845	18, 2,16←19, 1,19	Ground					27 192.		689
	1845	18, 3,15←19, 2,18	Ground					21 022.5		689
	1845	23, 2,22←22, 3,19	Ground					32 137.		689
$O^{18}O^{16}O^{18}$	1846	20, 3,17←21, 2,20	Ground					19 270.1		689
$O^bO^bO^b$	1847	Not Reported						9 225.8		689
	1847	Not Reported						9 641.7		689
	1847	Not Reported						9 823.		689
	1847	Not Reported						11 812.9		689
	1847	Not Reported						11 826.6		689
	1847	Not Reported						12 067.8		689
	1847	Not Reported						12 171.0		689
	1847	Not Reported						20 680.7		689
	1847	Not Reported						21 141.0		689
	1847	Not Reported						21 708.0		689
	1847	Not Reported						22 237.		689
	1847	Not Reported						23 421.		689
	1847	Not Reported						23 502.		689
	1847	Not Reported						23 552.		689
	1847	Not Reported						23 838.		689
	1847	Not Reported						24 932.6		689
	1847	Not Reported						25 651.		689
	1847	Not Reported						27 458.		689
	1847	Not Reported						27 912.		689
	1847	Not Reported						27 949.		689
	1847	Not Reported						28 116.		689

238-605 O-68—26

Isotopic Species	Id. No.	Rotational Quantum Nos.	Vib. State	Hyperfine F_1'	F'	F_1	F	Frequency MHz	Acc. ±MHz	Ref.
O^{b}O^{b}O^b	1847	Not Reported						28 239.		689
	1847	Not Reported						28 510.		689
	1847	Not Reported						29 111.		689
	1847	Not Reported						32 090.		689
	1847	Not Reported						32 254.		689
	1847	Not Reported						32 741.		689
	1847	Not Reported						32 800.		689
	1847	Not Reported						32 841.		689
	1847	Not Reported						33 009.		689
	1847	Not Reported						33 050.		689
	1847	Not Reported						33 248.		689
	1847	Not Reported						33 599.		689
	1847	Not Reported						33 631.		689
	1847	Not Reported						33 691.		689
	1847	Not Reported						33 781.		689
	1847	Not Reported						33 931.		689
	1847	Not Reported						34 622.		689
	1847	Not Reported						34 919.		689
	1847	Not Reported						34 967.		689
	1847	Not Reported						37 341.		689
	1847	Not Reported						37 608.		689
	1847	Not Reported						37 768.		689
	1847	Not Reported						37 970.		689
	1847	Not Reported						38 086.		689
	1847	Not Reported						38 270.		689
	1847	Not Reported						38 376.		689
	1847	Not Reported						40 080.		689
	1847	Not Reported						41 063.		689
	1847	Not Reported						42 332.		689
	1847	Not Reported						42 449.		689
	1847	Not Reported						44 575.		689
	1847	Not Reported						48 806.		689
	1847	Not Reported						48 806.		689

5. Index According to Empirical Formula

5.1. Inorganic

5.1. Inorganic (Continued)

5.2. Organic

6. Index According to Chemical Name

A

B

C

D

E

F

6. Index According to Chemical Name (Continued)

P

Q

R

S

T

6. Index According to Chemical Name (Continued)

T

7. Bibliography

1. Phys. Rev. **45,** 234 (1934) — Cleeton, C. E.; Williams, N. H.
2. J. Chem. Phys. **3,** 136 (1935) — Manning, M. F.
3. Phys. Rev. **52,** 1054 (1937) — Knerr, H. W.
4. J. Chem. Phys. **5,** 314 (1937) — Wall, F. T.; Glockler, G.
5. Physikalische Zeitschrift **40,** 1 (1939) — Scheffers, H.
6. Trans. Faraday Soc. **35,** 1373 (1939) — Sutherland, L.; Wu, C. S.
7. J. Chem. Phys. **8,** 403 (1940) — Langseth, A.; Bak, B.
8. Phys. Rev. **69,** 694 (1946) — Autler, S. H.; Becker, G. E.; Kellogg, J. M. B.
9. Phys. Rev. **70,** 300 (1946) — Becker, G. E.; Autler, S. H.
10. Repts. on Prog. in Phys. **11,** 178 (1946) — Bleaney, B.
11. Phys. Rev. **70,** 775L (1946) — Bleaney, B.; Penrose, R. P.
12. Nature **157,** 339 (1946) — Bleaney, B.; Penrose, R. P.

13. Phys. Rev. **70**, 979L (1946) — Coles, D. K.; Good, W. E.

14. Phys. Rev. **70**, 984L (1946) — Dailey, B. P.; Kyhl, R. L.; Strandberg, M. W. P.; Van Vleck, J. H.; Wilson, E. B., Jr.

15. Phys. Rev. **70**, 560L (1946) — Dakin, T. W.; Good, W. E.; Coles, D. K.

16. Phys. Rev. **69**, 539L (1946) — Good, W. E.

17. Phys. Rev. **70**, 109A (1946) — Good, W. E.

18. Phys. Rev. **70**, 213 (1946) — Good, W. E.

19. Phys. Rev. **70**, 108A (1946) — Hainer, R. M.; King, G. W.; Cross, P. C.

20. J. Appl. Phys. **17**, 495 (1946) — Hershberger, W. D.

21. Phys. Rev. **69**, 694A (1946) — Kyhl, R. L.; Dicke, R. H.; Beringer, R.

22. Phys. Rev. **70**, 109A (1946) — Townes, C. H.

23. Phys. Rev. **70**, 665 (1946) — Townes, C. H.

24. Phys. Rev. **70**, 558L (1946) — Townes, C. H.; Merritt, F. R.

25. Trans. Faraday Soc. **42A**, 114, 122 (1946) — Whiffen, D. H.; Thompson, H. W.

26. J. Chem. Phys. **15**, 762L (1947) — Beard, C. I.; Dailey, B. P.

27. Proc. Phys. Soc. **59**, 418 (1947) — Bleaney, B.; Penrose, R. P.

28. Proc. Roy. Soc. **A189**, 358 (1947) — Bleaney, B.; Penrose, R. P.

29. Tech. Rept., Gr. Brit. Admiralty, Dept. of Sci. Res. & Exp. (1947) — Bleaney, B.; Penrose, R. P.

30. Proc. Nat'l. Electronics Conf. **3**, 180 (1947) — Coles, D. K.

31. Phys. Rev. **72**, 973L (1947) — Coles, D. K.; Elyash, E. S.; Gorman, J. G.

32. Phys. Rev. **72**, 871L (1947) — Dailey, B. P.; Golden, S.; Wilson, E. B., Jr.

33. Phys. Rev. **72**, 522 (1947) — Dailey, B. P.; Wilson, E. B., Jr.

34. Phys. Rev. **71**, 640L (1947) — Dakin, T. W.; Good, W. E.; Coles, D. K.

35. Phys. Rev. **71**, 383L (1947) — Good, W. E.; Coles, D. K.

36. Phys. Rev. **72**, 157A (1947) — Good, W. E.; Coles, D. K.

37. Phys. Rev. **71**, 640L (1947) — Gordy, W.; Kessler, M.

38. Phys. Rev. **72**, 644L (1947) — Gordy, W.; Kessler, M.

39. Phys. Rev. **72**, 344 (1947) — Gordy, W.; Simmons, J. W.; Smith, A. G.

40. Phys. Rev. **71**, 917L (1947) — Gordy, W.; Smith, A. G.; Simmons, J. W.

41. Phys. Rev. **72**, 249 (1947) — Gordy, W.; Smith, A. G.; Simmons, J. W.

42. Phys. Rev. **72**, 259 (1947) — Gordy, W.; Smith, W. V.; Smith, A. G.; Ring, H.

43. Phys. Rev. **72**, 157A (1947) — Hillger, R. E.; Strandberg, M. W. P.; Wentink, T.; Kyhl, R. L.

44. Phys. Rev. **72**, 535A (1947) — Jauch, J. M.

45. Phys. Rev. **72**, 715 (1947) — Jauch, J. M.

46. Phys. Rev. **72**, 986L (1947) — Jen, C. K.

47. Phys. Rev. **72**, 158 (1947) — Johnson, R. C.; Weidler, R. C.; Williams, D.

48. Phys. Rev. **71**, 135A (1947) — King, G. W.; Hainer, R. M.

49. Phys. Rev. **71**, 433 (1947) — King, G. W.; Hainer, R. M.; Cross, P. C.

50. Phys. Rev. **72**, 86L (1947) — Nielsen, H. H.; Dennison, D. M.

51. Phys. Rev. **72**, 1101 (1947) — Nielsen, H. H.; Dennison, D. M.

52. Phys. Rev. **72**, 1121L (1947) — Pond, T. A.; Cannon, W. F.

53. Duke Univ. Rept. (1947) W28–099–ac–125 — Ring, H.; Edwards, H.; Kessler, M.

54. Phys. Rev. **72**, 1262 (1947) — Ring, H.; Edwards, H.; Kessler, M.; Gordy, W.

55. Phys. Rev. **71**, 126 (1947) — Smith W. V.

56. Phys. Rev. **72**, 638 (1947) — Smith, W. V.; Carter, R. L.

57. Phys. Rev. **71**, 326L (1947) — Strandberg, M. W. P.; Kyhl, R. L.; Wentink, T.; Hillger, R. E.

58. Phys. Rev. **71**, 639L (1947) — Strandberg, M. W. P.; Kyhl, R. L.; Wentink, T., Jr.; Hillger, R. E.

59. Phys. Rev. **71**, 909 (1947) — Townes, C. H.

60. Phys. Rev. **71**, 644L (1947) Errata **71**, 829L (1947) — Townes, C. H.; Holden, A. N.; Bardeen, J.; Merritt, F. R.

61. Phys. Rev. **71**, 64 (1947) — Townes, C. H.; Holden, A. N.; Merritt, F. R.

62. Phys. Rev. **71**, 479A (1947) — Townes, C. H.; Holden, A. N.; Merritt, F. R.

63. Phys. Rev. **72**, 513 (1947) — Townes, C. H.; Holden, A. N.; Merritt, F. R.

64. Phys. Rev. **72**, 740A (1947) — Townes, C. H.; Holden, A. N.; Merritt, F. R.

65. Phys. Rev. **71**, 425 (1947) — Van Vleck, J. H.

66. Phys. Rev. **71**, 639L (1947) — Watts, R. J.; Williams, D.

67. Phys. Rev. **72**, 157A (1947)   Watts, R. J.; Williams, D.

68. Phys. Rev. **72**, 975 (1947)   Weiss, P. R.; Whitmer, C. A.; Torry, H. C.; Hsiang, J. S.

69. Phys. Rev. **72**, 974L (1947)   Williams, D.

70. Phys. Rev. **73**, 627 (1948)   Bardeen, J.; Townes, C. H.

71. Nature **161**, 522L (1948)   Bleaney, B.; Loubser, J. H. N.

72. Proc. Phys. Soc. (London) **60**, 540 (1948)   Bleaney, B.; Penrose, R. P.

73. Proc. Phys. Soc. (London) **60**, 83 (1948)   Bleaney, B.; Penrose, R. P.

74. Phys. Rev. **74**, 533 (1948)   Bragg, J. K.

75. Phys. Rev. **73**, 1053 (1948)   Carter, R. L.; Smith, W. V.

76. Phys. Rev. **74**, 1537L (1948)   Cunningham, G. L.; LeVan, W. I.; Gwinn, W. D.

77. Phys. Rev. **74**, 1245A (1948)   Dailey, B. P.; Rusinow, K.; Shulman, R. G.; Townes, C. H.

78. J. Phys. Radium **9**, 29D (1948)   Freymann, R.; Freymann, M.; LeBot, J.

79. Phys. Rev. **73**, 635 (1948)   Gilliam, O. R.; Edwards, H. D.; Gordy, W.

80. Phys. Rev. **73**, 92L (1948)   Golden, S.; Wentink, T., Jr.; Hillger, R. E.; Strandberg, M. W. P.

81. Rev. Mod. Phys. **20**, 668 (1948)   Gordy, W.

82. Phys. Rev. **74**, 1191 (1948)   Gordy, W.; Ring, H.

83. Phys. Rev. **74**, 243 (1948)   Gordy, W.; Simmons, J. W.; Smith, A. G.

84. Phys. Rev. **74**, 107L (1948)
    Errata **74**, 626L (1948)   Henderson, R. S.

85. Phys. Rev. **74**, 106L (1948)   Henderson, R. S.; Van Vleck, J. H.

86. M.I.T. Progress Rept. (1948)   Hill, A. G.; Harvey, G. G.; Strandberg, M. W. P.; Hillger, R. E.; Lawrance, R. B.; Loomis, C. C.; Weiss, M.; Ingersoll, J. G.

87. Phys. Rev. **74**, 1262A (1948)   Jauch, J. M.

88. Phys. Rev. **73**, 1248A (1948)   Jen, C. K.

89. Phys. Rev. **74**, 1246A (1948)   Jen, C. K.

90. Phys. Rev. **74**, 1396 (1948)   Jen, C. K.

91. Phys. Rev. **73**, 1120L (1948)   Karplus, R.

92. Phys. Rev. **74**, 223 (1948)   Karplus, R.

93. Proc. Phys. Soc. **61**, 562 (1948)   Lamont, H. R. L.

94. Phys. Rev. **74**, 705L (1948)   Mizushima, M.

95. J. Chem. Phys. **16**, 310 (1948)   Newton, R. R.; Thomas, L. H.

96. Phys. Rev. **73**, 1405L (1948)   Roberts, A.

97. Phys. Rev. **74**, 1889A (1948)   Rogers, J. D.; Williams, D.

98. Phys. Rev. **74**, 1870L (1948)   Sharbaugh, A. H.

99. Phys. Rev. **74**, 846L (1948)   Shulman, R. G.; Dailey, B. P.; Townes, C. H.

100. Phys. Rev. **73**, 713 (1948)   Simmons, J. W.; Gordy, W.

101. Phys. Rev. **74**, 123A (1948)   Simmons, J. W.; Gordy, W.

102. Phys. Rev. **74**, 1246A (1948)   Simmons, J. W.; Gordy, W.; Smith, A. G.

103. J. Chem. Phys. **16**, 553L (1948)   Skinner, H. A.

104. Phys. Rev. **73**, 633L (1948)   Smith, A. G.; Ring, H.; Smith, W. V.; Gordy, W.

105. Phys. Rev. **74**, 123A (1948)   Smith, A. G.; Ring, H.; Smith, W. V.; Gordy, W.

106. Phys. Rev. **74**, 370 (1948)   Smith, A. G.; Ring, H.; Smith, W. V.; Gordy, W.

107. Phys. Rev. **74**, 506L (1948)   Smith, D. F.

108. MITRLE Rept. #85 (1948)   Strandberg, M. W. P.

109. Phys. Rev. **74**, 1245A (1948)   Strandberg, M. W. P.

110. Phys. Rev. **73**, 188L (1948)   Strandberg, M. W. P.; Wentink, T., Jr.; Hillger, R. E.; Wannier, G. H.; Deutsch, M. L.

111. Phys. Rev. **74**, 626L (1948)   Townes, C. H.; Geschwind, S.

112. Phys. Rev. **74**, 1113 (1948)   Townes, C. H.; Holden, A. N.; Merritt, F. R.

113. Thesis, McGill Univ. (1948)   Turner, T. E.

114. Thesis, Columbia Univ. (1948)   Weingarten, I. R.

115. Phys. Rev. **73**, 1249A (1948)   Wentink, T., Jr.; Strandberg, M. W. P.

116. Zh. Eksperim. i Teor. Fiz. **19**, 853 (1949)   Abrikosov, A. A.

117. Phys. Rev. **75**, 1622L (1949)   Bak, B.; Knudsen, E. S.; Madsen, E.

118. Phys. Rev. **75**, 1318A (1949) — Beard, C. I.; Dailey, B. P.

119. Phys. Rev. **76**, 473A (1949) — Bianco, D.; Matlack, G.; Roberts, A.

120. State Univ. of Iowa (1949) — Bianco, D. R.; Roberts, A.

121. Phys. Rev. **75**, 1774L (1949) — Bragg, J. K.; Sharbaugh, A. H.

122. Nuovo Cimento **6**, 552 (1949) — Carrara, N.; Lombardini, P.; Cine, R.; Sacconi, L.

123. Phys. Rev. **76**, 703L (1949) — Cohen, V. W.; Koski, W. S.; Wentink, T., Jr.

124. Phys. Rev. **76**, 178A (1949) — Coles, D. K.; Hughes, R. H.

125. Phys. Rev. **76**, 858L (1949) — Coles, D. K.; Hughes, R. H.

126. J. Chem. Phys. **17**, 211L (1949) — Cunningham, G. L.; Boyd, A. W.; Gwinn, W. D.; LeVan, W. I.

127. Phys. Rev. **76**, 136L (1949) — Dailey, B. P.; Mays, J. M.; Townes, C. H.

128. Phys. Rev. **76**, 472A (1949) — Dailey, B. P.; Mays, J. M.; Townes, C. H.

129. Phys. Rev. **75**, 1014 (1949) — Gilliam, O. R.; Edwards, H. D.; Gordy, W.

130. Phys. Rev. **76**, 195A (1949) — Gilliam, O. R.; Edwards, H. D.; Gordy, W.

131. Thesis, Harvard Univ. (1949) — Goldstein, J. H.

132. Phys. Rev. **75**, 1453L (1949) — Goldstein, J. H.; Bragg, J. K.

133. Phys. Rev. **76**, 443L (1949) — Gordy, W.; Gilliam, O. R.; Livingston, R.

134. Phys. Rev. **75**, 1325A (1949) — Gordy, W.; Ring, H.; Burg, A. B.

135. J. Chem. Phys. **17**, 972 (1949) — Gutowsky, H. S.; Kistiakowsky, G. B.; Pake, G. E.; Purcell, E. M.

136. Can. J. Res. **B27**, 332 (1949) — Herzberg, G.; Herzberg, L.

137. MITRLE Progress Rept. (1949) — Hillger, R. E.

138. Phys. Rev. **75**, 1319A (1949) — Jen, C. K.

139. Phys. Rev. **76**, 471A (1949) — Jen, C. K.

140. Phys. Rev. **76**, 1494 (1949) — Jen, C. K.

141. Phys. Rev. **75**, 889 (1949) — Karplus, R.; Sharbaugh, A. H.
    Errata **75**, 1449L (1949)

142. M.I.T. Progress Rept. (1949) — Lawrance, R. B.

143. Phys. Rev. **76**, 472A (1949) — Lawrance, R. B.; Kyhl, R. L.; Standberg, M. W. P.

144. Phys. Rev. **76**, 149L (1949) — Livingston, R.; Gilliam, O. R.; Gordy, W.

145. MITRLE Progress Rept. (1949) — Loomis, C. C.

146. Phys. Rev. **75**, 529L (1949) — Low, W.; Townes, C. H.

147. Phys. Rev. **75**, 1318A (1949) — Low, W.; Townes, C. H.

148. Phys. Rev. **76**, 121 (1949) — Margenau, H.

149. Phys. Rev. **76**, 585A (1949) — Margenau, H.

150. Phys. Rev. **76**, 1423 (1949) — Margenau, H.

151. J. Appl. Phys. **20**, 413L (1949) — Millman, G. H.; Raymond, R. C.

152. J. Phys. Soc. Japan **4**, 11 (1949) — Mizushima, M.

153. J. Phys. Soc. Japan **4**, 191 (1949) — Mizushima, M.

154. Phys. Rev. **75**, 1961L (1949) — Nielsen, H. H.

155. Phys. Rev. **76**, 690L (1949) — Pietenpol, W. J.; Rogers, J. D.

156. J. Chem. Phys. **17**, 742L (1949) — Roberts, A.; Edgell, W. F.

157. Phys. Rev. **76**, 178A (1949) — Roberts, A.; Edgell, W. F.

158. Nature **163**, 871 (1949) — Saxton, J. A.; Lane, J. A.

159. Phys. Rev. **76**, 1419L (1949) — Sharbaugh, A. H.; Bragg, J. K.; Madison, T. C.; Thomas, V. G.

160. Phys. Rev. **76**, 1529L (1949) — Sharbaugh, A. H.; Madison, T. C.; Bragg, J. K.

161. Phys. Rev. **75**, 1102L (1949) — Sharbaugh, A. H.; Mattern, J.

162. Phys. Rev. **76**, 472A (1949) — Shulman, R. G.; Dailey, B. P.; Townes, C. H.

163. Phys. Rev. **75**, 1318A (1949) — Shulman, R. G.; Townes, C. H.

164. Phys. Rev. **76**, 686L (1949) — Simmons, J. W.

165. Phys. Rev. **75**, 260 (1949) — Smith, A. G.; Gordy, W.; Simmons, J. W.; Smith, W. V.

166. J. Chem. Phys. **17**, 1348L (1949) — Smith, W. V.; Unterberger, R. R.

167. J. Chem. Phys. **17**, 901 (1949) — Strandberg, M. W. P.

168. J. Chem. Phys. **17**, 429L (1949) — Strandberg, M. W. P.; Pearsall, C. S.; Weiss, M. T.

169. Phys. Rev. **75**, 827 (1949) — Strandberg, M. W. P.; Wentink, T., Jr.; Hill, A. G.

170. Phys. Rev. **75**, 270 (1949) — Strandberg, M. W. P.; Wentink, T., Jr.; Kyhl, R. L.

171. Phys. Rev. **76**, 691L (1949)

172. Phys. Rev. **76**, 700L (1949)

173. Phys. Rev. **76**, 472 (1949)

174. J. Am. Chem. Soc. **71**, 797 (1949)

175. J. Chem. Phys. **17**, 1319 (1949)

176. Phys. Rev. **76**, 472A (1949)

177. Phys. Rev. **76**, 472 (1949)

178. J. Chem. Phys. **18**, 1422 (1950)

179. Phys. Rev. **80**, 511 (1950)

180. Phys. Rev. **79**, 190L (1950)

181. Phys. Rev. **80**, 101L (1950)

182. J. Chem. Phys. **18**, 1437 (1950)

183. J. Chem. Phys. **18**, 1514L (1950)

184. Phys. Rev. **77**, 144L (1950)

185. Proc. Phys. Soc. **63A**, 483 (1950)

186. Phys. Rev. **77**, 148L (1950)

187. Phys. Rev. **80**, 114L (1950)

188. Phys. Rev. **77**, 742A (1950)

189. Phys. Rev. **79**, 224A (1950)

190. Thesis, Harvard U. (1950)

191. J. Chem. Phys. **18**, 1118L (1950)

192. Naturwiss. **37**, 398 (1950)

193. Phys. Rev. **80**, 1106 (1950)

194. Phys. Rev. **78**, 174L (1950)

195. Phys. Rev. **79**, 226A (1950)

196. Phys. Rev. **78**, 140 (1950)

197. Phys. Rev. **78**, 347 (1950)

198. Phys. Rev. **78**, 512 (1950)

199. Phys. Rev. **79**, 224A (1950)

200. Duke Progress Rept. (1950)

201. MITRLE Progress Rept. (1950)

202. Phys. Rev. **78**, 339A (1950)

203. MITRLE Progress Rept. (1950)

204. J. Chem. Phys. **18**, 990L (1950)

205. Phys. Rev. **79**, 54 (1950)

206. J. Chem. Phys. **18**, 1109 (1950)

207. Phys. Rev. **78**, 347A (1950)

208. Experientia **VI**, 321 (1950)

209. M.I.T. Progress Rept. (1950)

210. Phys. Rev. **78**, 347 (1950)

211. M.I.T. Tech. Rept. No. 177 (1950)

212. Phys. Rev. **78**, 817 (1950)

213. M.I.T. Tech. Rept. No. 167 (1950)

214. Bull. Am. Phys. Soc. **25**, 1, 44 (1950)

215. Phys. Rev. **78**, 348A (1950)

216. Thesis, Columbia Univ. (1950)

217. Phys. Rev. **79**, 224A (1950)

218. J. Chem. Phys. **18**, 332 (1950)

219. Thesis, Columbia Univ. (1950)

220. Phys. Rev. **78**, 340A (1950)

221. Busseiron Kenkyu., No. 29, 25 (1950)

222. Phys. Rev. **77**, 130 (1950)

223. Phys. Rev. **78**, 296L (1950)

224. Phys. Rev. **80**, 910L (1950)

225. Phys. Rev. **77**, 741A (1950)

Townes, C. H.; Aamodt, L. C.

Townes, C. H.; Mays, J. M.; Dailey, B. P.

Townes, C. H.; Shulman, R. G.; Dailey, B. P.

Waddington, G.; Knowlton, J. W.; Scott, D. W.; Oliver, G. D.; Todd, S. S.; Hubbard, W. N.; Smith, J. C.; Huffman, H. M.

Westenberg, A. A.; Goldstein, J. H.; Wilson, E. B., Jr.

Westenberg, A. A.; Goldstein, J. H.; Wilson, E. B., Jr.

Wilson, M. K.; Badger, R. M.

Amble, E.; Dailey, B. P.

Anderson, P. W.

Bak, B.; Knudsen, E. S.; Madsen, E.; Rastrup-Andersen, J.

Bak, B.; Sloan, R.; Williams, D.

Beard, C. I.; Dailey, B. P.

Bernstein, H. J.

Birnbaum, G.

Bleaney, B.; Loubser, J. H. N.

Bragg, J. K.; Madison, T. C.; Sharbaugh, A. H.

Castle, J. G., Jr.; Beringer, R.

Cohen, V. W.; Koski, W. S.; Wentink, T. Jr.

Coles, D. K.; Good, W. E.; Hughes, R. H.

Cornwell, C. D.

Cornwell, C. D.

Dehmelt, H. G.

Eshbach, J. R.; Hillger, R. E.; Jen, C. K.

Geschwind, S.; Minden, H.; Townes, C. H.

Geschwind, S.; Minden, H.; Townes, C. H.

Gilliam, O. R.; Johnson, C. M.; Gordy, W.

Goldstein, J. H.; Bragg, J. K.

Gordy, W.; Ring, H.; Burg, A. B.

Gordy, W.; Sheridan, J.

Gordy, W.; Sheridan, J.; Williams, Q.

Hillger, R. E.

Jen, C. K.

Johnson, H. R.

Jones, L. H.; Shoolery, J. N.; Shulman, R. G.; Yost, D. M.

Kessler, M.; Ring, H.; Trambarulo, R.; Gordy, W.

Kisliuk, P.; Townes, C. H.

Kisliuk, P.; Townes, C. H.

Klages, V. G.

Lawrance, R. B.

Lawrance, R. B.

Lawrance, R. B.; Strandberg, M. W. P.

Lindström, G.

Loomis, C. C.; Strandberg, M. W. P.

Loubser, J. H. N.; Klein, J. A.

Loubser, J. H. N.; Klein, J. A.

Low, W.

Low, W.; Townes, C. H.

Matlack, G.; Glockler, G.; Bianco, D. R.; Roberts, A.

Mays, J. M.

McAfee, K. B., Jr.

Mizushima, M.

Nielsen, H. H.

Nielsen, H. H.

Okamura, T.; Torizuka, Y.; Kojima, Y.

Pietenpol, W. J.; Rogers, J. D.; Williams, D.

226. Phys. Rev. **78**, 480L (1950) — Pietenpol, W. J.; Rogers, J. D.; Williams, D.
227. Rev. Sci. Instr. **21**, 1014 (1950) — Rogers, J. D.; Cox, H. L.; Braunschweiger, P. G.
228. Phys. Rev. **78**, 293 (1950) — Senatore, S. J.
229. Phys. Rev. **77**, 302L (1950) — Sharbaugh, A. H.; Pritchard, B. S.; Madison, T. C.
230. Phys. Rev. **79**, 189L (1950) — Sharbaugh, A. H.; Pritchard, B. S.; Thomas, V. G.; Mays, J. M.; Dailey, B. P.
231. Phys. Rev. **78**, 64L (1950) — Sharbaugh, A. H.; Thomas, V. C.; Pritchard, B. S.
232. Phys. Rev. **77**, 292L (1950) — Sheridan, J.; Gordy, W.
233. Phys. Rev. **77**, 719L (1950) — Sheridan, J.; Gordy, W.
234. Phys. Rev. **79**, 224A (1950) — Sheridan, J.; Gordy, W.
235. Phys. Rev. **79**, 513 (1950) — Sheridan, J.; Gordy, W.
236. Phys. Rev. **78**, 145 (1950) — Shulman, R. G.; Dailey, B. P.; Townes, C. H.
237. Phys. Rev. **77**, 421L (1950) — Shulman, R. G.; Townes, C. H.
238. Phys. Rev. **77**, 500 (1950) — Shulman, R. G.; Townes, C. H.
239. Phys. Rev. **78**, 347A (1950) — Shulman, R. G.; Townes, C. H.
240. Phys. Rev. **80**, 338 (1950) — Simmons, J. W.; Anderson, W. E.
241. Phys. Rev. **77**, 77 (1950) — Simmons, J. W.; Anderson, W. E.; Gordy, W.
242. Phys. Rev. **80**, 289L (1950) — Simmons, J. W.; Swan, W. O.
243. J. Chem. Phys. **18**, 217 (1950) — Sinha, S. P.
244. Phys. Rev. **78**, 639A (1950) — Southern, A. L.; Morgan, H. W.; Keilholtz, G. W.; Smith, W. V.
245. Univ. of Kyoto, **A26**, Nos. 1 & 2, Article 12 (1950) — Takahashi, I.; Okaya, A.; Ogawa, T.; Hashi, T.
246. Nuovo Cimento **7**, 1 (1950) — Tomassini, M.
247. J. Chem. Phys. **18**, 1613 (1950) — Trambarulo, R.; Gordy, W.
248. Phys. Rev. **79**, 224A (1950) — Trambarulo, R.; Gordy, W.
249. J. Chem. Phys. **18**, 565L (1950) — Unterberger, R. R.; Trambarulo, R.; Smith, W. V.
250. Phys. Rev. **78**, 202 (1950) — Weiss, M. T.; Strandberg, M. W. P.; Lawrance, R. B.; Loomis, C. C.
251. J. Am. Chem. Soc. **72**, 199 (1950) — Westenberg, A. A.; Wilson, E. B., Jr.
252. Phys. Rev. **79**, 225A (1950) — Williams, Q.; Gordy, W.
253. Phys. Rev. **83**, 210 (1951) — Amble, E.
254. Phys. Rev. **82**, 328A (1951) — Amble, E.; Schawlow, A. L.
255. Phys. Rev. **81**, 819 (1951) — Anderson, W. E.; Sheridan, J.; Gordy, W.
256. Phys. Rev. **82**, 58 (1951) — Anderson, W. E.; Trambarulo, R.; Sheridan, J.; Gordy, W.
257. J. Chem. Phys. **19**, 975L (1951) — Beard, C. I.; Dailey, B. P.
258. Phys. Rev. **82**, 877 (1951) — Coles, D. K.; Good, W. E.; Bragg, J. K.; Sharbaugh, A. H.
259. O.N.R. Rept., Iowa State (1951) — Cornwell, C. D.
260. Phys. Rev. **82**, 108 (1951) — Costain, C. C.
261. Thesis, Duke Univ. (1951) — Crable, G. F.
262. J. Chem. Phys. **19**, 502 (1951) — Crable, G. F.; Smith, W. V.
263. J. Chem. Phys. **19**, 676 (1951) — Cunningham, G. L., Jr.; Boyd, A. W.; Myers, R. J.; Gwinn, W. D.; LeVan, W. I.
264. Phys. Rev. **82**, 342A (1951) — Eisenstein, J.
265. Phys. Rev. **82**, 327A (1951) — Eshbach, J. R.; Strandberg, M. W. P.
266. Compt. Rend. **232**, 401 (1951) — Freymann, M.; Freymann, R.
267. Compt. Rend. **232**, 1096 (1951) — Freymann, M.; Freymann, R.
268. Compt. Rend. **232**, 2312 (1951) — Freymann, M.; Rolland, M. T.; Freymann, R.
269. Phys. Rev. **81**, 882L (1951) — Geschwind, S.; Gunther-Mohr, G. R.
270. Phys. Rev. **82**, 346A (1951) — Geschwind, S.; Gunther-Mohr, G. R.
271. Phys. Rev. **81**, 288L (1951) — Geschwind, S.; Gunther-Mohr, G. R.; Townes, C. H.
272. Phys. Rev. **83**, 209A (1951) — Geschwind, S.; Gunther-Mohr, G. R.; Townes, C. H.
273. Phys. Rev. **83**, 881 (1951) — Gokhale, B. V.; Johnson, J. R.; Strandberg, M. W. P.

274. Phys. Rev. **83**, 880 (1951)

275. Phys. Rev. **82**, 343A (1951)

276. Phys. Rev. **83**, 741 (1951)
277. Phys. Rev. **82**, 327A (1951)
278. Phys. Rev. **83**, 575 (1951)
279. Phys. Rev. **83**, 488 (1951)
280. Naturwiss. **38**, 34 (1951)
281. Physica **17**, 378 (1951)
282. Phys. Rev. **84**, 1178 (1951)
283. M.I.T. Elect. Tech. Rept. #192 (1951)
284. Phys. Rev. **82**, 327A (1951)
285. Phys. Rev. **83**, 210A (1951)
286. Phys. Rev. **81**, 296A (1951)
287. Physica **17**, 446 (1951)
288. Phys. Rev. **83**, 363 (1951)
289. Phys. Rev. **81**, 798 (1951)
290. Phys. Rev. **81**, 297A (1951)

291. Phys. Rev. **81**, 630 (1951)

292. J. Chem. Phys. **19**, 1071 (1951)
293. Phys. Rev. **83**, 485A (1951)
294. Phys. Rev. **81**, 940 (1951)
295. Phys. Rev. **82**, 971 (1951)
296. Phys. Rev. **82**, 327 (1951)

297. J. Chem. Phys. **19**, 739 (1951)
298. Anal. Chem. **23**, 1000 (1951)

299. Thesis, Mich. Univ. (1951)
300. Phys. Rev. **83**, 1064L (1951)
301. Bull. Soc. Roy. Sci. Liege **20**, 439 (1951)
302. Physica **17**, 432 (1951)
303. Phys. Rev. **83**, 880A (1951)
304. Phys. Rev. **82**, 323A (1951)
305. Phys. Rev. **83**, 987 (1951)
306. O.N.R. Rept., Iowa State (1951)
307. Phys. Rev. **83**, 431 (1951)
308. Phys. Rev. **82**, 131A (1951)
309. Phys. Rev. **82**, 323A (1951)
310. M.I.T. Tech. Rept. **188** (1951)
311. J. Chem. Phys. **19**, 965 (1951)
312. Phys. Rev. **82**, 95 (1951)
313. Phys. Rev. **82**, 323A (1951)
314. J. Chem. Phys. **19**, 250L (1951)
315. J. Chem. Phys. **19**, 1364 (1951)

316. Phys. Rev. **83**, 485A (1951)
317. J. Chem. Phys. **19**, 938 (1951)
318. J. Chem. Phys. **19**, 1609L (1951)
319. Phys. Rev. **83**, 485A (1951)

320. Bull. Inst. Chem. Res. Kyoto Univ. **28**, 54 (1951)

321. Bull. Inst. Chem. Res. Kyoto Univ. **27**, 56 (1951)
322. J. Chem. Phys. **19**, 805 (1951)
323. Phys. Rev. **83**, 210A (1951)
324. Phys. Rev. **83**, 880 (1951)
325. Phys. Rev. **83**, 1058 (1951)
326. J. Chem. Phys. **19**, 381L (1951)

Good, W. E.; Coles, D. K.; Gunther-Mohr, G. R.; Schawlow, A. L.; Townes, C. H.
Gunther-Mohr, G. R.; Geschwind, S.; Townes, C. H.
deHeer, J.
Hillger, R. E.; Strandberg, M. W. P.
Hillger, R. E.; Strandberg, M. W. P.
Honer, R. E.
Honerjager, R.
Jen, C. K.
Johnson, C. M.; Trambarulo, R.; Gordy, W.
Johnson, H. R.; Strandberg, M. W. P.
Johnson, H. R.; Strandberg, M. W. P.
Kisliuk, P.; Townes, C. H.
Koski, W. S.; Wentink, T., Jr.; Cohen, V. W.
Lamont, H. R. L.
Lawrance, R. B.; Strandberg, M. W. P.
Loomis, C. C.; Strandberg, M. W. P.
Lyons, H.; Kessler, M.; Rueger, L. J.; Nuckolls, R. G.
Lyons, H.; Rueger, L. J.; Nuckolls, R. G.; Kessler, M.
Magnuson, D. W.
Magnuson, D. W.
Mays, J. M.; Townes, C. H.
McAfee, K. B., Jr.
Miller, S. L.; Kraitchman, J.; Dailey, B. P.; Townes, C. H.
Mizushima, M.; Ito, T.
Morgan, H. W.; Keilholtz, G. W.; Smith, W. V.; Southern, A. L.
Nethercot, A. H., Jr.
Newell, G., Jr.; Dicke, R. H.
Nielsen, H. H.
Nielsen, H. H.
Nuckolls, R. G.; Rueger, L. J.; Lyons, H.
Potter, C. A.; Bushkovitch, A. V.; Rouse, A. G.
Potter, C. A.; Bushkovitch, A. V.; Rouse, A. G.
Poynter, R. L.
Rogers, J. D.; Pietenpol, W. J.; Williams, D.
Rogers, J. D.; Williams, D.
Rogers, J. D.; Williams, D.
Sawyer, K. A.; Kierstead, J. D.
Sheridan, J.; Gordy, W.
Shoolery, J. N.; Sharbaugh, A. H.
Shoolery, J. N.; Shulman, R. G.
Shoolery, J. N.; Shulman, R. G.; Yost, D. M.
Shoolery, J. N.; Shulman, R. G.; Yost, D. M.; Sheehan, W. F., Jr.; Schomaker, V.
Simmons, J. W.; Goldstein, J. H.
Sirvetz, M. H.
Sirvetz, M. H.
Smith, D. F.; Tidwell, M.; Williams, D. V. P.; Senatore, S. J.
Takahashi, I.; Okaya, A.; Hashi, T.; Ogawa, T.; Ryuzan, O.
Takahashi, I.; Okaya, A.; Ogawa, T.
Talley, R. M.; Nielsen, A. H.
Ting, Y.; Weatherly, T. L.; Williams, D.
Van Vleck, J. H.
Weber, J.
Weber, J.; Laidler, K. J.

327. J. Chem. Phys. **19**, 1089 (1951) Weber, J.; Laidler, K. J.
328. Phys. Rev. **81**, 286 (1951) Weiss, M. T.; Strandberg, M. W. P.
329. Phys. Rev. **82**, 326A (1951) Weiss, M. T.; Strandberg, M. W. P.
330. Phys. Rev. **83**, 567 (1951) Weiss, M. T.; Strandberg, M. W. P.
331. Phys. Rev. **81**, 948 (1951) Wentink, T., Jr.; Koski, W. S.; Cohen, V. W.
332. J. Chem. Phys. **20**, 192L (1952) Amble, E.; Miller, S. L.; Schawlow, A. L.; Townes, C. H.

333. J. Chem. Phys. **20**, 1488L (1952) Beard, C. I.; Bianco, D. R.
334. Ann. N.Y. Acad. Sci. **55**, 814 (1952) Beringer, R.
335. Can. J. Chem. **30**, 963 (1952) Bernstein, H. J.; Pullin, A. D. E.
336. Phil. Mag. **43**, 995 (1952) Bleaney, B.; Scovil, H. E. D.; Trenam, R. S.
337. Nuovo Cimento **9**, 238 (1952) van den Bosch, J. C.; Bruin, F.
338. J. Phys. Chem. **56**, 321 (1952) Costain, C. C.; Sutherland, G. B. B. M.
339. Thesis, Univ. of Minn. (1952) Crawford, Harry D.
340. J. Chem. Phys. **20**, 101 (1952) DeHeer, J.; Nielsen, H. H.
341. J. Chem. Phys. **20**, 1804 (1952) Duchesne, J.
342. Nuovo Cimento **9**, 270 (1952) Duchesne, J.
343. J. Chem. Phys. **20**, 1708 (1952) Dunitz, J. D.; Feldman, H. G.; Schomaker, V.
344. Phys. Rev. **85**, 532 (1952) Eshbach, J. R.; Hillger, R. E.; Strandberg, M. W. P.

345. J. Chem. Phys. **20**, 1 (1952) Fristrom, R. M.
346. Phys. Rev. **85**, 717A (1952) Fristrom, R. M.
347. Ann. N.Y. Acad. Sci. **55**, 751 (1952) Geschwind, S.
348. Diss. Abstr. **12**, 88 (1952) Geschwind, S.
349. Phys. Rev. **85**, 474 (1952) Geschwind, S.; Gunther-Mohr, G. R.; Silvey, G.

350. J. Chem. Phys. **20**, 605 (1952) Ghosh, S. N.; Trambarulo, R.; Gordy, W.
351. Phys. Rev. **85**, 716A (1952) Gilbert, D. A.
352. Ann. N.Y. Acad. Sci. **55**, 774 (1952) Gordy, W.
353. J. Am. Chem. Soc. **74**, 4662 (1952) Guthrie, G. B., Jr.; Scott, D. W.; Hubbard, W. N.; Katz, C.; McCullough, J. P.; Gross, M. E.; Williamson, K. D.; Waddington, G.

354. Phys. Rev. **85**, 494L (1952) Hardy, W. A.; Silvey, G.; Townes, C. H.
355. Phys. Rev. **86**, 608A (1952) Hardy, W. A.; Silvey, G.; Townes, C. H.
356. J. Chem. Phys. **20**, 528L (1952) Hawkins, N. J.; Cohen, V. W.; Koski, W. S.
357. Nature **169**, 997 (1952) Herzberg, G.
358. J. Chem. Phys. **20**, 518 (1952) Hrostowski, H. J.; Myers, R. J.; Pimentel, G. C.

359. Phys. Rev. **85**, 717A (1952) Hughes, R. H.
360. Phys. Rev. **87**, 227A (1952) Javan, A.; Grosse, A. V.
361. Phys. Rev. **86**, 608A (1952) Javan, A.; Townes, C. H.
362. Ann. N.Y. Acad. Sci. **55**, 822 (1952) Jen, C. K.
363. Phys. Rev. **87**, 677 (1952) Johnson, C. M.; Slager, D. M.
364. J. Chem. Phys. **20**, 687 (1952) Johnson, H. R.; Strandberg, M. W. P.
365. Phys. Rev. **85**, 503L (1952) Johnson, H. R.; Strandberg, M. W. P.
366. Phys. Rev. **86**, 811L (1952) Johnson, H. R.; Strandberg, M. W. P.
367. Duke Univ. Progress Rept. AD–10 837 (1952) King, W. C.; Gordy, W.
368. J. Chem. Phys. **20**, 804 (1952) Kojima, S.; Tsukada, K.; Hagiwara, S.; Mizushima, M.; Ito, T.

369. Brit. J. Appl. Phys. **3**, 182 (1952) Lamont, H. R. L.; Hickin, E. M.
370. J. Am. Chem. Soc. **74**, 3548 (1952) Lide, D. R., Jr.
371. Phys. Rev. **87**, 227A (1952) Lide, D. R., Jr.
372. J. Chem. Phys. **20**, 1170 (1952) Livingston, R.
373. J. Chem. Phys. **20**, 229 (1952) Magnuson, D. W.
374. Ann. N.Y. Acad. Sci. **55**, 789 (1952) Mays, J. M.
375. J. Chem. Phys. **20**, 1695 (1952) Mays, J. M.; Dailey, B. P.
376. J. Chem. Phys. **20**, 1112 (1952) Miller, S. L.; Aamodt, L. C.; Dousmanis, G.; Townes, C. H.; Kraitchman, J.

377. Busseiron Kenkyu. No. 57, 9 (1952) Mizushima, M.; Venkateswarlu, P.
378. Phys. Rev. **87**, 172 (1952) Mockler, R.; Bailey, J. H.; Gordy, W.
379. J. Chem. Phys. **20**, 1981 (1952) Morgan, H. W.; Goldstein, J. H.
380. Phys. Rev. **87**, 172 (1952) Morgan, H. W.; Goldstein, J. H.

381. J. Chem. Phys. **20,** 1420 (1952)   Myers, R. J.; Gwinn, W. D.
382. Phys. Rev. **87,** 226A (1952)   Nethercot, A. H.; Javan, A.; Townes, C. H.
383. Nuovo Cimento **9,** 358 (1952)   Nethercot, A. H.; Klein, J. A.; Loubser, J. H. N.; Townes, C. H.

384. Phys. Rev. **86,** 798 (1952)   Nethercot, A. H., Jr.; Klein, J. A.; Townes, C. H.

385. Phys. Rev. **86,** 799 (1952)   Rank, D. H.; Ruth, B. P.; Vander Sluis, K. L.
386. Diss. Abstr. **12,** 653 (1952)   Robinson, G. W.
387. Phys. Rev. **86,** 654A (1952)   Rogers, J. D.; Williams, D.
388. Rev. Sci. Instr. **23,** 635 (1952)   Rueger, L. J.; Nuckolls, R. G.
389. Ann. N.Y. Acad. Sci. **55,** 955 (1952)   Schawlow, A. L.
390. Phys. Rev. **86,** 606 (1952)   Schwarz, R. F.
391. J. Chem. Phys. **20,** 591 (1952)   Sheridan, J.; Gordy, W.
392. J. Chem. Phys. **20,** 735 (1952)   Sheridan, J.; Gordy, W.
393. Phys. Rev. **87,** 236A (1952)   Silvey, G.; Hardy, W. A.; Townes, C. H.
394. Phys. Rev. **86,** 1055 (1952)   Simmons, J. W.; Anderson, W. E.; Gordy, W.
395. J. Chem. Phys. **20,** 122 (1952)   Simmons, J. W.; Goldstein, J. H.
396. Phys. Rev. **86,** 424 (1952)   Sinton, W. M.
397. Phys. Rev. **86,** 608A (1952)   Smith, D. F.
398. Phys. Rev. **87,** 226A (1952)   Smith, D. F.; Magnuson, D. W.
399. Ann. N.Y. Acad. Sci. **55,** 891 (1952)   Smith, W. V.
400. Bull. Inst. Chem. Res. Kyoto Univ. **28,** 54 (1952)   Takahashi, I.; Okaya, A.; Hashi, T.; Ogawa, T.; Ryuzan, O.

401. Phys. Rev. **86,** 440 (1952)   Tetenbaum, S. J.
402. Phys. Rev. **88,** 772 (1952)   Tetenbaum, S. J.
403. Duke Univ. Progress Rept. AD–10 837 (1952)   Trambarulo, R.; Ghosh, S. N.
404. Phys. Rev. **85,** 717A (1952)   Weatherly, T. L.; Manring, E. R.; Williams, D.

405. Phys. Rev. **87,** 517 (1952)   Weatherly, T. L.; Williams, D.
406. Can. J. Phys. **30,** 577 (1952)   Welsh, H. L.; Crawford, M. F.; Thomas, T. R.; Love, G. R.

407. J. Chem. Phys. **20,** 1820 (1952)   Weston, R. E., Jr.; Sirvetz, M. H.
408. J. Chem. Phys. **20,** 1656 (1952)   Wilcox, W. S.; Goldstein, J. H.
409. Phys. Rev. **87,** 172 (1952)   Wilcox, W. S.; Goldstein, J. H.; Simmons, J. W.

410. J. Chem. Phys. **20,** 1524 (1952)   Williams, Q.; Cox, J. T.; Gordy, W.
411. J. Chem. Phys. **20,** 164 (1952)   Williams, Q.; Sheridan, J.; Gordy, W.
412. J. Chem. Phys. **21,** 1898 (1953)   Arendale, W. F.; Fletcher, W. H.
413. J. Chem. Phys. **21,** 752L (1953)   Bak, B.; Bruhn, J.; Rastrup-Andersen, J.
414. J. Chem. Phys. **21,** 1305 (1953)   Bak, B.; Rastrup-Andersen, J.
415. Phys. Rev. **91,** 1014L (1953)   Beers, Y.; Weisbaum, S.
416. N.Y. Univ. Report No. 195.2 (1953)   Beers, Y.; Weisbaum, S.; Herrmann, G.; Sterzer, F.

417. J. Chem. Phys. **21,** 57 (1953)   Birnbaum, G.
418. J. Chem. Phys. **21,** 1774 (1953)   Birnbaum, G.; Maryott, A. A.
419. Phys. Rev. **89,** 895 (1953)   Birnbaum, G.; Maryott, A. A.
420. Phys. Rev. **92,** 270 (1953)   Birnbaum, G.; Maryott, A. A.
421. Nature **171,** 695L (1953)   Buchanan, T. J.
422. J. Opt. Soc. Am. **43,** 1058 (1953)   Burgess, J. S.; Bell, E. E.; Nielsen, H. H.
423. Phys. Rev. **90,** 303 (1953)   Burke, B. F.; Strandberg, M. W. P.
424. Phys. Rev. **90,** 338 (1953)   Burke, B. F.; Strandberg, M. W. P.
425. Phys. Rev. **92,** 274 (1953)   Burrus, C. A., Jr.; Gordy, W.
426. Phys. Rev. **91,** 222A (1953)   Cox, J. T.; Peyton, P. B., Jr.; Gordy, W.
427. J. Chem. Phys. **21,** 2099L (1953)   Crawford, H. D.
428. Phys. Rev. **90,** 338A (1953)   Dailey, B. P.
429. J. Chem. Phys. **21,** 1416L (1953)   Dousmanis, G. C.; Sanders, T. M., Jr.; Townes, C. H.; Zeiger, H. J.

430. Arkiv Fysik **7,** 189 (1953)   Erlandsson, G.
431. Arkiv Fysik **6,** 477 (1953)   Erlandsson, G.
432. Phys. Rev. **90,** 338A (1953)   Feeny, H.; Lackner, H.
433. Phys. Rev. **90,** 338A (1953)   Ferguson, R. C.; Wilson, E. B., Jr.
434. J. Chem. Phys. **21,** 308 (1953)   Ghosh, S. N.; Trambarulo, R.; Gordy, W.
435. J. Chimie Physique **50,** C114 (1953)   Gordy, W.

436. Phys. Rev. **92**, 1532 (1953) — Hardy, W. A.; Silvey, G.; Townes, C. H.; Burke, B. F.; Strandberg, M. W. P.; Parker, G. W.; Cohen, V. W.

437. Nature **172**, 774 (1953) — Harris, F. E.; Alder, B. J.
438. J. Chem. Phys. **21**, 1839 (1953) — Haynie, W. H.; Nielsen, H. H.
439. Nature **172**, 771 (1953) — Heath, G. A.; Thomas, L. F.; Sheridan, J.
440. Mem. Rept. 703, Ballistic Res. Lab. (1953) — Hicks, B. L.; Turner, T. E.
441. J. Chem. Phys. **21**, 564L (1953) — Hicks, B. L.; Turner, T. E.; Widule, W. W.
442. J. Chem. Phys. **21**, 959L (1953) — Hughes, R. H.
443. J. Chem. Phys. **21**, 520 (1953) — Jen, C. K.; Bianco, D. R.; Massey, J. T.
444. J. Chem. Phys. **21**, 1425L (1953) — Johnson, R. D.; Myers, R. J.; Gwinn, W. D.
445. Phys. Rev. **90**, 319 (1953) — King, W. C.; Gordy, W.
446. Diss. Abstr. **13**, 317 (1953) — Kisliuk, P.
447. J. Chem. Phys. **21**, 828 (1953) — Kisliuk, P.; Geschwind, S.
448. Private communication (1953) — Kivelson, D.
449. Phys. Rev. **90**, 338 (1953) — Kivelson, D.; Wilson, E. B., Jr.
450. Columbia Rad. Lab. Rept. (1953) — Klein, J. A.; Nethercot, A. H., Jr.
451. Am. J. Phys. **21**, 17 (1953) — Kraitchman, J.
452. J. Sci. Ind. Res. **12B**, 1 (1953) — Krishnaji; Swarup, P.
453. Z. Physik **136**, 374 (1953) — Krishnaji; Swarup, P.
454. J. Appl. Phys. **24** (1953) — Krishnaji; Swarup, P.
455. Phys. Rev. **92**, 1271 (1953) — Livingston, R.; Benjamin, B. M.; Cox, J. T.; Gordy, W.

456. NBS Circular 537 (1953) — Maryott, A. A.; Buckley, F.
457. J. Chem. Phys. **21**, 2082L (1953) — McCulloh, K. E.; Pollnow, G. F.
458. J. Chem. Phys. **21**, 539 (1953) — Mizushima, M.
459. J. Chem. Phys. **21**, 1222 (1953) — Mizushima, M.
460. Phys. Rev. **91**, 464A (1953) — Mizushima, M.
461. J. Chem. Phys. **21**, 705 (1953) — Mizushima, M.; Venkateswarlu, P.
462. J. Chem. Phys. **21** (1953) — Mockler, R. C.; Bailey, G. H.; Gordy, W.
463. J. Am. Chem. Soc. **75**, 860 (1953) — Muller, N.
464. J. Chem. Phys. **21**, 363 (1953) — Nethercot, A. H.; Javan, A.
465. J. Phys. Soc. Japan **8**, 426 (1953) — Nishokawa, T.; Shimoda, K.
466. Phys. Rev. **89**, 1101 (1953) — Nuckolls, R. G.; Rueger, L. J.; Lyons, H.
467. J. Opt. Soc. Am. **43**, 1065 (1953) — Overend, J.; Thompson, H. W.
468. M.I.T. Tech. Rept. #255 (1953) — Posener, D. W.
469. J. Chem. Phys. **21**, 1401L (1953) — Posener, D. W.; Strandberg, M. W. P.
470. J. Chem. Phys. **21**, 960 (1953) — Ramsay, D. A.
471. J. Chem. Phys. **21**, 1741 (1953) — Robinson, G. W.
472. J. Chem. Phys. **21**, 1436 (1953) — Robinson, G. W.; Cornwell, C. D.
473. J. Chimie Physique **50**, D42 (1953) — Roubine, E.
474. Nature **171**, 87 (1953) — Sharbaugh, A. H.; Heath, G. A.; Thomas, L. F.; Sheridan, J.

475. J. Chem. Phys. **21**, 1614 (1953) — Sirkar, S. C.; Ghosh, D. K.
476. J. Chem. Phys. **21**, 898 (1953) — Sirvetz, M. H.; Weston, R. E., Jr.
477. J. Chem. Phys. **21**, 609 (1953) — Smith, D. F.
478. J. Chem. Phys. **21**, 2072 (1953) — Stroup, R. E.; Oetjen, R. A.; Bell, E. E.
479. Bull. Inst. Chem. Res., Kyoto Univ. **31**, 125 (1953) — Takahashi, I.; Okaya, A.; Ogawa, T.
480. J. Chem. Phys. **21**, 851 (1953) — Trambarulo, R.; Ghosh, S. N.; Burrus, C. A., Jr.; Gordy, W.

481. Phys. Rev. **91**, 222A (1953) — Trambarulo, R.; Ghosh, S. N.; Burrus, C. A., Jr.; Gordy, W.

482. J. Chem. Phys. **21**, 564L (1953) — Turner, T. E.; Fiora, V. C.; Kendrick, W. M.; Hicks, B. L.

483. Phys. Rev. **90**, 338A (1953) — Turner, T. E.; Fiora, V. C.; Kendrick, W. M.; Hicks, B. L.

484. J. Chem. Phys. **21**, 1713 (1953) — Venkateswarlu, P.; Mockler, R. C.; Gordy, W.
485. Phys. Rev. **91**, 222A (1953) — Venkateswarlu, P.; Mockler, R. C.; Gordy, W.
486. J. Chem. Phys. **21**, 761 (1953) — Weatherly, T. L.; Davidson, E. H.; Williams, Q.

487. Phys. Rev. **90**, 338 (1953) — Weisbaum, S.; Beers, Y.; Herrmann, G.
488. Phys. Rev. **91**, 1014 (1953) — White, R. L.
489. Phys. Rev. **92**, 1256 (1953) — White, R. L.; Townes, C. H.

490. J. Chem. Phys. **21,** 563 (1953)

491. Phys. Rev. **94,** 789A (1954)

492. Rev. Sci. Instr. **25,** 319 (1954)
493. J. Phys. Radium **15,** 515 (1954)
494. Acta. Chem. Scand. **8,** 367 (1954)
495. J. Chem. Phys. **22,** 565L (1954)
496. J. Chem. Phys. **22,** 2013 (1954)
497. New York Univ. Rept. #289.2 (1954)
498. Phys. Rev. **95,** 1686L (1954)
499. Rev. Sci. Instr. **25,** 324 (1954)
500. Phys. Rev. **94,** 1203 (1954)
501. J. Chem. Phys. **22,** 1782L (1954)
502. Trans. Faraday Soc. **50,** 444 (1954)
503. Phys. Rev. **93,** 193 (1954)

504. Phys. Rev. **93,** 897L (1954)
505. Phys. Rev. **95,** 299 (1954)
506. Phys. Rev. **95,** 706 (1954)
507. Trans. Faraday Soc. **50,** 1027 (1954)
508. Phys. Rev. **95,** 1201 (1954)
509. J. Chem. Phys. **22,** 1257L (1954)
510. Phys. Rev. **95,** 299A (1954)
511. Columbia Univ. Rept. (1954)
512. J. Chem. Phys. **22,** 876 (1954)

513. Acta Acad. Aboensis **19,** 1 (1954)
514. J. Chem. Phys. **22,** 563 (1954)
515. J. Chem. Phys. **22,** 1152L (1954)
516. Arkiv Fysik **8,** 341 (1954)
517. J. Am. Chem. Soc. **76,** 850 (1954)
518. Z. Physik **139,** 578 (1954)
519. Rev. Mod. Phys. **26,** 444 (1954)

520. Phys. Rev. **95,** 282 (1954)
521. J. Phys. Radium **15,** 521 (1954)
522. J. Chem. Phys. **22,** 92 (1954)
523. Proc. Roy. Soc. **226,** 96 (1954)
524. Phys. Rev. **94,** 1191 (1954)

525. Phys. Rev. **94,** 1184 (1954)

526. Phys. Rev. **95,** 385 (1954)
527. Trans. Faraday Soc. **50,** 779 (1954)
528. J. Chem. Phys. **22,** 2093 (1954)
529. J. Chem. Phys. **22,** 262 (1954)
530. J. Phys. Soc. Japan **9,** 434 (1954)
531. Phys. Rev. **96,** 649 (1954)
532. J. Chem. Phys. **22,** 1473 (1954)
533. Phys. Rev. **93,** 407 (1954)
534. J. Chem. Phys. **22,** 86 (1954)
535. J. Chem. Phys. **22,** 904 (1954)
536. J. Chem. Phys. **22,** 2093L (1954)
537. J. Chem. Phys. **22,** 1477 (1954)
538. J. Chem. Phys. **22,** 568 (1954)
539. J. Chem. Phys. **22,** 1577 (1954)
540. Phys. Rev. **96,** 434 (1954)

541. Phys. Rev. **94,** 789A (1954)
542. J. Chem. Phys. **22,** 681 (1954)
543. J. Chem. Phys. **22,** 1144 (1954)
544. Ann. Rev. Phys. Chem. **5,** 385 (1954)

Wilcox, W. S.; Brannock, K. C.; DeMore, W.; Goldstein, J. H.

Aamodt, L.; Fletcher, P. C.; Silvey, G.; Townes, C. H.

Baird, D. H.; Bird, G. R.
Bak, B.; Bruhn, J.; Rastrup-Andersen, J.
Bak, B.; Bruhn, J.; Rastrup-Andersen, J.
Bak, B.; Hansen, L.; Rastrup-Andersen, J.
Bak, B.; Hansen, L.; Rastrup-Andersen, J.
Beers, Y.; Herrmann, G.; Sterzer, F.
Bird, G. R.
Bird, G. R.
Bird, G.; Townes, C. H.
Birnbaum, G.; Maryott, A. A.; Wacker, P. F.
Bowen, H. J. M.
Burke, B. F.; Strandberg, M. W. P.; Cohen, V. W.; Koski, S. W.

Burrus, C. A.; Gordy, W.
Burrus, C. A.; Jache, A.; Gordy, W.
Burrus, C. A.; Jache, A.; Gordy, W.
Christensen, M. T.; Thompson, H. W.
Collier, R. J.
Cornwell, C. D.; Poynter, R. L.
Cox, J.; Thomas, W. J.; Gordy, W.
Dailey, B. P.
DeMore, W. D.; Wilcox, W. S.; Goldstein, J. H.

Ekelund, B.
Erlandsson, G.
Erlandsson, G.
Erlandsson, G.
Ferguson, R. C.
Genzel, L.; Eckhardt, W.
Geschwind, S.; Gunther-Mohr, G. R.; Townes, C. H.

Gordon, J. P.; Zeiger, H. J.; Townes, C. H.
Gordy, W.
Gordy, W.; Sheridan, J.
Griffiths, J. H. E.; Owen, J.
Gunther-Mohr, G. R.; Townes, C. H.; Van Vleck, J. H.

Gunther-Mohr, G. R.; White, R. L.; Schawlow, A. L.; Good, W. E.; Coles, D. K.

Hardy, W. A.; Silvey, G.
Heath, G. A.; Thomas, L. F.; Sheridan, J.
Herrmann, G.
Hrostowski, H. J.; Myers, R. J.
Itoh, J.; Kusaka, R.
Javan, A.; Engelbrecht, A.
Katayama, M.
King, W. C.; Gordy, W.
Kisliuk, P.
Kivelson, D.
Kojima, S.; Tsukada, K.
Kraitchman, J.; Dailey, B. P.
Krishnaji; Swarup, P.
Lide, D. R., Jr.
Liuima, F. A.; Bushkovitch, A. V.; Rouse, A. G.

Lotspeich, J. F.; Javan, A.
McCulloh, K. E.; Pollnow, G. F.
McCulloh, K. E.; Pollnow, G. F.
Myers, R. J.; Gwinn, W. D.

545. Trans. Faraday Soc. **50,** 1270 (1954)   Mills, I. M.; Thompson, H. W.

546. Can. J. Phys. **32,** 635 (1954)   Moller, C. K.; Stoicheff, B. P.

547. J. Chem. Phys. **22,** 1275 (1954)   Monfils, A.; Duchesne, J.

548. J. Chem. Phys. **22,** 1427 (1954)   Morgan, H. W.; Goldstein, J. H.

549. J. Chem. Phys. **22,** 1383 (1954)   Nielsen, H. H.

550. J. Phys. Radium **15,** 601 (1954)   Nielsen, H. H.

551. J. Phys. Soc. Japan **9,** 135 (1954)   Okaya, A.

552. Zh. Eksperim. i Teor. Fiz. **27,** 115 (1954)   Osipov, B. D.

553. Phys. Rev. **95,** 622A (1954)   Peter, M.; Strandberg, M. W. P.

554. J. Phys. Radium **15,** 497 (1954)   Porter, G.

555. Phys. Rev. **95,** 374 (1954)   Posener, D. W.; Strandberg, M. W. P.

556. Diss. Abstr. **14,** 2193 (1954)   Poynter, R. L.

557. Phys. Rev. **93,** 248 (1954)   Rogers, T. F.

558. Phys. Rev. **95,** 622A (1954)   Rogers, T. F.

559. Nature **174,** 798L (1954)   Sheridan, J.; Thomas, L. F.

560. J. Phys. Soc. Japan **9,** 378 (1954)   Shimoda, K.

561. J. Phys. Soc. Japan **9,** 558 (1954)   Shimoda, K.

562. J. Phys. Soc. Japan **9,** 567 (1954)   Shimoda, K.

563. J. Chem. Phys. **22,** 1678 (1954)   Slayton, G. R.; Simmons, J. W.; Goldstein, J. H.

564. Phys. Rev. **95,** 299A (1954)   Slayton, G. R.; Simmons, J. W.; Goldstein, J. H.

565. J. Chem. Phys. **22,** 2094L (1954)   Sterzer, F.

566. J. Chem. Phys. **22,** 925 (1954)   Stone, S. A.

567. Phys. Rev. **94,** 1393 (1954)   Strandberg, M. W. P.; Dreicer, H.

568. J. Chem. Phys. **22,** 1718 (1954)   Thomas, W. J. O.; Cox, J. T.; Gordy, W.

569. Phys. Rev. **95,** 622A (1954)   Trambarulo, R.; Lackner, H.; Moser, P.; Feeny, H.

570. Phys. Rev. **93,** 716 (1954)   Westerkamp, J. F.

571. Phys. Rev. **94,** 789A (1954)   White, R. L.

572. Bull. Am. Phys. Soc. **29,** 10 (1954)   White, R. L.

573. J. Chem. Phys. **22,** 516 (1954)   Wilcox, W. S.; Goldstein, J. H.; Simmons, J. W.

574. Phys. Rev. **93,** 360A (1954)   Wolfe, P. N.; Williams, D.

575. Phys. Rev. **98,** 1224 (1955)   Aamodt, L. C.; Fletcher, P. C.

576. Phys. Rev. **98,** 1317 (1955)   Aamodt, L. C.; Fletcher, P. C.

577. Phys. Rev. **97,** 76 (1955)   Anderson, J. H.; Hutchison, C. A., Jr.

578. Phys. Rev. **97,** 1654 (1955)   Anderson, R. S.

579. Phys. Rev. **100,** 703 (1955)   Autler, S. H.; Townes, C. H.

580. Disc. Faraday Soc. **19,** 30 (1955)   Bak, B.; Hansen-Nygaard, L.; Rastrup-Andersen, J.

581. Microchimica Acta **2–3,** 512 (1955)   Bak, B.; Hansen, L.; Rastrup-Andersen, J.

582. Phys. Rev. **98,** 1160A (1955)   Bird, G. R.

583. Private communication (1955)   Bird, G.

584. Private communication (1955)   Bird, G. R.

585. Private communication (1955)   Bird, G. R.

586. Phys. Rev. **97,** 684 (1955)   Blevins, G. S.; Jache, A. W.; Gordy, W.

587. Phys. Rev. **98,** 1182A (1955)   Bray, P. J.; Barnes, R. G.

588. J. Chem. Phys. **23,** 1172 (1955)   Burkhalter, J. H.

589. Florence Univ. Rept., Contract AF 61(514)433 (1955)   Carrara, N.

590. Trans. Faraday Soc. **51,** 1 (1955)   Checkland, P. B.; Thompson, H. W.

591. J. Chem. Phys. **23,** 2037 (1955)   Costain, C. C.

592. Disc. Faraday Soc. **19,** 52 (1955)   Cox, J. T.; Gaumann, T.; Thomas, W. J. O.

593. J. Am. Chem. Soc. **77,** 774 (1955)   DiGiacomo, A.; Smyth, C. P.

594. Memoires Acad. Roy. de Belg. **28,** 1 (1955)   Duchesne, J.

595. Private communication (1955)   Erlandsson, G.

596. Arkiv Fysik **9,** 399 (1955)   Erlandsson, G.

597. Arkiv Fysik **9,** 341 (1955)   Erlandsson, G.

598. J. Chem. Phys. **23,** 758 (1955)   Fernandez, J.; Myers, R. J.; Gwinn, W. D.

599. Phys. Rev. **99,** 613A (1955)   Fletcher, P. C.; Aamodt, L. C.

600. J. Chem. Phys. **23,** 1557L (1955)   Friend, J. P.; Schneider, R. F.; Dailey, B. P.

601. J. Chem. Phys. **23,** 779 (1955)   Geller, S.; Schawlow, A. L.

602. Indian J. Sci. **29**, 161 (1955) — Ghosh, D. K.
603. J. Chem. Phys. **23**, 1540 (1955) — Gilliam, O. R.; Walter, R. I.; Cohen, V. W.
604. Phys. Rev. **99**, 666A (1955) — Gora, E. K.
605. Providence (R.I.) College Rept. 2 (1955) — Gora, E. K.
606. Providence College Rept. 9 (1955) AF–19(604)–831 — Gora, E. K.
607. Providence College QPR VIII (1955) — Gora, E. K.
608. Phys. Rev. **99**, 1253 (1955) — Gordon, J. P.
609. Disc. Faraday Soc. **19**, 43 (1955) — Gwinn, W. D.
610. Disc. Faraday Soc. **19**, 38 (1955) — Heath, G. A.; Thomas, L. F.; Sherrard, E. I.; Sheridan, J.

611. Nature **175**, 79 (1955) — Herzberg, G.; Stoicheff, B. P.
612. J. Chem. Phys. **23**, 1223 (1955) — Howe, J. A.; Goldstein, J. H.
613. J. Phys. Soc. Japan **10**, 56 (1955) — Itoh, T.
614. Phys. Rev. **97**, 680 (1955) — Jache, A. W.; Blevins, G. S.; Gordy, W.
615. Phys. Rev. **99**, 1302 (1955) — Javan, A.
616. J. Chem. Phys. **23**, 2450 (1955) — Kwak, N.; Simmons, J. W.; Goldstein, J. H.
617. Phys. Rev. **99**, 666A (1955) — Lin, C. C.
618. Phys. Rev. **97**, 1664 (1955) — Low, W.
619. Private communication (1955) — McLay, D.
620. Phys. Rev. **100**, 1267A (1955) — Meister, A. G.; Dowling, J. M.; Takata, A. N.; Bielecki, A. J.

621. Chem. & Ind., 538 (1955) — Millen, D. J.; Sinnott, K. M.
622. Private communication (1955) — Myers, R. J.; Gwinn, W. D.
623. J. Phys. Soc. Japan **10**, 89 (1955) — Nishikawa, T.; Shimoda, K.
624. Rev. Sci. Instr. **26**, 1024 (1955) — Okaya, A.
625. Dokl. Akad. Nauk. SSSR **102**, 933 (1955) — Osipov, B. D.; Prokhorov, A. M.
626. J. Chem. Phys. **23**, 2388 (1955) — Phillips, C. S. E.
627. J. Chem. Phys. **23**, 1728L (1955) — Posener, D. W.
628. Phys. Rev. **100**, 993 (1955) — Rank, D. H.; Bennett, H. E.; Bennett, J. M.
629. Phys. Rev. **99**, 532 (1955) — Romer, R. H.; Dicke, R. H.
630. Research **8**, 88 (1955) — Sheridan, J.
631. Private communication (1955) — Shimoda, K.; Miyahara, A.; Nishikawa, T.
632. Rev. Sci. Instr. **26**, 1148 (1955) — Shimoda, K.; Wang, T. C.
633. J. Chem. Phys. **23**, 1554 (1955) — Sidei, T.; Yano, S.
634. Thesis, Univ. of London (1955) — Sinnott, K. M.
635. Univ. of Delaware Tech. Note No. 4 OSR–TN–55–348 (1955) — Smith, W. V.
636. J. Chem. Phys. **23**, 389 (1955) — Smith, W. V.; Lackner, H. A.; Volkov, A. B.
637. J. Chem. Phys. **23**, 762 (1955) — Sterzer, F.
638. Phys. Rev. **100**, 1174 (1955) — Sterzer, F.; Beers, Y.
639. Trans. Faraday Soc. **51**, 619 (1955) — Thomas, L. F.; Sherrard, E. I.; Sheridan, J.
640. J. Phys. Soc. Japan **10**, 417 (1955) — Torizuka, Y.; Kojima, Y.; Okamura, T.; Kamiryo, K.

641. J. Phys. Soc. Japan **10**, 60 (1955) — Tsukada, K.
642. J. Chem. Phys. **23**, 1966L (1955) — Turner, T. E.; Fiora, V. C.; Kendrick, W. M.
643. J. Chem. Phys. **23**, 1601 (1955) — Weisbaum, S.; Beers, Y.; Herrmann, G.
644. J. Chem. Phys. **23**, 249 (1955) — White, R. L.
645. J. Chem. Phys. **23**, 253 (1955) — White, R. L.
646. Rev. Mod. Phys. **27**, 276 (1955) — White, R. L.
647. Georgia Inst. of Technology, Tech. Rept. #1 (1955) — Williams, Q.; Weatherly, T. L.
648. Proc. Inst. Elec. Engrs. (London) **102** (1955) — Willshaw, W. E.; Lamont, H. R. L.; Hickin, E. M.

649. J. Sci. Hiroshima Univ. **19**, 161 (1955) — Yamamura, H.; Kawano, T.; Murakami, I.
650. Busseiron Kenkyu **82**, 13 (1955) — Yoshizumi, H.; Ito, T.
651. Uspekhi Khim. **24**, 730 (1955) — Zhabotinskii, M. E.
652. J. Chem. Phys. **24**, 989 (1956) — Andersen, F. A.; Bak, B.; Brodersen, S.
653. J. Chem. Phys. **24**, 720 (1956) — Bak, B.; Christensen, D.; Hansen, L.; Rastrup-Andersen, J.

654. J. Chem. Phys. **25**, 892 (1956) — Bak, B.; Christensen, D.; Rastrup-Andersen, J.; Tannenbaum, E.

655. Arch. Sci. **9**, 68 (1956) — Battaglia, A.; Gozzini, A.; Bruin, F.
656. J. Chem. Phys. **24**, 468L (1956) — Bernstein, H. J.; Schneider, W. G.

657. J. Chem. Phys. **24**, 469L (1956) — Bernstein, H. J.; Schneider, W. G.

658. Private communication (1956) — Bird, G. R.; Rastrup-Andersen, J.; Baird, J. C.

659. J. Chem. Phys. **25**, 1040 (1956) — Bird, G. R.

660. Rev. Mod. Phys. **28**, 75 (1956) — Blin-Stoyle, R. J.

661. Helv. Phys. Acta **29**, 224 (1956) — Bonanomi, J.; Herrmann, J.

662. Helv. Phys. Acta **29**, 451 (1956) — Bonanomi, J.; Herrmann, J.

663. Phys. Rev. **101**, 599 (1956) — Burrus, C. A.; Gordy, W.

664. Bull. Am. Phys. Soc. Ser. II, **1**, 340 (1956) — Clayton, L.; Williams, Q.; Weatherly, T. L.

665. Duke Univ. Rept. #15 (1956) — Cowan, M.; Gordy, W.

666. Phys. Rev. **101**, 1298 (1956) — Cox, J. T.; Gordy, W.

667. Trans. Faraday Soc. **52**, 455 (1956) — Dagg, I. R.; Thompson, H. W.

668. Trans. Faraday Soc. **52**, 145 (1956) — Dodd, R. E.; Rolfe, J. A.; Woodward, L. A.

669. Georgia Inst. of Tech., Tech. Rept. #2 (1956) — Eagle, D. F.; Weatherly, T. L.; Williams, Q.

670. J. Am. Chem. Soc. **78**, 2358 (1956) — Edgell, W. F.; Parts, L.

671. J. Chem. Phys. **25**, 579 (1956) — Erlandsson, G.

672. ASTIA Doc. No. AD 110 334 (1956) — Erlandsson, G.; Cox, J.

573. J. Chem. Phys. **25**, 778 (1956) — Erlandsson, G.; Cox, J.

674. Arkiv Fysik **11**, 391 (1956) — Erlandsson, G.; Selén, H.

675. Private communication (1956) — Ferguson, R. C.

676. Columbia Rad. Lab. Rept. (1956) — Fletcher, P.

677. Z. Physik **144**, 311 (1956) — Genzel, L.

678. AF Surveys in Geophys. **82** (1956) — Ghosh, S. N.; Edwards, H. D.

679. J. Chem. Phys. **24**, 106 (1956) — Goldstein, J. H.

680. Cahiers de Phys. **69**, 44 (1956) — Grenier-Besson, M. L.

681. Bull. Am. Phys. Soc. Ser. II, **1**, 13 (1956) — Herrmann, G.

682. N.Y.U. Rept. #289.6 (1956) — Herrmann, G.

683. Helv. Phys. Acta **29**, 226 (1956) — Herrmann, J.

684. J. Phys. Soc. Japan **11**, 334 (1956) — Hirakawa, H.; Miyahara, A.; Shimoda, K.

685. J. Phys. Soc. Japan **11**, 1208 (1956) — Hirakawa, H.; Oka, T.; Shimoda, K.

686. J. Phys. Soc. Japan **11**, 1207 (1956) — Hirahawa, H.; Oka, T.; Shimoda, K.

687. Nature **178**, 1111 (1956) — Hoisington, R. W. R.; Kellner, C.; Pentz, M. J.

688. Emory Univ. Tech. Rept. #2 (1956) — Howe, J. A.; Goldstein, J. H.

689. J. Chem. Phys. **24**, 131 (1956) — Hughes, R. H.

690. J. Chem. Phys. **25**, 217 (1956) — Iguchi, K.

691. J. Chem. Phys. **25**, 209 (1956) — Jache, A. W.; Moser, P. W.; Gordy, W.

692. J. Chem. Phys. **25**, 779 (1956) — Kisliuk, P.

693. Opt. i Spektroskopiya **1**, 374 (1956) — Kolosov, A. A.; Myasnikov, L. L.

694. J. Chem. Phys. **25**, 1203 (1956) — Kwak, N.; Goldstein, J. H.; Simmons, J. W.

695. Emory Univ. Tech. Rept. #1 (1956) — Kwak, N.; Goldstein, J. H.; Simmons, J. W.

696. J. Chem. Phys. **24**, 635L (1956) — Laurie, V. W.

697. J. Chem. Phys. **25**, 595 (1956) — Lide, D. R., Jr.; Mann, D. E.

698. Bull. Am. Phys. Soc. Ser. II, **1**, 198A (1956) — Lide, D. R., Jr.; Mann, D. E.; Fristrom, R. M.

699. J. Chem. Phys. **26**, 734 (1956) — Lide, D. R., Jr.; Mann, D. E.; Fristrom, R. M.

700. Bull. Am. Phys. Soc. Ser. II, **1**, 13 (1956) — Lin, C. C.

701. J. Chem. Phys. **25**, 594 (1956) — Madden, R. P.; Benedict, W. S.

702. J. Chem. Phys. **24**, 344 (1956) — Magnuson, D. W.

703. J. Chem. Phys. **24**, 489 (1956) — Mashima, M.

704. J. Chem. Phys. **25**, 779 (1956) — Mashima, M.

705. J. Phys. Soc. Japan **11**, 1301 (1956) — Matsuura, K.; Sugiura, Y.; Hatoyama, G. M.

706. Private communication (1956) — McLay, D. B.

707. J. Chem. Phys. **24**, 814 (1956) — Miller, R. C.; Smyth, C. P.

708. J. Phys. Soc. Japan **11**, 335 (1956) — Miyahara, A.; Hirakawa, H.; Shimoda, K.

709. Private communication (1956) — Morgan, H. W.

710. Thesis, Ohio State Univ. (1956) — Nexsen, W. E., Jr.

711. Ohio State Univ. Sci. Rept. 1 (1956) — Nexsen, W. E., Jr.

712. J. Phys. Soc. Japan **11**, 258 (1956) — Okaya, A.

713. J. Phys. Soc. Japan **11**, 249 (1956) — Okaya, A.

714. Soviet Res. in Phys., Collection #5 (1956) — Osipov, B. D.; Prokhorov, A. M.

715. Bull. Am. Phys. Soc. Ser. II, **1**, 13A (1956) — Perkins, K. L.; Bushkovitch, A. V.; Kieffer, L. J.

716. J. Chem. Phys. **24**, 139 (1956) — Pierce, L.

717. Commonwealth Sci. & Ind. Res. Organiz., #7 (1956) — Posener, D. W.

718. J. Chem. Phys. 25, 413 (1956) — Rathmann, G. B.; Curtis, A. J.; McGeer, P. L.; Smyth, C. P.

719. Columbus Symposium 10A (1956) — Shearer, J. N.; Wiggins, T. A.; Guenther, A. H.; Rank, D. H.

720. Private communication (1956) — Sheridan, J.; Heath, G. A.; Thomas, L. F.

721. J. Chem. Phys. 25, 510 (1956) — Smith, W. V.

722. Phys. Rev. 104, 89 (1956) — Swarup, P.

723. Z. Physik 144, 632 (1956) — Swarup, P.

724. Private communication (1956) — Thomas, L. F.; Sherrard, E. I.; Heeks, J. S.; Sheridan, J.

725. J. Chem. Phys. 24, 924L (1956) — Turner, T. E.; Howe, J. A.

726. Private communication (1956) — Turner, T. E.; Kendrick, W. M.

727. J. Chem. Phys. 25, 717 (1956) — Weatherly, T. L.; Williams, Q.

728. Preprint (1956) — Weltner, W., Jr.

729. J. Chem. Phys. 25, 976 (1956) — Wolfe, P. N.

730. Bull. Am. Phys. Soc. Ser. II, 1, BA5 (1956) — Yarmus, L.

731. N.Y.U. Rept. #289.7 (1956) — Yarmus, L.

732. N.Y.U. Rept. #289.8 (1956) — Yarmus, L.

733. Phys. Rev. 104, 365 (1956) — Yarmus, L.

734. J. Phys. Soc. Japan 11, 85 (1956) — Yokozawa, Y.; Tatsuzaki, I.

735. Z. Elektrochem 60, 752 (1956) — Zeil, W. von

736. Bull. Am. Phys. Soc. Ser. II, 2, 99 (1957) — Baird, J. C.; Rastrup-Andersen, J.; Bird, G. R.

737. J. Chem. Phys. 26, 134 (1957) — Bak, B.; Christensen, D.; Hansen-Nygaard, L.; Tannenbaum, E.

738. Trans. Faraday Soc. 53, 1397 (1957) — Baker, J. G.; Jenkins, D. R.; Kenney, C. N.; Sugden, T. M.

739. Harvard Univ. Rept., AF 18(600)968 (1957) — Beers, Y.

740. Bull. Am. Phys. Soc. Ser. II, 2, 213 (1957) — Bhattacharya, B.; Gordy, W.; Fujii, O.

741. J. Chem. Phys. 27, 360 (1957) — Birnbaum, G.

742. Arch. Sci. 10, 187 (1957) — Bonanomi, J.; de Prins, J.; Herrmann, J.; Kartaschoff, P.; Rossel, J.

743. Rev. Mod. Phys. 29, 94 (1957) — Breene, R. G., Jr.

744. J. Chem. Phys. 27, 1108 (1957) — Brown, L. C.; Parker, P. M.

745. J. Chem. Phys. 26, 391 (1957) — Burrus, C. A.; Gordy, W.

746. Diss. Abstr. 17, 1359 (1957) — Dam, C. F.

747. J. Am. Chem. Soc. 79, 2691 (1957) — Edgell, W. F.; Kinsey, P. A.; Amy, J. W.

748. Bull. Am. Phys. Soc. Ser. II, 2, 212 (1957) — Erlandsson, G.; Gordy, W.

749. Phys. Rev. 106, 513 (1957) — Erlandsson, G.; Gordy, W.

750. Soviet Phys.-JETP 5, 501 (1957) — Fain, V. M.

751. J. Chem. Phys. 27, 898 (1957) — Feeny, H.; Modigowsky, W.; Winters, B.

752. Z. Physik 149, 471 (1957) — Fitzky, H. G.; Honerjäger, R.; Wilke, W.

753. Columbia Rad. Lab. Rept. (1957) — Fletcher, P.

754. Bull. Am. Phys. Soc. Ser. II, 2, 30 (1957) — Fletcher, P.; Amble, E.

755. Ann. Soc. Sci. Bruxelles 71, 128 (1957) — Goedertier, P.; Lee, K. L.

756. Ann. Soc. Sci. Bruzelles 71, 184 (1957) — Goedertier, P.; Lee, K. L.

757. J. Chem. Phys. 26, 1482 (1957) — Hadley, G. F.

758. Phys. Rev. 108, 291 (1957) — Hadley, G. F.

759. Z. Physik 147, 567 (1957) — Happ, H.

760. Bull. Am. Phys. Soc. Ser. II, 2, 209 (1957) — Javan, A.; Wang, T. C.

761. Bull. Am. Phys. Soc. Ser. II, 2, 318 (1957) — Kilpatrick, T. E.; Mizushima, M.

762. Opt. i Spektroskopiya 1, 374 (1957) — Kolosov, A. A.; Myasnikov, L. L.

763. J. Chem. Phys. 26, 1359 (1957) — Laurie, V. W.

764. J. Chem. Phys. 26, 734 (1957) — Lide, D. R., Jr.; Mann, D. E.; Fristrom, R. M.

765. J. Chem. Phys. 27, 223 (1957) — Magnuson, D. W.

766. Phys. Rev. 105, 1502 (1957) — Marshall, S. A.; Weber, J.

767. J. Chem. Phys. 27, 1221 (1957) — Maryott, A. A.; Kryder, S. J.

768. Denki Shikensho 21, 612 (1957) — Matsuura, K.; Sugiura, Y.; Hatoyama, G. M.

769. J. Phys. Soc. Japan 12, 835 (1957) — Matsuura, K.; Sugiura, Y.; Hatoyama, G. M.

770. J. Phys. Soc. Japan 12, 314 (1957) — Matsuura, K.; Sugiura, Y.; Hatoyama, G. M.

771. Phys. Rev. 106, 607 (1957) — Matsuura, K.; Sugiura, Y.; Hatoyama, G. M.

772. J. Am. Chem. Soc. 79, 4289 (1957) — McCullough, J. P.; Doushin, D. R.; Messerly, J. E.; Hossenlopp, I. A.; Kincheloe, T. C.; Waddington, G.

773. J. Phys. Soc. Japan **12**, 820 (1957) Oka, T.; Hirakawa, H.
774. J. Phys. Soc. Japan **12**, 39 (1957) Oka, T.; Hirakawa, H.; Miyahara, A.
775. Opt. i Spektroskopiya **3**, 94 (1957) Osipov, B. D.
776. J. Chem. Phys. **26**, 384 (1957) Pigott, M. T.; Rank, D. H.
777. Australian J. Phys. **10**, 276 (1957) Posener, D. W.
778. J. Phys. Soc. Japan **12**, 43 (1957) Sekino, S.; Nishikawa, T.
779. Arkiv Fysik **13**, 81 (1957) Selén, H.
780. Private communication (1957) Sharbaugh, A. H.
781. Private communication (1957) Sheridan, J.
782. J. Phys. Soc. Japan **12**, 558 (1957) Shimoda, K.
783. J. Phys. Soc. Japan **12**, 652 (1957) Shimomura, K.
784. J. Phys. Soc. Japan **12**, 657 (1957) Shimomura, K.
785. J. Phys. Soc. Japan **12**, 1356 (1957) Shimomura, K.
786. Nuovo Cimento **5**, 587 (1957) Simonetta, M.; Vaciago, A.
787. Bull. Am. Phys. Soc. Ser. II, **2**, 31 (1957) Stevenson, M. J.
788. Columbia Rad. Lab. Rept. (1957) Stevenson, M. J.
789. Phys. Rev. **107**, 635 (1957) Stevenson, M. J.; Townes, C. H.
790. Phys. Rev. **106**, 606 (1957) Takahashi, T.; Ogawa, T.; Yamano, M.; Hirai, A.

791. Arch. Sci. **10**, 180 (1957) Thomas, L. F.; Heeks, J. S.; Sheridan, J.
792. Private communication (1957) Thomas, L. F.; Sherrard, E. I.; Sheridan, J.
793. Univ. Kyoto Progress Rept. #6, 34 (1957) Tomita, K.
794. Progr. Theoret. Phys. **18**, 316 (1957) Tomita, K.
795. Zh. Eksperim. i Teor. Fiz. **32**, 620 (1957) Veselago, V. G.
796. Fiz. Sbornik III. 493 (1957) Veselago, V. G.; Prokhorov, A. M.
797. Zh. Eksperim. i Teor. Fiz. **31**, 731 (1957) Veselago, V. G.; Prokhorov, A. M.
798. Georgia Inst. of Tech. Rept. 1, Contract DA 01–009– ORD–465 (1957) Weatherly, T. L.; Williams, Q.; Clayton, L., Jr.
799. Compt. rend. **245**, 1793 (1957) Wertheimer, R.; Couard, M.
800. Phys. Rev. **105**, 928 (1957) Yarmus, L.
801. NBS Technical Report (1957) Zahn, C. T.
802. Z. Elektrochem. **61**, 938 (1957) Zeil, W. von; Pfrommer, J. F.
803. J. Mol. Spectr. **2**, 163 (1958) Amat, G.; Nielsen, H. H.
804. Spectrochimica Acta **13**, 120 (1958) Bak, B.; Christensen, D.; Hansen-Nygaard, L.; Rastrup-Andersen, J.

805. J. Mol. Spectr. **2**, 361 (1958) Bak, B.; Hansen-Nygaard, L.; Rastrup-Andersen, J.

806. Opt. i Spektroskopiya **4**, 532 (1958) Barchukov, A. I.; Basov, N. G.
807. Opt. i Spectroskopiya **4**, 795 (1958) Basov, N. G.; Osipov, B. D.
808. Astron. Zh. **35**, 656 (1958) Benediktov, E. A.
809. J. Chem. Phys. **28**, 738 (1958) Bird, G. R.; Baird, J. C.; Williams, R. B.
810. J. Chem. Phys. **28**, 992 (1958) Birnbaum, G.
811. Helv. Phys. Acta **31**, 285 (1958) Bonanomi, J.; dePrins, J.; Herrmann, J.; Kartaschoff, P.

812. Compt. rend. **246**, 730 (1958) Boudouris, G.; Demetre, I.
813. J. Chem. Phys. **29**, 427 (1958) Burrus, C. A.
814. J. Chem. Phys. **29**, 864 (1958) Costain, C. C.
815. Univ. of Birmingham Dept. of Chem. (1958) Cox, A. P.; Thomas, L. F.; Sheridan, J.
816. Nature **181**, 1000 (1958) Cox, A. P.; Thomas, L. F.; Sheridan, J.
817. Can. J. Phys. **36**, 1336 (1958) Dalby, F. W.
818. J. Chem. Phys. **29**, 577 (1958) Friend, J. P.; Dailey, B. P.
819. Diss. Abstr. **19**, 546 (1958) Garrison, A. K.
820. Phys. Rev. **108**, 291 (1958) Hadley, G. F.
821. Physica **24**, 589 (1958) Heineken, F. W.; Battaglia, A.
822. Varian Assoc. Tech. Rept. (1958) Helmer, J. C.; Weaver, H. E.
823. J. Chem. Phys. **29**, 875 (1958) Herrmann, G.
824. J. Chem. Phys. **29**, 444 (1958) Hirota, E.; Oka, T.; Morino, Y.
825. Helv. Phys. Acta **31**, 276 (1958) Kneubuhl, F.; Gaumann, T.; Ginsburg, T.; Gunthard, H. H.

826. Trudy Sibir. Lesotekh. Inst. **18**, 27 (1958) Korshunov, A. V.
827. Phys. Rev. **109**, 1560 (1958) Krishnaji; Srivastava, G. P.
828. J. Chem. Phys. **28**, 704 (1958) Laurie, V. W.

829. Diss. Abstr. **19,** 340 (1958) Lotspeich, J. F.
830. Rev. Sci. Instr. **29,** 717 (1958) Mattuck, R. D.; Strandberg, M. W. P.
831. Private communication Meschi, D. J.; Myers, R. J.
832. J. Chem. Soc., 350 (1958) Millen, D. J.; Sinnott, K. M.
833. Bull. Chem. Soc. Japan **31,** 423 (1958) Morino, Y.; Hirota, E.
834. J. Am. Chem. Soc. **80,** 3483 (1958) Muller, N.; Pritchard, D. E.
835. J. Chem. Phys. **28,** 691 (1958) Ohno, K.; Mizuno, Y.; Mizushima, M.
836. J. Phys. Radium **19,** 915 (1958) Ramadier, I. J.; Amat, G.
837. Bull. Am. Phys. Soc. Ser. II, **3,** 214 (1958) Rosenthal, J.
838. Boll. Sci. Fac. Chim. Ind. Bologna **16,** 71 (1958). Sheridan, J.
839. Australian J. Phys. **12,** 109 (1958) Smith, W. E.
840. Memoirs, Kyoto Univ. **29,** 57 (1958) Yamano, M.
841. Memoirs, Kyoto Univ. **29,** 67 (1958) Yamano, M.
842. Izvest. Akad. Nauk. SSSR **22,** 1145 (1958) Yaroslavskii, N. G.; Stanevich, A. E.
843. Opt. i Spektroskopiya **5,** 384 (1958) Yaroslavskii, N. G.; Stanevich, A. E.
844. J. Res., NBS **63A** (Phys. & Chem.) 145 (1959) Allen, H. C., Jr.; Plyler, E. K.
845. Bull. Am. Phys. Soc. Ser. II, **4,** 291 (1959) Beeson, E. L.; Williams, J. Q.; Weatherly, T. L.

846. Diss. Abstr. **19,** 3329 (1959) Burnside, P. B.
847. J. Chem. Phys. **30,** 976 (1959) Burrus, C. A.
848. J. Chem. Phys. **30,** 1328 (1959) Clayton, L.; Williams, Q.; Weatherly, T. L.
849. J. Chem. Phys. **31,** 554 (1959) Clayton, L.; Williams, Q.; Weatherly, T. L.
850. J. Chem. Phys. **31,** 389 (1959) Costain, C. C.; Morton, J. R.
851. J. Chem. Phys. **30,** 777 (1959) Costain, C. C.; Stoicheff, B. P.
852. J. Chem. Phys. **31,** 1678 (1959) DiGiorgio, V. E.; Robinson, G. W.
853. Can. J. Phys. **37,** 703 (1959) Dowling, J. M.; Stoicheff, B. P.
854. Physica **25,** 859 (1959) Dymanus, A.
855. Phys. Rev. **116,** 351 (1959) Dymanus, A.
856. J. Chem. Phys. **30,** 603 (1959) Eagle, D. F.; Weatherly, T. L.; Williams, Q.
857. Proc. Roy. Soc. **A250,** 39 (1959) Eaton, D. R.; Thompson, H. W.
858. J. Chem. Phys. **31,** 566 (1959) Favero, P.; Mirri, A. M.; Baker, J. G.
859. Phys. Rev. **114,** 1534 (1959) Favero, P.; Mirri, A. M.; Gordy, W.
860. Diss. Abstr. **19,** 1930 (1959) Goll, R. J.
861. J. Mol. Spectr. **3,** 78 (1959) Gora, E. K.
862. Bull. Am. Phys. Soc. Ser. II, **4,** 153 (1959) Gwinn, W. D.; Zinn, J.; Fernandez-Bertram, J.
863. Proc. Chem. Soc. **10** (1959) Jackson, R. H.; Millen, D. J.
864. Proc. Ind. Acad. Sci. **50A,** 108 (1959) Jaseja, T. S.
865. Trans. Faraday Soc. **55,** 1473 (1959) Jenkins, D. R.; Sugden, T. M.
866. J. Chem. Phys. **30,** 512 (1959) Kasai, P. H.; Myers, R. J.; Eggers, D. F., Jr.; Wiberg, K. B.

867. J. Chem. Phys. **31,** 1139 (1959) Kikuchi, Y.; Hirota, E.; Morino, Y.
868. J. Mol. Spectr. **3,** 349 (1959) Kneubuhl, F.; Gaumann, T.; Gunthard, H. H.
869. Arkiv Fysik **16,** 197 (1959) Kokeritz, P. G.; Selén, H.
870. J. Chem. Phys. **31,** 1438 (1959) Kowalewski, D. G.; Kokeritz, P.; Selén, H.
871. Phys. Rev. **116,** 903 (1959) Lin, C. C.
872. Rev. Mod. Phys. **31,** 841 (1959) Lin, C. C.; Swalen, J. D.
873. J. Chem. Phys. **31,** 633 (1959) Lotspeich, J. F.; Javan, A.; Englebrecht, A.
874. J. Mol. Spectr. **3,** 405 (1959) Meschi, D. J.; Myers, R. J.
875. J. Chem. Phys. **30,** 1025 (1959) Morgan, H. W.; Goldstein, J. H.
876. Spectrochim. Acta **15,** 313 (1959) Mould, H. M.; Price, W. C.; Wilkinson, G. R.
877. J. Mol. Spectr. **3,** 259 (1959) Palik, E. D.
878. Diss. Abstr. **20,** 336 (1959) Stroup, R. E.
879. Spectroscopia Mol. **8,** 51 (1959) Sundaram, S.
880. Opt. i Spektroskopiya **6,** 254 (1959) Svidzinskii, K. K.
881. J. Phys. Soc. Japan **14,** 1595 (1959) Takuma, H.; Shimizu, T.; Shimoda, K.
882. Nuovo Cimento **13,** 1060 (1959) Thaddeus, P.; Loubser, J.
883. J. Chem. Phys. **31,** 1677 (1959) Thaddeus, P.; Loubser, J.; Krisher, L.; Lecar, H.

884. Nature **183,** 1182 (1959) Tyler, J. K.; Cox, A. P.; Sheridan, J.
885. Proc. Chem. Soc., 155 (1959) Tyler, J. K.; Thomas, L. F.; Sheridan, J.
886. Bull. Classe Sci. Acad. Belg. **45,** 381 (1959) van Riet, R.; de Hemptinne, M.
887. J. Chem. Phys. **30,** 1372 (1959) Verdier, P. H.; Wilson, E. B., Jr.

888. Opt. and Spectr. (USSR) **6,** 286 (1959)

889. J. Chem. Phys. **30,** 1075 (1959)

890. Rev. Mod. Phys. **31,** 681 (1959)

891. Harvard Univ. Rept., Contract NONR−186614 (1959)

892. Compt. rend. **248,** 1640 (1959)

893. Opt. i Specktroskopiya **6,** 799 (1959)

894. Arch. Sci. **13,** 171 (1960)

895. Phys. Rev. **119,** 144 (1960)

896. J. Chem. Phys. **33,** 1643 (1960)

897. Bull. Am. Phys. Soc. Ser. II, **5,** 241 (1960)

898. Arch. Sci. (Sw.) **13,** N. Fasc. Spec., 143 (1960)

899. Nuovo Cimento **17,** 734 (1960)

900. Nuovo Cimento **17,** 740 (1960)

901. Nature **187,** 765 (1960)

902. Bull. Soc. Belge Phys. **4,** 273 (1960)

903. Bull. Acad. Roy. Belge **8,** 767 (1960)

904. Diss. Abstr. **16,** 668 (1956) publ. 1960

905. J. Chem. Phys. **32,** 1258 (1960)

906. Bull. Chem. Soc. Japan **33,** 158 (1960)

907. Proc. Chem. Soc., 220 (1960)

908. J. Chem. Phys. **33,** 949 (1960)

909. Trans. Faraday Soc. **56,** 1732 (1960)

910. J. Chem. Phys. **32,** 205 (1960)

911. Columbia Rad. Lab. Progr. Rept. #4

912. NBS Report (1960)

913. J. Chem. Phys. **33,** 598 (1960)

914. J. Chem. Phys. **33,** 508 (1960)

915. Spectrochimica Acta **16,** 1267 (1960)

916. J. Chem. Phys. **32,** 1577 (1960)

917. Univ. of Tokyo Progr. Rept. 4 (1960)

918. J. Chem. Phys. **32,** 1591 (1960)

919. J. Chem. Phys. **33,** 1256 (1960)

920. Australian J. Phys. **13,** 168 (1960)

921. Thesis, Purdue Univ. (1960)

922. J. Opt. Soc. Am. **50,** 421 (1960)

923. Diss. Abstr. **20,** 3536 (1960)

924. Padova Univ. (Italy) Tech. Rept. Contract DA 91−591−EUC−1050 (1960)

925. Proc. Chem. Soc. **21** (1960)

926. Nature **185,** 96 (1960)

927. J. Phys. Soc. Japan **15,** 1125 (1960)

928. J. Phys. Soc. Japan **15,** 646 (1960)

929. Emory Univ. Final Rept. (1960)

930. Geophysical Res. Papers #69 (1960)

931. Proc. Chem. Soc. 119 (1960)

932. Z. Naturforsch. **15a,** 1011 (1960)

933. J. Mol. Spectr. **4,** 93 (1960)

934. Diss. Abstr. **20,** 3343 (1960)

935. J. Mol. Spectr. **7,** 58 (1961)

936. Phys. Rev. **121,** 1119 (1961)

937. Ann. Soc. Sci. Brux. **75,** 174 (1961)

938. Ann. Soc. Sci. Brux. **75,** 194 (1961)

939. Private communication (1961)

940. Bull. Chem. Soc. Japan **34,** 341 (1961)

Veselago, V. G.

Weaver, J. R.; Heitsch, C. W.; Parry, R. W.

Weber, J.

Wilson, E. B., Jr.; Holton, J. R.

Wertheimer, R.

Yaroslavskii, N. G.; Stanevich, A. E.

Battaglia, A.; Gozzini, A.; Polacco, E.

Bhattacharya, B. N.; Gordy, W.

Chan, S. I.; Zinn, J.; Fernandez, J.; Gwinn, W. D.

Cowan, M.; Gordy, W.

DePrins, J.; Karteschoff, P.; Bonanomi, J.

Favero, P.; Baker, J. G.

Favero, P.; Mirri, A. M.; Baker, J. G.

Gebbie, H. A.; Stone, N. W. B.; Walshaw, C. D.

Goedertier, R. E.

Goedertier, R. E.

Graybeal, J. D.

Graybeal, J. D.

Hirota, E.; Morino, Y.

Jenkins, D. R.; Kewley, R.; Sugden, T. M.

Johnson, L. G.

Kewley, R.; Murty, K. S. R.; Sugden, T. M.

Kivelson, D.; Wilson, E. B., Jr.; Lide, D. R., Jr.

Kusch, P.

Laurie, V. W.

LeBlanc, O. H.; Laurie, V. W.; Gwinn, W. D.

Long, M. W.; Williams, Q.; Weatherly, T. L.

Moon, H.; Goldstein, J. H.; Simmons, J. W.

Muller, N.; Bracken, R. C.

Nishikawa, T.; Itoh, T.; Shimoda, K.

Norris, J. A.; Laurie, V. W.

Norris, J. A.; Laurie, V. W.

Posener, D. W.

Pritchard, D. E.

Rank, D. H.; Skorinko, G.; Eastman, D. P.; Wiggins, T. A.

Schneider, R. F.

Semerano, G.

Sheridan, J.; Turner, A. C.

Sheridan, J.; Tyler, J. K.; Aynsley, E. E.; Dodd, R. E.; Little, R.

Shimoda, K.; Kondo, K.

Shimuzu, T.; Takuma, H.

Simmons, J. W.; Goldstein, J. H.

Tsao, C. J.; Curnutte, B.

Tyler, J. K.; Sheridan, J.

Zeil, W. von; Winnewisser, M.; Bodenseh, H. K.; Bucheret, H.

Wait, S. C., Jr. Barnett, M. P.

Yarmus, L.

Bak, B.; Christensen, D.; Hansen-Nygaard, L.; Rastrup-Andersen, J.

Curl, R. F., Jr.; Kinsey, J. L.; Baker, J. G.; Baird, J. C.; Bird, G. R.; Heidelberg, R. F.; Sugden, T. M.; Jenkins, D. R.; Kenney, C. N.

Goedertier, R. E.

Hepcée, S. de

Herschbach, S.

Hirota, E.; Morino, Y.

941. J. Chem. Phys. **34**, 1247 (1961) Howe, J. A.
942. Nature **192**, 160 (1961) Job, B. E.; Sheridan, J.
943. Bull. Chem. Soc. Japan **34**, 348 (1961) Kikuchi, Y.; Hirota, E.; Morino, Y.
944. J. Chem. Phys. **34**, 291 (1961) Laurie, V. W.
945. Spectrochimica Acta **17**, 665 (1961) Lide, D. R., Jr.; Christensen, D.
946. J. Chem. Soc., 1322 (1961) Millin, D. J.; Pannell, J.
947. J. Chem. Phys. **34**, 1847 (1961) Miller, R. F.; Curl, R. F., Jr.
948. Nuovo Cimento **19**, 1189 (1961) Mirri, A. M.; Guarnieri, A.; Favero, P.
949. J. Chem. Phys. **35**, 2240L (1961) Pierce, L.; Jackson, R.; DiCianni, N.
950. Proc. Ind. Acad. Sci. **3A**, 89 (1961) Rajan, A. S.
951. Opt. and Spectr. (USSR) **10**, 288 (1961) Shipulo, G. P.
952. J. Chem. Phys. **35**, 750 (1961) Tanaka, I.; Carrington, T.; Broida, H. P.
953. Z. Physik **161**, 179 (1961) von Torring, T.
954. J. Mol. Spectr. **6**, 215 (1961) Venkateswarlu, P.; Baker, J. G.; Gordy, W.
955. J. Chem. Phys. **37**, 2027 (1962) Bak, B.; Christensen, D.; Dixon, W. B.; Hansen-Nygaard, L.; Rastrup-Andersen, J.

956. J. Mol. Spectr. **9**, 124 (1962) Bak, B.; Christensen, D.; Dixon, W. B.; Hansen-Nygaard, L.; Rastrup-Andersen, J.; Schottlander, M.

957. Adv. Mol. Spectr. (Bologna) (1962) Baker, J. G.; Jenkins, D. P.; Kenney, C. N.; Sugden, T. M.

958. J. Chem. Phys. **37**, 1133 (1962) Beaudet, R. A.; Wilson, E. B., Jr.
959. J. Mol. Spectr. **9**, 317 (1962) Costain, C. C.
960. Phys. Rev. **125**, 1993 (1962) Curl, R. F., Jr.; Heidelberg, R. F.; Kinsey, J. L.

961. J. Chem. Phys. **37**, 2503L (1962) Graybeal, J. D.; Roe, D. W.
962. Adv. Mol. Spectr. (Bologna) (1962) Ferronato, E.; Grifone, L.; Guarnieri, A.; Zuliani, G.

963. J. Chem. Phys. **36**, 440 (1962) Flygare, W. H.; Howe, J. A.
964. J. Chem. Phys. **36**, 200 (1962) Flygare, W. H.; Narath, A.; Gwinn, W. D.
965. J. Chem. Phys. **36**, 650 (1962) Howe, J. A.; Flygare, W. H.
966. Adv. Mol. Spectr. (Bologna) (1962) Jackson, R. H.; Millen, D. J.
967. J. Chem. Phys. **36**, 2525 (1962) Jen, M.; Lide, D. R., Jr.
968. Trans. Faraday Soc. **58**, 1284 (1962) Jenkins, D. R.; Kewley, R.; Sugden, T. M.
969. Private communication (1962) Job, B. E.; Sheridan, J.
970. Bull. Am. Phys. Soc. Ser. II, **7**, 272 (1962) Kewley, R.; Sastry, K. V. L. N.; Winnewisser, M.

971. J. Am. Chem. Soc. **84**, 334 (1962) Kirchhoff, W. H.; Wilson, E. B., Jr.
972. J. Chem. Phys. **37**, 2995 (1962) Laurie, V. W.; Pence, D. T.
973. J. Mol. Spectr. **9**, 366 (1962) Matsumura, C.; Hirota, E.; Oka, T.; Morino, Y.

974. Can. J. Phys. **40**, 61 (1962) McLay, D. B.; Mann, C. R.
975. J. Mol. Spectr. **8**, 153 (1962) Millen, D. J.; Topping, G.; Lide, D. R., Jr.
976. Nuovo Cimento **25**, 265 (1962) Mirri, A. M.; Guarnieri, A.; Favero, P.; Zuliani, G.

977. Proc. Int'l. Symp. Mol. Spectr. Japan (1962) Mirri, A. M.; Guarnieri, A.; Favero, P.; Zuliani, G.

978. J. Mol. Spectr. **8**, 300 (1962) Oka, T.; Morino, Y.
979. Proc. Int'l. Symp. Mol. Spectr. Japan (1962) Oka, T.; Morino, Y.
980. J. Chem. Phys. **37**, 2921 (1962) Pillai, M. G. K.; Curl, R. F., Jr.
981. Adv. Mol. Spectr. **1**, 139 (1962) Sheridan, J.
982. Opt. and Spectr. (USSR) **13**, 335 (1962) Shipulo, G. P.
983. Opt. and Spectr. (USSR) **12**, 78L (1962) Sobolev, G. A.; Shcherbakov, A. M.; Akishin, P. A.

984. Spectrochimica Acta **18**, 1368 (1962) Tolles, W. M.; Curl, R. F.; Heidelberg, R. F.
985. J. Chem. Phys. **36**, 1473 (1962) Treacy, E. B.; Beers, Y.
986. Opt. Soc. Am. **52**, 581 (1962) Tyler, J. K.; Thomas, L. F.; Sheridan, J.
987. Opt. and Spectr. (USSR) **12**, 250 (1962) Tyler, J. K.; Thomas, L. F.; Sheridan, J.
988. J. Chem. Phys. **38**, 1636 (1963) Cox, A. P.; Esbitt, A. S.
989. Private communication (1963) Esterowitz, L.
990. N.Y.U. Rept., Contract 285 (1963) Esterowitz, L.
991. J. Chem. Phys. **39**, 1133L (1963) Forman, R. A.; Lide, D. R., Jr.
992. Le J. de Phys. **24**, 633 (1963) Goedertier, R. E.

993. J. Mol. Spectr. **10**, 418 (1963) — Kewley, R.; Sastry, K. V. L. N.; Winnewisser, M.

994. J. Chem. Phys. **38**, 2329 (1963) — Larkin, D. M.; Gordy, W.

995. *Tables of Experimental Dipole Moments*, W. H. Freeman & Co., San Francisco & London (1963) — McClellan, A. L.

996. J. Chem. Phys. **38**, 730 (1963) — Pierce, L.; DiCianni, N.

997. J. Chem. Phys. **38**, 2029 (1963) — Pierce, L.; DiCianni, N.

998. Thesis, Univ. of Louvain (1963) — Windmolders, R.

999. J. Chem. Phys. **40**, 3378 (1964) — Bird, G.; Baird, J. C.; Jache, A. W.; Hodgeson, J. A.; Curl, R. F., Jr.; Kunkle, A. C.; Bransford, J. W.; Rastrup-Andersen, J.; Rosenthal, J.

1000. J. Mol. Spectr. **12**, 387 (1964) — Kewley, R.; Sastry, K. V. L. N.; Winnewisser, M.

1001. Ann. Soc. Sci. Brux. **78**, 200 (1964) — Savariraj, G. A.

1002. J. Chem. Phys. **40**, 257 (1964) — Thaddeus, P.; Krisher, L. C.; Loubser, J. H. N.

1003. J. Chem. Phys. **41**, 999 (1964) — Winnewisser, M.; Cook, R. L.

1004. Private communication (1965) — Smith, D. F.

1005. Private communication (1965) — Baird, J. C.; Rastrup-Andersen, J.; Bird, G. R.

1006. J. Chem. Phys. **37**, 2926 (1962) — Beeson, E. L.; Weatherly, T. L.; Williams, Q.

1007. Phys. Rev. **104**, 551 (1956) — Cowan, M.; Gordy, W.

1008. Private communication (1955) — Peter, M.; Strandberg, M. W. P.

1009. Private communication (1955) — Sheridan, J.; Thomas, L. F.; Heeks, J. S.

1010. Private communication (1956) — Simmons, J. W.; Harmon, G. L.

1011. Private communication (1957) — Sheridan, J.

1012. Private communication (1957) — Heeks, J. S.; Thomas, L. F.; Sheridan, J.

1013. Z. Electrochem. **61**, 935 (1957) — Thomas, L. F.; Heeks, J. S.; Sheridan, J.

1014. NYU Rept. 289.6 (1956) — Hermann, G.

1015. J. Appl. Phys. **17**, 814 (1946) — Walter, J. E.; Hershberger, W. D.

1016. Phys. Rev. **82**, 451 (1951) — Hill, R. M.; Smith, W. V.

1017. Phys. Rev. **96**, 434 (1954) — Liuima, F. A.; Bushkovitch, A. V.; Rouse, A. G.

1018. Princeton Univ. Rept. (1955) — Romer, R. H.; Wittke, J. P.

1019. Chim. Anal. **38**, 428 (1956) — Roubine, E.

1020. Molecular Microwave Spectra Tables, NBS Circular 518 (June 23, 1952) — Kisliuk, P.; Townes, C. H.

1021. Phys. Rev. **73**, 1249 (1948) — Merritt, F. R.; Townes, C. H.

1022. J. Chem. Phys. **21**, 1293 (1953) — Chang, T. S.; Dennison, D. M.

1023. CRPL Preprint 51–4 (1950) — Birnbaum, G.; Kryder, S. J.; Lyons, H.

1024. Phys. Rev. **102**, 1308 (1956) — Shimoda, K.; Wang, T. C.; Townes, C. H.

1025. J. Chem. Phys. **22**, 1456 (1954) — Krishnaji; Swarup, P.

1026. Univ. of Utrecht Diss. (1958) — Dymanus, A.

1027. Phys. Rev. **72**, 263 (1947) — Watts, R. J.; Williams, D.

1028. American Institute of Physics Handbook, McGraw-Hill Book Co., Inc., New York (1957) — Gray, D. E. (Coordinating Editor)

1029. Microwave Spectroscopy, McGraw-Hill Book Co., Inc., New York (1955) — Townes, C. H.; Schawlow, A. L.

1030. Table of Dielectric Constants and Electric Dipole Moments of Substances in the Gaseous State, NBS Circular 537 (June 25, 1953) — Maryott, A. A.; Buckley, F.

1031. Landolt-Bornstein: Zahlen-Werte und Funktionen aus Physik, Chemie, Astronomie, Geophysik und Tecknik (6th Edition) Springer-Verlag, Berlin (1951) — Euken, A. (Editor)

1032. Tables of Electric Dipole Moments, The Technology Press, M.I.T. (1947) — Wesson, L. G.

NATIONAL BUREAU OF STANDARDS

The National Bureau of Standards[1] was established by an act of Congress March 3, 1901. Today, in addition to serving as the Nation's central measurement laboratory, the Bureau is a principal focal point in the Federal Government for assuring maximum application of the physical and engineering sciences to the advancement of technology in industry and commerce. To this end the Bureau conducts research and provides central national services in three broad program areas and provides central national services in a fourth. These are: (1) basic measurements and standards, (2) materials measurements and standards, (3) technological measurements and standards, and (4) transfer of technology.

The Bureau comprises the Institute for Basic Standards, the Institute for Materials Research, the Institute for Applied Technology, and the Center for Radiation Research.

THE INSTITUTE FOR BASIC STANDARDS provides the central basis within the United States of a complete and consistent system of physical measurement, coordinates that system with the measurement systems of other nations, and furnishes essential services leading to accurate and uniform physical measurements throughout the Nation's scientific community, industry, and commerce. The Institute consists of an Office of Standard Reference Data and a group of divisions organized by the following areas of science and engineering:

> Applied Mathematics—Electricity—Metrology—Mechanics—Heat—Atomic Physics—Cryogenics[2]—Radio Physics[2]—Radio Engineering[2]—Astrophysics[2]—Time and Frequency.[2]

THE INSTITUTE FOR MATERIALS RESEARCH conducts materials research leading to methods, standards of measurement, and data needed by industry, commerce, educational institutions, and government. The Institute also provides advisory and research services to other government agencies. The Institute consists of an Office of Standard Reference Materials and a group of divisions organized by the following areas of materials research:

> Analytical Chemistry—Polymers—Metallurgy — Inorganic Materials — Physical Chemistry.

THE INSTITUTE FOR APPLIED TECHNOLOGY provides for the creation of appropriate opportunities for the use and application of technology within the Federal Government and within the civilian sector of American industry. The primary functions of the Institute may be broadly classified as programs relating to technological measurements and standards and techniques for the transfer of technology. The Institute consists of a Clearinghouse for Scientific and Technical Information,[3] a Center for Computer Sciences and Technology, and a group of technical divisions and offices organized by the following fields of technology:

> Building Research—Electronic Instrumentation — Technical Analysis — Product Evaluation—Invention and Innovation— Weights and Measures — Engineering Standards—Vehicle Systems Research.

THE CENTER FOR RADIATION RESEARCH engages in research, measurement, and application of radiation to the solution of Bureau mission problems and the problems of other agencies and institutions. The Center for Radiation Research consists of the following divisions:

> Reactor Radiation—Linac Radiation—Applied Radiation—Nuclear Radiation.

---

[1] Headquarters and Laboratories at Gaithersburg, Maryland, unless otherwise noted; mailing address Washington, D. C. 20234.
[2] Located at Boulder, Colorado 80302.
[3] Located at 5285 Port Royal Road, Springfield, Virginia 22151.

NBS TECHNICAL PUBLICATIONS

PERIODICALS

JOURNAL OF RESEARCH reports National Bureau of Standards research and development in physics, mathematics, chemistry, and engineering. Comprehensive scientific papers give complete details of the work, including laboratory data, experimental procedures, and theoretical and mathematical analyses. Illustrated with photographs, drawings, and charts.

Published in three sections, available separately:

● Physics and Chemistry

Papers of interest primarily to scientists working in these fields. This section covers a broad range of physical and chemical research, with major emphasis on standards of physical measurement, fundamental constants, and properties of matter. Issued six times a year. Annual subscription: Domestic, $5.00; foreign, $6.00*.

● Mathematical Sciences

Studies and compilations designed mainly for the mathematician and theoretical physicist. Topics in mathematical statistics, theory of experiment design, numerical analysis, theoretical physics and chemistry, logical design and programming of computers and computer systems. Short numerical tables. Issued quarterly. Annual subscription: Domestic, $2.25; foreign, $2.75*.

● Engineering and Instrumentation

Reporting results of interest chiefly to the engineer and the applied scientist. This section includes many of the new developments in instrumentation resulting from the Bureau's work in physical measurement, data processing, and development of test methods. It will also cover some of the work in acoustics, applied mechanics, building research, and cryogenic engineering. Issued quarterly. Annual subscription: Domestic, $2.75; foreign, $3.50*.

TECHNICAL NEWS BULLETIN

The best single source of information concerning the Bureau's research, developmental, cooperative and publication activities, this monthly publication is designed for the industry-oriented individual whose daily work involves intimate contact with science and technology—*for engineers, chemists, physicists, research managers, product-development managers, and company executives.* Annual subscription: Domestic, $1.50; foreign, $2.25*.

*Difference in price is due to extra cost of foreign mailing.

NONPERIODICALS

Applied Mathematics Series. Mathematical tables, manuals, and studies.

Building Science Series. Research results, test methods, and performance criteria of building materials, components, systems, and structures.

Handbooks. Recommended codes of engineering and industrial practice (including safety codes) developed in cooperation with interested industries, professional organizations, and regulatory bodies.

Special Publications. Proceedings of NBS conferences, bibliographies, annual reports, wall charts, pamphlets, etc.

Monographs. Major contributions to the technical literature on various subjects related to the Bureau's scientific and technical activities.

National Standard Reference Data Series. NSRDS provides quantitative data on the physical and chemical properties of materials, compiled from the world's literature and critically evaluated.

Product Standards. Provide requirements for sizes, types, quality and methods for testing various industrial products. These standards are developed cooperatively with interested Government and industry groups and provide the basis for common understanding of product characteristics for both buyers and sellers. Their use is voluntary.

Technical Notes. This series consists of communications and reports (covering both other agency and NBS-sponsored work) of limited or transitory interest.

CLEARINGHOUSE

The Clearinghouse for Federal Scientific and Technical Information, operated by NBS, supplies unclassified information related to Government-generated science and technology in defense, space, atomic energy, and other national programs. For further information on Clearinghouse services, write:

Clearinghouse
U.S. Department of Commerce
Springfield, Virginia 22151

Order NBS publications from:
Superintendent of Documents
Government Printing Office
Washington, D.C. 20402